Vector Partitions, Visible Points, and Ramanujan Functions

Vector Partitions, Visible Points and Ramanujan Functions offers a novel theory of Vector Partitions, though very much grounded in the long-established work of others, that could be developed as an extension to the existing theory of Integer Partitions. The book is suitable for graduate students in physics, applied mathematics, number theory and computational mathematics. It takes the reader up to research level, presenting new results alongside known classical results from integer partitions and areas of vector and multipartite partition theory. It also sets forth new directions for research for the more advanced reader.

Above all, the intention of the book is to bring new inspiration to others who study mathematics and related areas. It is hoped that some new ideas will be launched to add value and insight into many of the classical and new theories surrounding partitions. The book is an appreciation of the many gifted authors of research into partitions over the past century and before, in the hope that more may come of this for future generations.

Features

- Provides a step-by-step guide through the known literature on Integer and Vector Partitions, and a focus on the not so well-known Visible Point Vector identities
- Serves as a reference for graduate students and researchers in physics, applied mathematics, number theory and computational mathematics
- Offers a variety of practical examples as well as sets of exercises suitable for students and researchers

Geoffrey B. Campbell completed his PhD at Australian National University in 1998 under the esteemed physicist Professor Rodney Baxter. His affiliation with the Australian National University Mathematical Sciences Institute has continued for over 30 years. Within that time frame, Geoffrey also served eight years as an Honorary Research Fellow at LaTrobe University Mathematics and Statistics Department in Melbourne. Currently he writes ongoing articles for the Australian Mathematical Society Gazette. Within the international scope, Geoffrey currently serves as a PhD external committee member for a mathematics graduate student at Washington State University in America.

Geoffrey has built a career within Australian Commonwealth and State government departments, including as an Advisor at the Department of Prime Minister and Cabinet; as Analyst Researcher for a Royal Commission. Geoffrey specializes in complex data, machine learning including data analytics. He is also a published poet in Australian anthologies and literary magazines.

Vector Partitions, Visible Points, and Ramanujan Functions

Geoffrey B. Campbell

CRC Press
Taylor & Francis Group
Boca Raton London New York

CRC Press is an imprint of the
Taylor & Francis Group, an **informa** business
A CHAPMAN & HALL BOOK

First edition published 2024
by CRC Press
2385 NW Executive Center Drive, Suite 320, Boca Raton FL 33431

and by CRC Press
4 Park Square, Milton Park, Abingdon, Oxon, OX14 4RN

CRC Press is an imprint of Taylor & Francis Group, LLC

Library of Congress Cataloging-in-Publication Data

Names: Campbell, Geoffrey B., author.
Title: Vector partitions, visible points and Ramanujan functions / Geoffrey B. Campbell.
Description: First edition. | Boca Raton : C&H/CRC Press, 2024. | Includes bibliographical references and index.
Identifiers: LCCN 2023053940 (print) | LCCN 2023053941 (ebook) | ISBN 9781032004327 (hardback) | ISBN 9781032004327 (paperback) | ISBN 9781003174158 (ebook)
Subjects: LCSH: Partitions (Mathematics)
Classification: LCC QA165 .C238 2024 (print) | LCC QA165 (ebook) | DDC 512.7/3--dc23/eng/20240201
LC record available at https://lccn.loc.gov/2023053940
LC ebook record available at https://lccn.loc.gov/2023053941

ISBN: 978-1-032-00366-5 (hbk)
ISBN: 978-1-032-00432-7 (pbk)
ISBN: 978-1-003-17415-8 (ebk)

DOI: 10.1201/9781003174158

Typeset in Latin Modern Font
by KnowledgeWorks Global Ltd.

Publisher's note: This book has been prepared from camera-ready copy provided by the authors.

To Ashling Skye:
My reason to think of forests and mountains
on this beautiful planet we call Earth.

Contents

List of Figures

Foreword by Prof. Dr. Henk Koppelaar

I began my studies at the Utrecht University in 1971, graduating in Physics in 1976. In 1981 I started graduate study at the University of Amsterdam, where I obtained a PhD in Computer and Information Sciences and Support Services in 1984. My thesis covered the theory and new ways of data processing of modeling in the social sciences.

I was appointed Professor at the University of Amsterdam in 1986, but moved that year to Delft University of Technology, where I was Professor of Computer Science until 2004. In 1986 I founded the Dutch Artificial Intelligence Association, which merged into the Belgian Dutch Artificial Intelligence Association in 1988.

My early research focussed on automatic recognition of handprinted characters, where they developed a method of syntactic pattern recognition using fuzzy set theory. Their research had focussed on handwritten capitals. Later in the 1980s, my research had focussed the application of chaos theory in the social sciences. In 1994 I published a methodological analysis of these applications.

At various times I was interested in computing verifications of novel mathematical problems and results. So I believe that Campbell's book will provide numerous exercises for graduate students and researchers seeking to perform computer-assisted extensions of known areas such as in the multidimensional identities given in the present book by Campbell.

In many ways, this book is an homage to three luminaries; George E. Andrews in Mathematical Analysis, Rodney J. Baxter in Theoretical Physics and Jonathan M. Borwein in Computer Assisted Science. One could say, this book is a story of the legacy of Ramanujan, following several seemingly distinct journeys into new frontiers stemming from his work.

It is a story of the function introduced by Euler, for the number of partitions of a positive integer. The story starts out with properties of integer partitions, which becomes a story about applying the entire theory of integer partitions into two-dimensional, three-dimensional, and higher-dimensional space generalizations.

The first half of this book sets the scene. It brings forth the main knowledge base enabling the research work presented in the second half of the book.

If we look back to mid 1988, Campbell was not affiliated with any university. In a bid for clarity, he stuffed an envelope with his then current results mailing it to Professor Kurt Mahler at the Australian National University. Mahler wrote back a month before his passing, dismissing most of the handwritten results as unimportant, except for the unfamiliar Visible Point identity given as, for $|y| < 1$, $|xy| < 1$, and $x \neq 1$

$$(1-y)(1-xy^2)^{\frac{1}{2}}(1-xy^3)^{\frac{1}{3}}(1-x^2y^3)^{\frac{1}{3}}(1-xy^4)^{\frac{1}{4}}(1-x^3y^4)^{\frac{1}{4}}\cdots = \left(\frac{1-y}{1-xy}\right)^{\frac{1}{1-x}}$$

with coprime m and n left side terms $x^m y^n$ which he said "looks promising and may be worth following up". Inspired by Mahler's advice, Campbell wrote up a paper (Campbell [87]) on this result and related identities, then sending it to Professor George E. Andrews in 1991, who facilitated publication. Thankfully, Andrews broke with convention and shared with Campbell details of Professor Jonathan Borwein from Canada, who refereed the paper and was keen to hear more details from Campbell. This was the start of a 25-year time span of mathematical correspondence between Borwein and Campbell. Eventually, Borwein and Campbell met in Melbourne shuttling back and forth between La Trobe University and the University of Melbourne. Also of interest, these meetings facilitated an interesting discussion with M. Lawrence Glasser who headed up the Physics department of Clarkson University New York and happened to also be visiting Melbourne back then. Long-distance discussions on Visible Point Vector Identities between Borwein and Campbell occurred in the 1990s and the mid-first decade of the 2000s. In early 2007, Borwein offered to Campbell a teaching position at his university in Canada, at a professor's level. Unfortunately was Campbell at that time unable to accept the offer for personal reasons of the health of his then-wife. Later Borwein relocated to Newcastle, Australia and met up with Campbell again in Canberra Australia around 2013 at the Australian Academy of Science, then a few times before Borwein's unfortunate death in 2016.

Jon Borwein, as a computer science expert, wrote papers on experimental computations for various integrals and summations including so-called Euler Sums and Polylogarithms when he was at times in Australia eventually moving to University of Newcastle in New South Wales. Also most importantly he co-authored with David Bailey from Canada. In his lifetime Borwein was prolific in writing papers and whether Borwein knew it or not, these turned out to be applicable to the Visible Point identities, which in the 1990s Campbell had developed quite a way further than that initial paper. This situation is well covered in several places in the present book, including in chapter 22 where Campbell tabulates all of the known polylogarithm sum values as substitutions into the 2D, 3D, 4D, and 5D Visible Point Vector identities in the "first hyperquadrants" wherein the n-tuple co-ordinates are all positive (or non-negative) integers for the associated lattice sums. The 5D myriad tabulated cases are all instances of

If $|v| < 1, |w| < 1, |x| < 1, |y| < 1, |z| < 1$, and $q+r+s+t+u = 1$, then

$$\prod_{\substack{(a,b,c,d,e)=1 \\ a,b,c,d,e \geq 1}} \left(\frac{1}{1-v^a w^b x^c y^d z^e}\right)^{\frac{1}{a^q b^r c^s d^t e^u}}$$

$$= \exp\left\{\left(\sum_{g=1}^{\infty}\frac{v^g}{g^q}\right)\left(\sum_{h=1}^{\infty}\frac{w^h}{h^r}\right)\left(\sum_{i=1}^{\infty}\frac{x^i}{i^s}\right)\left(\sum_{j=1}^{\infty}\frac{y^j}{j^t}\right)\left(\sum_{k=1}^{\infty}\frac{z^k}{k^u}\right)\right\}.$$

This process in Chapter 22 is enabled by Borwein and his colleagues' work on polylogarithmic and Generalized Euler Sums, many of which were also published in the past decade by Sofo.

In many unpublished exchanges between Borwein and Campbell in the 1990s and 2000s decades, there were expressions like those in chapter 22 such as

For $|z| < 1$, $|yz| < 1$,

$$\prod_{\substack{m,n\geq 1\\ m\leq n;\,(m,n)=1}} \left(\frac{1}{1-y^m z^n}\right)^{\frac{m^1}{n^2}} = (1-yz)^{\frac{y}{1-y}} \exp\left\{\frac{y}{(1-y)^2}\left(Li_2(z)-Li_2(yz)\right)\right\},$$

and the reciprocal of this,

$$\prod_{\substack{m,n\geq 1\\ m\leq n;\,(m,n)=1}} (1-y^m z^n)^{\frac{m^1}{n^2}} = \left(\frac{1}{1-yz}\right)^{\frac{y}{1-y}} \exp\left\{-\frac{y}{(1-y)^2}\left(Li_2(z)-Li_2(yz)\right)\right\}.$$

So, we have digressed into the Visible Point Vector (VPV) identities. This is exclusively researched by Campbell in his 1990s papers and his occasional 21st-century papers. The combinatorics of VPV identities seems largely an open topic at present, but this may change given the emerging computer-assisted calculating technology.

Also, the book features some unusual combinatory identities, an example in Chapter 14 being,

$$\prod_{k=0}^{\infty}\left(\frac{1}{1-tq^k}\right)^{1/k!}$$

$$= 1 + e^q\frac{t}{1!} + \begin{vmatrix} e^q & -1 \\ e^{q^2} & e^q \end{vmatrix}\frac{t^2}{2!} + \begin{vmatrix} e^q & -1 & 0 \\ e^{q^2} & e^q & -2 \\ e^{q^3} & e^{q^2} & e^q \end{vmatrix}\frac{t^3}{3!} + \begin{vmatrix} e^q & -1 & 0 & 0 \\ e^{q^2} & e^q & -2 & 0 \\ e^{q^3} & e^{q^2} & e^q & -3 \\ e^{q^4} & e^{q^3} & e^{q^2} & e^q \end{vmatrix}\frac{t^4}{4!} + etc.$$

$$= 1 + e^q\frac{t}{1!} + \left(e^{q^2} + e^{2q}\right)\frac{t^2}{2!} + \left(2e^{q^3} + 3e^{q^2+q} + e^{3q}\right)\frac{t^3}{3!}$$
$$+ \left(6e^{q^4} + 8e^{q^3+q} + 3e^{2q^2} + 6e^{q^2+2q} + e^{4q}\right)\frac{t^4}{4!} + etc.$$

This result was also mentioned to Borwein by Campbell in the early 2000s, is not in the extant literature, but evidently corresponds someway to the well-known cases of the q-binomial theorem,

If $|t| < 1, |q| < 1$,

$$1 + \sum_{n=1}^{\infty}\frac{t^n}{(1-q)(1-q^2)\cdots(1-q^n)} = \prod_{n=0}^{\infty}\frac{1}{(1-tq^n)}.$$

$$1 + \sum_{n=1}^{\infty}\frac{t^n q^{\frac{1}{2}n(n-1)}}{(1-q)(1-q^2)\cdots(1-q^n)} = \prod_{n=0}^{\infty}(1+tq^n).$$

The book contains many novel extensions of partition-theoretic ideas related to continued fractions, congruence properties of coefficients in various offshoots of the original Ramanujan continued fractions and the Ramanujan partition congruence theories that were looked at by Freeman Dyson in the 1940s and settled in the late 1980s by Andrews and Garvan. Also, in the very different chapter 26, we have the generalization of Ramanujan's trigonometric function, for positive integers k, and n,

$$c_k(n) = \sum_{\substack{(j,k)=1 \\ 0<j<k}} \cos\left(\frac{2\pi nj}{k}\right) = \sum_{\substack{(j,k)=1 \\ 0<j<k}} e^{\frac{2\pi inj}{k}},$$

extended to the function

$$c_k(n_1, n_2, \ldots, n_m) = \sum_{\substack{(j_1, j_2, \ldots, j_m, k)=1 \\ 0 \le j_1 < k \\ \vdots \\ 0 \le j_m < k \\ j_1 + j_2 + \ldots + j_m \neq 0}} \cos\left(\frac{2\pi(j_1 n_1 + j_2 n_2 + \ldots + j_m n_m)}{k}\right)$$

$$= \sum_{\substack{(j_1, j_2, \ldots, j_m, k)=1 \\ 0 \le j_1 < k \\ \vdots \\ 0 \le j_m < k \\ j_1 + j_2 + \ldots + j_m \neq 0}} \exp\left(\frac{2\pi i(j_1 n_1 + j_2 n_2 + \ldots + j_m n_m}{k}\right).$$

Throughout the book, the extent of innovation and incidence of previously unpublished research is most pleasingly unusual.

It has to be mentioned also, that Campbell, in 1979, as a gifted student, was brought from La Trobe University in Melbourne to visit Professor George Szekeres at University of New South Wales; and during that month shared an office with Professor George E. Andrews, who was also visiting UNSW. It was Andrews who many years later in the 1990s contacted Professor Rodney J. Baxter at The Australian University and arranged for Campbell to complete a PhD under Baxter's Supervision. It was there that Campbell was exposed to the genius of the Statistical Mechanics exact solution of the Hard Hexagon Model and its relationship to the theory of partitions, hence providing the crossover from the Andrews Mathematical Analysis and the Baxter Statistical Mechanics as described in Chapter 11 of the present book. As a physicist myself with interests in mathematical analysis and also interest in the science of computation I am appreciative of the seamless interrelation between the theories underlying Campbell's Chapter 11, and the continued relevance of the Rogers-Ramanujan identities in Statistical Mechanics, even 44 years after Baxter found the initial connections.

However, an important new development given only in this book to my knowledge, is the generalization of integer partition theorems as applied to "radial from origin" straight lines in 2D, 3D, etc. and generally n-D cartesian integer lattice sums. We can illustrate a simple 2-dimensional example as follows, lifted from Chapter 12.

Example using 2D Vector Partitions whose parts are on two straight lines.

Consider the two lines with equations $y = z/2$ and $y = z/3$. The lattice point vectors along these lines in the first quadrant may be listed as

$$\begin{aligned} S_1 &= \{\langle 1,2\rangle, \langle 2,4\rangle, \langle 3,6\rangle, \langle 4,8\rangle, \langle 5,10\rangle, \ldots\}; \\ S_2 &= \{\langle 1,3\rangle, \langle 2,6\rangle, \langle 3,9\rangle, \langle 4,12\rangle, \langle 5,15\rangle, \ldots\}. \end{aligned}$$

Following the standard rationale outlined in Chapter 12 we see that the generating function for 2D vector partitions into parts contained in S_1 and S_2 is

$$\frac{1}{((1-yz^2)(1-y^2z^4)(1-y^3z^6)\cdots)((1-yz^3)(1-y^2z^6)(1-y^3z^9)\cdots)}$$

$$= \left(1+p(1)y^1z^2+p(2)y^2z^4+p(3)y^3z^6+\ldots\right)\left(1+p(1)y^1z^3+p(2)y^2z^6+p(3)y^3z^9+\ldots\right)$$

$$\begin{aligned} = \; & 1+p(1)yz^2+p(1)yz^3 \\ & +p(2)y^2z^4+p(1)^2y^2z^5+p(2)y^2z^6 \\ & +p(3)y^3z^6+p(1)p(2)y^3z^7+p(1)p(2)y^3z^8+p(3)y^3z^9 \\ & +\mathit{etc.} \end{aligned}$$

$$\begin{aligned} = \; & 1+yz^2+yz^3+2y^2z^4+y^2z^5+y^2(3y+2)z^6 \\ & +2y^3z^7+y^3(5y+2)z^8+3y^3(y+1)z^9+y^4(7y+4)z^{10} \\ & +y^4(5y+3)z^{11}+y^4(11y^2+6y+5)z^{12}+y^5(7y+6)z^{13} \\ & +y^5(14y^2+10y+5)z^{14}+y^5(11y^2+9y+7)z^{15} \\ & +\mathit{etc.} \end{aligned}$$

The coefficients here plot onto the grid

⋮	⋮	⋮	⋮	⋮	⋮	⋮	⋮	⋮	⋮
15						7	9	11	⋯
14						5	10	14	⋯
13						6	7		⋯
12					5	6	11		⋯
11					3	5			⋯
10					4	7			⋯
9				3	3				⋯
8				2	5				⋯
7				2					⋯
6			2	3					⋯
5			1						⋯
4			2						⋯
3		1							⋯
2		1							⋯
1									⋯
0	1								⋯
z/y	0	1	2	3	4	5	6	7	⋯

Example interpretations reading from this graph:

1) The number of partitions of vector $\langle 7,15\rangle$ into unrestricted number of parts from S_1 and S_2 is 11.

2) The number of partitions of vector $\langle 5,10\rangle$ into unrestricted number of parts from S_1 and S_2 is 7.

3) The number of partitions of vector $\langle 4,9\rangle$ into unrestricted number of parts from S_1 and S_2 is 3.

4) The number of partitions of vector $\langle 4,7\rangle$ into unrestricted number of parts from S_1 and S_2 is 0.

This concept is articulated for various extensions in 2D space, 3D space, etc. and for any finite number of radial from origin lines passing through those lattices. The theory underpinning the above result will no doubt be applicable to other types of partitions of vectors along other lines through the lattice spaces, and this will lead to new areas of vector partition research. Underpinning the many chapters on vector partitions and the early foundational chapters, is the indefatigable continuance over 60-plus years of ground-breaking research by George E. Andrews and his many colleagues.

I have known Geoffrey Campbell for 12 years and had many mathematical interactions with him when he managed a Number Theory Group on LinkedIn, guiding the membership from a few hundred up to 40,000 over a period of just a few years. He has a gift for original presentations of problems and general interest topics which no doubt has contributed to his being able to write the present volume.

As a Computer Science computation professional, I see the merit of Campbell's work in this book. Also, I share the opinion of Campbell's friend and mentor the late Jonathan Borwein, whom I think would have applied the new theories of vector partitions to many more aspects of research. In addition, I am sure the Visible Point Vector formulas will find a place in a variety of mathematical analysis applications, as well as in physics, computer science teaching and emerging higher dimensional analytics technology. Exciting and thought provoking at the least, at the most it can usher in new frontiers of research.

Henk Koppelaar
Delft University of Technology
January 2024

Preface

0.1 THE AUTHOR'S VISION AND MOTIVATIONS

The hope for this book is to contribute significantly in areas of graduate mathematics and physics education, and to provide a new framework for future research within the included topics, especially with respect to Visible Lattice Point identities and the associated theory.

The two main influences in my motivation, were:

1) The 1976 book by George E. Andrews entitled "The Theory of Partitions", [12] which is now considered a classical reference in the golden decades of partition research after the late 1960s.

2) The book by Rodney J. Baxter entitled "Exactly Solved Models in Statistical Mechanics", [52] which brings together many models from theoretical physics culminating with the famous solution of the Hard Hexagon Model that derived the Rogers-Ramanujan identities.

It is the intention of this book to formulate a theory of Vector Partitions. For a Conceptual Model placing Vector Partitions in a context of other related theories, see Figure 0.1. Some items in this conceptual model are awaiting development, such as *Dirichlet series analogues of q-series*, which appear to underlie contemporary ideas related to encoding *Quasicrystals.* (see Campbell [92], or Baake et al. [39], [40], [37]) The view is that *Vector Partitions* theory can be developed as an extension to the existing theory of *Integer Partitions.* Much of the vast current literature on Integer Partitions is applicable to Vector Partitions whose parts are on a line in n-dimensional space, as is explained in Chapter 12 of the present book. However, the material in this book may also contribute to research on Riemann Zeta functions, Dirichlet Series, Polylogarithms, Harmonic Numbers, Polygamma Functions, Multivariate Continuous Differentiable functions, Basic Hypergeometric Series, Lattice Sums in Physics and Chemistry, and Number Theory. It can provide material for exercises in computational verification of multivariable equations, and it may inspire researchers to new areas of suggested research.

The book will bring properties of higher dimensional space into practical focus via identities and theorems on vector partitions. We take steps to lay a foundation for some new branch of analysis that may inform many emerging technology applications, and provide new tools to examine, comprehend and maybe prove, longstanding Open Questions from a new multidimensional standpoint.

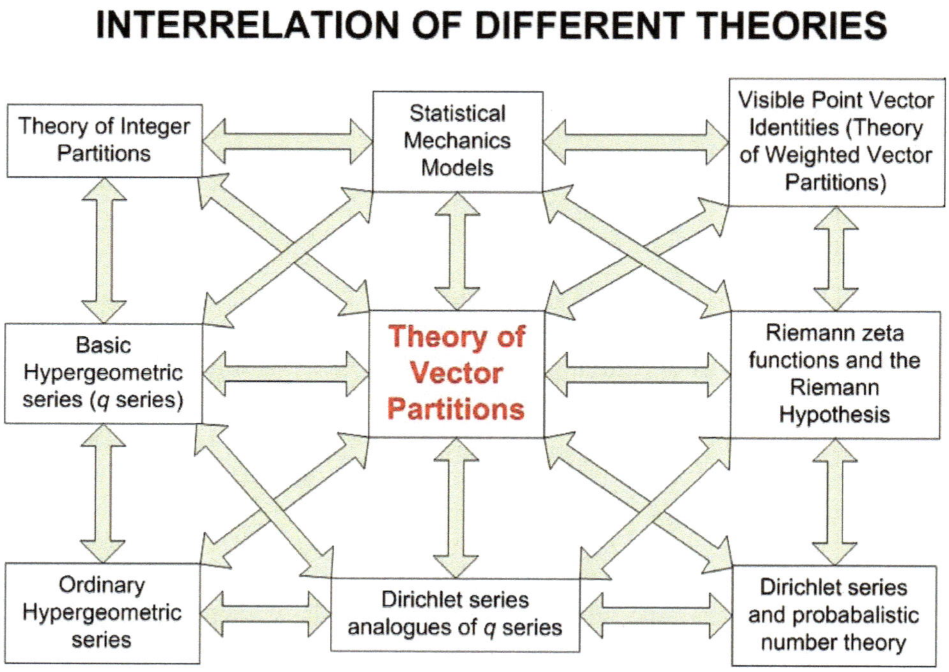

Figure 0.1 Conceptual Map Interrelating Theories

Some distinct features of intended value for the reader are:

1) To collect useful ideas underpinning Integer and Vector Partitions, from a variety of literature over the past 100 years, emphasizing the researches in the past 20 years, so that it will be readable to anyone with a basic level of understanding of undergraduate mathematics.

2) The book is leveled at number theorists, mathematical analysts, as well as for applied mathematicians, from a wide range of scientific fields, including theoretical physics, statistical mechanics, and lattice structures in chemistry.

3) To provide a step-by-step guide through the known literature on Integer and Vector Partitions, and a focus on the not so well-known Visible Point Vector identities.

4) To provide a variety of practical examples as well as sets of exercises suitable for students and researchers.

5) To encourage and include computational experimentation leading to proofs, such as occurred with modern-day Euler sums and Mordell-Tornheim-Witten (MTW) Zeta Functions which are now seen as applicable in areas of Vector Partitions.

The book can be used by Masters and/or PhD level graduate students, and therefore by researchers in any field of pure and applied mathematics, theoretical physics, and physical sciences. The basic contents of the book can be taught to students. Researchers can use it as a handbook for a starting level in investigations. As a graduate text:

1) This book may serve as a useful textbook for a course on *Theory of Integer Partitions.*

2) It can add value for courses in *Number Theory, Theory of Vector Partitions and the Visible Points*, *Statistical Mechanics Exactly Solved Models*, and *Lattice Sums in Chemistry.*

3) The book may also contribute value for courses in *Combinatorics*, *Structural Properties of Polylogarithms*, *Basic Hypergeometric Series*, and *Random Walks in 2D and 3D scenarios.*

The existing books on partitions listed here provide basic background material on the topic of Integer Partitions or in the case of Statistical Mechanics, on the novel way partitions have arisen in the calculation of free energy. The books by Andrews are the classical accounts that are cited by all the researchers. However, years have passed since original publications, so the researcher must account for literature since then of between 16 years and up to 40 years ago in order to "cover the bases" for asserting new results. The proposed book may alleviate much of the research gap, including a large amount of relevant updates. Similarly, the work in Statistical Mechanics since Baxter's *Exactly Solved Models* book in 1980 has vastly diversified since, the epitome of this seen in the 2020 Australian National University *Baxter 80th birthday conference on Integrable Functions.* This conference featured research on topics evolved from Baxter's work such as:

1) Ising correlation C(M,N) for $v = -k$.

2) Spin chain overlaps and the AdS/CFT correspondence.

3) Quantum separation of variables.

4) Exact solution of the spin$-\frac{1}{2}$ XXX chain with off-diagonal boundary fields.

5) Generalized hydrodynamics in box-ball system.

6) New developments for integrability and the star-triangle relations.

7) Baxter's Q operator for open quantum spin chains.

8) A loop model for the superintegrable chiral Potts model.

9) Painlevé transcendents in quantum mechanics.

10) Wronskians of Hermite polynomials, anharmonic oscillators, and Painlevé IV.

0.2 PLANS FOR THE BOOK AND ITS ACADEMIC CONTEXT

This book should be suitable for graduate students in physics, applied mathematics, number theory and computational mathematics. It should take the reader up to the research level, presenting new results alongside known classical results from integer partitions and areas of vector and multipartite partition theory. However, as with most topics in mathematics, we have tried to include the main features, and unfortunately, it is not possible to include everything that may be of value to the future researchers.

The theory of Integer Partitions has been well developed and applied in the latter half of the 20th century, and the first decades of the 21st century, with signs that it will be developed further as the 21st century progresses. Partitions are a research area in their own right, and the classical identities and theorems going back to Euler and Gauss are ample testimony to this. However, the coming of Ramanujan, Hardy and MacMahon early 20th century, set a firm path into many diverse areas of fruitful research going into Statistical Mechanics, String Theory and q-series or Basic Hypergeometric Series. In terms of the theory of special functions in mathematical physics and chemistry, partitions have arisen in a broad-ranging context that has developed in different and applicable ways. Today this includes polylogarithms, Mordell-Wittam-Tornheim functions, Random Walks, and physics topics in Integrable Functions.

Andrews' inspiring papers from the late 1960s and 1970s culminated in his *The Theory of Partitions.* This then led on to his works on Mock Theta Functions and Rogers-Ramanujan type summations, q-series, flowering into a golden age of several decades of discovery. Amongst all this came Baxter in 1980 with a Statistical Mechanics derivation of the Rogers-Ramanujan partition identities and similar results. This has evolved into applications in theoretical physics areas such as Integrable Function theory, the finding of many Exactly Solvable Models, the wide occurrences of Yang-Baxter star-triangle relationships, and far-reaching branches of physics. These are now a seemingly self-contained branch of Statistical Mechanics research.

So this book includes results from the theory of integer partitions, generalized circa 2020 into two, three, and higher dimensions involving vectors, in the context of known theories relating the established integer partition tools:

- Generating functions, Ferrers graphs, and Durfee squares
- Basic Hypergeometric Series or q-series
- Circle Method and Saddle Point Method of asymptotics
- Polylogarithms and computational research related results
- Partition theory in Statistical Mechanics and other Theoretical Physics
- Extension of known Vector Partitions generating function identities

The reader/student will be led far enough into the above places to start research of their own, and the book has a relevant bibliography of papers and books, both historical and contemporary, intended as a *kick-off* place into new research.

An additional hoped-for value of this book is that in the 1990s, the author introduced Visible Point Vector (VPV) partitions and their generating functions; a departure from known works on higher-dimensional identities. However, these new equations back then were not given a thorough detailed derivation or context, for various reasons. The new VPV identities were never main stream and understandably were not studied by others other than the present author. Back in the 1990s, it was possible but awkward to calculate and verify the results. Despite that difficulty, the author had enthusiastic conversations with Jonathan Borwein on perspectives to publish the new VPV equations. Borwein around the year 2000 at the computing forefront, realized that the Visible Point Vector and similar identities would soon be in the realm of easy to verify, having been a bit out-of-range up until the 21st century. So it is, that today, these previously *under the radar* VPV identities can be verified term-by-term online at WolframAlpha.com, or on Mathematica or Maple software quite easily.

It also has emerged over the past two decades, that the VPV identities host the polylogarithm function occurring often and naturally along with Mordell-Tornheim-Witten (MTW) functions; both areas where new summations have been published related to Riemann Zeta functions and digamma function particular values and integrals. Borwein and his colleagues experimentally found calculated MTW value conjectures that verified to 100s of decimal places first, before then seeking and finding the actual proofs. Many of these can be substituted into VPV equations as particular cases of theorems on weighted vector partitions.

In addition to the work from Lattice Sums in Chemical Molecular Structures, we can include the very important work by Borwein on Polylogarithms and experimental equalities found by computer calculations in our chapters on 2D, 3D, 4D etc. Vector Partitions. This work extends papers by Borwein et al., with the 21st-century version of "Euler Sums" which are derived from so-named Eulerian Log-Gamma Integrals and Mordell-Tornheim-Witten (MTW) Zeta Functions.

The results on MTW Sums along with about thirty papers over the past ten years by Sofo et al., are also applicable for particular VPV identity cases. These modern-day Euler Sums featuring Harmonic Numbers, sum to exact linear combinations of Riemann zeta function values applicable to the VPV infinite products and weighted vector partition results in this book.

Some of the recent VPV identities seem to cast some new interpretations on well-known conjectures like the *Riemann Hypothesis* and the *abc Conjecture.* At this early stage of research, it seems there may well be new approaches to open questions via work in the book. Related snapshot images from the current draft of the book are given here.

A sample snapshot image of new results using cases of Euler Sums recent research.

If $|y| < 1$,

$$\prod_{\substack{a,b\geq 1\\(a,b)=1}} \left(1 - \frac{y^b}{2^a}\right)^{-\frac{1}{a}} = 2^{\frac{y}{1-y}}, \tag{0.2.1}$$

$$\prod_{\substack{a,b\geq 1\\(a,b)=1}} \left(1-\frac{y^b}{2^a}\right)^{-\frac{b}{a^2}} = \exp\left[\left(\frac{\pi^2}{12}-\frac{1}{2}(\ln 2)^2\right)\frac{y}{(1-y)^2}\right], \tag{0.2.2}$$

$$\prod_{\substack{a,b\geq 1\\(a,b)=1}} \left(1-\frac{y^b}{2^a}\right)^{-\frac{b^2}{a^3}} = \exp\left[\left(\frac{1}{6}(\ln 2)^3-\frac{\pi^2}{12}(\ln 2)^2-\frac{7}{8}\zeta(3)\right)\frac{y(1+y)}{(1-y)^3}\right], \tag{0.2.3}$$

$$\prod_{\substack{a,b\geq 1\\(a,b)=1}} \left(1-\frac{y^b}{2^a}\right)^{-\frac{b^3}{a^4}} = \exp\left[\left(\frac{\pi^4}{360}-\frac{1}{24}(\ln 2)^4-\frac{\pi^2}{24}(\ln 2)^4-\frac{1}{2}\zeta(\overline{3},\overline{1})\right)\frac{y(1+4y+y^2)}{(1-y)^4}\right], \tag{0.2.4}$$

where the alternating double sum $\zeta(\overline{3},\overline{1}) = \sum_{m>n>0}(-1)^{m+n}m^{-3}n^{-1}$, is given from

$$Li_4(1/2) = \frac{\pi^4}{360}-\frac{1}{24}(\ln 2)^4-\frac{\pi^2}{24}(\ln 2)^4-\frac{1}{2}\zeta(\overline{3},\overline{1}). \tag{0.2.5}$$

A snapshot with *new to literature* results including (0.2.7) related to the *abc Conjecture.*

For each of $|v|,|w|,|x|,|y|,|z|<1$,

$$\prod_{\substack{a,b\geq 1\\(a,b)=1}} \left(1-y^a z^b\right)^{\frac{1}{a}} = (1-y)^{\frac{z}{1-z}}, \tag{0.2.6}$$

$$\prod_{\substack{a,b,c\geq 1\\(a,b,c)=1}} \left(1-x^a y^b z^c\right)^{\frac{c}{ab}} = ((1-x)(1-y))^{\frac{z}{(1-z)^2}}, \tag{0.2.7}$$

$$\prod_{\substack{a,b,c,d\geq 1\\(a,b,c,d)=1}} \left(1-w^a x^b y^c z^d\right)^{\frac{d^2}{abc}} = ((1-w)(1-x)(1-y))^{\frac{z(1+z)}{(1-z)^3}}, \tag{0.2.8}$$

$$\prod_{\substack{a,b,c,d,e\geq 1\\(a,b,c,d,e)=1}} \left(1-v^a w^b x^c y^d z^e\right)^{\frac{e^3}{abcd}} = ((1-v)(1-w)(1-x)(1-y))^{\frac{z(1+4z+z^2)}{(1-z)^4}}. \tag{0.2.9}$$

The identities (0.2.7) to (0.2.9) are new to the literature. The index functions on the right sides will be recognized by many as the well-known values of the non-positive integer polylogarithms, $Li_0(z), Li_{-1}(z), Li_{-2}(z), Li_{-3}(z)$. The partition-theoretic asymptotics are applicable to special cases of these identities and may give us insights into the *abc Conjecture*, especially suggestive of bounds on the behaviors of coefficients in the product expansions.

Above all, the intention of the book is to bring new inspiration to others who study mathematics and related areas. It is hoped that some new ideas will be launched to add value and insight into many of the classical and new theories surrounding partitions. The book is an appreciation of the many gifted authors of research into partitions over the past century and before, in the hope that more may come of this for future generations.

GEOFFREY B. CAMPBELL
The Australian National University
September 2023

Author Biographic Note

Geoffrey Campbell completed his PhD at the Australian National University in 1998 under the esteemed physicist Professor Rodney Baxter. His affiliation with the Australian National University Mathematical Sciences Institute has continued for over 30 years. Within that time frame, Geoffrey also served eight years as an Honorary Research Fellow at LaTrobe University Mathematics and Statistics Department in Melbourne. Currently, he writes ongoing articles for the Australian Mathematical Society Gazette. Within the international scope, Geoffrey currently serves as a PhD external committee member for a mathematics graduate student at Washington State University in America.

Geoffrey has built a career within Australian Commonwealth and State government departments, including as an Advisor at the Department of Prime Minister and Cabinet and as an Analyst Researcher for a Royal Commission. Geoffrey specializes in complex data, and machine learning including data analytics. He is also a published poet in Australian anthologies and literary magazines.

Contributor

Henk Koppelaar

Delft University
Mekelweg 5, 2628 CD Delft, Netherlands

I

Background and History

CHAPTER 1

Historical Background – The Range of Literature

1.1 DISCOVERING THE RAMANUJAN MIRACLE

IT WOULD BE DIFFICULT NOT TO BE IMPRESSED by the exceptional and seamless mathematical ideas of George E. Andrews, Rodney J. Baxter, and Jonathan M. Borwein, each of whom gave, in their own way, a profound and durable push into the 21st century from known works on partitions of integers.

When starting university, I first learned of Ramanujan and his life. I recall my first delving into the problems he posed and sensing that Ramanujan was able to do mathematics like no one else. My fascination with Ramanujan grew through the years in differing ways, as I realized the longevity of his works and the subsequent fullness of application and diversity of thought.

The theory of integer partitions goes back to Euler, added to by Gauss, Cauchy, and others. We cannot ignore the foundation they built, and we rely upon. Rogers, Ramanujan, MacMahon, Hardy, and Littlewood, ushered in the 20th century, while the mid and latter part of the 20th century saw developments in q-series and q-analogues.

In 1980 we experienced the entrance of *exactly solved models* of statistical mechanics by Baxter, connecting partitions famously to physics by finding the Rogers-Ramanujan identities solution in the hard hexagon model. Andrews, with his classical book in the late 1970s, progressed with his 1980s work on Special Functions and q-series with Askey. In the late 1980s, the Borweins' *Pi and the AGM* [66] kicked off what became computer-assisted computations in the late 1990s and into the 2000s leading into experimental computed conjectures. All this set the scene for celebrated research associated with partitions for the 21st century, especially for congruences and plane partitions.

The rise of integer partition theory in the 20th century was based on works of Ramanujan. In this book, we cover material from the Ramanujan ideas on partitions associated with continued fractions, congruence properties, the circle method and asymptotics for integer partitions; also many q-series generating function identities. The Ramanujan trigonometric functions also generalize to new Visible Point Vector

DOI: 10.1201/9781003174158-1

(VPV) identities. Ramanujan is truly the basis of the sophisticated 20th- and 21st-century developments underpinning the current book.

Figure 1.1 Srinivasa Ramanujan in India, circa 1913.

1.2 THREE NOTABLE GEORGE E. ANDREWS TEXTS ON PARTITIONS

1.2.1 *The Theory of Partitions* (1976) by G.E. Andrews

Although the history of partitions of numbers goes back hundreds of years, we are drawn to this now classical book (see Andrews [12]) as being the first on the topic to bring together all major aspects of the theory, up to the end of the 1970s. Andrews' book covers a wide range of territory from the elementary up to the most analytically sophisticated approaches to partitions of numbers.

1.2.2 *Integer Partitions* (2004) by G.E. Andrews with K. Eriksson

As a natural way forward from the 1976 classic, the book by Andrews and Eriksson [29] entitled *Integer Partitions* published in 2004 gives many simplified updates of results. It has chapters about Lecture Hall partitions, Growing Ferrars Boards, Domino

Figure 1.2 Professor George E. Andrews circa 1979.

Tilings, and The Arctic Circle Theorem. This book shifts the historical emphasis of proof by generating functions of the past 100 years, into the proof by *bijection*, or *near bijection* in the style of Franklin's 19th-century proof of Euler's Pentagonal Number Theorem. The new emphasis came after Bressoud's brilliant bijective proof of the Rogers-Ramanujan identities in the 1980s, followed by simple proofs of previously difficult partition theorems. (See Bressoud [72, 74], and Chapman [107].) While this is a welcome development in the theory of Integer Partitions, maybe the method of bijection may need innovation to apply successfully in Vector Partition theory, as given in this book.

1.2.3 *G.E. Andrews 80 Years of Combinatory Analysis* (2021)

This book collects 37 articles related to research by George Andrews, plus two original papers and three personal contribution articles. (See Alladi et al. [9].) So, to get a *near complete* picture of the combinatory analysis surrounding integer partitions, one can equip ones' self for research by mastering the two above-mentioned books along with the 2021 tribute book. It is broadly accepted that Andrews is the foremost contributor to the modern-day theory of partitions. In our present book we try to summarize significant features of the theory as a lead-up to the considerations of research in the areas of Vector Partitions. The three books by Andrews et al. briefly mentioned here provide a vantage place for the 21st-century frontier into a New World of Partitions embracing Physics and Higher Dimensional mathematics.

1.3 BASIC HYPERGEOMETRIC SERIES

Because most of the partition theory identities arise from q-series, it is important that we have a solid resource book to cover off *Basic Hypergeometric Series.* Incidentally that is the title of the book by Gasper and Rahman [134] published in 1990, and this contains numerous generating function hypergeometric q-series. The results presented, again relevant to integer partitions, and also to vector partitions, cover historical results right up to the research level as it was in 1990 (see [134]). Of course since then, many new q-series identities have arisen, but Gasper and Rahman's book is now considered a primary source for basic hypergeometric series.

1.4 LATTICE SUMS IN CHEMISTRY MOLECULAR STRUCTURES

A further important resource is the book by Borwein et al., entitled *Lattice Sums Then and Now* (see [69]). This book, published in 2013, covers many lattice point summations that arise in physical situations notably molecular structures that occur in researches associated with chemistry. Although this book at first appears to not be about partitions, our research on vector partitions leads us into lattice structures in Cartesian space. These arise naturally in proofs and encode many of the multi-variable identities that underpin Vector Partitions generally as defined by Andrews above, and especially for weighted vector partitions, such as those arising from multi-sums of Visible Point Vectors (VPVs).

1.5 POLYLOGARITHMS AND COMPUTATIONAL RESEARCH-RELATED RESULTS

An important reference for Polylogarithms, which frequently occur in our vector partition summations, is the book by L. Lewin [185]. Lewin gives a complete historical account of relevant functional equations and particular values known up to 1991. He also introduces the concept of polylogarithm ladders and there are numerous examples of these that may find a place in our researches on vector partitions. Polylogarithms arise as features in the 2D, 3D, and nD VPV identities. In the past two decades many cases of Polylogarithmic sums involving Harmonic Numbers and variants on these have been found in terms of Riemann zeta functions and polygamma functions. These are in the present book applied to give nice examples of vector partition identities.

In addition to the work from the 2013 text *Lattice Sums Then and Now*, we bring in important work by Borwein and his collaborators on Polylogarithms and experimental equalities found by computer calculations often to several 100 significant figures. This work extends papers by Borwein et al., [68] with the 21st-century version of "Euler Sums" which are derived from *Eulerian Log-Gamma Integrals* and *Mordell-Tornheim-Witten (MTW) Zeta Functions*. The results on *MTW Sums* along with the many papers over the past ten years by Sofo (see for example: [238, 239, 240, 279]) on Euler Sums, Harmonic Numbers and Riemann zeta function values summed to polylogarithm values apply to finding new results on the VPV infinite products and weighted vector partitions as given in the current book.

1.6 PARTITION THEORY IN STATISTICAL MECHANICS AND THEORETICAL PHYSICS

Finally, there is a wealth of literature on *Exactly solved models in statistical mechanics*. My primary reference for this material is the classical book by Rodney Baxter first published in 1982 but reprinted in 2007. See Baxter [52] and Wikipedia [300], the latter having an excellent partition-theoretic account of the Hard Hexagon Model.

Theoretical Physics interprets partitions in a, seemingly at first, different manner to the earlier-mentioned references. A partition is here considered as a sum over particle states. There is no doubt that the partition equations arising in the calculation of free energy across differing states of molecules occurring in different physical models is in fact a relevant and yet seemingly distinct area of research that relates to partitions at least in two-dimensional space.

In early 2020 at the Australian National University Mathematical Sciences Institute, an international conference was held featuring the latest developments in integrable functions and the Yang-Baxter equation exhibiting Star-Triangle relationships. This is a prolific growth area that has developed and spread across many related research categories in Theoretical Physics. Central to the significance of this fascinating branch of partition theory is the completely different kind of proof found by Baxter in solving the so-called hard hexagon model.

It is well known that the solution found by Baxter gave a physical application for the celebrated Rogers-Ramanujan identities previously thought to be part of the most non-applicable kinds of research. The remarkable part of this discovery is that the hard hexagon model represents the physical adsorption of helium into graphite. Since this discovery in the early 1980s, Baxter [50] and Andrews [21] followed up with several interesting papers that solved statistical mechanics models for the Gordon-Gollnitz identities, which are of a type similar to the Rogers-Ramanujan identities. From that point on, physicists researching statistical mechanics became interested in finding two- and three-dimensional molecular models and indeed other kinds of models that satisfied the Yang-Baxter equation.

In a pleasing way, the *Integrable Functions* conference papers presented in early 2020 at The Australian National University included recent work upon *string theory* and other branches of physics previously not mindful by mathematicians as a theory of partitions. We mention the theoretical physics approach here as it seems to progress along separate independent paths of research compared with the *Lattice Sums* in Chemistry (see Borwein et al., [69]) and even the vector partition results presented in this book. It does seem that our work on *weighted vector partitions* associated with VPVs may be relevant to a model similar to those in Baxter [52] or Berkovich, McCoy, and Schilling [60, 61], for example.

The partition function concept in Baxter's statistical mechanics book is as follows. We find a way to calculate thermodynamic functions such as $M(H,T)$ from the microscopic forces between the components of the system. The method for this was given by John Willard Gibbs in 1902 (see [136]). Consider a system with states s and Hamiltonian $E(s)$. Form the partition function

$$Z = \sum_{s} \exp\left(\frac{-E(s)}{kT}\right), \tag{1.6.1}$$

where k is Boltzmann's constant, and the summation is over all allowed states s of the system. Then the free energy F is given by

$$F = -kTlog(Z). \tag{1.6.2}$$

Also, the probability of the system being in a state s is

$$Z^{-1} \exp\left(\frac{-E(s)}{kT}\right), \tag{1.6.3}$$

so if X is some observable property of the system, such as its total energy or magnetization, with value $X(s)$ for state s, then its observed average thermodynamic value is

$$\langle X \rangle = Z^{-1} \sum_{s} X(s) \exp\left(\frac{-E(s)}{kT}\right). \tag{1.6.4}$$

In particular, the internal energy is

$$U = \langle E \rangle \tag{1.6.5}$$

$$= Z^{-1} \sum_{s} E(s) \exp\left(\frac{-E(s)}{kT}\right), \tag{1.6.6}$$

and by using the above definitions we can verify that

$$U = kT^2 \frac{\delta}{\delta T} log(Z) \tag{1.6.7}$$

$$= -T^2 \frac{\delta}{\delta T}(F/T), \tag{1.6.8}$$

in agreement with standard thermodynamics. The basic problem of equilibrium statistical mechanics is therefore to calculate the sum-over-states in the partition (for continuum systems this sum becomes an integral, for quantum mechanical ones a trace). This will give Z and F as functions of T and of any variables that occur in $E(s)$, such as a magnetic field. The thermodynamic properties can then be obtained by differentiation. Unfortunately, for any realistic interacting system of macroscopic size, including the examples mentioned above, the evaluation of Z is hopelessly difficult. One is therefore forced to do one or both of the following:

A. Replace the real system by some simple idealization of it: this idealization is known as a model. Mathematically it consists of specifying the states s and the energy Hamiltonian function $E(s)$.

B. Make some approximation to evaluate the partition sum-over-states.

The so-named *Spherical Model* in statistical mechanics introduces a constraint coupling all lattice point molecular components equally, no matter how far apart they are. Thus it is *unphysical* in that it involves infinite range interactions. It may also lend itself to VPV (visible from the origin) concepts present in our research in the present volume. However, Stanley (1968) [282] has shown that it can be regarded as a limiting case of a system with only nearest neighbor interactions. The model is discussed in Chapter 5 of Baxter's book. It is interesting in that its exponents are not classical in three dimensions.

Think About It...

Ramanujan, starving in 1910, sought work at the government revenue department, where he met an official named Ramaswamy Aiyer. Ramanujan did not have a resume to show; only his notebooks – the results of his mathematical work.

Ramanujan was in luck, as Ramaswamy Aiyer was a mathematician. He had in 1907 founded the Indian Mathematical Society, and his jaw dropped when he saw Ramanujan's work.

"I was struck by the extraordinary mathematical results contained in it. I had no mind to smother his genius by an appointment in the lowest rungs of the revenue department".

V. RAMASWAMY AIYER
Indian Mathematician, 1871–1936

GLOSSARY

Bijective Proof: In combinatorics, bijective proof is a proof technique for proving that two sets have equally many elements, or that the sets in two combinatorial classes have equal size.

Boltzmann's Constant: The Boltzmann constant is defined to be $1.380649 \times 10^{-23} m^2 kg^{-2} K^{-1}$. It is the proportionality factor that relates the average relative kinetic energy of particles in a gas with the thermodynamic temperature of the gas.

Euler Sum: The double Euler sum is

$$\zeta(s,t) = \sum_{n>m>0} \frac{1}{n^s m^t} = \sum_{n=1}^{\infty} \frac{1}{n^s} \sum_{m=1}^{n-1} \frac{1}{m^t}.$$

Euler Sum of Depth N: The two-place function $\zeta(s,t)$ has a natural generalization

$$\zeta(s_1, s_2, s_3, \ldots, s_N) = \sum_{n_1>n_2>n_3>\cdots>n_N} \frac{1}{n_1^{s_1} n_2^{s_2} n_3^{s_3} \cdots n_N^{s_N}}, \tag{1.6.9}$$

which defines an Euler sum of depth N and weight $\sum_i s_i$. Some authors reverse the order of the variables.

Exactly Solved Model: In statistical mechanics this is a lattice point model exhibiting spin charges at each point subject to certain positioning rules. The feature of an exactly solvable model is that there are exact formulas to calculate the free energy and the transition phase temperature. Examples of exactly solved models are Six-Vertex Model, Eight-Vertex Model, and Hard Hexagon Model.

Generating Function: Suppose there is a sequence of numbers or of functions $a_1, a_2, a_3, \ldots$. The generating function for that sequence is the power series $1 + a_1 z + a_2 z^2 + a_3 z^3 + \ldots$. Often the power series is a Taylor series of some closed form function, say $f(z)$.

Integer Partition: An integer partition of a positive integer n is a sum of non-increasing positive integers, that sum equal to n. The function $p(n)$ is the number of possible ways this sum can occur for n.

Integrable Function: Yang-Baxter integrable models in condensed matter have long been the traditional setting in which a range of exactly solved models have been realized in experiments. These include the early experiments related to the two-dimensional Ising model. Other integrable (exactly solved) models in statistical mechanics are the hard hexagon model, the Heisenberg spin chain, the quantum Ising chain with transverse and longitudinal fields, and a spin ladder model.

Polylogarithm: The polylogarithm function is defined by a power series in z, which is also a Dirichlet series in s:

$$Li_s(z) = \sum_{k=1}^{\infty} \frac{z^k}{k^s} = z + \frac{z^2}{2^s} + \frac{z^3}{3^s} + \ldots$$

This definition is valid for arbitrary complex order s and for all complex arguments z with $|z| < 1$; it can be extended to $|z|$ greater than unity by the process of analytic continuation. Note $Li_s(1)$ is the Riemann zeta function $\zeta(s)$.

Visible Point Vectors (VPVs): A VPV is a vector $\langle a_1, a_2, a_3, \ldots, a_n \rangle$ with each a_i a positive integer in n-dimensional Cartesian space where the greatest common divisor of the co-ordinates $gcd(a_1, a_2, a_3, \ldots)$ is equal to unity. Conceptually this can be thought of as a vector arrow starting at the origin $(0, 0, 0, \ldots, 0)$ with destination point $(a_1, a_2, a_3, \ldots, a_n)$. Since the co-ordinates are n-wise coprime, the arrow goes unimpeded from the origin to the point, so the point is said to be *visible from the origin*. It is well known that the probability that any n-space point is visible-from-origin, is $1/\zeta(n)$.

Yang-Baxter Equation: In physics, the Yang-Baxter equation (or star-triangle relation) is a consistency equation which was first introduced in the field of statistical mechanics. It depends on the idea that in some scattering situations, particles may preserve their momentum while changing their quantum internal states. The Yang-Baxter equation has long been recognized as the masterkey to integrability, providing the basis for exactly solved models which capture the fundamental physics of a number of realistic classical and quantum systems.

FURTHER READING

ALLADI, K., BERNDT, B.C., PAULE, P., SELLARS, J. and YEE, A.J. (Editors). *George E. Andrews 80 Years of Combinatory Analysis.* Springer Nature Birkhäuser, Trends in Mathematics, Switzerland, 2021.

ANDREWS, G.E. *The Theory of Partitions*, volume 2 of *Encyclopedia of Mathematics and its Applications.* Addison-Wesley Publishing Company, Advanced Book Program, Reading, Massachusetts, USA, 1976.

ANDREWS, G.E. The hard-hexagon model and Rogers-Ramanujan type identities. *Proc Nat. Acad. Sci. USA*, 78:5290–5292, 1981.

ANDREWS, G.E. and ERIKSSON, K. *Integer Partitions.* Cambridge University Press, Cambridge, UK, New York, USA, Port Melbourne, Australia, Madrid, Spain, Cape Town, South Africa, 2004.

BAXTER, R.J. Hard hexagons: exact solution. *J. Phys. A: Math. Gen.*, 13(3), 1980. L61-L70, Bibcode 1980JPhA...13L..61B.

BAXTER, R.J. *Exactly Solved Models in Statistical Mechanics.* Academic Press, New York, 1982. Reprinted 1989 and 2007.

BERKOVICH, A. and MCCOY, B.M. Rogers-Ramanujan Identities: A Century of Progress from Mathematics to Physics. *Documenta Mathematica*, Extra Volume ICM 1998 - III:163–172, 1998.

BERKOVICH, A., MCCOY, B.M. and SCHILLING, A. Rogers-Schur-Ramanujan type identities for M(p, p') minimal models of conformal field theory. *Comm. Math. Phys.*, 191(325), 1998.

BORWEIN, J.M. and BORWEIN, P. *Pi and the AGM.* John Wiley and Sons, New York, 1986.

BORWEIN, J.M., BRADLEY, D.M., BROADHURST, D.J. and LISONEK, P. Special values of multiple polylogarithms. *Trans. Am. Math. Soc.*, 353:907–941, 2001.

BORWEIN, J.M., GLASSER, M.L., MCPHEDRAN, R.C., WAN, J.G. and ZUCKER, I.J. *Lattice Sums then and now*, volume 150 of *Encyclopedia of Mathematics and Its Applications.* Cambridge University Press, 2013.

BRESSOUD, D.M. A new family of partition identities. *Pacific J. Math.*, 77(3):71–74, 1978.

BRESSOUD, D.M. Some identities for terminating q-series. *Math. Proc. Cambridge Phil. Soc.*, 89:211–223, 1981.

CHAPMAN, R. A new proof of some identities of Bressoud. *Int. J. Math. and Math. Sci.*, 32:627–633, 2002.

GASPER, G. and RAHMAN, M. *Basic Hypergeometric Series*, volume 35 of *Encyclopedia of Mathematics and its Applications.* Cambridge University Press, (Cambridge - New York - Port Chester - Melbourne - Sydney), 1990.

GIBBS, J.W. *Elementary Principles in Statistical Mechanics.* (Reprinted by Dover, New York 1960), 1902.

LEWIN, L. (Ed.). *Structural Properties of Polylogarithms.* Amer. Math. Soc. - Providence, RI., 1991. ISBN 0-8218-1634-9.

SOFO, A. Integral identities for sums. *Math. Commun.*, 13:303–309, 2008.

SOFO, A. Double Integral Representation of Sums. *Int. J. Math. Anal.*, 3(10):483–489, 2009.

SOFO, A. Harmonic numbers and double binomial coefficients. *Integral Transforms Spec. Funct.*, 20(11):847–857, 2009.

SOFO, A. and NIMBRAN, A. Euler sums and integral connections. *Mathematics*, 7:833, 09 2019.

STANLEY, H.E. Spherical Model as the Limit of Infinite Spin Dimensionality. *Phys. Rev.*, 176(2):718–722, 1968.

WIKIPEDIA. Hard Hexagon Model. On-line encyclopedia article, July 2020. Brief statistical mechanics article on the model and its solution. *https://en.wikipedia.org/wiki/Hard hexagon model.*

CHAPTER 2

A Brief History Timeline for Partitions

2.1 LEONHARD EULER – 18th CENTURY

THE IDEA of partitions of positive integers goes way back to Leonhard Euler [123] in the 18th century. Originally Euler considered breakups of positive integers into sums of smaller positive integers. For example, the number 50 may be written as 49 + 1 or 1 + 1 + 1 +... etc. fifty times. So the number of ways of writing the number 50 as a sum of positive integers is called the number of partitions of 50 written as $p(50) = 204226$.

Euler's original approach discovered many formulas that would soon become classical in the theory of integer partitions. Way back in the 18th century, all this was very new, and his results were a departure from known types of combinatorial identities. Euler was the first to find formulas for the number of unrestricted partitions of a positive integer. He also found formulas and generating functions and equivalent for different kinds of partitions of positive integers such as partitions into distinct numbers or partitions into the odd numbers or partitions into the even numbers. Leonhard Euler lived from 1707 until 1783 and his works embraced eighty volumes of revolutionary thinking in mathematics. His theory of partitions was the breaking of new ground for the new series that were to follow in subsequent centuries.

2.2 GAUSS, CAUCHY, AND HEINE – 19th CENTURY

The work of Carl Friedrich Gauss in 1813 [135] signaled the next major breakthrough for the theory of partitions. Gauss introduced his theory of ordinary hypergeometric series, which was revolutionary and set the scene for the follow-up work in the mid-19th century by Heine (see [154] and [155]) who was the first to introduce the formal idea of basic hypergeometric series which have become known as q-series in 20th- and 21st-century analysis. Gauss, however, also introduced many identities that were relevant to the theory of partitions, including the celebrated results,

$$\sum_{k\in Z}(-1)^k q^{k^2} = 1-2q+2q^4-2q^9+2q^{16}-2q^{25}+\cdots = \prod_{k=1}^{\infty}\left(\frac{1-q^k}{1+q^k}\right), \qquad (2.2.1)$$

DOI: 10.1201/9781003174158-2

$$\sum_{k=0}^{\infty} q^{\frac{k(k+1)}{2}} = 1+q+q^3+q^6+q^{10}+q^{15}+\cdots = \prod_{k=1}^{\infty}\left(\frac{1-q^{2k}}{1+q^{2k-1}}\right). \tag{2.2.2}$$

Cauchy was the first to state what is today called the q-binomial theorem, so for that alone, he would figure as an important contributor to 20th century partition generating functions. Also, Cauchy introduced clear thinking on functions of a complex variable, including the Residue Theorem and so, in the context of the Hardy-Ramanujan Circle Method, and the 1950s Theorem of Meinardus, the retrospective view of Cauchy's contribution to asymptotics of partitions is that it was indispensable.

The other stand-out 19th-century contributor to 20th-century partitions generating functions was Heine. He introduced the first q-series identities and transformations, so many of which are routinely cited in 20th and 21st century research in mathematical partition theory and in physics.

2.3 ROGERS AND RAMANUJAN – 1890s TO 1920s

In 1894 Rogers [225] wrote his proof of the *later-named* Rogers-Ramanujan identities and the continued fractions associated with them. This paper went largely unnoticed until it was found by Ramanujan in Cambridge when he was reading journals. At that time, nobody else was able to prove these identities, given in the letters from Ramanujan to Hardy prior to 1913. Rogers and Ramanujan were soon brought together to publish further proofs of the identities, and in MacMahon's book *Combinatory Analysis* [193] published in 1916, the identities appeared without proof. Although these two identities became jointly attributed to both Rogers and Ramanujan in the 20th century, Roger's proof preceded Ramanujan's involvement for over two decades, and meanwhile in Germany in 1917 during World War I, Schur [230] also proved the identities. They are

$$\sum_{k=0}^{\infty} \frac{q^{k^2}}{(q;q)_k} = \prod_{k=0}^{\infty} \frac{1}{(1-q^{1+5k})(1-q^{4+5k})}, \tag{2.3.1}$$

$$\sum_{k=0}^{\infty} \frac{q^{k^2+k}}{(q;q)_k} = \prod_{k=0}^{\infty} \frac{1}{(1-q^{2+5k})(1-q^{3+5k})}. \tag{2.3.2}$$

Schur's proof was very different from that of Roger's original paper 23 years earlier. Partly, the mystique of these identities and their place in the Hardy-Ramanujan correspondence, along with their continued fractions of which particular cases (see Chapter 4) were first evaluated by Ramanujan, has led to them being known as the Rogers-Ramanujan identities, rather than the Rogers-Schur or simply the Rogers identities.

The Rogers-Ramanujan identities famously appeared in Baxter's solution [50] of the hard hexagon model in statistical mechanics (see also [52]).

2.4 MAJOR PERCY MACMAHON – 1896 TO CIRCA 1920

Another important contributor in the 20th century was Major Percy MacMahon, (see [193]) whose papers from about 1900 through until the book on combinatorial analysis published in 1916 introduced ideas that led to multipartite numbers including concepts associated with chess boards and vector partitions. This book laid foundations for much of the work associated with modern-day combinatorics. However, MacMahon's lasting contributions were related to the so-named *Plane Partitions*, especially his famous generating function formula for $PL(n)$ given by (see MacMahon [193]),

$$\sum_{n=0}^{\infty} PL(n)x^n = \prod_{k=1}^{\infty} \frac{1}{(1-x^k)^k} = 1 + x + 3x^2 + 6x^3 + 13x^4 + 24x^5 + \ldots \tag{2.4.1}$$

This is often called the *MacMahon function*. There are several remarkable proofs in the literature, and I refer the reader to G.E. Andrews' eloquent descriptions and proofs in [12]. It seems the matrix definition is good and precise, but the cube pile model is probably the most intuitive way to describe $\pi_{i,j}$. Formula (2.4.1) may be viewed as the 2-dimensional analogue of Euler's product formula for the number of integer partitions of n. There is no known analogs formula known for partitions in higher dimensions (i.e., for the extension to partitions stacked as 4D tesseracts instead of 3D cubes).

In 1896 Percy A. MacMahon published a generating function of plane partitions that are subsets of the $i \times j \times k$ box $\mathcal{B}(r,s,t)$ in his first paper on plane partitions. The formula is given by

$$\sum_{\pi \in \mathcal{B}(r,s,t)} q^{|\pi|} = \prod_{i=1}^{r}\prod_{j=1}^{s} \frac{1-q^{i+j+t-1}}{1-q^{i+j-1}}.$$

A proof of this formula is in the book *Combinatory Analysis* written by MacMahon. [193] He shows how this involves the generating function of plane partitions of the cube representations inside an $r \times s \times t$ box in the 3D space rectangular prism from the origin $(0,0,0)$ to the point (r,s,t). The formula for the generating function can be written another way, given by

$$\sum_{\pi \in \mathcal{B}(r,s,t)} q^{|\pi|} = \prod_{i=1}^{r}\prod_{j=1}^{s}\prod_{k=1}^{t} \frac{1-q^{i+j+k-1}}{1-q^{i+j+k-2}}.$$

Plane partitions have continued relevance into the 21st century where they have found applications in mathematical physics.

2.5 HARDY AND RAMANUJAN – CIRCA 1918s TO THE 1940s

In the early 20th century, works by Hardy and Ramanujan were significant developments associated with the asymptotic behavior of partitions, introducing the Circle Method to the analysis (see [152]). For example, $p(50) = 204226$ but the early 20th century asymptotic estimates showed that $p(50) \approx 217590$ when given by the simple formula $p(n) \sim e^{\pi\sqrt{2n/3}}/(4n\sqrt{3})$. Ramanujan also brought in the significant work associated with congruence properties of partitions and the Mock Theta functions, which have had a deeper impact almost a century later in previously unimagined applications such as String Theory.

The asymptotics of the Circle Method techniques were refined by others such as Rademacher in the 1930s and Szekeres with his *Saddle Point method* approaches in the 1950s. (See Rademacher [212], [213]; also Szekeres [289, 290], Roth and Szekeres [227]; and Richmond [219, 220]) This opened the way for the significant works of George E Andrews culminating in his book *The Theory of Partitions* first published in 1976, but inspiring the work from numerous contributors in the latter part of the 20th century and the early decades of the 21st century. (see Andrews [11], [17], [22], [23], for example Andrews and Askey [25])

2.6 ANDREWS – EARLY 1960s TO CIRCA 2020s

Andrews began writing numerous partition theoretic papers in the 1960s, culminating in 1976 with his classical book *The Theory of Partitions.* He then was instrumental in popularizing the Statistical Mechanics derivations of Rogers-Ramanujan type identities that are now a main topic of research in mathematical physics.

In Chapter 12 of Andrews' book on the theory of partitions, we find the basis for our research pertaining to the theory of partitions of vectors in n-dimensional space. Andrews states how problems often arise concerning the additive decomposition of nonzero vectors with non-negative integral coordinates (also called multipartite numbers). Andrews' book chapter on vector partitions has a good introduction to analysis of the partition functions of the multipartite number $(n_1, n_2, n_3, ..., n_r)$,

$$P(\mathbf{n}) = P(n_1, n_2, n_3, ..., n_r) \tag{2.6.1}$$

$$Q(\mathbf{n}) = Q(n_1, n_2, n_3, ..., n_r), \tag{2.6.2}$$

for *unrestricted* and *distinct* r-space partitions respectively. As is usual, *compositions* take the order of summands into account whereas *partitions* do not. In our present book treating weighted vector partitions, this feature is again adopted for our work.

The elementary theory of multipartite partitions resembles greatly the elementary theory of ordinary partitions presented in the earlier known theory and related to infinite product-generating functions going back to those introduced by Euler. Unfortunately, there is not known any truly simple type of infinite series that is useful in treating multipartite partitions as the basic hypergeometric series (introduced in earlier years) in treating ordinary partitions. Thus, we only in the 20th century knew of comparatively simple identities related to multipartite partition functions and we did not have any device like Euler's pentagonal number theorem for the rapid calculation of any multipartite partition functions. The most useful theorems for the actual evaluation of multipartite partition functions are presented in Chapter 12 of Andrews' book where he mentions vector or multipartite partitions.

Leonard Carlitz [102] and others have considered problems of restricted multipartite partitions in the 1960s. In particular, they looked at k-partite partitions. These k-partite partitions were subject to decreasing conditions.

Andrew's book gives r-space partition-generating functions identical to those we consider, as follows

$$\sum_{a_1,a_2,\ldots,a_r\geq 0} P(\mathbf{n}){x_1}^{n_1}{x_2}^{n_2}\cdots{x_r}^{n_r} = \prod_{\substack{n_1,n_2,\ldots,n_r\geq 0\\ \textit{not all zero}}} (1-{x_1}^{n_1}{x_2}^{n_2}\cdots{x_r}^{n_r})^{-1}, \quad (2.6.3)$$

$$\sum_{a_1,a_2,\ldots,a_r\geq 0} Q(\mathbf{n}){x_1}^{n_1}{x_2}^{n_2}\cdots{x_r}^{n_r} = \prod_{n_1,n_2,\ldots,n_r\geq 0} (1+{x_1}^{n_1}{x_2}^{n_2}\cdots{x_r}^{n_r}), \quad (2.6.4)$$

where for absolute convergence, we have $|x_i| \leq 1$ for $1 \leq i \leq r$.

In the present book, we consider a generalization of many of the concepts and classical results from the theory of integer partitions, as they apply to vectors in higher dimensional space.

This includes a generalized q-binomial theorem, and some cases of it offering a new context for the theory of vector partitions. We also give some new q series that are illustrative of hybrid forms in n-space of the new vector identities and variations on the traditional q series integer partition generating functions.

We consider the set of vector partitions in the Cartesian plane, interpreting them as the number of stepping stone combinations in integer coordinate steps from the origin. Although we do not fully solve many of the issues that present themselves naturally from generalizing integer partition concepts, clearly, we are introducing a theoretical basis for further research in this generalized approach to partitions.

The new identities enumerate weighted vector partitions using the analogy of n-space stepping stones. We touch upon concepts first given by Ramanujan, with respect to arithmetical functions that become applicable as tools for evaluating many of the new vector partition generating functions. We state new non-weighted n-space vector partition-related functional equations, and theorems that indicate ways to graph n-space vector partitions by combining Cartesian Lattice points with an extended Ferrars graph concept.

2.7 RODNEY BAXTER – 1980 TO 2023 AND BEYOND

In the early 1980s, there was an undercurrent of novelty, shock and euphoric elation when the area of Statistical Mechanics suddenly had a connection to the famous Rogers-Ramanujan identities. So then, Rodney Baxter forty years later remarked that

Think About It...

"I wondered often if Hardy had turned in his grave; Hardy the pure mathematician, who says proudly in one of his books that his mathematical works have no use in physical reality at all. This contrasted with his colleague Littlewood, who was doing war work, including in ballistics.

So, I don't know that Hardy would have been very pleased to find that Rogers-Ramanujan identities (which Hardy knew well) cropped up in a practical, physical way."

Interview with Rodney James Baxter)
Australian Mathematical Society Gazette, 2020 (see [94]).

There is something "larger than life" about Baxter solving the "Hard Hexagon and related Models". The gravity of the discovery and its connecting theories of partitions to physics strongly and definitely drew in the big personalities. To start with, it happened that the great Kurt Mahler whose research history on partitions went back 50 years to the 1930s, happened to work on the same campus as Baxter. Then the worldwide announcements brought George E. Andrews to Canberra for a year, where he, Baxter and Forrester found new classes of related theoretical identities. (see Andrews, Baxter and Forrester [31] for example) The physics theories "broke new ground" for the mathematics, and vice versa. Now, 40 years after Baxter's original discovery, physicists are churning out papers on Yang-Baxter equation solutions and an abundance of new "exactly solvable models" are being revealed. Despite the overlaps between the mathematics and the physics leading to prolific researches, there is yet a tendency for physics and math researchers to "stay in their lanes". The present book hopefully will realign some of that separation.

It seems Russians especially, led by Ludwig Faddeev (see [125]); and also the Kyoto Japan Group [116], has taken up the Yang-Baxter relation which determines whether a Statistical Mechanics model is exactly solvable. There are now many papers on Yang-Baxter and Star-Triangle relations, forging new trails into 21st century theoretical physics.

FURTHER READING

ANDREWS, G.E. *Problems and prospects for basic hypergeometric functions. The theory and application of Special Functions. (R Askey, Editor).* Academic Press, New York, 1975.

ANDREWS, G.E. *The Theory of Partitions*, volume 2 of *Encyclopedia of Mathematics and its Applications.* Addison-Wesley Publishing Company, Advanced Book Program, Reading, Massachusetts, USA, 1976.

ANDREWS, G.E. *Partitions: yesterday and today.* New Zealand Math. Soc., Wellington, 1979.

ANDREWS, G.E. q-Series: Ten Lectures. In *NSF-CBMS Regional Conference*, pages 175–182. Arizona State University, 1985.

ANDREWS, G.E. *q-series: Their development and application in analysis, number theory, combinatorics, physics, and computer algebra.* Number 66 in Regional conference series in mathematics. Amer. Math. Soc., Providence, RI, 1986.

ANDREWS, G.E. and ASKEY, R. Enumeration of Partitions: The Role of Eulerian Series and q-Orthogonal Polynomials. In Martin Aigner, editor, *Higher Combinatorics*, pages 3–26, Dordrecht, 1977. Springer Netherlands.

ANDREWS, G.E., BAXTER, R.J. and FORRESTER, P.J. Eightvertex SOS model and generalized Rogers-Ramanujan-type identities. *J. Stat. Phys.*, 35:193–266, 1984.

BAXTER, R.J. Hard hexagons: exact solution. *Journal of Physics A: Mathematical and General*, 13(3), 1980. L61-L70, Bibcode 1980JPhA...13L..61B.

BAXTER, R.J. *Exactly Solved Models in Statistical Mechanics*. Academic Press, New York, 1982. Reprinted 1989 and 2007.

CAMPBELL, G.B. An interview with Rodney James Baxter. *Aust. Math. Soc. Gazette*, 47(1):24–32, March 2020.

CARLITZ, L. Rectangular arrays and plane partitions. *Acta Arith.*, 13:29–47, 1967.

DATA, E., JIMBO, M., KUNIBA, A., MIWA, T. and OKADO, M. Exactly solvable SOS models: local height probabilities and theta function identities. *Nucl. Phys. B*, 290(231), 1987.

EULER, L. Introductio in analysin infinitorum. In *Opera Omnia*, volume 1. Marcum-Michaelum, Brousquet, Lausannae, 1748. Chapter 16.

FADDEEV, L.D. How Algebraic Bethe Ansatz works for integrable model, May 1996. https://arxiv.org/abs/hep-th/9605187.

GAUSS, C.F. Disquisitiones generales circa seriem infinitam ... . In *Werke 3*, volume II, pages 123–162. Comm. soc. reg. sci. Gött. rec., 1876. article first printed 1813; reprinted in Werke 3 1876.

HARDY, G.H. and RAMANUJAN, S. Asymptotic formulae in combinatory analysis. *Proc. London. Math. Soc.*, 2(17):75–115, 1918. Also in Collected papers of S. Ramanujan, pp. 276-309, Cambridge Univ. Press, Cambridge, 1927; reprinted by Chelsea, New York, 1962.

HEINE, E. Untersuchungen uber die Reihe ... . *J. Reine angew. Math.*, 34:285–328, 1847.

HEINE, E. *Handbuch der Kugelfunctionen*, Theorie und Andwendungen, volume 1. Reimer, Berlin, 1878.

MACMAHON, P.A. *Combinatory Analysis*, volume 2. Cambridge Univ. Press, London and New York (reprinted by Chelsea, New York, 1960), 1916.

RADEMACHER, H. On the partition function $p(n)$. *Proc. London Math. Soc.*, 2(43):241–254, 1937.

RADEMACHER, H. On the expansion of the partition function in a series. *Ann. Of Math.*, 44:416–422, 1943.

RICHMOND, B. Asymptotic relations for partitions. *J. Number Theory*, 7:389–405, 1975.

RICHMOND, B. Asymptotic relations for partitions. *Canadian J. Math.*, 27:1083–1091, 1975.

ROGERS, L.J. Second memoir on the expansion of certain infinite products. *Proc. London Math. Soc.*, 25:318–343, 1894.

ROTH, K.F., and SZEKERES, G. Some asymptotic formulae in the theory of partitions. *Q. J. Math. Oxf. Ser(2)*, 5:241–259, 1954.

SCHUR, I. Ein Beitrag zur additive Zahlentheorie und zue Theorie der Kettenbruche. *S.-B. Preuss, Akad. Wies. Phys.-Math, Kl.*, pages 302–321, 1973. First printed 1917: Proceedings of the Royal Prussian Academy of Sciences in Berlin; Reprinted 1973: Ges. Abhandlungen, Vol. 2, pp. 117-136.

SZEKERES, G. An asymptotic formula in the theory of partitions, I. *Quart. J. Math. Oxford Ser.* 2(2):85–108, 1951.

SZEKERES, G. An asymptotic formula in the theory of partitions, II. *Quart. J. Math. Oxford Ser.(2)*, 4:96–111, 1951.

II

Integer Partition Theory

CHAPTER 3

Integer Partitions and Their Generating Functions

3.1 PREVIEW

THIS CHAPTER exists thanks to the efforts of many researchers over the past 270 years since Euler's pioneering work. Historically, the research was carried forward into the nineteenth century by Gauss, Jacobi, Heine, and Franklin. In the early 20th century Hardy, MacMahon, Ramanujan, Watson, and Schur found a large body of identities and theorems that broadened the literature greatly. Finally in the late 20th century the lucid, culminating work of Carlitz, Baxter, Andrews, Bressoud, and Askey found the new theory of partitions generating functions and bijective proofs. The bijections now rival the generating function combinatorial identities approach, often providing visual insights helping us to understand a combinatorial identity. This chapter covers a few of the well-known results in the integer theory of partitions based on generating functions.

3.2 EULER'S APPROACH TO PARTITIONS

Euler recognized that partitions of integers could be shown to be identical to the act of multiplying out algebraic terms. Hence he was able to show that for $|x| < 1$,

$$\prod_{k=1}^{\infty} \frac{1}{1-x^k} = \sum_{n=0}^{\infty} p(n)x^n, \tag{3.2.1}$$

where $p(n)$ is the number of unrestricted partitions into integer summands of positive integer n. The right side series in (3.2.1) is

$$1 + x + 2x^2 + 3x^3 + 5x^4 + 7x^5 + 11x^6 + 15x^7 + 22x^8 + 30x^9 + 42x^{10} + 56x^{11}$$

$$+77x^{12} + 101x^{13} + 135x^{14} + 176x^{15} + 231x^{16} + 297x^{17} + 385x^{18} + O(x^{19}).$$

DOI: 10.1201/9781003174158-3

For example, the seven unrestricted partitions of 5 are:

$$5, 4+1, 3+2, 3+1+1, 2+2+1, 2+1+1+1, \textit{ and } 1+1+1+1+1,$$

so demonstrating that $p(5)=7$.

3.3 EULER'S PARTITION IDENTITY

Euler also noted that for $|x|<1$,

$$\prod_{k=1}^{\infty}(1+x^k)=\sum_{n=1}^{\infty}\mathcal{D}(n)x^n, \tag{3.3.1}$$

where $\mathcal{D}(n)$ is the number of partitions of positive integer n into distinct integer parts.

$$\prod_{k=1}^{\infty}\frac{1}{1-x^{2k-1}}=\sum_{n=1}^{\infty}\mathcal{O}(n)x^n, \tag{3.3.2}$$

where $\mathcal{O}(n)$ is the number of partitions of positive integer n into odd number parts.

The 270 years old partition identity statement due to Euler [123] is,

Theorem 3.1 *For each $n\geq 1$,*

$$\mathcal{O}(n)=p(n\,|\,\textit{odd parts}\in\mathbb{N})=\mathcal{D}(n)=p(n\,|\,\textit{distinct parts}\in\mathbb{N}). \tag{3.3.3}$$

For example, the partitions of 9 into distinct parts are

$$9, 8+1,\ 7+2,\ 6+3,\ 6+2+1,\ 5+4,\ 5+3+1,\ 4+3+2,$$

so there are 8 such partitions, and the partitions of 9 into odd parts are

$$9,\ 7+1+1,\ 5+3+1,\ 5+1+1+1+1,\ 3+3+3,\ 3+3+1+1+1,$$

$$3+1+1+1+1+1+1,\ 1+1+1+1+1+1+1+1+1,$$

so there are also 8 partitions of 9 into odd parts.

Theorem 3.1 in generating function form, is the so-named *Euler's Identity,*

$$\prod_{n=1}^{\infty}\frac{1}{1-q^{2n-1}}=\prod_{n=1}^{\infty}(1+q^n),\ |q|<1. \tag{3.3.4}$$

$$=1+x+x^2+2x^3+2x^4+3x^5+4x^6+5x^7+6x^8+8x^9+10x^{10}+\ldots.$$

whence we have the sequence $1,1,1,2,2,3,4,5,6,8,10,\ldots$ (OEIS A000009, see [234]).

3.4 EULER'S PENTAGONAL NUMBER THEOREM

Euler also studied the generating function for the difference between the number of partitions into an even number of unequal parts and the number of partitions in an odd number of unequal parts as given by his identity,

$$\prod_{k=1}^{\infty}(1-z^k) = 1-z-z^2+z^5+z^7-z^{12}-z^{15}+\ldots = 1+\sum_{k=1}^{\infty} e(k)z^k, \qquad (3.4.1)$$

where

$$e(k) = \begin{cases} (-1)^j, & if\ k = \frac{j(3j-1)}{2}; \\ 0, & \text{otherwise.} \end{cases}$$

The numbers $\frac{j(3j-1)}{2} \in \mathbb{Z}$ are the *generalized pentagonal numbers*, which is the Sloane OEIS integer sequence, A000326 at [235]. The identity (3.4.1) encodes the

Theorem 3.2 *(Euler [123]) For each integer* $n \geq 1$,

$$p(n\,|\,\textit{even number of distinct parts}) = e(n) + p(n\,|\,\textit{odd number of distinct parts}). \qquad (3.4.2)$$

The above theorem was early on proved by way of a particular case of the Jacobi Triple Product identity, which we derive in a later section from the q-binomial theorem. However, the often-cited and presented, different proof of the Euler pentagonal number theorem was the near bijection proof of Franklin, first given in 1881. (See Franklin [130], and Hardy and Wright [153, Franklin's proof, pp. 286–287].)

3.5 THE q-BINOMIAL THEOREM

In his classical account of the theory of partitions, Andrews [12, chapter 2] shows that many of the time-honored partition identities first given by Euler [123], Gauss [135], Heine [154], and Jacobi [160] derive from the q-binomial theorem originally given by Cauchy [104],

Theorem 3.3 ***The q-binomial theorem.*** *If* $|q|<1, |t|<1$, *for all complex* a,

$$1+\sum_{k=1}^{\infty}\frac{(1-a)(1-aq)(1-aq^2)\ldots(1-aq^{k-1})t^k}{(1-q)(1-q^2)(1-q^3)\ldots(1-q^k)} = \prod_{k=0}^{\infty}\frac{1-atq^k}{1-tq^k}. \qquad (3.5.1)$$

Before giving proof, we mention the notation commonly used for these types of identities. We will often use the terminology:

$$(a)_n = (a;q)_n = (1-a)(1-aq)\cdots(1-aq^{n-1}),$$
$$(a)_\infty = (a;q)_\infty = \lim_{n\to\infty}(a;q)_n,$$

$$(a)_0 = 1.$$

For all real numbers n we see also that

$$(a)_n = \frac{(a)_\infty}{(aq^n)_\infty}.$$

Theorem 3.3 has become well known as the q-analogue of the binomial series, since if we write $a = q^\alpha$ where α is a nonnegative integer, then (3.5.1) approaches formally

$$1 + \sum_{k=1}^{\infty} \binom{\alpha+n-1}{n} t^k = (1-t)^{-\alpha}, \quad as \quad q \to 1^-.$$

Proof of Theorem 3.3. We start with

$$F(t) = \prod_{n=0}^{\infty} \frac{(1-atq^n)}{(1-tq^n)} = \sum_{n=0}^{\infty} A_n t^n \tag{3.5.2}$$

where $A_n = A_n(a,q)$ for $|t| < 1$. We have then

$$(1-t)F(t) = (1-at) \prod_{n=1}^{\infty} \frac{(1-atq^n)}{(1-tq^n)} \tag{3.5.3}$$

$$= (1-at) \prod_{n=0}^{\infty} \frac{(1-atq^{n+1})}{(1-tq^{n+1})} = (1-at)F(tq). \tag{3.5.4}$$

So $A_0 = F(0) = 1$ and equating coefficients of powers of t we have

$$A_n = \frac{(1-aq^{n-1})}{(1-q^n)} A_{n-1}. \tag{3.5.5}$$

Repeated substitution of this tells us that

$$A_n = \frac{(1-aq^{n-1})(1-aq^{n-2})\cdots(1-a)A_0}{(1-q^n)(1-q^{n-1})\cdots(1-q)} \tag{3.5.6}$$

$$= \frac{(a)_n}{(q)_n}. \tag{3.5.7}$$

Substituting this A_n into (3.5.2) proves the theorem. ■

Following the approach of Andrews in his book [12, chapter 2] we see immediate consequences of Theorem 3.3 in the

Theorem 3.4 *If* $|t| < 1, |q| < 1,$

$$1 + \sum_{n=1}^{\infty} \frac{t^n}{(1-q)(1-q^2)\cdots(1-q^n)} = \prod_{n=0}^{\infty} \frac{1}{(1-tq^n)}. \tag{3.5.8}$$

$$1 + \sum_{n=1}^{\infty} \frac{t^n q^{\frac{1}{2}n(n-1)}}{(1-q)(1-q^2)\cdots(1-q^n)} = \prod_{n=0}^{\infty} (1+tq^n). \tag{3.5.9}$$

These were first found by Euler [123].

Proof: (3.5.8) is a trivial consequence of (3.5.1) with $a = 0$, and (3.5.9) also follows simply from (3.5.1) by setting $a \to a/b$ with $t = bz$ yielding

$$1+\sum_{k=1}^{\infty} \frac{(b-a)(b-aq)(b-aq^2)...(b-aq^{k-1})z^k}{(1-q)(1-q^2)(1-q^3)...(1-q^k)} = \prod_{k=0}^{\infty} \frac{1-azq^k}{1-bzq^k}.$$

Setting $a = -1, b = 0$ in this gives us (3.5.9). ■

3.6 THE JACOBI TRIPLE PRODUCT

In 1829 Jacobi, in his masterwork on elliptic functions, *Fundamenta nova theoriae functionum ellipticarum*, (see Jacobi [160]) first proved the classical result today named after him as the Jacobi triple product, enabling many of the now well-known partition identities to be proven.

Theorem 3.5 ***Jacobi Triple Product.*** *For* $z \neq 0, |q| < 1$,

$$\sum_{n=-\infty}^{\infty} z^n q^{n^2} = \prod_{n=0}^{\infty} (1-q^{2n+2})(1+q^{2n+1}z)(1+q^{2n+1}z^{-1}). \tag{3.6.1}$$

Proof. For $|z| \neq 0, |q| < 1$,

$$\begin{aligned}\prod_{n=0}^{\infty}(1+zq^{2n+1}) &= \sum_{m=0}^{\infty} \frac{z^m q^{m^2}}{(q^2;q^2)_m} \quad \text{(by (3.6.1))}\\ &= \frac{1}{(q^2;q^2)} \sum_{m=0}^{\infty} z^m q^{m^2} (q^{2m+2};q^2)_\infty\\ &= \frac{1}{(q^2;q^2)} \sum_{m=-\infty}^{\infty} z^m q^{m^2} (q^{2m+2};q^2)_\infty\end{aligned}$$

(since $(q^{2m+2};q^2)_\infty$ vanishes for $m < 0$)

$$\begin{aligned}&= \frac{1}{(q^2;q^2)_\infty} \sum_{m=-\infty}^{\infty} z^m q^{m^2} \sum_{r=0}^{\infty} \frac{(-1)^r q^{r^2+2mr+r}}{(q^2;q^2)_r}\\ &= \frac{1}{(q^2;q^2)_\infty} \sum_{r=0}^{\infty} \frac{(-1)^r z^{-r} q^r}{(q^2;q^2)_r} \sum_{m=-\infty}^{\infty} q^{(m+r)^2} z^{m+r}\\ &= \frac{1}{(q^2;q^2)_\infty} \sum_{r=0}^{\infty} \frac{(-q/z)^r}{(q^2;q^2)_r} \sum_{m=-\infty}^{\infty} q^{m^2} z^m\\ &= \frac{1}{(q^2;q^2)_\infty(-q/z;q^2)_\infty} \sum_{m=-\infty}^{\infty} q^{m^2} z^m.\end{aligned}$$

This is essentially the required result. Absolute convergence holds everywhere provided $|z| > |q|, |q| < 1$. The full result of the theorem comes from analytic continuation or by carrying through the same rationale again using z^{-1} in place of z. ■

3.7 TWO IDENTITIES FROM GAUSS

Theorem 3.6 ***Two Gauss Identities.*** *If* $q < 1$

$$\sum_{n=-\infty}^{\infty} (-1)^n q^{n^2} = \prod_{n=1}^{\infty} \frac{(1-q^n)}{(1+q^n)}, \tag{3.7.1}$$

$$\sum_{n=0}^{\infty} q^{\frac{n(n+1)}{2}} = \prod_{n=1}^{\infty} \frac{(1-q^{2n})}{(1+q^{2n-1})}. \tag{3.7.2}$$

Proof. By (3.6.1) with $z = -1$,

$$\begin{aligned}\sum_{n=-\infty}^{\infty} (-1)^n q^{n^2} &= (q^2;q^2)_\infty (q;q^2)_\infty (q;q^2)_\infty \\ &= (q)_\infty (q;q^2)_\infty = \frac{(q)_\infty}{(-q)_\infty}\end{aligned}$$

where simplifying with *Euler's Identity* (3.3.4) leads to (3.7.1). Next, we have

$$\sum_{n=0}^{\infty} q^{\frac{n(n+1)}{2}} = \; = \frac{1}{2} \sum_{n=-\infty}^{\infty} q^{\frac{n(n+1)}{2}} \tag{3.7.3}$$

$$= \frac{1}{2}(q)_\infty (-q)_\infty (-1)_\infty \tag{3.7.4}$$

$$= (q)_\infty (-q)_\infty (-q)_\infty = (q^2;q^2)_\infty (-q)_\infty = \frac{(q^2;q^2)_\infty}{(q:q^2)_\infty} \tag{3.7.5}$$

where again simplifying with *Euler's Identity* (3.3.4) leads to (3.7.2). ■

3.8 THE WATSON QUINTUPLE PRODUCT

In 1929 G. N. Watson, in a paper on Ramanujan's Notebooks, proved the so-named quintuple product identity. Evidently, the identity arose in the course of verifying unpublished *Notebook* continued fraction results of Ramanujan.

Theorem 3.7 ***The Watson Quintuple Product.*** *[296] For* $z \neq 0, |q| < 1$,

$$\sum_{n=-\infty}^{\infty} \left(z^{3n} - z^{-3n-1}\right) q^{\frac{3n^2+n}{2}} \tag{3.8.1}$$

$$= \prod_{n=1}^{\infty} (1-q^n)(1-q^n z)(1-q^{n-1}z^{-1})(1-q^{2n-1}z^2)(1-q^{2n-1}z^{-2}).$$

This identity, whose origin may be traced to an elliptic sigma formula of Weierstrass, has a very interesting history. Now, however, we derive this identity as a simple and natural theorem of Jacobi's triple product identity.

Proof of theorem 3.7. The identity (3.8.1) may be written in the form

$$\prod_{n=1}^{\infty}(1-q^{2n})(1-q^{2n}z)(1-q^{2n-2}z^{-1})(1-q^{4n-2}z^{2})(1-q^{4n-2}z^{-2}). \tag{3.8.2}$$

$$= \sum_{n=-\infty}^{\infty}\left(z^{3n}-z^{-3n-1}\right)q^{3n^2+n}$$

Let $A(s,t)$ denote $\prod_{n=1}^{\infty}(1-q^{4n})$ times the bilateral summation of (3.8.1). Applying (3.6.1) twice, we have

$$\begin{aligned} A(q,z) &= \sum_{j=1}^{\infty}(-1)^j q^{j^2+j}z^j\sum_{k=1}^{\infty}(-1)^k q^{2k^2}z^{2k} \\ &= \sum_{j,k\geq 1}^{\infty}(-1)^{j+k}q^{j^2+2k^2+j}z^{j+2k}. \end{aligned} \tag{3.8.3}$$

Writing $j+2k=n$, we get

$$\begin{aligned} A(q,z) &= \sum_{n=1}^{\infty}(-1)^n z^n\sum_{k=1}^{\infty}(-1)^k q^{(n-2k)^2+2k^2+n-2k} \\ &= \sum_{n=1}^{\infty}(-1)^n z^n\sum_{k=1}^{\infty}(-1)^k q^{n^2+6k^2-4nk+n-2k} \\ &= \sum_{n=1}^{\infty}(-1)^n z^n q^{n^2+n}\sum_{k=1}^{\infty}(-1)^k q^{6k^2-4nk-2k}. \end{aligned} \tag{3.8.4}$$

Now, for all integers p,

$$\sum_{k=1}^{\infty}(-1)^k q^{6k^2+6k(2p+1)}=0.$$

This follows on replacing k by $-k-2p-1$. Hence the inner sum in (3.8.4) vanishes unless $-4n-2\equiv 0 \pmod 6$, that is, $2n+1\equiv 0 \pmod 3$; so that we can assume in (3.8.4) that $n\equiv 0$or $-1 \pmod 3$. Let $A_1(q,z)$ and $A_2(q,z)$ denote respectively the parts of $A(q,z)$ corresponding to $n\equiv 0 \pmod 3$ and $n\equiv -1 \pmod 3$. Then

$$A_1(q,z)=\sum_{n=1}^{\infty}(-1)^n z^{3n}q^{9n^2+3n}\sum_{k=1}^{\infty}(-1)^k q^{6k^2-12nk-2k}.$$

Write $m=k-n$ in the inner series. We get, after a routine simplification,

$$A_1(q,z)=\sum_{n=1}^{\infty}z^{3n}q^{3n^2+n}\sum_{n=1}^{\infty}(-1)^m q^{6m^2-2m}=Q\sum_{n=1}^{\infty}z^{3n}q^{3n^2+n} \tag{3.8.5}$$

where $Q = \prod_{m=1}^{\infty}(1-q^{4m}) = \sum_{m=1}^{\infty}(-1)^m q^{6m^2-2m}$. Similarly, we get

$$\begin{aligned} A_2(q,z) &= Q\sum_{n=1}^{\infty} z^{3n-1}q^{3n^2-n} \\ &= Q\sum_{n=1}^{\infty} z^{-3n-1}q^{3n^2+n} \end{aligned} \tag{3.8.6}$$

Combining (3.8.5) and (3.8.6), we obtain (3.8.2) and hence (3.8.1). ■

3.9 THE HEINE FUNDAMENTAL TRANSFORMATION

In 1847 Eduard Heine wrote his introductory work [154] on what later became known as the *basic hypergeometric series.* In this, he first gave the following result known today as Heine's fundamental transformation, which leads us to corollaries relevant to the theory of integer partitions.

Theorem 3.8 *For* $|q| < 1, |t| < 1, |b| < 1,$

$$\begin{aligned} &1+\sum_{n=1}^{\infty}\frac{(1-a)(1-aq)\cdots(1-aq^{n-1})(1-b)(1-bq)\cdots(1-bq^{n-1})t^n}{(1-q)(1-q^2)\cdots(1-q^n)(1-c)(1-cq)\cdots(1-cq^{n-1})} \\ &= \prod_{m=0}^{\infty}\frac{(1-bq^m)(1-atq^m)}{(1-cq^m)(1-tq^m)} \\ &\times\left(1+\sum_{n=1}^{\infty}\frac{(1-\frac{c}{b})(1-\frac{cq}{b})\cdots(1-\frac{cq^{n-1}}{b})(1-t)(1-tq)\cdots(1-tq^{n-1})b^n}{(1-q)(1-q^2)\cdots(1-q^n)(1-at)(1-atq)\cdots(1-atq^{n-1})}\right). \end{aligned} \tag{3.9.1}$$

Proof. In the workings here we apply the q-binomial theorem.

$$\begin{aligned} \sum_{n=0}^{\infty}\frac{(a)_n(b)_n t^n}{(q)_n(c)_n} &= \frac{(b)_\infty}{(c)_\infty}\sum_{n=0}^{\infty}\frac{(a)_n}{(q)_n}\frac{(cq^n)_\infty}{(bq^n)_\infty}t^n \\ &= \frac{(b)_\infty}{(c)_\infty}\sum_{n=0}^{\infty}\sum_{m=0}^{\infty}\frac{(a)_n t^n}{(q)_n}\frac{(c/b)_m b^m q^{nm}}{(q^m)_\infty} \\ &= \frac{(b)_\infty}{(c)_\infty}\sum_{m=0}^{\infty}\frac{(c/b)_m b^m}{(q)_m}\frac{(atq^m)_\infty}{(tq^m)_\infty} \\ &= \frac{(b)_\infty(at)_\infty}{(c)_\infty(t)_\infty}\sum_{m=0}^{\infty}\frac{(c/b)_m(t)_m b^m}{(q)_m(at)_m}. \end{aligned}$$

■

3.10 THE q-ANALOGUES OF GAUSS'S AND KUMMER'S THEOREMS

Also in his 1847 treatise [154], Eduard Heine proved the *basic hypergeometric series* q-analog for the hypergeometric series Gauss Theorem. 100 years later, Bailey [47] proved the hypergeometric Kummer Theorem q-analogue. The two q-analogues are now here given as theorems.

Theorem 3.9 *(Gauss q-analogue, by Heine [154])*
For $|q|<1$, $|t|<1$, $0<|a|<1$, $0<|b|<1$, $c\neq 1$,

$$1+\sum_{n=1}^{\infty}\frac{(1-a)(1-aq)\cdots(1-aq^{n-1})(1-b)(1-bq)\cdots(1-bq^{n-1})}{(1-q)(1-q^2)\cdots(1-q^n)(1-c)(1-cq)\cdots(1-cq^{n-1})}\left(\frac{c}{ab}\right)^n \tag{3.10.1}$$

$$=\prod_{m=0}^{\infty}\frac{(1-\frac{c}{a}q^m)(1-\frac{c}{b}q^m)}{(1-cq^m)(1-\frac{c}{ab}q^m)}.$$

Theorem 3.10 *(Kummer q-analogue, by Bailey [47])*
For $|q|<\min(1,|b|)$, $|t|<1$, $0<|a|<1$, $0<|b|<1$, $a\neq b$,

$$1+\sum_{n=1}^{\infty}\frac{(1-a)(1-aq)\cdots(1-aq^{n-1})(1-b)(1-bq)\cdots(1-bq^{n-1})}{(1-q)(1-q^2)\cdots(1-q^n)\left(1-\frac{a}{b}\right)\left(1-\frac{a}{b}q\right)\cdots\left(1-\frac{a}{b}q^{n-1}\right)}\left(\frac{-q}{b}\right)^n \tag{3.10.2}$$

$$=\prod_{m=0}^{\infty}\frac{(1-aq^{2m+1})(1+q^{m+1})\left(1-\frac{a}{b^2}q^{2m+2}\right)}{\left(1-\frac{a}{b}q^{m+1}\right)\left(1-\frac{1}{b}q^{m+1}\right)}.$$

Before proving these two theorems, we note that the Gauss (3.10.3) and the Kummer sum (3.10.4) are limiting cases of the two theorems (3.10.1) and (3.10.2) respectively.

$${}_2F_1(a,b;c;1)=\frac{\Gamma(c)\Gamma(c-a-b)}{\Gamma(c-a)\Gamma(c-b)} \tag{3.10.3}$$

$$=1+\frac{ab}{c1!}+\frac{a(a+1)b(b+1)}{c(c+1)2!}+\frac{a(a+1)(a+2)b(b+1)(b+2)}{c(c+1)(c+2)3!}+\dots,$$

$${}_2F_1(a,b;1+a-b;-1)=\frac{\Gamma(1+a-b)\Gamma(1+\frac{1}{2}a)}{\Gamma(1+a)\Gamma(1+\frac{1}{2}a-b)}=1-\frac{ab}{(a+1-b)1!} \tag{3.10.4}$$

$$+\frac{a(a+1)b(b+1)}{(a+1-b)(a+1-b+1)2!}-\frac{a(a+1)(a+2)b(b+1)(b+2)}{(a+1-b)(a+1-b+1)(a+1-b+2)3!}+\dots.$$

Proof of theorem 3.9. Starting from theorem 3.8,

$$\begin{aligned}\sum_{n=0}^{\infty}\frac{(a)_n(b)_n(c/ab)^n}{(q)_n(c)_n} &= \frac{(b)_\infty(c/b)_\infty}{(c)_\infty(c/ab)_\infty}\sum_{n=0}^{\infty}\frac{(c/ab)_n b^n}{(q)_n}\\ &= \frac{(b)_\infty(c/b)_\infty}{(c)_\infty(c/ab)_\infty}\times\frac{(c/a)_\infty}{(b)_\infty}\\ &= \frac{(c/a)_\infty(c/b)_\infty}{(c)_\infty(c/ab)_\infty}.\end{aligned}$$

■

Proof of theorem 3.10. Starting from (3.10.4) interchanging a and b,

$$\begin{aligned}\sum_{n=0}^{\infty}\frac{(b)_n(a)_n(-q/b)^n t^n}{(q)_n(aq/b)_n} &= \frac{(a)_\infty(-q)_\infty}{(aq/b)_\infty(-q/b)_\infty}\sum_{n=0}^{\infty}\frac{(q/b)_n(-q/b)_n a^n}{(q)_n(-q)_n}\\ &= \frac{(a)_\infty(-q)_\infty}{(aq/b)_\infty(-q/b)_\infty}\sum_{n=0}^{\infty}\frac{(q^2/b^2;q^2)_n a^n}{(q^2;q^2)_n}\\ &= \frac{(a)_\infty(-q)_\infty(aq^2/b^2;q^2)_\infty}{(aq/b)_\infty(-q/b)_\infty(a;q^2)_\infty}\\ &= \frac{(aq;q^2)_\infty(-q)_\infty(aq^2/b^2;q^2)_\infty}{(a/b)_\infty(-q/b)_\infty}. \qquad ■\end{aligned}$$

We follow these two theorems up with two corollaries that each are applicable as generating functions for classes of integer partitions. The first of these is due to Cauchy [104] and the second one due to Euler [123].

Theorem 3.11 *If* $|q|<1$,

$$1+\sum_{n=1}^{\infty}\frac{q^{n^2-n}z^n}{(1-q)(1-q^2)\cdots(1-q^n)(1-z)(1-zq)\cdots(1-zq^{n-1})} = \prod_{n=0}^{\infty}\frac{1}{1-zq^n}, \tag{3.10.5}$$

$$1+\sum_{n=1}^{\infty}\frac{q^{n^2}}{((1-q)(1-q^2)\cdots(1-q^n))^2} = \prod_{n=1}^{\infty}\frac{1}{1-q^n}. \tag{3.10.6}$$

Proof of theorem 3.11. First we see that (3.10.6) comes from (3.10.5) with $z=q$. In theorem 3.8.6, set $a=\alpha^{-1}, b=\beta^{-1}, c=z$. Then

$$1+\sum_{n=1}^{\infty}\frac{(\alpha-1)(\alpha-q)\cdots(\alpha-q^{n-1}(\beta-1)(\beta-q)\cdots(\beta-q^{n-1}))z^n}{(q)_n(z)_n} = \frac{(z\alpha)_\infty(z\beta)_\infty}{(z)_\infty(z\alpha\beta)_\infty},$$

and if we let $\alpha=\beta=0$ here we obtain (3.10.5). ■

Theorem 3.12 *If* $|q|<1$,

$$1+\sum_{n=1}^{\infty}\frac{(1-a)(1-aq)\cdots(1-aq^{n-1})q^{\frac{n(n+1)}{2}}}{(1-q)(1-q^2)\cdots(1-q^n)} = \prod_{n=0}^{\infty}(1-aq^{2n-1})(1+q^n). \tag{3.10.7}$$

Proof of theorem 3.12. In theorem 3.10.4, set $b=\beta^{-1}$. Then

$$1+\sum_{n=1}^{\infty}\frac{(\alpha)_n(\beta-1)(\beta-q)\cdots(\beta-q^{n-1})(-q)^n}{(q)_n(aq\beta)_n} = \frac{(z\alpha)_\infty(z\beta)_\infty}{(z)_\infty(z\alpha\beta)_\infty}.$$

Now set $\beta=0$ and we have our proof. ■

3.11 EULER-PAIR THEOREM

Euler proved that the partitions of a natural number n into distinct parts are equinumerous with the partitions of n into odd parts [[153], p. 277]. A second theorem due to Euler states that every natural number is uniquely representable as a sum of distinct powers of 2 [see again [153], p. 277]. If S_1 and S_2 are subsets of the natural numbers $\mathbb{N}$, let us call (S_1, S_2) an Euler-pair if for all natural numbers, n, the number of partitions of n into distinct parts taken from S_1 equals the number of partitions of n into parts taken from S_2.

Theorem 3.13 *(Schur [230]) For each $n \geq 1$,*

$$p(n \,|\, \textit{parts} \in \mathbb{L}) = p(n \,|\, \textit{distinct parts} \in \mathbb{M}) + e(n), \tag{3.11.1}$$

where $\mathbb{L}$ is any set of positive integers such that the ratio of two elements of $\mathbb{L}$ is never a power of two, and $\mathbb{M}$ is the set of all elements of $\mathbb{L}$ and all their multiples by powers of two.

3.12 ROGERS-RAMANUJAN IDENTITIES

In 1894 Rogers [225] wrote his significant proof of the *later-named* Rogers-Ramanujan identities and related results that proved to be ahead of their time. The Rogers-Ramanujan identities are

$$\prod_{k=0}^{\infty} \frac{1}{(1-q^{1+5k})(1-q^{4+5k})} = \sum_{k=0}^{\infty} \frac{q^{k^2}}{(q;q)_k}, \tag{3.12.1}$$

$$\prod_{k=0}^{\infty} \frac{1}{(1-q^{2+5k})(1-q^{3+5k})} = \sum_{k=0}^{\infty} \frac{q^{k^2+k}}{(q;q)_k}. \tag{3.12.2}$$

The combinatorial interpretation for these two identities in terms of integer partitions is in the following statements:

1. The number of partitions of n such that with k parts the smallest part is at least k is the same as the number of partitions of n such that each part is congruent to either 1 or 4 modulo 5.

2. The number of partitions of n such that with k parts the smallest part is at least $k+1$ is the same as the number of partitions of n such that each part is congruent to either 2 or 3 modulo 5.

The Rogers-Ramanujan identities famously and unexpectedly were derived in Baxter's 1980 solution [50] of the hard hexagon model in statistical mechanics (see also [52]).

Although the literature over the past century has numerous proofs of the Rogers-Ramanujan identities, there still does not seem to be any simple proof *from The Book* as Erdös put it. There are polynomial proofs, generating function recurrence proofs, bijection proofs, basic hypergeometric function proofs, and even the Hard Hexagon

Model statistical mechanics derivation, but nothing simple that makes it all obvious. The simplest proof seems to be the bijective one by Bressoud and Zeilberger [75], which we give in a later chapter on successive Durfee squares.

EXERCISES

3.1 *Prove the following two identities.* If $|q| < 1$,

$$1+\sum_{n=1}^{\infty}\frac{q^n}{(1-q)(1-q^2)\cdots(1-q^n)} = \prod_{n=1}^{\infty}\frac{1}{(1-q^n)}.$$

$$1+\sum_{n=1}^{\infty}\frac{q^{\frac{1}{2}n(n+1)}}{(1-q)(1-q^2)\cdots(1-q^n)} = \prod_{n=1}^{\infty}(1+q^n).$$

3.2 *Interpretation.* Show that from the previous exercise, the number of partitions of n is equal to $p_1(n)+p_2(n)+...+p_n(n)$, where $p_i(n)$ is the number of partitions of n into parts at least i.

Think About It...

"Mathematics knows no races or geographic boundaries; for mathematics, the cultural world is one country.

One can measure the importance of a scientific work by the number of earlier publications rendered superfluous by it."

David Hilbert
German Mathematician, 1862–1943

FURTHER READING

ANDREWS, G.E. *The Theory of Partitions*, volume 2 of *Encyclopedia of Mathematics and its Applications.* Addison-Wesley Publishing Company, Advanced Book Program, Reading, Massachusetts, USA, 1976.

BAILEY, W.N. Some identities in combinatory analysis. *Proc. Lond. Math. Soc. Second Ser.*, 49:421–435, 1947.

BAXTER, R.J. Hard hexagons: exact solution. *Journal of Physics A: Mathematical and General*, 13(3), 1980. L61-L70, Bibcode 1980JPhA...13L..61B.

BAXTER, R.J. *Exactly Solved Models in Statistical Mechanics.* Academic Press, New York, 1982. Reprinted 1989 and 2007.

BRESSOUD, D.M. and ZEILBERGER, D. A short Rogers-Ramanujan bijection. *Discrete Math.*, 38:313–315, 1982.

CAUCHY, A. Mémoire sur les fonctions dont plusieurs . . . . In *Oeuvres de Cauchy*, pages 42–50. C. R. Acad. Sci. Paris, Gauthier-Villars, Paris, 1893. T. XVII, p. 523, 1re série, T. VIII.

EULER, L. Introductio in analysin infinitorum. In *Opera Omnia*, volume 1. Marcum-Michaelum, Brousquet, Lausannae, 1748. Chapter 16.

FRANKLIN, F. Sur les developpement du produit infiniti $(1-x)(1-x^2)(1-x^3)\ldots$,. *Comptes Rendus*, 82:448–450, 1881.

GAUSS, C.F. Disquisitiones generales circa seriem infinitam . . .. In *Werke 3*, volume II, pages 123–162. Comm. soc. reg. sci. Gött. rec., 1876. article first printed 1813; reprinted in Werke 3 1876.

HARDY, G.H. and WRIGHT, E.M. *An Introduction to the Theory of Numbers.* Oxford University Press, Clarendon, London, 1971.

HEINE, E. Untersuchungen uber die Reihe ... . *J. Reine angew. Math.*, 34:285–328, 1847.

JACOBI, C.G.J. *Fundamenta nova theoriae functionum ellipticarum*, volume 1. Regiomonte fratrum Bornträger., 1829. (reprinted in *Gesammelte Werke*, Vol. 1, pp 49-239; Reimer, Berlin, 1881).

ROGERS, L.J. Second memoir on the expansion of certain infinite products. *Proc. London Math. Soc.*, 25:318–343, 1894.

SCHUR, I. Ein Beitrag zur additive Zahlentheorie und zue Theorie der Kettenbruche. *S.-B. Preuss, Akad. Wies. Phys.-Math, Kl.*, pages 302–321, 1973. First printed 1917: Proceedings of the Royal Prussian Academy of Sciences in Berlin; Reprinted 1973: Ges. Abhandlungen, Vol. 2, pp. 117–136.

SLOANE, N.J.A. Expansion of $\prod_{m\geq 1}(1+x^m)$; number of partitions of n into distinct parts; number of partitions of n into odd parts. The On-Line Encyclopedia of Integer Sequences (OEIS), 2020. sequence A000009 : *https : //oeis.org/A000009.*

SLOANE, N.J.A. Generalized Pentagonal Numbers. The On-Line Encyclopedia of Integer Sequences (OEIS), 2020. sequence A000326 : *https : //oeis.org/A000326.*

WATSON, G.N. Theorems stated by Ramanujan. VII: Theorems on continued fractions. *J. London Math. Soc.*, 4:37–48, 1929.

CHAPTER 4

Continued Fractions for Partition Generating Functions

WE DERIVE CONTINUED FRACTIONS for partition-generating functions, utilizing both Euler's techniques and Ramanujan's techniques. Although our results are for integer partitions there is scope to extend this work to vector partitions, including for binary and n-ary partitions. Most of this chapter's material is in a recent paper, Campbell [96].

4.1 EULER'S CONTINUED FRACTION

Almost 290 years ago in 1737, Leonhard Euler wrote *De fractionibus continuis dissertatio*, which gave mathematics a first-ever comprehensive account of the properties of continued fractions, and included the first proof that the number e is irrational. (See Sandifer [229]) Later, but still 275 years ago in 1748, Euler, in his *Introductio in analysin infinitorum Vol. I, Chapter 18* [123], proved:

(a) The equivalence of his continued fraction to a generalized infinite series.

(b) Every rational number can be written as a finite continued fraction.

(c) The continued fraction of an irrational number is infinite.

Euler's continued fraction is the very nice identity, whose first few cases are:

$$\begin{aligned} a_0 + a_0 a_1 &= a_0/(1 - a_1/(1 + a_1)) \\ &= \cfrac{a_0}{1 - \cfrac{a_1}{1 + a_1}}; \end{aligned} \tag{4.1.1}$$

DOI: 10.1201/9781003174158-4

$$a_0+a_0a_1+a_0a_1a_2 = a_0/(1-a_1/(1+a_1-a_2/(1+a_2))) = \cfrac{a_0}{1-\cfrac{a_1}{1+a_1-\cfrac{a_2}{1+a_2}}}; \tag{4.1.2}$$

$$a_0+a_0a_1+a_0a_1a_2+a_0a_1a_2a_3 = a_0/(1-a_1/(1+a_1-a_2/(1+a_2-a_3/(1+a_3)))) = \cfrac{a_0}{1-\cfrac{a_1}{1+a_1-\cfrac{a_2}{1+a_2-\cfrac{a_3}{1+a_3}}}}. \tag{4.1.3}$$

Hence, we can state Euler's Continued Fraction in the following

Theorem 4.1 *If a_0, a_1, a_3, ... a_n are defined functions such that no denominator is zero in the following equations then*

$$\sum_{k=0}^{n}\prod_{j=0}^{k} a_j = a_0+a_0a_1+a_0a_1a_2+...+a_0a_1...a_n \tag{4.1.4}$$

$$= a_0/(1-a_1/(1+a_1-a_2/(1+a_2-a_3/(1+...a_{n-1}/(1+a_{n-1}-a_n/(1+a_n))))).$$

$$= \cfrac{a_0}{1-\cfrac{a_1}{1+a_1-\cfrac{a_2}{1+a_2-\cfrac{a_3}{1+a_3-\cfrac{\ddots}{\ddots\cfrac{a_{n-1}}{1+a_{n-1}-\cfrac{a_n}{1+a_n}}}}}}}.$$

Obviously, this lends itself to many of the elementary series that arise in school and university analysis. However, we shall put this to good use in applying it to partition generating functions. The fact that this theorem involves a finite sum allows us to incrementally extend the number of terms until we can infer the infinite versions of the theorem.

Example 1: The exponential function is

$$\exp(z) = 1+\frac{z}{1!}+\frac{z^2}{2!}+\frac{z^3}{3!}+... = 1+\left(\frac{z}{1}\right)+\left(\frac{z}{1}\right)\left(\frac{z}{2}\right)+\left(\frac{z}{1}\right)\left(\frac{z}{2}\right)\left(\frac{z}{3}\right)+... \tag{4.1.5}$$

$$= 1/\left(1-z/\left(1+z-\left(\frac{z}{2}\right)/\left(1+\left(\frac{z}{2}\right)-\left(\frac{z}{3}\right)/\left(1+\left(\frac{z}{3}\right)-\left(\frac{z}{4}\right)/\left(1+\left(\frac{z}{4}\right)-...\right)\right)\right)\right)\right).$$

Applying an "equivalence transformation" that consists of clearing the fractions, this example is simplified to

$$\exp(z) = 1/(1-z/(1+z-z/(2+z-2z/(3+z-3z/(4+z-\ldots))))),$$

or the equivalent statement

$$\exp(z) = \cfrac{1}{1-\cfrac{z}{1+z-\cfrac{z}{2+z-\cfrac{2z}{3+z-\cfrac{3z}{4+z-\ldots}}}}}$$

and we know this continued fraction converges uniformly on every bounded domain in the complex plane because it is equivalent to the power series for $\exp(z)$.

Example 2: There is the well-known logarithmic function series

$$\log\left(\frac{1+z}{1-z}\right) = 2z\left(\frac{1}{1}+\frac{z^2}{3}+\frac{z^4}{5}+\ldots\right) \tag{4.1.6}$$

$$= 2z\left(1+\left(\frac{z^2}{3}\right)+\left(\frac{z^2}{3}\right)\left(\frac{3z^2}{5}\right)+\left(\frac{z^2}{3}\right)\left(\frac{3z^2}{5}\right)\left(\frac{5z^2}{7}\right)+\ldots\right).$$

Applying Euler's continued fraction formula to this expression shows that:

$$\log\left(\frac{1+z}{1-z}\right) = 2z/\left(1-\left(\frac{z^2}{3}\right)/\left(1+\left(\frac{z^2}{3}\right)-\left(\frac{3z^2}{5}\right)/\left(1+\left(\frac{3z^2}{5}\right)\right.\right.\right.$$
$$\left.\left.\left.\left.\left.-\left(\frac{5z^2}{7}\right)/\left(1+\left(\frac{5z^2}{7}\right)-\left(\frac{7z^2}{9}\right)/\left(1+\left(\frac{7z^2}{9}\right)-\ldots\right)\right)\right)\right)\right).$$

Applying the "equivalence transformation" this example is simplified to

$$\log\left(\frac{1+z}{1-z}\right) = 2z/(1-z^2/(z^2+3-(3z)^2/(3z^2+5$$
$$-(5z)^2/(5z^2+7-(7z)^2/(7z^2+9-\ldots)))))$$

$$= \cfrac{2z}{1-\cfrac{z^2}{z^2+3-\cfrac{(3z)^2}{3z^2+5-\cfrac{(5z)^2}{5z^2+7-\cfrac{(7z)^2}{7z^2+9-\ldots}}}}}$$

Example 3: A continued fraction for π. We can use the previous example involving the principal branch of the natural logarithm function to construct a continued fraction representation of π. First, we note that

$$(i+1)/(i-1) = i, \quad \textit{so then} \quad \log((i+1)/(i-1)) = i\pi/2.$$

Setting $z=i$ in the previous result, and remembering that $i^2=-1$, we obtain immediately

$$\pi=\cfrac{4}{1+\cfrac{1^2}{2+\cfrac{3^2}{2+\cfrac{5^2}{2+\cfrac{7^2}{2+\dots}}}}}$$

4.2 EULER'S CONTINUED FRACTION APPLIED TO PARTITIONS

In this section, we will technically do no more than apply the previous section. However, the theory of partitions is full of generating functions that are amenable to the Euler continued fraction. In a subsequent section, we will examine Ramanujan-type continued fractions, but firstly we will gather some *low hanging fruit* from some elementary series-product identities.

We begin with the well-known telescoping identities:

If a_1, a_2, a_3, ... , a_n, are functions chosen for nonzero denominators, then

$$1+\frac{a_1}{1-a_1}+\frac{a_2}{(1-a_1)(1-a_2)}+...+\frac{a_n}{(1-a_1)(1-a_2)...(1-a_n)} \tag{4.2.1}$$
$$=\frac{1}{(1-a_1)(1-a_2)(1-a_3)...(1-a_n)};$$

and

$$1+a_1+a_2(1+a_1)+a_3(1+a_1)(1+a_2)+...+a_n(1+a_1)(1+a_2)...(1+a_{n-1}) \tag{4.2.2}$$
$$=(1+a_1)(1+a_2)(1+a_3)...(1+a_n).$$

The series in (4.2.1) and (4.2.2) are already close to being in the required form to apply the Euler continued fraction since

$$1+\frac{a_1}{1-a_1}+\frac{a_2}{(1-a_1)(1-a_2)}+...+\frac{a_n}{(1-a_1)(1-a_2)...(1-a_n)} \tag{4.2.3}$$
$$=1+\frac{a_1}{1-a_1}+\frac{a_1}{1-a_1}\frac{a_2(1-a_1)}{a_1(1-a_2)}+...+\frac{a_1}{1-a_1}\frac{a_2(1-a_1)}{a_1(1-a_2)}\cdots\frac{a_n(1-a_{n-1})}{a_{n-1}(1-a_n)};$$

and

$$1+a_1+a_2(1+a_1)+a_3(1+a_1)(1+a_2)+...+a_n(1+a_1)(1+a_2)...(1+a_{n-1}) \tag{4.2.4}$$
$$=1+a_1+a_1\frac{a_2(1+a_1)}{a_1}+a_1\frac{a_2(1+a_1)}{a_1}\frac{a_3(1+a_2)}{a_2}$$
$$+...+a_1\frac{a_2(1+a_1)}{a_1}\frac{a_3(1+a_2)}{a_2}\cdots\frac{a_n(1+a_{n-1})}{a_{n-1}}.$$

Hence combining (4.2.1) with (4.2.3) and then (4.2.2) with (4.2.4) respectively, we obtain

$$\frac{1}{(1-a_1)(1-a_2)(1-a_3)...(1-a_n)} \tag{4.2.5}$$

$$= \cfrac{1}{1-\cfrac{\frac{a_1}{1-a_1}}{1+\frac{a_1}{1-a_1}-\cfrac{\frac{a_2(1-a_1)}{a_1(1-a_2)}}{1+\frac{a_2(1-a_1)}{a_1(1-a_2)}-\cfrac{\frac{a_3(1-a_2)}{a_2(1-a_3)}}{1+\frac{a_3(1-a_2)}{a_2(1-a_3)}-\cfrac{\ddots}{\ddots\cfrac{\frac{a_{n-1}(1-a_{n-2})}{a_{n-2}(1-a_{n-1})}}{1+\frac{a_{n-1}(1-a_{n-2})}{a_{n-2}(1-a_{n-1})}-\cfrac{\frac{a_n(1-a_{n-1})}{a_{n-1}(1-a_n)}}{1+\frac{a_n(1-a_{n-1})}{a_{n-1}(1-a_n)}}}}}}};$$

and

$$(1+a_1)(1+a_2)(1+a_3)...(1+a_n) \tag{4.2.6}$$

$$= \cfrac{1}{1-\cfrac{a_1}{1+a_1-\cfrac{\frac{a_2(1+a_1)}{a_1}}{1+\frac{a_2(1+a_1)}{a_1}-\cfrac{\frac{a_3(1+a_2)}{a_2}}{1+\frac{a_3(1+a_2)}{a_2}-\cfrac{\ddots}{\ddots\cfrac{\frac{a_{n-1}(1+a_{n-2})}{a_{n-2}}}{1+\frac{a_{n-1}(1+a_{n-2})}{a_{n-2}}-\cfrac{\frac{a_n(1+a_{n-1})}{a_{n-1}}}{1+\frac{a_n(1+a_{n-1})}{a_{n-1}}}}}}}}.$$

After applying the "equivalence transformation" to both (4.2.5) and then (4.2.6) to eliminate denominator terms, each continued fraction is simplified giving us the following two theorems.

Theorem 4.2 *If* a_1, a_2, a_3, ... , a_n, *are functions chosen for nonzero denominators, then*

$$\frac{1}{(1-a_1)(1-a_2)(1-a_3)...(1-a_n)} \tag{4.2.7}$$

$$= \cfrac{1}{1-\cfrac{a_1}{1-\cfrac{a_2}{a_1+a_2-2a_1a_2-\cfrac{a_1a_3}{a_2+a_3-2a_2a_3-\cfrac{\ddots}{\ddots\cfrac{a_{n-2}a_n}{a_{n-1}+a_n-2a_{n-1}a_n}}}}}}.$$

At first glance, we can see this theorem as being applicable to generating functions for unrestricted partitions of various kinds. Similarly the next theorem applies for partitions of various sorts into distinct parts.

Theorem 4.3 *If a_1, a_2, a_3, ... , a_n, are functions chosen for nonzero denominators, then*

$$(1+a_1)(1+a_2)(1+a_3)...(1+a_n) \tag{4.2.8}$$

$$= \cfrac{1}{1-\cfrac{a_1}{1+a_1-\cfrac{(1+a_1)a_2}{a_1+a_2+a_1a_2-\cfrac{(1+a_2)a_3}{a_2+a_3+a_2a_3-\cfrac{\ddots}{\ddots\, \cfrac{(1+a_{n-1})a_n}{a_{n-1}+a_n+a_{n-1}a_n}}}}}}.$$

There are many examples we could choose for substitution into theorems 4.2 and 4.3. So, let's start with the generating functions for unrestricted partitions, and for distinct partitions as follows.

Corollary 4.1 *If $p_n(k)$, is the number of unrestricted partitions of k into integers no greater than n, then*

$$\frac{1}{(1-q^1)(1-q^2)(1-q^3)...(1-q^n)} = \sum_{k=0}^{\infty} p_n(k)q^k \tag{4.2.9}$$

$$= \cfrac{1}{1-\cfrac{q^1}{1-\cfrac{q^2}{q^1+q^2-2q^1q^2-\cfrac{q^1q^3}{q^2+q^3-2q^2q^3-\cfrac{\ddots}{\ddots\, \cfrac{q^{n-2}q^n}{q^{n-1}+q^n-2q^{n-1}q^n}}}}}}.$$

Corollary 4.2 *If $p_n(\mathfrak{D},k)$, is the number of distinct partitions of k into integers no greater than n, then*

$$(1+q^1)(1+q^2)(1+q^3)...(1+q^n) = \sum_{k=0}^{\infty} p_n(\mathfrak{D},k)q^k \tag{4.2.10}$$

$$= \cfrac{1}{1-\cfrac{q^1}{1+q^1-\cfrac{(1+q^1)q^2}{q^1+q^2+q^1q^2-\cfrac{(1+q^2)q^3}{q^2+q^3+q^2q^3-\cfrac{\ddots}{\ddots\, \cfrac{(1+q^{n-1})q^n}{q^{n-1}+q^n+q^{n-1}q^n}}}}}}.$$

Next we choose the odd integer powers substituted into the two theorems.

Corollary 4.3 *If $p_n(\mathfrak{O},k)$, is the number of unrestricted partitions of k into odd integers no greater than $2n-1$, then*

$$\frac{1}{(1-q^1)(1-q^3)(1-q^5)...(1-q^{2n-1})} = \sum_{k=0}^{\infty} p_n(\mathfrak{O},k)q^k \qquad (4.2.11)$$

$$= \cfrac{1}{1-\cfrac{q^1}{1-\cfrac{q^3}{q^1+q^3-2q^1q^3-\cfrac{q^1q^5}{q^3+q^5-2q^3q^5-\cfrac{\ddots}{\ddots\, \cfrac{q^{n-2}q^n}{q^{2n-3}+q^{2n-1}-2q^{2n-3}q^{2n-1}}}}}}}.$$

Corollary 4.4 *If $p_n(\mathfrak{D}\mathfrak{O},k)$, is the number of distinct partitions of k into odd integers no greater than $2n-1$, then*

$$(1+q^1)(1+q^3)(1+q^5)...(1+q^{2n-1}) = \sum_{k=0}^{\infty} p_n(\mathfrak{D}\mathfrak{O},k)q^k \qquad (4.2.12)$$

$$= \cfrac{1}{1-\cfrac{q^1}{1+q^1-\cfrac{(1+q^1)q^3}{q^1+q^3+q^1q^3-\cfrac{(1+q^3)q^5}{q^3+q^5+q^3q^5-\cfrac{\ddots}{\ddots\, \cfrac{(1+q^{2n-3})q^{2n-1}}{q^{2n-3}+q^{2n-1}+q^{2n-3}q^{2n-1}}}}}}}.$$

It is a well-known result due to Euler that $p_\infty(\mathfrak{D}\mathfrak{O},k) = p_\infty(\mathfrak{O},k)$. Explicitly, as $n \to \infty$ equations (4.2.12) and (4.2.11) are equal to each other.

Next, let us give the cases covering binary partitions.

Corollary 4.5 *If $b_n(y,k)$, is the number of unrestricted binary partitions of k into non-negative powers of two no greater than 2^n, then*

$$\frac{1}{(1-q^1)(1-q^2)(1-q^4)...(1-q^{2^n})} = \sum_{k=0}^{\infty} b_n(y,k)q^k \tag{4.2.13}$$

$$= \cfrac{1}{1-\cfrac{q^1}{1-\cfrac{q^2}{q^1+q^2-2q^1q^2-\cfrac{q^1q^4}{q^2+q^4-2q^2q^4-\cfrac{\ddots}{\ddots\ \cfrac{q^{2^{n-2}}q^{2^n}}{q^{2^{n-1}}+q^{2^n}-2q^{2^{n-1}}q^{2^n}}}}}}}.$$

The following distinct binary partitions example is completely solvable.

Corollary 4.6 *If $p_n(y\mathfrak{D},k)$, is the number of binary partitions of k into distinct non-negative powers of two no greater than 2^n, then*

$$(1+q^1)(1+q^2)(1+q^4)...(1+q^{2^n}) = \frac{1-q^{2^{n+1}}}{1-q} = \sum_{k=0}^{2^{n+1}-1} p_n(y\mathfrak{D},k)q^k \tag{4.2.14}$$

$$= \cfrac{1}{1-\cfrac{q^1}{1+q^1-\cfrac{(1+q^1)q^2}{q^1+q^2+q^1q^2-\cfrac{(1+q^2)q^4}{q^2+q^4+q^2q^4-\cfrac{\ddots}{\ddots\ \cfrac{(1+q^{2^{n-1}})q^{2^n}}{q^{2^{n-1}}+q^{2^n}+q^{2^{n-1}}q^{2^n}}}}}}}.$$

Note that from (4.2.14) we have directly that

$$p_n(y\mathfrak{D},k) = \begin{cases} 1, & \text{when } 0 \le k < 2^{n+1}; \\ 0, & \text{when } k \ge 2^{n+1}. \end{cases}$$

The following distinct ternary partitions example is easily stated.

Corollary 4.7 *If $p_n(v\mathfrak{D},k)$, is the number of ternary partitions of k into distinct non-negative powers of three no greater than 3^n, then*

$$(1+q^1)(1+q^3)(1+q^9)...(1+q^{3^n}) = \sum_{k=0}^{3^n-1} p_n(v\mathfrak{D},k)q^k \tag{4.2.15}$$

$$= \cfrac{1}{1 - \cfrac{q^1}{1+q^1 - \cfrac{(1+q^1)q^3}{q^1+q^3+q^1q^3 - \cfrac{(1+q^3)q^9}{\ddots \, q^3+q^9+q^3q^9 - \cfrac{\ddots}{\ddots \cfrac{(1+q^{3^{n-1}})q^{3^n}}{q^{3^{n-1}}+q^{3^n}+q^{3^{n-1}}q^{3^n}}}}}}}.$$

Note that from (4.2.15) we have directly that

$$p_n(v\mathfrak{D},k) = \begin{cases} 1, & \text{for } 0 \le k < 3^{n+1};\ k \text{ is a sum of distinct powers of 3.} \\ 0, & \text{for } 0 \le k < 3^{n+1};\ k \text{ not a sum of distinct powers of 3.} \\ 0, & \text{for } k \ge 3^{n+1}. \end{cases}$$

In our next example, the property of Fibonacci numbers that $f(n)+f(n+1) = f(n+2)$ can be usefully applied to our work here. The Fibonacci number are famously the sequence starting with $0,1,1,2,3,5,8,13,21,34,55,89,144,233....$

Corollary 4.8 *If $p_n(\mathfrak{f},k)$, is the number of partitions of k into distinct Fibonacci numbers no greater than $f(n)$, then*

$$(1+q^0)(1+q^1)(1+q^1)(1+q^2)...(1+q^{f(n)}) = \sum_{k=0}^{\infty} p_n(\mathfrak{f},k)q^k \tag{4.2.16}$$

$$= \cfrac{1}{1 - \cfrac{q^0}{1+q^0 - \cfrac{q^1+q^1}{q^0+q^1+q^1 - \cfrac{q^1+q^2}{q^1+q^1+q^2 - \cfrac{\ddots}{\ddots \cfrac{q^{f(n)}+q^{f(n+1)}}{q^{f(n-1)}+q^{f(n)}+q^{f(n+1)}}}}}}}.$$

Clearly, this topic of Euler Continued Fractions applied to partition-generating functions is an interesting elementary study for students, and a possible tool for researchers. The above results are no doubt old, and have probably been well-worked over time.

4.3 ROGERS-RAMANUJAN CONTINUED FRACTIONS FOR PARTITION FUNCTIONS

The fraction given here was mentioned by Ramanujan in his second letter to Hardy (see Adiga et al. [2, p. xxviii]); namely

$$R(a,b) = 1 + \cfrac{bq}{1+aq+\cfrac{bq^2}{1+aq^2+\cfrac{bq^3}{1+aq^3+\cfrac{bq^4}{\ddots}}}}. \tag{4.3.1}$$

However, these now famous continued fractions, as with the Rogers-Ramanujan identities, were first discovered in 1894 by Rogers (see [225]). We define the functions $G(q)$ and $H(q)$ in the context of the Rogers-Ramanujan identities,

$$G(q) = \sum_{n=0}^{\infty} \frac{q^{n^2}}{(1-q)(1-q^2)\cdots(1-q^n)} = \sum_{n=0}^{\infty} \frac{q^{n^2}}{(q:q)_n} \tag{4.3.2}$$

$$= \frac{1}{(q;q^5)(q^4;q^5)} = \prod_{n=1}^{\infty} \frac{1}{(1-q^{5n-4})(1-q^{5n-1})},$$

and

$$H(q) = \sum_{n=0}^{\infty} \frac{q^{n^2+n}}{(1-q)(1-q^2)\cdots(1-q^n)} = \sum_{n=0}^{\infty} \frac{q^{n^2+n}}{(q:q)_n} \tag{4.3.3}$$

$$= \frac{1}{(q^2;q^5)(q^3;q^5)} = \prod_{n=1}^{\infty} \frac{1}{(1-q^{5n-3})(1-q^{5n-2})}.$$

The Rogers-Ramanujan continued fraction is then,

$$R(q) = \frac{q^{\frac{11}{60}}H(q)}{q^{\frac{-1}{60}}G(q)} = q^{\frac{1}{5}} \prod_{n=1}^{\infty} \frac{(1-q^{5n-4})(1-q^{5n-1})}{(1-q^{5n-3})(1-q^{5n-2})} \tag{4.3.4}$$

$$= 1 + \cfrac{q^{\frac{1}{5}}}{1+\cfrac{q}{1+\cfrac{q^2}{1+\cfrac{q^3}{\ddots}}}}.$$

So, we note that $R(0,1)$ leads us to the celebrated Rogers-Ramanujan continued fraction, which has been researched by many (see Andrews [12, Chapter 7], for example). In the course of analyzing identities from Ramanujan's Lost Notebook [27], Andrews and Berndt have discussed the fraction $R(a,b)$, but mainly from the viewpoint of transformation formulas. Our emphasis here is on using (4.3.1) in a generalized approach to several partition identities, but there is a whole adjacent theory on particular values of these continued fractions determined from applying the theory of modular forms. Hence the examples, using φ as the golden ratio $(\sqrt{5}+1)/2$,

$$\cfrac{e^{-\frac{-\pi}{5}}}{1+\cfrac{e^{-\pi}}{1+\cfrac{e^{-2\pi}}{1+\cfrac{e^{-3\pi}}{1+\ddots}}}} = \frac{1}{2}\varphi(\sqrt{5}-\varphi^{3/2})(\sqrt[4]{5}+\varphi^{3/2}), \tag{4.3.5}$$

$$\cfrac{e^{-\frac{-2\pi}{5}}}{1+\cfrac{e^{-2\pi}}{1+\cfrac{e^{-4\pi}}{1+\cfrac{e^{-6\pi}}{1+\ddots}}}} = \sqrt[4]{5}\varphi^{1/2}-\varphi, \tag{4.3.6}$$

$$\cfrac{e^{-\frac{-4\pi}{5}}}{1+\cfrac{e^{-4\pi}}{1+\cfrac{e^{-8\pi}}{1+\cfrac{e^{-12\pi}}{1+\ddots}}}} = \frac{1}{2}\varphi(\sqrt{5}-\varphi^{3/2})(-\sqrt[4]{5}+\varphi^{3/2}). \tag{4.3.7}$$

So next we examine the continued fraction $R(a,b)$ of Ramanujan and consider various restricted partition functions. For further reading, a good reference is Alladi and Gordon [8]. We use the continued fraction to give results for several partition identities, some of which generalize results of Bressoud [73] and Göllnitz [140]. We also give a combinatorial interpretation for the coefficients in the power series expansion of the reciprocal $\frac{1}{R(-a,-b)}$, extending a result of Odlyzko and Wilf [205]. The full description of this approach would add several more pages to our work, but [8] covers all of this very nicely.

It turns out that **Lebesgue's identity** plays a major role in our analysis with respect to the numerators and denominators of the finite continued fractions we consider.

$$\sum_{k\geq 0} \frac{q^{k(k+1)/2}\prod_{j=1}^{k}(1+bq^j)}{(1-q)(1-q^2)...(1-q^k)} = \prod_{m\geq 1}(1+bq^{2m})(1+q^m). \tag{4.3.8}$$

It is known that Lebesgue's identity implies Ramanujan's fraction $R(a,b)$ has a product representation when $a=1$. More precisely (4.3.14) and (4.3.15) (see below) yield

$$1+\cfrac{bq}{1+q+\cfrac{bq^2}{1+q^2+\cfrac{bq^3}{1+q^3+\cfrac{bq^4}{\ddots}}}} = \prod_{m=1}^{\infty}\frac{(1+bq^{2m-1})}{(1+bq^{2m})}. \tag{4.3.9}$$

A neat case of (4.3.9) is obtained from $q \mapsto q^2$ and $b \mapsto bq^{-1}$ so then

$$1+\cfrac{bq}{1+q^2+\cfrac{bq^3}{1+q^4+\cfrac{bq^5}{1+q^6+\cfrac{bq^7}{\ddots}}}} = \prod_{m=1}^{\infty} \frac{(1+bq^{4m-3})}{(1+bq^{4m-1})}. \tag{4.3.10}$$

For a continued fraction F, let P_n/Q_n denote its nth convergent, and suppose that $\lim_{n\to\infty} P_n = P$, $\lim_{n\to\infty} Q_n = Q$ in a suitable topology. We then say that F has numerator P and denominator Q, and write $P = F^N$, $Q = F^D$. Consider the fraction

$$F(a,c) = 1+a+\cfrac{acq}{1+aq+\cfrac{acq^2}{1+aq^2+\cfrac{acq^3}{1+aq^3+\cfrac{acq^4}{\ddots}}}}.$$

This can be written in the form

$$F(a,c) = \frac{f(a,c)}{f(aq,c)}, \quad \text{where} \quad f(a,c) = \sum_{k\geq 0} A_k q^k.$$

We now compute the coefficients $A_k = A_k(c,q)$, observing that $f(a,c)$ satisfies the recurrence

$$f(a,c) = (1+a)f(aq,c) + acq\, f(aq^2,c).$$

Therefore the coefficients A_k satisfy

$$A_k = q^k A_k + q^{k-1} A_{k-1} q - cq^{2k-1} A_{k-1},$$

which is the same as

$$A_k = \frac{q^{k-1}(1+cq^k)}{(1-q^k)} A_{k-1}.$$

By iteration this yields

$$F(a,c) = \sum_{k\geq 0} \frac{a^k q^{\frac{k(k-1)}{2}} (-cq)_k}{(q)_k}.$$

Let $c = a^{-1}b$. Then

$$R(a,b) = \frac{f(a,a^{-1}b)}{f(aq,a^{-1}b)} - a$$

is Ramanujan's fraction (4.3.1).

Lemma 4.1 *For the fraction $R(a,b)$, the numerator is*

$$R^N(a,b) = \sum_{k\geq 0} \frac{a^k q^{k(k+1)/2}(-a^{-1}b)_k}{(q)_k}, \tag{4.3.11}$$

and the denominator is

$$R^D(a,b) = \sum_{k\geq 0} \frac{a^k q^{k(k+1)/2}(-a^{-1}bq)_k}{(q)_k}. \tag{4.3.12}$$

Proof: The expansion (4.3.12) is an immediate consequence of

$$R^D(a,b) = f(aq, a^{-1}b). \tag{4.3.13}$$

The expansion (4.3.11) is more complicated. To obtain it, observe that

$$\begin{aligned}
R^N(a,b) &= f(a,a^{-1}b) - a f(aq,a^{-1}b) \\
&= \sum_{k\geq 0} \frac{a^k q^{k(k-1)/2}(-a^{-1}bq)_k}{(q)_k} - \sum_{k\geq 0} \frac{a^{k+1} q^{k(k+1)/2}(-a^{-1}bq)_k}{(q)_k} \\
&= 1 + \sum_{k\geq 0} \frac{a^{k+1} q^{k(k+1)/2}(-a^{-1}bq)_k}{(q)_k} \left(\frac{1+a{-}1bq^{k+1}}{1-q^{k+1}} - 1 \right) \\
&= 1 + \sum_{k\geq 0} \frac{a^{k+1} q^{(k+1)(k+2)/2}(-a^{-1}bq)_k (1-a^{-1}b)}{(q)_{k+1}} \\
&= \sum_{k\geq 0} \frac{a^k q^{k(k+1)/2}(-a^{-1}b)_k}{(q)_k}
\end{aligned}$$

as required. ■

Andrews (see [15] and [20]) considered the expansions in lemma 4.1 while discussing a transformation formula of Ramanujan [218] for $R(a,b)$. Our emphasis here is on the partition theorems that can be derived using $R(a,b)$, and for this the following lemma is crucial.

Lemma 4.2 *For the fraction $R(a,b)$, we also have the expansions*

$$R^N(a,b) = \sum_{i,j\geq 0} \frac{a^i b^j q^{(i^2+i)/2+ij+j^2}}{(q)_i (q)_j}, \tag{4.3.14}$$

and the denominator is

$$R^D(a,b) = \sum_{i,j\geq 0} \frac{a^i b^j q^{(i^2+i)/2+ij+j^2+j}}{(q)_i (q)_j}. \tag{4.3.15}$$

Proof: To obtain (4.3.14) and (4.3.15) from (4.3.12) and (4.3.13) we use the q-binomial theorem,

$$(-z)_k = \sum_{j=0}^{k} z^j q^{j(j-1)/2} \binom{k}{j}_q$$

with $z=a^{-1}b$ and $z=a^{-1}bq$. (See Campbell [93] for the n-space q-binomial theorem.) Therefore

$$\begin{aligned} R^N(a,b) &= \sum_{k\geq 0}\frac{a^k q^{k(k+1)/2}}{(q)_k}\sum_{j=0}^{k}\frac{a^{-j}b^j q^{j(j-1)/2}(q)_k}{(q)_j(q)_{j-k}} \\ &= \sum_{i,j\geq 0}\frac{a^i b^j q^{(i+j)(i+j+1)/2}}{(q)_i(q)_j}, \end{aligned}$$

where $i=k-j$; this is equivalent to (4.3.12). To obtain (4.3.13), observe that

$$R^D(a,b)=R^N(a,bq) \tag{4.3.16}$$

by comparing (4.3.14) and (4.3.15).

The following two theorems relate successively to the numerator and the denominator of the fraction (4.3.1), so then to (4.3.14) and (4.3.15). For a proof of these see Alladi and Gordon [8].

Theorem 4.4 ***(Numerator)***

Let $A^N(n;i,j)$ be the number of partitions of n into $i+j$ distinct red parts and j distinct blue parts such that one of the blue parts may be zero and every blue part is $\leq i+j-1$.

Let $B^N(n;i,j)$ be the number of partitions of n into i distinct red parts and j distinct non-consecutive blue parts such that every red part is $>j$.

Let $C^N(n;i,j)$ be the number of partitions of n into i red parts and j blue parts such that all parts are distinct and after each blue part there is a gap of at least 2. Then

$$A^N(n;i,j)=B^N(n;i,j)=C^N(n;i,j).$$

Theorem 4.5 ***(Denominator)***

Let $A^D(n;i,j)$ be as in $A^N(n;i,j)$ except that every blue part is >0 and $\leq i+j$.
Let $B^D(n;i,j)$ be as in $B^N(n;i,j)$ except that part 1 cannot be blue.
Let $C^D(n;i,j)$ be as in $C^N(n;i,j)$ except that part 1 cannot be blue. Then

$$A^D(n;i,j)=B^D(n;i,j)=C^D(n;i,j).$$

So reprising (4.3.10) namely

$$1+\cfrac{bq}{1+q^2+\cfrac{bq^3}{1+q^4+\cfrac{bq^5}{1+q^6+\cfrac{bq^7}{\ddots}}}}=\prod_{m=1}^{\infty}\frac{(1+bq^{4m-3})}{(1+bq^{4m-1})}),$$

we have interesting cancellations in numerator-denominator equations. That is, the numerator is given by

$$\begin{aligned}\sum_{k\geq 0}\frac{q^{k(k+1)}(-bq^{-1};q^2)_k}{(q^2;q^2)_k} &= \prod_{m=1}^{\infty}(1+bq^{4m-3})(1+q^{2m})\\ &= \prod_{m=1}^{\infty}(1+bq^{4m-3})(1+q^{4m-2})(1+q^{4m})\end{aligned}$$

and the denominator is given by

$$\sum_{k\geq 0}\frac{q^{k(k+1)}(-bq;q^2)_k}{(q^2;q^2)_k} = \prod_{m=1}^{\infty}(1+bq^{4m-1})(1+q^{4m-2})(1+q^{4m})$$

with right sides having common factors that eliminate.

This leads in particular to the continued fraction identity

$$1+\cfrac{q}{1+q^2+\cfrac{q^3}{1+q^4+\cfrac{q^5}{1+q^6+\cfrac{q^7}{\ddots}}}} = \frac{\prod_{j\equiv 2,3,7\,(\mathrm{mod} 8)}(1-q^j)}{\prod_{j\equiv 1,5,6\,(\mathrm{mod} 8)}(1-q^j)}. \tag{4.3.17}$$

Göllnitz [140] states similar results, but (4.4) seems to have escaped attention. There is a continued fraction identity due to Gordon [142] and Göllnitz [140] which looks very similar to (4.3.17), namely

$$1+q+\cfrac{q^2}{1+q^3+\cfrac{q^4}{1+q^5+\cfrac{q^4}{1+q^7+\cfrac{bq^6}{\ddots}}}} = \frac{\prod_{j\equiv 3,4,5\,(\mathrm{mod} 8)}(1-q^j)}{\prod_{j\equiv 1,4,7\,(\mathrm{mod} 8)}(1-q^j)}. \tag{4.3.18}$$

However, this result first appears in Alladi and Gordon [8] almost 30 years after (4.4).

4.4 RAMANUJAN'S THREE PARAMETER CONTINUED FRACTION

Ramanujan [216] obtained in addition to (4.3.1), the following continued fraction with three parameters a, b, q which has also a product representation

$$1-ab+\cfrac{(a-bq)(b-aq)}{(1-ab)(1+q^2)+\cfrac{(a-bq^3)(b-aq^3)}{(1-ab)(1+q^4)+\cfrac{(a-bq^5)(b-aq^5)}{(1-ab)(1+q^6)+\cfrac{(a-bq^7)(b-aq^7)}{\ddots}}}} \tag{4.4.1}$$

$$= \prod_{m=1}^{\infty} \frac{(1+a^2q^{4m-3})(1+b^2q^{4m-3})}{(1+a^2q^{4m-1})(1+b^2q^{4m-1})}. \tag{4.4.2}$$

This was proved only in 1985 by the reviewers of Chapter 16 of Ramanujan's Second Notebook [2], 65 years after Ramanujan's death. If we put $a = 0$ and replace b^2 by $-b$ in (4.4.1), we get (4.3.10). It seems there is still scope to study the combinatorial properties of the coefficients in the power series expansion of this fraction.

4.5 THE RAMANUJAN MACHINE AND COMPUTER ASSISTED RESEARCH

Well-known constants like π, e, and γ arise in physics, biology, chemistry, engineering, and mathematical analysis. And yet, over time new mathematical formulas relating different constants are scarce and are found at times by accident or serendipity. The *Ramanujan Machine* (see [214]) uses the combined computer power of thousands of individual PCs to make new discoveries. This is sometimes achieved by brute force but mostly by use of clever algorithms that are capable of self-adjustments to find interesting formulas experimentally, that, though unproven, are very likely true as they are verified to hundreds of decimal places.

Continued fractions are an area that has been fruitfully applied to this computer assisted research methodology. A really good recent example of the successful application of this approach is in the examples given below due to Elimelech et al. (see [121]). The sheer number of formulas discovered by the algorithm unveils a novel mathematical structure that is called the conservative matrix field. Such matrix fields:

(a) Unify thousands of existing formulas.

(b) Generate infinitely many new formulas.

(c) Reveal unexpected relations between different mathematical constants, including multiple integer values of the Riemann zeta function.

Conservative matrix fields also enable new mathematical proofs of irrationality. For example, they were applied to generalize the celebrated proof by Apéry of the irrationality of $\zeta(3)$. Linking thousands of personal computers worldwide, the computer-supported research strategy shows the power of experimental mathematics, and brings large-scale computations into longstanding open problems and finds unexpected ties across diverse fields of research.

4.5.1 Computer Discovered Continued Fraction Values of Polylogarithms

In our book from Chapter 20 onward, we use polylogarithm particular values to derive n-space Visible Point Vector identities. A computer-assisted algorithm was recently applied to polylogarithms and came up with the following new and unproven continued fractions.

We employ the notation

$$\mathbf{F}(a_n, b_n) = a_0 + \cfrac{b_1}{a_1 + \cfrac{b_2}{a_2 + \cfrac{b_3}{a_3 + \cfrac{b_4}{\ddots}}}}. \tag{4.5.1}$$

This enabled the team of seven computer-assisted researchers, Elimelech et al. [121] to find the remarkable experimental result for the polylogarithm,

$$\mathbf{F}(n^d + c(n+1)^d, -cn^{2d}) = Li_d\left(\frac{1}{c}\right). \tag{4.5.2}$$

A few of the results listed below are called *formula candidates* since they are unproven conjectures that evidently verify to hundreds of decimal places.

$\mathbf{a_n}$	$\mathbf{b_n}$	**formula**
$n^4 + (n+1)^4 + 2(n^2 + (n+1)^2)$	$-n^8$	$\frac{1}{-\zeta(4)-2\zeta(2)+8}$
$n^5 + (n+1)^5 + 6(n^3 + (n+1)^3)$	$-n^{10}$	$\frac{2}{2\zeta(5)+6\zeta(3)-9}$
$n^5 + (n+1)^5 + 6(n^3 + (n+1)^3) - 4(2n+1)$	$-n^{10}$	$\frac{2}{2\zeta(5)-2\zeta(3)-1}$
$n^5 + (n+1)^5 + 16(n^3 + (n+1)^3) - 4(2n+1)$	$-n^{10}$	$\frac{64}{64\zeta(5)+176\zeta(3)-273}$
$n^7 + (n+1)^7 + 8(n^5 + (n+1)^5) -$ $8(n^3 + (n+1)^3) - 4(2n+1)$	$-n^{14}$	$\frac{1}{\zeta(7)-4\zeta(3)+4}$

These results come from the *Ramanujan Machine* project [214] computer algorithm, which has discovered an unprecedented number of continued fraction formulas for fundamental mathematical constants. So, our message here is that the *Ramanujan Machine* project and similar distributed systems assisted research, may benefit by participation of our readers. Also, researchers may benefit from checking *Ramanujan Machine* discoveries for useful *formula candidates* giving rise to research projects to find formal proofs of new results related to current studies.

Think About It...

A man is like a fraction whose numerator is what he is and whose denominator is what he thinks of himself.

The larger the denominator, the smaller the fraction.

Leo Tolstoy
Russian Author; 1828 - 1910

FURTHER READING

ADIGA, C., BERNDT, B.C., BHARGAVA, S., and WATSON, G.N. Chapter 16 of Ramanujan's Second Notebook: Theta Functions and q-Series. In *Memoirs of the American Mathematical Society*, volume 315. Amer. Math. Soc., Providence, RI, 1985.

ALLADI, K. and GORDON H. Partition Identities and a Continued Fraction of Ramanujan. *J. Comb. Theory*, Series A, 63:275–300, 1993.

ANDREWS, G.E. *The Theory of Partitions*, volume 2 of *Encyclopedia of Mathematics and its Applications*. Addison-Wesley Publishing Company, Advanced Book Program, Reading, Massachusetts, USA, 1976.

ANDREWS, G.E. An introduction to Ramanujan's "lost" notebook. *Amer. Math. Monthly*, 86:89–108, 1979.

ANDREWS, G.E. Ramanujan's "Lost" Notebook. III. The Rogers-Ramanujan continued fraction. *Adv. Math.*, 41:186–208, 1981.

ANDREWS, G.E. and BERNDT, B.C. *Ramanujan's Lost Notebook: Part V Paperback*. Ramanujan's Notebooks. Springer-Verlag, New York, 2018. ISBN-13: 978-3030085506.

BRESSOUD, D.M. On a partition theorem of Göllnitz. *J. Reine Angew. Math.*, 305:215–217, 1979.

CAMPBELL, G.B. Some n-space q-binomial theorem extensions and similar identities. arxiv preprint, June 2019. arXiv:1906.07526v1 [math.NT]: https://arxiv.org/abs/1906.07526.

CAMPBELL, G.B. Continued Fractions for partition generating functions. arxiv preprint, Jan 2023. arXiv:2301.12945 [math.CO], https://arxiv.org/abs/2301.12945.

ELIMELECH, R., DAVID, O., De la CRUZ MENGUAL, C., KALISCH, R., BERNDT, W., SHALYT, M., SILBERSTEIN, M., HADAD, Y., and KAMINER, I. Algorithm-assisted discovery of an intrinsic order among mathematical constants, Oct 2023. https://arxiv.org/abs/2308.11829.

EULER, L. Introductio in analysin infinitorum. In *Opera Omnia*, volume 1. Marcum-Michaelum, Brousquet, Lausannae, 1748. Chapter 16.

GÖLLNITZ, H. Partitionen mit Differenzenbedingungen. *J. Reine Angew. Math.*, 225:154–190, 1967.

GORDON, B. Some continued fractions of the Rogers-Ramanujan type. *Duke Math. J.*, 32:741–748, 1965.

ODLYZKO, A.M. and WILF, H.S. n coins in a fountain. *Amer. Math. Monthly*, 95:840–843, 1988.

The RAMANUJAN MACHINE. The Ramanujan Machine parallel computing project. (web page), 2023. [Online; accessed 22 October 2023]. www.ramanujanmachine.com.

RAMANUJAN, S. *Notebooks (Two Volumes)*. Tata Institute, Bombay, 1957.

RAMANUJAN, S. *The Lost Notebook, and Other Unpublished Papers*. Narosa, New Delhi, 1988.

ROGERS, L.J. Second memoir on the expansion of certain infinite products. *Proc. London Math. Soc.*, 25:318–343, 1894.

SANDIFER, C.E. Chapter 32: Who proved e is irrational?. How Euler Did It. *Mathematical Association of America*, pages 185–190, 2006. ISBN: 978-0-88385-563-8. LCCN 2007927658.

CHAPTER 5

Congruence Properties of Partitions

5.1 RAMANUJAN'S PARTITION CONGRUENCES

We now discuss Ramanujan's striking divisibility identities, suggested when Ramanujan examined MacMahon's table of partitions from $p(1)$ to $p(200)$, which was derived by manual calculation circa 1917.

Hardy states three of Ramanujan's congruence equations, (see [215]) specifically,

$$p(5m+4) \equiv 0, \pmod 5 \tag{5.1.1}$$

$$p(7m+5) \equiv 0, \pmod 7 \tag{5.1.2}$$

$$p(11m+6) \equiv 0, \pmod{11} \tag{5.1.3}$$

5.1.1 Ramanujan's Most Beautiful Identity

Inferred from (5.1.1) and (5.1.2), Hardy then gives the two connected identities (5.1.4) and (5.1.5) from Ramanujan. The first of these Hardy famously rates, in his personal opinion, as Ramanujan's most beautiful identity. That's a significant statement when you consider the hundreds of charming and astonishingly attractive identities Ramanujan produced in his lifetime.

$$\sum_{n=0}^{\infty} p(5n+4)q^n = 5\frac{(q^5;q^5)_\infty^5}{(q;q)_\infty^6}, \tag{5.1.4}$$

$$\sum_{n=0}^{\infty} p(7n+5)q^n = 7\frac{(q^7;q^7)_\infty^3}{(q;q)_\infty^4} + 49\frac{(q^7;q^7)_\infty^7}{(q;q)_\infty^8}, \tag{5.1.5}$$

where as usual

$$(a;q)_\infty = \prod_{n=1}^{\infty}(1-q^n) \quad \text{and} \quad (a_1,a_2,...,a_m;q) = (a_1;q)_\infty(a_2;q)_\infty \dots (a_m;q)_\infty. \tag{5.1.6}$$

In describing 21st-century updates of this theory we adopt the abbreviation $E(n) = (q^n;q^n)_\infty$.

DOI: 10.1201/9781003174158-5

It is well known that the two identities (5.1.4) and (5.1.5) remarkably encode the congruences (5.1.1) and (5.1.2) shown above.

In 1967 Atkin [34] proved the modified conjecture of Ramanujan that

Theorem 5.1 *If* $\delta = 5^a 7^b 11^c$ *and* $24\lambda \equiv 1 (mod\, \delta)$ *then*

$$p(\lambda) \equiv 0, \quad (mod\, 5^a 7^{[(b+2)/2]} 11^c). \tag{5.1.7}$$

In terms of congruences, $p(n)$ has served as a testing ground for fundamental constructions in the theory of modular forms. Some of the deepest results on partition congruences have been obtained by making use of modular equations, Hecke operators, Shimura's correspondence, and the Deligne-Serre theory of ℓ-adic Galois representations.

Here we revisit the theory of Ramanujan's celebrated congruences, which assert that

$$p(5^m n + \delta_5(m)) \equiv 0, \quad (\text{mod } 5^m) \tag{5.1.8}$$

$$p(7m^n + \delta_7(m)) \equiv 0, \quad (\text{mod } 7^{\lfloor m/2 \rfloor + 1}) \tag{5.1.9}$$

$$p(11^m n + \delta_{11}(m)) \equiv 0, \quad (\text{mod } 11^m) \tag{5.1.10}$$

where $0 < \delta_\ell(m) < \ell^m$ satisfies the congruence $24\delta_\ell(m) \equiv 1\ (mod\, \ell^m)$. To prove these congruences, Atkin, Ramanujan, and also Watson [6, 29, 30, 35] made use of special modular equations to produce ℓ-adic expansions of the generating functions

$$P_\ell(b; z) = \sum_{n=0}^{\infty} p\left(\frac{\ell^b n + 1}{24}\right) q^{\frac{n}{24}} \tag{5.1.11}$$

where $q = e^{2\pi i z}$ throughout, $p(0) = 1$, and $p(\alpha) = 0$ if $\alpha < 0$ or $\alpha \in \mathbb{Z}$.

5.2 1944, FREEMAN DYSON AND HIS HYPOTHESIZED CRANK

The crank of a partition of an integer is a certain integer mapped to from the partition. The term *crank* derives from the similarity to the *rank* of a partition, with the added benefit of congruence information about that partition. It was first introduced without a definition by Freeman Dyson in a 1944 paper published in Eureka, a journal of the Mathematics Society of Cambridge University. [119] Dyson then gave a list of properties this not-yet-defined quantity should have. In 1988, George E. Andrews and Frank Garvan [30] discovered a definition for the crank satisfying the properties hypothesized for it by Dyson 44 years earlier. It is possible that Ramanujan in the last year of his life already knew about the crank partition function, as the *Lost Notebook* has related draft formulas.

5.2.1 Dyson's Definition of Rank

In [119] Freeman Dyson defined the rank of a partition as the largest part minus the number of parts. He let $N(m,t,n)$ denote the number of partitions of n of rank congruent to m modulo t, and he conjectured that

$$N(m,5,5n+4) = \frac{p(5n+4)}{5} \quad 0 \le m \le 4, \tag{5.2.1}$$

$$N(m,7,7n+5) = \frac{p(7n+5)}{7} \quad 0 \le m \le 6. \tag{5.2.2}$$

These conjectures were later proved by Atkin and Swinnerton-Dyer [35]. Dyson gave several interesting equations that he couldn't prove in 1944, that later turned out to be true. For example, Dyson asserted that: In the power series generated from the product of the series

$$\left(\sum_{m=0}^{\infty} p(m)x^m\right)\left(\sum_{r=1}^{\infty}(-1)^r \frac{\{x^{\frac{r(3r-1)}{2}} - x^{\frac{r(3r+1)}{2}}\}\{x^r - x^{3r} - x^{4r} + x^{6r}\}}{(1-x^{7r})}\right) \tag{5.2.3}$$

the coefficients of x^{7n+1}, x^{7n+3}, x^{7n+4}, x^{7n+5}, vanish identically.

5.3 1988, GEORGE ANDREWS AND FRANK GARVAN FIND DYSON'S CRANK

The *crank* whose properties were stated and proposed to exist by Dyson in 1944 was *sought and found* 44 years later by Andrews and Garvan [30]. (See also Garvan [132, 133].) Andrews and Garvan developed insights from Dyson's note that the rank did not separate the partition of $11n+6$ into 11 equal classes, even though Ramanujan's congruence

$$p(11n+6) \equiv 0 \pmod{11} \tag{5.3.1}$$

holds. Dyson had conjectured the existence of some other function of a partition (which he called the *crank*); this new function to give a combinatorial interpretation of $\frac{1}{11}p(lln+6)$ in the same way that (5.2.1) and (5.2.2) do for primes 5 and 7.

5.3.1 Andrews and Garvan's Definition of Crank

George Andrews and Frank Garvan in 1988 defined the crank of a partition as follows:

Definition 5.1 *For a partition λ, let $l(\lambda)$ denote the largest part of λ, $\omega(\lambda)$ denote the number of 1s in λ, and $\mu(\lambda)$ denote the number of parts of λ larger than $\mu(\lambda)$. The crank $c(\lambda)$ is given by*

$$c(\lambda) = \begin{cases} l(\lambda), & \text{if } \omega(\lambda) = 0; \\ \mu(\lambda) - \omega(\lambda), & \text{if } \omega(\lambda) > 0. \end{cases} \tag{5.3.2}$$

With this definition we get the following verifications to show how Ramanujan's congruences (5.1.1), (5.1.2), and (5.1.3) separate partitions of:

(a) $5n+4$ into 5 equal classes, the first case mapping the 5 partitions of 4;

partitions of 4	$l(\lambda)$	$\omega(\lambda)$	$\mu(\lambda)$	$c(\lambda)$
4	4	0	1	4
3+1	3	1	1	0
2+2	2	0	2	2
2+1+1	2	2	0	−2
1+1+1+1	1	4	0	−4

(b) $7n+5$ into 7 equal classes, the first case mapping the 7 partitions of 5;

partitions of 5	$l(\lambda)$	$\omega(\lambda)$	$\mu(\lambda)$	$c(\lambda)$
5	5	0	1	5
4+1	4	1	1	0
3+2	3	0	2	3
3+1+1	3	2	1	−1
2+2+1	2	1	2	1
2+1+1+1	2	3	0	−3
1+1+1+1+1	1	5	0	−5

(c) $11n+6$ into 11 equal classes, the first case mapping the 11 partitions of 6.

partitions of 6	$l(\lambda)$	$\omega(\lambda)$	$\mu(\lambda)$	$c(\lambda)$
6	6	0	1	6
5+1	5	1	1	0
4+2	4	0	2	4
4+1+1	4	2	1	−1
3+3	3	0	2	3
3+2+1	3	1	2	1
3+1+1+1	3	3	0	−3
2+2+2	2	0	3	2
2+2+1+1	2	2	0	−2
2+1+1+1+1	2	4	0	−4
1+1+1+1+1+1	1	6	0	−6

For all integers $n \geq 0$ and all integers m, the number of partitions of n with crank equal to m is denoted by $M(m,n)$ except for $n=1$ where $M(-1,1) = -M(0,1) = M(1,1) = 1$ as given by the following generating function. The number of partitions of n with crank equal to m modulo q is denoted by $M(m,q,n)$.

The generating function for $M(m,n)$ is given as

$$\sum_{n=0}^{\infty} \sum_{m=-\infty}^{\infty} M(m,n) z^m q^n = \prod_{n=1}^{\infty} \frac{(1-q^n)}{(1-zq^n)(1-z^{-1}q^n)}. \tag{5.3.3}$$

5.4 FURTHER CONGRUENCE IDENTITIES FOR PARTITIONS

In the years since Atkin's 1960s proofs and the 2013 proofs of Bruinier and Ono [81] and Bruinier et al. [80, 82] from 2006, we have also the 2012 work of Toh [294], as well as Zhao and Zhong [305] which is worth consideration, as the ideas may bear fruit in extensions to vector partitions. Using elementary methods, Toh establishes several new Ramanujan type identities and congruences for certain pairs of partition functions. Toh uses the $E(n) = (q^n; q^n)_\infty$ as an adopted notation, so with this notation the partition function $p(n)$ for example, has generating function $\sum_{n=0}^{\infty} p(n)q^n = 1/E(1)$. Toh next defines the following seven partition functions intended for later usage:

1) $p_o(n) = p(n \mid \text{odd parts})$, where

$$\sum_{n=0}^{\infty} p_o(n)q^n = \frac{1}{(q;q^2)_\infty} = \frac{E(2)}{E(1)}.$$

2) $p_e(n) = p(n \mid \text{even parts})$, where

$$\sum_{n=0}^{\infty} p_e(n)q^n = \frac{1}{(q^2;q^2)_\infty} = \frac{1}{E(2)}.$$

3) $\overline{p}(n) = p(n \mid \text{overpartitions})$, where

$$\sum_{n=0}^{\infty} \overline{p}(n)q^n = \frac{(-q;q)_\infty}{(q;q)_\infty} = \frac{E(2)}{E(1)^2}.$$

4) $\overline{p}_o(n) = p(n \mid \text{overpartitions into odd parts})$, where

$$\sum_{n=0}^{\infty} \overline{p}_o(n)q^n = \frac{(-q;q^2)_\infty}{(q;q^2)_\infty} = \frac{E(2)^3}{E(1)^2E(4)}.$$

5) $\overline{p}_e(n) = p(n \mid \text{overpartitions into even parts})$, where

$$\sum_{n=0}^{\infty} \overline{p}_e(n)q^n = \frac{(-q^2;q^2)_\infty}{(q^2;q^2)_\infty} = \frac{E(4)}{E(2)^2}.$$

6) $p_{od}(n) = p(n \mid \text{odd parts are distinct})$, where

$$\sum_{n=0}^{\infty} p_{od}(n)q^n = \frac{(-q;q^2)_\infty}{(q^2;q^2)_\infty} = \frac{E(2)}{E(1)E(4)}.$$

7) $p_{ed}(n) = p(n \mid \text{even parts are distinct})$, where

$$\sum_{n=0}^{\infty} p_{ed}(n)q^n = \frac{(-q^2;q^2)_\infty}{(q;q^2)_\infty} = \frac{E(4)}{E(1)}.$$

Properties of these partition functions, as well as identities and congruences satisfied by them, have been the subject of many 21st century studies See for example

Hammond and Lewis [147], Bringmann and Lovejoy [77],or Hirschhorn and Sellers [156] the latter proving that the overpartition function $\overline{p}(n)$ satisfies the following analogue of (5.1.4) and (5.1.5),

$$\sum_{n=0}^{\infty} \overline{p}(2n+1)q^n = 2\frac{E(2)^2E(8)^2}{E(1)^4E(4)}. \tag{5.4.1}$$

It emerged that instead of studying a single partition function, one may also consider a partition pair formed from partition functions $p_1(n)$ and $p_2(n)$. Let $Q(p_1,p_2)(n)$ be the number of partitions of n into two colors (say, red and blue), where the parts colored red satisfy restrictions of partitions counted by $p_1(n)$, while the parts colored blue satisfy restrictions of partitions counted by $p_2(n)$. The generating function for $Q(p_1,p_2)(n)$ is then the product of the generating functions of $p_1(n)$ and $p_2(n)$. For example, partitions into two colors with no restrictions on the red parts, but where the blue parts must be even, are counted by $Q_{(p,pe)}(n)$. Chan [105] over a decade ago, showed that such partitions satisfy the following remarkable identity

$$\sum_{n=0}^{\infty} Q_{(p,pe)}(3n+2)q^n = 3\frac{E(3)^3E(6)^3}{E(1)^4E(2)^4}. \tag{5.4.2}$$

When expressed without so much abbreviation, we see that this reads as

$$\sum_{n=0}^{\infty} Q_{(p,pe)}(3n+2)q^n = 3\prod_{n=1}^{\infty} \frac{(q^3;q^3)_\infty^3(q^6;q^6)_\infty^3}{(q;q)_\infty^4(q^2;q^2)_\infty^4} \tag{5.4.3}$$

$$= 3\prod_{n=1}^{\infty} \frac{\{(1-q^3)(1-q^6)(1-q^9)\cdots\}^3\{(1-q^6)(1-q^{12})(1-q^{18})\cdots\}^3}{\{(1-q)(1-q^2)(1-q^3)\cdots\}^4\{(1-q^2)(1-q^4)(1-q^6)\cdots\}^4} \tag{5.4.4}$$

$$= 3\prod_{n=1}^{\infty} \frac{\{(1-q^3)(1-q^6)^2(1-q^9)(1-q^{12})^2(1-q^{15})(1-q^{18})^2(1-q^{21})(1-q^{24})^2\cdots\}^3}{\{(1-q)(1-q^2)^2(1-q^3)(1-q^4)^2(1-q^5)(1-q^6)^2(1-q^7)(1-q^8)^2\cdots\}^4}. \tag{5.4.5}$$

Pairs of partition functions with assigned colors seem *a priori* to be not too far an intellectual stretch from examining two variable 2D vector partition generating function theory, which interests us in later chapters. Recently in 2023, Dicks [118] has studied the congruence relations for r-colored partitions; a concept that can perhaps be in some way related to r-dimensional vector partitions.

Let $\ell \geq 5$ be prime. For the partition function $p(n)$ and $5 \leq \ell \leq 31$, Atkin [34] found a number of examples of primes $Q \geq 5$ such that there exist congruences of the form $p(\ell Q^3 n + \beta) \equiv 0 \pmod{\ell}$. Recently, Ahlgren, Allen, and Tang [3] proved that there are infinitely many such congruences for every ℓ. (Also see Ahlgren et al. [4] in a paper to appear.) In the 2023 work by Dicks [118], for a wide range of $c \in F\ell$, he proves congruences of the form

$$p(\ell Q^3 n + \beta_0) \equiv c \cdot p(\ell Q n + \beta_1) \pmod{\ell}$$

for infinitely many primes Q. For a positive integer r, let $p_r(n)$ be the r-colored partition function. Dicks' methods yield similar congruences for $p_r(n)$. In particular, if r is an odd positive integer for which $\ell > 5r+19$ and $2^{r+2} \not\equiv 2^{\pm 1} (\mathrm{mod}\, \ell)$, then we show that there are infinitely many congruences of the form $pr(\ell Q^3 n+\beta) \equiv 0\, (\mathrm{mod}\, \ell)$. Dicks' approaches involve the theory of modular Galois representations.

5.5 RÖDSETH-GUPTA THEOREM ON BINARY PARTITIONS

We consider partitions of the natural number n into the powers of two. The partitions which differ in the order of summands are considered to be identical. Let the number of partitions, that are the number of solutions of the equation

$$n = m_s 2^s + m_{s-1} 2^{s-1} + \cdots + m_0$$

in nonnegative integers be $b(n)$. This function was investigated by Euler (1750) [123], Tanturi (1918), Mahler (1940) [194] and many others. A more contemporary account with generalizations is given by Alkauskas and Mendel-Strasse [7]. Let, for simplicity, $b(0) = 1$. The generating power series of the sequence $b(n)$ is

$$F(x) = \sum_{n=0}^{\infty} b(n) x^n = \prod_{n=0}^{\infty} \frac{1}{1 - x^{2^n}}.$$

This function satisfies the functional equation $(1-x)F(x) = F(x^2)$ and, by comparing coefficients of like powers of x, we get

$$b(2n+1) = b(2n), \quad b(2n) = b(2n-1) + b(n). \tag{5.5.1}$$

These two equations have a fairly straightforward combinatorial meaning:

(a) Every partition of the number $2n+1$ has at least one summand equal to 1, and this gives the first formula of (5.5.1).

(b) The number of binary partitions of the number $2n$ with $m_0 \neq 0$ is $b(2n-1)$. The summands of the rest of each partition are divisible by 2; hence the number of such partitions is $b(n)$.

It is well known that $2^{n^2/4} < b(2n) < 2^{n^2/2}$ for $n \geq 3$. This statement can be proved using the recurrence relations (5.5.1). It can be deduced from here that $b(2h) \leq hb(h)$ and $b(4h) > 2hb(h)$ for $h \geq 2$. Mahler in 1940 investigated the difference equation $(f(z+\omega) - f(z))\omega^{-1} = f(qz)$ and derived the following asymptotic formula for $b(n)$:

$$b(2n) = e^{O(1)} \sum_{v=0}^{\infty} \frac{2^{-v(v-1)} n^v}{v!}, \quad n \to \infty.$$

This gives the exact asymptotic for $\log b(n)$. Related results and generalizations of de Bruin are given in Pennington [208] who sharpened the Mahler results by using a Tauberian theorem. He extracted from the remainder term $O(1)$ some periodical function, thus getting an $o(1)$ term. The first asymptotic term of the logarithm

$$\log b(n) \sim \frac{(\log n)^2}{2\log 2} \tag{5.5.2}$$

can be easily calculated using the geometrical method of Alkauskas [5]. This result was proved in an elementary (combinatorial) way by Eckstein [120].

In 1969, Churchhouse [110] noted some arithmetical properties of the sequence $b(n)$. He proved some theorems and made a conjecture which now is known as the Rödseth-Gupta theorem.

Theorem 5.2 *(Rödseth-Gupta theorem.) If n is an odd positive integer, then for any integer $s \geq 1$ we have*

$$b(2^{s+2}n) \equiv b(2^s n) \ (mod\, 2^{\mu(s)}), \quad where \quad \mu(s) = \left\lfloor \frac{3s+4}{2} \right\rfloor; \tag{5.5.3}$$

and this congruence is exact.

This was independently proved by Rödseth [224] and also by Gupta [146]. A third proof was given in Andrews' classical book on the Theory of Partitions [12].

5.6 THE 2D AND 3D BINARY PARTITIONS CONGRUENCE THEORY

There seems to be no existing 2D or 3D, etc. congruence theory in the present literature for vector partitions $b_2(j,k)$ = the number of partitions of vector $\langle j,k\rangle$ having into unrestricted vectors of form $\langle 2^a, 2^b\rangle$ where $a,b \geq 0$. However, an approach can probably be made into this as a research area by way of the functional equations presented in chapter 15. In that chapter, the 2D, 3D, and higher binary partition functional equations are given; and these appear to be a good starting place to discover more about congruences for higher dimensional vector partitions with parts n-space vectors with binary components. The 2D version we give here.

Theorem 5.3 *If for $|x| < 1, |y| < 1, |z| < 1$, and all complex numbers a and t,*

$$B_2(y,z) = \prod_{j,k\geq 0} \frac{1}{1-y^{2^j}z^{2^k}} = \sum_{\alpha,\beta\geq 0}^{\infty} b_2(\alpha,\beta) y^\alpha z^\beta,$$

then

$$\frac{1}{1-yz} = \frac{B_2(y,z)B_2(y^2,z^2)}{B_2(y^2,z)B_2(y,z^2)}.$$

For a simple proof of this see chapter 15. Theorem 5.3 implies that

$$B_2(y^2,z)B_2(y,z^2) = (1-yz)B_2(y,z)B_2(y^2,z^2).$$

Substituting in this from theorem 5.3 shows it is equivalent to

$$\sum_{\alpha,\beta,\gamma,\delta\geq 0}^{\infty} b_2(\alpha,\beta)b_2(\gamma,\delta)y^{2\alpha+\gamma}z^{\beta+2\delta} = (1-yz)\sum_{\alpha,\beta,\gamma,\delta\geq 0}^{\infty} b_2(\alpha,\beta)b_2(\gamma,\delta)y^{\alpha+2\gamma}z^{\beta+2\delta} \tag{5.6.1}$$

$$= \sum_{\alpha,\beta,\gamma,\delta\geq 0}^{\infty} b_2(\alpha,\beta)b_2(\gamma,\delta)\left(y^{\alpha+2\gamma}z^{\beta+2\delta} - y^{1+\alpha+2\gamma}z^{1+\beta+2\delta}\right)$$

$$= \sum_{\alpha,\beta,\gamma,\delta\geq 0}^{\infty} \left[b_2(\alpha,\beta)b_2(\gamma,\delta) - b_2(\alpha-1,\beta)b_2(\gamma-1,\delta)\right] y^{\alpha+2\gamma}z^{\beta+2\delta}, \tag{5.6.2}$$

defining $b_2(-1,\beta) = 0$ for any $\beta \geq 0$. Comparing (5.6.1) to (5.6.2), we can find a functional equation for the 2D binary partitions $b_2(\alpha,\beta)$ by equating coefficients of like powers of $y^m z^n$. This is achieved by first equating coefficients of powers of y^m and then coefficients of powers of z^n. So, to start, we equate coefficients of like powers of y by solving a system of simultaneous equations, for positive integers m,

$$2\alpha+\gamma = m, \quad \alpha+2\gamma = m, \tag{5.6.3}$$

obtaining $\alpha = \frac{m}{3}$ and $\gamma = \frac{m}{3}$.

Hence, comparing (5.6.1) with (5.6.2) we obtain, for $\alpha,\beta \geq 0$,

$$b_2(3\alpha,\beta)b_2(\gamma,3\delta) = \left[b_2(3\alpha,\beta)b_2(3\gamma,\delta) - b_2(3\alpha-1,\beta)b_2(3\gamma-1,\delta)\right].$$

so then

$$b_2(3\alpha-1,\beta)b_2(3\gamma-1,\delta) = 0. \tag{5.6.4}$$

In the 3D and 4D situations, also there is scope to research the functional equations and to infer congruence information. We here preview the results on this up to the 4D version from our future chapter 15.

Theorem 5.4 *If for all complex numbers w, x, y, and z, such that $|w| < 1$,*

$|x| < 1, |y| < 1, |z| < 1,$

$$\begin{aligned}
B_1(z) &= \prod_{k\geq 0}\frac{1}{1-z^{2^k}} = \sum_{\alpha\geq 0}^{\infty} b(\alpha)z^{\alpha}, \\
B_2(y,z) &= \prod_{j,k\geq 0}\frac{1}{1-y^{2^j}z^{2^k}} = \sum_{\alpha,\beta\geq 0}^{\infty} b_2(\alpha,\beta)y^{\alpha}z^{\beta}, \\
B_3(x,y,z) &= \prod_{j,k,l\geq 0}\frac{1}{1-x^{2^j}y^{2^k}z^{2^l}} = \sum_{\alpha,\beta,\gamma\geq 0}^{\infty} b_3(\alpha,\beta,\gamma)x^{\alpha}y^{\beta}z^{\gamma}, \\
B_4(w,x,y,z) &= \prod_{j,k,l,m\geq 0}\frac{1}{1-w^{2^j}x^{2^k}y^{2^l}z^{2^m}} = \sum_{\alpha,\beta,\gamma,\delta\geq 0}^{\infty} b_4(\alpha,\beta,\gamma,\delta)w^{\alpha}x^{\beta}y^{\gamma}z^{\delta},
\end{aligned}$$

etc.

then

$$\begin{aligned}
\frac{1}{1-z} &= \frac{B_1(z)}{B_1(z^2)}, \\
\frac{1}{1-yz} &= \frac{B_2(y,z)B_2(y^2,z^2)}{B_2(y^2,z)B_2(y,z^2)}, \\
\frac{1}{1-xyz} &= \frac{B_3(x,y,z)B_3(x,y^2,z^2)B_3(x^2,y^2,z)B_3(x^2,y,z^2)}{B_3(x^2,y,z)B_3(x,y^2,z)B_3(x,y,z^2)B_3(x^2,y^2,z^2)}, \\
\frac{1}{1-wxyz} &= \frac{B_4(w,x,y,z)B_4(w,x,y^2,z^2)B_4(w,x^2,y^2,z)B_4(w,x^2,y,z^2)}{B_4(w^2,x,y,z)B_4(w,x^2,y,z)B_4(w,x,y^2,z)B_4(w,x,y,z^2)} \\
&\quad \times \frac{B_4(w^2,x^2,y,z)B_4(w^2,x,y^2,z)B_4(w^2,x,y,z^2)B_4(w^2,x^2,y^2,z^2)}{B_4(w^2,x^2,y^2,z)B_4(w^2,x^2,y,z^2)B_4(w^2,x,y^2,z^2,)B_4(w,x^2,y^2,z^2)}, \\
&\textit{etc.}
\end{aligned}$$

where the numerator and denominator combinations are derived from the even and the odd symmetric functions respectively.

Again, for a simple proof of this see Chapter 15. Theorem 5.4 implies that

$$\begin{aligned}
&B_3(x^2,y,z)B_3(x,y^2,z)B_3(x,y,z^2)B_3(x^2,y^2,z^2) \\
&= (1-xyz)B_3(x,y,z)B_3(x,y^2,z^2)B_3(x^2,y^2,z)B_3(x^2,y,z^2).
\end{aligned}$$

Substituting in this from theorem 5.4 shows it is equivalent to the statement,

$$\begin{aligned}
&\sum_{a,b,c,d,e,f,g,h,i,j,k,l\geq 0}^{\infty} b_3(a,b,c)b_3(d,e,f)b_3(g,h,i)b_3(j,k,l)x^{2a+d+g+2j}y^{b+2e+h+2k}z^{c+f+2i+2l} \\
&= (1-yz)\times \\
&\sum_{a,b,c,d,e,f,g,h,i,j,k,l\geq 0}^{\infty} b_3(a,b,c)b_3(d,e,f)b_3(g,h,i)b_3(j,k,l)x^{a+d+2g+2j}y^{b+2e+2h+k}z^{c+2f+i+2l}.
\end{aligned} \tag{5.6.5}$$

From this, it seems plausible to find functional equations for $b_3(a,b,c)$. Evidently inferences arising from this section regarding congruence properties of binary 2D and 3D vector partitions may provide topics for further research.

5.7 CONGRUENCES ARISING FROM EXPONENTIAL GENERATING FUNCTIONS

From Chapter 21 we borrow some of our derived bivariate, trivariate and several-variable infinite products that involve exponential generating functions, such as for $|y|<1$, $|z|<1$,

$$\prod_{\substack{j,k\geq 1 \\ j\leq k;\,(j,k)=1}} \left(1-y^j z^k\right)^{\frac{1}{k}} = \left(\frac{1-z}{1-yz}\right)^{\frac{y}{1-y}}. \tag{5.7.1}$$

From this we derive, again for $|y| < 1$, $|z| < 1$,

$$\prod_{\substack{j,k \geq 1; j \leq k \\ gcd(j,k)=1}} \left(1 + y^j z^k\right)^{\frac{1}{k}} = \left(\frac{1 - yz}{1 - z}\right)^{\frac{y}{1-y}} \left(\frac{1 - z^2}{1 - y^2 z^2}\right)^{\frac{y^2}{1-y^2}}. \tag{5.7.2}$$

Particular cases arising from these are, firstly, for $\varphi(k)$ the Euler totient function, the number of positive integers less than and coprime to k,

$$\prod_{k=1}^{\infty} \left(1 - z^k\right)^{\frac{\varphi(k)}{k}} = e^{\frac{z}{z-1}} = \sum_{k=0}^{\infty} \frac{\alpha(k)}{k!} \tag{5.7.3}$$

$$= 1 - \frac{z}{1!} - \frac{z^2}{2!} - \frac{z^3}{3!} + \frac{z^4}{4!} + \frac{19z^5}{5!} + \frac{151z^6}{6!} + \frac{1091z^7}{7!}$$

$$+ \frac{7841z^8}{8!} + \frac{56519z^9}{9!} + \frac{396271z^{10}}{10!} + O(z^{11}),$$

demonstrating sequence $\alpha(k)$ has the exponential generating function $e^{\frac{z}{z-1}}$. It is worth noting that in for most values of k, we have $\gcd(\alpha(k), k!) = 1$, (the first break with this is at k=34) and also there is a recurrence relation

$$\alpha(n) + (n-1)(n-2)\,\alpha(n-2) = (2n-3)\,\alpha(n-1)$$

holds. (See OEIS sequence A293116 [237]) This recurrence relation implies continued fractions for the ratios $\alpha(n+1)/\alpha(n)$.

In a superficial observation of the exponential generating function coefficients for (5.7.3), one notices the sequence $\alpha(k)$ in base 10 always ends in digit 1 or 9, so, one easily conjectures that $\alpha(k) \equiv 1 \; or \; 9 \; (mod\ 10)$. In fact, evidently we have the congruence $\alpha(n+k) \equiv \alpha(n) \; (mod\ k)$ true for all positive integers n and k.

$\mathbf{e^{\frac{z}{z-1}}}$ coefficients showing 1 or 9 modulo 10 feature.

n	α(n) from (5.7.3)
0	1
1	−1
2	−1
3	−1
4	1
5	19
6	151
7	1091
8	7841
9	56519
10	396271
11	2442439
12	7701409
13	−145269541
14	−4833158329
15	−104056218421
16	−2002667085119
17	−37109187217649
18	−679877731030049
19	−12440309297451121
20	−227773259993414719
21	−4155839606711748061
22	−74724654677947488521
23	−1293162252850914402221
24	−20381626111249718908319
25	−244110863655032038665001
26	267543347653261450406351
27	172316772106087159102974551
28	8944973491570029894272392801
29	361702062324149751903132843499
30	13353699077321671584329389125031

The second particular case arising again in Chapter 21 derived from (5.7.3) is

$$\prod_{k=1}^{\infty}\left(1+z^k\right)^{\frac{\varphi(k)}{k}} = e^{\frac{z}{1-z^2}} = \sum_{k=0}^{\infty}\frac{\beta(k)}{k!} \tag{5.7.4}$$

$$= 1+\frac{z}{1!}+\frac{z^2}{2!}+\frac{7z^3}{3!}+\frac{25z^4}{4!}+\frac{181z^5}{5!}+\frac{1201z^6}{6!}$$

$$+\frac{10291z^7}{7!}+\frac{97777z^8}{8!}+\frac{202709z^9}{9!}+O(z^{10}).$$

Upon similar examining of the sequence $\beta(k)$ in (5.7.4), we see remarkably that $\gcd(\beta(k),k!)=1$ for most cases so far calculated.

The first 30 coefficients for this exponential generating function are given by

n	β(n) from (5.7.4)
0	1
1	1
2	1
3	7
4	25
5	181
6	1201
7	10291
8	97777
9	**202709** $\times$ **5**
10	**1743223** $\times$ **7**
11	151573951
12	2173233481
13	31758579997
14	**104811403133** $\times$ **5**
15	8838296029291
16	164416415570401
17	**449336774160103** $\times$ **7**
18	65057767274601217
19	**278248694109978827** $\times$ **5**
20	31671795881695430521
21	747996624368605997701
22	**1430133056253846489397** $\times$ **13**
23	480126276655495820797987
24	**3697872284538200564519 63** $\times$ **5** $\times$ **7**
25	3625075061022493145870094 01
26	105371977207333078536961680 01
27	317940307036989922598772811951
28	9915242854364117895813434738857
29	**2563767360580175587315966478129** $\times$ **5**^{3}
30	1067453725131060215237041288037 2081

where the **bolded** numbers are shown here to be "not coprime" to their factorial denominator $n!$ in the exponential generating function. We end this chapter with the suggestion that the above table highlights further congruence properties affiliated with the e.g.f. coefficients, and our enunciating properties evident from the above coefficient tables of this section provide an opening into new theories and understanding of these generating functions that are limiting cases of Visible Point Vector identities derived later in the present book.

Think About It...

Ramanujan wrote, upon seeing the table of $p(1), p(2), p(3), \ldots, p(200)$ manually calculated by Major MacMahon, that:

$$\begin{array}{llllll l}
(1) & p(4), & p(9), & p(14), & p(19), & \ldots & \equiv 0 (mod 5) \\
(2) & p(5), & p(12), & p(19), & p(26), & \ldots & \equiv 0 (mod 7) \\
(3) & p(6), & p(17), & p(28), & p(39), & \ldots & \equiv 0 (mod 11) \\
(4) & p(24), & p(49), & p(74), & p(99), & \ldots & \equiv 0 (mod 25) \\
(5) & p(19), & p(54), & p(89), & p(124), & \ldots & \equiv 0 (mod 35) \\
(6) & p(47), & p(96), & p(145), & p(194), & \ldots & \equiv 0 (mod 49) \\
(7) & p(39), & p(94), & p(149), & \ldots & & \equiv 0 (mod 55) \\
(8) & p(61), & p(138), & \ldots & & & \equiv 0 (mod 77) \\
(9) & p(116), & \ldots & & & & \equiv 0 (mod 121) \\
(10) & p(99), & \ldots & & & & \equiv 0 (mod 125)
\end{array}$$

From these data I conjectured the truth of the following theorem: If $\delta = 5^a 7^b 11^c$ and $24\lambda \equiv 1 (mod \delta)$, then

$$p(\lambda),\ p(\lambda+\delta),\ p(\lambda+2\delta),\ \ldots \equiv 1\ (mod \delta).$$

The theorem is supported by all the available evidence; but I have not yet been able to find a general proof.

Srinivasa Ramanujan
circa 1917

FURTHER READING

AHLGREN, S., ALLEN, P., and TANG, S. Congruences like Atkin's for the partition function. *Trans. Amer. Math. Soc. Ser. B*, to appear, 2023. arXiv:2112.09481v2 [math.NT].

AHLGREN, S., BECKWITH, O., and RAUM, M. Scarcity of congruences for the partition function. *Am. J. Math.*, to appear, 2023. arXiv:2006.07645v3 [math.NT].

ALKAUSKAS, G. The beginnings of significant mathematical results in the school children competitions (in Lithuanian). *Alpha Plus Omega*, 1(4):94–98, 1997.

ALKAUSKAS, G. and MENDEL-STRASSE, G. Congruence Properties of the Function that Counts Compositions into Powers of 2. *Journal of Integer Sequences*, 13, 2010. Article 10.5.3.

ANDREWS, G.E. *The Theory of Partitions*, volume 2 of *Encyclopedia of Mathematics and its Applications.* Addison-Wesley Publishing Company, Advanced Book Program, Reading, Massachusetts, USA, 1976.

ANDREWS, G.E. and GARVAN, F.G. Dyson's crank of a partition. *Bulletin of the American Mathematical Society. New Series.*, 18(2):6 pages, April 1988. PDF:www.ams.org/journals/bull/1988-18-02/S0273-0979-1988- 15637-6/S0273-0979-1988-15637-6.pdf.

ATKIN, A.O.L. Proof of a conjecture of Ramanujan. *Glasgow Math. J.*, 8:14–32, 1967.

ATKIN, A.O.L. and SWINNERTON-DYER, P. Some properties of partitions. *Proc. London Math. Soc.*, 3(4):84–106, 1954.

BRINGMANN, K. and LOVEJOY , J. Ranks and congruences for overpartition pairs. *Int. J. Number Theory*, 4:303–322, 2008.

BRUINIER, J.H. and FUNKE, J. Traces of CM-values of modular functions. *J. Reine Angew. Math.*, 594:1–33, 2006.

BRUINIER, J.H., and ONO, K. Algebraic formulas for the coefficients of halfintegral weight harmonic weak Maass forms. *Adv. Math.*, 246:198–219, 2013.

BRUINIER, J.H., JENKINS, P., and ONO, K. Hilbert class polynomials and traces of singular moduli. *Math. Ann.*, 334:373–393, 2006.

CHAN, H.C. Ramanujan's cubic continued fraction and an analog of his "most beautiful identity". *Int. J. Number Theory*, 6:673–680, 2010.

CHURCHHOUSE, R.F. Congruence properties of the binary partition function. *Proc. Cambridge Philos. Soc.*, 66:371–376, 1969.

DICKS, R. Congruence relations for r-colored partitions. *J. Number Theory*, 249:377–399, 2023.

DYSON, F.J. Some Guesses in The Theory of Partitions. *Eureka (Cambridge)*, 8:10–15, 1944. ISBN 9780821805619. PDF:www.archim.org.uk/eureka/archive/Eureka-8.pdf.

ECKSTEIN, G. On an Olympiad problem. *Romanian Math. Mag.*, 7:15–16, 1998.

EULER, L. Introductio in analysin infinitorum. In *Opera Omnia*, volume 1. Marcum-Michaelum, Brousquet, Lausannae, 1748. Chapter 16.

GARVAN, F.G. Generalizations of Dyson's rank. PhD Thesis, Pennsylvania State Univ., May 1986. https://dokumen.pub/generalizations-of-dysons-rank.html.

GARVAN, F.G. New combinatorial interpretations of Ramanujan's partition congruences mod 5, 7 and 11. *Trans. Amer. Math. Soc.*, 305(1):31 pages, Jan 1988.

GUPTA, H. Proof of the Churchhouse conjecture concerning binary partitions. *Proc. Cambridge Philos. Soc.*, 70:53–56, 1971.

HAMMOND, P. and LEWIS, R. Congruences in ordered pairs of partitions. *Int. J. Math. Math. Sci.*, 47:2509–2512, 2004.

HIRSCHHORN, M.D. and SELLERS, J.A. Arithmetic relations for overpartitions. *J. Comb. Math. Comb. Comp.*, 53:65–73, 2005.

MAHLER, K. On a Special Functional Equation. *J. London Math. Soc.*, 15:115– 123, 1940.

PENNINGTON, W.B. On Mahler partition problem. *Ann. of Math.*, 57:531– 546, 1952.

RAMANUJAN, S. Some properties of p(n), the number of partitions of n. *Proc. Cambridge Philos. Soc.*, 19:207–210, Nov 1919.

RÖDSETH, Ö. Some arithmetical properties of m-ary partitions. *Proc. Cambridge Philos. Soc.*, 68:447–453, 1970.

SLOANE, N.J.A. Expansion of e.g.f. exp(x/(x-1)). The On-Line Encyclopedia of Integer Sequences (OEIS), 2023. sequence A293116. (www.oeis.org/A293116).

TOH, P.C. Ramanujan type identities and congruences for partition pairs. *Discrete Mathematics*, 312:1244–1250, 2012.

ZHAO, H. and ZHONG, Z. Ramanujan type congruences for a partition function. *Electron. J. Combin.*, 18:P58, 2011.

CHAPTER 6

Ferrers Graphs and Ferrers Boards

6.1 FERRERS GRAPHS AND CONJUGATE GRAPHS

MANY RESULTS on partitions can be obtained by the use of a Ferrers diagram. The diagram of a partition is obtained by putting down a row of dots equal in number to the largest part, then immediately below it a row of dots equal in number to the next part, and so on. Such a diagram for $16 = 5+3+3+2+1+1+1$ is shown here

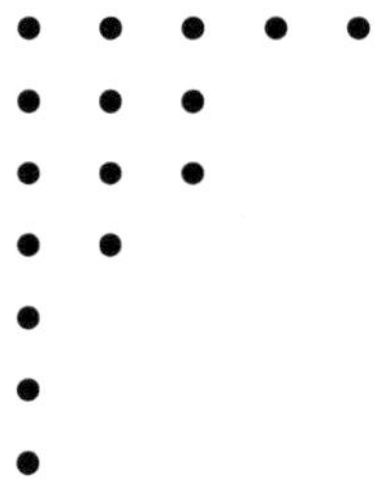

By rotating the Ferrers diagram of the partition about the diagonal, we obtain the **conjugate**, so the above example becomes the partition $16 = 7+4+3+1+1$, having the Ferrers diagram,

6.2 FERRERS BOARDS

The two most frequently met types of graphs for partitions are using either dots, the **Ferrers graph**; or using squares, the **Ferrers board**. Another name in common usage for the Ferrers board is the **Young diagram**. As an illustration, we give the Ferrers board for $16 = 5+3+3+2+1+1+1$, shown here

DOI: 10.1201/9781003174158-6

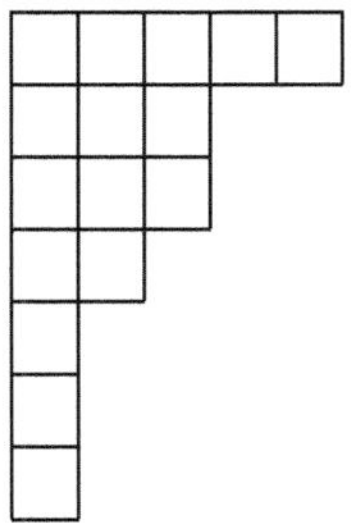

In many integer partition representations, the dots form is more simply applied, but there are many situations where the Ferrers board is more convenient for the purposes of illustrating partitions. Also, the Ferrers board may have numbers or vectors in each square making the concept useful for higher dimensional partitions.

6.3 BIJECTION PROOFS USING FERRERS GRAPHS

In the very significant *Integer Partitions* book by Andrews and Eriksson [29] there are several chapters applying beautiful and simple Ferrers graph diagrams that imply bijection equivalence statements on various kinds of partitions. In many ways, this seems the natural approach to proving partition identities. However, it's noted that while often Ferrers diagrams may easily infer a partition identity, the converse situation is generally not as easy to achieve. ie. A partition identity for which we have a generating function representation may not suggest *a priori* a suitable Ferrers diagram bijective proof. We present a summary of these now, each stated as a theorem, despite our indicative proof being not rigorous, but evident from the Ferrers diagrams provided.

Theorem 6.1

$$p(n \mid \textit{greatest part is } r) = p(n-r \mid \textit{all parts } \leq r). \tag{6.3.1}$$

Proof outline: Consider the Ferrers graph transform,

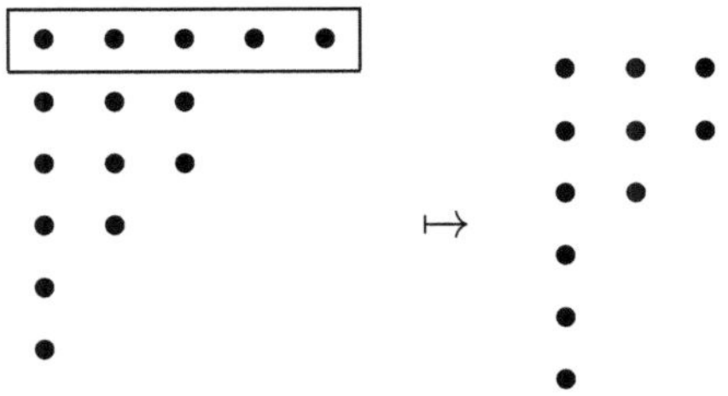

where the operation for transform is the *removal of the top row.* This diagram illustrates one case of the theorem, which applies to all cases. If the removed row has length r, then all the new rows have length less than or equal to r. Hence we arrive at the bijection that proves the theorem. ■

A similar approach removing instead the first column would imply a similar theorem, namely

Theorem 6.2

$$p(n \mid m \text{ parts}) = p(n - m \mid \text{at most } m \text{ parts}). \tag{6.3.2}$$

Next we consider the Ferrers graph transform,

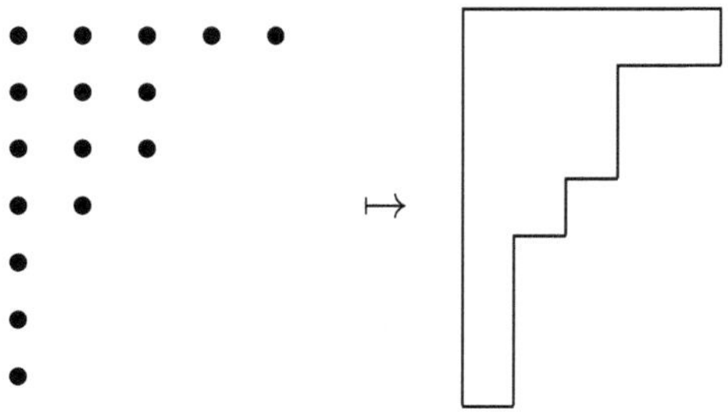

which maps the Ferrers graph to its outline perimeter. This outline has a top line, a left-side edge line and a jagged rim semi-diagonal line. The partition is encoded in this jagged rim, which comprises *outer corners* ⌋; and *inner corners* ⌈. The two end-points of the rim are counted as inner corners. The outer corners are the squares that can be removed so that what remains is still a Ferrers board. Likewise, the inner corners are where squares can be added in, forming a new Ferrers board. There is an elegant theory associated with Ferrers boards growing this way.

Theorem 6.3

$$p(n \mid m \text{ parts}) = p(n \mid \text{greatest part is } m). \tag{6.3.3}$$

Proof outline: Consider the Ferrers graph *conjugation* transform.

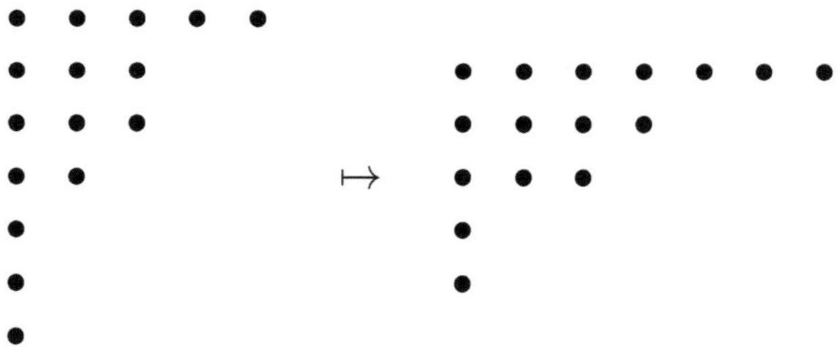

(6.3.4)

Here, the rows of the first graph map to the columns of the second graph, and vice versa. Hence, the transform arrives at a partition having the *greatest part* equal to the *number of parts* of the original partition. This gives a bijective proof of the theorem 6.3.

Theorem 6.4

$$p(n \mid \leq m \text{ parts}) = p(n \mid \text{all parts } \leq m). \tag{6.3.5}$$

Proof outline: Consider the Ferrers graph conjugate map (6.3.4). Theorem 6.4 is almost immediate *a priori* from this diagram mapping; the *m parts* of the left side first column Ferrers diagram mapping to the top row on the right side being, therefore, m with the rows after that all obviously less than or equal to m.

Partition identities often impose limitations on the *sizes of parts* and also on the *number of parts*. Another common restriction on partitions are that the parts be *distinct*. Clearly, the idea of conjugation is relevant to these three important features of partitions. With this in mind, we can easily arrive at the following

Theorem 6.5

$$p(n \mid \text{ distinct parts}) = p(n \mid \text{ parts of each size from 1 to largest part}). \tag{6.3.6}$$

Proof outline: Consider again the conjugate map (6.3.4). An example given is $n = 6$, there are four distinct partitions, 6, $5+1$, $4+2$, and $3+2+1$, while the right side of (6.3.6) gives us the four partitions $1+1+1+1+1+1$, $2+1+1+1+1$, $2+2+1+1$, $3+2+1$. Of course a proof is not a partial verification. However, the principle underpinning this verification leads to the proof.

Theorem 6.6

$$p(n \mid \text{ self-conjugate}) = p(n \mid \text{ distinct odd parts}). \tag{6.3.7}$$

Proof outline: Consider the Ferrers graph transform of a self-conjugate partition; that is, one symmetric about the 45° diagonal.

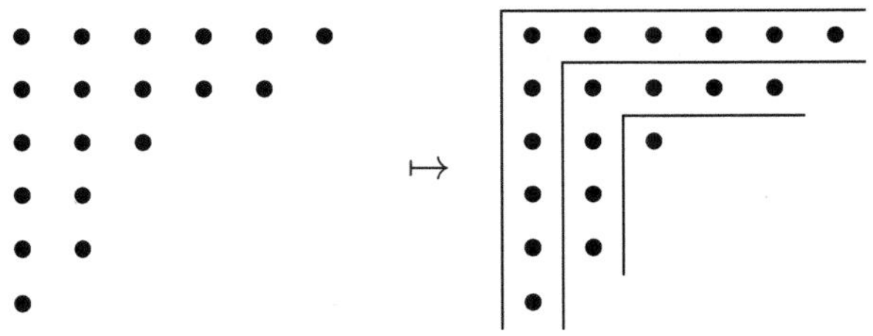

(6.3.8)

Clearly, in this instance, the left side of (6.3.8) is self-conjugate, and the right side of (6.3.8) dots inside each $\ulcorner$ shaped column-and-row of dots represent a distinct odd part of the partition. Following this convention for every possible partition will verify the theorem, but the proof of the theorem is in showing that the diagram for this given case is representative of every possible case of the theorem.

Next, we look into the remarkable aspect that the sequence $p(n)$ is upper-bounded by the Fibonacci number sequence F_n.

Theorem 6.7 *(see Andrews and Eriksson [29, Chapter 3, page 22])*
For all $n \geq 0$, the partition function $p(n)$ is less than or equal to the $(n+1)$st Fibonacci number F_{n+1}.

Proof outline: The proof is achieved in several steps. Steps (b) and (d) are inferred from bijections. The steps are as follows:

(a) $p(n) > p(n-1)$ for all $n \geq 2$.

(b) $p(n) = p(n-1) + p(n \mid \text{no 1-part} > p(n-1))$ for all $n \geq 2$.

(c) Fibonacci numbers $F_0, F_1, F_2, \ldots$ where $F_0 = 0$, $F_1 = 1$, and $F_n = F_{n-1} + F_{n-2}$.

(d) $p(n-2) = p(n\,|\text{no 1-part}) + p(n-2\,|\,\text{smallest non-1-part} < 2 + \text{number of 1-parts})$.

(e) $p(n) \le p(n-1) + p(n-2)$ for all $n \ge 2$.

(f) Applying induction to the above points we have: Since $p(0) = F_1 = p(1) = F_2 = 1$, the theorem holds for $n = 0$ and $n = 1$. Assuming it's true for all $n < k$ for some $k \ge 2$, gives $p(k) \le p(k-1) + p(k-2) \le F_k + F_{k-1} = F_{k+1}$. ■

6.4 FRANKLIN'S NEAR-BIJECTION PROOF OF EULER'S PENTAGONAL NUMBER THEOREM

We now reproduce the classical nineteenth century proof by Franklin of the Euler Pentagonal Number Theorem. We start by stating the

Theorem 6.8 *(Euler [123]) For each integer $n \ge 1$,*

$$p_e(\mathfrak{D}, n) = e(n) + p_o(\mathfrak{D}, n), \tag{6.4.1}$$

where

$$e(k) = \begin{cases} (-1)^j, & \text{if } k = \frac{j(3j-1)}{2}; \\ 0, & \text{otherwise;} \end{cases}$$

$$\begin{aligned} p_e(\mathfrak{D}, n) &= p(n\,|\,\text{even number of distinct parts}), \\ p_o(\mathfrak{D}, n) &= p(n\,|\,\text{odd number of distinct parts}). \end{aligned}$$

Proof: (Franklin [130]) We attempt to establish a one-to-one correspondence between the partitions $p_e(\mathfrak{D}, n)$ and the partitions $p_o(\mathfrak{D}, n)$. For most integers n our attempt will be successful. But, whenever n is one of the pentagonal numbers $\frac{1}{2}m(3m \pm 1$, a single, exceptional case will arise.

To start, we note that each partition $\lambda = (\lambda_1, \lambda_2, \ldots, \lambda_r)$ of n has a smallest part $s_\lambda = \lambda_r$; and we note that the largest part λ_1 of $\lambda = (\lambda_1, \lambda_2, \ldots, \lambda_r)$ is the first part of a sequence of say, $\sigma(\lambda)$ consecutive integers that are parts of λ (that is, $\sigma(\lambda)$ is the largest j such that $\lambda_j = \lambda_1 - j + 1$). Graphically the terms $s(\lambda)$ and $\sigma(\lambda)$ are shown as

odd distinct $\lambda = (76432)$ *even distinct* $\lambda = (8765)$

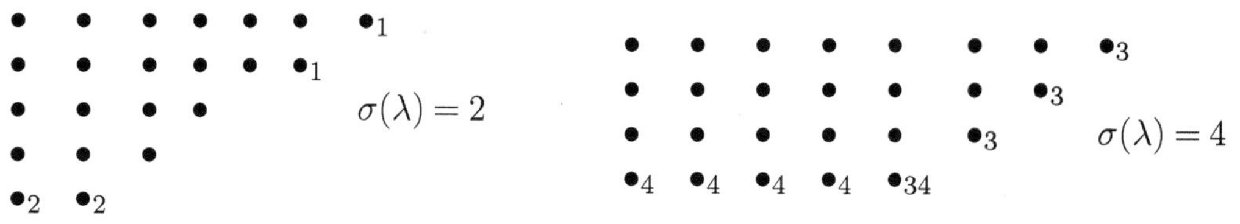

$s(\lambda) = 2$ $s(\lambda) = 5$

We transform partitions as follows. Two scenarios emerge, which we call Case 1 and Case 2.

Case 1: $s(\lambda) \le \sigma(\lambda)$. In this event we add one to each of the $s(\lambda)$ largest parts of λ and we delete the smallest part. Hence

$$\lambda = (76432) \quad \mapsto \quad \lambda' = (8743);$$

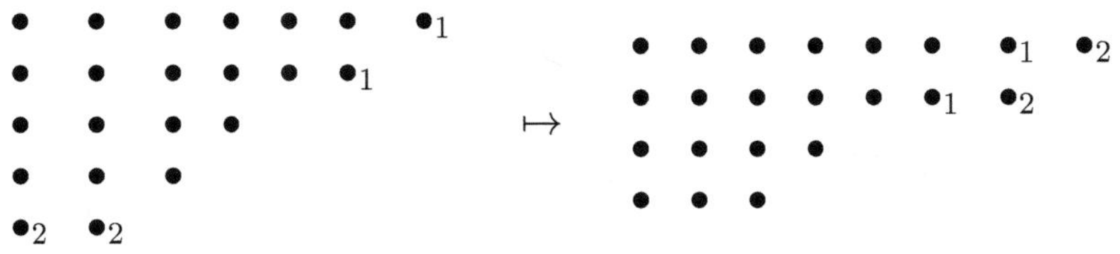

Case 2: $s(\lambda) > \sigma(\lambda)$. In this event, we subtract one from each of the $\sigma(\lambda)$ largest parts of λ and we insert a new smallest part of size $\sigma(\lambda)$. Hence

$$\lambda = (8743) \quad \mapsto \quad \lambda' = (76432);$$

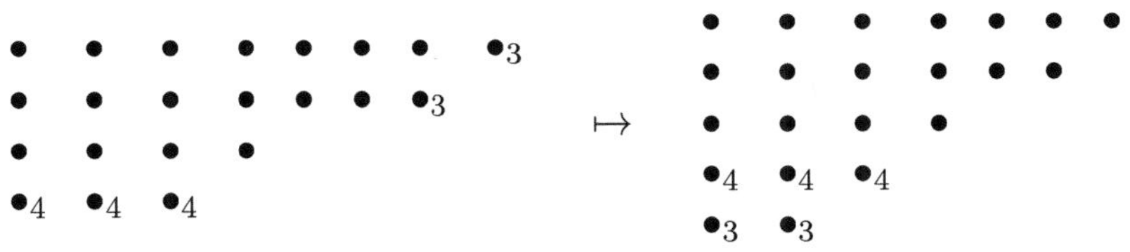

The above two exceptional case scenarios change the parity from plus to minus and vice versa for the number of parts of the partition, and since exactly one case is applicable to any partition λ, we note our mapping preserves a one-to-one correspondence. And yet for some partitions, this mapping fails. An example $\lambda = (8765)$ illustrates this. Case 2 should apply to it, but the image partition is no longer one with distinct parts. For Case 2 fails in cases where the partition has r parts, $\sigma(\lambda) = r$ and $s(\lambda) = r+1$. In exactly those cases the number being partitioned is

$$(r+1)+(r+2)+\cdots+2r = \frac{1}{2}r(3r+1).$$

The other scenario, Case 1 fails in exactly those cases when the partition has r parts, $\sigma(\lambda) = r$ and $s(\lambda) = r$, so the number being partitioned is

$$r+(r+1)+\cdots+(2r-1) = \frac{1}{2}r(3r-1).$$

As a consequence, if n is not a pentagonal number, $p_e(\mathfrak{D},n) = p_o(\mathfrak{D},n)$. Otherwise for $n = \frac{1}{2}r(3r \pm 1)$ then $p_e(\mathfrak{D},n) = p_o(\mathfrak{D},n) + (-1)^r$. ■

For more on this chapter's topic see Andrews [16], with also Bressoud and Zeilberger ([75] and [76]).

EXERCISES

6.1 *Towards a 2D Vector Partition extension of the Euler pentagonal number theorem.* The generating function we could examine for a 2D extension of the Euler pentagonal number theorem is

$$\prod_{j,k\geq 1}^{\infty}(1-x^j y^k) = 1+\sum_{j,k\geq 1}^{\infty}(p_e(\mathfrak{D};j,k)-p_o(\mathfrak{D};j,k))x^j y^k$$

where $p_e(\mathfrak{D};j,k)$ is the number of partitions into distinct even number of vector parts $\langle a,b\rangle$ for $1\leq a$ and $1\leq b$; and $p_o(\mathfrak{D};j,k)$ is the number of partitions into distinct odd number of vector parts $\langle a,b\rangle$ for $1\leq a$ and $1\leq b$.

The partial grid for this is

⋮	⋮	⋮	⋮	⋮	⋮	⋮	⋮	⋮	⋮	⋱
8	0	−1	2	4	0	−5	−9	−11	**−14**	⋯
7	0	−1	2	4	2	−3	−5	**−7**	−11	⋯
6	0	−1	1	3	2	1	**−3**	−5	−9	⋯
5	0	−1	1	3	3	**3**	1	−3	−5	⋯
4	0	−1	0	2	**2**	3	2	2	0	⋯
3	0	−1	0	**−1**	2	3	3	4	4	⋯
2	0	−1	**−1**	0	0	1	1	2	2	⋯
1	0	**−1**	−1	−1	−1	−1	−1	−1	−1	⋯
0	**1**	0	0	0	0	0	0	0	0	⋯
$z\uparrow y\rightarrow$	0	1	2	3	4	5	6	7	8	⋯

So, the research question involves finding a formula for the generated 2D coefficients as begun in the above grid.

6.2 The *Largest is k cf exactly k parts.* The number of partitions of n with the largest part k equals the number of partitions of n with exactly k parts.

6.3 *2s and singletons.* The total number of 2s in all partitions of n is equal to the total number of singletons in all partitions of $n-1$. A singleton is a part with multiplicity one.

6.4 *Odd parts cf Distinct parts.* The number of partitions of n with odd parts equals the number of partitions of n with distinct parts.

6.5 *Odd parts up to 9 times cf No parts divisible by 10.* The number of partitions of n for which no part occurs more than 9 times is equal to the number of partitions of n with no parts divisible by 10.

6.6 *Self-conjugate cf Distinct odd parts.* The number of self-conjugate partitions of n equals the number of partitions of n into distinct odd parts.

Think About It...

Geometric diagrams are to geometers what board and pieces are to chessmasters: visual aids, helpful but not indispensable.

Richard J. Trudeau
Introduction to Graph Theory

FURTHER READING

ANDREWS, G.E. Partitions and Durfee dissection. *Amer: J. Math.*, 101:735–742, 1979.

ANDREWS, G.E. and ERIKSSON, K. *Integer Partitions.* Cambridge University Press, Cambridge, UK, New York, USA, Port Melbourne, Australia, Madrid, Spain, Cape Town, South Africa, 2004.

BRESSOUD, D.M. and ZEILBERGER, D. A short Rogers-Ramanujan bijection. *Discrete Math.*, 38:313–315, 1982.

BRESSOUD, D.M. and ZEILBERGER, D. Generalized Rogers-Ramanujan Bijections. *Advances in Math.*, 78:42–75, 1989.

EULER, L. Introductio in analysin infinitorum. In *Opera Omnia*, volume 1. Marcum-Michaelum, Brousquet, Lausannae, 1748. Chapter 16.

FRANKLIN, F. Sur les developpement du produit infiniti $(1-x)(1-x^2)(1-x^3)...$,. *Comptes Rendus*, 82:448–450, 1881.

CHAPTER 7

Durfee Squares

7.1 WHAT IS A DURFEE SQUARE?

IN NUMBER THEORY, a Durfee square is an attribute of an integer partition. A partition of n has a Durfee square of side k if k is the largest number such that the partition contains at least k parts with values $\geq k$. An equivalent, but more visual, definition is that the Durfee square is the largest square that is contained within a partition's Ferrers diagram. The side-length of the Durfee square is known as the rank of the partition.

Any Durfee Square has an associated Frobenius Symbol, which we will discuss later in this chapter.

The partition $18 = 6+5+3+2+1+1$:

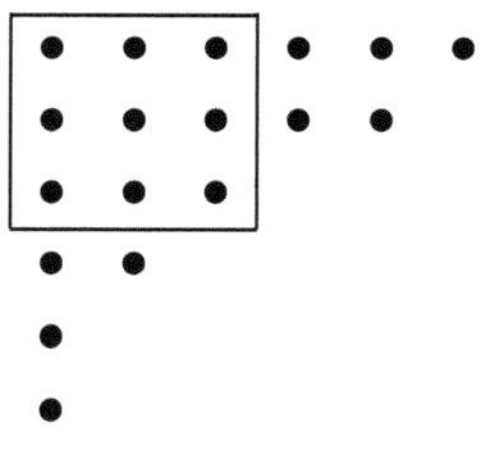

has a Durfee square of side 3 (3×3 square of dots) because it contains three parts that are ≥ 3, but does not contain four parts that are ≥ 4.

It is clear from the visual definition that the Durfee square of a partition and its conjugate partition have the same size. Or, equivalently, *The size of the Durfee square is unchanged between a partition and its conjugate partition.* The partitions of an integer n contain Durfee squares with sides up to and including $\lfloor \sqrt{n} \rfloor$.

7.2 DURFEE SQUARES APPLIED TO GENERATING FUNCTIONS

Suppose we consider a generic $k \times k$ Durfee Square in its generic partition,

DOI: 10.1201/9781003174158-7

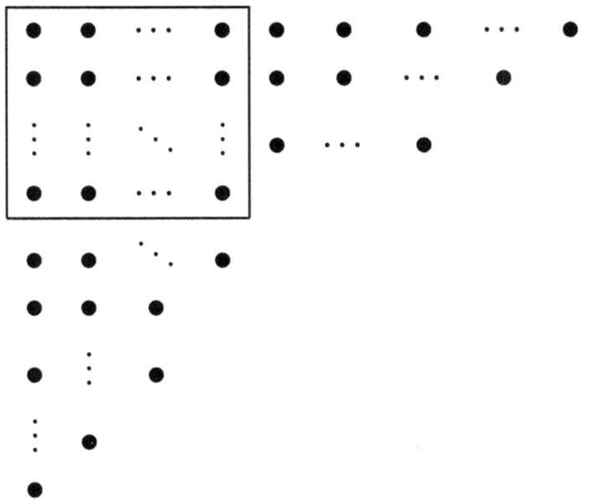

The part of this partition below the *Durfee Square whose side is* k, say; is a partition into $\leq k$ parts. Likewise, the portion of this generic partition to the right of the *Durfee Square whose side is* k, is a partition into $\leq k$ parts. So, for an $k \times k$ Durfee Square thought of as a partition itself, is generated by the term q^{k^2}. From the definition of q binomial numbers, we know the *right of the Durfee Square portion* is generated by

$$\begin{bmatrix} n \\ k \end{bmatrix}_q,$$

and similarly, the portion beneath the $k \times k$ Durfee Square is generated by

$$\begin{bmatrix} n \\ n-k \end{bmatrix}_q,$$

so the generating function for partitions into at most n parts each less than or equal to n, with Durfee Square of side k is

$$q^{k^2} \begin{bmatrix} n \\ k \end{bmatrix}_q \begin{bmatrix} n \\ n-k \end{bmatrix}_q = q^{k^2} \begin{bmatrix} n \\ k \end{bmatrix}_q^2.$$

But we know already that the generating function for all partitions with at most n parts each less than or equal to n is

$$\begin{bmatrix} 2n \\ n \end{bmatrix}_q.$$

Hence we have shown that

Theorem 7.1

$$\sum_{k=0}^{n} q^{k^2} \begin{bmatrix} n \\ k \end{bmatrix}_q^2 = \begin{bmatrix} 2n \\ n \end{bmatrix}_q. \tag{7.2.1}$$

If we take the case of this with $n \to \infty$ we obtain Jacobi's well-known identity

$$\sum_{k=0}^{\infty} \frac{q^{k^2}}{(1-q)^2(1-q^2)^2 \ldots (1-q^k)^2} = \prod_{n=1}^{\infty} \frac{1}{(1-q^n)}. \tag{7.2.2}$$

If we reconsider our rationale to study all partitions with parts less than or equal to m, while accounting for both the size of the Durfee Square and the number of parts, we obtain the extended version of (7.2.1),

$$\sum_{k=0}^{m} \begin{bmatrix} m \\ k \end{bmatrix}_q \frac{z^k q^{k^2}}{(1-zq)(1-zq^2)\ldots(1-zq^k)} = \prod_{n=1}^{m} \frac{1}{(1-zq^n)}. \tag{7.2.3}$$

7.3 SUCCESSIVE DURFEE SQUARES

We extend the idea of a Durfee Square of a partition, to having successive Durfee Squares in a partition.

Consider for example the partition $33 = 7+5+4+4+3+3+3+2+1+1$ of Ferrers graph

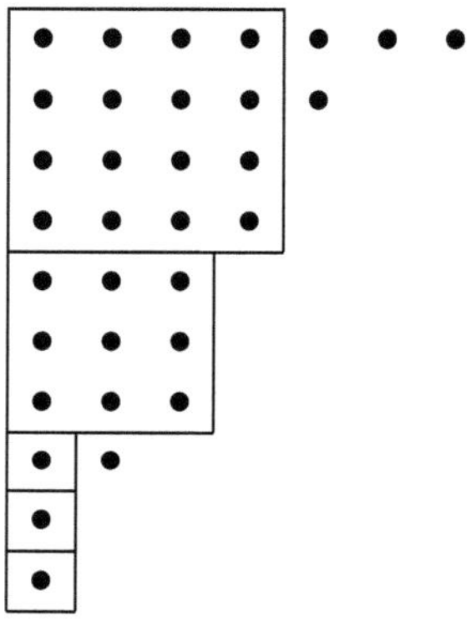

having successive Durfee squares of sides $4, 3, 1, 1, 1$.

So, adopting this generalization from one Durfee Square to many successive Durfee Squares, we can then find the generating function for partitions that have at most k successive Durfee Squares. For an account of the below process see Andrews and Eriksson [29]. Recall that we saw earlier that if $k = 1$ then

$$s_n(q) = \sum_{j=0}^{n} q^{j^2} \begin{bmatrix} m \\ j \end{bmatrix}_q .$$

So for general k, we take account of k possible squares $j_1^2 \geq j_2^2 \geq j_3^2 \geq \cdots \geq j_k^2$ and we see how the portions of the partition to the right side of each Durfee Square contribute to the partition in this form. Hence the generating function for partitions with at most k successive Durfee Squares, $s_{k,n}(q)$ is given by

$$\begin{aligned} s_{k,n}(q) &= \sum_{n \geq j_1 \geq j_2 \geq \cdots \geq j_k \geq 0} q^{j_1^2+j_2^2+\cdots+j_k^2} \begin{bmatrix} n \\ j_1 \end{bmatrix}_q \begin{bmatrix} j_1 \\ j_2 \end{bmatrix}_q \begin{bmatrix} j_2 \\ j_3 \end{bmatrix}_q \cdots \begin{bmatrix} j_{k-1} \\ j_k \end{bmatrix}_q \\ &= \sum_{n \geq j_1 \geq j_2 \geq \cdots \geq j_k \geq 0} \frac{q^{j_1^2+j_2^2+\cdots+j_k^2}(q;q)_n}{(q;q)_{n-j_1}(q;q)_{n-j_2}(q;q)_{n-j_3}\cdots(q;q)_{n-j_{k-1}-j_k}(q;q)_{j_k}}. \end{aligned}$$

We next take $n \to \infty$, so the generating function for all partitions with at most k successive Durfee Squares is

$$s_{k,\infty}(q) = \sum_{n \geq j_1 \geq j_2 \geq \cdots \geq j_k \geq 0} \frac{q^{j_1^2+j_2^2+\cdots+j_k^2}(q;q)_n}{(q;q)_{n-j_1}(q;q)_{n-j_2}(q;q)_{n-j_3}\cdots(q;q)_{n-j_{k-1}-j_k}(q;q)_{j_k}}.$$

This enables us to state the

Theorem 7.2 *(see Andrews [16]) The number of partitions of n, each with at most k successive Durfee Squares, equals the number of partitions of n into parts not congruent to* $0, \pm(k+1)\,(mod 2k+3)$. *The generating function encoding this is*

$$\sum_{n \geq j_1 \geq j_2 \geq \cdots \geq j_k \geq 0} \frac{q^{j_1^2+j_2^2+\cdots+j_k^2}(q;q)_n}{(q;q)_{n-j_1}(q;q)_{n-j_2}(q;q)_{n-j_3}\cdots(q;q)_{n-j_{k-1}-j_k}(q;q)_{j_k}} \tag{7.3.1}$$

$$= \prod_{\substack{n=1 \\ n \not\equiv 0, \pm(k+1)\,(mod 2k+3);}}^{\infty} \frac{1}{(1-q^n)}.$$

The case of theorem 7.2 with $k=1$ is the first Rogers-Ramanujan identity. This remarkable proof was given by Bressoud and Zeilberger in [75] and a generalized approach using bijections by these authors is given in the next section.

7.4 BIJECTION PROOFS OF ROGERS-RAMANUJAN TYPE IDENTITIES

The successive Durfee square proof of the first Rogers-Ramanujan identity was given by Bressoud and Zeilberger in 1982 [75] and a generalized Rogers-Ramanujan bijections sequel appeared in 1989. (See [76].) We now give a few significant bijection results from this topic, as exercises making use of Durfee squares.

EXERCISES

7.1 *Number of 2s cf singletons.* The total number of 2s in all partitions of n is equal to the total number of singletons in all partitions of $n-1$. A *singleton* is a part with multiplicity one. For instance, the partition $(8,8,7,5,4,4,4,2,2,1)$ has two 2s and three singletons.

7.2 *Partitions into powers of 2.* The number of partitions of $n \geq 2$ into powers of 2 is even. For instance, when $n=4$ there are the four partitions $4 = 2+2 = 2+1+1 = 1+1+1+1$.

7.3 *Number of Odd parts = Number of Distinct parts.* The number of partitions of n with odd parts equals the number of partitions of n with distinct parts.

7.4 *k parts and k distinct parts.* The number of partitions of n with k parts equals the number of partitions of $n + \binom{k}{2}$ with k distinct parts.

7.5 *Parts occurring less than 9 times.* The number of partitions of n for which no part occurs more than 9 times is equal to the number of partitions of n with no parts divisible by 10.

7.6 *Self-conjugate partitions.* The number of self-conjugate partitions of n equals the number of partitions of n into distinct odd parts.

7.7 *Parts even/odd and self-conjugate.* Let $e(n)$, $o(n)$, and $k(n)$ denote, respectively, the number of partitions of n with an even number of even parts, with an odd number of even parts, and that are self-conjugate. Then $e(n) - o(n) = k(n)$.

Think About It...

A mathematician, like a painter or poet, is a maker of patterns.
If his patterns are more permanent than theirs, it is because they are made with ideas.

G.H. Hardy
A Mathematician's Apology, Chapter 10 (1940)

FURTHER READING

ANDREWS, G.E. Partitions and Durfee dissection. *Amer: J. Math.*, 101:735–742, 1979.

ANDREWS, G.E. and ERIKSSON, K. *Integer Partitions.* Cambridge University Press, Cambridge, UK, New York, USA, Port Melbourne, Australia, Madrid, Spain, Cape Town, South Africa, 2004.

BRESSOUD, D.M. and ZEILBERGER, D. A short Rogers-Ramanujan bijection. *Discrete Math.*, 38:313–315, 1982.

BRESSOUD, D.M. and ZEILBERGER, D. Generalized Rogers-Ramanujan Bijections. *Advances in Math.*, 78:42–75, 1989.

CHAPTER 8

Gaussian Polynomials

8.1 DEFINITION AND PARTICULAR CASES

THE GAUSSIAN binomial coefficients are defined by:

$$\begin{bmatrix} m \\ r \end{bmatrix}_q = \frac{(1-q^m)(1-q^{m-1})...(1-q^{m-r+1})}{(1-q)(1-q^2)...(1-q^r)} \tag{8.1.1}$$

where m and r are non-negative integers. If $r > m$, this evaluates to 0. For $r = 0$, the value is 1 since both the numerator and denominator are empty products.

Although the formula at first appears to be a rational function, it actually is a polynomial, because the division is exact in $Z[q]$.

All of the factors in the numerator and denominator are divisible by $1-q$, and the quotient is the q number:

$$[k]_q = \sum_{0 \le i < k} q^i = 1+q+q^2+...+q^{k-1} = \frac{1-q^k}{1-q} \text{ for } q \neq 1 \text{ and } k \text{ for } q = 1. \tag{8.1.2}$$

Dividing out these factors gives the equivalent formula

$$\begin{bmatrix} m \\ r \end{bmatrix}_q = \frac{[m]_q![m-1]_q!...[m-r+1]_q!}{[1]_q![2]_q!...[r]_q!}, \text{ for } r \le m. \tag{8.1.3}$$

Substituting $q = 1$ into $\begin{bmatrix} m \\ r \end{bmatrix}_q$ gives the ordinary binomial coefficient $\binom{m}{r}$. Gaussian polynomials were introduced primarily as they encoded many profound q-series identities by way of an analog extension of the simple binomial coefficient. For good accounts of this concept see Andrews [12], Andrews and Eriksson [29], or Gasper and Rahman [134] where the concept is fully developed into a complete 1990 contemporary volume of q-series identities and relationships. The Gaussian polynomial is therefore a useful concept applicable to the theory of integer partitions, which

DOI: 10.1201/9781003174158-8

has successfully been also applied in physics, especially in statistical mechanics partition applications. (See Baxter [52] or Bazhnov et al. [53].)

The Gaussian binomial coefficient (8.1.3) has finite values as $m \to \infty$:

$$\begin{bmatrix} \infty \\ r \end{bmatrix}_q = \lim_{m\to\infty} \begin{bmatrix} m \\ r \end{bmatrix}_q = \frac{1}{(1-q)(1-q^2)...(1-q^r)} = \frac{1}{[r]_q!(1-q)^r} \tag{8.1.4}$$

Examples

$$\begin{bmatrix} 0 \\ 0 \end{bmatrix}_q = \begin{bmatrix} 1 \\ 0 \end{bmatrix}_q = 1,$$

$$\begin{bmatrix} 1 \\ 1 \end{bmatrix}_q = \frac{1-q}{1-q} = 1,$$

$$\begin{bmatrix} 2 \\ 1 \end{bmatrix}_q = \frac{1-q^2}{1-q} = 1+q,$$

$$\begin{bmatrix} 3 \\ 1 \end{bmatrix}_q = \frac{1-q^3}{1-q} = 1+q+q^2,$$

$$\begin{bmatrix} 3 \\ 2 \end{bmatrix}_q = \frac{(1-q^3)(1-q^2)}{(1-q)(1-q^2)} = 1+q+q^2,$$

$$\begin{bmatrix} 4 \\ 2 \end{bmatrix}_q = \frac{(1-q^4)(1-q^3)}{(1-q)(1-q^2)} = (1+q^2)(1+q+q^2) = 1+q+2q^2+q^3+q^4,$$

$$\begin{bmatrix} 6 \\ 3 \end{bmatrix}_q = \frac{(1-q^6)(1-q^5)(1-q^4)}{(1-q)(1-q^2)(1-q^3)} = (1+q^2)(1+q^3)(1+q+q^2+q^3+q^4)$$
$$= 1+q+2q^2+3q^3+3q^4+3q^5+3q^6+2q^7+q^8+q^9.$$

8.2 COMBINATORIAL INTERPRETATIONS

8.2.1 Inversions

One combinatorial description of Gaussian binomial coefficients involves inversions. The ordinary binomial coefficient $\binom{m}{r}$ counts the r-combinations chosen from an m-element set. If one takes those m elements to be the different character positions in a word of length m, then each r-combination corresponds to a word of length m using an alphabet of two letters, say $0, 1$, with r copies of the letter 1 (indicating the positions in the chosen combination) and $m-r$ letters 0 (for the remaining positions). So, for example, the $\binom{4}{2} = 6$ words using 0s and 1s are $0011, 0101, 0110, 1001, 1010, 1100$.

To obtain the Gaussian binomial coefficient $\begin{bmatrix} m \\ r \end{bmatrix}_q$, each word is associated with a factor q^d where d is the number of inversions of the word, where, in this case, an inversion is a pair of positions where the left of the pair holds the letter 1 and the right position holds the letter 0.

With the example above, there is one word with 0 inversions, 0011, one word with 1 inversion, 0101, two words with 2 inversions, 0110, 1001, one word with 3 inversions, 1010, and one word with 4 inversions, 1100. This is also the number of left-shifts of the 1s from the initial position. These correspond to the coefficients in

$$\begin{bmatrix} 4 \\ 2 \end{bmatrix}_q = \frac{(1-q^4)(1-q^3)}{(1-q)(1-q^2)} = 1+q+2q^2+q^3+q^4.$$

Another way to see this is to associate each word with a path across a rectangular grid with height r and width $m-r$, going from the bottom left corner to the top right corner. The path takes a step right for each 0 and a step up for each 1. An inversion switches the directions of a step (*right and up* becomes *up and right* and vice versa), hence the number of inversions equals the area under the path.

8.2.2 Balls into Bins

Let $B(n,m,r)$ be the number of ways of throwing r indistinguishable balls into m indistinguishable bins, where each bin can contain up to n balls. The Gaussian binomial coefficient can be used to characterize $B(n,m,r)$. We have then

$$B(n,m,r) = [q^r]\begin{bmatrix} n+m \\ m \end{bmatrix}_q,$$

where $[q^r]P$ denotes the coefficient of q^r in polynomial P.

8.2.3 Reflection

Like the ordinary binomial coefficients, the Gaussian binomial coefficients are center-symmetric, i.e., invariant under the reflection $r \to m-r$,

$$\begin{bmatrix} m \\ r \end{bmatrix}_q = \begin{bmatrix} m \\ m-r \end{bmatrix}_q.$$

In particular,

$$\begin{bmatrix} m \\ 0 \end{bmatrix}_q = \begin{bmatrix} m \\ m \end{bmatrix}_q = 1,$$

$$\begin{bmatrix} m \\ 1 \end{bmatrix}_q = \begin{bmatrix} m \\ m-1 \end{bmatrix}_q = \frac{1-q^m}{1-q} = 1+q+\ldots+q^{m-1} \text{ for } m \geq 1.$$

The name Gaussian binomial coefficient stems from the fact that their evaluation at $q=1$ is

$$\lim_{q\to 1}\begin{bmatrix} m \\ r \end{bmatrix}_q = \binom{m}{r}$$

for all m and r.

8.3 ANALOGS OF PASCAL'S IDENTITY

The analogs of Pascal's identity for the Gaussian binomial coefficients are:

$$\begin{bmatrix} m \\ r \end{bmatrix}_q = q^r \begin{bmatrix} m-1 \\ r \end{bmatrix}_q + \begin{bmatrix} m-1 \\ r-1 \end{bmatrix}_q$$

and

$$\begin{bmatrix} m \\ r \end{bmatrix}_q = \begin{bmatrix} m-1 \\ r \end{bmatrix}_q + q^{m-r} \begin{bmatrix} m-1 \\ r-1 \end{bmatrix}_q .$$

When $q = 1$, these both give the usual binomial identity. We can see that as $m \to \infty$, both equations remain valid. The first Pascal identity allows one to compute the Gaussian binomial coefficients recursively (with respect to m) using the initial values

$$\begin{bmatrix} m \\ 0 \end{bmatrix}_q = \begin{bmatrix} m \\ m \end{bmatrix}_q = 1,$$

and also incidentally shows that the Gaussian binomial coefficients are indeed polynomials (in q).

The second Pascal identity follows from the first using the substitution $r \to m - r$ and the invariance of the Gaussian binomial coefficients under the reflection $m - r \to r$.

Both Pascal identities can be proved by first noting the definitions

$$\begin{bmatrix} m \\ r \end{bmatrix}_q = \frac{1-q^m}{1-q^{m-r}} \begin{bmatrix} m-1 \\ r \end{bmatrix}_q \tag{8.3.1}$$

and

$$\begin{bmatrix} m \\ r \end{bmatrix}_q = \frac{1-q^m}{1-q^r} \begin{bmatrix} m-1 \\ r-1 \end{bmatrix}_q . \tag{8.3.2}$$

Now,

$$\frac{1-q^m}{1-q^{m-r}} = q^r + \frac{1-q^r}{1-q^{m-r}}$$

and directly equating (8.3.2) and (8.4.1), we get:

$$\frac{1-q^r}{1-q^{m-r}} \begin{bmatrix} m-1 \\ r \end{bmatrix}_q = \begin{bmatrix} m-1 \\ r-1 \end{bmatrix}_q$$

Similarly for (8.3.2).

8.4 q-BINOMIAL THEOREM ANALOGUE

There is an analog of the binomial theorem for q-binomial coefficients. In fact it is the well known q-binomial theorem disguised in the notation of q-binomial coefficients.

$$\prod_{k=0}^{n-1}(1-q^k t)=\sum_{k=0}^{n} q^{\frac{k(k-1)}{2}}\begin{bmatrix} n \\ k \end{bmatrix}_q t^k. \tag{8.4.1}$$

Like the usual binomial theorem, this formula has numerous generalizations and extensions. One such extension, corresponding to Newton's generalized binomial theorem for negative powers, is

$$\prod_{k=0}^{n-1}\frac{1}{1-q^k t}=\sum_{k=0}^{\infty}\begin{bmatrix} n+k-1 \\ k \end{bmatrix}_q t^k. \tag{8.4.2}$$

It is not hard to see that setting $t=q$ in (8.4.2) and (8.4.3) and allowing $n\to\infty$ leads respectively to the generating functions for *distinct* partitions and *unrestricted* partitions.

There are many extensions and analogues of binomial coefficient sums and identities. A typical example is for the so-named *Central Binomial Coefficients*, in which we see that

$$\sum_{k=0}^{n} q^{k^2}\begin{bmatrix} n \\ k \end{bmatrix}_q^2=\begin{bmatrix} 2n \\ n \end{bmatrix}_q. \tag{8.4.3}$$

Clearly this is the extended q-analogue of the well-known sum

$$\sum_{k=0}^{n}\binom{n}{k}^2=\binom{2n}{n}.$$

8.5 GAUSSIAN POLYNOMIALS IN THE THEORY OF PARTITIONS

Gaussian binomial coefficients occur in the counting of symmetric polynomials and in the theory of partitions. The coefficient of q^r in

$$\begin{bmatrix} n+m \\ m \end{bmatrix}_q$$

is the number of partitions of r with m or fewer parts each less than or equal to n. Equivalently, it is also the number of partitions of r with n or fewer parts each less than or equal to m.

Gaussian binomial coefficients also play an important role in the enumerative theory of projective spaces defined over a finite field. In particular, for every finite field F_q with q elements, the Gaussian binomial coefficient

$$\begin{bmatrix} n \\ k \end{bmatrix}_q$$

counts the number of k-dimensional vector subspaces of an n-dimensional vector space over F_q (a Grassmannian). When expanded as a polynomial in q, it yields the well-known decomposition of the Grassmannian into Schubert cells. For example, the Gaussian binomial coefficient

$$\begin{bmatrix} n \\ 1 \end{bmatrix}_q = 1+q+q^2+...+q^{n-1}$$

is the number of one-dimensional subspaces in $(F_q)^n$ (equivalently, the number of points in the associated projective space). Furthermore, when q is 1 (respectively -1), the Gaussian binomial coefficient yields the Euler characteristic of the corresponding complex (respectively real) Grassmannian.

The number of k-dimensional affine subspaces of F_q^n is equal to

$$q^{n-k}\begin{bmatrix} n \\ k \end{bmatrix}_q.$$

This allows another interpretation of the identity

$$\begin{bmatrix} m \\ r \end{bmatrix}_q = \begin{bmatrix} m-1 \\ k \end{bmatrix}_q + q^{m-r}\begin{bmatrix} m-1 \\ r-1 \end{bmatrix}_q.$$

as counting the $(r-1)$-dimensional subspaces of $(m-1)$-dimensional projective space by fixing a hyperplane, counting such subspaces contained in that hyperplane, and then counting the subspaces not contained in the hyperplane; these latter subspaces are in bijective correspondence with the $(r-1)$-dimensional affine subspaces of the space obtained by treating this fixed hyperplane as the hyperplane at infinity.

In the conventions common in applications to quantum groups, a slightly different definition is used. The quantum binomial coefficient there is

$$q^{k^2-nk}\begin{bmatrix} n \\ k \end{bmatrix}_{q^2}.$$

This version of the quantum binomial coefficient is symmetric under exchange of q and q^{-1}.

Finally, the generating function for partitions bounded in number by j and in magnitude by k is given by

$$\begin{bmatrix} j+k \\ k \end{bmatrix}_q = \begin{bmatrix} j+k \\ j \end{bmatrix}_q.$$

Think About It...

We must admit with humility that, while number is purely a product of our minds, space has a reality outside our minds, so that we cannot completely prescribe its properties *a priori*.

C.F. Gauss (1777–1855)
in an 1830 letter to Bessel

FURTHER READING

ANDREWS, G.E. *The Theory of Partitions*, volume 2 of *Encyclopedia of Mathematics and its Applications*. Addison-Wesley Publishing Company, Advanced Book Program, Reading, Massachusetts, USA, 1976.

ANDREWS, G.E. and ERIKSSON, K. *Integer Partitions*. Cambridge University Press, Cambridge, UK, New York, USA, Port Melbourne, Australia, Madrid, Spain, Cape Town, South Africa, 2004.

BAXTER, R.J. *Exactly Solved Models in Statistical Mechanics*. Academic Press, New York, 1982. Reprinted 1989 and 2007.

BAZHANOV, V.V., KELS, A.P. and SERGEEV, S.M. Quasi-classical expansion of the star-triangle relation and integrable systems on quad-graphs. *J. Phys. A: Math. Theor.*, 49:44pp., October 2016.

GASPER, G. and RAHMAN, M. *Basic Hypergeometric Series*, volume 35 of *Encyclopedia of Mathematics and its Applications*. Cambridge University Press, (Cambridge - New York - Port Chester - Melbourne - Sydney), 1990.

CHAPTER 9

Plane Partitions from MacMahon to Andrews

9.1 DEFINITIONS OF PLANE PARTITIONS

In mathematics and especially in combinatorics, a *plane partition* is a two-dimensional array of nonnegative integers $\pi_{i,j}$ (with positive integer subscripts i and j) that are nonincreasing in both subscripts. This means that

$\pi_{i,j} \geq \pi_{i,j+1}$ and $\pi_{i,j} \geq \pi_{i+1,j}$ for all i and j.

Moreover, only finitely many of the $\pi_{i,j}$ are nonzero. A plane partition may be represented visually by the placement of a stack of $\pi_{i,j}$ unit cubes above the point (i,j) in the plane, giving a three-dimensional solid as shown in the picture.

The sum of a plane partition is

$n = \sum_{i,j} \pi_{i,j}$.

The sum describes the number of cubes of which the plane partition consists. The number of plane partitions with sum n is denoted $PL(n)$.

For example, there are six plane partitions with sum 3.

The matrix index representation of this and the cube pile representation of this are shown in Figure 2, so it is seen that $PL(3) = 6$. (Here the plane partitions are drawn using *matrix indexing* for the coordinates and the entries equal to 0 are suppressed for readability.) Let $\mathcal{N}_i(r,s,t)$ be the total number of plane partitions in which r is the number of rows that are unequal to zero, s is the number of columns that are non-zero, and t is the largest integer of the matrix. Plane partitions are often described by the positions of the unit cubes. Therefore, a plane partition is defined as a finite

DOI: 10.1201/9781003174158-9

Figure 9.1 A plane partition of 34 shown as a stack of cubes.

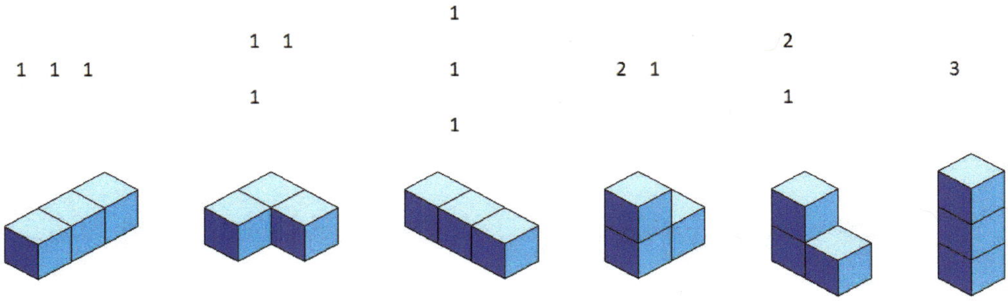

Figure 9.2 The six plane partitions of 3 shown in matrix form, and as a stack of cubes.

subset $\mathcal{P}$ of positive integer lattice points (i,j,k) in $\mathbb{N}^3$, such that if (r,s,t) lies in $\mathcal{P}$ and if (i,j,k) satisfies $1 \leq i \leq r$, $1 \leq j \leq s$ and $1 \leq k \leq t$, then (i,j,k) also lies in $\mathcal{P}$.

$$\mathcal{B}(r,s,t) = \{(i,j,k) | 1 \leq i \leq r, 1 \leq j \leq s, 1 \leq k \leq t\}$$

This is basically a $i \times j \times k$ box, and it is often called that.

9.2 LOZENGE TILINGS OF A HEXAGON

A lozenge tiling is a tiling of a planar domain with three types of lozenges which we refer to as *left*, *top* and *right* tiles. These correspond respectively to the dark blue, light blue and the middle blue tiles in the Figures 9.1 and 9.2 diagrams that look like cubes. Let $LT(a,b,c)$ be the set of lozenge tilings of a hexagon with sides (a,b,c,a,b,c) as in Figure 9.1. Although much is written on lozenge tilings elsewhere, we shall not specifically focus upon that aspect of plane partitions here; preferring to follow the cubes stack version in the context of the ten symmetry classes outlined in the subsequent sections here.

9.3 GENERATING FUNCTIONS FOR PLANE PARTITIONS

A much celebrated result of Percy A. MacMahon [193], is the generating function for $PL(n)$ given by

$$\sum_{n=0}^{\infty} PL(n)x^n = \prod_{k=1}^{\infty} \frac{1}{(1-x^k)^k} = 1+x+3x^2+6x^3+13x^4+24x^5+... \tag{9.3.1}$$

This is often called the *MacMahon function*. There are several remarkable proofs in the literature, and I refer the reader to G.E. Andrews' eloquent descriptions and proofs in [12]. Although the matrix definition is good and precise, the cube pile model may be the most intuitive way to describe $\pi_{i,j}$. It is said that formula (9.3.1) may be viewed as the 2-dimensional analog of Euler's product formula for the number of integer partitions of n. There is no known analogous formula known for partitions in higher dimensions (i.e., for solid partitions). However, 2D vector partitions may also fit that supposition. The asymptotics of plane partitions applying the circle method was published by E. M. Wright [302]. He obtained for large n:

$$PL(n) \sim \frac{\zeta(3)^{7/36}}{\sqrt{12\pi}} \left(\frac{n}{2}\right)^{-25/36} \exp\left(3\,\zeta(3)^{1/3}\left(\frac{n}{2}\right)^{2/3} + \zeta'(-1)\right). \tag{9.3.2}$$

Here the typographical error (in Wright's paper) has been corrected. Evaluating numerically yields

$$\log(PL(n)) \sim 2.00945n^{2/3} - 0.69444\log n - 1.4631. \tag{9.3.3}$$

In 1896 Percy A. MacMahon published a generating function of plane partitions that are subsets of the $i \times j \times k$ box $\mathcal{B}(r,s,t)$ in his first paper on plane partitions. The formula is given by

$$\sum_{\pi \in \mathcal{B}(r,s,t)} q^{|\pi|} = \prod_{i=1}^{r}\prod_{j=1}^{s} \frac{1-q^{i+j+t-1}}{1-q^{i+j-1}}. \tag{9.3.4}$$

A proof of this formula is in MacMahon's book *Combinatory Analysis* written in 1916 [193]. He shows how this involves the generating function of plane partitions of the cube representations inside an $r \times s \times t$ box in the 3D space rectangular prism from the origin $(0,0,0)$ to the point (r,s,t). The formula for the generating function can be written another way, given by

$$\sum_{\pi \in \mathcal{B}(r,s,t)} q^{|\pi|} = \prod_{i=1}^{r}\prod_{j=1}^{s}\prod_{k=1}^{t} \frac{1-q^{i+j+k-1}}{1-q^{i+j+k-2}}. \tag{9.3.5}$$

This now brings us to consider the beautiful work since MacMahon's ground-breaking book of 1916 [193]. This important work lay dormant until Leonard Carlitz [102], [103]; and Basil Gordon and Lorne Houten [143], [144], in the 1960s, relaunched the interest in plane partitions. Continued revitalization in researches from 1971 to present times, started with Richard Stanley (see [283], [285], and [286]) and was continued by George E. Andrews (see [14] and [13]), Ian Macdonald [191] and others. There are now ten symmetry classes of Plane Partitions that now show productive applications and beautiful theories. We now give a brief outline of each of these ten classes.

9.4 THE TEN SYMMETRY CLASSES OF PLANE PARTITIONS

In this section, we summarise the types of enumeration of symmetry classes of plane partitions. We begin by defining the symmetry operations which underpin these 10 symmetry classes. The classification and most of the theory behind these 10 classes goes back to Richard Stanley circa 1970 (see Stanley [283] to [286]). This way of classifying symmetries for plane partitions was neatly summarized by Kattenthaler [175] (see also [171, 173]) in a survey article we largely follow in this chapter. For each symmetry class, we give its enumeration formula, and cite who had proved that formula, mentioning some open questions that follow from this. The first operation is **reflection.** To reflect a plane partition π means to reflect it along the main diagonal, that is, to map $\pi = (\pi_{i,j})_{i,j}$ to $\pi = (\pi_{j,i})_{i,j}$. In the representation of a plane partition as a rhombus tiling as in Figure 9.1, reflection is along the vertical symmetry axis of the hexagon (if there is one).

Rotation is the second operation. Given a plane partition π, viewed as a pile of cubes as in Figure 9.1, to rotate it means to rotate it by 120° with the rotation axis the set of (t,t,t) for $-1 < t < 1$ (this rotation leaves the origin of the coordinate system invariant and maps the coordinate axes to each other; so for us it is irrelevant to favor a "left" or "right" rotation). In the representation of a plane partition as a rhombus tiling, this rotation corresponds to a rotation of the hexagon (together with the tiling) by 120°.

Intuitively, the **complement** of a plane partition in a given $a \times b \times c$ box is what the name says, namely the pile of cubes which remains (suitably rotated and reflected) after the plane partition is removed from the box. To be precise, if the cubes of a plane partition π contained in an $a \times b \times c$ box are given co-ordinates in the obvious way by (i,j,k) with $1 \le i \le a$, $1 \le j \le b$, and $1 \le k \le c$, then the complement π^c of π is defined by

$$\pi^c = \{(a+1-i, b+1-j, c+1-k) : (i,j,k) \notin \pi\}.$$

In the representation of a plane partition as a rhombus tiling as in Figure 9.1, to take the **complement** of a plane partition means to rotate the hexagon (together with the tiling) by 180°.

If we combine the three operations—**reflection, rotation, complementation**—each possible way, then there are ten symmetry classes of plane partitions. Of interest, for each class, there is a closed-form product formula for the number of plane partitions contained in a given box, and also generating functions for each class. The first complete presentation of the corresponding results and conjectures was given by Stanley in [285], together with the state of the art at the time, and the proof of the conjectured formulas for Class 5.

In order to have a convenient notation to write down some of the formulas, we use the symbol $\mathcal{N}_d(a, b, c)$ for the number of all plane partitions in symmetry class d which are contained in an $a \times b \times c$ box.

9.4.1 Class 1: Unrestricted Plane Partitions

We know MacMahon proved as per (9.3.5) in the earlier section, that the generating function for all plane partitions contained in an $a \times b \times c$ box is given by

$$\sum_{\pi \in \mathcal{N}_1(a,b,c)} q^{|\pi|} = \prod_{i=1}^{a} \prod_{j=1}^{b} \prod_{k=1}^{c} \frac{1 - q^{i+j+k-1}}{1 - q^{i+j+k-2}}, \tag{9.4.1}$$

where the sum is over all plane partitions π contained in an $a \times b \times c$ box.

9.4.2 Class 2: Symmetric Plane Partitions

A plane partition π is called symmetric if it is invariant under reflection along the main diagonal. In the representation of a plane partition as a rhombus tiling as in Figure 9.1, to be symmetric means to be invariant under reflection along the vertical symmetry axis of the hexagon (i.e., two of the side lengths of the hexagons must be equal).

For this class, two different weights lead to generating functions which have closed form product formulae. The first weight is the usual "size" of a plane partition, and it leads to MacMahon's conjecture [192] that

$$\sum_{\pi \in \mathcal{B}_2(a,a,c)} q^{|\pi|} = \prod_{i=1}^{a} \frac{1 - q^{c+2i-1}}{1 - q^{2i-1}} = \prod_{1 \le i \le a} \frac{1 - q^{2(c+i+j-1)}}{1 - q^{2(i+j-1)}}, \tag{9.4.2}$$

where the sum is over all symmetric plane partitions π that are contained in an $a \times a \times c$ box.

MacMahon's conjecture was first proved by Andrews [14] and Macdonald [191, Ex. 16 and 17, pp. 83–85]. Since then, several other proofs and refinements were given. (See for example Krattenhaler [171]).

The second weight is roughly "half" the size of a plane partition, and it leads to a conjecture of Bender and Knuth [57, pp. 50]. Given a plane partition π, the weight π_0 of π is defined as the sum $\sum_{1 \le j \le i} \pi_{i,j}$. Then

$$\sum_{\pi \in \mathcal{B}_2(a,a,c)} q^{|\pi|} = \prod_{1 \le i \le a} \frac{1 - q^{c+i+j-1}}{1 - q^{i+j-1}}, \tag{9.4.3}$$

where the sum is over all symmetric plane partitions π that are contained in an $a \times a \times c$ box. The first published proofs are due to Andrews [14] and Macdonald [191, Ex. 19, pp. 86].

9.4.3 Class 3: Cyclically Symmetric Plane Partitions

A plane partition π is called *cyclically symmetric* if, viewed as a pile of cubes as in Figure 9.1, it is invariant under rotation by 120° as described above. In the representation of a plane partition as a rhombus tiling as in Figure 9.1, to be cyclically symmetric means to be invariant under rotation of the hexagon (together with the tiling) by 120° (all sides of the hexagon must have the same length). For this class, Macdonald [191, Ex. 18, p.85] conjectured

$$\sum_{\pi\in\mathcal{B}_3(a,a,a)} q^{|\pi|} = \prod_{i=1}^{a} \frac{1-q^{3i-1}}{1-q^{3i-2}} \prod_{1\le i<j\le a} \frac{1-q^{3(2i+j-1)}}{1-q^{3(2i+j-2)}} \prod_{1\le i<j,k\le a} \frac{1-q^{2(i+j+k-1)}}{1-q^{2(i+j+k-2)}}, \tag{9.4.4}$$

where the sum is over all cyclically symmetric plane partitions π that are contained in an $a\times a\times a$ box.

Andrews [18, Theorem 4] had found a determinant for the generating function in question. In retrospect, this determinant can be explained by non-intersecting lattice paths. The conjecture was proved by Mills, Robbins, and Rumsey [202] by evaluating this determinant using some elegant linear algebra, and it is still the only proof.

9.4.4 Class 4: Totally Symmetric Plane Partitions

A plane partition is called *totally symmetric* if it is at the same time symmetric and cyclically symmetric. For this class, Andrews (see [19]) conjectured

$$\sum_{\pi\in\mathcal{B}_2(a,a,a)} q^{|\pi|_0} = \prod_{1\le i\le j\le k\le a} \frac{1-q^{i+j+k-1}}{1-q^{i+j+k-2}}, \tag{9.4.5}$$

where the sum is over all symmetric plane partitions π that are contained in an $a\times a\times a$ box.

A determinant was found by Okada [207] for the generating function here, but for many years no-one could evaluate it, and so this was the longest standing conjecture of all plane partition research.

Then, nearly 30 years after Andrews' initial assertion of (9.4.5), in 2011, Koutschan, Kauers, and Zeilberger [170] evaluated Okada's determinant, by a computer-assisted approach.

9.4.5 Class 5: Self-Complementary Plane Partitions

A plane partition is called *self-complementary* if it is equal to its complement. In the representation of a plane partition as a rhombus tiling as in Figure 9.1, is to be self-complementary means to be invariant under rotation of the hexagon (together with the tiling) by 120°. In other words, to be self-complementary in the rhombus tiling picture means to be centrally symmetric. For this class, Robbins and Stanley independently conjectured that

$$\mathcal{N}_5(2a,2b,2c) = (\mathcal{N}_1(a,b,c))^2, \tag{9.4.6}$$

$$\mathcal{N}_5(2a+1,2b,2c) = \mathcal{N}_1(a,b,c)\mathcal{N}_1(a+1,b,c), \tag{9.4.7}$$

$$\mathcal{N}_5(2a+1,2b+1,2c) = \mathcal{N}_1(a+1,b,c)\mathcal{N}_1(a,b+1,c). \tag{9.4.8}$$

Stanley proved this conjecture in [285, Section 3].

9.4.6 Class 6: Transpose-Complementary Plane Partitions

A plane partition is called *transpose-complementary* if it is equal to the reflection of its complement. In the representation of a plane partition as a rhombus tiling as in Figure 9.1, to be transpose-complementary means to be invariant under reflection of the hexagon (together with the tiling) in its horizontal symmetry axis (two of the side lengths of the hexagon must be equal).

The number of all transpose-complementary plane partitions contained in an $a \times a \times c$ box is equal to

$$\mathcal{N}_6(a,a,c) = \binom{c+a-1}{a-1} \prod_{1 \le i \le j \le a-2} \frac{2c+i+j+1}{i+j+1}, \tag{9.4.9}$$

This was proved by Proctor [211, para before Cor. 1, Cases 'CG'] by essentially relating these plane partitions to the symplectic tableaux of King and El-Sharkaway [168].

9.4.7 Class 7: Symmetric Self-Complementary Plane Partitions

A plane partition is called *symmetric self-complementary* if it is equal to its complement and invariant under reflection in the main diagonal. In the representation of a plane partition as a rhombus tiling as in Figure 9.1, to be symmetric self-complementary means to be invariant under reflection of the hexagon (together with the tiling) in its vertical symmetry axis and under reflection in its horizontal symmetry axis (all side lengths of the hexagon must be equal). For this class, we have

$$\mathcal{N}_7(2a,2a,2c) = \mathcal{N}_1(a,a,c), \tag{9.4.10}$$

$$\mathcal{N}_7(2a+1,2a+1,2c) = \mathcal{N}_1(a+1,a,c). \tag{9.4.11}$$

This was proved by Proctor [210] by again using the representation theory of symplectic groups.

9.4.8 Class 8: Cyclically Symmetric Transpose-Complementary Plane Partitions

A plane partition is called *cyclically symmetric transpose-complementary* if it is equal to the transpose of its complement and invariant under rotation by 120°. In the representation of a plane partition as a rhombus tiling as in Figure 9.1, to be cyclically symmetric transpose-complementary means to be invariant under rotation of the hexagon (together with the tiling) by 120° and under reflection in its horizontal symmetry axis (all side lengths of the hexagon must be equal). The number of all cyclically symmetric transpose-complementary plane partitions contained in a $2a \times 2a \times 2a$ box is equal to

$$\mathcal{N}_8(2a,2a,2a) = \prod_{i=0}^{a-1} \frac{(3i+1)!(6i)!(2i)!}{(4i+1)!(4i)!}. \tag{9.4.12}$$

This was proved by Mills, Robbins, and Rumsey [201] by an interesting determinant evaluation.

For a different proof, using Ciucu's matchings factorization theorem, see Ciucu [111], and also Ciucu and Krattenthaler [113, Sec. 4].

9.4.9 Class 9: Cyclically Symmetric Self-Complementary Plane Partitions

A plane partition is called *cyclically symmetric self-complementary* if it is equal to its complement and invariant under rotation by 120°. In the representation of a plane partition as a rhombus tiling as in Figure 9.1, to be cyclically symmetric self-complementary means to be invariant under rotation of the hexagon (together with the tiling) by 60° (all side lengths of the hexagon must be equal).

The number of all cyclically symmetric self-complementary plane partitions contained in a $2a \times 2a \times 2a$ box is equal to

$$\mathcal{N}_9(2a,2a,2a) = \prod_{i=0}^{a-1} \frac{(3i+1)!^2}{(a+i)!^2}. \tag{9.4.13}$$

This was first proved by Kuperberg [177] by using Kasteleyn's method (see [163] and [164]). Ciucu [112] found a combinatorial explanation of the factorization

$$\mathcal{N}_9(2a,2a,2a) = \mathcal{N}_{10}(2a,2a,2a)^2 \tag{9.4.14}$$

that results from a comparison of (9.4.13) and (9.4.15), based on his matching factorisation theorem [111].

9.4.10 Class 10: Totally Symmetric Self-Complementary Plane Partitions

A plane partition is called *totally symmetric self-complementary* if it is equal to its complement, is invariant under rotation by 120°, and also under reflection in the main diagonal. In the representation of a plane partition as a rhombus tiling as in Figure 9.1, to be totally symmetric self-complementary means to be invariant under all symmetry operations of the hexagon, that is, under reflection of the hexagon (together with the tiling) in its horizontal symmetry axis and under rotation by 60° (all side lengths of the hexagon must be equal). The number of all totally symmetric self-complementary plane partitions contained in a $2a \times 2a \times 2a$ box is equal to

$$\mathcal{N}_{10}(2a,2a,2a) = \prod_{i=0}^{a-1} \frac{(3i+1)!}{(a+i)!}. \tag{9.4.15}$$

Stembridge [287, Theorem 8.3] found a determinant for this number using non-intersecting lattice paths, and Andrews [24] succeeded in evaluating this determinant. See Andrews and Burge [28] for a much simpler proof of that determinant evaluation, and Krattenthaler [172] for a generalization containing an additional parameter.

9.5 THE STATE OF PLAY CIRCA 2023 AND OPEN QUESTIONS

Is this everything there is to know? Or is there more to research? Yes, as reported in the above sections, all conjectures on the enumeration of the ten symmetry classes of plane partitions have now been proven. However, although significant advances have been made since Richard Stanley studied plane partitions, one cannot claim that we have a good understanding of the formulas. An immediate question which poses itself is whether there are insightful explanations why, for all ten symmetry classes, there are closed form product formulae for the number or the generating function of plane partitions in the class which are contained in a given box?

Representation theory, in the form of the observation that certain plane partitions index bases of representations of classical groups and the fact that there are closed form product formulae for the dimensions of such representations provide explanations for several symmetry classes, namely for Classes 1, 2, 3 (only for $q = 1$), then for 5, 6, 7, 8 (see [176, 178], [211] and the references therein). Not only is it not one idea that works for all these classes but several, each of which works for a subset, for the remaining cases, no such explanation is available. The "worst" case is certainly the class of totally symmetric plane partitions, where the proof consists of a heavily computer-assisted verification of a (complicated) determinant evaluation.

A combinatorial demigod would aspire to find a bijective methodology to explain the product formulas. However, we seem far from achieving that dream. Only in the simplest case, namely the case of MacMahon's theorem for the generating function for all plane partitions in a given box — presented here in (9.4.1), there exists a bijective proof by Krattenthaler [174]. In all other cases, finding a bijective proof is an open problem. As mentioned earlier, there exists at least a combinatorial explanation of the factorization (9.4.13) occurring for cyclically symmetric self-complementary plane partitions, see Ciucu [112].

As a matter of fact, already Mills, Robbins and Rumsey looked deeper into this mysterious triple occurrence of the same fascinating number sequence. They attempted to find parametric refinements and variations of these "coincidences." Their idea was that, if the numerical equality took place even at a finer level, then this may be the guide to the sought for conceptual explanation(s) and, in the best case, to bijections between these objects.

The result of their search for refinements was several conjectures, one [199, Conjecture 3] connecting refined enumerations of descending plane partitions and of alternating sign matrices with each other, and several [200, Conjectures 2-7] connecting refined enumerations of totally symmetric self-complementary plane partitions and of alternating sign matrices with each other. The first conjecture has been recently demonstrated by Behrend, Di Francesco and Zinn-Justin [54] then continued and refined in [55]) in a remarkable combination of the new ideas developed by the second and third author in their (so far, unsuccessful) attempt to prove the so-called Razumov-Stroganov conjecture (see Cantini and Sportiello [101] for comparison) with a skillful determinant calculation. As this description indicates, this proof is again far from being bijective or illuminating as to why the numerical equality occurs.

The second group of conjectures is still open, although Zeilberger's *magnum opus* [304] actually proves a weak version of [200, Conjecture 7]. Again, the proof is not bijective and not transparent. In 1996, Krattenthaler is said to have found a generalization of this second conjecture of Mills, Robbins and Rumsey. It is also still open, and in particular it has so far not led to the sought-for bijection between alternating sign matrices and totally symmetric self-complementary plane partitions. Krattenthaler's conjecture has not yet been published anywhere.

9.6 THE GOG AND MAGOG TRAPEZOIDS AND ALTERNATING SIGN MATRICES

However, the biggest unsolved mystery seems to be the question *What has plane partitions to do with alternating sign matrices?* (see below, for their definition). This question was first raised by Stanley in [284], and there is still no answer. Mills, Robbins and Rumsey asserted [199, Conjecture 1] and then Zeilberger [304] and also Kuperberg [179] proved, that the number of $n \times n$ alternating sign matrices is given by (9.4.15) (with a replaced by n), which also counts totally symmetric self-complementary plane partitions contained in a $2n \times 2n \times 2n$ box. Moreover, as Andrews [18] had shown, the same numbers also arise as the numbers of "size n" so-called *descending plane partitions*. Can these be coincidences? Certainly not. However, so far nobody has found any conceptual explanation.

Definition 9.1 *An* (m,n,k)*-Magog trapezoid is an array of positive integers consisting of the first* k *rows of an array*

$$\begin{array}{ccccc} b_{11} & b_{12} & \dots & \dots & b_{1n} \\ b_{21} & b_{22} & \dots & b_{2,n-1} & \\ \dots & \dots & \dots & \dots & \\ b_{n1} & & & & \end{array} \tag{9.6.1}$$

such that entries along rows are weakly increasing, entries along columns are weakly decreasing, and such that the entries in the first row are bounded by $b_{11} \le m+1$, $b_{12} \le m+2, \dots, b_{in} \le m+n$. *So here we have an alternating sign matrix turned into a monotone triangle.*

$$\begin{pmatrix} 0 & 0 & 1 & 0 & 0 & 0 \\ 1 & 0 & -1 & 1 & 0 & 0 \\ 0 & 0 & 1 & -1 & 0 & 1 \\ 0 & 1 & -1 & 1 & 0 & 0 \\ 0 & 0 & 1 & -1 & 1 & 0 \\ 0 & 0 & 0 & 1 & 0 & 0 \end{pmatrix} \to \begin{pmatrix} 1 & 1 & 1 & 1 & 1 & 1 \\ 1 & 1 & 0 & 1 & 1 & 1 \\ 0 & 1 & 1 & 0 & 1 & 1 \\ 0 & 1 & 0 & 1 & 1 & 0 \\ 0 & 0 & 1 & 0 & 1 & 0 \\ 0 & 0 & 0 & 1 & 0 & 0 \end{pmatrix} \to \begin{array}{cccccc} 1 & 2 & 3 & 4 & 5 & 6 \\ 1 & 2 & 4 & 5 & 6 & \\ 2 & 3 & 5 & 6 & & \\ 2 & 4 & 5 & & & \\ 3 & 5 & & & & \\ 4 & & & & & \end{array} \tag{9.6.2}$$

Likewise, we may transform an $n \times n$ alternating sign matrix into a triangular array. This is done by replacing the i-th row of the matrix with the sum of rows $i, i+1, \dots, n$, for $i = 1, 2, \dots, n$. The result is a matrix with only 0's and 1's, see the middle of (9.6.2). To obtain the corresponding triangular array, for each row we record the positions of the 1's in that row; see again (9.6.2). The entries in the triangular array are positive integers. they have the property that they are strictly increasing along rows, weakly increasing along columns, weakly increasing along diagonals in the direction north-east, and the entries in the first row are $1, 2, \dots, n$. Again, it is easy to see that the above defines a bijection. The triangular arrays which one obtains here are most often called *monotone triangles*. The "Gog trapezoids" (terminology again following [304]) below generalise these monotone triangles. In view of the above correspondence, $n \dots n$ alternating sign matrices correspond to $(0,n,n)$-Gog trapezoids.

Definition 9.2 *An (m,n,k)-Gog trapezoid is an array of positive integers consisting of the first k rows of an array*

$$\begin{matrix} b_{11} & b_{12} & \dots & \dots & b_{1n} \\ b_{21} & b_{22} & \dots & b_{2,n-1} & \\ \dots & \dots & \dots & \dots & \\ b_{n1} & & & & \end{matrix} \tag{9.6.3}$$

such that entries along rows are strictly increasing, entries along columns are weakly increasing, and entries along diagonals from lower-left to upper-right are weakly increasing, and such that the entries in the right-most column are bounded by $a_{1k} \le m+k$, $a_{2k} \le m+k+1$, …, $a_{n+1-k,k} \le m+n$.

The conjecture associated with this is

Conjecture 9.1 *The number of (m,n,k)-Magog trapezoids with s Maxima in the first row and t Minima in the last row equals the number of (m,n,k)-Gog trapezoids with t Maxima in the right-most column and s Minima in the left-most column. Here, a Maximum is an entry that is equal to its upper bound, whereas a Minimum is an entry that is 1.*

This conjecture has been unproven for several years now.

What is known about Conjecture 9.1?

(1) If $m = 0$ and $k = n$, and if we forget about Maxima and Minima, then the conjecture reduces to the equality of the number of $n \times n$ alternating sign matrices and the number of totally symmetric self-complementary plane partitions contained in a $2n \times 2n \times 2n$ box. As was said earlier, this is a theorem, thanks to Andrews [24] and Zeilberger [304], but no bijection is known.

(2) If $m = 0$ and if we forget about Maxima and Minima, then this is Zeilberger's main theorem in [304, Lemma 1] from 1996, but, again, his proof is not bijective.

(3) If $m = 0$, then the above conjecture reduces to Mills, Robbins and Rumsey [200, Conjecture 7] from 1986.

(4) If $k = 1$ then we want to show that the number of arrays of positive integers

$$a_1 \quad a_2 \quad \dots \quad a_n$$

with $a_k \le m+k$ and with s Maxima and t Minima is exactly the same as the number of arrays of the same type but with t Maxima and s Minima. It is easy to show that the number in question is

$$\binom{m+2n-s-t-2}{m+n-2} - \binom{m+2n-s-t-2}{m+n-1},$$

which is symmetric in s and t. For this special case, it is easy to construct a bijection.

For further conjectures around totally symmetric self-complementary plane partitions see Robbins [222], and Ishikawa [158] to [159]. Some of these have been proved in Fonseca and Zinn-Justin [128].

So, interest in plane partitions is still very much alive, yet evolved from the times of MacMahon in 1916 or when Stanley studied them and devised innovations for them in 1971. There has also been a long term contribution from George E Andrews dating from the 1960s through to recent times over an almost 60 year period.

Aside from the notorious problem of connecting plane partitions and alternating sign matrices in a conceptual way, it is rhombus tilings and, more generally, perfect matchings of bipartite graphs which are primarily in the focus of current research. On the one hand, unexpected closed-form product formulae for rhombus tilings of particular regions are constantly discovered, see, for example, Gilmore [137], Lai and Rohatgi [180]; and Cook and Nagel [114] and the references cited there. On the other hand, a deep theory of asymptotic properties of perfect matchings (and thus, in particular, of rhombus tilings and plane partitions) has been developed in the past decade, noting the excellent survey by Kenyon [167, see Section 4] on Kasteleyn-Percus matrices, are the starting point of this theory. We go into more detail of the subsequent asymptotic analysis when we consider the circle method results culminating in the Rademacher Theorem; and the Theorem of Meinardus in another chapter.

9.7 INFERENCES FROM PLANE PARTITIONS THAT MAY APPLY TO VECTOR PARTITIONS

One of the key results starting way back with MacMahon's ground-breaking work, was the consideration of plane partitions as cube stacks that fitted inside an $a \times b \times c$ box.

Of course, MacMahon proved as per (9.3.5) in the earlier section, that the generating function for all plane partitions contained in an $a \times b \times c$ box is given by

$$\sum_{\pi \in \mathcal{N}_1(a,b,c)} q^{|\pi|} = \prod_{i=1}^{a} \prod_{j=1}^{b} \prod_{k=1}^{c} \frac{1-q^{i+j+k-1}}{1-q^{i+j+k-2}}, \tag{9.7.1}$$

where the sum is over all plane partitions π contained in an $a \times b \times c$ box.

Suppose, in the context of 3D vector partitions, where the corner of our $a \times b \times c$ box has co-ordinates (a, b, c), that we consider the vector $\langle a, b, c \rangle$ which is an arrow from the origin $(0, 0, 0)$ to the point (a, b, c).

We know that the generating function for 3D unrestricted vector partitions is given by

$$\sum_{a,b,c \geq 0} p(a,b,c) x^a y^b z^c = \prod_{a,b,c \geq 0} \frac{1}{1 - x^a y^b z^c}, \tag{9.7.2}$$

excluding the case $a = b = c = 0$ from sum and product, and where $p(a, b, c)$ is the number of 3D vector partitions of $\langle a, b, c \rangle$. That is, vector partitions inside the box $a \times b \times c$.

It is clear that the definitive operations that led us to consider the ten classes of plane partitions may also be relevant for the analysis of vector partitions. So recall, that in defining the ten classes:

(a) The first operation is reflection.

(b) Rotation is the second operation.

(c) The complement is the third operation.

If one combines the three operations - reflection, rotation, complementation - in all possible ways, then there result ten symmetry classes of plane partitions. A similar thing can be applied for vector partitions. Already in the 1990s papers by the author on VPV (Visible Point Vector) partition identities, some examples of reflection and rotation have been given for 2D, 3D and even 4D and 5D VPV identities. (See Campbell [84, 85, 87, 88] for example.) However, VPV identities are always *weighted* partitions, and to develop our theories it will be prudent to examine first the less complicated *non-weighted* partitions. Obviously, for vector partitions, we are not limited to a set of cubes piled in a box. In fact is turns out we can define sets of or collections of vector that grouped together enable easier analysis than examining vector partitions singly. We call such clusters of vectors *constellations*, as they are points in space not unlike the star constellations in our sky.

EXERCISES

9.1 *Plane Partition research problem.* A plane partition of n is an array $\pi = (\pi_{ij})_{i,j\geq 1}$ of nonnegative integers whose rows and columns are weakly decreasing and whose entries sum to n. When writing π, the entries equal to 0 are often omitted. Hence the plane partitions of the integers $0 \leq n \leq 3$ are given by

$$\begin{array}{ccccccccccc} \emptyset & 1 & 2 & 11 & 1 & 3 & 21 & 111 & 11 & 2 & 1 \\ & & & & 1 & & & & 1 & 1 & 1 \\ & & & & 1 & & & & & & 1 \end{array}$$

If π is a plane partition of n, then we write $|\pi| = n$. Let $a_{rs}(n)$ denote the number of plane partitions of n with at most r rows and at most s columns (of nonzero entries). Then

$$\sum_{n\geq 0} a_{rs}(n)x^n = \prod_{i=1}^{r}\prod_{j=1}^{s}(1-x^{i+j-1})^{-1}. \tag{P9.1}$$

In particular, let $a(n)$ denote the total number of plane partitions of n. If we let r, $s \to \infty$ in (P9.1) we see that

$$\sum_{n\geq 0} a(n)x^n = \prod_{i=1}^{\infty}(1-x^i)^{-i},$$

which is MacMahon's famous formula. At this point it's natural to consider *three-dimensional* (and higher-dimensional) partitions, but almost nothing is known about them, and a "reasonable" enumeration of them was long believed to be hopeless. However, there seems scope to develop such a theory using rationale from the current book. This is the challenge.

9.2 *Plane partitions with largest parts.* Fix r, s, $t > 0$. Let $P(r,s,t)$ denote the set of plane partitions with at most r rows, at most s columns, and with the largest part at most t. Then

$$\sum_{\pi \in P(r,s,t)} a_{rs}(n)x^{|\pi|} = \prod_{i=1}^{r}\prod_{j=1}^{s}\prod_{k=1}^{t} \frac{(1-x^{i+j+k-1})}{(1-x^{i+j+k-2})}. \qquad \text{(P9.2)}$$

Note that the previous exercise is equivalent to the case $t \to \infty$.

9.3 *Symmetric Plane Partition generating function.* A plane partition $\pi = (\pi_{ij})$ is symmetric if $\pi_{ij} = \pi_{ji}$ for all i, j. Let $b(n)$ denote the number of symmetric plane partitions of n. Then

$$\sum_{n \geq 0} b(n)x^n = \prod_{i \geq 1} \frac{1}{(1-x^{2i-1})(1-x^{2i})^{\lfloor i/2 \rfloor}}. \qquad \text{(P9.3)}$$

9.4 *Plane Partitions with trace n.* Let $f_{rs}(n)$ denote the number of plane partitions $\pi = (\pi_{ij})$ with at most r rows, at most s columns, and with *trace* $tr(\pi) := \pi_{11} + \pi_{22} + \ldots = n$. Then

$$f_{rs}(n) = \binom{rs+n-1}{rs-1}.$$

9.5 *Monotone Triangle Plane Partitions and Alternating Matrices.* A *monotone triangle* of length n is a triangular array of integers whose first row is $1, 2, \ldots, n$, every row is strictly increasing, and each entry is (weakly) between its two neighbors above. This definition is clarified by the following example:

$$\begin{array}{ccccccccccc}
1 & & 2 & & 3 & & 4 & & 5 & & 6 \\
 & 1 & & 2 & & 3 & & 4 & & 6 & \\
 & & 1 & & 3 & & 4 & & 5 & & \\
 & & & 2 & & 4 & & 5 & & & \\
 & & & & 2 & & 5 & & & & \\
 & & & & & 3 & & & & &
\end{array}.$$

To show this, there are seven monotone triangles of length 3, given by

$$\begin{array}{ccccccc}
123 & 123 & 123 & 123 & 123 & 123 & 123 \\
12 & 12 & 13 & 13 & 13 & 23 & 23 \\
1 & 2 & 1 & 2 & 3 & 2 & 3
\end{array}.$$

An alternating sign matrix is a square matrix with entries $0, \pm 1$, such that the nonzero entries in every row and column alternate $1, -1, 1, -1, \ldots, 1, -1, 1$. (So every row and column sum is 1.) An example is

$$\begin{pmatrix} 0 & 0 & 0 & 1 & 0 & 0 \\ 0 & 0 & 1 & 0 & 0 & 0 \\ 0 & 1 & -1 & 0 & 1 & 0 \\ 1 & -1 & 1 & -1 & 0 & 1 \\ 0 & 0 & 0 & 1 & 0 & 0 \\ 0 & 1 & 0 & 0 & 0 & 0 \end{pmatrix}.$$

The number of monotone triangles of length n is equal to the number of $n \times n$ alternating sign matrices.

9.6 *Plane Partitions with trace n.* An $n \times n$ Totally Symmetric Self-Complementary (TSSC) plane partition is a plane partition $\pi = (\pi_{ij})_{i,j=1}^{n}$ satisfying: (i) π is symmetric, (ii) every row (and hence every column by symmetry) of π is a self-conjugate partition and (iii) π is invariant under the operation of replacing each entry i with $n-i$ and rotating 180°. We observe that n must be even. The 4×4 TSSC plane partitions are given by:

$$\begin{pmatrix} 4 & 4 & 2 & 2 \\ 4 & 4 & 2 & 2 \\ 2 & 2 & 0 & 0 \\ 2 & 2 & 0 & 0 \end{pmatrix} \quad \begin{pmatrix} 4 & 4 & 3 & 2 \\ 4 & 3 & 2 & 1 \\ 3 & 2 & 1 & 0 \\ 2 & 1 & 0 & 0 \end{pmatrix}$$

A *descending plane partition* is an array of positive integers satisfying:

(i) Each row after the first contains fewer elements than the row above,

(ii) Each row is indented one space to the right from the row above,

(iii) The entries weakly decrease in each row,

(iv) The entries strictly decrease in each column,

(v) The first entry in each row (except the first) does not exceed the number of entries in the preceding row, and

(vi) The first entry in each row is greater than the number of entries in its own row.

The descending partitions with largest part at most three are given by

$$\emptyset \quad 2 \quad 3 \quad 31 \quad 32 \quad 33 \quad \begin{array}{c} 33 \\ \;\;2 \end{array}.$$

The following four numbers are all equal:

(a)

$$\prod_{i=0}^{n-1} \frac{(3i+1)!}{(n+i)!};$$

(b) the number of monotone triangles of length n;

(c) the number of $2n \times 2n$ TSSC plane partitions;

(d) the number of descending plane partitions with largest part at most n.

Open Problem Challenge: There are $\binom{4}{2} = 6$ pairs of equal numbers above. None of these six pairs is known to be equal by a bijective proof! Yet, all are known to be equal by complicated indirect arguments. This is one of the most interesting open problems in the topic of bijective proofs. There are certain refinements of the numbers (a) to (d) which may be useful in finding bijections. For example, it appears that the number of descending plane partitions with largest part at most n and with exactly k parts equal to n is equal to the number of monotone triangles of length n and bottom element $k+1$. Similarly, it seems that the number of descending plane partitions with the largest part at most n and with exactly k parts is equal to the number of monotone triangles of length n with exactly k entries which are greater than the entry to the upper left.

9.7 *Alternating sign matrix sum.* If A is an alternating sign matrix, let $s(A)$ denote the number of -1's in A. Then

$$\sum_A 2^{S(A)} = 2^{\binom{n}{2}},$$

where A ranges over all $n \times n$ alternating sign matrices.

Think About It...

George E. Andrews is generally recognized as the world's leading authority on partitions and is the author of the foremost treatise on the subject. [Bruce Berndt and Robert Rankin, 1995]

Andrews' awards and honors include: In 2003 elected a Member of the National Academy of Sciences. Elected a Fellow of the American Academy of Arts and Sciences in 1997. In 1998 he was an Invited Speaker at the International Congress of Mathematicians in Berlin. In 2012 he became a Fellow of the American Mathematical Society. President of the American Mathematical Society 2009-2010.

Andrews was given honorary doctorates from the University of Parma in 1998, the University of Florida in 2002, the University of Waterloo in 2004, SASTRA University in Kumbakonam, India in 2012, and University of Illinois at Urbana-Champaign in 2014. His Award from the Department of Mathematics, the University of Pennsylvania says "... in recognition of contributions to pure mathematics and [...] mathematical education ..."

[Andrews in September 2006 said he was]... reviving MacMahon's "Partition Analysis", collaborating on further applications of partitions to statistical mechanics and computer science, and completing his study of the relationship between Ramanujan's enigmatic identities and quadratic forms.

Quotes in honor of George E. Andrews
The Modern Father of the Theory of Partitions

GLOSSARY

Cube Stack: A plane partition may be represented visually by the placement of a stack of $\pi_{i,j}$ unit cubes above the point (i, j) in the plane, giving a three-dimensional solid as shown in the picture at the start of this chapter.

Abnormal Variation: Changes in process performance that cannot be accounted for by typical day-to-day variation. Also referred to as non-random variation.

Matrix Index representation: The minimum number of parts that must comply with quality standards, usually stated as a percentage.

Lozenge tilings of a hexagon: A lozenge tiling is a tiling of a planar domain with three types of lozenges which we refer to as *left*, *top* and *right* tiles. These correspond respectively to the dark blue, light blue and middle blue tiles in the Figures 9.1 and 9.2 diagrams that look like cubes.

Plane Partition: In mathematics and especially in combinatorics, a *plane partition* is a two-dimensional array of nonnegative integers $\pi_{i,j}$ (with positive integer subscripts i and j) that are nonincreasing in both subscripts.

FURTHER READING

ANDREWS, G.E. *The Theory of Partitions*, volume 2 of *Encyclopedia of Mathematics and its Applications.* Addison-Wesley Publishing Company, Advanced Book Program, Reading, Massachusetts, USA, 1976.

ANDREWS, G.E. Plane partitions II: The equivalence of the Bender-Knuth and MacMahon conjectures. *Pacific J. Math.*, 72:283–291, 1977.

ANDREWS, G.E. Plane partitions I: The MacMahon conjecture. In G.-C. Rota, editor, *Studies in foundations and combinatorics*, volume 1 of Adv. in *Math. Suppl. Studies*, pages 131–150, 1978.

ANDREWS, G.E. Macdonald's conjecture and descending plane partitions. In J. G. Williams T. V. Narayana, R. M. Mathsen, editor, *Combinatorics, representation theory and statistical methods in groups*, volume 57 of Lecture Notes in *Pure Math.*, pages 91–106. Marcel Dekker, New York, Basel, 1980. Young Day Proceedings.

ANDREWS, G.E. Totally symmetric plane partitions. *Abstracts Amer. Math. Soc.*, 1:415, 1980.

ANDREWS, G.E. Plane partitions (V): The t.s.s.c.p.p. conjecture. *J. Combin. Theory Ser. A*, 66:28–39, 1994.

ANDREWS, G.E. and BURGE, W.H. Determinant identities. *Pacific J. Math.*, 158:1–14, 1993.

BEHREND, R.E., DI FRANCESCO, P.D. and ZINN-JUSTIN, P. On the weighted enumeration of alternating sign matrices and descending plane partitions. *J. Combin. Theory Ser. A*, 119:331–363, 2012.

BEHREND, R.E., DI FRANCESCO, P.D. and ZINN-JUSTIN, P. A doublyrefined enumeration of alternating sign matrices and descending plane partitions. *J. Combin. Theory Ser. A*, 120:409–432, 2013.

BENDER E.A. and KNUTH D.E. Enumeration of plane partitions. *J. Combin. Theory Ser. A*, 13:40–54, 1972.

CAMPBELL, G.B. Multiplicative functions over Riemann zeta function products. *J. Ramanujan Soc.*, 7(1):52–63, 1992.

CAMPBELL, G.B. Dirichlet summations and products over primes. *Int. J. Math. Math. Sci.*, 16(2):359–372, 1993.

CAMPBELL, G.B. A new class of infinite products, and Euler's totient. *Int. J. Math. and Math. Sci.*, 17(3):417–422, 1994.

CAMPBELL, G.B. Infinite products over visible lattice points. *Int. J. Math. and Math. Sci.*, 17(4):637–654, 1994.

CANTINI, L. and SPORTIELLO, A. Proof of the Razumov-Stroganov conjecture. *J. Combin. Theory Ser. A*, 118:1549–1574, 2011.

CARLITZ, L. Rectangular arrays and plane partitions. *Acta Arith.*, 13:29–47, 1967.

CARLITZ, L. Some determinants of q-binomial coefficients. *J. Reine Angew. Math.*, 226:216–220, 1967.

CIUCU, M. Enumeration of perfect matchings in graphs with reflective symmetry. *J. Combin. Theory Ser. A*, 77:67–97, 1997.

CIUCU, M. The equivalence between enumerating cyclically symmetric, selfcomplementary and totally symmetric, self-complementary plane partitions. *J. Combin. Theory Ser. A*, 86:382–389, 1999.

CIUCU, M. and KRATTENTHALER, C. Plane partitions II: 512 symmetry classes. In M. Kashiwara, K. Koike, S. Okada, I. Terada, H. Yamada, editor, *Combinatorial Methods in Representation Theory*, volume 28 of *Advanced Studies in Pure Mathematics*, pages 83–103. RIMS, Kyoto, 2000.

COOK II, D. and NAGEL, U. Enumerations of lozenge tilings, lattice paths, and perfect matchings and the weak Lefschetz property. arxiv preprint, May 2013. website: www.arxiv.org/abs/1305.1314.

FONSECA, T. and ZINN-JUSTIN, P. On the doubly refined enumeration of alternating sign matrices and totally symmetric self-complementary plane partitions. *Electron. J. Combin.*, 15, 2008. Article Number R81, 35 pp.

GILMORE, T. Three interactions of holes in two dimensional dimer systems. arxiv preprint, Mar 2015. preprint last revised 16 Mar 2015. doi.org/10.48550/arXiv.1501.05772.

GORGON, B. and HOUTEN, L. Notes on plane partitions I. *J. Combin. Theory Ser. A*, 4:72–80, 1968.

GORGON, B. and HOUTEN, L. Notes on plane partitions III. *Duke Math. J.*, 36:801–824, 1969.

ISHIKAWA, M. On refined enumerations of totally symmetric selfcomplementary plane partitions, I. arxiv preprint, June 2006. preprint 12 Jun 2006.

ISHIKAWA, M. On refined enumerations of totally symmetric selfcomplementary plane partitions, II. arxiv preprint, June 2006. preprint 6 Jun 2006: www.arxiv.org/abs/math/0606082.

KASTELEYN, P.W. Dimer statistics and phase transitions. *J. Math. Phys.*, 4:287–293, 1963.

KASTELEYN, P.W. Graph theory and crystal physics. In *Graph Theory and Theoretical Physics*, pages 43–110. Academic Press, London, 1967.

KENYON, R. *An introduction to the dimer model*, volume 16 of *Statistical mechanics*, IAS/Park City Math. Ser. Amer. Math. Soc., Providence, R.I., 2009. pp. 191-230.

KING, R.C. and EL-SHARKAWAY, N.G.I. Standard Young tableaux and weight multiplicities of the classical Lie groups. *J. Physics A*, 16:3153–3178, 1983.

KOUTSCHAN, C., KAUERS, M. and ZEILBERGER, D. Proof of George Andrews's and David Robbins's q-TSPP conjecture. *Proceedings of the National Academy of Sciences (PNAS)*, 108(6):2196–2199, February 2011.

KRATTENTHALER, C. *The major counting of nonintersecting lattice paths and generating functions for tableaux*, volume 115 of *Memoirs of the American Mathematical Society. Amer. Math. Soc.* Providence, R.I., 1995. (fourth of five numbers).

KRATTENTHALER, C. Determinant identities and a generalization of the number of totally symmetric self-complementary plane partitions. *Electron. J. Combin.*, 4(1), 1997. Article R27, 62 pp.

KRATTENTHALER, C. Advanced Determinant Calculus. arxiv preprint, 1999. 67 pages arXiv:math/9902004v3 : www.arxiv.org/pdf/math/9902004.pdf.

KRATTENTHALER, C. Another involution principle-free bijective proof of Stanley's hook-content formula. *J. Combin. Theory Ser. A*, 88:66–92, 1999.

KRATTENTHALER, C. Plane Partitions in the work of Richard Stanley and his School. arxiv preprint, May 2015. 33 pages arXiv:1503.05934v2 [math.CO]. www.doi.org/10.48550/arXiv.1503.05934.

KUPERBERG, G. Self-complementary plane partitions by Proctor's minuscule method. *Europ. J. Combin.*, 15:545–553, 1994.

KUPERBERG, G. Symmetries of plane partitions and the permanent determinant method. *J. Combin. Theory Ser. A*, 68:115–151, 1994.

KUPERBERG, G. Four symmetry classes of plane partitions under one roof. *J. Combin. Theory Ser. A*, 75:295–315, 1996.

KUPERBERG, G. An exploration of the permanent-determinant method. *Electron. J. Combin.*, 5, 1998. Article Number R46, 34 pp.

LAI, T. and ROHATGI, R. Enumeration of lozenge tilings of a hexagon with a shamrock missing on the symmetry axis: www.arxiv.org/abs/1711.02818. arxiv preprint, July 2018. preprint last revised 31 Jul 2018.

MACDONALD, I.G. *Symmetric Functions And Hall Polynomials.* Oxford:Clarendon Press; New York/London: Oxford University Press, 2nd edition, 1995.

MACMAHON, P.A. Partitions of numbers whose graphs possess symmetry. *Trans. Cambridge Philos. Soc.*, 17:149–170, 1899.

MACMAHON, P.A. *Combinatory Analysis*, volume 2. Cambridge Univ. Press, London and New York (reprinted by Chelsea, New York, 1960), 1916.

MILLS, W.H., ROBBINS, D.H. and RUMSEY, H. Alternating sign matrices and descending plane partitions. *J. Combin. Theory Ser. A*, 34:340–359, 1983.

MILLS, W.H., ROBBINS, D.H. and RUMSEY, H. Self-complementary totally symmetric plane partitions. *J. Combin. Theory A*, 42:277–292, 1986.

MILLS, W.H., ROBBINS, D.H. and RUMSEY, H. Enumeration of a symmetry class of plane partitions. *Discrete Math.*, 67:43–55, 1987.

MILLS, W.H., ROBBINS, D.H. and RUMSEY,H. Proof of the Macdonald conjecture. *Invent. Math.*, 66:73–87, 1982.

OKADA, S. On the generating functions for certain classes of plane partitions. *J. Combin. Theory Ser. A*, 51:1–23, 1989.

PROCTOR, R.A. Shifted plane partitions of trapezoidal shape. *Proc. Amer. Math. Soc.*, 89:553–559, 1983.

PROCTOR, R.A. New symmetric plane partition identities from invariant theory work of De Concini and Procesi. *Europ. J. Combin.*, 11:289–300, 1990.

ROBBINS, D.P. The story of 1,2,7,42,429,7436, ... *Math. Intelligencer*, 13:12–19, 1991.

STANLEY, R.P. Theory and applications of plane partitions: Part 1,2. *Stud. Appl. Math.*, 50:167–188, 259–279, 1971.

STANLEY, R.P. A baker's dozen of conjectures concerning plane partitions. In P. Leroux (eds.) G. Labelle, editor, *Combinatoire énumérative*, volume 1234 of *Lecture Notes in Math.*, pages 285–293. Springer-Verlag, Berlin-New York, 1986.

STANLEY, R.P. Symmetries of plane partitions. *J. Combin. Theory A*, 43:103–113, 1986. Erratum 44 (1987), 310.

STANLEY, R.P. *Enumerative Combinatorics, volume 2.* Cambridge University Press, Cambridge, 1999.

STEMBRIDGE, J.R. Nonintersecting paths, pfaffians and plane partitions. *Adv. Math.*, 83:96–131, 1990.

WRIGHT, E.M. Partitions of multipartite numbers. *Proc. Amer. Math. Soc.*, 28:880–890, 1956.

ZEILBERGER, D. Proof of the alternating sign matrix conjecture. In *Electron. J. Combin. 3 (no. 2, "The Foata Festschrift")*, 1996. number R13, 84 pp.

CHAPTER 10

Asymptotics for Partition Functions

10.1 RAMANUJAN'S EARLY PARTITION ASYMPTOTIC CONJECTURE

Around 1913, when Ramanujan was still in India, one of his early conjectures concerned an asymptotic estimate for $p(n)$. This result resurfaced when Ramanujan was in England a few years later and led to Hardy's involvement in a joint research project that led to the famous 1918 paper. (See Hardy and Ramanujan [152])

So, in 1918 Hardy and Ramanujan co-authored their ground-breaking paper that gave an asymptotic formula for the number of unrestricted partitions of an integer n, introducing the now famous *Circle Method.* The formula arising from this research was said to represent the very best mathematical creativity and skill of two standout gifted mathematicians of the early 20th century. Ramanujan's intuition led him to assert that a formula with error less than 1 existed, and Hardy's penchant for creative hard analysis of functions of a complex variable combined perfectly to bring out the best and most profound aspects of these two uniquely talented men.

In 1934 E. M. Wright [301] published the asymptotic formula for plane partitions of a number $PR(n)$ using a variation on the circle method.

Then, in 1937, Hans Rademacher [212] was able to refine the work of Hardy and Ramanujan, and produced the convergent formula for the number of unrestricted partitions $p(n)$. Rademacher later supervised the PhD studies of George Andrews, who has since become the main creative force in bringing the theory of partitions from an interesting topic in number theory, to a subject of broad-ranging and profound implications in mathematics and physics. This he did in an exciting, eloquent and sustained but seemingly effortless authoritative approach across half a century of creative inspiration that saw the developments in q-series and important discoveries in Statistical Mechanics, Mock Theta Functions and importantly, the growing appreciation of the works of Ramanujan.

In the 1950s the asymptotic formulas received a further boost with the saddle point approach taken in Theorem of Meinardus. [198] Then, in subsequent decades, came extensions of the related theories by foremost mathematicians of the day, like Erdös, Mahler [194],[195]; Roth and Szekeres [227], [289], and [290]; Erdös and

DOI: 10.1201/9781003174158-10

Richmond [122]. These individuals brought many different approaches grown from various statistical limit theorems, extended circle methods, saddle point variations, or by judicious application of Mellin inversions.

In the period from around 2000 up to near 2020, further extensions of these asymptotics have arisen looking into polynomial coefficients of 2D functions, or examining contour integral expressions with multiple singularities.

So the present chapter aims to give the reader a taste of many ideas, and a practical insight on the relevant topics already researched, but with a view to future possible extensions of the methods. The current book has numerous examples of multivariate and so multidimensional weighted partition results. Some of these results have been described in our text with open ends suggesting possible new directions for research.

From about 2015 to the early 2020s, there have been some developments that may be applicable to our current analyses in vector partitions and to Visible Point Vector identities.

We shall present a few different approaches taken for deriving asymptotic formulas for various kinds of partitions.

10.2 THE CIRCLE METHOD OF HARDY AND RAMANUJAN

Our topic began when Ramanujan wrote, in one of his early Indian conjectures, that

$$p(n) \sim \frac{1}{4n\sqrt{3}} e^{\pi\sqrt{2n/3}} \tag{10.2.1}$$

provides a very good approximation to $p(n)$. This was independently discovered by Uspensky in 1920. Back then around 1917, Major MacMahon using Euler's recurrence calculated a table of $p(1)$, $p(2)$, up to $p(200)$, so it was known that

$$p(200) = 3972999029388.$$

In 1918 Hardy and Ramanujan [152] devised and then used the circle method and modular functions to obtain an $o(1)$ asymptotic solution. (ie. An asymptotic approximation valid to the nearest integer.) That formula asserts $p(n)$ is the integer nearest

$$\frac{1}{2\sqrt{2}} \sum_{q=1}^{\nu} \sqrt{q} A_q(n) \psi_q(n), \tag{10.2.2}$$

where $A_q(n) = \sum \omega_{p,q} e^{-2np\pi i/q}$, the sum being over all p less than and prime to q, with $\omega_{p,q}$ a certain 24th root of unity, ν is of the order of $\sqrt{n}$, and

$$\psi_q(n) = \frac{d}{dn}\left(\exp\left\{\frac{C\sqrt{\left(n-\frac{1}{24}\right)}}{q}\right\}\right), \quad C = \pi\sqrt{\frac{2}{3}}.$$

The Hardy and Ramanujan method can be summarized in a few steps as follows:

Step 1. Use Cauchy's formula to transform coefficients of generating functions into integrals. For example, Cauchy's integral theorem implies

$$p(n) = \frac{1}{2\pi i} \oint_C \frac{P(x)}{x^{n+1}} dx$$

where

$$P(x) = \prod_{n=1}^{\infty} \frac{1}{1-x^n} = \sum_{n=0}^{\infty} p(n)x^n.$$

Step 2. Integrate over an appropriate circle that will allow easier calculations.
Step 3. Bound the integrals over the major and minor arcs.
Step 4. Manipulate the integral contours to avoid or shift around simple poles.
Step 5. Incorporate the modular inversion formula of the Dedekind Eta function

$$(q;q)_\infty = \frac{1}{\sqrt{-i\tau}} \exp\left(\frac{-\pi i\tau}{12} - \frac{\pi i}{12\tau}\right)\left(1+O\left(e^{\frac{-2\pi i}{\tau}}\right)\right);$$

the Dedekind-Eta function being $\eta(\tau) = q^{1/24}(q;q)_\infty$ with $q = e^{2\pi i\tau}$.

As stated above, $p(200) = 3972999029388$, and in fact the first eight terms of (10.2.2) produce

$$\begin{array}{r} +\ 3{,}972{,}998{,}993{,}185{,}896 \\ +\ 36{,}282{,}978 \\ -\ 87.555 \\ +\ 5.147 \\ +\ 1.424 \\ +\ 0.071 \\ +\ 0.000 \\ +\ 0.043 \\ \hline 3{,}972{,}999{,}029{,}388.004 \end{array}$$

which is the correct value of $p(200)$ to within 0.004. So, after inspiration from Hardy and Ramanujan's first sophisticated formula, the methodologies for partition asymptotic formulas kept coming over the next century.

In 1942, Erdös showed that the formula of Hardy and Ramanujan could be derived by elementary means (Hoffman 1998, p. 91).

In 2016 Václav Kotěšovec (see [169]) showed that the Hardy and Ramanujan formula is equivalent to

$$p(n) \sim \frac{1}{4n\sqrt{3}} \exp\left(\pi\sqrt{\frac{2n}{3} + c_0 + \frac{c_1}{n^{\frac{1}{2}}} + \frac{c_2}{n} + \frac{c_3}{n^{\frac{3}{2}}} + \frac{c_4}{n^2} + \ldots}\right) \tag{10.2.3}$$

where the coefficients c_0 to c_4 are approximately

$$c_0 = -0.2304201450624533206655367041972\,33... = -\frac{1}{36} - \frac{2}{\pi^2},$$

$$c_1 = -0.0178416569128570889795021353499\,49... = \frac{1}{6\sqrt{6\pi}} - \frac{\sqrt{3}}{\sqrt{2}\pi^3},$$

$$c_2 = 0.0051329911273421675945763916335\,59... = \frac{1}{2\pi^4},$$

$$c_3 = -0.0011129404895597609082366028434\,97... = \frac{3\sqrt{3}}{4\sqrt{2}\pi^5} - \frac{5}{16\sqrt{6}\pi^3},$$

$$c_4 = 0.0009573432848069729589686943491\,96... = \frac{1}{576\pi^2} - \frac{1}{24\pi^4} + \frac{93}{80\pi^6}.$$

10.3 THE RADEMACHER EXACT FORMULA

The next major development was when Rademacher [212] in 1937 subsequently obtained an exact convergent series solution which yields the Ramanujan formula (10.2.1) as the first term. Rademacher's exact infinite series formula is

$$p(n) = \frac{1}{\pi\sqrt{2}} \sum_{k=1}^{\infty} A_k(n)\sqrt{k}\frac{d}{dn}\left[\frac{\sinh\left(\frac{\pi}{k}\sqrt{\frac{2}{3}(n-\frac{1}{24})}\right)}{\sqrt{n-\frac{1}{24}}}\right] \tag{10.3.1}$$

in which

$$A_k(n) = \sum_{h=1}^{k} \delta_{\gcd(h,k),1} \exp\left[\pi i \sum_{j=1}^{k-1} \frac{j}{k}\left(\frac{hj}{k} - \left\lfloor\frac{hj}{k}\right\rfloor - \frac{1}{2}\right) - \frac{2\pi i h n}{k}\right], \tag{10.3.2}$$

where $\delta_{m,n}$ is the Kronecker delta, and $\lfloor x \rfloor$ is the floor function. The remainder after N terms is

$$R(N) < CN^{-1/2} + D\sqrt{\frac{N}{n}}\sinh\left(\frac{K\sqrt{n}}{N}\right), \tag{10.3.3}$$

where C and D are fixed constants (see Apostol [32]). The contour for the integral used by Rademacher involves Farey sequences and Ford circles.

10.4 THE THEOREM OF MEINARDUS

We know that there are many partition functions generated by infinite products. Many researchers have studied the asymptotics of the generated coefficients from these infinite product generating functions. In 1954 Meinardus [198] published what has become a very useful asymptotic formula applicable to weighted integer partitions. In recent years, extensions of this theorem have taken us to asymptotics for some weighted partition vector generating function coefficients. This will definitely be useful to us in some of our work on 2D and 3D partitions.

So, we examine the infinite product

$$f(\tau) = \prod_{n=1}^{\infty} \left(\frac{1}{1-q^n}\right)^{a_n} \tag{10.4.1}$$

$$= 1+\sum_{n=1}^{\infty} r(n)q^n, \tag{10.4.2}$$

where $q = e^{-t}$ with $\Re\tau > 0$, which is $|q| < 1$. Traditionally a_n are taken to be non-negative real integers, but in later chapters, we will use a_n when rational numbers. In our work, we need the function

$$D(s) = \sum_{n=1}^{\infty} \frac{a_n}{n^s}, \quad \textit{with } s = \sigma + it, \tag{10.4.3}$$

assuming convergence for $\sigma > \alpha$ being positive real. Also, we assume $D(s)$ has analytic continuation in the region $\sigma \geq C_0$ for $0 < C_0 < 1$ with $D(s)$ analytic in the region except for a simple pole at $s = \alpha$ with residue A. Next, we assume

$$D(s) = O(|t|^{C_1} \tag{10.4.4}$$

uniformly in $\sigma \geq -C_0$ as $|t| \to \infty$, where C_1 is a fixed positive real number.

Last of all we shall use the function

$$g(\tau) = \sum_{n=1}^{\infty} a_n q^n, \quad \textit{for } q = e^{-\tau}. \tag{10.4.5}$$

Also, if $\tau = y + 2\pi i x$ (for real x and y), we need next to assume that for $|\arg\tau| > \frac{\pi}{4}$, $|x| < \frac{1}{2}$,

$$\Re(g(\tau)) - g(y) \leq -C_2 y^{-\varepsilon} \tag{10.4.6}$$

for sufficiently small y, where ε is an arbitrary fixed positive number and C_2 is a suitably chosen positive real number dependent on ε. So, having covered the above-required definitions and functions, we can make sense of the theorem following here.

While the proof of Theorem of Meinardus is interesting and involves contour integration and Mellin inversion, in this book we shall simply outline the approach taken and refer the reader to Andrews [12, Chapter 6], where the proof is described over several pages. So, the result we will use and may apply to many generating function coefficients in our present book is Meinardus' main theorem:

Theorem 10.1 *(Meinardus, 1954)*

As $n \to \infty$,

$$r(n) = Cn^{\kappa} \exp\left(n^{\frac{\alpha}{\alpha+1}} \left(\frac{\alpha+1}{\alpha}\right) (A\Gamma(\alpha+1)\zeta(\alpha+1))^{\frac{1}{\alpha+1}}\right) (1+O(n^{-\kappa_1})) \tag{10.4.7}$$

where $\zeta(s) = \sum_{m=1}^{\infty} m^{-s}$ is the Riemann zeta function, and

$$C = e^{D'(0)}\sqrt{\frac{(A\Gamma(\alpha+1)\zeta(\alpha+1))^{\frac{1-2D(0)}{\alpha+1}}}{2\pi(\alpha+1)}}, \tag{10.4.8}$$

$$\kappa = \frac{D(0)-1-\frac{1}{2}\alpha}{\alpha+1}, \tag{10.4.9}$$

$$\kappa_1 = \frac{\alpha}{\alpha+1}\min\left(\frac{C_0}{\alpha}-\frac{\delta}{4}, \frac{1}{2}-\delta\right), \tag{10.4.10}$$

for δ an arbitrary real number.

This theorem depends on the saddle point method. In order to neatly apply this theorem it is important to consider some associated features of this theorem. To start with, we use the formula stated as a lemma:

Lemma 10.1 *Under the assumptions on $f(\tau)$, $D(s)$, and $g(s)$ as given earlier in this section with $\tau = y+2\pi x$,*

$$f(\tau) = \exp\left(A\Gamma(\alpha)\zeta(\alpha+1))\tau^{-\alpha} - D(0)\log\tau + D'(0) + O(y^{C_0})\right) \tag{10.4.11}$$

uniformly in x as $y \to 0$, provided $|\arg\tau \le \frac{\pi}{4}$, $|x| \le \frac{1}{2}$; there exists a positive ε_1 such that

$$f(y+2\pi ix) = O(\exp(A\Gamma(\alpha)\zeta(\alpha+1)y^{-s} - C_3 y^{-\varepsilon_1}) \tag{10.4.12}$$

uniformly in x with $y^{\beta} \le |x| \le \frac{1}{2}$, as $y \to 0$, where

$$\beta = 1+\frac{\alpha}{2}\left(1-\frac{\delta}{2}\right) \quad \textit{with } 0 \le \delta \le \frac{2}{3}, \tag{10.4.13}$$

and C_3 a fixed real number.

This early 1950s version of Meinardus and/or variations on the 1917 Hardy-Ramanujan Circle Method are useful for the integer partition generating functions related to:

(a) Partitions into unrestricted parts.

(b) Partitions into odd parts, even parts and modular congruent parts.

(c) Partitions where the parts are of minimal integer difference restriction.

(d) Classes of Plane Partitions of various kinds with known generating functions.

(e) Binary partitions and more generally, those related to parts that are sums of powers of integer m.

The theorem of Meinardus can supply us in our current book, with asymptotic behaviors of coefficients generated by functions that arise in our analysis in various chapters. There are particular cases of our higher dimensional formulas that are examinable by the Meinardus Saddle Point Methods and Hardy-Ramanujan Circle Methods.

Some examples from the present book will evidently be applicable to one asymptotic method or an other. Clearly this is an opening to further research.

As a brief case in point, there is the so-called hybrid example of a q-binomial identity. None of the known methods seem to be directly applicable to this kind of identity, but the known methodologies might be a good starting place for the 2D and 3D results we bring in other chapters of the present book. So, as a moment's aside look at the following example hybrid infinite product identity.

$$G_1(q;0,t) = \prod_{k=0}^{\infty} \left(\frac{1}{1-tq^k}\right)^{1/k!} = \sum_{k=0}^{\infty} g_{(t,k)}(q)t^k \tag{10.4.14}$$

$$= 1 + e^q \frac{t}{1!} + \begin{vmatrix} e^q & -1 \\ e^{q^2} & e^q \end{vmatrix} \frac{t^2}{2!} + \begin{vmatrix} e^q & -1 & 0 \\ e^{q^2} & e^q & -2 \\ e^{q^3} & e^{q^2} & e^q \end{vmatrix} \frac{t^3}{3!} + \begin{vmatrix} e^q & -1 & 0 & 0 \\ e^{q^2} & e^q & -2 & 0 \\ e^{q^3} & e^{q^2} & e^q & -3 \\ e^{q^4} & e^{q^3} & e^{q^2} & e^q \end{vmatrix} \frac{t^4}{4!} + O(t^5)$$

$$= 1 + e^q \frac{t}{1!} + \left(e^{q^2} + e^{2q}\right) \frac{t^2}{2!} + \left(2e^{q^3} + 3e^{q^2+q} + e^{3q}\right) \frac{t^3}{3!}$$

$$+ \left(6e^{q^4} + 8e^{q^3+q} + 3e^{2q^2} + 6e^{q^2+2q} + e^{4q}\right) \frac{t^4}{4!} + O(t^5).$$

Although we give this as a power series in t, it could also be given as a power series in q. In that situation, we would have

$$G_1(q;0,t) = \prod_{k=0}^{\infty} \left(\frac{1}{1-tq^k}\right)^{1/k!} = \sum_{k=0}^{\infty} h_{(q,k)}(t)q^k \tag{10.4.15}$$

$$= 1 + t\frac{q}{1!} + (2t^2+t)\frac{q^2}{2!} + (6t^3+3t^2+t)\frac{q^3}{3!} + (24t^4+12t^3+13t^2+t)\frac{q^4}{4!}$$

$$+ (120t^5+60t^4+65t^3+15t^2+t)\frac{q^5}{5!} + O(q^6).$$

In recent years the theorem of Meinardus has been extended to polynomial coefficients, somewhat like $h_{(q,k)}$ which is a polynomial in t. A good resource for fairly recent updates on this topic may be found in the 2017 Ph.D. Dissertation by Richard Alexander Frnka (see [131]). So, next we present a few, more recent versions and extensions of the Meinardus and Circle Method results, which of themselves can lead to many interesting asymptotic estimates that are not yet in the known literature. In particular, some *nice-to-have* results may involve the 2D and 3D extensions of the theory of integer partitions into the new areas of vector partitions.

10.5 A POLYNOMIAL ANALOGUE OF MEINARDUS' THEOREM

Recent authors have developed a polynomial analog of the Meinardus Theorem for bivariate Euler products and applied it to the study of complex multiplicatively weighted partitions. The work in those researches explores what appear to be *simply weighted* partitions of a particular kind, but may be applicable to integer and vector partitions weighted more broadly. So, the particular approach given in this section may be a good starting place for further multidimensional vector partitions asymptotics. We follow the terminology of Parry [209].

As with the theorem of Meinardus, there are several preliminary conditions that must be met in order to apply the methodology for obtaining the coefficient estimates. We start with a statement of the theorem, then list the conditions afterward, as then we see upfront the usefulness of the theorem.

Theorem 10.2 *Consider the equation*

$$P(z,q) = 1 + \sum_{n=1}^{\infty} Q_n(z) q^n = \prod_{m=1}^{\infty} \frac{1}{(1-zq^m)^{a_m}}, \tag{10.5.1}$$

valid for z *in the unit disk* $\mathbb{D}$. *Then*

$$Q_n(z) = \sum_{\{(h,k):L_{h,k}(z)\}\in[L_{p,q}(z)]} w_{h,k,n}(z)\bar{I}_{h,k,n}(z) + O_X\left(\frac{1}{n}\exp(m_{h,k,n}(z))\right), \tag{10.5.2}$$

where the following statements 1 to 3 conditions hold.

Statement 1 *Define the family of twisted Dirichlet series* $D_{r,h,k}(s)$ *with* $r,h \leq k$ *positive integers, and* $gcd(h,k) = 1$, *by*

$$D_{r,h,k}(s) = \sum_{m=1}^{\infty} \frac{e^{2\pi i \frac{hrm}{k}} a_m}{m^s} \tag{10.5.3}$$

and assume

(a) Each Dirichlet series $D_{r,h,k}(s)$ *converges uniformly and absolutely in some half plane* $\sigma = c > 0$.

(b) For some $\sigma_0 \in (-1,0), D_{r,h,k}(s)$ *has either an analytic continuation to* $\Re s = \sigma = \sigma_0$ *or has a meromorphic continuation to* $\sigma = \sigma_0$ *with a simple pole at* $0 < s_0 < c$ *with residue* $A_{r,h,k}$.

(c) There is a $T > 0$ *so that for* $|t| \geq T$ *and* $\sigma \in [-\sigma_0; c]$ *there exists a* $C > 0$ *so that* $|D_{r,h,k}(s)| = O_{h,k,T}(|t|^C)$.

(d) There exists a $C > 0$ *so that* $D_{r,h,k}(\sigma_0 + it) = O(\max(1;|t|^C)k^{s_0-\sigma_0})$*:*

(e) $D_{r,h,k}(0) = O(k^{s_0})$.

We set $A_{r,h,k} = 0$ *if* $D_{r,h,k}(s)$ *is analytic and not meromorphic as Residue* $(s_0, D_{r,h,k}(s)) = 0$*: This makes the notation more uniform. Define the operator*

$$J[D_{r,h,k}(s)](z,\alpha) = \Gamma(s)\Phi(z^k, s+1, \frac{r}{k})D_{r,h,k}(s)(k\alpha)^{-s}$$

where $\Phi(z,s,\nu)$ *is the Lerch phi function*

$$\Phi(z,s,\nu) = \sum_{n=0}^{\infty} \frac{z^n}{(n+\nu)^s}.$$

Statement 2 *Let* $z \in \mathbb{D}$ *be the open unit disk,* h, k *be relatively prime positive integers and* $\Re\alpha > 0$. *Then*

$$\log P(z, e^{2\pi i \frac{h}{k} - \alpha}) = \Psi_{h,k}(z,\alpha) + \log \omega_{h,k,n}(z) + g_{h,k}(z,\alpha) + 2\pi i n \frac{h}{k},$$

where

$$\begin{aligned}
\Psi_{h,k}(z,\alpha) &= \frac{\Gamma(s_0+1)}{s_0 k^{s_0+1}\alpha^{s_0}} \left(\sum_{r=1}^{k} z^r \Phi(z^k, s_0+1, \frac{r}{k}) A_{r,h,k} \right), \\
\omega_{h,k,n}(z) &= \exp\left(\sum_{r=1}^{k} \frac{z^r}{k} \Phi(z^k, 1, \frac{r}{k}) D_{r,h,k}(0) - 2\pi i \frac{h}{k} \right), \\
g_{h,k}(z,\alpha) &= \frac{1}{k} \sum_{r=1}^{k} z^r \frac{1}{2\pi i} \int_{\sigma_0 - i\infty}^{\sigma_0 + i\infty} J[D_{r,h,k}(s)](z) ds.
\end{aligned}$$

Statement 3 *For* $h, k \in \mathbb{N}$; $(h,k) = 1$; *and* $0 < h \le k$ *define* $L_{h,k}(z) = \sqrt[s_0+1]{\Phi_{h,k}(z)}$ *where we write*

$$\Phi_{h,k}(z) = \Gamma(s_0+1) \sum_{r=1}^{k} z^r \Phi(z^k, 1, \frac{r}{k}) A_{r,h,k},$$

and $D \subset \mathbb{D}$. *We define* $[L_{p,q}(z)]$ *as the equivalence class of* $\{L_{p,q}(z)\}$ *whose real component is identical to* $L_{p,q}(z)$ *on* $\mathbb{D}$*: The set* $R(p,q)$ *we call the* (p,q)*-th phase (or simply phase* (p,q)*) is defined by*

$$R(p,q) = \{z \in D : L_{h,k}(z) \notin [L_{p,q}(z)] \Rightarrow \Re L_{p,q}(z) > \Re L_{h,k}(z)\}.$$

10.6 THE CONVOLUTION METHOD FOR PRODUCTS OF TWO SERIES

In 2015 Kočšovec [169] gave a method of finding the asymptotics of q-series based on the convolution of generating functions. It involves what is essentially a saddle point method. The technique yields many new results and may be applicable to 2D, 3D, 4D, and higher vector partition-generating functions. He starts with a

Theorem 10.3 *Convolution of two generating functions with subexponential growth. Let r_1 and $r_2 > 0$, and $0 < p < 1$. If*

$$g_1(x) = \sum_{n=0}^{\infty} \alpha_n x^n \quad with \quad \alpha_n \sim \nu_1 \frac{\exp(r_1 n^p)}{n^{b_1}}, \tag{10.6.1}$$

$$g_2(x) = \sum_{n=0}^{\infty} \beta_n x^n \quad with \quad \beta_n \sim \nu_2 \frac{\exp(r_2 n^p)}{n^{b_2}} \tag{10.6.2}$$

and

$$g(x) = g_1(x)\, g_2(x) = \sum_{n=0}^{\infty} a_n x^n, \tag{10.6.3}$$

then we have

$$a_n \sim \frac{\sqrt{2\pi}\nu_1\nu_2 \left(r_1^{\frac{1}{1-p}} + r_2^{\frac{1}{1-p}}\right)^{b_1+b_2-\frac{3-p}{2}} \exp\left(n^p \left(r_1^{\frac{1}{1-p}} + r_2^{\frac{1}{1-p}}\right)^{1-p}\right)}{\sqrt{p(1-p)}\, r_1^{\frac{b_1-1/2}{1-p}}\, r_2^{\frac{b_2-1/2}{1-p}}\, n^{b_1+b_2+\frac{p}{2}-1}}. \tag{10.6.4}$$

The proof of this by Koěšovec [169] inspired several others to continue this kind of asymptotic analysis. However, we now reproduce that proof.

PROOF OF THEOREM 10.3.
From the product of two power series we know that

$$g(x) = g_1(x) g_2(x) = \left(\sum_{n=0}^{\infty} \alpha_n x^n\right)\left(\sum_{n=0}^{\infty} \beta_n x^n\right) = \sum_{n=0}^{\infty}\left(\sum_{k=0}^{n} \alpha_{n-k}\beta_k\right) x^n.$$

We set $k = m\, n$, and find the maximum of the term

$$\begin{aligned} \alpha_{n-k}\beta_k \;&\sim\; v_1 v_2 k^{-b_2}(n-k)^{-b_1} \exp(r_2 k^p + r_1 (n-k)^p) \\ &\sim\; v_1 v_2 (mn)^{-b_2}((1-m)n)^{-b_1} \exp(r_1((1-m)n)^p + r_2(mn)^p). \end{aligned}$$

The derivative with respect to m is, for $m, n, 1-m \neq 0$,

$$\frac{v_1 v_2 (mn)^{-b_2} \exp(r_1((1-m)n)^p + r_2(mn)^p)}{(1-m)n(mn)^{b_2}((1-m)n)^{b_2}}$$

$$\times \frac{((1-m)(b_2 - pr_2(mn)^p) - b_1 m + mpr_1((1-mn)n)^p)}{(1-m)n(mn)^{b_2}((1-m)n)^{b_2}},$$

so for the required maximum we then must solve the equation

$$(1-m)(b_2 - pr_2(mn)^p) - b_1 m + mpr_1((1-mn)n)^p = 0.$$

However, observe that

$$\lim_{n\to\infty} \frac{-b_1 m - b_2 m + b_2 + p r_2 m^{p+1} n^p - p r_2 m^p n^p + m p r_1 (1-m)^p n^p}{n^p}$$

$$= p \times ((1-m) r_2 m^p - m r_1 (1-m)(1-m)^p)$$

and the equation

$$(1-m)\, r_1^{\frac{1}{p-1}} = m\, r_2^{\frac{1}{p-1}}$$

has the solution

$$k = \frac{r_2^{\frac{1}{1-p}}}{r_1^{\frac{1}{1-p}} + r_2^{\frac{1}{1-p}}}.$$

We then set

$$k = \left(\frac{r_2^{\frac{1}{1-p}}}{r_1^{\frac{1}{1-p}} + r_2^{\frac{1}{1-p}}} \right) n + x.$$

This enables us to write the desired maximum value as

$$v_1 v_2 A_1^{-b_1} A_2^{-b_2} \exp\left(r_1 A_1^p + r_2 A_2^p\right)$$

$$\text{where} \quad A_1 = \left(\frac{n r_1^{\frac{1}{1-p}}}{r_2^{\frac{1}{1-p}} + r_1^{\frac{1}{1-p}}} - x \right), \quad A_2 = \left(\frac{n r_2^{\frac{1}{1-p}}}{r_1^{\frac{1}{1-p}} - r_2^{\frac{1}{1-p}}} + x \right).$$

We approximate the exponent with the Maclaurin series method given by Auluck [36, p. 683] in 1951. So then

$$r_1 A_1^p + r_2 A_2^p \sim n^p \left(r_1^{\frac{1}{1-p}} + r_2^{\frac{1}{1-p}} \right)^{1-p} \left(1 - \frac{(1-p) p\, r_1^{\frac{1}{1-p}}\, r_2^{\frac{1}{1-p}} \left(r_1^{\frac{1}{p-1}}\, r_2^{\frac{1}{1-p}} + 1 \right)^2 x^2}{2n^2} \right)$$

and near $x = 0$ we have

$$a_n \sim \frac{v_1\, v_2 \int_{-\infty}^{\infty} \exp\left\{ n^p (r_1^{\frac{1}{1-p}} + r_2^{\frac{1}{1-p}})^{1-p} \left(1 - \frac{(1-p) p r_1^{\frac{1}{1-p}}\, (r_1/r_2)^{\frac{1}{1-p}}\, (r_2^{\frac{1}{1-p}} + 1)^2 x^2}{2n^2} \right) \right\} dx}{(A_1 + x)^{b_1} (A_2 - x)^{b_2}}.$$

Finally, after evaluating the integral and simplifying, we obtain

$$a_n \sim \frac{\sqrt{2\pi} \nu_1 \nu_2 \left(r_1{}^{\frac{1}{1-p}} + r_2{}^{\frac{1}{1-p}} \right)^{b_1 + b_2 - \frac{3-p}{2}} \exp\left(n^p \left(r_1{}^{\frac{1}{1-p}} + r_2{}^{\frac{1}{1-p}} \right)^{1-p} \right)}{\sqrt{p(1-p)}\; r_1{}^{\frac{b_1 - 1/2}{1-p}}\; r_2{}^{\frac{b_2 - 1/2}{1-p}}\; n^{b_1 + b_2 + \frac{p}{2} - 1}}.$$

This proves the desired result. ■

We can write two immediate corollaries as cases of theorem 10.3 as follows, from firstly where $p = \frac{1}{2}$ and secondly where $p = \frac{2}{3}$.

Corollary 10.1 *(see Dewar and Ram Murty [117]) Let $g_1(x)$ and $g_1(x)$ be generating functions with subexponential growth in the indices of x and the following known asymptotic behaviours. If r_1 and $r_2 > 0$,*

$$g_1(x) = \sum_{n=0}^{\infty} \alpha_n x^n \quad with \quad \alpha_n \sim \nu_1 \frac{\exp(r_1 \sqrt{n})}{n^{b_1}}, \tag{10.6.5}$$

$$g_2(x) = \sum_{n=0}^{\infty} \beta_n x^n \quad with \quad \beta_n \sim \nu_2 \frac{\exp(r_2 \sqrt{n})}{n^{b_2}} \tag{10.6.6}$$

and

$$g(x) = g_1(x)\, g_2(x) = \sum_{n=0}^{\infty} a_n x^n, \tag{10.6.7}$$

then we have

$$a_n \sim \frac{2\, \sqrt{2\pi}\, v_1\, v_2\, (r_1^2 + r_2^2)^{b_1+b_2-5/4} \exp(\sqrt{(r_1^2 + r_2^2)n})}{n^{b_1+b_2-3/4}\, r_1^{2b_1-1}\, r_1^{2b_2-1}}. \tag{10.6.8}$$

There are many known asymptotic generating functions for partitions that satisfy the conditions of corollary 10.1. Next, we look at the second inferred corollary 10.2 corresponding to case $p = \frac{2}{3}$ of theorem 10.3. Similarly, there are known examples generating functions for partitions associated with this result.

Corollary 10.2 *Let $g_1(x)$ and $g_1(x)$ be generating functions with subexponential growth in the indices of x and the following known asymptotic behaviours. If r_1 and $r_2 > 0$,*

$$g_1(x) = \sum_{n=0}^{\infty} \alpha_n x^n \quad with \quad \alpha_n \sim \nu_1 \frac{\exp(r_1 \sqrt[3]{n^2})}{n^{b_1}}, \tag{10.6.9}$$

$$g_2(x) = \sum_{n=0}^{\infty} \beta_n x^n \quad with \quad \beta_n \sim \nu_2 \frac{\exp(r_2 \sqrt[3]{n^2})}{n^{b_2}} \tag{10.6.10}$$

and

$$g(x) = g_1(x)\, g_2(x) = \sum_{n=0}^{\infty} a_n x^n, \tag{10.6.11}$$

then we have

$$a_n \sim \frac{3\, \sqrt{\pi}\, v_1\, v_2\, (r_1^3 + r_2^3)^{b_1+b_2-7/6} \exp(\sqrt[3]{(r_1^3 + r_2^3)n^2})}{n^{b_1+b_2-2/3}\, r_1^{3b_1-3/2}\, r_2^{3b_2-3/2}}. \tag{10.6.12}$$

10.7 BRUINIER AND ONO'S EXACT FORMULA FOR p(n)

In 2013 Jan Bruinier and Ken Ono published a finite algebraic form for $p(n)$; their account of which we follow in this section (see [81]). They spectacularly obtained a new formula for $p(n)$, which expressed $p(n)$ as a finite sum of algebraic numbers. These numbers are singular moduli for a weak Maass form which they described using Dedekind's eta-function

$$\eta(z) = q^{\frac{1}{24}} \prod_{n=1}^{\infty} (1-q^n)$$

with $q = e^{2\pi i z}$ by definition and the quasimodular Eisenstein series

$$E_2(z) := 1 - \sum_{n=1}^{\infty} \sum_{d|n} dq^n. \tag{10.7.1}$$

With this, they define the $\Gamma_0(6)$ weight -2 meromorphic modular form $F(z)$ by

$$F(z) := \frac{1}{2} \frac{E_2(z) - 2E_2(2z) - 3E_2(3z) - 6E_2(6z)}{\eta(z)^2\eta(2z)^2\eta(3z)^2\eta(6z)^2} = q^{-1} - 10 - 29q - \ldots. \tag{10.7.2}$$

Setting $z = x + iy$, with $x, y \in \Re$, they defined the weak Maass form

$$P(z) := -\left(\frac{1}{2\pi i}\frac{d}{dz} + \frac{1}{2\pi y}\right) F(z) = \left(1 - \frac{1}{2\pi y}\right) q^{-1} + \frac{5}{\pi y} + \left(29 + \frac{29}{2\pi y}\right) q + \ldots. \tag{10.7.3}$$

This nonholomorphic form has weight 0 and is a *weak Maass form.* For a more complete account of the related context of this topic of weak Maass forms, see firstly the monograph by Bruinier in 2002 [78], then Bruinier and Funke [79] and [80], then the 2006 paper by Bruinier, Jenkins, and Ono [82]. This weak Maass form has eigenvalue -2 with respect to the hyperbolic Laplacian

$$\Delta := -y^2 \left(\frac{\partial^2}{\partial x^2} + \frac{\partial^2}{\partial y^2}\right). \tag{10.7.4}$$

To describe the new $p(n)$ formula, we use discriminant $-24n+1 = b^2 - 4ac$ positive definite integral binary quadratic forms $Q(x,y) = ax^2 + bxy + cy^2$ with the property that $6\,|\,a$. The group $\Gamma_0(6)$ acts on such forms, and we let $\mathcal{Q}_n$ be any set of representatives of those equivalence classes with $a > 0$ and $b \equiv 1 \pmod{12}$. For each $Q(x,y)$, we let α_Q be the CM point in $\mathbb{H}$, the upper half of the complex plane, for which $Q(\alpha_Q, 1) = 0$. We then define the "trace"

$$Tr(n) := \sum_{Q \in \mathcal{Q}_n} P(\alpha_Q). \tag{10.7.5}$$

The following theorem then gives the finite algebraic formula for $p(n)$.

Theorem 10.4 *If n is a positive integer, then we have that*

$$p(n) = \frac{1}{24n-1} Tr(n). \tag{10.7.6}$$

The numbers $P(\alpha_Q)$*, as* Q *varies over* $\mathcal{Q}_n$*, form a multiset of algebraic numbers which is the union of Galois orbits for the discriminant* $-24n+1$ *ring class field. Moreover, for each* $Q \in \mathcal{Q}_n$ *we have that* $6(24n-1)P(\alpha_Q)$ *is an algebraic integer.*

The claim in theorem 10.4 that $p(n) = Tr(n)/(24n-1)$ is an example of a general theorem on "traces" of CM values of certain weak Maass forms. This result pertains to weight 0 weak Maass forms which are images under the Maass raising operator of weight -2 harmonic Maass forms.

To complete the proof of theorem 10.4, we must show that the values $P(\alpha_Q)$ are algebraic numbers with bounded denominators. To prove these claims, we require the classical theory of complex multiplication, as well as new results which bound denominators of suitable singular moduli.

Since theorem 10.4 is a culmination of decades of refinements and improvements of the original exact formula due to Rademacher, we refer the reader to Zagier's seminal paper [303] on "traces" of singular moduli published in 2002 which inspired and led to many subsequent resulting papers by others.

There are many known identities that are susceptible to application of the *Theorem of Meinardus*, the *Hardy-Ramanujan Circle Method*, or to polynomial extensions of these. Also, there is the *Convolution method* involving a product of two or more infinite series whose asymptotic coefficient behaviors are already known. In the following exercises, we cover quite a lot of cases that relate to these various asymptotic methods applied to partition generating functions. The intention is not to be exhaustive, but to indicate plausible directions for research on similar asymptotic behaviors of 2D, 3D, 4D, etc. generating function coefficients. This approach may embrace many identities given in other chapters of the present book.

EXERCISES

10.1 *Unrestricted Integer Partitions.* The number of integer partitions $p(n)$ has generating function

$$\sum_{n=0}^{\infty} p(n)q^n = \prod_{n=1}^{\infty} \frac{1}{1-q^k}. \tag{P10.1}$$

Show that by applying either the Circle Method or the Theorem of Meinardus, this leads to the asymptotic formula,

$$p(n) \sim \frac{1}{4\sqrt{3}n} e^{\pi\sqrt{\frac{2n}{3}}}. \tag{P10.2}$$

10.2 *Unrestricted Odd Partitions and Distinct Partitions.* We know the number of integer partitions into odd parts $p_o(n)$, is also the number of integer partitions into distinct parts $p_d(n)$. The generating function is

$$\sum_{n=0}^{\infty} p_o(n)q^n = \prod_{n=1}^{\infty} \frac{1}{1-q^{2n-1}} = \prod_{n=1}^{\infty}(1+q^n) = \sum_{n=0}^{\infty} p_d(n)q^n. \tag{P10.3}$$

Show that, the asymptotic formula derived from the Meinardus Theorem is

$$p_o(n) = p_d(n) \sim \frac{1}{4\pi \sqrt[4]{6n^3}} e^{2\pi \sqrt{\frac{n}{6}}}.$$

10.3 *Partitions with Congruences in the Parts.* Show that, for the number of integers partitions into parts so that each part is congruent to $r \,(\mathrm{mod}\, m)$ and letting $p_{(r,m)}(n)$ count these partitions, has an asymptotic formula given by Meinardus [198] as

$$p_{(r,m)}(n) \sim \Gamma\left(\frac{r}{m}\right) \pi^{\left(\frac{r}{m}-1\right)} 2^{\left(-\frac{3}{2}-\frac{r}{2m}\right)} 3^{\left(-\frac{r}{2m}\right)} m^{\left(-\frac{1}{2}+\frac{r}{2m}\right)} n^{-\frac{1}{2}\left(1+\frac{r}{m}\right)} e^{\pi \sqrt{\frac{2n}{3m}}}. \tag{P10.4}$$

With the parts allowed to be $\pm r \,(\mathrm{mod}\, m)$ and letting $p_{(\pm r,m)}(n)$ be the number of partitions, show, as did Livingood [187] that for

$$\sum_{n=0}^{\infty} p_{(\pm r,m)}(n) q^n = \prod_{1 \le n \equiv \pm r \,(\mathrm{mod}\, m)}^{\infty} \frac{1}{1-q^n}, \tag{P10.5}$$

we then have

$$p_{(\pm r,m)}(n) \sim \frac{\csc\left(\frac{\pi r}{m}\right)}{4\pi \sqrt[4]{3mn^3}} e^{2\pi \sqrt{\frac{n}{3m}}}. \tag{P10.6}$$

Show also that the particular case of this with $r = 1$, $m = 2$ gives us,

$$p_o(n) = p_d(n) \sim \frac{1}{4\pi \sqrt[4]{6n^3}} e^{2\pi \sqrt{\frac{n}{6}}}. \tag{P10.7}$$

10.4 *Plane Partitions.* In an earlier chapter we examined plane partitions, and it was seen that the much celebrated result of Percy A. MacMahon [193], is the generating function for $PL(n)$ given by

$$\sum_{n=0}^{\infty} PL(n) q^n = \prod_{k=1}^{\infty} \frac{1}{(1-q^k)^k} \tag{P10.8}$$

$$= 1 + q + 3q^2 + 6q^3 + 13q^4 + 24q^5 + 48q^6 + 86q^7 + 160q^8$$
$$+ 282q^9 + 500q^{10} + 859q^{11} + 1479q^{12} + O(q^{13}).$$

We know this as the *MacMahon function*, and the sequence $1, 1, 3, 6, 13, 24, \ldots$ is sequence A000219 in the OEIS. (see Sloane [236]) The asymptotics for plane partitions were first calculated by E. M. Wright in 1931, using the circle method. (see [301]) Show, using methods of the present chapter, that for large n,

$$PL(n) \sim \frac{\zeta(3)^{7/36}}{\sqrt{12\pi}} \left(\frac{n}{2}\right)^{-25/36} \exp\left(3\zeta(3)^{1/3} \left(\frac{n}{2}\right)^{2/3} + \zeta'(-1)\right). \tag{P10.9}$$

Incidentally, $\zeta'(-1) = 1/12 - \log(A)$ where A is the Glaisher-Kinkelin Constant. Show that for practical purposes, the asymptotic formula reduces to

$$\log(PL(n)) \sim 2.00945n^{2/3} - 0.69444\log n - 1.4631. \tag{P10.10}$$

10.5 *Partitions into primes.* We in this case have

$$\sum_{n=0}^{\infty} Pr(n)q^n = \prod_{k=1}^{\infty} \frac{1}{1-q^{p_k}} \tag{P10.11}$$

$$= 1 + q^2 + q^3 + q^4 + 2q^5 + 2q^6 + 3q^7 + 3q^8 + 4q^9 + 5q^{10} + 6q^{11} + 7q^{12}$$
$$+ 9q^{13} + 10q^{14} + 12q^{15} + 14q^{16} + 17q^{17} + 19q^{18} + O(q^{19}),$$

where p_k denotes the kth prime number and $Pr(n)$ the number of unrestricted partitions into prime numbers. The following formula is due to Roth and Szekeres [227] in 1954. The proof is very intricate and beautiful, and it involves the Riemann Hypothesis, or more accurately stated, it depends upon the positioning of the real parts of the complex zeroes of the Riemann zeta function. After all of the analysis, Roth and Szekeres prove the following statement within a theorem. So, the exercise here is to follow their proof. Are there similar results for other sequences than the prime numbers that this method can apply to?

Let θ be the *least upper bound* of the real parts of the imaginary roots of the Riemann zeta function. Then

$$\log Pr(n) - F(\alpha) = \begin{cases} O(\alpha^{-\theta+\varepsilon}), & \forall \varepsilon > 0; \\ O(\alpha^{-\psi}), & \forall \theta \leq \psi \text{ otherwise}; \end{cases} \tag{P10.12}$$

where

$$\alpha \sim \frac{\pi}{\sqrt{3n\log n}},$$

and we have that

$$F(\alpha) = \left(\frac{1}{\alpha}\right) \sum_{j=0}^{N} \frac{C_j\, j!}{\log^{j+1}\left(\frac{1}{\alpha}\right)} + O\left(\left(\frac{1}{\alpha}\right) \log^{-N+2}\left(\frac{1}{\alpha}\right)\right),$$

with

$$(2+u)\Gamma(1+u)\zeta(2+u) = \sum_{j=0}^{\infty} C_j u^j, \qquad C_0 = \frac{\pi^2}{3};$$

and there exists a constant $C > 0$ such that

$$\log Pr(n) = F(\alpha) + O\left(\frac{1}{\alpha} \exp\left(-C \sqrt[5]{\frac{\log^3\left(\frac{1}{\alpha}\right)}{\log\log\left(\frac{1}{\alpha}\right)}}\right)\right). \tag{P10.13}$$

10.6 *Partitions into r-th powers.* We also have asymptotic formulas for partitions into powers of a fixed positive integer $r > 1$, due originally to Mahler in 1940 (see [194]) and extended by Erdös and Richmond in 1976 (see [122]).

Suppose $A = \{r^0, r^1, r^2, \dots\}$ and have $P_A(n)$ be the number of unrestricted partitions of n with summands chosen from A, so that

$$\sum_{n=0}^{\infty} P_A(n)q^n = \prod_{k=0}^{\infty} \frac{1}{(1-q^{r^k})}. \tag{P10.14}$$

In 1940 Mahler [194] showed quite remarkably that

$$\log(P_A(rn)) = \frac{1}{2\log r}(\log(n\log n))^2 + \left(\frac{1}{2} + \frac{1}{\log r} + \frac{\log\log r}{\log r}\right)\log n \tag{P10.15}$$
$$-\left(1 + \frac{\log\log r}{\log r}\right)\log\log n + O(1).$$

In subsequent papers the $O(1)$ order term was shown to be of the form

$$U\left(\frac{\log n - \log\log n + \log\log r}{\log r}\right) + O\left(\frac{(\log\log n)^2}{\log n}\right),$$

where U is a periodic function with period 1 having an exact Fourier expansion. This property is known as the *de Bruijn-Mahler phenomenon.* To read an ingenious account of this see the beautiful Erdös and Richmond 1976 paper [122].

So, the exercise question is here to write down and get a sense of the periodic nature of the error term in the asymptotic approximation above. In later chapters, we examine vector partition extensions of this kind of generating function. Write down any research topics that seem to arise from these asymptotic estimates having the *de Bruijn-Mahler phenomenon.*

10.7 *Partitions into positive integers that are not r-th powers.* Let A be the set of integers which are not powers of the fixed integer r. Show that the number $p_A(n)$ of partitions of n into summands from A satisfies conditions given by a theorem of Roth and Szekeres (1954) (see [227]) and allows us to assert that

$$\log(p_A(rn)) = (2\pi A_2)^{-\frac{1}{2}} \exp\left(\alpha n - \sum_{a\in A} a^2 \frac{e^{\alpha a}}{(e^{\alpha a}-1)^2}\right) \tag{P10.16}$$

with

$$A_2 = \sum_{a\in A} a^2 \frac{e^{\alpha a}}{(e^{\alpha a}-1)^2} \tag{P10.17}$$

in which α is defined by

$$n = \sum_{a \in A} \frac{e^a}{e^{\alpha a} - 1}. \tag{P10.18}$$

Show that we can derive our asymptotic formula from considering the Mellin transformation formula

$$\sum_{a \in A} \frac{e^a}{e^{\alpha a} - 1} = \frac{1}{2\pi i} \int_{\alpha - i\infty}^{\alpha + i\infty} \alpha^{-t} \Gamma(t) \zeta(t) \zeta_A(t-1) dt \quad (\alpha > 2) \tag{P10.19}$$

where $\Gamma(t)$ denotes the usual gamma function, $\zeta(t)$ denotes the Riemann zeta function and

$$\zeta_A(t) = \sum_{a \in A} a^{-t}.$$

Now

$$\zeta_A(t) = \sum_{n=1}^{\infty} n^{-t} - \sum_{j=1}^{\infty} r^{-jt} = \zeta(t) - \frac{1}{r^t - 1},$$

so $\zeta_A(t)$ is defined in the entire plane except for simple poles at $t = 1$ and $t = 2\pi ik/\log r \, (k = 0, 1, 2, ...)$. It is known that we may shift the contour of integration to a line $\alpha < 0$ and obtain (using $\zeta(0) = -\frac{1}{2}$) that

$$n = \frac{1}{\alpha^2} \frac{\pi^2}{6} - \frac{1}{\alpha} \frac{\log(1/\alpha)}{\log r}$$
$$- \frac{1}{\alpha \log r} \sum_{\nu}{}' \Gamma\left(1 + \frac{2\pi i\nu}{\log r}\right) \zeta\left(1 + \frac{2\pi i\nu}{\log r}\right) \exp\left(-\frac{2\pi i\nu \log \alpha}{\log r}\right) \tag{P10.20}$$
$$+ O(1),$$

where $\sum_{\nu}'$ denotes summation over nonzero ν. We may solve this for α to obtain

$$\alpha = \frac{\pi}{\sqrt{6n}} \left[1 - \frac{\sqrt{6}}{2\pi \sqrt{n} \log r} \left\{\log \frac{\sqrt{6}}{\pi} + \Sigma \right\}\right] + O(n^{-2/3} \log^2 n), \tag{P10.21}$$

where

$$\Sigma = \sum_{\nu}{}' \Gamma\left(1 + \frac{2\pi i\nu}{\log r}\right) \zeta\left(1 + \frac{2\pi i\nu}{\log r}\right) \exp\left(2\pi i\nu \log \frac{\sqrt{6n}}{\pi}\right). \tag{P10.22}$$

From Mellin's transformation formula

$$-\sum_{a \in A} \log(1 - e^{-\alpha a}) = \frac{1}{2\pi i} \int_{\sigma - i\infty}^{\sigma + i\infty} \alpha^{-t} \zeta(t+1) \Gamma(t) \zeta_A(t) dt \quad (\sigma > 1). \tag{P10.23}$$

Again shifting the contour of integration to $\sigma < 0$, using $\zeta'(0) = \frac{1}{2}\log(2\pi)$ and the series $z\Gamma(z+1) = 1 + \gamma z + \gamma_2 z^2 + \gamma_2 z^2 + \gamma_3 z^3 + ...$ leading us to assert

$$\zeta(t+1)\Gamma(t)\zeta_A(t) = \frac{-1}{t^3 \log r} - \left\{\frac{\log^2 r}{12} - \frac{1}{2}\log(2\pi) + \frac{\gamma_2 + \frac{\gamma^2}{2} + \frac{\pi^2}{12}}{\log r}\right\} \frac{1}{t} + O(1), \tag{P10.24}$$

we arrive at

$$-\sum_{a\in A}\log(1-e^{-\alpha a})=\frac{\pi^2}{6\alpha}-\frac{\log^2 r}{12}-\frac{\log^2(1/\alpha)}{\log r}+\frac{\log r}{12}+\frac{\log(2\pi)}{2}+\frac{\gamma_2+\frac{\gamma^2}{2}+\frac{\pi^2}{12}}{\log r} \tag{P10.25}$$

$$+\frac{1}{\log r}\sum_{\nu}{}'\Gamma\left(\frac{2\pi i\nu}{\log r}\right)\zeta\left(1+\frac{2\pi i\nu}{\log r}\right)\exp\left(\frac{2\pi i\nu\log\alpha}{\log r}\right)+O(\alpha).$$

From (P10.20) and (P10.25) we obtain

$$\alpha n-\sum_{a\in A}\log(1-e^{-\alpha a})=\frac{\pi^2}{3\alpha}-\frac{\log^2(1/\alpha)}{\log r}-\frac{\log(1/\alpha)}{\log r}+\frac{\log(2\pi)}{2}-\frac{\log r}{12} \tag{P10.26}$$

$$+\frac{\gamma_2+\frac{\gamma^2}{2}+\frac{\pi^2}{12}}{\log r}-\frac{1}{\log r}\sum_{\nu}{}'\Gamma\left(\frac{2\pi i\nu}{\log r}\right)\zeta\left(1+\frac{2\pi i\nu}{\log r}\right)$$
$$\times\left(1+\frac{2\pi i\nu}{\log r}\right)\exp\left(\frac{2\pi i\nu\log\alpha}{\log r}\right)+O(\alpha).$$

We note that

$$A_2=\frac{d}{\alpha a}\left[\frac{a}{e^{\alpha a}-1}\right]=\frac{\pi^2}{6\alpha^3}+O(\alpha^{-2}). \tag{P10.27}$$

Show that from (P10.26), (P10.27) and (P10.20) applied to (P10.16) we obtain

Theorem 10.5 *Let A be the set of integers which are not positive powers of the fixed integer r. Then*

$$\log(p_A(n))=\sqrt{\frac{2}{3n}}\pi-\frac{\log^2(\sqrt{6n}/\pi)}{\log r}+\frac{3}{2}\log\left(\frac{\sqrt{6n}}{\pi}\right) \tag{P10.28}$$

$$-\frac{\log r}{12}-\frac{1}{2}\log\left(\frac{\pi^2}{3}\right)-\left(\gamma_2+\frac{\gamma^2}{2}+\frac{\pi^2}{12}\right)\frac{1}{\log r}$$

$$-\frac{1}{\log r}\sum_{\nu}{}'\Gamma\left(\frac{2\pi i\nu}{\log r}\right)\zeta\left(1+\frac{2\pi i\nu}{\log r}\right)\left(\frac{\sqrt{6n}}{\pi}\right)^{\frac{2\pi i\nu}{\log r}}+O(n^{-1/2}\log^2 n).$$

10.8 *Some known asymptotic coefficients for partition functions.* The following remaining exercises for this chapter are some useful known asymptotic formulas for various kinds of partitions. These formulas were arrived at via one of the methods:

a. A variation on the 1918 Hardy and Ramanujan Circle Method.

b. Theorem of Meinardus 1954.

c. Series convolution saddle point method.

d. Meinardus adapted with finite number of simple poles.

e. Circle Method or Meinardus approach for polynomial coefficients.

The latter three methods appear in several papers dating from about 2000 up to 2022. The results given below include a few classical asymptotic formulas for integer partitions of various kinds. It seems that there is scope to research the corresponding formulas that may arise in studying the generating functions for the vector partitions. We have drawn from the 2015 paper by Václav Kotěšovec [169]. However, there is also the paper from 2017 by Andrews on *stacks* related to *unimodal* partitions.

10.9 *Unrestricted integer partitions (revisited)* have generating function

$$\prod_{k=1}^{\infty} \frac{1}{1-q^k} = \sum_{n=0}^{\infty} p(n)q^n \tag{P10.29}$$

and the asymptotic formula

$$p(n) \sim \frac{\exp\left(\pi\sqrt{\frac{2n}{3}}\right)}{4n\sqrt{3}}. \tag{P10.30}$$

Prove this using the theorem of Meinardus.

10.10 *Partitions into distinct parts (odd parts)* have generating function

$$\prod_{k=1}^{\infty} (1+q^k) = \sum_{n=0}^{\infty} p(\mathfrak{D},n)q^n \tag{P10.31}$$

and the asymptotic formula

$$p(\mathfrak{D},n) \sim \frac{\exp\left(\pi\sqrt{\frac{n}{3}}\right)}{4\sqrt[4]{3n^3}}. \tag{P10.32}$$

Apply theorem of Meinardus to prove this.

10.11 *Number of overpartitions of n* have generating function

$$\prod_{k=1}^{\infty} \frac{1+q^k}{1-q^k} = \sum_{n=0}^{\infty} \bar{p}(n)q^n \tag{P10.33}$$

and the asymptotic formula

$$\bar{p}(n) \sim \frac{\exp(\pi\sqrt{n})}{8n}. \tag{P10.34}$$

Prove this by the convolution method.

10.12 *Unrestricted integer partitions with multiplicities.* Show that this result can be found using either a convolution approach or the Meinardus method. Here we have $m = 1, 2, 3, 4, \ldots$. Also the Online Encyclopedia of Integer Sequences, (commonly referred to as the OEIS) has the following cases associated at OEIS numbers A000041 (m=1), A000712 (m=2), A000716 (m=3), A023003 (m=4), A144064 (m=5), etc. The generating function for these is

$$\prod_{k=1}^{\infty} \frac{1}{(1-q^k)^m} = \sum_{n=0}^{\infty} p_m(n) q^n \tag{P10.35}$$

and the asymptotic formula

$$p_m(n) \sim \sqrt[4]{\frac{m^{m+1}}{2^{3m+5} 3^{m+1} n^{m+3}}} \exp\left(\pi \sqrt{\frac{2mn}{3}}\right). \tag{P10.36}$$

10.13 *Distinct parts integer partitions with multiplicities.* Show that, as with the previous result, this can be found using either a convolution approach or the Meinardus method. Again we have $m = 1, 2, 3, 4, \ldots$. Also the Online Encyclopedia of Integer Sequences has the following cases associated with OEIS numbers A000009 (m=1), A022567 (m=2), A022568 (m=3), A022569 (m=4), A022570 (m=5), etc. The generating function for these is

$$\prod_{k=1}^{\infty} (1+q^k)^m = \sum_{n=0}^{\infty} p_m(\mathfrak{D}, n) q^n \tag{P10.37}$$

and the asymptotic formula

$$p_m(\mathfrak{D}, n) \sim \sqrt[4]{\frac{m}{4^{m+3} 3 n^3}} \exp\left(\pi \sqrt{\frac{mn}{3}}\right). \tag{P10.38}$$

10.14 A *number of overpartitions with multiplicities.* Derive the following using the convolution method. These types of partitions have the following OEIS entries at A015128 (m=1), A001934 (m=2), A004404 (m=3, alternating). Show that the generating function is

$$\prod_{k=1}^{\infty} \left(\frac{1+q^k}{1-q^k}\right)^m = \sum_{n=0}^{\infty} \bar{p}_m(n) q^n \tag{P10.39}$$

and the asymptotic formula is

$$\bar{p}_m(n) \sim \sqrt[4]{\frac{m^{m+1}}{4^{3m+3} n^{m+3}}} \exp\left(\pi \sqrt{mn}\right). \tag{P10.40}$$

10.15 *Number of distinct partitions into certain arithmetic progressions.* Use theorem of Meinardus or the convolution method to prove the following.

For any fixed $s>0$ along with any fixed $t>0$ such that $\gcd(s,t)=1$, we consider the generating function

$$\prod_{k=1}^{\infty}\left(\frac{1}{1-q^{sk+t}}\right)=\sum_{n=0}^{\infty}\mathbf{p}_{s,t}(n)q^{n}. \tag{P10.41}$$

The coefficients $\mathbf{p}_{s,t}(n)$ have the asymptotic formula

$$\mathbf{p}_{s,t}(n)\sim\Gamma\left(\frac{t}{s}\right)\pi^{A}2^{B}3^{C}s^{D}n^{E}\exp\left(\pi\sqrt{\frac{2n}{3s}}\right), \tag{P10.42}$$

where $A=\frac{t}{s}-1$, $B=-\frac{3}{2}-\frac{t}{2s}$, $C=-\frac{t}{2s}$, $D=-\frac{1}{2}+\frac{t}{2s}$, and $E=-\frac{1}{2}-\frac{t}{2s}$. This is a *Tauberian formula* that was found by Ingham in 1941. (See Ingham [157], and more recently Kotěšovec [169].)

10.16 *Number of distinct partitions into certain arithmetic progressions.* (See Meinardus [198]) Use the theorem of Meinardus to prove the following.

For any fixed $s>0$ along with any fixed $t>0$ such that $\gcd(s,t)=1$, we consider the generating function

$$\prod_{k=1}^{\infty}\left(1+q^{sk+t}\right)=\sum_{n=0}^{\infty}\mathbf{q}_{s,t}(n)q^{n}. \tag{P10.43}$$

The coefficients $\mathbf{q}_{s,t}(n)$ have the asymptotic formula

$$\mathbf{q}_{s,t}(n)\sim\frac{\exp\left(\pi\sqrt{\frac{n}{3s}}\right)}{2^{1+\frac{t}{s}}(3s)^{\frac{1}{4}}n^{\frac{3}{4}}}. \tag{P10.44}$$

Think About It...

Ramanujan independently discovered results of Gauss, Kummer and others on hypergeometric series. Ramanujan's own work on partial sums and products of hypergeometric series have led to major developments in the topic. Perhaps his most famous work was on the number $p(n)$ of partitions of an integer n into summands.

In a joint paper with Hardy, Ramanujan gave an asymptotic formula for $p(n)$ to less than an integer, as later proven by Rademacher.

J.E. Littlewood
review of Collected Papers of Srinivasa Ramanujan, 1929 (see [186])

FURTHER READING

ANDREWS, G.E. *The Theory of Partitions*, volume 2 of *Encyclopedia of Mathematics and its Applications.* Addison-Wesley Publishing Company, Advanced Book Program, Reading, Massachusetts, USA, 1976.

APOSTOL, T. *Introduction to Analytic Number Theory.* Springer-Verlag, New York, 1976.

AULUCK, F.C. On some new types of partitions associated with generalized Ferrers graphs. *Proc. Cambridge Philos. Soc.*, 47:679–686, 1951.

BRUINIER, J.H. *Borcherds products on O(2, l) and Chern classes of Heegner divisors*, volume 1780 of *Springer Lecture Notes in Mathematics.* Springer-Verlag, 2002.

BRUINIER, J.H. and FUNKE, J. On two geometric theta lifts. *Duke Math. J.*, 125:45–90, 2004.

BRUINIER, J.H. and FUNKE, J. Traces of CM-values of modular functions. *J. Reine Angew. Math.*, 594:1–33, 2006.

BRUINIER, J.H., and ONO, K. Algebraic formulas for the coefficients of halfintegral weight harmonic weak Maass forms. *Adv. Math.*, 246:198–219, 2013.

BRUINIER, J.H., JENKINS, P., and ONO, K. Hilbert class polynomials and traces of singular moduli. *Math. Ann.*, 334:373–393, 2006.

DEWAR, M. and RAM MURTY, M. An asymptotic formula for the coefficients of j(z). *International Journal of Number Theory*, 9(3):1–12, 2013.

ERDÖS P. and RICHMOND B. Concerning Periodicity in the Asymptotic Behaviour of Partition Functions. *J. Austral. Math. Soc. (Series A)*, 21:447–456, 1976.

FRNKA, R.A. *Asymptotic Formulae for Restricted Unimodal Sequences.* Phd thesis, Louisiana State University and Agricultural and Mechanical College, April 2017. LSU Doctoral Dissertations. DOI: 10.31390/gradschool dissertations.4460.

HARDY, G.H. and RAMANUJAN, S. Asymptotic formulae in combinatory analysis. *Proc. London. Math. Soc.*, 2(17):75–115, 1918. Also in Collected papers of S. Ramanujan, pp. 276-309, Cambridge Univ. Press, Cambridge, 1927; reprinted by Chelsea, New York, 1962.

INGHAM, A.E. A Tauberian theorem for partitions. *Annals of Mathematics*, 42:1075–1090, 1941.

KOTĚŠOVEC, V. A method of finding the asymptotics of q-series based on the convolution of generating functions. arxiv preprint, Sep 2015. arXiv:1509.08708 [math.CO]. doi.org/10.48550/arXiv.1509.08708.

LITTLEWOOD, J.E. Review of the Collected Papers of Srinivasa Ramanujan. *Mathematical Gazette*, 14:427–428, 1929.

LIVINGOOD, J,. A partition function with the prime modulus $p > 3$. *American Journal of Mathematics*, 67:194–208, 1945.

MACMAHON, P.A. *Combinatory Analysis*, volume 2. Cambridge Univ. Press, London and New York (reprinted by Chelsea, New York, 1960), 1916.

MAHLER, K. On a Special Functional Equation. *J. London Math. Soc.*, 15:115–123, 1940.

MAHLER, K. On a Special Function. *J. Number Theory*, 12:20–26, 1980.

MEINARDUS, G. Asymptotische Aussagen über Partitionen. *Math. Z.*, 59:388– 398, 1954.

PERRY, D. A polynomial variation of Meinardus' theorem. *International Journal of Number Theory*, 11(01):251–268, 2015.

RADEMACHER, H. On the partition function $p(n)$. *Proc. London Math. Soc.*, 2(43):241–254, 1937.

ROTH, K.F., and SZEKERES, G. Some asymptotic formulae in the theory of partitions. *Q.J.Math.Oxf.Ser(2)*, 5:241–259, 1954.

SLOANE, N.J.A. Number of planar partitions (or plane partitions) of n. The On-Line Encyclopedia of Integer Sequences (OEIS), 2022. sequence A000219. (www.oeis.org/A000219).

SZEKERES, G. An asymptotic formula in the theory of partitions, I. *Quart. J. Math. Oxford Ser., 2(2)*:85–108, 1951.

SZEKERES, G. An asymptotic formula in the theory of partitions, II. *Quart. J. Math. Oxford Ser.(2)*, 4:96–111, 1951.

WRIGHT, E.M. Asymptotic partition formulae I. Plane partitions. *The Quarterly Journal of Mathematics*, 1:177–189, 1931.

ZAGIER, D. Traces of singular moduli, Motives, polylogarithms and Hodge theory, Part I. In *Int. Press Lect. Ser., 3*, pages 211–244. Int. Press, Somerville, MA, 2002. first printed Irvine, CA,1998.

CHAPTER 11

Rogers-Ramanujan functions in Statistical Mechanics

11.1 BAXTER'S HARD HEXAGON MODEL SOLVED EXACTLY BY ROGERS-RAMANUJAN IDENTITIES

THE YEARS 1980 AND 1981 were amazing in the theory of partitions; with significance arisen from a remarkable discovery in theoretical physics. In 1980 Baxter [50] found real-world physical application of the Rogers-Ramanujan identities coming directly from his solving the Hard Hexagon Model in Statistical Mechanics. This opened up a new world of research in both theoretical physics and partition-theoretic mathematics that is still evolving, over forty years later. This discovery has given rise to thousands of mathematics and physics research papers and promises continued relevance into future decades.

Some of this excitement can be gleaned from the important paper by Andrews in 1981 (see [21]) from quoting his opening paragraph here.

Think About It...

"In 1980, Baxter found his beautiful solution to the hard-hexagon model of statistical mechanics. His treatment of this model is naturally divided into six regimes that depend on values taken by various parameters associated with the model. Then in truly astounding fashion it turns out that eight Rogers-Ramanujan-type identities, all essentially known to Rogers ([225], [226]), are the fundamental keys for finding infinite product representations of the related statistical mechanics partition functions in **regimes I, III, and IV**."

George E. Andrews
Proceedings of the National Academy of Sciences, 1981

DOI: 10.1201/9781003174158-11

Figure 11.1 Professor Rodney J. Baxter around 1970.
(Courtesy, Australian National University Archives, reference number ANUA225-67)

11.1.1 A Few Words about the Types of Lattice Models Studied Here

Baxter formulates his approach based on the *hard hexagon* diagram (see next page Figure 11.2) which has constraints on the particles such that they can only be in certain places at certain moments.

The hard hexagon model is a two-dimensional lattice model of a gas of hard (i.e. non-overlapping) molecules. In it, particles are placed on the sites of the triangular lattice so that no two particles are together or adjacent. A typical allowed arrangement of particles is shown in Figure 11.2. If we regard each particle as the center of a hexagon covering the six adjacent faces (such hexagons are shown shaded in the figure), then the rule only allows hexagons that do not overlap: hence the name of the model. For a lattice of N sites, the grand partition function is

$$Z(z) = \sum_{n=0}^{n/3} z^n g(n, N) \qquad (11.1.1)$$

where $g(n, N)$ is the allowed number of ways of placing n particles on the lattice, and the sum is over all possible values of n. (Since no more than $\frac{1}{3}$ of the sites can be occupied, n takes values from 0 to $\frac{N}{3}$.)

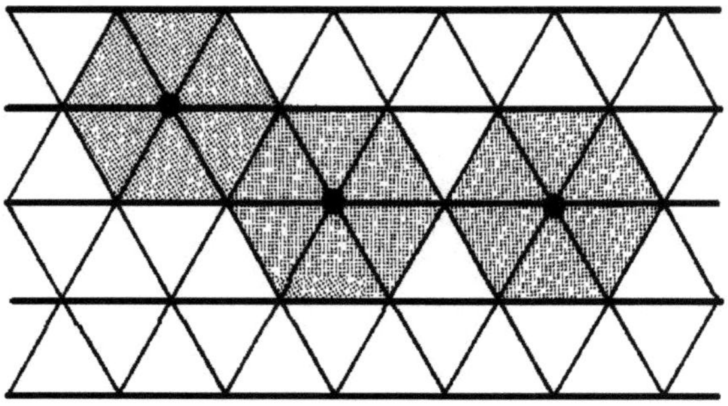

A typical arrangement of particles (black circles) on the triangular lattice, such that no two particles are together or adjacent. The six faces round each particle are shaded: they form non-overlapping (ie. "hard") hexagons.

Figure 11.2 The particle arrangement for the Hard Hexagon Model.

Clearly, from the mathematical 'non-physicist' perspective, that this schematic the diagram could lead to the enumeration of well defined types of integer partitions was (back in 1980 at least) seen as a major departure from the more typical Ferrers Graphs, Durfee Squares and Plane Partition cube stack representations adopted in the literature by Andrews and others over the contemporary mathematical landscapes. As a mathematician already versed in the ideas of our earlier chapters, it becomes natural to ask how say, Bijection equivalences, q-series partition generating functions, and such, may become the same world as the statistical mechanics models involving assigned values of *spin charges* along a vertex or edge of a lattice of a certain structure, shape and form. We put such questions aside for now, as we want to present Baxter's extraordinary findings here. However, these questions of equivalences of theories and analyses are worthy of deeper understanding it seems.

In seeking to calculate the free energy in the hard hexagon model as depicted in Figure 11.2, Baxter was led to consider a range of logical options or *Regimes*, which are here depicted as Baxter described in 1980.

The three sub-lattices of the triangular lattice: sub-lattice 1 consists of all sites of type 1, and similarly for sub-lattices 2 and 3. Adjacent sites lie on different sub-lattices. a close-packed arrangement of particles (black circles) is shown: all sites of one sub-lattice (in this case sub-lattice 1) are occupied, the rest are empty. (See Figure 11.3)

11.1.2 What Baxter found in 1980

So, continuing our discussion of equation (11.1.1), we want to calculate Z, or rather the partition-function per site of the infinite lattice

$$\kappa = \lim_{N\to\infty} Z^{\frac{1}{N}} \tag{11.1.2}$$

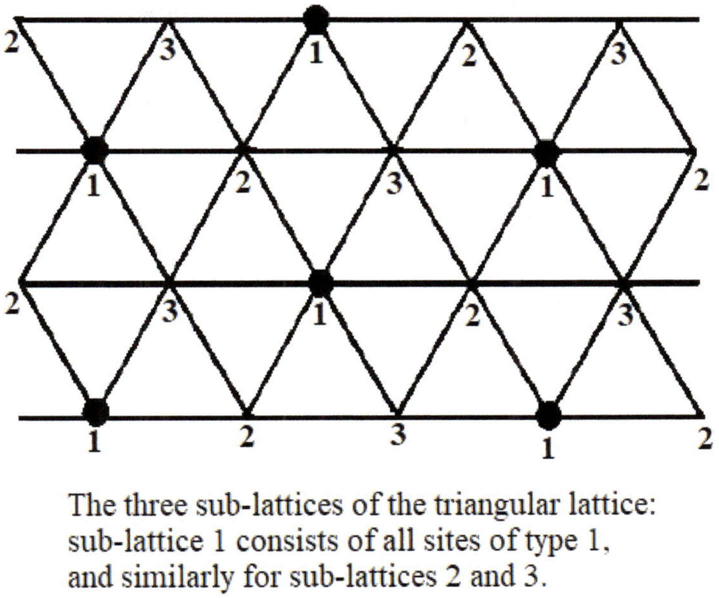

Figure 11.3 Baxter's three sub-lattices of the triangular lattice.

as a function of the positive real variable z. This z is known as the 'activity'. This problem can be put into what is called 'spin'-type language by associating with each site i a variable σ_i. However, instead of the usual approach in statistical mechanics models of letting σ_i take values 1 and -1, we take the values 0 and 1: if the site (lattice point) is empty, then $\sigma_i = 0$; if it is full then $\sigma_i = 1$. Thus σ_i is the number of particles at site i: the 'occupation number'. Then (11.1.1) can be written as

$$Z = \sum_{\sigma} z^{(\sigma_1+\sigma_2+\ldots+\sigma_N)} \prod_{i,j}(1-\sigma_i\sigma_j), \tag{11.1.3}$$

where the product is over all edges (i,j) of the triangular lattice, and the sum is over all values (0 and 1) of all the occupation numbers $\sigma_1, \sigma_2, \ldots, \sigma_N$.

We expect this model to undergo a phase transition from a homogeneous fluid state at low activity z to an inhomogeneous solid state at high activity z. To see this, divide the lattice into three sub-lattices $1,2,3$ so that no two sites of the same type are adjacent. Then there are three possible close-packed configurations of particles on the lattice: either all sites of type 1 are occupied, or all sites of type 2, or all sites of type 3. Suppose we fix the boundary sites as in the first possibility, i.e. all boundary sites of type 1 are full, and all other boundary sites are empty. Then for an infinite lattice the second and third possibilities give a negligible contribution to the sum-over-states in (11.1.3). Clearly, sites on different sub-lattices are no longer equivalent. Let ρ_r be the local density at a site of type r, given by,

$$\rho_r = \langle \sigma_l \rangle = Z^{-1} \sum_{\sigma} \sigma_l z^{(\sigma_1+\sigma_2+\ldots+\sigma_N)} \prod_{i,j}(1-\sigma_i\sigma_j), \tag{11.1.4}$$

where l is a site of type r.

When z is infinite, the system is close-packed with all sites of type 1 occupied, so $\rho_1, = 1$, $\rho_2 = \rho_3 = 0$. We can expand each ρ_r in inverse powers of z by considering successive perturbations of the close-packed state. For a site l deep inside a large lattice, this gives

$$\begin{aligned} \rho_1 &= 1 - z^{-1} - 5z^{-2} - 34z^{-3} - 267z^{-4} - 2037z^{-5} - \ldots \\ \rho_2 &= \rho_3 = z^{-2} + 9z^{-3} + 80z^{-4} + 965z^{-5} + \ldots \end{aligned} \tag{11.1.5}$$

The system is therefore not homogeneous, since ρ_1, ρ_2, ρ_3, are not all equal. This contrasts with the low-activity situation: starting from the state with all sites empty and successively introducing particles, we obtain

$$\rho_1 = \rho_2 = \rho_3 = z^1 - 7z^2 - 58z^3 - 519z^4 + 4856z^5 - \ldots \tag{11.1.6}$$

To all orders in this expansion, it is true that $\rho_1 = \rho_2 = \rho_3$.

The system is therefore inhomogeneous for sufficiently large z, and homogeneous for sufficiently small z if the series converges. There must be a critical value z_c, of z above which the system ceases to be homogeneous. Since the homogeneous phase is typical of a fluid, and the ordered inhomogeneous phase is typical of a solid, the model can be said to undergo a fluid-to-solid transition at $z = z_c$.

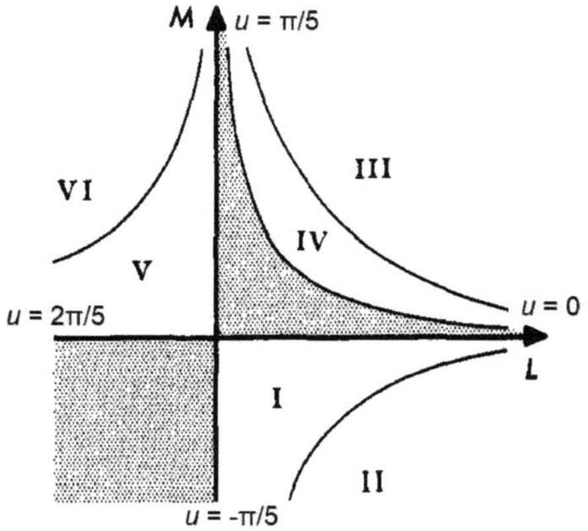

THE SIX REGIMES

I: $\Delta > \Delta_c$, $q^2 < 0$, $-\pi/5 < u < 0$,

II: $0 < \Delta < \Delta_c$, $q^2 > 0$, $-\pi/5 < u < 0$,

III: $-\Delta_c < \Delta < 0$, $q^2 > 0$, $0 < u < \pi/5$,

IV: $\Delta < -\Delta_c$, $q^2 < 0$, $0 < u < \pi/5$,

V: $\Delta > \Delta_c$, $q^2 < 0$, $\pi/5 < u < 2\pi/5$,

VI: $0 < \Delta < \Delta_c$, $q^2 > 0$, $\pi/5 < u < 2\pi/5$.

The six regimes in the (L, M) plane, as listed at right. Shaded areas are unphysical, since $z = (1 - e^{-L})(1 - e^{-M})/(e^{L+M} - e^L - e^M)$ and $\Delta = z^{-1/2}(1 - z\, e^{L+M})$ for $\Delta = \Delta_1$ gives z therein to be negative. Regimes I, III, V are disordered, II and VI have triangular ordering, IV has square ordering. The system is critical on the (I, II), (III, IV) and (V, VI) boundaries, where $|\Delta| = \Delta_c$ the values of u on the (L, M) axes are indicated.

Figure 11.4 Baxter's six Regimes for the Hard Hexagon Model. The six regimes in the (L, M) plane. Shaded areas are not physical. Regimes I, III and V are disordered, and regimes II and VI have triangular ordering, while Regime IV has square ordering.

11.1.3 Baxter's 1980 numerical preliminaries

Baxter then applied the Corner Transfer Matrix approach to the matrix eigenvalues $a_1, \ldots, a_{10}$, for the hard hexagon model with $z = 1$. The values were approximate, being calculated from finite truncations of the triangular lattice analog of the matrix equation related to this. The eigenvalues occur in groups of comparable magnitude, and it is sensible to include all members of a group. For this reason the truncations used were 2×2, 3×3, 5×5, 7×7 and 10×10. Each a_i was given for successively larger truncations, and clearly, each tended rapidly to a limit. This limit is its exact value for the infinite-dimensional corner transfer matrix. Refer to Baxter [52, Chapter 13] for how the Corner Transfer Matrix context applies.

Baxter was able to determine the required coefficients up to 30 terms of the power series *exact solution.* Equipped with knowledge from his famous eight-vertex model exact solution from a few years earlier, Baxter was able to put the hard hexagon proposed solution expansion into the form

$$z = -x \prod_{n=1}^{\infty} (1 - x^n)^{c_n}. \tag{11.1.7}$$

With this in mind he discovered that

$$\begin{aligned} c_1, c_2, \ldots, c_{29} \quad = \quad & 5, -5, -5, 5, 0, 5, -5, -5, 5, 0, \\ & 5, -5, -5, 5, 0, 5, -5, -5, 5, 0, \\ & 5, -5, -5, 5, 0, 5, -5, -5, 5. \end{aligned} \tag{11.1.8}$$

Baxter then was able to infer that

$$z = -x \left(\frac{H(x)}{G(x)} \right)^5, \tag{11.1.9}$$

where

$$G(x) = \prod_{n=1}^{\infty} \frac{1}{(1 - x^{5n-4})(1 - x^{5n-1})}, \tag{11.1.10}$$

$$H(x) = \prod_{n=1}^{\infty} \frac{1}{(1 - x^{5n-3})(1 - x^{5n-2})}. \tag{11.1.11}$$

The mathematical partition theorist will will recognise the products $G(x)$ and $H(x)$ as from the Rogers-Ramanujan identities.

So, Baxter followed through this Corner Transfer Matrix approach to calculate the coefficients and determine a range of identities for each of the Hard Hexagon Regime types of allowable spin configurations. What he derived, is what Andrews subsequently was able to prove on his visit to Baxter in Australia the following year. So, we have set the scene for Andrews' proofs in 1981.

11.1.4 What Andrews Found in 1981

So, after Baxter's solution to the Hard Hexagon Model in 1980, the opening to a new world of research in both theoretical physics and partition-theoretic mathematics was about to happen. George E. Andrews visited Baxter in 1981, and proved the identities arisen from the Regimes of Baxter's Hard Hexagon exact solutions.

In a landmark paper, Andrews [21] goes on then to say that Baxter found the following identities occurring for the six regimes, conspicuously missing (but later returning to) the more complicated Regime II, as follows:

Regime I

$$\sum_{n=0}^{\infty} \frac{q^{n^2}}{(q)_n} = \frac{1}{\Pi(q, q^4; q^5)}; \tag{11.1.12}$$

$$\sum_{n=0}^{\infty} \frac{q^{n(n+1)}}{(q)_n} = \frac{1}{\Pi(q^2, q^3; q^5)}; \tag{11.1.13}$$

Regime III

$$\sum_{n=0}^{\infty} \frac{q^{\frac{n(3n-1)}{2}}}{(q)_n (q; q^2)_n} = \frac{\Pi(q^4, q^6, q^{10}; q^{10})}{(q)_\infty}; \tag{11.1.14}$$

$$\sum_{n=0}^{\infty} \frac{q^{\frac{3n(n+1)}{2}}}{(q)_n (q; q^2)_{n+1}} = \frac{\Pi(q^2, q^8, q^{10}; q^{10})}{(q)_\infty}; \tag{11.1.15}$$

Regime IV

$$\sum_{n=0}^{\infty} \frac{q^{n(n+1)}}{(q)_{2n+1}} = \frac{\Pi(q^3, q^7, q^{10}; q^{10})\Pi(q^4, q^{16}; q^{20})}{(q)_\infty}; \tag{11.1.16}$$

$$\sum_{n=0}^{\infty} \frac{q^{n(n+1)}}{(q)_{2n}} = \frac{\Pi(q, q^9, q^{10}; q^{10})\Pi(q^8, q^{12}; q^{20})}{(q)_\infty}; \tag{11.1.17}$$

$$\sum_{n=0}^{\infty} \frac{q^{n^2}}{(q)_{2n}} = \frac{1}{\Pi(q^4, q^{16}; q^{20})(q; q^2)_\infty}; \tag{11.1.18}$$

$$\sum_{n=1}^{\infty} \frac{q^{n^2}}{(q)_{2n-1}} = \frac{q}{\Pi(q^8, q^{12}; q^{20})(q; q^2)_\infty}. \tag{11.1.19}$$

These use the standard notation of Slater [233],

$$(a)_n = (a; q)_n = \prod_{j=0}^{n-1} (1 - aq^j); \tag{11.1.20}$$

$$(a)_\infty = (a; q)_\infty = \prod_{j=0}^{\infty} (1 - aq^j); \tag{11.1.21}$$

$$\Pi(a_1, a_2, ..., a_r; q) = (a_1)_\infty (a_2)_\infty ... (a_r)_\infty. \tag{11.1.22}$$

These results either are given explicitly by Rogers ([225], [226]) or are immediate consequences of his work: (11.1.12) is equation 1 on page 328 of [225]; (11.1.13) is equation 2 on page 329 of [225]. (11.1.12) and (11.1.13) are of course the now-famous Rogers-Ramanujan identities.

(11.1.14) is equation 2, line 3, on page 330 of [226]; (11.1.15) is equation 2, line 2, on page 330 of [226]; (11.1.16) is implicit in the identity of equations 2 and 3 on page 330 of [225] when $\lambda = 1$ and q is replaced by $-q$ (explicitly given by Slater [233], equation 94); (11.1.17) is equation 13 on page 332 of [225]; (11.1.18) is the second equation of page 331 of [225]; (11.1.19) is equation 3, line 2, on page 330 of [226].

For **Regime II**, however, it turns out that one must consider the following rather complicated one-dimensional partition function:

$$F_k(\sigma_1) = \lim_{m\to\infty} \sum_{\sigma_2,\sigma_3,\ldots,\sigma_m} q^{i(\sigma_i\sigma_{i+2}-\sigma_{i+1}+\overline{\sigma}_{i+1})}. \tag{11.1.23}$$

in which the summation runs over all possible $(m-1)$-tuples $(\sigma_2,\sigma_3,\ldots,\sigma_m)$ subject to the fairly complicated and stringent conditions:

$$\begin{cases} \sigma_i = 0\, or\, 1, & 1 \le j \le m; \\ \sigma_i + \sigma_{i+1} \le 1, & 1 \le j \le m-1; \\ \sigma_m = \sigma_{m+1} = 0, & 1; \\ \overline{\sigma}_i = 1, & i \equiv k(mod 3); \\ \overline{\sigma}_i = 0, & i \not\equiv k(mod 3). \end{cases}$$

Baxter obtains recurrence relations for refinements of these functions $F_k(\sigma_1)$; however, the techniques that he applies successfully to solve the recurrence relations in the other three regimes fail here. For this reason, he is unable to find counterparts of the infinite series in (11.1.12) to (11.1.19). By direct expansion, he obtains overwhelming evidence to conjecture that each of $F_1(0), F_1(1), F_2(0), F_2(1), F_3(0)$, and $F_3(1)$ are identical with elegant combinations of infinite products.

Next, we shall give double series expansions for the $F_k(\sigma_1)$ that establish all six of Baxter's conjectures. Apart from their contribution to Baxter's solution of the hard-hexagon model, these results are also surprising mathematically. They are not apparently limiting cases of known basic hypergeometric series identities; this is in contradistinction to the fact that the place of equations (11.1.12) to (11.1.19) in the hierarchy of basic hypergeometric series was back then well known (see Slater [233], Bailey [47] and [48]). Here later, we shall describe the results and techniques required to establish these theorems.

11.2 HARD HEXAGON MODEL REGIME II IDENTITIES FROM BAXTER

So, the 1980 conjectures by Baxter for Regime II were rigorously proven by Andrews in 1981 (see [21]), although, as Andrews says, the results could have been known to, or derived by, Rogers [226] or Schur [230]) in 1917, or by Slater [233] originally in 1952. However, these results were not explicitly given back then, and Baxter's derivation was after all, innovative and a significant step forward in the theory. So we state the relevant identities in the following six equations.

Regime II

$$F_1(0)=\sum_{n=0}^{\infty}\sum_{0\le r\le\frac{3n+1}{2}}\frac{q^{\frac{3n(n+1)}{2}-r}}{(q^2;q^2)_r(q)_{3n-2r+1}}=\frac{\Pi(q^4,q^{11},q^{15};q^{15})+q\Pi(q,q^{14},q^{15};q^{15})}{(q)_\infty}; \tag{11.2.1}$$

$$F_1(1)=\sum_{n=0}^{\infty}\sum_{0\le r\le\frac{3n}{2}}\frac{q^{\frac{3n(n+1)}{2}-r}}{(q^2;q^2)_r(q)_{3n-2r}}=\frac{\Pi(q^7,q^8,q^{15};q^{15})+q\Pi(q^2,q^{13},q^{15};q^{15})}{(q)_\infty}; \tag{11.2.2}$$

$$F_2(0)=\sum_{n=1}^{\infty}\sum_{0\le r\le\frac{3n-1}{2}}\frac{q^{\frac{n(3n-1)}{2}-r}}{(q^2;q^2)_r(q)_{3n-2r-1}}=\frac{\Pi(q^6,q^9,q^{15};q^{15})}{(q)_\infty} \tag{11.2.3}$$

$$=1+q+2q^2+3q^3+5q^4+7q^5+10q^6+14q^7+20q^8+26q^9+36q^{10}+47q^{11}+63q^{12}$$
$$+81q^{13}+106q^{14}+135q^{15}+174q^{16}+219q^{17}+278q^{18}+347q^{19}+436q^{20}+O(q^{21});$$

$$F_2(1)=\sum_{n=0}^{\infty}\sum_{0\le r\le\frac{3n+1}{2}}\frac{q^{\frac{n(3n+5)}{2}+1-r}}{(q^2;q^2)_r(q)_{3n-2r+1}}=\frac{q\Pi(q^3,q^{12},q^{15};q^{15})}{(q)_\infty} \tag{11.2.4}$$

$$=q+q^2+2q^3+2q^4+4q^5+5q^6+8q^7+10q^8+15q^9+19q^{10}+27q^{11}+34q^{12}+46q^{13}$$
$$+58q^{14}+77q^{15}+96q^{16}+125q^{17}+155q^{18}+198q^{19}+244q^{20}+O(q^{21});$$

$$F_3(0)=F_2(0)=\sum_{n=0}^{\infty}\sum_{0\le r\le\frac{3n}{2}}\frac{q^{\frac{n(3n+1)}{2}-r}}{(q^2;q^2)_r(q)_{3n-2r}}=\frac{\Pi(q^6,q^9,q^{15};q^{15})}{(q)_\infty}; \tag{11.2.5}$$

$$F_3(1)=F_2(1)=\sum_{n=1}^{\infty}\sum_{0\le r\le\frac{3n-1}{2}}\frac{q^{\frac{n(3n+1)}{2}-r}}{(q^2;q^2)_r(q)_{3n-2r-1}}=\frac{q\Pi(q^3,q^{12},q^{15};q^{15})}{(q)_\infty}. \tag{11.2.6}$$

So, Baxter conjectured the identity of each of the $F_k(\sigma_i)$ with the corresponding infinite products given above. His approach involved firstly, methods for the treatment of the expressions given in (11.1.22) so that the double series representations given above can be found. Secondly, a set of transformations was required to allow the identification of the double series with the appropriate infinite product expression.

11.3 OUTLINE OF PROOFS OF BAXTER'S REGIME II CONJECTURES

Our methodology differs from that of Baxter immediately. Baxter's development of series-product identities relied on the taking of the limit as m tends to ∞ in (11.1.22). We instead find representations for the partition functions arising in **Regime III** with m remaining fixed and finite. We then make use of the significant fact that when m is finite we can map from **Regime III** to **Regime II** by the transformation $q\mapsto 1/q$. Our solution of **Regime III** (on the way proving (11.1.1) to (11.1.16) relies on the two following polynomial identities:

$$\sum_{n,r\geq 0} q^{\frac{n(3n+1}{2}} \begin{bmatrix} N-2n-2r \\ n \end{bmatrix}_q \begin{bmatrix} r+n \\ r \end{bmatrix}_{q^2} q^r = \sum_{\lambda=-\infty}^{\infty} (-1)^\lambda q^{\lambda(5\lambda+1)} \begin{bmatrix} N \\ \left[\frac{n-5\lambda}{2}\right] \end{bmatrix}_q; \tag{11.3.1}$$

$$\sum_{n,r\geq 0} q^{\frac{n(3n+1}{2}} \begin{bmatrix} N-2n-2r-1 \\ n \end{bmatrix}_q \begin{bmatrix} r+n \\ r \end{bmatrix}_{q^2} q^r = \sum_{\lambda=-\infty}^{\infty} (-1)^\lambda q^{\lambda(5\lambda-3)} \begin{bmatrix} N \\ \left[\frac{n-5\lambda}{2}\right]+1 \end{bmatrix}_q; \tag{11.3.2}$$

in which

$$\begin{bmatrix} N \\ M \end{bmatrix}_q = \begin{cases} \frac{(1-q^N)(1-q^{N-1})\cdots(1-q^{N-M+1})}{(1-q^M)(1-q^{M-1})\cdots(1-q)}, & \text{for } M \geq 0; \\ 0, & \text{for } M < 0, N \geq 0, \end{cases} \tag{11.3.3}$$

and $[x]$ = the largest integer not exceeding x. If in (11.3.1) we let $N \to \infty$, the first result required for **Regime III**, that being (11.3.3), is obtained. Similarly. if $N \to \infty$ in (11.4.1), we obtain (11.1.14).

To obtain (11.2.1) to (11.2.6), we replace N by $3N + a$ ($a = 0, \pm 1$) in (11.3.1) and (11.3.2), and then replace q by $1/q$. Next multiply by the minimal power of q necessary to produce polynomials in q, and then let $N \to \infty$. This process produces the identities of series and products described in these (11.2.5) to (11.2.6), and the relationship between **Regimes II** and **III** that follows from the replacement of q by $1/q$ resulting in the identity with the various $F_k(\sigma_l)$.

11.4 TRANSFORMS BETWEEN BAXTER'S REGIME II AND REGIME III

The results described here belie the fact that the Rogers-Ramanujan type identities for **regime II** of the hard-hexagon model are now rigorously established. On the other hand, there are numerous interesting long range questions more of interest in the theory of partitions and q-series that have been extensively explored under three approaches:

Approach (i): Suppose the Rogers-Ramanujan partition ideal (see Andrews [25, chapter 8] for a detailed discussion of partition ideals) is replaced by another classical partition ideal; what happens in **Regimes II**, **III**, and **IV** appropriately modified?

Approach (ii): The $q \to 1/q$ duality of **Regimes II** and **III** also exists between **Regimes I** and **IV**. In fact, the relevant polynomial identities for this latter relationship are

$$\sum_{j\geq 0} q^{j^2} \begin{bmatrix} N-j \\ j \end{bmatrix}_q = \sum_{\lambda=-\infty}^{\infty} (-1)^\lambda q^{\frac{\lambda(5\lambda+1)}{2}} \begin{bmatrix} N \\ \left[\frac{N-5\lambda}{2}\right] \end{bmatrix}_q; \tag{11.4.1}$$

$$\sum_{j\geq 0} q^{j(j+1)} \begin{bmatrix} N-j \\ j \end{bmatrix}_q = \sum_{\lambda=-\infty}^{\infty} (-1)^\lambda q^{\frac{\lambda(5\lambda-3)}{2}} \begin{bmatrix} N \\ \left[\frac{N+1-5\lambda}{2}\right]+1 \end{bmatrix}_q. \tag{11.4.2}$$

These identities were completely stated in Andrews [10] and have their origin in the work of Schur [231]. The arguments used to obtain (11.2.1) to (11.2.6) from (11.3.1) and (11.3.2) may now be turned on (11.4.1) and (11.4.2) to obtain (11.1.16) to (11.1.19), a relationship previously unnoticed. Thus a *duality theory* between various sets of identities of the Rogers-Ramanujan type presents for exploration.

Approach (iii): The analytic duality described above has a corresponding manifestation in the partition-theoretic interpretations of the various identities considered. Thus the well-known combinatorial interpretations of (11.1.12) and (11.1.13) are *dual* to the combinatorial interpretations of (11.1.16) to (11.1.19). Since the 1980s this duality has been developed in two distinct streams, via statistical mechanics Integrable Systems and in the theory of mathematical integer partitions.

11.5 THE PARTITION FUNCTION OF THE HARD HEXAGON MODEL

The hard hexagon model occurs within the framework of the grand canonical ensemble, where the total number of particles (the *hexagons*) is allowed to vary naturally and is fixed by a chemical potential. In the hard hexagon model, all valid states have zero energy, and so the only important thermodynamic control variable is the ratio of chemical potential to temperature $\frac{\mu}{\kappa T}$. The exponential of this ratio, $z = \exp\left(\frac{\mu}{\kappa T}\right)$ is called the activity and larger values correspond roughly to denser configurations.

For a triangular lattice with N sites, the grand partition function is

$$Z(z) = \sum_{n} z^n g(n,N) - 1 + Nz + \frac{1}{2}N(N-7)z^2 + \cdots \tag{11.5.1}$$

where $g(n,N)$ is the number of ways of placing n particles on distinct lattice sites such that no two are adjacent. The function κ is defined by

$$\kappa(z) = \lim Z(z)^{1/N} = 1 + z - 3z^2 + \cdots \tag{11.5.2}$$

so that $\log(\kappa)$ is the free energy per unit site. Solving the hard hexagon model means (roughly) finding an exact expression for κ as a function of z.

The **mean density** ρ is given for small z by

$$\rho = z\frac{d\log(\kappa)}{dz} = z - 7z^2 + 58z^3 - 519z^4 + 4856z^5 - \cdots. \tag{11.5.3}$$

The vertices of the lattice fall into three classes numbered 1, 2, and 3, given by the 3 different ways to fill space with hard hexagons. There are 3 local densities ρ_1, ρ_2, ρ_3, corresponding to the 3 classes of sites. When the activity is large the system approximates one of these 3 packings, so the local densities differ, but when the activity is below a critical point the three local densities are the same. The critical point separating the low-activity homogeneous phase from the high-activity ordered phase is

$$z_c = \frac{(11+5\sqrt{5})}{2} = \phi^5 = 11.09017\ldots$$

with golden ratio $\phi = \frac{1+\sqrt(5)}{2}$. Above the critical point the local densities differ and in the phase where most hexagons are on sites of type 1 can be expanded as

$$\rho_1 = 1 - z^{-1} - 5z^{-2} - 34z^{-3} - 267z^{-4} - 2037z^{-5} - \cdots,$$

$$\rho_2 = z^{-2} - 9z^{-3} + 80z^{-4} + 965z^{-5} - \cdots.$$

11.5.1 Solution

The solution is given for small values of $z < z_c$ by

$$z = \frac{-xH(x)^5}{G(x)^5},$$

$$\kappa = \frac{H(x)^3 Q(x^5)^2}{G(x)^2} \prod_{n\geq 1} \frac{(1-x^{6n-4})(1-x^{6n-3})^2(1-x^{6n-2})}{(1-x^{6n-5})(1-x^{6n-1})(1-x^{6n})^2},$$

$$\rho = \rho_1 = \rho_2 = \rho_3 = \frac{-xG(x)H(x^6)P(x^3)}{P(x)},$$

where

$$G(x) = \prod_{n\geq 1} \frac{1}{(1-x^{5n-4})(1-x^{5n-1})},$$

$$H(x) = \prod_{n\geq 1} \frac{1}{(1-x^{5n-3})(1-x^{5n-2})},$$

$$P(x) = \prod_{n\geq 1} (1-x^{2n-1}) = \frac{Q(x)}{Q(x)^2},$$

$$Q(x) = \prod_{n\geq 1} (1-x^n).$$

For large $z > z_c$ the solution (in the phase where most occupied sites have type 1) is given by

$$z = \frac{G(x)^5}{xH(x)^5},$$

$$\kappa = x^{\frac{1}{3}} \frac{G(x)^3 Q(x^5)^2}{H(x)^2} \prod_{n\geq 1} \frac{(1-x^{3n-2})(1-x^{3n-1})}{(1-x^{3n})^2},$$

$$\rho_1 = \frac{H(x)Q(x)(G(x)Q(x) + x^2 H(x^9)Q(x^9))}{Q(x^3)^2},$$

$$\rho_2 = \rho_3 = \frac{x^2 H(x)Q(x)H(x^9)Q(x^9)}{Q(x^3)^2},$$

$$R = \rho_1 - \rho_2 = \frac{Q(x)Q(x^5)}{Q(x^3)^2}.$$

The functions G and H turn up in the Rogers-Ramanujan identities, and the function Q is the Euler function, which is closely related to the Dedekind eta function. If $x = e^{2\pi i\tau}$, then $q^{-1/60}G(x)$, $x^{11/60}H(x)$, $x^{-1/24}P(x)$, $z, \kappa, \rho, \rho_1, \rho_2$, and ρ_3 are modular functions of τ, while $x^{1/24}Q(x)$ is a modular form of weight 1/2. Since any two modular functions are related by an algebraic relation, this implies that the functions κ, z, R, ρ are all algebraic functions of each other of quite high degree.

11.6 ROGERS-RAMANUJAN SHIFT FROM MATHEMATICS TO PHYSICS

In 1998 the physicists Alexander Berkovich and Barry M. McCoy from the State University of New York wrote an expository paper [60] explaining how the story of the Rogers-Ramanujan identities transitioned from mathematics to physics. Put simply, in 1894 L.J. Rogers [225] proved the following identities for $a = 0, 1$ between infinite series and products valid for $|q| < 1$,

$$\sum_{n=0}^{\infty} \frac{q^{n(n+a)}}{(q)_n} = \prod_{n=1}^{\infty} \frac{1}{(1-q^{5n-1-a})(1-q^{5n-4+a})} \tag{11.6.1}$$

$$= \frac{1}{(q)_\infty} \sum_{m=-\infty}^{\infty} (q^{n(10n+1+2a)} - q^{(5n+2-a)(2n+1)}),$$

where $(q)_n = \prod_{j=0}^{n}(1-q^j)$.

A fascinating quote from Berkovich and McCoy on Rogers-Ramanujan identities:

Think About It...

"For about the first 85 years after their discovery interest in these identities and their generalizations was confined to mathematicians. Many ingenious proofs and relations applying combinatorics, basic hypergeometric functions and Lie algebras were discovered by MacMahon, Rogers, Schur, Ramanujan, Watson, Bailey, Slater, Gordon, Göllnitz, Andrews, Bressoud, Lepowsky and Wilson; so by 1980 there were over 130 isolated identities and several infinite families of identities known.

The entry of these identities into physics occurred in the early 1980s when Baxter [51], Andrews, Baxter and Forrester (see [31] and [129]), and the Kyoto group [116] encountered (11.6.1) and various generalizations in the computation of order parameters of certain lattice models of statistical mechanics.

A further glimpse of the relation to physics is seen in the development of conformal field theory by Belavin, Polyakov and Zamolodchikov [56] and the form of computation of characters of representations of Virasoro algebra by Kac [161], Feigin and Fuchs [126] and Rocha-Caridi [223]. The occurrence of (11.6.1) in this context led Kac [162] to suggest that *every modular invariant representation of Virasoro should produce a Rogers-Ramanujan type identity.*"

Alexander Berkovich and Barry M. McCoy
State University of New York, 1998

The full relation between physics and Rogers-Ramanujan identities is more extensive than we may think from these first indications. Starting in 1993 both Berkovich and McCoy (see for example [165, 166], [58, 59, 60, 61], and [26]) have fused the physical insight of solvable lattice models in statistical mechanics with the classical work of the first 85 years and the recent developments in conformal field theory to greatly enlarge the theory of Rogers-Ramanujan identities. We can summarize the results of this work and present some of the current results. Our point of view for the rest of this chapter will be influenced by the statistical mechanics, but we will try to indicate where alternative viewpoints exist. Hopefully some language differences between physicists and mathematicians can be understood.

The work of the last 30 years arising in physics problems has given a new context and point of view in the study of Rogers-Ramanujan identities. The emphasis is not the same as in the earlier mathematical investigations and thus it is good to discuss this before giving detailed results.

In mathematics, Rogers-Ramanujan identities always were seen as *partitions of integers equivalence theorems* or as *generating function algebraic identities for integer partitions*. This latter mostly took the form of q-series identities where one side was a sum of $(q)_n$ terms and the other side was an infinite product of $(q)_n$ terms. In physics the discussion still centered around partitions, but now defined by terms of *statistical mechanical and conformal field theory applications*. In physics, nowadays in most cases where we have generalizations of the identities between the two sums, a product form is not known. So in physics, a Rogers-Ramanujan identity means the equality of the sums without further reference to possible product forms.

What is interesting about the two diverging approaches to this topic, is that underlying mathematics and physics is a universality that embraces both approaches. However, for now, the two approaches not just to Rogers-Ramanujan identities, but to the theory of partitions, are distinct, but equally valid.

The full range of Rogers-Ramanujan identities is by no means yet understood and it is anticipated that both in the related mathematics and in the related physics there is much still left to be discovered. The physics has grown into innumerable research areas due to the now well-established fact that the Yang-Baxter equation has long been recognised as the masterkey to integrability, providing the basis for exactly solved models which capture the fundamental physics of a number of realistic classical and quantum systems. The theory of partitions has therefore become an essential tool for physicists involved with exactly solvable models, of which there is now, including the hard-hexagon model, the Heisenberg spin chain, the transverse quantum Ising chain, a spin ladder model, the Lieb-Liniger Bose gas, the Gaudin-Yang Fermi gas and the two-site Bose-Hubbard model. For an eloquent review of this in the context of *condensed matter to ultracold atoms* see Batchelor and Foerster [49].

11.7 PARTITION GENERATING FUNCTIONS IN PHYSICS ARE POLYNOMIAL GENERALIZATIONS

The second insight which is also present in the very first papers on the connection of Rogers-Ramanujan identities with physics, (see Baxter [51], Andrews, Baxter and Forrester [31], and Forrester and Baxter [129]) is the fact that the physics will often lead to polynomial identities, with an order depending on an integer L, which yield infinite series identities as $L \to \infty$.

The polynomial generalization of (11.6.1) is the identity first proven by Andrews [10] in 1970,

$$F_a(L,q) = B_a(L,q) \tag{11.7.1}$$

where

$$F_a(L,q) = \sum_{n=0}^{\infty} q^{n(n+a)} \begin{bmatrix} L-n-a \\ n \end{bmatrix}_q \tag{11.7.2}$$

and

$$B_a(L,q) = \sum_{n=-\infty}^{\infty} (-1)^n q^{n(5n+1+2a)/2} \begin{bmatrix} L \\ \lfloor \frac{1}{2}(L-5n-a) \rfloor \end{bmatrix}_q \tag{11.7.3}$$

where $\lfloor x \rfloor$ denotes the integer part of x and the Gaussian polynomials (q-binomial coefficients) are defined for integer m, n by

$$\begin{bmatrix} n \\ m \end{bmatrix}_q = \begin{cases} \frac{(q)_n}{(q)_m (q)_{n-m}}, & 0 \leq m \leq n; \\ 0, & \text{otherwise.} \end{cases} \tag{11.7.4}$$

The identity (11.7.1) is obtained by using

$$\lim_{n \to \infty} \begin{bmatrix} n \\ m \end{bmatrix}_q = \frac{1}{(q)_m}.$$

So (11.7.1) is a generalization of the identity (11.6.1) which we will call a Rogers-Ramanujan identity.

There are many known generalizations of $F_a(L,q)$ that are written in terms of the following function due to Kedem et al. [166]

$$f = \sum_{restrictions} q^{\frac{1}{2}\mathbf{mBm} - \frac{1}{2}\mathbf{Am}} \prod_{\alpha=1}^{n} \begin{bmatrix} ((1-\mathbf{B})\mathbf{m} + \frac{\mathbf{u}}{2})_\alpha \\ m_\alpha \end{bmatrix}_q \tag{11.7.5}$$

where $\mathbf{m}$, $\mathbf{u}$ and $\mathbf{A}$ are $\mathbf{n}$ dimensional vectors and $\mathbf{B}$ is an $n \times n$ dimensional matrix and the sum is over all values of the variables m_α possibly subject to some restrictions (such as being even or odd). In many cases the q-binomials are defined by (11.7.4) but there do occur cases with an extended definition

$$\begin{bmatrix} m+n \\ m \end{bmatrix}_q = \begin{cases} (q^{n+1})_m, & \text{for } m \geq 0,\ n \text{ integers;} \\ 0, & \text{otherwise.} \end{cases} \tag{11.7.6}$$

This brings (11.7.4) to cover negative n.

The function (11.7.5) is regarded as the partition function for a collection of n different species of free massless (right moving) fermions with a linear energy momentum relation $e(P_j^\alpha)_\alpha = vP_j^\alpha$ where the momenta are quantized in units of $2\pi/M$ and are chosen from the sets

$$P_j^\alpha \in \left\{ P_{\text{min}}^\alpha, P_{\text{min}}^\alpha + \frac{2\pi}{M}, P_{\text{min}}^\alpha + \frac{4\pi}{M}, P_{\text{min}}^\alpha + \frac{6\pi}{M}, \ldots, P_{\text{max}}^\alpha \right\} \tag{11.7.7}$$

where

$$P^{\alpha}_{\text{min}}(\mathbf{m}) = \frac{\pi}{M}\left[((\mathbf{B}-1)\mathbf{m})_\alpha - \mathbf{A}_\alpha + 1\right]$$

and

$$P^{\alpha}_{\text{max}}(\mathbf{m}) = -P^{\alpha}_{\text{min}}(\mathbf{m}) + \frac{\pi}{M}\left(\frac{\mathbf{u}}{2} - \mathbf{A}\right)_\alpha$$

with the Fermi exclusion rule $P_j^\alpha \neq P_k^\alpha$ for $j \neq k$ and all $\alpha = 1, 2, \ldots, n$.

11.8 THE ELLIPTIC q-GAMMA FUNCTION

The elliptic gamma function is a generalization of the q-gamma function, which is itself the q-analog of the usual gamma function. It was first examined in detail by Ruijsenaars in 1997 as an integrable function (see [228]), and can be expressed in terms of the triplegamma function. It is defined by

$$\Gamma(z;p,q) = \prod_{m=0}^{\infty}\prod_{n=0}^{\infty} \frac{1 - p^{m+1}q^{n+1}/z}{1 - p^m q^n z}. \tag{11.8.1}$$

The following relation is easily seen to apply

$$\Gamma(z;p,q)\,\Gamma(pq/z;p,q) = 1. \tag{11.8.2}$$

Using the q-theta function $\theta(z;q) = \prod_{n=0}^{\infty}(1-q^n z)(1-q^{n+1}/z)$, it is well-known that

$$\begin{aligned}\Gamma(pz;p,q) &= \theta(z;q)\Gamma(z;p,q),\\ \Gamma(qz;p,q) &= \theta(z;p)\Gamma(z;p,q).\end{aligned}$$

At $p = 0$ we have $\Gamma(z;0,q)\,(z,q)_\infty = 1$ using the infinite q-Pochhammer notation.

As with the usual gamma function, there is a duplication, a triplication and a multiplication formula. We need firstly to define the function

$$\widetilde{\Gamma}(z;p,q) := \frac{(q,q)_\infty}{(p,p)_\infty}(\theta(q,p))^{1-z} \prod_{m=0}^{\infty}\prod_{n=0}^{\infty} \frac{1 - p^{m+1}q^{n+1-z}}{1 - p^m q^{n+z}}. \tag{11.8.3}$$

We follow the approach by Felder and Varchenko [127] who proved the

Theorem 11.1 *If $r = q^n$ then*

$$\begin{aligned}&\widetilde{\Gamma}(nz;p,q)\widetilde{\Gamma}(\frac{1}{n};p,r)\widetilde{\Gamma}(\frac{2}{n};p,r)\cdots\widetilde{\Gamma}(\frac{n-1}{n};p,r) \\ &= \left(\frac{\theta(r;p)}{\theta(q;p)}\right)^{nz-1}\widetilde{\Gamma}(z;p,q)\widetilde{\Gamma}(z+\frac{1}{n};p,r)\cdots\widetilde{\Gamma}(z+\frac{n-1}{n};p,r).\end{aligned} \tag{11.8.4}$$

The two most significant corollaries of theorem 11.1 are as follows:

Corollary 11.1 *The Duplication Formula.*

$$\widetilde{\Gamma}(2z;p,q)\widetilde{\Gamma}(\frac{1}{2};p,q^2) = \left(\frac{\theta(q^2;p)}{\theta(q;p)}\right)^{2z-1}\widetilde{\Gamma}(z;p,q^2)\widetilde{\Gamma}(z+\frac{1}{2};p,q^2). \tag{11.8.5}$$

Corollary 11.2 *The Triplication Formula.*

$$\widetilde{\Gamma}(3z;p,q)\widetilde{\Gamma}(\frac{1}{3};p,q^3)\widetilde{\Gamma}(\frac{2}{3};p,q^3) = \left(\frac{\theta(q^3;p)}{\theta(q;p)}\right)^{3z-1} \widetilde{\Gamma}(z;p,q^3)\widetilde{\Gamma}(z+\frac{1}{3};p,q^3)\widetilde{\Gamma}(z+\frac{2}{3};p,q^3). \tag{11.8.6}$$

The above theorem with the Duplication and Triplication formula cases may be worthwhile examining and comparing with the functional equations given in Chapter 15. For example, is there a connection applicable between the Multiplication Formula of the Elliptic Gamma Function and the functions from Chapter 15, namely,

Definition 11.1 *Define the function $F_n(t)$ for all complex numbers a,t with $|a|,|t| < 1$, and for $n \geq 1$ with $|q_1|,|q_2|,...,|q_n| < 1$ by the sequence of functions $F_n(t)$ with*

$$F_0(-;a,t) = \frac{1-at}{1-t} \tag{11.8.7}$$

and for all of $|q_1|,|q_2|,|q_3|,...,|q_n| < 1$, by

$$F_n(q_1,q_2,q_3,...,q_n;a,t) = \prod_{\alpha_1,\alpha_2,\alpha_3,...,\alpha_n \geq 0} \frac{1-q_1^{\alpha_1}q_2^{\alpha_2}...q_n^{\alpha_n}at}{1-q_1^{\alpha_1}q_2^{\alpha_2}...q_n^{\alpha_n}t} \equiv \sum_{k=1}^{\infty} {}_nA_k t^k \tag{11.8.8}$$

where $\alpha_1,\alpha_2,\alpha_3,...,\alpha_n$ may be all zero.

We now state a functional equation theorem for $F_n(t)$, next highlighting the 2D, 3D and 4D cases as examples in a corollary.

Theorem 11.2 *(see chapter 15) If the left side of (11.8.8) is for the moment considered only as a function, $F_n(t)$, of t, then the following finite product holds,*

$$\frac{1-at}{1-t} = \frac{F_n(t)\left(\prod_{S_2} F_n(q_{i_1}q_{i_2}t)\right)\left(\prod_{S_4} F_n(q_{i_1}q_{i_2}q_{i_3}q_{i_4}t)\right)...}{\left(\prod_{S_1} F_n(q_{i_1}t)\right)\left(\prod_{S_3} F_n(q_{i_1}q_{i_2}q_{i_3}t)\right)\left(\prod_{S_5} F_n(q_{i_1}q_{i_2}q_{i_3}q_{i_4}q_{i_5}t)\right)...} \tag{11.8.9}$$

the numerator products each being over the nth order even symmetric combinations whilst the denominator products are over the nth order odd symmetric combinations of the variables. We have used the notation $S_1,S_2,S_3,S_4,...$ to denote the respective nth order symmetric function product of terms.

It is worth noting that we have used the abbreviation $F(t) = F_n(q_1,q_2,q_3,...,q_n;a,t)$ and the right side of (11.8.9) is a finite product of such functions going up to the nth order symmetric function product of terms. It is also seen here that (11.8.9) is a functional equation for a *generalized right side of (11.8.8)* form of the q-binomial product. As an illustration we next give the $n = 1$ up to $n = 4$ examples of the functional equation (11.8.8), with $q_1 \to x$, $q_2 \to y$, $q_3 \to z$. With these variables, we assert

Corollary 11.3 *(see chapter 15) If for $|w|<1, |x|<1, |y|<1, |z|<1$, and all complex numbers a and t,*

$$\begin{aligned}
F_1(t) = F_1(z;a,t) &= \prod_{j\geq 0} \frac{1-z^j at}{1-z^j t}, \\
F_2(t) = F_2(y,z;a,t) &= \prod_{j,k\geq 0} \frac{1-y^j z^k at}{1-y^j z^k t}, \\
F_3(t) = F_3(x,y,z;a,t) &= \prod_{j,k,l\geq 0} \frac{1-x^j y^k z^l at}{1-x^j y^k z^l t}, \\
F_4(t) = F_4(w,x,y,z;a,t) &= \prod_{j,k,l,m\geq 0} \frac{1-w^j x^k y^l z^m at}{1-w^j x^k y^l z^m t}, \\
\text{etc.} & \qquad\qquad (11.8.10)
\end{aligned}$$

then

$$\begin{aligned}
\frac{1-at}{1-t} &= \frac{F_1(t)}{F_1(zt)}, \\
\frac{1-at}{1-t} &= \frac{F_2(t)F_2(yzt)}{F_2(yt)F_2(zt)}, \\
\frac{1-at}{1-t} &= \frac{F_3(t)F_3(yzt)F_3(xyt)F_3(xzt)}{F_3(xt)F_3(yt)F_3(zt)F_3(xyzt)}, \\
\frac{1-at}{1-t} &= \frac{F_4(t)F_4(yzt)F_4(xyt)F_4(xzt)F_4(wxt)F_4(wyt)F_4(wzt)F_4(wxyzt)}{F_4(wt)F_4(xt)F_4(yt)F_4(zt)F_4(wxyt)F_4(wxzt)F_4(wyzt)F_4(xyzt)}, \\
\text{etc.} & \qquad\qquad (11.8.11)
\end{aligned}$$

The above equations are suggestive of three as yet undeveloped theories:

1) The functional equations for generating functions imply recurrence relations for the nD vector partitions themselves. These are higher dimensional extensions of the Ramanujan integer partition congruences. See Bruinier [78], and Bruinier et al. [79] to [81] for unrestricted partition congruences.

2) An nD m-ary vector partition congruence theory; extending the binary partition congruences of Rödseth and Gupta. See Alkauskas [6] for an account of binary integer congruences.

3) The nD generating functions suggest their own versions of Duplication Formulas, Triplication formulas, and Multiplication formulas, similar to those stated here for the Elliptic Gamma function. This function is *integrable* and therefore related to existing theories in Statistical Mechanics, Yang-Baxter equations, Boltzmann weights and phase transitions for Lattice Models. (See the original 1997 account by Ruijsenaars [228], and more recently the 2016 Physics in Bazhanov et al. [53] for example.

This narrative adjourns to future discussions on topics of functional equations, integrable functions, vector partition congruences and nD Lattice Models. The Elliptic Gamma Function is a 2D example of an nD function that may *play out* in a similar way in higher Euclidean spaces. The Elliptic Gamma Function seems useful for vector partition concepts not yet articulated. Also, our chapter 15 functions and functional equations may be related to solvable models associated with Yang-Baxter equations. This topic brings us deeply into the physics of Statistical Mechanics but is beyond the present scope of our book. It may bring future researchers to consider cases of single or multiple spin degrees of freedom at each site of a higher dimensional lattice. The Yang-Baxter equation for such models may reduce to particular simple forms called Star-Triangle relations. For a 2023 contemporary appraisal of Yang-Baxter equations related to Brauer algebras see Cañada [83].

Think About It...

The 1982 book by Rodney Baxter *Exactly solved models in statistical mechanics* has over 7,180 citations as at early 2023.

These citations are probably split approximately evenly between physics and mathematics research articles and books; the breakup as follows: Highly Influential Citations = 919; Background Citations = 2,656; Methods Citations = 1,857; Results Citations = 125.

Semantic Scholar
semanticscholar.org, February 2023.

FURTHER READING

ALKAUSKAS, G. Generalization of the Rödseth-Gupta Theorem on Binary Partitions. *Lith. Math. J.*, April 2003.

ANDREWS, G.E. A polynomial identity which implies the Rogers-Ramanujan identities. *Scr. Math.*, 28:297–305, 1970.

ANDREWS, G.E. The hard-hexagon model and Rogers-Ramanujan type identities. *Proc Nat. Acad. Sci. USA*, 78:5290–5292, 1981.

ANDREWS, G.E. and ASKEY, R. Enumeration of Partitions: The Role of Eulerian Series and q-Orthogonal Polynomials. In Martin Aigner, editor, *Higher Combinatorics*, pages 3–26, Dordrecht, 1977. Springer Netherlands.

ANDREWS, G.E. and BERKOVICH, A. A trinomial analogue of Bailey's lemma and $N = 2$ superconformal invariance. *Comm. Math. Phys.*, 192:245–260, 1998.

ANDREWS, G.E., BAXTER, R.J. and FORRESTER, P.J. Eightvertex SOS model and generalized Rogers-Ramanujan-type identities. *J. Stat. Phys.*, 35:193– 266, 1984.

BAILEY, W.N. Some identities in combinatory analysis. *Proc. Lond. Math. Soc. Second Ser.*, 49:421–435, 1947.

BAILEY, W.N. Identities of the Rogers-Ramanujan type. *Proc. Lond. Math. Soc. Second Ser.*, 50:1–10, 1949.

BATCHELOR, M.T. and FOERSTER, A. Yang-Baxter integrable models in experiments: From condensed matter to ultracold atoms. *J. Phys. A: Math. Theor.*, 49(17), March 2016.

BAXTER, R.J. Hard hexagons: exact solution. *Journal of Physics A: Mathematical and General*, 13(3), 1980. L61-L70, Bibcode 1980JPhA...13L..61B.

BAXTER, R.J. Rogers-Ramanujan identities in the hard hexagon model. *J. Stat. Phys.*, 26:427, 1981.

BAXTER, R.J. *Exactly Solved Models in Statistical Mechanics.* Academic Press, New York, 1982. Reprinted 1989 and 2007.

BAZHANOV, V.V., KELS, A.P. and SERGEEV, S.M. Quasi-classical expansion of the star-triangle relation and integrable systems on quad-graphs. *J. Phys. A: Math. Theor.*, 49:44pp., October 2016.

BELAVIN, A.A., POLYAKOV, A.M. and ZAMOLODCHIKOV, A.B. Infinite conformal symmetry in two-dimensional quantum field theory. *Nucl. Phys. B*, 241:333, 1984.

BERKOVICH, A. Fermionic counting of RSOS states and Virasoro character formulae for the unitary minimal series $M(\upsilon, \upsilon+1)$: Exact results. *Nucl. Phys. B*, 431(1-2):315–348, 1994.

BERKOVICH, A. and MCCOY, B.M. Continued fractions and fermionic representations for characters of $M(p,p')$ minimal models. *Lett. Math. Phys.*, 37(49), 1996.

BERKOVICH, A. and MCCOY, B.M. Rogers-Ramanujan Identities: A Century of Progress from Mathematics to Physics. *Documenta Mathematica*, Extra Volume ICM 1998 - III:163–172, 1998.

BERKOVICH, A., MCCOY, B.M. and SCHILLING, A. Rogers-Schur- Ramanujan type identities for $M(p,p')$ minimal models of conformal field theory. *Comm. Math. Phys.*, 191(325), 1998.

BRUINIER, J.H. *Borcherds products on $O(2,l)$ and Chern classes of Heegner divisors*, volume 1780 of *Springer Lecture Notes in Mathematics.* Springer-Verlag, 2002.

BRUINIER, J.H. and FUNKE, J. On two geometric theta lifts. *Duke Math. J.*, 125:45–90, 2004.

BRUINIER, J.H., and ONO, K. Algebraic formulas for the coefficients of halfintegral weight harmonic weak Maass forms. *Adv. Math.*, 246:198–219, 2013.

CAÑADAS, A.M.; BALLESTER-BOLINCHES, A.; and GAVIRIA, I.D.M. Solutions of the Yang-Baxter Equation Arising from Brauer Configuration Algebras. *Computation*, 11(1), 2023.

DATA, E., JIMBO, M., KUNIBA, A., MIWA, T. and OKADO, M. Exactly solvable SOS models: local height probabilities and theta function identities. *Nucl. Phys. B*, 290(231), 1987.

FEIGIN, B.L. and FUCHS,D.B. Verma modules over the Virasoro algebra. *Funct. Anal. Appl.*, 17(3):241–242, 1983.

FELDER, G. and VARCHENKO, A. Multiplication Formulas for the Elliptic Gamma Function. arxiv preprint, Dec 2002. arXiv:math/0212155v1 [math.QA]. www.doi.org/10.48550/arXiv.math/0212155.

FORRESTER,P.J. and BAXTER,R.J. Further exact solutions of the eightvertex SOS model and generalizations of the Rogers-Ramanujan identities. *J. Stat. Phys.*, 38(435), 1985.

KAC, V.G. Contravariant form for infinite-dimensional Lie algebras and superalgebras. *Lect. Notes in Phys.*, 94:441, 1979.

KAC, V.G. Modular invariance in mathematics and physics. In *Proceedings of the AMS Centennial Symposium*, volume 337. American Mathematical Society, 1992.

KEDEM, R., KLASSEN, T.R., MCCOY, B.M. and MELZER, E. Fermionic quasi-particle representations for characters of $(G^{(1)})_1 \times (G^{(1)})_1/G^{(1)})_2$. *Phys. Lett. B*, 304:263–270, 1993.

KEDEM, R., KLASSEN, T.R., MCCOY, B.M. and MELZER, E. Fermionic sum representations for conformal field theory characters. *Phys. Lett. B*, 307:68–76, 1993.

ROCHA-CARIDI, A. *Vertex Operators in Mathematics and Physics.* Springer, Berlin, 1985. ed. J. Lepowsky, S. Mandelstam and I.M. Singer.

ROGERS, L.J. Second memoir on the expansion of certain infinite products. *Proc. London Math. Soc.*, 25:318–343, 1894.

ROGERS, L.J. On Two Theorems of Combinatory Analysis and Some Allied Identities. *Proc. Lond. Math. Soc. Second Ser.*, 16:315–336, 1917.

RUIJSENAARS, S.N.M. First order analytic difference equations and integrable quantum systems. *J. Math. Phys.*, 38(2):1069–1146, February 1997.

SCHUR, I. Ein Beitrag zur additive Zahlentheorie und zue Theorie der Kettenbruche. *S.-B. Preuss, Akad. Wies. Phys.-Math, Kl.*, pages 302–321, 1973. First printed 1917: Proceedings of the Royal Prussian Academy of Sciences in Berlin; Reprinted 1973: Ges. Abhandlungen, Vol. 2, pp. 117–136.

SCHUR, I. Zur additive Zahlentheorie und zue Theorie der Kettenbruche. *S.-B. Preuss, Akad. Wies. Phys.-Math, Kl.*, pages 488–495, 1973. First printed 1917: Proceedings of the Royal Prussian Academy of Sciences in Berlin; Reprinted 1973: Ges. Abhandlungen, Vol. 3, pp. 43-50.

SLATER, L.J. *Generalized Hypergeometric Series.* Cambridge University Press, 1966.

III

Vector Partition Theory

CHAPTER 12

Vector Partitions and Their Generating Function Identities

12.1 DEFINING VECTOR PARTITIONS

WITH RESPECT TO VECTOR PARTITIONS in this book, we study the additive decomposition of vectors $v_n = \langle a_1, a_2, \ldots, a_n \rangle$ where each of the components $a_1, a_2, \ldots, a_n$ are an integer, usually a non-negative integer, but this may vary by situation. A partition, whether it be for *integer partitions* or it be for *vector partitions*, is a partially ordered set or *poset*. For integer partitions, this partial ordering is always a sum of weakly decreasing integers.

That is, an integer partition of integer n is just a one-dimensional vector partition that is a sum, equal to n, of weakly decreasing integers greater than or equal to zero.

For example, the following are all integer partitions:

$$10+9+9+9,\ 12+12+11+5+2+1+1+1+1,\ 1+1+1+1+1,$$

$$3+3+3,\ 100+98+98+90+2+1,$$

whereas the following are not integer partitions

$$5+7+10+9+9+9,\ 3+2+1+12+12+11+5+2+1+1+1+1,$$

$$1+1+1+1+1+5,\ 3+3+6+2+3,\ 9+100+98+98+90+2+1.$$

So, for a partition of integer n, we have $n = \lambda_1 + \lambda_2 + \ldots + \lambda_r$ under the rule that $\lambda_1 \geq \lambda_2 \geq \ldots \geq \lambda_r \geq 0$ for λ as integers.

For a vector partition we generalize the weakly decreasing concept to apply to the component integers of the vector under partitioning. So for an nD (n-dimensional, or n-component) vector $v_n = \langle a_1, a_2, a_3, \ldots, a_n \rangle$ we say an nD vector partition of v_n has the form $\lambda_1 + \lambda_2 + \ldots + \lambda_r = v$ where we impose that $\|\lambda_1\| \geq \|\lambda_2\| \geq \ldots \geq \|\lambda_r\|$ for each λ using some poset definition applied to each nD vector λ. The poset rule may be defined via a vector norm, or by some other pre-ordering that allows a weak

DOI: 10.1201/9781003174158-12

ordering of the parts of the vector partition sum. Our definition supplied next is but one of many possible acceptable poset definitions for a vector partition.

Definition 12.1 ***Vector Partition.***

A partition of an n-dimensional vector $v_n = \langle a_1, a_2, \ldots, a_n \rangle$ *is a finite sum of n-dimensional (or nD) vectors each having a weakly decreasing sequence of 1st vector components, 2nd vector components, 3rd vector components, etc. under vector addition. The vectors that comprise a single sum adding to* v_n *are called parts of the vector partition. A suitable partial ordering, is for example, a set of weakly decreasing integer components that sum to each component of the vector* v_n*, such as:*

$$\begin{aligned} v_n = \langle a_1, a_2, a_3, \ldots, a_n \rangle &= \langle b_1, c_1, d_1, \ldots, \alpha_1 \rangle \\ &+ \langle b_2, c_2, d_2, \ldots, \alpha_2 \rangle \\ &+ \langle b_3, c_3, d_3, \ldots, \alpha_3 \rangle \\ &+ \langle b_4, c_3, d_3, \ldots, \alpha_4 \rangle \\ &\vdots \quad \vdots \\ &= \langle b_m, c_m, d_m, \ldots, \alpha_m \rangle \end{aligned}$$

so we have then clearly that

$$\begin{aligned} a_1 &= b_1 + b_2 + b_3 + \cdots + b_m, \quad \textit{for} \quad b_1 \geq b_2 \geq b_3 \geq \cdots \geq b_m \geq 0; \\ a_2 &= c_1 + c_2 + c_3 + \cdots + c_m, \quad \textit{for} \quad c_1 \geq c_2 \geq c_3 \geq \cdots \geq c_m \geq 0; \\ a_3 &= d_1 + d_2 + d_3 + \cdots + d_m, \quad \textit{for} \quad d_1 \geq d_2 \geq d_3 \geq \cdots \geq d_m \geq 0; \\ \vdots \quad &\vdots \quad \vdots \\ a_n &= \alpha_1 + \alpha_2 + \alpha_3 + \cdots + \alpha_m, \quad \textit{for} \quad \alpha_1 \geq \alpha_2 \geq \alpha_3 \geq \cdots \geq \alpha_m \geq 0. \end{aligned}$$

$\langle a_1, a_2, a_3, \ldots, a_n \rangle$ *may be construed as the arrow starting at the nD origin ending at the nD first hyperquadrant Cartesian lattice point* $(a_1, a_2, a_3, \ldots, a_n)$*, where each* a_i $(1 \leq i \leq n)$ *co-ordinate is a non-negative integer.*

We will work to obtain theorems and conjectures with regard to nD vectors generally with positive integer n, but mostly we shall begin with or attempt to illustrate or justify these theorems and conjectures by first considering the 2D and 3D situations, since they are most easily visualized in the Cartesian co-ordinate settings. We say the $p_n(\langle a_1, a_2, \ldots, a_n \rangle)$ abbreviated to $p_n(a_1, a_2, \ldots, a_n)$ or simply $p_n(v)$ is the number of nD partitions of vector v. Here are a few 2D examples:

$$\begin{aligned} p_2(1,1) = 2 &\equiv \langle 1,0 \rangle + \langle 0,1 \rangle = \langle 1,1 \rangle; \\ p_2(2,1) = 4 &\equiv \langle 2,0 \rangle + \langle 0,1 \rangle = \langle 1,1 \rangle + \langle 1,0 \rangle \\ &\quad = \langle 1,0 \rangle + \langle 1,0 \rangle + \langle 0,1 \rangle = \langle 2,1 \rangle. \end{aligned}$$

12.2 VECTOR OR MULTIPARTITE PARTITIONS FROM ANDREWS

Andrews [12, Chapter 12, page 203] defines $P(\mathbf{n}) = P(n_1, n_2, \ldots, n_r)$ as the number of unrestricted partitions of the *r-partite* or *multipartite* number $(n_1, n_2, \ldots, n_r)$ as an ordered r-tuple of non-negative integers not all zero. Similarly, the $Q(\mathbf{n})$ function is the number of these partitions of $\mathbf{n}$ into distinct parts. Andrews states the two generating functions

$$\sum_{n_1,n_2,\ldots,n_r\geq 0} P(\mathbf{n})x_1^{n_1}\cdots x_r^{n_r} = \prod_{\substack{n_1,n_2,\ldots,n_r\geq 0\\ not\,all\,zero}} \frac{1}{1-x_1^{n_1}\cdots x_r^{n_r}} \tag{12.2.1}$$

$$\sum_{n_1,n_2,\ldots,n_r\geq 0} Q(\mathbf{n})x_1^{n_1}\cdots x_r^{n_r} = \prod_{n_1,n_2,\ldots,n_r\geq 0} (1+x_1^{n_1}\cdots x_r^{n_r}) \tag{12.2.2}$$

and also gives the result due to Cheema and Motzkin [109] extending the well-known Euler theorem that the number of integer partitions into *odd parts* is the same as the number of integer partitions into *distinct parts*. In our work however, we often replace $n_i \geq 0$ by $n_i \geq 1$ or even apply mixed versions of these.

12.3 DEFINING VECTOR GRIDS AND RESULTING PARTITION GRIDS

The set of first quadrant cartesian 2D lattice point vectors is a countable set. The lattice points radial from the origin in the radial region from $y = x$ to the positive y axis can be replicated in the region between $y = x$ and the positive x axis. Therefore, the set of all 2D lattice point vectors in the first quadrant are also countable. They are shown as follows.

DEFINITION The first quadrant 2D vector grid, denoted as $V_2(y \in [1,\infty); z \in [1,\infty))$ is the set of vectors $\langle a,b\rangle$ such that a and b are positive integers. We say that a vector grid generally is a collection of vectors in a defined region. For example, in the following table, the **first quadrant 2D vector grid** is split into an $\overbrace{\textit{overbraced Upper}}$ vector grid, a **Bold Diagonal** vector grid and an $\underbrace{\textit{underbraced Lower}}$ vector grid.

First quadrant 2D vector grid

$\vdots$	$\vdots$	$\vdots$	$\vdots$	$\vdots$	$\vdots$	$\vdots$	$\vdots$	$\vdots$	$\ddots$
8	$\overbrace{\langle 1,8\rangle}$	$\overbrace{\langle 2,8\rangle}$	$\overbrace{\langle 3,8\rangle}$	$\overbrace{\langle 4,8\rangle}$	$\overbrace{\langle 5,8\rangle}$	$\overbrace{\langle 6,8\rangle}$	$\overbrace{\langle 7,8\rangle}$	$\langle \mathbf{8},\mathbf{8}\rangle$	$\cdots$
7	$\overbrace{\langle 1,7\rangle}$	$\overbrace{\langle 2,7\rangle}$	$\overbrace{\langle 3,7\rangle}$	$\overbrace{\langle 4,7\rangle}$	$\overbrace{\langle 5,7\rangle}$	$\overbrace{\langle 6,7\rangle}$	$\langle \mathbf{7},\mathbf{7}\rangle$	$\underbrace{\langle 8,7\rangle}$	$\cdots$
6	$\overbrace{\langle 1,6\rangle}$	$\overbrace{\langle 2,6\rangle}$	$\overbrace{\langle 3,6\rangle}$	$\overbrace{\langle 4,6\rangle}$	$\overbrace{\langle 5,6\rangle}$	$\langle \mathbf{6},\mathbf{6}\rangle$	$\underbrace{\langle 7,6\rangle}$	$\underbrace{\langle 8,6\rangle}$	$\cdots$
5	$\overbrace{\langle 1,5\rangle}$	$\overbrace{\langle 2,5\rangle}$	$\overbrace{\langle 3,5\rangle}$	$\overbrace{\langle 4,5\rangle}$	$\langle \mathbf{5},\mathbf{5}\rangle$	$\underbrace{\langle 6,5\rangle}$	$\underbrace{\langle 7,5\rangle}$	$\underbrace{\langle 8,5\rangle}$	$\cdots$
4	$\overbrace{\langle 1,4\rangle}$	$\overbrace{\langle 2,4\rangle}$	$\overbrace{\langle 3,4\rangle}$	$\langle \mathbf{4},\mathbf{4}\rangle$	$\underbrace{\langle 5,4\rangle}$	$\underbrace{\langle 6,4\rangle}$	$\underbrace{\langle 7,4\rangle}$	$\underbrace{\langle 8,4\rangle}$	$\cdots$
3	$\overbrace{\langle 1,3\rangle}$	$\overbrace{\langle 2,3\rangle}$	$\langle \mathbf{3},\mathbf{3}\rangle$	$\underbrace{\langle 4,3\rangle}$	$\underbrace{\langle 5,3\rangle}$	$\underbrace{\langle 6,3\rangle}$	$\underbrace{\langle 7,3\rangle}$	$\underbrace{\langle 8,3\rangle}$	$\cdots$
2	$\overbrace{\langle 1,2\rangle}$	$\langle \mathbf{2},\mathbf{2}\rangle$	$\underbrace{\langle 3,2\rangle}$	$\underbrace{\langle 4,2\rangle}$	$\underbrace{\langle 5,2\rangle}$	$\underbrace{\langle 6,2\rangle}$	$\underbrace{\langle 7,2\rangle}$	$\underbrace{\langle 8,2\rangle}$	$\cdots$
1	$\langle \mathbf{1},\mathbf{1}\rangle$	$\underbrace{\langle 2,1\rangle}$	$\underbrace{\langle 3,1\rangle}$	$\underbrace{\langle 4,1\rangle}$	$\underbrace{\langle 5,1\rangle}$	$\underbrace{\langle 6,1\rangle}$	$\underbrace{\langle 7,1\rangle}$	$\underbrace{\langle 8,1\rangle}$	$\cdots$
z/y	1	2	3	4	5	6	7	8	$\cdots$

In certain situations, the set of all overlined *upper* vectors can be mapped to the set of all underlined *lower* vectors. The bold diagonal is defined by $y = z$. We have in fact

Statement 4 *For any positive integers such that $a < b$,*

$$\overbrace{\langle a,b\rangle} \mapsto \underbrace{\langle b,a\rangle},$$

and

$$\underbrace{\langle b,a\rangle} \mapsto \overbrace{\langle a,b\rangle}.$$

Hence, the Upper grid can map onto the Lower grid, and the Lower grid can map onto the Upper grid.

While the consideration of the **upper vectors** and the **lower vectors** may seem unnecessary in the present discussion, in later chapters we shall make good use of the symmetries and possible mappings, especially when we come to discuss the Visible Point Vector (VPV) partitions wherein these symmetries can be applied to practical examples.

DEFINITION A finite 2D vector grid is a collection of vectors in a defined finite rectangular region. $V_2(y \in [a,b]; z \in [c,d])$ is notation for

$$\begin{array}{c|cccccc}
d & \langle a,d\rangle & \langle a+1,d\rangle & \ldots & \langle b-1,d\rangle & \langle b,d\rangle \\
d-1 & \langle a,d-1\rangle & \langle a+1,d-1\rangle & \ldots & \langle b-1,d-1\rangle & \langle b,d-1\rangle \\
d-2 & \langle a,d-2\rangle & \langle a+1,d-2\rangle & \ldots & \langle b-1,d-2\rangle & \langle b,d-2\rangle \\
\vdots & \vdots & \vdots & \ddots & \vdots & \vdots \\
c+2 & \langle a,c+2\rangle & \langle a+1,c+2\rangle & \ldots & \langle b-1,c+2\rangle & \langle b,c+2\rangle \\
c+1 & \langle a,c+1\rangle & \langle a+1,c+1\rangle & \ldots & \langle b-1,c+1\rangle & \langle b,c+1\rangle \\
c & \langle a,c\rangle & \langle a+1,c\rangle & \ldots & \langle b-1,c\rangle & \langle b,c\rangle \\
\hline
z/y & a & a+1 & \ldots & b-1 & b
\end{array}$$

and we see from this, that the first 2D quadrant finite grid starting at $\langle a,c\rangle$ can have b and d extended *to infinity*, which could then be written $V_2(y \in [a,\infty]; z \in [c,\infty])$.

We give some examples now.

$$V_2(y \in [-1,1]; z \in [-1,1]) =$$

$$\begin{array}{c|ccc}
1 & \langle -1,1\rangle & \langle 0,1\rangle & \langle 1,1\rangle \\
0 & \langle -1,0\rangle & \langle 0,0\rangle & \langle 1,0\rangle \\
-1 & \langle -1,-1\rangle & \langle 0,-1\rangle & \langle 1,-1\rangle \\
\hline
z/y & -1 & 0 & 1
\end{array}$$

$$V_2(y \in [-2,2]; z \in [-2,2]) =$$

$$\begin{array}{c|ccccc}
2 & \langle -2,2\rangle & \langle -1,2\rangle & \langle 0,2\rangle & \langle 1,2\rangle & \langle 2,2\rangle \\
1 & \langle -2,1\rangle & \langle -1,1\rangle & \langle 0,1\rangle & \langle 1,1\rangle & \langle 2,1\rangle \\
0 & \langle -2,0\rangle & \langle -1,0\rangle & \langle 0,0\rangle & \langle 1,0\rangle & \langle 2,0\rangle \\
-1 & \langle -2,-1\rangle & \langle -1,-1\rangle & \langle 0,-1\rangle & \langle 1,-1\rangle & \langle 2,-1\rangle \\
-2 & \langle -2,-2\rangle & \langle -1,-2\rangle & \langle 0,-2\rangle & \langle 1,-2\rangle & \langle 2,-2\rangle \\
\hline
z/y & -2 & -1 & 0 & 1 & 2
\end{array}$$

In our work on vector partitions, it will be a standard approach that we specify the vector grid, and then based on the type of partition function applied to the grid, specify a corresponding partition grid. Hence our partition grid definition is as follows.

DEFINITION The first quadrant 2D **partition grid** is the set of 2D partitions denoted

$$p_2(V_2(\langle m,n\rangle \subseteq \{y \in [a,b] \cup z \in [c,d]\}))$$

of $\langle a,b\rangle$ such that a and b are positive integers. We say that a **partition grid** is a range of a partition function applied to a **vector grid** domain in a defined region. For example, in the following picture, the first quadrant 2D partition grid is split into an $\overbrace{\textit{overlined Upper}}$ partition grid, a *Diagonal upslope* grid, and an $\underbrace{\textit{underlined Lower}}$ partition grid.

First quadrant 2D partition grid.

$\vdots$	$\vdots$	$\vdots$	$\vdots$	$\vdots$	$\vdots$	$\vdots$	$\vdots$	$\vdots$	$\vdots$	$\ddots$
8	$\overbrace{p_2(0,8)}$	$\overbrace{p_2(1,8)}$	$\overbrace{p_2(2,8)}$	$\overbrace{p_2(3,8)}$	$\overbrace{p_2(4,8)}$	$\overbrace{p_2(5,8)}$	$\overbrace{p_2(6,8)}$	$\overbrace{p_2(7,8)}$	$p_2(8,8)$	$\cdots$
7	$\overbrace{p_2(0,7)}$	$\overbrace{p_2(1,7)}$	$\overbrace{p_2(2,7)}$	$\overbrace{p_2(3,7)}$	$\overbrace{p_2(4,7)}$	$\overbrace{p_2(5,7)}$	$\overbrace{p_2(6,7)}$	$p_2(7,7)$	$\underbrace{p_2(8,7)}$	$\cdots$
6	$\overbrace{p_2(0,6)}$	$\overbrace{p_2(1,6)}$	$\overbrace{p_2(2,6)}$	$\overbrace{p_2(3,6)}$	$\overbrace{p_2(4,6)}$	$\overbrace{p_2(5,6)}$	$p_2(6,6)$	$\underbrace{p_2(7,6)}$	$\underbrace{p_2(8,6)}$	$\cdots$
5	$\overbrace{p_2(0,5)}$	$\overbrace{p_2(1,5)}$	$\overbrace{p_2(2,5)}$	$\overbrace{p_2(3,5)}$	$\overbrace{p_2(4,5)}$	$p_2(5,5)$	$\underbrace{p_2(6,5)}$	$\underbrace{p_2(7,5)}$	$\underbrace{p_2(8,5)}$	$\cdots$
4	$\overbrace{p_2(0,4)}$	$\overbrace{p_2(1,4)}$	$\overbrace{p_2(2,4)}$	$\overbrace{p_2(3,4)}$	$p_2(4,4)$	$\underbrace{p_2(5,4)}$	$\underbrace{p_2(6,4)}$	$\underbrace{p_2(7,4)}$	$\underbrace{p_2(8,4)}$	$\cdots$
3	$\overbrace{p_2(0,3)}$	$\overbrace{p_2(1,3)}$	$\overbrace{p_2(2,3)}$	$p_2(3,3)$	$\underbrace{p_2(4,3)}$	$\underbrace{p_2(5,3)}$	$\underbrace{p_2(6,3)}$	$\underbrace{p_2(7,3)}$	$\underbrace{p_2(8,3)}$	$\cdots$
2	$\overbrace{p_2(0,2)}$	$\overbrace{p_2(1,2)}$	$p_2(2,2)$	$\underbrace{p_2(3,2)}$	$\underbrace{p_2(4,2)}$	$\underbrace{p_2(5,2)}$	$\underbrace{p_2(6,2)}$	$\underbrace{p_2(7,2)}$	$\underbrace{p_2(8,2)}$	$\cdots$
1	$\overbrace{p_2(0,1)}$	$p_2(1,1)$	$\underbrace{p_2(2,1)}$	$\underbrace{p_2(3,1)}$	$\underbrace{p_2(4,1)}$	$\underbrace{p_2(5,1)}$	$\underbrace{p_2(6,1)}$	$\underbrace{p_2(7,1)}$	$\underbrace{p_2(8,1)}$	$\cdots$
0	$p_2(0,0)$	$\underbrace{p_2(1,0)}$	$\underbrace{p_2(2,0)}$	$\underbrace{p_2(3,0)}$	$\underbrace{p_2(4,0)}$	$\underbrace{p_2(5,0)}$	$\underbrace{p_2(6,0)}$	$\underbrace{p_2(7,0)}$	$\underbrace{p_2(8,0)}$	$\cdots$
z/y	0	1	2	3	4	5	6	7	8	$\cdots$

Depending on how the partition function is defined, it may be that the set of all *Upper* partitions can be mapped to the set of all *Lower* partitions. The *Diagonal* upslope is defined here by $y = z$. We may often seek out or observe these features and apply them to generating functions for vector partitions in 2D, or generalized to 3D, 4D, and so on. As an analogy with partitions into integers having Ferrers graphs, vector partitions can have generating functions over different grids, yielding theorems on vector partitions that arise simply from different equivalences of algebraic rational functions of several variables. What we often call weighted vector partitions are in fact functions mapped from the vector partition sums.

For example, a 2D vector sum may map to a weighted value as in,

$$\langle 1,2\rangle + \langle 1,3\rangle + \langle 2,3\rangle = \langle 4,8\rangle \quad \mapsto \quad f_2(4,8) = \binom{2}{1}\binom{3}{1}\binom{3}{2},$$

so a vector partition sum may map to a *function of the sum of that partition* $\langle 4,8\rangle$, the right side showing the function as a product of binomial coefficients in this instance.

We now try to lay the foundation for the language of vector partitions needed to progress our work in this book.

⋮	⋮	⋮	⋮	⋮	⋮	⋮	⋮	⋮	⋮	⋱
8	1	1	1	1	1	1	1	1	1	⋯
7	1	1	1	1	1	1	1	1	1	⋯
6	1	1	1	1	1	1	1	1	1	⋯
5	1	1	1	1	1	1	1	1	1	⋯
4	1	1	1	1	1	1	1	1	1	⋯
3	1	1	1	1	1	1	1	1	1	⋯
2	1	1	1	1	1	1	1	1	1	⋯
1	1	1	1	1	1	1	1	1	1	⋯
0	1	1	1	1	1	1	1	1	1	⋯
z/y	0	1	2	3	4	5	6	7	8	⋯

and the generating function for this, say $\hat{o}_2(a,b)$, for $|y|$ and $|z|$ both less than 1, is

$$\sum_{a,b\geq 0} \hat{o}_2(a,b) y^a z^b = \sum_{a,b\geqq 0} y^a z^b = \frac{1}{(1-y)(1-z)}. \tag{12.3.1}$$

So we have described the simplest vector partition scenario, where there is only one part in each partition.

The next simplest case is then the scenario where each partition $\langle a,b\rangle$ is into exactly two parts, based on the vector grid for a and b both greater than or equal to 1. An arbitrary example where the vector partition is into two parts is for $\langle 5,2\rangle$,

$$\langle 5,2\rangle = \langle 4,1\rangle + \langle 1,1\rangle = \langle 3,1\rangle + \langle 2,1\rangle.$$

A reminder here, that *compositions* count different orders of summands whereas *partitions* do not. We recall that partitions are partially ordered sets, so these two vector equations are the only two partitions into two parts using vectors from the 2D first quadrant grid.

12.4 PARTITIONS INTO EXACTLY TWO PARTS, AND EXACTLY THREE PARTS

It is easy to find the number of partitions of 2D first quadrant vectors into exactly two parts, by a process of observation, then by formalizing this. However, a much neater approach is given from generating functions in chapter 13 leading to a determinant form solution given here.

Theorem 12.1 *The number of first quadrant 2D partitions of vectors into two parts* $\spadesuit_2(a,b)$*, and the number of first quadrant 2D partitions of vectors into three parts* $\clubsuit_2(a,b)$*, are generated by the following equations valid for* $|y| < 1$, $|z| < 1$,

$$\sum_{a,b\geq 0}^{\infty} \spadesuit_2(a,b) y^a z^b = \frac{1}{2!} \begin{vmatrix} \frac{1}{(1-y)(1-z)} & -1 \\ \frac{1}{(1-y^2)(1-z^2)} & \frac{1}{(1-y)(1-z)} \end{vmatrix}$$

$$= \frac{(1+yz)}{2(1+y)(1+z)(1-y^2)^2(1-z^2)^2};$$

and

$$\sum_{a,b\geq 0}^{\infty} \clubsuit_2(a,b)y^a z^b = \frac{1}{3!}\begin{vmatrix} \frac{1}{(1-y)(1-z)} & -1 & 0 \\ \frac{1}{(1-y^2)(1-z^2)} & \frac{1}{(1-y)(1-z)} & -2 \\ \frac{1}{(1-y^3)(1-z^3)} & \frac{1}{(1-y^2)(1-z^2)} & \frac{1}{(1-y)(1-z)} \end{vmatrix}$$

$$= \frac{y^3z^3+y^2z^2+y^2z+yz^2+yz+1}{(1-y)^3(1+y)(y^2+y+1)(1-z)^3(1+z)(z^2+z+1)}.$$

The two related partition grids for these are:

for $\spadesuit_2(a,b)$,

First quadrant 2D partitions of vectors into two parts $\spadesuit_{\mathbf{2}}(\mathbf{a},\mathbf{b})$

⋮	⋮	⋮	⋮	⋮	⋮	⋮	⋮	⋮	⋮	⋱
8	5	9	14	18	23	27	32	36	**41**	⋯
7	4	8	12	16	20	24	28	**32**	36	⋯
6	4	7	11	14	18	21	**25**	28	32	⋯
5	3	6	9	12	15	**18**	21	24	27	⋯
4	3	5	8	10	**13**	15	18	20	23	⋯
3	2	4	6	**8**	10	12	14	16	18	⋯
2	2	3	**5**	6	8	9	11	12	14	⋯
1	1	**2**	3	4	5	6	7	8	9	⋯
0	**1**	1	2	2	3	3	4	4	5	⋯
$z\uparrow y\rightarrow$	0	1	2	3	4	5	6	7	8	⋯

and for $\clubsuit_2(a,b)$,

First quadrant 2D partitions of vectors into three parts $\clubsuit_{\mathbf{2}}(\mathbf{a},\mathbf{b})$

⋮	⋮	⋮	⋮	⋮	⋮	⋮	⋮	⋮	⋮	⋱
8	10	25	50	80	120	165	220	280	**350**	⋯
7	8	20	40	64	96	132	176	**224**	280	⋯
6	7	16	32	51	76	104	**139**	176	220	⋯
5	5	12	24	38	57	**78**	104	132	165	⋯
4	4	9	18	28	**42**	57	76	96	120	⋯
3	3	6	12	**19**	28	38	51	64	80	⋯
2	2	4	**8**	12	18	24	32	12	40	⋯
1	1	**2**	4	6	9	12	16	20	25	⋯
0	**1**	1	2	3	4	5	7	8	10	⋯
$z\uparrow y\rightarrow$	0	1	2	3	4	5	6	7	8	⋯

Some notes on the partition grid sequences in theorem 12.1
Observations on the diagonal sequences, and symmetries:

Observation 1: The **bold** diagonal sequence 1,2,5,8,13,18,25,... in the $\spadesuit_2(a,b)$ grid is easily found in the OEIS as sequence number A000982 and is given by the ceiling function, so we have the exact formula $\spadesuit_2(n,n) = \lceil \frac{(n+1)^2}{2} \rceil$. In plain words this means "The number of 2D vector partitions of $\langle n,n \rangle$ into exactly two parts where n is a positive integer is equal to $\lceil \frac{(n+1)^2}{2} \rceil$."

Observation 2: The $\spadesuit_2(a,b)$ grid diagonal has generating function valid for $|q| < 1$,

$$\frac{1+q^2}{(1+q)(1-q)^3} = 1+2q+5q^2+8q^3+13q^4+18q^5+25q^6+\dots.$$

Observation 3: The sequence $\spadesuit_2(n,n)$ is also given as a *floor* function by $\lfloor \frac{n^2+1}{2} \rfloor$

Observation 4: If we partition n in two parts, say r and s so that r^2+s^2 is minimal, then $\spadesuit_2(n,n) = r^2+s^2$. Geometrical significance: folding a rod with length n units at right angles in such a way that the endpoints are at the least distance, which is given by $\sqrt{\spadesuit_2(n,n)}$ as the hypotenuse of a right triangle with the sum of the base and height $= n$ units.

Observation 5: *A Dirichlet summation.* With respect to the diagonal sequence $\spadesuit_2(n,n)$ it is known that

$$\sum_{n>0} \frac{1}{\lfloor \frac{n^2+1}{2} \rfloor} = \sum_{n>0} \frac{1}{2n^2} + \sum_{n\geq 0} \frac{1}{2n^2+2n+1}$$

$$1+\frac{1}{2}+\frac{1}{5}+\frac{1}{8}+\frac{1}{13}+\frac{1}{18}+\frac{1}{25}+\dots = \frac{1}{2}\left(\frac{\pi^2}{6}+\pi \tanh\left(\frac{\pi}{2}\right)\right) = 2.26312655....$$

Observation 6: The **bold** diagonal sequence 1,2,8,19,42,78,139,224,350,... in the $\clubsuit_2(a,b)$ grid is found in the OEIS sequence number A101427, and is given by the following cases,

$$\clubsuit_2(n,n) = \begin{cases} ((n+2)^2(n+1)^2+12(\lfloor n/2 \rfloor+1)^2+8)/24, & \text{for } n \text{ divisible by 3;} \\ ((n+2)^2(n+1)^2+12(\lfloor n/2 \rfloor+1)^2)/24, & \text{otherwise.} \end{cases}$$

Therefore we have an exact formula for $\clubsuit_2(n,n)$; the number of 2D vector partitions of $\langle n,n \rangle$ into exactly three parts; for n any positive integer. The description of this sequence A101427 from OEIS is "Number of different cuboids with volume $(pq)^n$, where p, q are distinct prime numbers".

As an illustrative example of the formula, the first ten terms of $((n+2)^2(n+1)^2+12(\lfloor n/2 \rfloor+1)^2+8)/24$ are $1, \frac{7}{3}, \frac{25}{3}, 19, \frac{127}{3}, \frac{235}{3}, 139, \frac{673}{3}, \frac{1051}{3}, 517$, and the first ten terms of $((n+2)^2(n+1)^2+12(\lfloor n/2 \rfloor+1)^2)/24$ are $\frac{2}{3}, 2, 8, \frac{56}{3}, 42, 78, \frac{416}{3}, 224, 350, \frac{1550}{3}$.

Observation 7: The $\clubsuit_2(a,b)$ grid diagonal has generating function valid for $|q|<1$,

$$\frac{q^6+3q^4+4q^3+3q^2+1}{(1-q^3)(1-q^2)^2(1-q)^2} = 1+2q+8q^2+19q^3+42q^4+78q^5+139q^6+\ldots.$$

Observation 8: *Horizontal rows and Vertical columns sequences generating functions.*

Row and column sequences for $\spadesuit_2(a,b)$ and $\clubsuit_2(a,b)$ have certain observable properties worth mentioning:

(a) The jth row sequence is the same as the jth column sequence.

(b) The kth term in the jth row sequence is the same as k term in the jth column sequence.

(c) That is, $\spadesuit_2(a,b) = \spadesuit_2(b,a)$ and $\clubsuit_2(a,b) = \clubsuit_2(b,a)$.

(d) The generating function for the jth row sequence is the same as the generating function for the jth column sequence.

PROOF OF THEOREM 12.1.

In Chapter 13, we have the following 2D corollary of the main nD extension of Cauchy's q-binomial theorem, which is equation (13.3.3) in that chapter. The proof of this nD theorem used Cramer's Rule and gave us the determinants as coefficients. So, setting $a=0$ in identity (13.3.3) of Chapter 13 we obtain, for y and z both less than unity,

$$F_2(y,z;0,t) = \prod_{j,k\geq 0} \frac{1}{1-y^j z^k t} \tag{12.4.1}$$

$$= 1 + \frac{1}{(1-y)(1-z)}\frac{t}{1!} + \begin{vmatrix} \frac{1}{(1-y)(1-z)} & -1 \\ \frac{1}{(1-y^2)(1-z^2)} & \frac{1}{(1-y)(1-z)} \end{vmatrix} \frac{t^2}{2!}$$

$$+ \begin{vmatrix} \frac{1}{(1-y)(1-z)} & -1 & 0 \\ \frac{1}{(1-y^2)(1-z^2)} & \frac{1}{(1-y)(1-z)} & -2 \\ \frac{1}{(1-y^3)(1-z^3)} & \frac{1}{(1-y^2)(1-z^2)} & \frac{1}{(1-y)(1-z)} \end{vmatrix} \frac{t^3}{3!}$$

$$+ \begin{vmatrix} \frac{1}{(1-y)(1-z)} & -1 & 0 & 0 \\ \frac{1}{(1-y^2)(1-z^2)} & \frac{1}{(1-y)(1-z)} & -2 & 0 \\ \frac{1}{(1-y^3)(1-z^3)} & \frac{1}{(1-y^2)(1-z^2)} & \frac{1}{(1-y)(1-z)} & -3 \\ \frac{1}{(1-y^4)(1-z^4)} & \frac{1}{(1-y^3)(1-z^3)} & \frac{1}{(1-y^2)(1-z^2)} & \frac{1}{(1-y)(1-z)} \end{vmatrix} \frac{t^4}{4!}$$

$$+ \ etc.$$

The proof of the theorem comes from recognizing that $♠_2(b,a)$ and $♣_2(a,b)$ are the coefficients of t^2 and t^3 respectively in (12.4.1). The determinants are simplified by applying from Chapter 31, identities (31.1.1) and (31.1.2) there, setting

$$a_1 = \frac{1}{1-q}, \quad a_2 = \frac{1}{1-q^2}, \quad a_3 = \frac{1}{1-q^3},$$

substituted into

$$\begin{vmatrix} a_1 & -1 \\ a_2 & a_1 \end{vmatrix} = a_1{}^2 + a_2;$$

and

$$\begin{vmatrix} a_1 & -1 & 0 \\ a_2 & a_1 & -2 \\ a_3 & a_2 & a_1 \end{vmatrix} = a_2{}^3 + 3a_1a_2 + 2a_3$$

respectively. This gives us the appropriate functions generating the partitions $♠_2(b,a)$ and $♣_2(a,b)$. Expanding these generating functions allows us to fill in the relevant partition grids. ■

12.5 FIRST 3D HYPERQUADRANT TABLEAU REDUCED TO A 2D TABLEAU

The set of all lattice points in 3D space first hyperquadrant (extended cube) are similarly a countable set. The lattice point vectors in nD space first hyperquadrant comprised of points $\langle a_1, a_2, a_3, ..., a_n\rangle$ with each a_i a positive integer is a countable set.

In 3D the set of first hyperquadrant points can be portrayed as shown below, as 2D horizontal layers or lamina, each one as follows.

First hyperquadrant 3D vector grid as 2D layers

$\vdots$	$\vdots$	$\vdots$	$\vdots$	$\vdots$	$\vdots$	$\vdots$	$\cdot^{\cdot^{\cdot}}$
6	$\langle 1,6,a\rangle$	$\langle 2,6,a\rangle$	$\langle 3,6,a\rangle$	$\langle 4,6,a\rangle$	$\langle 5,6,a\rangle$	$\langle 6,6,a\rangle$	$\cdots$
5	$\langle 1,5,a\rangle$	$\langle 2,5,a\rangle$	$\langle 3,5,a\rangle$	$\langle 4,5,a\rangle$	$\langle 5,5,a\rangle$	$\langle 6,5,a\rangle$	$\cdots$
4	$\langle 1,4,a\rangle$	$\langle 2,4,a\rangle$	$\langle 3,4,a\rangle$	$\langle 4,4,a\rangle$	$\langle 5,4,a\rangle$	$\langle 6,4,a\rangle$	$\cdots$
3	$\langle 1,3,a\rangle$	$\langle 2,3,a\rangle$	$\langle 3,3,a\rangle$	$\langle 4,3,a\rangle$	$\langle 5,3,a\rangle$	$\langle 6,3,a\rangle$	$\cdots$
2	$\langle 1,2,a\rangle$	$\langle 2,2,a\rangle$	$\langle 3,2,a\rangle$	$\langle 4,2,a\rangle$	$\langle 5,2,a\rangle$	$\langle 6,2,a\rangle$	$\cdots$
1	$\langle 1,1,a\rangle$	$\langle 2,1,a\rangle$	$\langle 3,1,a\rangle$	$\langle 4,1,a\rangle$	$\langle 5,1,a\rangle$	$\langle 6,1,a\rangle$	$\cdots$
$y\uparrow, x\rightarrow, z=a$	1	2	3	4	5	6	$\cdots$

(12.5.1)

where layer $a \in \{1,2,3,\ldots\}$.

In order to count each lattice point vector in the entire first 3D hyperquadrant, it suffices to count the following vectors $\langle a,b,c\rangle$ such that $0 < a < b < c$ and apply symmetries. We can start with an arrangement like

$$\begin{array}{llllllll}
\text{etc.} & & & & & & & \\
\langle 1,3,6\rangle & \langle 2,3,7\rangle & \langle 3,6,7\rangle & \langle 4,6,8\rangle & \langle 5,6,9\rangle & \langle 6,8,9\rangle & \langle 7,8,10\rangle & \langle 8,9,10\rangle \\
\langle 1,2,6\rangle & \langle 2,5,6\rangle & \langle 3,5,7\rangle & \langle 4,5,8\rangle & \langle 5,7,8\rangle & \langle 6,7,9\rangle & \langle 7,8,9\rangle & \\
\langle 1,4,5\rangle & \langle 2,4,6\rangle & \langle 3,4,7\rangle & \langle 4,6,7\rangle & \langle 5,6,8\rangle & \langle 6,7,8\rangle & & \\
\langle 1,3,5\rangle & \langle 2,3,6\rangle & \langle 3,5,6\rangle & \langle 4,5,7\rangle & \langle 5,6,7\rangle & & & \\
\langle 1,2,5\rangle & \langle 2,4,5\rangle & \langle 3,4,6\rangle & \langle 4,5,6\rangle & & & & \\
\langle 1,3,4\rangle & \langle 2,3,5\rangle & \langle 3,4,5\rangle & & & & & \\
\langle 1,2,4\rangle & \langle 2,3,4\rangle & & & & & & \\
\langle 1,2,3\rangle & & & & & & &
\end{array} \tag{12.5.2}$$

where the bottom entry for "column a" is $\langle a,a+1,a+2\rangle$ and each column has terms $\bigcup_{a<j<k}\langle a,j,k\rangle$ in an order that exhausts successive k integer terms with respect to the j terms between a and k. The above arrangement of 3D vectors enables us to write the generating function

Theorem 12.2

$$\prod_{0<a<b<c}(1+x^a y^b z^c) = 1 + \sum_{0<i<j<k} p_3(\mathfrak{D};i,j,k)x^i y^j z^k \tag{12.5.3}$$

$$\begin{aligned}
= {} & (1+x^1y^2z^3) \\
& (1+x^1y^2z^4)(1+x^2y^3z^4) \\
& (1+x^1y^3z^4)(1+x^2y^3z^5)(1+x^3y^4z^5) \\
& (1+x^1y^2z^5)(1+x^2y^4z^5)(1+x^3y^4z^6)(1+x^4y^5z^6) \\
& (1+x^1y^3z^5)(1+x^2y^3z^6)(1+x^3y^5z^6)(1+x^4y^5z^7)(1+x^5y^6z^7) \\
& (1+x^1y^4z^5)(1+x^2y^4z^6)(1+x^3y^4z^7)(1+x^4y^6z^7)(1+x^5y^6z^8)(1+x^6y^7z^8) \\
& (1+x^1y^2z^6)(1+x^2y^5z^6)(1+x^3y^5z^7)(1+x^4y^5z^8)(1+x^5y^7z^8)(1+x^6y^7z^9)(1+x^7y^8z^9) \\
& \textit{etc.}
\end{aligned} \tag{12.5.4}$$

where $p_3(\mathfrak{D};i,j,k)$ *is the number of 3D partitions of* $\langle i,j,k\rangle$ *into distinct vector parts of type* $\langle a,b,c\rangle$ *such that* $0<a<b<c$ *with positive integer values of* a*,* b*, and* c*. In other words,* $p_3(\mathfrak{D};i,j,k)$ *is the number of 3D partitions into distinct parts from (12.5.2).*

As a logical companion to theorem 12.2, also easily inferred from our reduction of 3D tableau to 2D tableau (12.5.2) is the following

Theorem 12.3

$$\prod_{0<a<b<c}\frac{1}{(1-x^a y^b z^c)} = 1 + \sum_{0<i<j<k} p_3(\mathfrak{U};i,j,k)x^i y^j z^k \tag{12.5.5}$$

$$= \cfrac{1}{(1-x^1y^2z^3)} \\ \cfrac{1}{(1-x^1y^2z^4)(1-x^2y^3z^4)} \\ \cfrac{1}{(1-x^1y^3z^4)(1-x^2y^3z^5)(1-x^3y^4z^5)} \\ \cfrac{1}{(1-x^1y^2z^5)(1-x^2y^4z^5)(1-x^3y^4z^6)(1-x^4y^5z^6)} \\ \cfrac{1}{(1-x^1y^3z^5)(1-x^2y^3z^6)(1-x^3y^5z^6)(1-x^4y^5z^7)(1-x^5y^6z^7)} \\ \cfrac{1}{(1-x^1y^4z^5)(1-x^2y^4z^6)(1-x^3y^4z^7)(1-x^4y^6z^7)(1-x^5y^6z^8)(1-x^6y^7z^8)} \\ \cfrac{1}{(1-x^1y^2z^6)(1-x^2y^5z^6)(1-x^3y^5z^7)(1-x^4y^5z^8)(1-x^5y^7z^8)(1-x^6y^7z^9)(1-x^7y^8z^9)} \\ \text{etc.} \tag{12.5.6}$$

where $p_3(\mathfrak{U}; i,j,k)$ is the number of 3D partitions of $\langle i,j,k\rangle$ into unrestricted vector parts of type $\langle a,b,c\rangle$ such that $0<a<b<c$ with positive integer values of a, b, and c. In other words, $p_3(\mathfrak{U}; i,j,k)$ is the number of 3D partitions into unrestricted parts from (12.5.2).

An interesting exercise here is to also note the relationships between

$$\Phi(x,y,z) = \prod_{0<a<b<c} \frac{1}{(1-x^ay^bz^c)}$$

and

$$\Psi(x,y,z) = \prod_{\substack{0<a<b<c \\ gcd(a,b,c)=1}} \frac{1}{(1-x^ay^bz^c)}$$

evidently bringing a connection with the Mordell-Tornheim-Witten (MTW) zeta-function.

On the one hand, we have

$$\prod_{k=1}^{\infty} \Psi(x^k,y^k,z^k) = \Phi(x,y,z).$$

There are some as yet unpublished 3D identities that would relate to this $\Psi(x^k,y^k,z^k)$ to $\Phi(x,y,z)$ relationship. A good place to start with the underpinning theory for this is, for example, the skewed hyperpyramid theorem from Chapter 24 given briefly here as

Theorem 12.4 ***The skewed hyperpyramid VPV identity.*** *If $i = 1,2,3,...,n$ then for each $x_i \in \mathbb{C}$ such that $|x_i| < 1$ and $b_i \in \mathbb{C}$ such that $\sum_{i=1}^{n} b_i = 1$, and in*

addition for $\alpha_i \in \Re^+$, we have

$$\exp\left\{\left(\sum_{k=1}^{\infty}\frac{{x_n}^k}{k^{b_n}}\right)\right\} \prod_{\substack{gcd(a_1,a_2,\ldots,a_n)=1\\ a_1\alpha_1,a_2\alpha_2,\ldots,a_{n-1}\alpha_{n-1}<a_n\\ a_1,a_2,\ldots,a_{n-1}>0\\ a_n>1}} \left(\frac{1}{1-{x_1}^{a_1}{x_2}^{a_2}{x_3}^{a_3}\cdots{x_n}^{a_n}}\right)^{\frac{1}{{a_1}^{b_1}{a_2}^{b_2}{a_3}^{b_3}\cdots{a_n}^{b_n}}} \tag{12.5.7}$$

$$= \exp\left\{\sum_{k=1}^{\infty}\prod_{i=1}^{n-1}\left(\sum_{j=1}^{F(k)}\frac{{x_i}^j}{j^{b_i}}\right)\frac{{x_n}^k}{k^{b_n}}\right\},$$

where

$$F(k) = \begin{cases} \left[\frac{k-1}{\alpha_i}\right], & \alpha_i > 1; \\ k-1, & \alpha_i = 1; \\ \left[\frac{k}{\alpha_i} - 1\right], & \alpha_i < 1. \end{cases}$$

A 3D version example identity is given by

Corollary 12.1 *For each of $|x|, |y|, |z| < 1$,*

$$\prod_{\substack{(a,b,c)=1\\ \frac{a}{5},\frac{b}{7}<c\\ a,b\geq 0, c>0}} \left(\frac{1}{1-x^a y^b z^c}\right)^{\frac{1}{c}} = \left(\frac{(1-x^5 z)(1-y^7 z)}{(1-z)(1-x^5 y^7 z)}\right)^{\frac{1}{(1-x)(1-y)}} \tag{12.5.8}$$

$$= 1 + \frac{(1-x^5)(1-y^7)}{(1-x)(1-y)}\frac{z}{1!} + \begin{vmatrix} \frac{(1-x^5)(1-y^7)}{(1-x)(1-y)} & -1 \\ \frac{(1-x^{10})(1-y^{14})}{(1-x)(1-y)} & \frac{(1-x^5)(1-y^7)}{(1-x)(1-y)} \end{vmatrix} \frac{z^2}{2!}$$

$$+ \begin{vmatrix} \frac{(1-x^5)(1-y^7)}{(1-x)(1-y)} & -1 & 0 \\ \frac{(1-x^{10})(1-y^{14})}{(1-x)(1-y)} & \frac{(1-x^5)(1-y^7)}{(1-x)(1-y)} & -2 \\ \frac{(1-x^{30})(1-y^{21})}{(1-x)(1-y)} & \frac{(1-x^{10})(1-y^{14})}{(1-x)(1-y)} & \frac{(1-x^5)(1-y^7)}{(1-x)(1-y)} \end{vmatrix} \frac{z^3}{3!}$$

$$+ \begin{vmatrix} \frac{(1-x^5)(1-y^7)}{(1-x)(1-y)} & -1 & 0 & 0 \\ \frac{(1-x^{10})(1-y^{14})}{(1-x)(1-y)} & \frac{(1-x^5)(1-y^7)}{(1-x)(1-y)} & -2 & 0 \\ \frac{(1-x^{30})(1-y^{21})}{(1-x)(1-y)} & \frac{(1-x^{10})(1-y^{14})}{(1-x)(1-y)} & \frac{(1-x^5)(1-y^7)}{(1-x)(1-y)} & -3 \\ \frac{(1-x^{40})(1-y^{28})}{(1-x)(1-y)} & \frac{(1-x^{30})(1-y^{21})}{(1-x)(1-y)} & \frac{(1-x^{10})(1-y^{14})}{(1-x)(1-y)} & \frac{(1-x^5)(1-y^7)}{(1-x)(1-y)} \end{vmatrix} \frac{z^4}{4!} + etc.$$

In this identity, the determinant coefficients are found by applying Cramer's Rule to the coefficient recurrence relation arrived at by logarithmic differentiation of both sides of (12.5.8).

12.6 VECTOR PARTITIONS WHOSE PARTS ARE ON LINES IN nD.

In this section, we show that integer partition generating functions map one-to-one onto the number of "n-dimensional space" vector partitions whose parts are in a straight line radial from the origin. For example, from the familiar unrestricted integer partitions generating function, for $|q| < 1$

$$\prod_{n=1}^{\infty} \frac{1}{1-q^n} = \prod_{n=1}^{\infty} \left(1+q^{1n}+q^{2n}+q^{3n}+\ldots\right) = \sum_{n=0}^{\infty} p(n)q^n$$

we see that replacing q by $y^a z^b$ having $|y^a z^b| < 1$ where a and b are coprime positive integers, gives the equation

$$\prod_{n=1}^{\infty} \frac{1}{1-(y^a z^b)^n} = \prod_{n=1}^{\infty} \left(1+(y^a z^b)^{1n}+(y^a z^b)^{2n}+(y^a z^b)^{3n}+\ldots\right) \tag{12.6.1}$$

$$= 1+p(1)(y^a z^b)^1+p(2)(y^a z^b)^2+p(3)(y^a z^b)^3+\ldots.$$

Of course, this is just a thinly disguised version of the generating function for unrestricted integer partitions. ie. one-dimensional partitions. In the 2D case, we are saying that the number of partitions of an integer lattice point vector $\langle A, B\rangle$ on the line $z = ay/b$ for $gcd(a,b) = 1$ into vector parts also on this line, is equal to $p(gcd(A,B))$.

So (12.6.1), enumerating the number of vector partitions for the lattice points along a 1D line in 2D space, applies equally to a 1D line in 3D space as follows, where a, b and c are positive integers with $gcd(a,b,c) = 1$.

$$\prod_{n=1}^{\infty} \frac{1}{1-(x^a y^b z^c)^n} = \prod_{n=1}^{\infty} \left(1+(x^a y^b z^c)^{1n}+(x^a y^b z^c)^{2n}+(x^a y^b z^c)^{3n}+\ldots\right) \tag{12.6.2}$$

$$= 1+p(1)(x^a y^b z^c)^1+p(2)(x^a y^b z^c)^2+p(3)(x^a y^b z^c)^3+\ldots.$$

Similarly, in the 4D case extension corresponding to the 3D equation (12.6.2) we have, where a, b, c and d are positive integers with gcd(a,b,c,d)=1,

$$\prod_{n=1}^{\infty} \frac{1}{1-(w^a x^b y^c z^d)^n} \tag{12.6.3}$$

$$= \prod_{n=1}^{\infty} \left(1+(w^a x^b y^c z^d)^{1n}+(w^a x^b y^c z^d)^{2n}+(w^a x^b y^c z^d)^{3n}+\ldots\right)$$

$$= 1+p(1)(w^a x^b y^c z^d)^1+p(2)(w^a x^b y^c z^d)^2+p(3)(w^a x^b y^c z^d)^3+\ldots.$$

From considering the above analysis of the vector partitions into parts contained on a 2D, 3D, or 4D straight line through the multidimensional cartesian space, we can state the following

Theorem 12.5 *Take an nD lattice point vector of fixed non-negative integers* $\langle a_1, a_2, ..., a_n \rangle$ *(not all zero) where* $\gcd(a_1, a_2, ..., a_n)$*, and this defines a straight line from the origin through the n-space. Suppose then, that all lattice point vectors on that line are given by the notation*

$$\mathbb{L}_n(a_1, a_2, ..., a_n) = \{\langle a_1, a_2, ..., a_n \rangle, \langle 2a_1, 2a_2, ..., 2a_n \rangle, \langle 3a_1, 3a_2, ..., 3a_n \rangle, \ldots\}.$$

Similarly define for "n-tuple coprime" the lattice point vector $\langle b_1, b_2, ..., b_n \rangle$*; and for "n-tuple coprime"* $\langle c_1, c_2, ..., c_n \rangle$*; integer lattice point vectors on the two lines through the n-space, as follows:*

$$\mathbb{L}_n(b_1, b_2, ..., b_n) = \{\langle b_1, b_2, ..., b_n \rangle, \langle 2b_1, 2b_2, ..., 2b_n \rangle, \langle 3b_1, 3b_2, ..., 3b_n \rangle, \ldots\};$$

$$\mathbb{L}_n(c_1, c_2, ..., c_n) = \{\langle c_1, c_2, ..., c_n \rangle, \langle 2c_1, 2c_2, ..., 2c_n \rangle, \langle 3c_1, 3c_2, ..., 3c_n \rangle, \ldots\}.$$

Let $\mathbb{V}_n(A_1, A_2, ..., A_n)$ *be the number of vector partitions of the n-space vector* $\langle A_1, A_2, ..., A_n \rangle$ *with* $A_1, A_2, ..., A_{n-1}$ *each a non-negative integer and* A_n *a positive integer, into unrestricted parts included only from the lines* $\mathbb{L}_n(a_1, a_2, ..., a_n)$*,* $\mathbb{L}_n(b_1, b_2, ..., b_n)$*, and* $\mathbb{L}_n(c_1, c_2, ..., c_n)$*.*

Then $\mathbb{V}_n(A_1, A_2, ..., A_n)$ *is the coefficient of* ${q_1}^{A_1}{q_2}^{A_2}...{q_n}^{A_n}$ *in either side of the expression,*

$$\prod_{m=1}^{\infty} \frac{1}{(1-({q_1}^{a_1}{q_2}^{a_2}...{q_n}^{a_n})^m)(1-({q_1}^{b_1}{q_2}^{b_2}...{q_n}^{b_n})^m)(1-({q_1}^{c_1}{q_2}^{c_2}...{q_n}^{c_n})^m)} \tag{12.6.4}$$

$$= \left(\sum_{m\geq 0} p(m)({q_1}^{a_1}{q_2}^{a_2}\cdots{q_n}^{a_n})^m\right) \times \left(\sum_{m\geq 0} p(m)({q_1}^{b_1}{q_2}^{b_2}\cdots{q_n}^{b_n})^m\right)$$

$$\times \left(\sum_{m\geq 0} p(m)({q_1}^{c_1}{q_2}^{c_2}\cdots{q_n}^{c_n})^m\right).$$

That is, $\mathbb{V}_n(A_1, A_2, ..., A_n)$ *can be found exactly as a combination of ordinary unrestricted integer partition functions.*

We now proceed to give several examples of the theorem firstly applying unrestricted integer partitions in 2D and 3D situations. After that we give examples with partitions into distinct parts. After this, we show that there are a wide range of integer partition types that can be applied to the parts of the vector partition, such as the various partition types applied in our earlier chapter on congruence properties of partitions; the theory of which goes back to Ramanujan.

2D Vector Partitions whose parts are on two straight lines.

The above analysis is with respect to vector partitions in the upper half first quadrant above and including the line $y = z$. The simplest version of this theory departing from the single radial from the origin line of lattice points would be to state the result if dealing with 2D partitions from two such radial lines of lattice points. For example, consider the two lines with equations $y = z/2$ and $y = z/3$. The lattice point vectors along these lines in the first quadrant may be listed as

$$\begin{aligned} S_1 &= \{\langle 1,2\rangle, \langle 2,4\rangle, \langle 3,6\rangle, \langle 4,8\rangle, \langle 5,10\rangle, \ldots\}; \\ S_2 &= \{\langle 1,3\rangle, \langle 2,6\rangle, \langle 3,9\rangle, \langle 4,12\rangle, \langle 5,15\rangle, \ldots\}. \end{aligned}$$

Following the above rationale we see that the generating function for 2D vector partitions into parts contained in S_1 and S_2 is

$$\frac{1}{((1-yz^2)(1-y^2z^4)(1-y^3z^6)\cdots)(1-yz^3)(1-y^2z^6)(1-y^3z^9)\cdots)}$$

$$= \left(1+p(1)y^1z^2+p(2)y^2z^4+p(3)y^3z^6+\ldots\right)\left(1+p(1)y^1z^3+p(2)y^2z^6+p(3)y^3z^9+\ldots\right)$$

$$\begin{aligned} = \;& 1+p(1)yz^2+p(1)yz^3 \\ & +p(2)y^2z^4+p(1)^2y^2z^5+p(2)y^2z^6 \\ & +p(3)y^3z^6+p(1)p(2)y^3z^7+p(1)p(2)y^3z^8+p(3)y^3z^9 \\ & + \textit{etc.} \end{aligned}$$

$$\begin{aligned} = \;& 1+yz^2+yz^3+2y^2z^4+y^2z^5+y^2(3y+2)z^6 \\ & +2y^3z^7+y^3(5y+2)z^8+3y^3(y+1)z^9+y^4(7y+4)z^{10} \\ & +y^4(5y+3)z^{11}+y^4(11y^2+6y+5)z^{12}+y^5(7y+6)z^{13} \\ & +y^5(14y^2+10y+5)z^{14}+y^5(11y^2+9y+7)z^{15} \\ & + \textit{etc.} \end{aligned}$$

The coefficients here plot onto the grid

⋮	⋮	⋮	⋮	⋮	⋮	⋮	⋮	⋮	⋮
15						7	9	11	⋯
14						5	10	14	⋯
13						6	7		⋯
12					5	6	11		⋯
11					3	5			⋯
10					4	7			⋯
9				3	3				⋯
8				2	5				⋯
7				2					⋯
6			2	3					⋯
5			1						⋯
4			2						⋯
3		1							⋯
2		1							⋯
1									⋯
0	1								⋯
z/y	0	1	2	3	4	5	6	7	⋯

Example interpretations reading from this graph:

1) The number of partitions of vector $\langle 7,15\rangle$ into an unrestricted number of parts from S_1 and S_2 is 11.

2) The number of partitions of vector $\langle 5,10\rangle$ into an unrestricted number of parts from S_1 and S_2 is 7.

3) The number of partitions of vector $\langle 4,9\rangle$ into an unrestricted number of parts from S_1 and S_2 is 3.

4) The number of partitions of vector $\langle 4,7\rangle$ into an unrestricted number of parts from S_1 and S_2 is 0.

3D Vector Partitions whose parts are on two straight lines.

A further simple version of the above approach allows us to work out exactly the number of partitions into vector parts that lie upon "radial from the origin" lines of lattice points. We state the following example in 3D partitions in the first 3D hyperquadrant (ie. with lattice points whose co-ordinates are triples of positive integers.) For example, consider the first line through 3D space with equation $x = y/2 = z/3$; then the second line with equation $x = y/3 = z/4$. The lattice point vectors along these lines in the first quadrant may be listed as

$$\begin{aligned} S_3 &= \{\langle 1,2,3\rangle,\langle 2,4,6\rangle,\langle 3,6,9\rangle,\langle 4,8,12\rangle,\langle 5,10,15\rangle,\ldots\}; \\ S_4 &= \{\langle 1,3,4\rangle,\langle 2,6,8\rangle,\langle 3,9,12\rangle,\langle 4,12,16\rangle,\langle 5,15,20\rangle,\ldots\}. \end{aligned}$$

Applying our rationale we see that the generating function for 3D vector partitions into parts contained in S_3 and S_4 is

$$\frac{1}{((1-xy^2z^3)(1-x^2y^4z^6)(1-x^3y^6z^9)\cdots)(1-xy^3z^4)(1-x^2y^6z^8)(1-x^3y^9z^{12})\cdots)}$$

$$= \left(1+p(1)xy^2z^3+p(2)x^2y^4z^6+p(3)x^3y^6z^9+\ldots\right)$$
$$\times\left(1+p(1)xy^3z^4+p(2)x^2y^6z^8+p(3)x^3y^9z^{12}+\ldots\right)$$

$$\begin{aligned} = \quad & 1+p(1)xy^3z^4+p(1)xy^2z^3 \\ & +p(2)x^2y^6z^8+p(1)^2x^2y^5z^7+p(2)x^2y^4z^6 \\ & +p(3)x^3y^6z^9+p(1)p(2)x^3y^7z^{10}+p(1)p(2)x^3y^8z^11+p(3)x^3y^9z^{12} \\ & +\mathit{etc}. \end{aligned}$$

There are many extended possibilities for the above enumerations of partitions in higher cartesian space sets of lattice point vectors. There is no reason other than a formal complexity why we can't create exact enumerative formulas for "nD Vector Partitions whose parts are on m straight lines" for m and n both arbitrary fixed positive integers.

Distinct Vector Partitions along an n-space line.

Recall that Euler, in addition to giving us the "unrestricted" integer partitions generating function, also noted that for $|x| < 1$,

$$\prod_{k=1}^{\infty}(1+x^k) = \sum_{n=1}^{\infty} \mathcal{D}(n)x^n, \tag{12.6.3}$$

where $\mathcal{D}(n)$ is the number of partitions of positive integer n into distinct integer parts.

From equation (12.6.3) we see that replacing x by $y^a z^b$ for $|y^a z^b| < 1$ where a and b are coprime positive integers, gives the equation

$$\prod_{n=1}^{\infty}(1+(y^a z^b)^n) = 1+\mathcal{D}(1)(y^a z^b)^1+\mathcal{D}(2)(y^a z^b)^2+\mathcal{D}(3)(y^a z^b)^3+\dots. \tag{12.6.4}$$

Similarly to the previous rationale using "unrestricted" partitions, this is a version of the generating function for integer partitions into distinct vector parts. ie. one-dimensional partitions into distinct parts. In the 2D case, we are saying that the number of partitions of an integer lattice point vector $\langle A, B\rangle$ on the line $z = ay/b$ for $gcd(a,b) = 1$ into "distinct" vector parts also on this line, is equal to $\mathcal{D}(gcd(A,B))$.

In a further example, consider again the two lines with equations $y = z/2$ and $y = z/3$. The lattice point vectors along these lines in the first quadrant we list again as

$$\begin{aligned} S_1 &= \{\langle 1,2\rangle, \langle 2,4\rangle, \langle 3,6\rangle, \langle 4,8\rangle, \langle 5,10\rangle, \ldots\}; \\ S_2 &= \{\langle 1,3\rangle, \langle 2,6\rangle, \langle 3,9\rangle, \langle 4,12\rangle, \langle 5,15\rangle, \ldots\}. \end{aligned}$$

Following the above rationale we see that the generating function for 2D vector partitions into distinct parts contained in S_1 and S_2 is

$$\begin{aligned} &((1+yz^2)(1+y^2z^4)(1+y^3z^6)\cdots)((1+yz^3)(1+y^2z^6)(1+y^3z^9)\cdots) \\ &\quad = \left(1+\mathcal{D}(1)y^1z^2+\mathcal{D}(2)y^2z^4+\mathcal{D}(3)y^3z^6+\ldots\right) \\ &\quad\quad \times \left(1+\mathcal{D}(1)y^1z^3+\mathcal{D}(2)y^2z^6+\mathcal{D}(3)y^3z^9+\ldots\right) \end{aligned}$$

$$\begin{aligned} = \; & 1+\mathcal{D}(1)yz^2+\mathcal{D}(1)yz^3 \\ & +\mathcal{D}(2)y^2z^4+\mathcal{D}(1)^2y^2z^5+\mathcal{D}(2)y^2z^6 \\ & +\mathcal{D}(3)y^3z^6+\mathcal{D}(1)\mathcal{D}(2)y^3z^7+\mathcal{D}(1)\mathcal{D}(2)y^3z^8+\mathcal{D}(3)y^3z^9 \\ & +\mathit{etc.} \end{aligned}$$

$$\begin{aligned} = \; & 1+yz^2+yz^3+y^2z^4+y^2z^5+y^2(2y+1)z^6 \\ & +y^3z^7+y^3(2y+1)z^8+2y^3(y+1)z^9+y^4(3y+1)z^{10} \\ & +2y^4(y+1)z^{11}+2y^4(2y^2+y+1)z^{12}+y^5(3y+2)z^13 \\ & +2y^5(2y^2+y+1)z^{14}+y^5(4y^2+4y+3)z^{15} \\ & +\mathit{etc.} \end{aligned}$$

We can easily plot the coefficients here onto the grid

z/y	0	1	2	3	4	5	6	7	$\cdots$
$\vdots$	$\vdots$	$\vdots$	$\vdots$	$\vdots$	$\vdots$	$\vdots$	$\vdots$	$\vdots$	$\vdots$
15						3	4	4	$\cdots$
14						2	2	4	$\cdots$
13						2	3		$\cdots$
12					2	2	4		$\cdots$
11					2	2			$\cdots$
10					1	3			$\cdots$
9				2	2				$\cdots$
8				1	2				$\cdots$
7				1					$\cdots$
6			1	2					$\cdots$
5			1						$\cdots$
4			1						$\cdots$
3		1							$\cdots$
2		1							$\cdots$
1									$\cdots$
0	1								$\cdots$

Example interpretations reading from this graph.

1) The number of partitions of vector $\langle 7, 15\rangle$ using distinct parts from S_1 and S_2 is 11.

2) The number of partitions of vector $\langle 5, 10\rangle$ using distinct parts from S_1 and S_2 is 7.

3) The number of partitions of vector $\langle 4, 9\rangle$ using distinct parts from S_1 and S_2 is 3.

4) The number of partitions of vector $\langle 4, 7\rangle$ using distinct parts from S_1 and S_2 is 0.

12.6.1 Any Integer Partition Theorem Applies to Lines in n-Space

The partition function $p(n)$ for the number of partitions of integer n is generated from the product of infinite series:

$$\begin{aligned}
&\prod_{k=1}^{\infty} \frac{1}{1-x^k} \\
= \quad & (1+x^1+x^2+x^3+x^4+x^5+x^6+x^7+\ldots) \\
\times \quad & (1+x^2+x^4+x^6+x^8+x^{10}+x^{12}+x^{14}+\ldots) \\
\times \quad & (1+x^3+x^6+x^9+x^{12}+x^{15}+x^{18}+x^{21}+\ldots)
\end{aligned}$$

$$
\begin{aligned}
&\times \quad (1+x^4+x^8+x^{12}+x^{16}+x^{20}+x^{24}+x^{28}+\ldots) \\
&\times \quad (1+x^5+x^{10}+x^{15}+x^{20}+x^{25}+x^{30}+x^{35}+\ldots) \\
&\times \quad etc. \\
&= \quad \sum_{n=0}^{\infty} x^{(each\, partition\, of\, n)} \\
&= \quad \sum_{n=0}^{\infty} (number\ of\ partitions\ of\ n) \times x^n \\
&= \quad \sum_{n=0}^{\infty} p(n)x^n.
\end{aligned}
$$

Crucially, any partition of n contributes 1 to $p(n)$. For example the partition $10+9+9+7+2+2+2+1$ of 42 contributes 1 to $p(42)$. By the way, $p(42)=53174$. So, the familiar partition function $p(n)$ maps each partition one-to-one onto its sum, and in the case of $p(42)$ this happens 53174 times, with each partition sum counted once only. The above partition of 42 can be represented as jumps along a line of length 42, as follows:

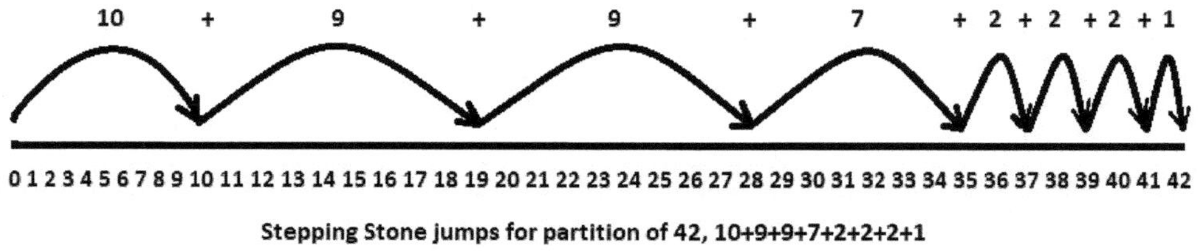

Figure 12.1 Stepping Stone jumps for an Integer Partition.

In the chapter on Congruences for partitions (Chapter 5), the following seven commonly used integer partition functions were stated:

1) $p_o(n) = p(n \,|\, \text{odd parts})$, where

$$\sum_{n=0}^{\infty} p_o(n)q^n = \frac{1}{(q;q^2)_\infty} = \frac{E(2)}{E(1)}. \tag{12.6.5}$$

2) $p_e(n) = p(n \,|\, \text{even parts})$, where

$$\sum_{n=0}^{\infty} p_e(n)q^n = \frac{1}{(q^2;q^2)_\infty} = \frac{1}{E(2)}. \tag{12.6.6}$$

3) $\overline{p}(n) = p(n \,|\, \text{overpartitions})$, where

$$\sum_{n=0}^{\infty} \overline{p}(n)q^n = \frac{(-q;q)_\infty}{(q;q)_\infty} = \frac{E(2)}{E(1)^2}. \tag{12.6.7}$$

4) $\overline{p}_o(n) = p(n\,|$ overpartitions into odd parts), where

$$\sum_{n=0}^{\infty} \overline{p}_o(n) q^n = \frac{(-q;q^2)_\infty}{(q;q^2)_\infty} = \frac{E(2)^3}{E(1)^2 E(4)}. \tag{12.6.8}$$

5) $\overline{p}_e(n) = p(n\,|$ overpartitions into even parts), where

$$\sum_{n=0}^{\infty} \overline{p}_e(n) q^n = \frac{(-q^2;q^2)_\infty}{(q^2;q^2)_\infty} = \frac{E(4)}{E(2)^2}. \tag{12.6.9}$$

6) $p_{od}(n) = p(n\,|$ odd parts are distinct), where

$$\sum_{n=0}^{\infty} p_{od}(n) q^n = \frac{(-q;q^2)_\infty}{(q^2;q^2)_\infty} = \frac{E(2)}{E(1)E(4)}. \tag{12.6.10}$$

7) $p_{ed}(n) = p(n\,|$ even parts are distinct), where

$$\sum_{n=0}^{\infty} p_{ed}(n) q^n = \frac{(-q^2;q^2)_\infty}{(q;q^2)_\infty} = \frac{E(4)}{E(1)}. \tag{12.6.11}$$

where we use the $E(n) = (q^n;q^n)_\infty$ adopted notation.

The above functions each being integer partitions, exhibit partitions with parts as 1D vectors along a straight line. Hence they can each be applied to vector partitions along 1D lines in 2D, 3D, etc. Euclidean space where the line "*from origin to a visible lattice point*" optimizes the respective gradients of the straight line with respect to each axis in the region being considered.

Hence we can easily write a variation on our theorem 12.5 given earlier pertaining then to $p(n)$, but here applying to any "*mixed-and-matched*" combination from $p_o(n), p_e(n), \overline{p}(n), \overline{p}_o(n), \overline{p}_e(n), p_{od}(n), p_{ed}(n)$, provided we apply the correct generating function from the above equations (12.6.5) to (12.6.11) to a particular associated 1D "*radial from origin line*" in the higher dimensional space of lattice point vectors. So we have appropriate generating functions in nD vector partitions into parts contained on any straight line through the multidimensional cartesian space, utilizing whichever of the seven integer partition functions we please associate the chosen line. We therefore state the example of this just arbitrarily choosing the overpartition function from among equations (12.6.5) to (12.6.11). Note, in the following theorem we reproduce the preamble words of definition from theorem 12.5 defining the three lines.

Theorem 12.6 *Take an nD lattice point vector of fixed non-negative integers $\langle a_1, a_2, \ldots, a_n\rangle$ (not all zero) where* $\gcd(a_1, a_2, \ldots, a_n)$, *and this defines a straight line from the origin through the n-space. Suppose then, that all lattice point vectors on that line are given by the notation*

$$\mathbb{L}_n(a_1, a_2, \ldots, a_n) = \{\langle a_1, a_2, \ldots, a_n\rangle, \langle 2a_1, 2a_2, \ldots, 2a_n\rangle, \langle 3a_1, 3a_2, \ldots, 3a_n\rangle, \ldots\}.$$

Similarly define for "n-tuple coprime" the lattice point vector $\langle b_1, b_2, \ldots, b_n\rangle$; and for *"n-tuple coprime" $\langle c_1, c_2, ..., c_n\rangle$; integer lattice point vectors on the two lines through the n-space, as follows:*

$$\mathbb{L}_n(b_1, b_2, ..., b_n) = \{\langle b_1, b_2, ..., b_n\rangle, \langle 2b_1, 2b_2, ..., 2b_n\rangle, \langle 3b_1, 3b_2, ..., 3b_n\rangle, \ldots\};$$

$$\mathbb{L}_n(c_1, c_2, ..., c_n) = \{\langle c_1, c_2, ..., c_n\rangle, \langle 2c_1, 2c_2, ..., 2c_n\rangle, \langle 3c_1, 3c_2, ..., 3c_n\rangle, \ldots\}.$$

*Let $\mathbb{V}_n(A_1, A_2, ..., A_n)$ be the number of vector partitions of the n-space vector $\langle A_1, A_2, ..., A_n\rangle$ with $A_1, A_2, ..., A_{n-1}$ each a non-negative integer and A_n a positive integer, into **overpartition parts** from the line $\mathbb{L}_n(a_1, a_2, ..., a_n)$; **overpartition parts** from the line $\mathbb{L}_n(b_1, b_2, ..., b_n)$, and **overpartition parts** from the line $\mathbb{L}_n(c_1, c_2, ..., c_n)$.*

Then $\mathbb{V}_n(A_1, A_2, ..., A_n)$ is the coefficient of ${q_1}^{A_1}{q_2}^{A_2}...{q_n}^{A_n}$ in either side of the expression,

$$\frac{(-({q_1}^{a_1}{q_2}^{a_2}...{q_n}^{a_n}); ({q_1}^{a_1}{q_2}^{a_2}...{q_n}^{a_n}))_\infty}{(({q_1}^{a_1}{q_2}^{a_2}...{q_n}^{a_n}); ({q_1}^{a_1}{q_2}^{a_2}...{q_n}^{a_n}))_\infty} \times \frac{(-({q_1}^{b_1}{q_2}^{b_2}...{q_n}^{b_n}); ({q_1}^{b_1}{q_2}^{b_2}...{q_n}^{b_n}))_\infty}{(({q_1}^{b_1}{q_2}^{b_2}...{q_n}^{b_n}); ({q_1}^{b_1}{q_2}^{b_2}...{q_n}^{b_n}))_\infty} \quad (12.6.12)$$

$$\times \frac{(-({q_1}^{c_1}{q_2}^{c_2}...{q_n}^{c_n}); ({q_1}^{c_1}{q_2}^{c_2}...{q_n}^{c_n}))_\infty}{(({q_1}^{c_1}{q_2}^{c_2}...{q_n}^{c_n}); ({q_1}^{c_1}{q_2}^{c_2}...{q_n}^{c_n}))_\infty}$$

$$= \left(\sum_{m\geq 0} \overline{p}(m)({q_1}^{a_1}{q_2}^{a_2}\cdots{q_n}^{a_n})^m\right) \times \left(\sum_{m\geq 0} \overline{p}(m)({q_1}^{b_1}{q_2}^{b_2}\cdots{q_n}^{b_n})^m\right)$$

$$\times \left(\sum_{m\geq 0} \overline{p}(m)({q_1}^{c_1}{q_2}^{c_2}\cdots{q_n}^{c_n})^m\right).$$

*That is, $\mathbb{V}_n(A_1, A_2, ..., A_n)$ can be found exactly as a combination of **number of overpartitions** integer partition functions.*

Obviously, we could have chosen a different partition function for a similar theorem to the above. We instead of using overpartitions $\overline{p}(m)$ for example we could have used instead say $p_{od}(n) = p(n|$ odd parts are distinct$)$, or $\overline{p}_o(n) = p(n|$ overpartitions into odd parts$)$, or we could have been innovative and justified use of plane partitions $PL(n)$ along selected 1D lines through the nD-space.

12.6.2 Weighted integer partitions with function of partition function, f(p(n))

The partition function $p(n)$ for the number of partitions of integer n represents a model where each partition is of weight 1. That is, each partition, as with the above example for 42 contributes a one-for-one value to the $p(n)$. So in our example above, a mapping like $10+9+9+7+2+2+2+1 \mapsto 1$ holds for each partition of 42 to allow $p(42)$ to work out as 53174. However, in the next example, we look into situations where each partition contributes a value that is different from unity.

The partition function $PL(n)$ for the number of plane partitions of integer n is generated from the product of infinite series:

$$
\begin{aligned}
&\prod_{k=1}^{\infty} \frac{1}{(1-x^k)^k} \\
= \;& (1+x^1+x^2+x^3+x^4+x^5+x^6+x^7+\ldots)^1 \\
\times \;& (1+x^2+x^4+x^6+x^8+x^{10}+x^{12}+x^{14}+\ldots)^2 \\
\times \;& (1+x^3+x^6+x^9+x^{12}+x^{15}+x^{18}+x^{21}+\ldots)^3 \\
\times \;& (1+x^4+x^8+x^{12}+x^{16}+x^{20}+x^{24}+x^{28}+\ldots)^4 \\
\times \;& (1+x^5+x^{10}+x^{15}+x^{20}+x^{25}+x^{30}+x^{35}+\ldots)^5 \\
\times \;& etc.
\end{aligned}
$$

$$
\begin{aligned}
= \;& (\sum_{{}_1n_1 \geq 0} x^{1\,{}_1n_1}) \\
\times \;& (\sum_{{}_2n_1 \geq 0} x^{2\,{}_2n_1})(\sum_{{}_2n_2 \geq 0} x^{2\,{}_2n_2}) \\
\times \;& (\sum_{{}_3n_1 \geq 0} x^{3\,{}_3n_1})(\sum_{{}_3n_2 \geq 0} x^{3\,{}_3n_2})(\sum_{{}_3n_3 \geq 0} x^{3\,{}_3n_3}) \\
\times \;& (\sum_{{}_4n_1 \geq 0} x^{4\,{}_4n_1})(\sum_{{}_4n_2 \geq 0} x^{4\,{}_4n_2})(\sum_{{}_4n_3 \geq 0} x^{4\,{}_4n_3})(\sum_{{}_4n_4 \geq 0} x^{4\,{}_4n_4})
\end{aligned}
$$

$$
\begin{aligned}
\times \;& (\sum_{{}_5n_1 \geq 0} x^{5\,{}_5n_1})(\sum_{{}_5n_2 \geq 0} x^{5\,{}_5n_2})(\sum_{{}_5n_3 \geq 0} x^{5\,{}_5n_3})(\sum_{{}_5n_4 \geq 0} x^{5\,{}_5n_4})(\sum_{{}_5n_5 \geq 0} x^{5\,{}_5n_5}) \\
\times \;& etc. \\
= \;& \sum_{\substack{n \geq 0 \\ {}_1n_1, {}_2n_1, {}_2n_2, {}_3n_1, {}_3n_2, {}_3n_3, \ldots \geq 0)}} x^{\{{}_1n_1+2({}_2n_1+{}_2n_2)+3({}_3n_1+{}_3n_2+{}_3n_3)+4({}_4n_1+{}_4n_2+{}_4n_3+{}_4n_4)+\ldots\}} \\
& for\ each\ {}_1n_1+2({}_2n_1+{}_2n_2)+3({}_3n_1+{}_3n_2+{}_3n_3)+4({}_4n_1+{}_4n_2+{}_4n_3+{}_4n_4)+\ldots = n \\
= \;& \sum_{n=0}^{\infty} PL(n)x^n.
\end{aligned}
$$

Crucially, any weighted partition of n of the form

$$ {}_1n_1+2({}_2n_1+{}_2n_2)+3({}_3n_1+{}_3n_2+{}_3n_3)+4({}_4n_1+{}_4n_2+{}_4n_3+{}_4n_4)+\ldots $$

contributes a plane partition to $PL(n)$.

12.7 2D AND 3D *UPPER RADIAL REGIONS* OF VECTORS FOR PARTITION IDENTITIES

In the next sections, we study 2D and 3D sets of vectors in regions radial from the origin. A 2D or 3D vector partition is essentially a single entity with simultaneous features:

(a) a finite quiver of vector arrows,

(b) laid out and joined head to tail,

(c) each with non-increasing positive rational gradients in 2D or in 3D with respect to the relevant axes,

(d) arranged in non-increasing length,

(e) starting at the origin,

(f) with ending destination an integer co-ordinate point,

(g) for example, $\langle a,b\rangle$ with a and b both non-negative integers in 2D, or perhaps $\langle a,b,c\rangle$ with a, b and c all non-negative integers in 3D.

The number of arrows in this vector partition (or in the quiver) is the number of parts of the partition.

Since a vector partition is a partially ordered set summed in a defined way, this means there are potentially many ways to define the vector partition. Rules used in defining a vector partition are possibly one or more of:

- non-increasing arrow length (or non-decreasing),
- non-increasing coordinate numbering (or non-decreasing) or co-ordinate orderings, say X then Y; or Y- then X-axes,
- non-increasing (respectively non-decreasing) angular measure from an axis as polar co-ordinate systemic,
- non-increasing (respectively non-decreasing) color gradations.

We consider 2D vectors $\langle a,b\rangle$ where $0 < a < b$. We mainly look into two types of such vectors for our vector partition aggregations; both types being in "radial from the origin" regions:

(a) All of the 2D vectors $\langle a,b\rangle$ where $0 \leq a < b$ are in the infinitely extended region radial from the origin. This on occasion means an examination of defined "subset finite patches" of that radial region.

(b) All VPV 2D vectors are $\langle a,b\rangle$ where $0 < a < b$ and $\gcd(a,b) = 1$ are in the infinitely extended region radial from the origin. This on occasion means an examination of defined "subset finite patches" of that radial region.

Using "All 2D first quadrant vectors" we have the examples

$$\prod_{j,k\geq 1} \frac{1}{1-x^j y^k} = \prod_{j,k\geq 1} \left\{1 + x^j y^k + x^{2j} y^{2k} + x^{3j} y^{3k} + \ldots\right\} = \sum_{a,b\geq 0} \mathbf{p}_2(a,b) x^a y^b. \tag{12.7.1}$$

Here we generate $\mathbf{p}_2(a,b)$ as the number of partitions of $\langle j,k\rangle$ into unrestricted parts

$\langle a,b\rangle$ with $a,b \geq 0$. In this case, each partition has a value of 1, so the number of partitions is unsurprisingly just a count of each partition.

$$\prod_{j,k\geq 1}\left(1+\frac{x^j y^k}{(1-x^j y^k)^2}\right) = \prod_{j,k\geq 1}\left\{1+x^j y^k+2x^{2j}y^{2k}+3x^{3j}y^{3k}+\ldots\right\} = \sum_{a,b\geq 0} \mathbf{q}_2(a,b)x^a y^b. \tag{12.7.2}$$

This generates $\mathbf{q}_2(a,b)$ as a weighted sum of partitions of $\langle j,k\rangle$ into unrestricted parts $\langle a,b\rangle$ with $a,b \geq 0$. This is a *function of a partition* applied across all possible partitions of $\langle j,k\rangle$. In this case, each partition is assigned a value (called also a weight) of

$$n_1^{a_1} n_2^{a_2} \ldots n_r^{a_r}$$

for a partition of $\langle j,k\rangle$

$$(\langle n_1 j, n_1 k\rangle^{a_1} \langle n_2 j, n_2 k\rangle^{a_2} \ldots \langle n_r j, n_r k\rangle^{a_r}),$$

so the function of partitions (as distinct from a number of partitions) on $\langle j,k\rangle$ is derived as a combined function of each partition. So $\mathbf{q}_2(a,b)$ is a sum of products $\sum n_1^{a_1} n_2^{a_2} \ldots n_r^{a_r}$ taken over every partition of $\langle j,k\rangle$.

$$\prod_{j,k\geq 1}\left(1+\frac{x^j y^k(1+x^j y^k)}{(1-x^j y^k)^3}\right) = \prod_{j,k\geq 1}\left\{1+1^2 x^j y^k+2^2 x^{2j}y^{2k}+3^2 x^{3j}y^{3k}+\ldots\right\} \tag{12.7.3}$$

$$= \sum_{a,b\geq 0} \mathbf{r}_2(a,b)x^a y^b.$$

In this instance, $\mathbf{r}_2(a,b)$ is a sum of products $\sum n_1^{2a_1} n_2^{2a_2} \ldots n_r^{2a_r}$ taken over every partition $(\langle n_1 j, n_1 k\rangle^{a_1} \langle n_2 j, n_2 k\rangle^{a_2} \ldots \langle n_r j, n_r k\rangle^{a_r})$ of $\langle j,k\rangle$.

$$\prod_{j,k\geq 1}\left(1+\frac{x^j y^k}{1-x^{2j}y^{2k}}\right) = \prod_{j,k\geq 1}\left\{1+x^j y^k+3x^{3j}y^{3k}+5x^{5j}y^{5k}+\ldots\right\} \tag{12.7.4}$$

$$= \sum_{a,b\geq 0} \mathbf{s}_2(a,b)x^a y^b.$$

In this case $\mathbf{x}_2(a,b)$ is a sum of products $\sum(2n_1+1)^{a_1}(2n_2+1)^{a_2}\ldots(2n_r+1)^{a_r}$ taken over every partition $(\langle n_1 j, n_1 k\rangle^{a_1} \langle n_3 j, n_3 k\rangle^{a_2} \ldots \langle n_{2r+1} j, n_{2r+1} k\rangle^{a_r})$ of $\langle j,k\rangle$.

In the examples that follow here we use $\gcd(j,k) = 1$ abbreviated to $(j,k) = 1$. It will become evident that when applying these generating function forms to radial-from-origin regions over vector partitions, the theory of $\langle j,k\rangle$ where $(j,k) = 1$ simplifies things.

$$\prod_{\substack{j,k\geq 1;\\(j,k)=1}} \frac{1}{1-x^j y^k} = \prod_{\substack{j,k\geq 1;\\(j,k)=1}} \left\{1+x^j y^k+x^{2j}y^{2k}+x^{3j}y^{3k}+\ldots\right\} = \sum_{a,b\geq 0} \mathbf{v}_2(a,b)x^a y^b. \tag{12.7.5}$$

Note that the curly-bracketed term in (12.7.5) contributes vector parts of rational gradient j/k with $(j,k)=1$ in the first 2D quadrant exactly one part per partition. So $\mathbf{v}_2(a,b)$ is the number of partitions of $\langle a,b\rangle$ into vectors each part of distinct positive rational gradient. In traditional integer partition parlance this would say $\mathbf{v}_2(a,b)$ is the number of partitions into unrestricted Visible Point Vector (VPV) parts.

$$\prod_{\substack{j,k\geq 1;\\(j,k)=1}} \left(1+\frac{x^j y^k}{(1-x^j y^k)^2}\right) = \prod_{\substack{j,k\geq 1;\\(j,k)=1}} \left\{1+x^j y^k + 2x^{2j}y^{2k} + 3x^{3j}y^{3k} + \ldots\right\}$$
$$= \sum_{a,b\geq 0} \mathbf{w}_2(a,b)x^a y^b. \tag{12.7.6}$$

$$\prod_{\substack{j,k\geq 1;\\(j,k)=1}} \left(1+\frac{x^j y^k(1+x^j y^k)}{(1-x^j y^k)^3}\right) = \prod_{\substack{j,k\geq 1;\\(j,k)=1}} \left\{1+1^2x^j y^k + 2^2x^{2j}y^{2k} + 3^2x^{3j}y^{3k} + \ldots\right\}$$
$$= \sum_{a,b\geq 0} \mathbf{x}_2(a,b)x^a y^b. \tag{12.7.7}$$

$$\prod_{\substack{j,k\geq 1;\\gcd(j,k)=1}} \left(1+\frac{x^j y^k}{1-x^{2j} y^{2k}}\right) = \prod_{\substack{j,k\geq 1;\\gcd(j,k)=1}} \left\{1+x^j y^k + 3x^{3j}y^{3k} + 5x^{5j}y^{5k} + \ldots\right\} \tag{12.7.8}$$
$$= \sum_{a,b\geq 0} \mathbf{y}_2(a,b)x^a y^b.$$

We also consider 3D vectors $\langle a,b,c\rangle$ where $0<a<b<c$. Again we mainly look into two types of such vectors for our vector partition aggregations:

(a) All of the 3D vectors $\langle a,b,c\rangle$ where $0<a<b<c$ are in the infinitely extended region radial from the origin. This on occasion means an examination of defined "finite patches" of that radial region.

(b) All of the 3D vectors $\langle a,b,c\rangle$ where $0<a<b<c$ and $gcd(a,b,c)=1$ are in the infinitely extended region radial from the origin. This on occasion means an examination of defined "finite patches" of that radial region.

12.8 DEFINING 2D *UPPER VISIBLE POINT VECTORS* IN ORIGIN-RADIAL REGIONS

These lattice point vectors *count* the visible from the origin points in the infinite radial region of the first quadrant bounded by the positive y-axis and the line $y=x$. The visible lattice points in 3D space are similarly a countable set, and in fact in nD space with n a positive integer greater than unity. In 3D the set of visible points can be counted as shown below. In order to count the entire first 3D hyperquadrant it suffices to count the following vectors $\langle a,b,c\rangle$ such that $0<a<b<c$ has

A set of identities involving n-dimensional visible lattice points was discovered by Campbell in 1994, and the 2D cases were published by him in 1992. (see Campbell [84, 85, 86, 87]. Also, see the 1997 thesis by the author [89] and follow-up papers from

1998, 2000 and 2019. (see Campbell [90, 91, 93]) The visible lattice points in the 2 dimensional first quadrant (see Weisstein, E [299], [298]) correspond to the countable set of rational numbers between 0 and 1, listed here as

$$\begin{array}{ccccccccc}
\frac{1}{2} & & & & & & & & \\
\frac{1}{3} & \frac{2}{3} & & & & & & & \\
\frac{1}{4} & & \frac{3}{4} & & & & & & \\
\frac{1}{5} & \frac{2}{5} & \frac{3}{5} & \frac{4}{5} & & & & & \\
\frac{1}{6} & & & & \frac{5}{6} & & & & \\
\frac{1}{7} & \frac{2}{7} & \frac{3}{7} & \frac{4}{7} & \frac{5}{7} & \frac{6}{7} & & & \\
\frac{1}{8} & & \frac{3}{8} & & \frac{5}{8} & & \frac{7}{8} & & \\
\frac{1}{9} & \frac{2}{9} & & \frac{4}{9} & \frac{5}{9} & & \frac{7}{9} & \frac{8}{9} & \\
\frac{1}{10} & & \frac{3}{10} & & & & \frac{7}{10} & & \frac{9}{10}
\end{array}$$

etc.

giving rise to the *visible from the origin vector* (VPV) lattice points $\langle x,y\rangle$ with $x<y$ and $\gcd(x,y)=1$ in the first 2D quadrant.

etc.

$\langle 1,10\rangle$		$\langle 3,10\rangle$				$\langle 7,10\rangle$		$\langle 9,10\rangle$
$\langle 1,9\rangle$	$\langle 2,9\rangle$		$\langle 4,9\rangle$	$\langle 5,9\rangle$		$\langle 7,9\rangle$	$\langle 8,9\rangle$	
$\langle 1,8\rangle$		$\langle 3,8\rangle$		$\langle 5,8\rangle$		$\langle 7,8\rangle$		
$\langle 1,7\rangle$	$\langle 2,7\rangle$	$\langle 3,7\rangle$	$\langle 4,7\rangle$	$\langle 5,7\rangle$	$\langle 6,7\rangle$			
$\langle 1,6\rangle$				$\langle 5,6\rangle$				
$\langle 1,5\rangle$	$\langle 2,5\rangle$	$\langle 3,5\rangle$	$\langle 4,5\rangle$					
$\langle 1,4\rangle$		$\langle 3,4\rangle$						
$\langle 1,3\rangle$	$\langle 2,3\rangle$							
$\langle 1,2\rangle$								

These are the countable list of *upper first quadrant* 2D Visible Point Vectors. That is, the VPVs in the infinitely extended radial region of the 2D first quadrant above the line $y=x$.

12.9 EXAMPLES OF 2D VPV FINITE GENERATING FUNCTIONS

12.9.1 2D Distinct *Upper VPV* Coefficients – Order 2

$$(1+xy^2) \qquad \textit{and} \qquad (1-xy^2) \tag{12.9.1}$$

y/x	0	1	2	$\sum r$
3				
2		1		1
1				0
0	1			1
$\sum c$	1	1		

y/x	0	1	2	$\sum r$
3				
2		−1		−1
1				0
0	1			1
$\sum c$	1	−1		

12.9.2 2D Distinct *Upper VPV* Coefficients – Order 3

$$(1+xy^2)(1+xy^3)(1+x^2y^3) \qquad \textit{and} \qquad (1-xy^2)(1-xy^3)(1-x^2y^3) \tag{12.9.2}$$

y/x	0	1	2	3	4	$\sum r$
8					1	1
7						0
6				1		1
5			1	1		2
4						0
3		1	1			2
2		1				1
1						0
0	1					1
$\sum c$	1	2	2	2	1	

y/x	0	1	2	3	4	$\sum r$
8					−1	−1
7						0
6				1		1
5			1	1		2
4						0
3		−1	−1			−2
2		−1				−1
1						0
0	1					1
$\sum c$	1	−2	0	2	−1	

Note that row totals in the left side grid generate from

$$(1+xy^2)(1+xy^3)(1+x^2y^3) \quad \text{where} \quad x=1,$$

expanding to

$$y^8+y^6+2y^5+2y^3+y^2+1.$$

Also row totals in the right side grid generate from

$$(1-xy^2)(1-xy^3)(1-x^2y^3) \quad \text{where} \quad x=1,$$

expanding to

$$-y^8+y^6+2y^5-2y^3-y^2+1.$$

Likewise, the left side grid column totals generate from

$$(1+xy^2)(1+xy^3)(1+x^2y^3) \quad \text{where} \quad y=1,$$

expanding to

$$x^4+2x^3+2x^2+2x+1;$$

whilst the right side grid column totals generate from

$$(1-xy^2)(1-xy^3)(1-x^2y^3) \quad \text{where} \quad y=1,$$

expanding to

$$-x^4+2x^3-2x+1.$$

Combinatorial interpretation: Consider the three vectors

$$\begin{array}{ll} \langle 1,3\rangle & \langle 2,3\rangle \\ \langle 1,2\rangle & \end{array} \tag{12.9.3}$$

Let $p_{2,\mathrm{up}}(\mathfrak{D}_{vpv},\langle a,b\rangle)$, denote the number of partitions of $\langle a,b\rangle$ into distinct parts from (12.9.3). Then

$$\prod_{k=2}^{3}\prod_{\substack{j=1;\\ \gcd(j,k)=1}}^{k-1}(1+x^jy^k)=\sum_{b=2}^{8}\sum_{a=1}^{b-1}p_{2,\mathrm{up}}(\mathfrak{D}_{vpv},\langle a,b\rangle)x^ay^b, \tag{12.9.4}$$

and each entry in the left side grid after (12.9.2) gives each numerical value.

Let $p_{2,\mathrm{up}}(\mathfrak{D}_{vpv,\mathrm{odd}},\langle a,b\rangle)$, denote the number of partitions of $\langle a,b\rangle$ into an odd number of distinct parts from (12.9.3). Let $p_{2,\mathrm{up}}(\mathfrak{D}_{vpv,\mathrm{even}},\langle a,b\rangle)$, denote the number of partitions of $\langle a,b\rangle$ into an even number of distinct parts from (12.9.3).Then

$$\prod_{k=2}^{3}\prod_{\substack{j=1;\\ \gcd(j,k)=1}}^{k-1}(1+x^jy^k)=\sum_{b=2}^{8}\sum_{a=1}^{b-1}[p_{2,\mathrm{up}}(\mathfrak{D}_{vpv,\mathrm{odd}},\langle a,b\rangle)-p_{2,\mathrm{up}}(\mathfrak{D}_{vpv,\mathrm{even}},\langle a,b\rangle)]x^ay^b, \tag{12.9.5}$$

and each entry in the right side grid after (12.9.2) gives each numerical value.

12.9.3 2D Distinct *Upper VPV* Coefficients – Order 4

The product

$$(1+xy^2)(1+xy^3)(1+x^2y^3)(1+xy^4)(1+x^3y^4) \tag{12.9.6}$$

generates Upper VPV 2D first quadrant partitions of Order 4 having $\langle a,b\rangle$ with $0<1\leq a<b\leq 4$ and $gcd(a,b)=1$.

First quadrant 2D distinct Upper VPV partitions – Order 4

y/x	0	1	2	3	4	5	6	7	8	$\sum r$
16									1	1
15										0
14								1		1
13							1	1		2
12						1		1		2
11						1	1			2
10					1	1	1			3
9				1	1	1	1			4
8					2					2
7			1	1	1	1				4
6			1	1	1					3
5			1	1						2
4		1		1						2
3		1	1							2
2		1								1
1										0
0	1									1
$\sum c$	1	3	4	5	6	5	4	3	1	

Note that column totals generate from

$$(1+xy^2)(1+xy^3)(1+x^2y^3)(1+xy^4)(1+x^3y^4) \quad \text{where} \quad y=1,$$

expanding to

$$x^8+3x^7+4x^6+5x^5+6x^4+5x^3+4x^2+3x+1.$$

Likewise, row totals generate from

$$(1+xy^2)(1+xy^3)(1+x^2y^3)(1+xy^4)(1+x^3y^4) \quad \text{where} \quad x=1,$$

expanding to

$$y^{16}+y^{14}+2y^{13}+2y^{12}+2y^{11}+3y^{10}+4y^9$$
$$+2y^8+4y^7+3y^6+2y^5+2y^4+2y^3+y^2+1.$$

Combinatorial interpretation: Consider the aggregate of vectors

$$\begin{array}{lll} \langle 1,4\rangle & & \langle 3,4\rangle \\ \langle 1,3\rangle & \langle 2,3\rangle & \\ \langle 1,2\rangle & & \end{array} \tag{12.9.7}$$

Let $p_{2,\mathrm{up}}(\mathfrak{D}_{vpv},\langle a,b\rangle)$, denote the number of partitions of $\langle a,b\rangle$ into distinct parts from (12.9.7). Then

$$\prod_{k=2}^{4}\prod_{\substack{j=1;\\ \gcd(j,k)=1}}^{k-1}(1+x^jy^k)=\sum_{b=2}^{20}\sum_{a=1}^{b-1}p_{2,\mathrm{up}}(\mathfrak{D}_{vpv},\langle a,b\rangle)x^ay^b, \tag{12.9.8}$$

and each entry in the grid after (12.9.6) above gives each numerical value.

We continue in this way by considering the generating function

$$(1-xy^2)(1-xy^3)(1-x^2y^3)(1-xy^4)(1-x^3y^4) \tag{12.9.9}$$

which expanded gives us the grid,

First quadrant 2D $p_2(e,v)-p_2(o,v)$ distinct Upper VPV partitions – Order 4

y/x	0	1	2	3	4	5	6	7	8	$\sum r$
16									-1	-1
15										0
14								1		1
13							1	1		2
12						1		1		2
11						-1	-1			-2
10					-1	-1	-1			-3
9				-1	-1	-1	-1			-4
8					2					2
7			1	1	1	1				4
6			1	1	1					3
5			1	1						2
4		-1		-1						-2
3		-1	-1							-2
2		-1								-1
1										0
0	1									1
$\sum c$	1	-3	2	1	2	-1	-2	3	-1	

Note that column totals generate from

$$(1-xy^2)(1-xy^3)(1-x^2y^3)(1-xy^4)(1-x^3y^4) \quad \text{where} \quad y=1,$$

expanding to

$$-x^8+3x^7-2x^6-x^5+x^3+2x^2-3x+1.$$

Likewise, row totals generate from

$$(1-xy^2)(1-xy^3)(1-x^2y^3)(1-xy^4)(1-x^3y^4) \quad \text{where} \quad x=1,$$

expanding to

$$-y^{16}+y^{14}+2y^{13}+2y^{12}-2y^{11}-3y^{10}-4y^9$$
$$+4y^7+3y^6+2y^5-2y^4-2y^3-y^2+1.$$

Combinatorial interpretation: Consider the aggregate of vectors

$$\begin{array}{lll} \langle 1,4\rangle & & \langle 3,4\rangle \\ \langle 1,3\rangle & \langle 2,3\rangle & \\ \langle 1,2\rangle & & \end{array} \qquad (12.9.10)$$

Let $p_{2,\text{up}}(\mathfrak{D}_{(vpv,\text{odd})};\langle a,b\rangle)$, denote the number of partitions of $\langle a,b\rangle$ into distinct odd number of parts from (12.9.7). Also let $p_{2,\text{up}}(\mathfrak{D}_{(vpv,\text{even})};\langle a,b\rangle)$, denote the number of partitions of $\langle a,b\rangle$ into distinct even number of parts from (12.9.7).Then

$$\prod_{k=2}^{4}\prod_{\substack{j=1;\\ \gcd(j,k)=1}}^{k-1}(1-x^j y^k)=\sum_{b=2}^{20}\sum_{a=1}^{b-1}[p_{2,\text{up}}(\mathfrak{D}_{(vpv,\text{odd})};\langle a,b\rangle)-p_{2,\text{up}}(\mathfrak{D}_{(vpv,\text{even})};\langle a,b\rangle)]x^a y^b, \tag{12.9.11}$$

and each entry in the grid after (12.9.9) above gives each numerical value.

12.9.4 2D Distinct *Upper VPV* Coefficients – Order 5

$(1+xy^2)(1+xy^3)(1+x^2y^3)(1+xy^4)(1+x^3y^4)(1+xy^5)(1+x^2y^5)(1+x^3y^5)(1+x^4y^5)$ generates the grid shown on the next page. Note that column totals for that grid generate from

$(1+xy^2)(1+xy^3)(1+x^2y^3)(1+xy^4)(1+x^3y^4)(1+xy^5)(1+x^2y^5)(1+x^3y^5)(1+x^4y^5)$ where $y=1$, expanding to

$$1+4x+8x^2+14x^3+23x^4+32x^5+41x^6+50x^7+55x^8+56x^9+55x^{10}+50x^{11}$$
$$+41x^{12}+32x^{13}+23x^{14}+14x^{15}+8x^{16}+4x^{17}+x^{18}.$$

Likewise, row totals generate from

$(1+xy^2)(1+xy^3)(1+x^2y^3)(1+xy^4)(1+x^3y^4)(1+xy^5)(1+x^2y^5)(1+x^3y^5)(1+x^4y^5)$ where $x=1$, expanding to

$$1+y^2+2y^3+2y^4+6y^5+3y^6+8y^7+10y^8+12y^9+17y^{10}+14y^{11}$$
$$+24y^{12}+22y^{13}+29y^{14}+28y^{15}+27y^{16}+36y^{17}+28y^{18}+36y^{19}$$
$$+27y^{20}+28y^{21}+29y^{22}+22y^{23}+24y^{24}+14y^{25}+17y^{26}+12y^{27}$$
$$+10y^{28}+8y^{29}+3y^{30}+6y^{31}+2y^{32}+2y^{33}+y^{34}+y^{36}.$$

Combinatorial interpretation: Consider the aggregate of vectors

$$\begin{matrix}\langle 1,5\rangle & \langle 2,5\rangle & \langle 3,5\rangle & \langle 4,5\rangle\\ \langle 1,4\rangle & & \langle 3,4\rangle & \\ \langle 1,3\rangle & \langle 2,3\rangle & & \\ \langle 1,2\rangle & & & \end{matrix} \tag{12.9.12}$$

Let $p_{2,\text{vpv}}(\mathfrak{D}, \langle a,b\rangle)$, denote the number of partitions of $\langle a,b\rangle$ into distinct parts from (12.9.12). Then we have the generating function

$$\prod_{k=2}^{5} \prod_{\substack{j=1; \\ \gcd(j,k)=1}}^{k-1} (1+x^j y^k) = \sum_{b=2}^{36} \sum_{a=1}^{b-1} p_{2,\text{vpv}}(\mathfrak{D}, \langle a,b\rangle) x^a y^b, \tag{12.9.13}$$

encoding all possible partitions of this kind, so each entry on the grid shown next gives all numerical values of $p_2(\mathfrak{D}, \langle a,b\rangle)$.

Distinct 2D Upper VPV partitions – Order 5

y/x	0	1	2	3	4	5	6	7	8	9	10	11	12	13	14	15	16	17	18	$\sum r$
36																			1	1
35																				0
34																		1		1
33																	1	1		2
32																1		1		2
31															1	2	2	1		6
30															1	1	1			3
29														2	2	2	2			8
28													1	2	4	2	1			10
27												1	2	3	3	2	1			12
26												2	4	5	4	2				17
25											1	2	4	4	2	1				14
24										1	3	5	6	5	3	1				24
23										1	4	6	6	4	1					22
22									2	3	6	7	6	3	2					29
21									2	5	7	7	5	2						28
20								1	3	6	7	6	3	1						27
19							1	3	6	8	8	6	3	1						36
18								4	6	8	6	4								28
17						1	3	6	8	8	6	3	1							36
16						1	3	6	7	6	3	1								27
15						2	5	7	7	5	2									28
14					2	3	6	7	6	3	2									29
13					1	4	6	6	4	1										22
12				1	3	5	6	5	3	1										24
11				1	2	4	4	2	1											14
10				2	4	5	4	2												17
9			1	2	3	3	2	1												12
8			1	2	4	2	1													10
7			2	2	2	2														8
6			1	1	1															3
5		1	2	2	1															6
4		1		1																2
3		1	1																	2
2		1																		1
1																				0
0	1																			1
$\sum c$	1	4	8	14	23	32	41	50	55	56	55	50	41	32	23	14	8	4	1	

Next, we again consider the cluster of vectors (12.9.12), where the number of VPV partitions $\langle a,b\rangle, gcd(a,b) = 1, 1 \leq a < b \leq 5$ into the number of distinct even number of parts minus the number of partitions into a distinct odd number of parts is generated by the product

$$(1-xy^2)(1-xy^3)(1-x^2y^3)(1-xy^4)(1-x^3y^4)(1-xy^5)(1-x^2y^5)(1-x^3y^5)(1-x^4y^5) \tag{12.9.14}$$

is a generating function for the grid

First quadrant 2D $p_2(e,v)-p_2(o,v)$ distinct Upper VPV partitions – Order 5

y/x	0	1	2	3	4	5	6	7	8	9	10	11	12	13	14	15	16	17	18	$\sum r$
36																			−1	−1
35																				0
34																		1		1
33																	1	1		2
32																1		1		2
31															1			1		2
30															−1	−1	−1			−3
29														−2	−2	−2	−2			−8
28													−1	−2	−2	−2	−1			−8
27												−1		−1	−1		−1			−4
26													2	1	2					5
25											1	2	4	4	2	1				14
24										1	3	3	6	3	3	1				20
23										1	2	2	2	2	1					10
22									−1		−3		−1							−5
21									−2	−3	−5	−5	−3	−2						−20
20								−1	−3	−6	−5	−6	−3	−1						−25
19							−1	−1	4	−4	−4	−4	−1	−1						−20
18																				0
17						1	1	4	4	4	4	1	1							20
16						1	3	6	5	6	3	1								25
15						2	3	5	5	3	2									20
14						1		3		1										5
13					−1	−2	−2	−2	−2	−1										−10
12				−1	−3	−3	−6	−3	−3	−1										−20
11				−1	−2	−4	−4	−2	−1											−14
10					−2	−1	−2													−5
9			1		1	1		1												4
8			1	2	2	2	1													8
7			2	2	2	2														8
6			1	1	1															3
5		−1			−1															−2
4		−1		−1																−2
3		−1	−1																	−2
2		−1																		−1
1																				0
0	1																			1
$\sum c$	1	−4	4	2	−3	0	−7	10	−1	0	1	−10	7	0	3	−2	−4	4	−1	

Note that column totals generate from

$$(1-xy^2)(1-xy^3)(1-x^2y^3)(1-xy^4)(1-x^3y^4)(1-xy^5)(1-x^2y^5)(1-x^3y^5)(1-x^4y^5),$$

where $y=1$ expanding to

$$1-4x+4x^2+2x^3-3x^4-7x^6+10x^7-x^8+x^{10}-10x^{11}$$
$$+7x^{12}+3x^{14}-2x^{15}-4x^{16}+4x^{17}-x^{18}.$$

Likewise, row totals generate from

$$(1-xy^2)(1-xy^3)(1-x^2y^3)(1-xy^4)(1-x^3y^4)(1-xy^5)(1-x^2y^5)(1-x^3y^5)(1-x^4y^5),$$

where $x=1$ expanding to

$$1-y^2-2y^3-2y^4-2y^5+3y^6+8y^7+8y^8+4y^9-5y^{10}-14y^{11}$$
$$-20y^{12}-10y^{13}+5y^{14}+20y^{15}+25y^{16}+20y^{17}-20y^{19}-25y^{20}$$
$$-20y^{21}-5y^{22}+10y^{23}+20y^{24}+14y^{25}+5y^{26}-4y^{27}-8y^{28}$$
$$-8y^{29}-3y^{30}+2y^{31}+2y^{32}+2y^{33}+y^{34}-y^{36}.$$

Combinatorial interpretation: Consider the aggregate of vectors

$$\begin{array}{cccc} \langle 1,5\rangle & \langle 2,5\rangle & \langle 3,5\rangle & \langle 4,5\rangle \\ \langle 1,4\rangle & & \langle 3,4\rangle & \\ \langle 1,3\rangle & \langle 2,3\rangle & & \\ \langle 1,2\rangle & & & \end{array} \tag{12.9.15}$$

Let $p_{2,\mathrm{up}}(\mathfrak{D}_{(vpv,\mathrm{odd})};\langle a,b\rangle)$, denote the number of partitions of $\langle a,b\rangle$ into a distinct odd number of parts from (12.9.15). Also let $p_{2,\mathrm{up}}(\mathfrak{D}_{(vpv,\mathrm{even})};\langle a,b\rangle)$, denote the number of partitions of $\langle a,b\rangle$ into distinct even numbers of parts from (12.9.15).Then

$$\prod_{k=2}^{5}\prod_{\substack{j=1;\\ \gcd(j,k)=1}}^{k-1}(1-x^j y^k) = \sum_{b=2}^{36}\sum_{a=1}^{b-1}[p_{2,\mathrm{up}}(\mathfrak{D}_{(vpv,\mathrm{odd})};\langle a,b\rangle) - p_{2,\mathrm{up}}(\mathfrak{D}_{(vpv,\mathrm{even})};\langle a,b\rangle)]x^a y^b, \tag{12.9.16}$$

and each entry in the grid after (12.9.14) above gives each numerical value.

12.9.5 2D weighted *Upper VPV* Coefficients – Order 5

The following product is five of the factors in the known VPV infinite product for $\left(\frac{1-y}{1-xy}\right)^{\frac{1}{1-x}}$.

$$\begin{gathered}(1-xy^2)^{1/2} \\ (1-xy^3)^{1/3}(1-x^2y^3)^{1/3} \\ (1-xy^4)^{1/4}(1-x^3y^4)^{1/4} \\ (1-xy^5)^{1/5}(1-x^2y^5)^{1/5}(1-x^3y^5)^{1/5}(1-x^4y^5)^{1/5}.\end{gathered} \tag{12.9.17}$$

This product is encoded by the grid

First quadrant 2D weighted distinct Upper VPV partitions – Order 5

y/x	0	1	2	3	4	5	6	7	8	9	$\sum r$
13				⋮	⋮	⋮	⋮	⋮	⋮	⋮	⋮
12				$\frac{-301}{9600}$	$\frac{-163279}{1555200}$	$\frac{86923}{3110400}$	$\frac{360869}{2073600}$	$\frac{86923}{3110400}$	$\frac{-163279}{1555200}$	$\frac{-301}{9600}$	$\frac{-2431243}{6220800}$
11				$\frac{41}{1440}$	$\frac{127}{3240}$	$\frac{-899}{17280}$	$\frac{-899}{17280}$	$\frac{127}{3240}$	$\frac{41}{1440}$		$\frac{811}{25920}$
10			$\frac{-2}{25}$	$\frac{1171}{14400}$	$\frac{-1511}{14400}$	$\frac{23}{6400}$	$\frac{-1511}{14400}$	$\frac{1171}{14400}$	$\frac{-2}{25}$		$\frac{-12143}{57600}$
9			$\frac{1}{20}$	$\frac{-23}{810}$	$\frac{61}{432}$	$\frac{61}{432}$	$\frac{-23}{810}$	$\frac{1}{20}$			211/648
8			$\frac{-13}{480}$	$\frac{317}{1440}$	$\frac{583}{5760}$	$\frac{317}{1440}$	$\frac{-13}{480}$				2807/5760
7			$\frac{11}{60}$	$\frac{9}{40}$	$\frac{9}{40}$	$\frac{11}{60}$					49/60
6			$\frac{1}{72}$	$\frac{1}{144}$	$\frac{1}{72}$						11/144
5		$\frac{-1}{5}$	$\frac{-1}{30}$	$\frac{-1}{30}$	$\frac{-1}{5}$						−7/15
4		$\frac{1}{4}$	$\frac{1}{8}$	$\frac{1}{4}$							−5/8
3		$\frac{-1}{3}$	$\frac{-1}{3}$								−2/3
2		$\frac{-1}{2}$									−1/2
1											0
0	1										1
$\sum c$	1	$\frac{-77}{60}$	$\frac{-2531}{7200}$	…	*see*	*below*					

Note that column totals generate from

$$(1-xy^2)^{1/2}(1-xy^3)^{1/3}(1-x^2y^3)^{1/3}(1-xy^4)^{1/4}(1-x^2y^4)^{1/4}(1-x^3y^4)^{1/4}$$

$$(1-xy^5)^{1/5}(1-x^2y^5)^{1/5}(1-x^3y^5)^{1/5}(1-x^4y^5)^{1/5} \quad \text{where} \quad y=1,$$

expanding to

$$1-\frac{77}{60}x-\frac{2531}{7200}x^2+\frac{360127}{1296000}x^3+\frac{54348601}{311040000}x^4+\frac{52396725643}{93312000000}x^5-\frac{15725249008811}{33592320000000}x^6$$
$$+\frac{536095858573681}{2015539200000000}x^7-\frac{221227534655582777}{9674588160000000000}x^8-\frac{15078454659730017851}{522427760640000000000}x^9$$
$$+\frac{5521836728167586290170 7}{31345665638400000000000}x^{10}+O(x^{11}).$$

However, only the first three grid columns are complete. Likewise, row totals generate from

$$(1-xy^2)^{1/2}(1-xy^3)^{1/3}(1-x^2y^3)^{1/3}(1-xy^4)^{1/4}(1-x^2y^4)^{1/4}(1-x^3y^4)^{1/4}$$
$$(1-xy^5)^{1/5}(1-x^2y^5)^{1/5}(1-x^3y^5)^{1/5}(1-x^4y^5)^{1/5} \quad \text{where} \quad x=1,$$

expanding to

$$1-\frac{1}{2}y^2-\frac{2}{3}y^3-\frac{5}{8}y^4-\frac{7}{15}y^5+\frac{11}{144}y^6+\frac{49}{60}y^7+\frac{2807}{5760}y^8+\frac{211}{648}y^9-\frac{12143}{57600}y^{10}$$
$$+\frac{811}{25920}y^{11}-\frac{2431243}{6220800}y^{12}+\frac{19889}{259200}y^{13}+\frac{626597}{2488320}y^{14}-\frac{7647377}{46656000}y^{15}+O(y^{15}).$$

Combinatorial interpretation: Consider the sub-aggregate of the countable list of *upper first quadrant* 2D Visible Point Vectors.

$$\begin{array}{llll} \langle 1,5\rangle & \langle 2,5\rangle & \langle 3,5\rangle & \langle 4,5\rangle \\ \langle 1,4\rangle & & \langle 3,4\rangle & \\ \langle 1,3\rangle & \langle 2,3\rangle & & \\ \langle 1,2\rangle & & & \end{array} \tag{12.9.18}$$

Let $p_{2,\text{up}}(\mathfrak{W},\langle a,b\rangle)$, denote the *weighted sum of partitions* of $\langle a,b\rangle$ into parts of the form $\binom{1/k}{n}\langle j,k\rangle$ for integers $n\geq 0$, with $1\leq j<k\leq 5$, $\gcd(j,k)=1$, so that each $\langle j,k\rangle$ from (12.9.18). Then

$$\prod_{k=2}^{5}\prod_{\substack{j=1,\\ \gcd(j,k)=1}}^{k-1}(1-x^jy^k)^{1/k}=\sum_{b=2}^{\infty}\sum_{a=1}^{b-1}p_{2,\text{up}}(\mathfrak{W},\langle a,b\rangle)x^ay^b, \tag{12.9.19}$$

and each entry in the grid after (12.9.17) gives each numerical value of $p_{2,\text{up}}(\mathfrak{W},\langle a,b\rangle)$.

While the above argument of partitions of vectors looks a bit disheveled and onerous, *a priori* there is another interpretation which appears more natural. That is the light diffusion model, where each visible point from the origin receives a light ray from the origin, and in turn, sends a relayed-on further light ray onto the points visible from there, and so on.

12.10 DEFINING RADIAL FROM ORIGIN REGION 2D *UPPER ALL VECTORS* AGGREGATES

These are *all lattice point vectors counted* in the infinite radial region of the first quadrant bounded by the positive y-axis and the line $y = x$. The visible lattice points in 3D space are similarly a countable set, and so also are the n-component Vectors in nD space with n a positive integer greater than unity. In order to count the entire first 3D hyperquadrant comprising all a, b, and c positive integer components of $\langle a, b, c\rangle$ it suffices to count the following vectors $\langle a, b, c\rangle$ such that $0 < a < b < c$ and apply symmetries in ways to be shown later in this book.

A set of identities involving n-dimensional *visible lattice points* was discovered by Campbell (1994). However, there has not been any emphasis on possible research projects based on radial-from-origin n-dimensional extended regions.

The *Upper All Vector* (UAV) lattice points $\langle x, y\rangle$ with $x < y$ in the first 2D quadrant are:

First quadrant 2D Upper UAV vector grid

etc.

$\langle 1,10\rangle$	$\langle 2,10\rangle$	$\langle 3,10\rangle$	$\langle 4,10\rangle$	$\langle 5,10\rangle$	$\langle 6,10\rangle$	$\langle 7,10\rangle$	$\langle 8,10\rangle$	$\langle 9,10\rangle$
$\langle 1,9\rangle$	$\langle 2,9\rangle$	$\langle 3,9\rangle$	$\langle 4,9\rangle$	$\langle 5,9\rangle$	$\langle 6,9\rangle$	$\langle 7,9\rangle$	$\langle 8,9\rangle$	
$\langle 1,8\rangle$	$\langle 2,8\rangle$	$\langle 3,8\rangle$	$\langle 4,8\rangle$	$\langle 5,8\rangle$	$\langle 6,8\rangle$	$\langle 7,8\rangle$		
$\langle 1,7\rangle$	$\langle 2,7\rangle$	$\langle 3,7\rangle$	$\langle 4,7\rangle$	$\langle 5,7\rangle$	$\langle 6,7\rangle$			
$\langle 1,6\rangle$	$\langle 2,6\rangle$	$\langle 3,6\rangle$	$\langle 4,6\rangle$	$\langle 5,6\rangle$				
$\langle 1,5\rangle$	$\langle 2,5\rangle$	$\langle 3,5\rangle$	$\langle 4,5\rangle$					
$\langle 1,4\rangle$	$\langle 2,4\rangle$	$\langle 3,4\rangle$						
$\langle 1,3\rangle$	$\langle 2,3\rangle$							
$\langle 1,2\rangle$								

These are the countable list of *upper first quadrant* 2D Vectors. That is, the vectors in the infinitely extended radial region of the 2D first quadrant between the y-axis and the line $y = x$.

12.11 EXAMPLES OF 2D *UPPER ALL VECTORS* FINITE GENERATING FUNCTIONS

12.11.1 2D Distinct *Upper All Vectors* Coefficients – Order 2

Note that for Orders 2 and 3 the *Upper All Vectors* set is identical to the *Upper VPSs.*

$$(1 + xy^2) \qquad \textit{and} \qquad (1 - xy^2) \tag{12.11.1}$$

y/x	0	1	2	$\sum r$
3				
2		1		1
1				0
0	1			1
$\sum c$	1	1		

y/x	0	1	2	$\sum r$
3				
2		−1		−1
1				0
0	1			1
$\sum c$	1	−1		

12.11.2 2D Distinct *Upper All Vectors* Coefficients – Order 3

Note that for Order 3 the *Upper All Vectors* set is identical to the *Upper VPVs* set.

$$(1+xy^2)(1+xy^3)(1+x^2y^3) \qquad \textit{and} \qquad (1-xy^2)(1-xy^3)(1-x^2y^3) \tag{12.11.2}$$

y/x	0	1	2	3	4	$\sum r$
8					1	1
7						0
6				1		1
5			1	1		2
4						0
3		1	1			2
2		1				1
1						0
0	1					1
$\sum c$	1	2	2	2	1	

y/x	0	1	2	3	4	$\sum r$
8					−1	−1
7						0
6				1		1
5			1	1		2
4						0
3		−1	−1			−2
2		−1				1
1						0
0	1					1
$\sum c$	1	−2	0	2	−1	

12.11.3 2D Distinct *Upper All Vectors* Coefficients – Order 4

$$(1+xy^2)(1+xy^3)(1+x^2y^3)(1+xy^4)(1+x^2y^4)(1+x^3y^4) \tag{12.11.3}$$

2D Upper All Vectors partitions into distinct parts – Order 4

y/x	0	1	2	3	4	5	6	7	8	9	10	$\sum r$
20											1	1
19												0
18										1		1
17									1	1		2
16								1	1	1		3
15								1	1			2
14							1	2	1			4
13						1	2	2	1			6
12						1	2	1				4
11					1	2	2	1				6
10					2	2	2					6
9				1	2	2	1					6
8				1	2	1						4
7			1	2	2	1						6
6			1	2	1							4
5			1	1								2
4		1	1	1								3
3		1	1									2
2		1										1
1												0
0	1											1
$\sum c$	1	3	5	8	10	10	10	8	5	3	1	

Note that column totals generate from

$$(1+xy^2)(1+xy^3)(1+x^2y^3)(1+xy^4)(1+x^2y^4)(1+x^3y^4) \quad \text{where} \quad y=1,$$

expanding to

$$x^{10}+3x^9+5x^8+8x^7+10x^6+10x^5+10x^4+8x^3+5x^2+3x+1.$$

Likewise, row totals generate from

$$(1+xy^2)(1+xy^3)(1+x^2y^3)(1+xy^4)(1+x^2y^4)(1+x^3y^4) \quad \text{where} \quad x=1,$$

expanding to

$$y^{20}+y^{18}+2y^{17}+3y^{16}+2y^{15}+4y^{14}+6y^{13}+4y^{12}+6y^{11}$$
$$+6y^{10}+6y^9+4y^8+6y^7+4y^6+2y^5+3y^4+2y^3+y^2+1.$$

Combinatorial interpretation: Consider the aggregate of vectors

$$\begin{matrix} \langle 1,4\rangle & \langle 2,4\rangle & \langle 3,4\rangle \\ \langle 1,3\rangle & \langle 2,3\rangle & \\ \langle 1,2\rangle & & \end{matrix} \tag{12.11.4}$$

Let $p_2(\mathfrak{D},\langle a,b\rangle)$, denote the number of partitions of $\langle a,b\rangle$ into distinct parts from (12.11.4). Then

$$\prod_{k=2}^{4}\prod_{j=1}^{k-1}(1+x^j y^k) = \sum_{b=2}^{20}\sum_{a=1}^{b-1} p_2(\mathfrak{D},\langle a,b\rangle)x^a y^b, \tag{12.11.5}$$

and each entry in the grid after (12.11.3) above gives each numerical value.

Next, we consider the following product generating an even number of distinct partitions minus the odd number of distinct partitions into vectors from the cluster $\langle a,b\langle$ with $0<1\leq a<b\leq 4$,

$$(1-xy^2)(1-xy^3)(1-x^2y^3)(1-xy^4)(1-x^2y^4)(1-x^3y^4) \tag{12.11.6}$$

which is the generating function for the grid

2D Upper All Vector partitions $\mathbf{p_2(e,v)-p_2(o,v)}$ distinct parts – Order 4

y/x	0	1	2	3	4	5	6	7	8	9	10	$\sum_r$
20											1	1
19												0
18										−1		−1
17									−1	−1		−2
16								−1	−1	1		−3
15								1	1			2
14							1	2	1			4
13						1	2	2	1			6
12						1		1				2
11					−1	−2	−2	−1				−6
10					−2	−2	−2					−6
9				−1	−2	−2	−1					−6
8				1		1						2
7			1	2	2	1						6
6			1	2	1							4
5			1	1								2
4		−1	−1	−1								−3
3		−1	−1									−2
2		−1										−1
1												0
0	1											1
$\sum_c$	1	−3	1	4	−2	−2	−2	4	1	−3	1	

We see that column totals generate from

$$(1-xy^2)(1-xy^3)(1-x^2y^3)(1-xy^4)(1-x^2y^4)(1-x^3y^4) \quad \text{where} \quad y=1,$$

expanding to

$$x^{10}-3x^9+x^8+4x^7-2x^6-2x^5-2x^4+4x^3+x^2-3x+1.$$

Likewise, row totals generate from

$$(1-xy^2)(1-xy^3)(1-x^2y^3)(1-xy^4)(1-x^2y^4)(1-x^3y^4) \quad \text{where} \quad x=1,$$

expanding to

$$y^{20}-y^{18}-2y^{17}-3y^{16}+2y^{15}+4y^{14}+6y^{13}+2y^{12}-6y^{11}$$
$$-6y^{10}-6y^{9}+2y^{8}+6y^{7}+4y^{6}+2y^{5}-3y^{4}-2y^{3}-y^{2}+1.$$

Combinatorial interpretation: Consider the aggregate of vectors

$$\begin{array}{lll} \langle 1,4\rangle & \langle 2,4\rangle & \langle 3,4\rangle \\ \langle 1,3\rangle & \langle 2,3\rangle & \\ \langle 1,2\rangle & & \end{array} \tag{12.11.7}$$

Let $p_{2,e}(\mathfrak{D},\langle a,b\rangle)$, respectively $p_{2,o}(\mathfrak{D},\langle a,b\rangle)$, denote the number of partitions of $\langle a,b\rangle$ from (12.11.7) into an *even* respectively *odd* number of distinct parts. Then

$$\prod_{k=2}^{4}\prod_{j=1}^{k-1}(1-x^{j}y^{k}) = \sum_{b=2}^{20}\sum_{a=1}^{b-1}\left[p_{2,e}(\mathfrak{D},\langle a,b\rangle)-p_{2,o}(\mathfrak{D},\langle a,b\rangle)\right]x^{a}y^{b}, \tag{12.11.8}$$

and each entry in the grid after (12.11.6) above gives each numerical value.

12.11.4 2D Unrestricted *Upper All Vectors* Partitions – Order 4

$$\frac{1}{(1-xy^2)(1-xy^3)(1-x^2y^3)(1-xy^4)(1-x^2y^4)(1-x^3y^4)} \tag{12.11.9}$$

encodes the following part sector of the infinitely extended radial grid

2D Upper All Vectors partitions into unrestricted parts – Order 4

y/x	0	1	2	3	4	5	6	7	8	9	10	11	12	13	14	$\sum r$
							⋮	⋮	⋮	⋮	⋮	⋮	⋮	⋮	⋮	
20						1	4	11	21	31	37	31	21	11	4	173
19						1	1	5	12	21	28	21	12	5	1	134
18						2	7	15	23	28	23	15	7	2		122
17						2	7	14	20	20	14	7	2			86
16					1	4	10	17	21	17	10	4	1			85
15					1	5	10	15	15	10	5	1				62
14					2	6	12	15	12	6	2					55
13					2	6	10	10	6	2						36
12				1	4	8	11	8	4	1						37
11				1	4	7	7	4	1							24
10				2	5	8	5	2								22
9				2	4	4	2									12
8			1	3	5	3	1									13
7			1	3	3	1										8
6			2	3	2											7
5			1	1												2
4		1	2	1												4
3		1	1													2
2		1														1
1																0
0	1															1
$\sum c$	1	3	8	17	33	58	97	153	233	342	489	681	930	1245	1641	

Note that column totals generate from

$$\frac{1}{(1-xy^2)(1-xy^3)(1-x^2y^3)(1-xy^4)(1-x^2y^4)(1-x^3y^4)} \quad \text{where} \quad y=1,$$

expanding to

$$1+3x+8x^2+17x^3+33x^4+58x^5+97x^6+153x^7+233x^8+342x^9$$
$$+489x^{10}+681x^{11}+930x^{12}+1245x^{13}+1641x^{14}+2130x^{15}+2730x^{16}$$
$$+3456x^{17}+4330x^{18}+O(x^{19})$$

Likewise, row totals generate from

$$\frac{1}{(1-xy^2)(1-xy^3)(1-x^2y^3)(1-xy^4)(1-x^2y^4)(1-x^3y^4)} \quad \text{where} \quad x=1,$$

expanding to

$$1+y^2+2y^3+4y^4+2y^5+7y^6+8y^7+13y^8+12y^9+22y^{10}+24y^{11}+37y^{12}$$
$$+36y^{13}+55y^{14}+62y^{15}+85y^{16}+86y^{17}+122y^{18}+134y^{19}+173y^{20}+182y^{21}$$
$$+239y^{22}+260y^{23}+326y^{24}+344y^{25}+431y^{26}+470y^{27}+569y^{28}+602y^{29}$$
$$+734y^{30}+794y^{31}+938y^{32}+O(y^{33})$$

Combinatorial interpretation: Consider the aggregate of vectors

$$\begin{array}{lll} \langle 1,4\rangle & \langle 2,4\rangle & \langle 3,4\rangle \\ \langle 1,3\rangle & \langle 2,3\rangle & \\ \langle 1,2\rangle & & \end{array} \tag{12.11.10}$$

Let $p_{2,\text{up}}(\mathfrak{U}, \langle a,b\rangle)$, denote the number of partitions of $\langle a,b\rangle$ into unrestricted parts from (12.11.10). Then

$$\prod_{k=2}^{4}\prod_{j=1}^{k-1}\frac{1}{(1-x^jy^k)} = \sum_{b=2}^{\infty}\sum_{a=1}^{b-1} p_{2,\text{up}}(\mathfrak{U}, \langle a,b\rangle)x^ay^b, \tag{12.11.11}$$

and each entry in the grid after (12.11.9) above gives each numerical value.

Observable Features. Just quickly inspecting the grid suggests the following points for the 2D Unrestricted *Upper All Vectors* Partitions of Order 4:

(a) The grid table extends indefinitely and therefore is not simply a finite patch as are most of the distinct partitions grids so far given.

(b) The grid bounds are best defined by the radial region between the lines $y = 4x$ and $y = x$ with some of the vectors actually on $y = 4x$.

(c) Each row of number entries is a symmetric finite sequence.

12.11.5 2D Distinct *Upper All Vectors* Coefficients – Order 5

The product

$$(1+xy^2)(1+xy^3)(1+x^2y^3)(1+xy^4)(1+x^2y^4) \qquad (12.11.12)$$

$$(1+x^3y^4)(1+xy^5)(1+x^2y^5)(1+x^3y^5)(1+x^4y^5)$$

generating partitions from the vectors $\langle m,n\rangle$ for $0<m<n\leq 5$, into distinct parts, is encoded by the grid

2D Upper All Vectors partitions into distinct parts – Order 5

y/x	0	1	2	3	4	5	6	7	8	9	10	11	12	13	14	15	16	17	18	19	20	$\sum r$
40																					1	1
39																						0
38																				1		1
37																			1	1		2
36																		1	1	1		3
35																	1	2	2	1		6
34																	1	2	1			4
33																2	3	3	2			10
32															1	3	4	3	1			12
31														1	3	5	5	3	1			18
30														2	5	6	5	2				20
29													1	4	6	6	4	1				8
28												1	4	7	10	7	4	1				34
27												2	6	9	9	6	2					34
26											2	5	10	12	10	5	2					46
25											3	7	11	11	7	3						42
24										2	6	11	13	11	6	2						51
23									1	4	10	14	14	10	4	1						58
22									2	7	12	15	12	7	2							57
21								1	5	11	15	15	11	5	1							64
20								2	6	12	14	12	6	2								54
19							1	5	11	15	15	11	5	1								64
18							2	7	12	15	12	7	2									57
17						1	4	10	14	14	10	4	1									58
16						2	6	11	13	11	6	2										51
15						3	7	11	11	7	3											42
14					2	5	10	12	10	5	2											46
13					2	6	9	9	6	2												34
12				1	4	7	10	7	4	1												34
11				1	4	6	6	4	1													22
10				2	5	6	5	2														20
9			1	3	5	5	3	1														18
8			1	3	4	3	1															12
7			2	3	3	2																10
6			1	2	1																	4
5		1	2	2	1																	6
4		1	2	1																		4
3		1	1																			2
2		1																				1
1																						0
0	1																					1
$\sum c$	1	4	9	18	31	46	64	82	96	106	110	106	96	82	64	46	31	18	9	4	1	

Note that column totals generate from

$$(1+xy^2)(1+xy^3)(1+x^2y^3)(1+xy^4)(1+x^2y^4)(1+x^3y^4)$$

$$(1+xy^5)(1+x^2y^5)(1+x^3y^5)(1+x^4y^5)$$

where $y=1$, expanding to

$$x^{20}+4x^{19}+9x^{18}+18x^{17}+31x^{16}+46x^{15}+64x^{14}+82x^{13}+96x^{12}+106x^{11}$$

$$+110x^{10}+106x^9+96x^8+82x^7+64x^6+46x^5+31x^4+18x^3+9x^2+4x+1.$$

Likewise, row totals generate from again

$$(1+xy^2)(1+xy^3)(1+x^2y^3)(1+xy^4)(1+x^2y^4)(1+x^3y^4)$$
$$(1+xy^5)(1+x^2y^5)(1+x^3y^5)(1+x^4y^5)$$

where $x=1$, expanding to

$$1+y^2+2y^3+3y^4+6y^5+4y^6+10y^7+12y^8+18y^9+20y^{10}+22y^{11}+34y^{12}$$
$$+34y^{13}+46y^{14}+42y^{15}+51y^{16}+58y^{17}+57y^{18}+64y^{19}+54y^{20}+64y^{21}$$
$$+57y^{22}+58y^{23}+51y^{24}+42y^{25}+46y^{26}+34y^{27}+34y^{28}+22y^{29}+20y^{30}$$
$$+18y^{31}+12y^{32}+10y^{33}+4y^{34}+6y^{35}+3y^{36}+2y^{37}+y^{38}+y^{40}.$$

Combinatorial interpretation: Consider the aggregate of vectors corresponding to (12.11.12),

$$\begin{array}{llll} \langle 1,5\rangle & \langle 2,5\rangle & \langle 3,5\rangle & \langle 4,5\rangle \\ \langle 1,4\rangle & \langle 2,4\rangle & \langle 3,4\rangle & \\ \langle 1,3\rangle & \langle 2,3\rangle & & \\ \langle 1,2\rangle & & & \end{array} \tag{12.11.13}$$

Let $p_2(\mathfrak{D},\langle a,b\rangle)$, denote the number of partitions of $\langle a,b\rangle$ into distinct parts from (12.11.13). Then we have the generating function

$$\prod_{k=2}^{4}\prod_{j=1}^{k-1}(1+x^jy^k)=\sum_{b=2}^{20}\sum_{a=1}^{b-1}p_2(\mathfrak{D},\langle a,b\rangle)x^ay^b, \tag{12.11.14}$$

encoding all possible partitions of this kind, so each entry in the previous page 20×40 grid gives all numerical values of $p_2(\mathfrak{D},\langle a,b\rangle)$.

Next, we again consider the cluster of vectors (12.11.13), where the number of partitions into distinct even numbers of parts minus the number of partitions into distinct odd number of parts is generated by the product

$$(1-xy^2)(1-xy^3)(1-x^2y^3)(1-xy^4)(1-x^2y^4) \tag{12.11.15}$$
$$(1-x^3y^4)(1-xy^5)(1-x^2y^5)(1-x^3y^5)(1-x^4y^5).$$

This result is encoded by the grid

2D Upper All Vector partitions $p_2(e,v) - p_2(o,v)$ distinct parts – Order 5

y/x	0	1	2	3	4	5	6	7	8	9	10	11	12	13	14	15	16	17	18	19	20	$\sum r$
40																					1	1
39																						0
38																				−1		−1
37																			−1	−1		−2
36																		−1	−1	−1		−3
35																	−1			−1		−2
34																	1	2	1			4
33															2	3	3	3	2			10
32															1	3	2	3	1			10
31														1	1	1	1	1	1			6
30															−2	−2	−3					−8
29													−1	−4	−6	−6	−4	−1				−22
28												−1	−4	−5	−8	−5	−4	−1				−28
27												−2	−2	−3	−3	−2	−2					−14
26												1	2	4	2	1						10
25											3	5	9	9	5	3						34
24										2	6	9	11	9	6	2						45
23									1	2	6	6	6	6	2	1						30
22										−1		−3		−1								−5
21								−1	−3	−7	−9	−9	−7	−3	−1							−40
20								−2	−6	−12	−10	−12	−6	−2								−50
19							−1	−3	−7	−9	−9	−7	−3	−1								−40
18								−1		−3		−1										−5
17						1	2	6	6	6	6	6	2	1								30
16						2	6	9	11	9	6	2										45
15						3	5	9		9	5	3										34
14						1	2	4	2	1												10
13					−2	−2	−3	−3	−2	−2												−14
12				−1	−4	−5	−8	−5	−4	−1												−28
11				−1	−4	−6	−6	−4	−1													−22
10					−3	−2	−2															−8
9			1	1	1	1	1	1														6
8			1	3	2	3	1															10
7			2	3	3	2																10
6			1	2	1																	4
5		−1			−1																	−2
4		−1	−1	−1																		−3
3		−1	−1																			−2
2		−1																				−1
1																						0
0	1																					1
$\sum c$	1	−4	3	6	−7	−2	−4	10	6	−10	2	−10	6	10	−4	−2	−7	6	3	−4	1	

Note that column totals generate from

$$(1-xy^2)(1-xy^3)(1-x^2y^3)(1-xy^4)(1-x^2y^4)(1-x^3y^4)$$
$$(1-xy^5)(1-x^2y^5)(1-x^3y^5)(1-x^4y^5)$$

where $y=1$, expanding to

$$1-4x+3x^2+6x^3-7x^4-2x^5-4x^6+10x^7+6x^8-10x^9+2x^{10}-10x^{11}$$
$$+6x^{12}+10x^{13}-4x^{14}-2x^{15}-7x^{16}+6x^{17}+3x^{18}-4x^{19}+x^{20}.$$

Likewise, row totals generate from again

$$(1-xy^2)(1-xy^3)(1-x^2y^3)(1-xy^4)(1-x^2y^4)(1-x^3y^4)$$
$$(1-xy^5)(1-x^2y^5)(1-x^3y^5)(1-x^4y^5)$$

where $x=1$, expanding to

$$1-y^2-2y^3-3y^4-2y^5+4y^6+10y^7+10y^8+6y^9-8y^{10}-22y^{11}-28y^{12}$$

$$-14y^{13}+10y^{14}+34y^{15}+45y^{16}+30y^{17}-5y^{18}-40y^{19}-50y^{20}-40y^{21}$$
$$-5y^{22}+30y^{23}+45y^{24}+34y^{25}+10y^{26}-14y^{27}-28y^{28}-22y^{29}-8y^{30}$$
$$+6y^{31}+10y^{32}+10y^{33}+4y^{34}-2y^{35}-3y^{36}-2y^{37}-y^{38}+y^{40}.$$

Combinatorial interpretation: Consider the aggregate of vectors

$$\begin{array}{llll}\langle 1,5\rangle & \langle 2,5\rangle & \langle 3,5\rangle & \langle 4,5\rangle \\ \langle 1,4\rangle & \langle 2,4\rangle & \langle 3,4\rangle & \\ \langle 1,3\rangle & \langle 2,3\rangle & & \\ \langle 1,2\rangle & & & \end{array} \tag{12.11.16}$$

Let $p_{(2,e)}(\mathfrak{D},\langle a,b\rangle)$, denote the number of partitions of $\langle a,b\rangle$ into an even number of distinct parts from (12.11.16). Let $p_{(2,o)}(\mathfrak{D},\langle a,b\rangle)$, denote the number of partitions of $\langle a,b\rangle$ into an odd number of distinct parts from (12.11.16). Then we have the generating function

$$\prod_{k=2}^{4}\prod_{j=1}^{k-1}(1-x^jy^k)=\sum_{b=2}^{20}\sum_{a=1}^{b-1}[p_{(2,e)}(\mathfrak{D},\langle a,b\rangle)-p_{(2,o)}(\mathfrak{D},\langle a,b\rangle)]x^ay^b, \tag{12.11.17}$$

encoding all 20×40 grid numerical values of $p_{(2,e)}(\mathfrak{D},\langle a,b\rangle)-p_{(2,o)}(\mathfrak{D},\langle a,b\rangle)$, with respect to $\langle a,b\rangle$).

Think About It...

I paint objects as I think them, not as I see them.

Pablo Picasso
(1881–1973)

FURTHER READING

ANDREWS, G.E. *The Theory of Partitions*, volume 2 of *Encyclopedia of Mathematics and its Applications*. Addison-Wesley Publishing Company, Advanced Book Program, Reading, Massachusetts, USA, 1976.

CAMPBELL, G.B. Multiplicative functions over Riemann zeta function products. *J. Ramanujan Soc.*, 7(1):52–63, 1992.

CAMPBELL, G.B. Dirichlet summations and products over primes. *Int. J. Math. Math. Sci.*, 16(2):359–372, 1993.

CAMPBELL, G.B. A generalized formula of Hardy. *Int. J. Math. Math. Sci.*, 17(2):369–378, 1994.

CAMPBELL, G.B. A new class of infinite products, and Euler's totient. *Int. J. Math. and Math. Sci.*, 17(3):417–422, 1994.

CAMPBELL, G.B. *Combinatorial identities in number theory related to q-series and arithmetical functions.* Doctor of philosophy thesis, School of Mathematical Sciences, The Australian National University, October 1997.

CAMPBELL, G.B. A closer look at some new identities. *Int. J. Math. and Math. Sci.*, 21(3):581–586, 1998.

CAMPBELL, G.B. Infinite products over hyperpyramid lattices. *Int. J. Math. and Math. Sci.*, 23(4):271–277, 2000.

CAMPBELL, G.B. Some n-space q-binomial theorem extensions and similar identities. arxiv preprint, June 2019. arXiv:1906.07526v1 [math.NT]: https://arxiv.org/abs/1906.07526.

CHEEMA, M.S. and MOTZKIN, T.S. Multipartitions and multipermutations. *Proc. Symp. Pure Math.*, 19:37–39, 1971.

WEISSTEIN, Eric W. Visible Point Vector Identity. From MathWorld– A Wolfram Web Resource., July 2012. Article with 2D and 3D identities (https://mathworld.wolfram.com/VisiblePointVectorIdentity.html).

WEISSTEIN, Eric W. Visible Point. From MathWorld–A Wolfram Web Resource., July 2020. Article with references (https://mathworld.wolfram.com/VisiblePoint.html).

CHAPTER 13

Integer Partitions Generalized to Vector Partitions

13.1 USE OF THE q-BINOMIAL THEOREM

THE LITERATURE on q-series goes way back to the nineteenth century, starting with Heine [154] and [155]. This generalized the classical hypergeometric series work introduced by Gauss [135]. A beautiful account of these q-series identities and their proof as well as their application to integer partitions is given in the excellent book by Andrews [12]. Such q-series identities, including the celebrated Rogers-Ramanujan identities, were shown in the 1980s by Baxter [52] to be applicable in the calculation of free energy in statistical mechanics models, such as in his exact solution of [52]. The results in this chapter mostly follow the author's ([93] from 2019, but some date back 22 years earlier to the author's PhD thesis [89], or beyond that to the early 1990s papers Campbell [84, 85, 86, 87], and circa 2000 papers [90, 91].

We recount our introduction in chapter 3, of the very useful q-binomial theorem.

Theorem 13.1 *If* $|q| < 1, |t| < 1$, *for all complex* a,

$$1+\sum_{k=1}^{\infty} \frac{(1-a)(1-aq)(1-aq^2)...(1-aq^{k-1})t^k}{(1-q)(1-q^2)(1-q^3)...(1-q^k)} = \prod_{k=0}^{\infty} \frac{1-atq^k}{1-tq^k}. \tag{13.1.1}$$

We extend the approach in Andrews [12] which derives many generating functions of integer partition identities from cases of the q-binomial theorem. We do so by moving the logic into n-space generalizations of the q-binomial theorem, as well as applying the methodology to the newer, and not so well-known, *visible point vector identities* that appeared in the 1990s. We therefore, give some new identities that are similar to traditional q-series pertaining to *stepping stone* type vector partitions in Euclidean n-space regions for the lattice point summations. Along the way, we use the standard notation for q-shifted factorials,

DOI: 10.1201/9781003174158-13

$$(a;q)_k = \begin{cases} 1, & k=0; \\ (1-a)(1-aq)...(1-aq^{k-1}), & k=1,2,..., \end{cases}$$

so then $(a;q)_\infty$ is understood as the limit $k \to \infty$, with occasional abbreviated form $(a)_k = (a;q)_k$.

13.2 A HIGHER DIMENSIONAL KIND OF q-BINOMIAL THEOREM

Note, the generalization of the q-binomial series is no longer a q-series; having transitioned the q variable into $q_1, q_2, \ldots, q_n$ variables. However, we begin with the

Definition 13.1 *Define the function $F_n(t)$ for all complex numbers a,t with $|a|,|t| < 1$, and for $n \geq 1$ with $|q_1|,|q_2|,...,|q_n| < 1$ by the sequence of functions $F_n(t)$ with*

$$F_0(-;a,t) = \frac{1-at}{1-t} \tag{13.2.1}$$

and for all of $|q_1|,|q_2|,|q_3|,...,|q_n| < 1$, by

$$F_n(q_1,q_2,q_3,...,q_n;a,t) = \prod_{\alpha_1,\alpha_2,\alpha_3,...,\alpha_n \geq 0} \frac{1-q_1^{\alpha_1}q_2^{\alpha_2}...q_n^{\alpha_n}at}{1-q_1^{\alpha_1}q_2^{\alpha_2}...q_n^{\alpha_n}t} \equiv \sum_{k=0}^{\infty} {}_nA_k t^k \tag{13.2.2}$$

where $\alpha_1,\alpha_2,\alpha_3,...,\alpha_n$ may be all zero.

Also, note that by this definition the right side of (13.1.1) is $F_1(q;a,t)$, so it is evident we are working on a generalized q-binomial expression, despite having used q_i to replace q. Next, based on this definition we can assert the

Theorem 13.2 *The n-space q-binomial theorem.*

$${}_nA_k(q_1,q_2,q_3,...,q_n;a) = \det(a_{ij})/k! \tag{13.2.3}$$

where the determinant is of order k, and

$$a_{ij} = \begin{cases} \frac{1-a^{i-j+1}}{(1-q_1{}^{i-j+1})(1-q_2{}^{i-j+1})(1-q_3{}^{i-j+1})...(1-q_n{}^{i-j+1})}, & i \geq j; \\ -i, & i=j-1; \\ 0, & otherwise. \end{cases}$$

We follow this with the

Definition 13.2 *Define the function $G_n(t)$ for all $|t| < 1$, and for $n \geq 1$ with $|q_1|,|q_2|,...,|q_n| < 1$ by the sequence of functions $G_n(t)$ with*

$$G_0(-;t) = 1+t \tag{13.2.4}$$

and for all of $|q_1|,|q_2|,|q_3|,...,|q_n| < 1$, by

$$G_n(q_1,q_2,q_3,...,q_n;t) = \prod_{\alpha_1,\alpha_2,\alpha_3,...,\alpha_n \geq 0} (1+q_1^{\alpha_1}q_2^{\alpha_2}...q_n^{\alpha_n}t) \equiv \sum_{k=0}^{\infty} {}_nC_k t^k \tag{13.2.5}$$

where $\alpha_1,\alpha_2,\alpha_3,...,\alpha_n$ may be all zero.

Next, based on this definition we can assert the

Theorem 13.3 *The n-space product generating distinct partitions.*

$$ {}_nC_k(q_1, q_2, q_3, \ldots, q_n; t) = \det(a_{ij})/k! \tag{13.2.6}$$

where the determinant is of order k, and

$$a_{ij} = \begin{cases} \frac{(-1)^{i-j+1}}{(1+q_1{}^{i-j+1})(1+q_2{}^{i-j+1})(1+q_3{}^{i-j+1})\ldots(1+q_n{}^{i-j+1})}, & i \geq j; \\ -i, & i = j-1; \\ 0, & otherwise. \end{cases}$$

13.3 PROVING THE n-SPACE VERSION OF q-BINOMIAL THEOREM

Before giving the full n-space proof, we mention that the cases $n=2$ and $n=3$ of theorem 13.2.1 are not in the literature. Also, for $n=1$, the q-binomial theorem itself, there is no proof using determinants in the literature. However, we first prove the n-space identity, before then giving a diverse range of examples.

Proof 1 *We see first of all that*

$$\begin{aligned} F_n(q_1, q_2, q_3, \ldots, q_n; a, t) &= \exp\left(\sum_{\alpha_1, \alpha_2, \alpha_3, \ldots, \alpha_n \geq 0} \log\left(\frac{1 - q_1^{\alpha_1} q_2^{\alpha_2} \ldots q_n^{\alpha_n} at}{1 - q_1^{\alpha_1} q_2^{\alpha_2} \ldots q_n^{\alpha_n} t} \right)\right) \\ &= \exp\left(\sum_{k=1}^{\infty} \frac{1-a^k}{(1-q_1{}^k)(1-q_2{}^k)(1-q_3{}^k)\cdots(1-q_n{}^k)} \frac{t^k}{k} \right). \end{aligned}$$

Since $F_n(q_1, q_2, q_3, \ldots, q_n; a, t) = F_n(t) = \sum_{k=0}^{\infty} {}_nA_k t^k$, *differentiating both sides of the equation*

$$\log\left(\sum_{k=0}^{\infty} {}_nA_k t^k \right) = \sum_{k=1}^{\infty} \frac{1-a^k}{(1-q_1{}^k)(1-q_1{}^k)(1-q_2{}^k)\cdots(1-q_n{}^k)} \frac{t^k}{k}$$

with respect to t, then equating like powers of t, leads to the set of simultaneous equations for any positive integer k

$$1 \times {}_nA_1 = \frac{1-a}{(1-q_1)(1-q_2)\cdots(1-q_n)},$$

$$2 \times {}_nA_2 = \frac{1-a^2}{(1-q_1{}^2)(1-q_2{}^2)\cdots(1-q_n{}^2)} + {}_nA_1 \frac{1-a}{(1-q_1)(1-q_2)\cdots(1-q_n)},$$

$$\begin{aligned} 3 \times {}_nA_3 = &\frac{1-a^3}{(1-q_1{}^3)(1-q_2{}^3)\cdots(1-q_n{}^3)} + {}_nA_1 \frac{1-a^2}{(1-q_1{}^2)(1-q_2{}^2)\cdots(1-q_n{}^2)} \\ &+ {}_nA_2 \frac{1-a}{(1-q_1)(1-q_2)\cdots(1-q_n)}, \end{aligned}$$

$$4 \times {}_nA_4 = \frac{1-a^4}{(1-{q_1}^4)(1-{q_2}^4)\cdots(1-{q_n}^4)} + {}_nA_1 \frac{1-a^3}{(1-{q_1}^3)(1-{q_2}^3)\cdots(1-{q_n}^3)}$$

$$+{}_nA_2 \frac{1-a^2}{(1-{q_1}^2)(1-{q_2}^2)\cdots(1-{q_n}^2)} + {}_nA_3 \frac{1-a}{(1-q_1)(1-q_2)\cdots(1-q_n)},$$

etc., down to the kth equation

$$k \times {}_nA_k = \frac{1-a^k}{(1-{q_1}^k)(1-{q_2}^k)\cdots(1-{q_n}^k)} + {}_nA_1 \frac{1-a^{k-1}}{(1-{q_1}^{k-1})(1-{q_2}^{k-1})\cdots(1-{q_n}^{k-1})} + \ldots$$

$$\ldots + {}_nA_{k-2} \frac{1-a^2}{(1-{q_1}^2)(1-{q_2}^2)\cdots(1-{q_n}^2)} + {}_nA_{k-1} \frac{1-a}{(1-q_1)(1-q_2)\cdots(1-q_n)}.$$

We note that for the above, the first equation has one unknown, the first and second equations together are two equations in two unknowns, the first, second and third equations are three equations in three unknowns and so on. The set of k equations here is a solvable set of k equations in k unknowns, and is neatly solved by Cramer's Rule involving determinants. Applying this rule in a straightforward manner yields the theorem. ■

The above proof is essentially the same as that given in Macdonald [191] when discussing such combinatorial objects as Schur functions and Jack polynomials. Logarithmic derivatives are taken with respect to t, then the recurrence for the coefficients is solved by Cramer's rule.

We write theorem 13.2 as a single equation identity as follows and rate it as a theorem.

Theorem 13.4 *The n-space generalization of the q-binomial product.*

$$F_n(q_1, q_2, q_3, \ldots, q_n; a, t) = \prod_{\alpha_1, \alpha_2, \alpha_3, \ldots, \alpha_n \geq 0} \frac{1 - q_1^{\alpha_1} q_2^{\alpha_2} \ldots q_n^{\alpha_n} at}{1 - q_1^{\alpha_1} q_2^{\alpha_2} \ldots q_n^{\alpha_n} t} \tag{13.3.1}$$

$$= 1 + \frac{1-a}{(1-q_1)(1-q_2)\cdots(1-q_n)} \frac{t}{1!}$$

$$+ \begin{vmatrix} \frac{1-a}{(1-q_1)(1-q_2)\cdots(1-q_n)} & -1 \\ \frac{1-a^2}{(1-{q_1}^2)(1-{q_2}^2)\cdots(1-{q_n}^2)} & \frac{1-a}{(1-q_1)(1-q_2)\cdots(1-q_n)} \end{vmatrix} \frac{t^2}{2!}$$

$$+ \begin{vmatrix} \frac{1-a}{(1-q_1)(1-q_2)\cdots(1-q_n)} & -1 & 0 \\ \frac{1-a^2}{(1-{q_1}^2)(1-{q_2}^2)\cdots(1-{q_n}^2)} & \frac{1-a}{(1-q_1)(1-q_2)\cdots(1-q_n)} & -2 \\ \frac{1-a^3}{(1-{q_1}^3)(1-{q_2}^3)\cdots(1-{q_n}^3)} & \frac{1-a^2}{(1-{q_1}^2)(1-{q_2}^2)\cdots(1-{q_n}^2)} & \frac{1-a}{(1-q_1)(1-q_2)\cdots(1-q_n)} \end{vmatrix} \frac{t^3}{3!}$$

$$+ \begin{vmatrix} \frac{1-a}{(1-q_1)(1-q_2)\cdots(1-q_n)} & -1 & 0 & 0 \\ \frac{1-a^2}{(1-{q_1}^2)(1-{q_2}^2)\cdots(1-{q_n}^2)} & \frac{1-a}{(1-q_1)(1-q_2)\cdots(1-q_n)} & -2 & 0 \\ \frac{1-a^3}{(1-{q_1}^3)(1-{q_2}^3)\cdots(1-{q_n}^3)} & \frac{1-a^2}{(1-{q_1}^2)(1-{q_2}^2)\cdots(1-{q_r}^2)} & \frac{1-a}{(1-q_1)(1-q_2)\cdots(1-q_n)} & -3 \\ \frac{1-a^4}{(1-{q_1}^4)(1-{q_2}^4)\cdots(1-{q_n}^4)} & \frac{1-a^3}{(1-{q_1}^3)(1-{q_2}^3)\cdots(1-{q_n}^3)} & \frac{1-a^2}{(1-{q_1}^2)(1-{q_2}^2)\cdots(1-{q_n}^2)} & \frac{1-a}{(1-q_1)(1-q_2)\cdots(1-q_n)} \end{vmatrix} \frac{t^4}{4!}$$

$$+etc.$$

Following though equation (13.2.5) in complete analogy to the above process that led us to theorem 13.4 with its equation (13.3.1) brings us similarly to state the following.

Theorem 13.5 *An n-space generalization of the q-product for distinct partitions.*

$$G_n(q_1,q_2,q_3,...,q_n;t) = \prod_{\alpha_1,\alpha_2,\alpha_3,...,\alpha_n \geq 0} (1+q_1^{\alpha_1}q_2^{\alpha_2}...q_n^{\alpha_n}t) \tag{13.3.2}$$

$$= 1 + \frac{1}{(1-q_1)(1-q_2)\cdots(1-q_n)}\frac{t}{1!}$$

$$+ \begin{vmatrix} \frac{1}{(1-q_1)(1-q_2)\cdots(1-q_n)} & -1 \\ \frac{-1}{(1-q_1{}^2)(1-q_2{}^2)\cdots(1-q_n{}^2)} & \frac{1}{(1-q_1)(1-q_2)\cdots(1-q_n)} \end{vmatrix} \frac{t^2}{2!}$$

$$+ \begin{vmatrix} \frac{1}{(1-q_1)(1-q_2)\cdots(1-q_n)} & -1 & 0 \\ \frac{-1}{(1-q_1{}^2)(1-q_2{}^2)\cdots(1-q_n{}^2)} & \frac{1}{(1-q_1)(1-q_2)\cdots(1-q_n)} & -2 \\ \frac{1}{(1-q_1{}^3)(1-q_2{}^3)\cdots(1-q_n{}^3)} & \frac{-1}{(1-q_1{}^2)(1-q_2{}^2)\cdots(1-q_n{}^2)} & \frac{1}{(1-q_1)(1-q_2)\cdots(1-q_n)} \end{vmatrix} \frac{t^3}{3!}$$

$$+ \begin{vmatrix} \frac{1}{(1-q_1)(1-q_2)\cdots(1-q_n)} & -1 & 0 & 0 \\ \frac{-1}{(1-q_1{}^2)(1-q_2{}^2)\cdots(1-q_n{}^2)} & \frac{1}{(1-q_1)(1-q_2)\cdots(1-q_n)} & -2 & 0 \\ \frac{1}{(1-q_1{}^3)(1-q_2{}^3)\cdots(1-q_n{}^3)} & \frac{-1}{(1-q_1{}^2)(1-q_2{}^2)\cdots(1-q_r{}^2)} & \frac{1}{(1-q_1)(1-q_2)\cdots(1-q_n)} & -3 \\ \frac{-1}{(1-q_1{}^4)(1-q_2{}^4)\cdots(1-q_n{}^4)} & \frac{1}{(1-q_1{}^3)(1-q_2{}^3)\cdots(1-q_n{}^3)} & \frac{-1}{(1-q_1{}^2)(1-q_2{}^2)\cdots(1-q_n{}^2)} & \frac{1}{(1-q_1)(1-q_2)\cdots(1-q_n)} \end{vmatrix} \frac{t^4}{4!}$$

$$+ \ etc. \ .$$

The 1-space example of equation (13.1.1) is the q-binomial theorem itself, and the 1D case of (13.3.2) is equation (3.5.9) of Chapter 3. The following derived form of the q-binomial theorem is proven by applying the famous Faà de Bruno's formula (see Abramowitz and Stegun [1, chapter 24, pp824]). This form of the q-binomial theorem seems to not be in the literature.

Corollary 13.1

$$F_1(q;a,t) = \prod_{k=0}^{\infty} \frac{1-atq^k}{1-tq^k} = 1 + \frac{1-a}{1-q}\frac{t}{1!} + \begin{vmatrix} \frac{1-a}{1-q} & -1 \\ \frac{1-a^2}{1-q^2} & \frac{1-a}{1-q} \end{vmatrix} \frac{t^2}{2!} \tag{13.3.3}$$

$$+ \begin{vmatrix} \frac{1-a}{1-q} & -1 & 0 \\ \frac{1-a^2}{1-q^2} & \frac{1-a}{1-q} & -2 \\ \frac{1-a^3}{1-q^3} & \frac{1-a^2}{1-q^2} & \frac{1-a}{1-q} \end{vmatrix} \frac{t^3}{3!} + \begin{vmatrix} \frac{1-a}{1-q} & -1 & 0 & 0 \\ \frac{1-a^2}{1-q^2} & \frac{1-a}{1-q} & -2 & 0 \\ \frac{1-a^3}{1-q^3} & \frac{1-a^2}{1-q^2} & \frac{1-a}{1-q} & -3 \\ \frac{1-a^4}{1-q^4} & \frac{1-a^3}{1-q^3} & \frac{1-a^2}{1-q^2} & \frac{1-a}{1-q} \end{vmatrix} \frac{t^4}{4!} + etc.$$

An interesting related aside is that in Mathematica or WolframAlpha, the code

$$Det[\{(1-a)/(1-q),-1,0\},\{(1-a^2)/(1-q^2),(1-a)/(1-q),-2\},$$
$$\{(1-a^3)/(1-q^3),(1-a^2)/(1-q^2),(1-a)/(1-q)\}]$$

$$= 6((1-a)(1-aq)(1-aq^2))/((1-q)(1-q^2)(1-q^3))$$

yields the particular verification

$$\begin{vmatrix} \frac{1-a}{1-q} & -1 & 0 \\ \frac{1-a^2}{1-q^2} & \frac{1-a}{1-q} & -2 \\ \frac{1-a^3}{1-q^3} & \frac{1-a^2}{1-q^2} & \frac{1-a}{1-q} \end{vmatrix} = 6 \times \frac{(1-a)(1-aq)(1-aq^2)}{(1-q)(1-q^2)(1-q^3)},$$

and the direct application of Faà De Bruno's formula to the general determinant coefficient in (13.3.3) will complete this new proof of the q-binomial theorem. ■

Corollary 13.2 *The 2-space example of (13.3.3) is*

$$F_2(x,y;a,t) = \prod_{j,k\geq 0} \frac{1-x^j y^k at}{1-x^j y^k t} \tag{13.3.4}$$

$$= 1 + \frac{1-a}{(1-x)(1-y)} \frac{t}{1!} + \begin{vmatrix} \frac{1-a}{(1-x)(1-y)} & -1 \\ \frac{1-a^2}{(1-x^2)(1-y^2)} & \frac{1-a}{(1-x)(1-y)} \end{vmatrix} \frac{t^2}{2!}$$

$$+ \begin{vmatrix} \frac{1-a}{(1-x)(1-y)} & -1 & 0 \\ \frac{1-a^2}{(1-x^2)(1-y^2)} & \frac{1-a}{(1-x)(1-y)} & -2 \\ \frac{1-a^3}{(1-x^3)(1-y^3)} & \frac{1-a^2}{(1-x^2)(1-y^2)} & \frac{1-a}{(1-x)(1-y)} \end{vmatrix} \frac{t^3}{3!}$$

$$+ \begin{vmatrix} \frac{1-a}{(1-x)(1-y)} & -1 & 0 & 0 \\ \frac{1-a^2}{(1-x^2)(1-y^2)} & \frac{1-a}{(1-x)(1-y)} & -2 & 0 \\ \frac{1-a^3}{(1-x^3)(1-y^3)} & \frac{1-a^2}{(1-x^2)(1-y^2)} & \frac{1-a}{(1-x)(1-y)} & -3 \\ \frac{1-a^4}{(1-x^4)(1-y^4)} & \frac{1-a^3}{(1-x^3)(1-y^3)} & \frac{1-a^2}{(1-x^2)(1-y^2)} & \frac{1-a}{(1-x)(1-y)} \end{vmatrix} \frac{t^4}{4!}$$

$$+ \textit{etc.}$$

The function $F_2(x,y;a,t)$ is interesting for a variety of reasons. Firstly, $F_2(0,q;a,t)$ and $F_2(q,0;a,t)$ are the same as the right side of the equation (13.1.1), and therefore corollary 13.2 is an extension of the q-binomial formula. Secondly, particular cases such as for $|x| < 1$

$$F_2(x,x;a,t) = \prod_{k\geq 0} \left(\frac{1-x^k at}{1-x^k t}\right)^k$$

lead easily to functions like

$$F_2(q,q;0,qa) = \prod_{k\geq 1} \left(\frac{1}{1-q^k a}\right)^k$$

whose coefficient of $a^m q^n$ is the number of plane partitions of n for which the sum of the diagonal parts is m. (see Andrews [12, pages 189 and 199].

That is, we have

$$\prod_{k\geq 1}\left(\frac{1}{1-q^k a}\right)^k = 1 + \frac{1}{(1-q)^2}\frac{aq}{1!} + \begin{vmatrix} \frac{1}{(1-q)^2} & -1 \\ \frac{1}{(1-q^2)^2} & \frac{1}{(1-q)^2} \end{vmatrix} \frac{a^2q^2}{2!}$$

$$+ \begin{vmatrix} \frac{1}{(1-q)^2} & -1 & 0 \\ \frac{1}{(1-q^2)^2} & \frac{1}{(1-q)^2} & -2 \\ \frac{1}{(1-q^3)^2} & \frac{1}{(1-q^2)^2} & \frac{1}{(1-q)^2} \end{vmatrix} \frac{a^3q^3}{3!}$$

$$+ \begin{vmatrix} \frac{1}{(1-q)^2} & -1 & 0 & 0 \\ \frac{1}{(1-q^2)^2} & \frac{1}{(1-q)^2} & -2 & 0 \\ \frac{1}{(1-q^3)^2} & \frac{1}{(1-q^2)^2} & \frac{1}{(1-q)^2} & -3 \\ \frac{1}{(1-q^4)^2} & \frac{1}{(1-q^3)^2} & \frac{1}{(1-q^2)^2} & \frac{1}{(1-q)^2} \end{vmatrix} \frac{a^4q^4}{4!}$$

$$+ etc.$$

In this instance we have derived the determinants from $F_2(q, q; 0, qa)$ substituted into (13.3.4). The beginning patch of it's 2D coefficient grid is as follows

Number of plane partitions of n where the sum of the diagonal parts is m

⋮																⋮
15															1	⋯
14														1	2	⋯
13													1	2	6	⋯
12												1	2	6	14	⋯
11											1	2	6	14	33	⋯
10										1	2	6	14	33	70	⋯
9									1	2	6	14	33	70	149	⋯
8								1	2	6	14	33	70	149	298	⋯
7							1	2	6	33	33	70	149	298	582	⋯
6						1	2	6	14	70	70	149	290	561	1034	⋯
5					1	2	6	14	33	70	142	272	507	904	1568	⋯
4				1	2	6	14	33	64	127	228	404	672	1100	1724	⋯
3			1	2	6	14	28	52	93	152	242	370	546	784	1103	⋯
2		1	2	6	10	19	28	44	60	85	110	146	182	231	280	⋯
1	1	2	3	4	5	6	7	8	9	10	11	12	13	14	15	⋯
m/n	1	2	3	4	5	6	7	8	9	10	11	12	13	14	15	⋯

Further esoteric restrictions on (13.3.4) yield the celebrated MacMahon generating functions (see Andrews [12, page 184]),

$$F_2(q, q; q^k, q) = \prod_{j\geq 1}\left(\frac{1}{1-q^j}\right)^{\min(k,j)},$$

$$F_2(q,q;0,q) = \prod_{j \geq 1} \left(\frac{1}{1-q^j} \right)^j,$$

whose coefficients of q^j enumerate respectively:-

(a) The number of k *rowed* partitions of j.

(b) The number of *unlimited row* plane partitions of j.

Equation (13.2.3) can also lead to vector partition extensions of the formulae given in Andrews [12, chapter 12, page 203] for the n-space vector partition generating functions

$$F_n(q_1,q_2,q_3,...,q_n;0,1)$$

$$= \prod_{\alpha_1,\alpha_2,\alpha_3,...,\alpha_n \geq 0} \frac{1}{1-q_1^{\alpha_1} q_2^{\alpha_2} ... q_n^{\alpha_n}} = \sum_{\alpha_1,\alpha_2,\alpha_3,...,\alpha_n \geq 0} P(\mathbf{n}) q_1^{\alpha_1} q_2^{\alpha_2} ... q_n^{\alpha_n},$$

with left side product boundary condition that $\alpha_1,\alpha_2,\alpha_3,...,\alpha_n$ cannot be all zero.

$$F_n(q_1,q_2,q_3,...,q_n;0,1)/F_n({q_1}^2,{q_2}^2,{q_3}^2,...,{q_n}^2;0,1)$$

$$= G_n(q_1,q_2,q_3,...,q_n;1)$$

$$= \prod_{\alpha_1,\alpha_2,\alpha_3,...,\alpha_n \geq 0} (1+q_1^{\alpha_1} q_2^{\alpha_2} \cdots q_n^{\alpha_n}) = \sum_{\alpha_1,\alpha_2,\alpha_3,...,\alpha_n \geq 0} Q(\mathbf{n}) q_1^{\alpha_1} q_2^{\alpha_2} \cdots q_n^{\alpha_n},$$

where the left side product includes the boundary condition that $\alpha_1,\alpha_2,\alpha_3,...,\alpha_n$ may be all zero.

For example in the 2D scenario of this, setting $q_1 = x$ and $q_2 = y$, we see that for $|x| < 1$, $|y| < 1$,

$$\prod_{j,k \geq 0; j+k \neq 0} \left(\frac{1}{1-x^j y^k} \right),$$

$$\prod_{j,k \geq 0} \left(1+x^j y^k \right),$$

whose coefficients of $x^m y^n$ generate two-dimensional first quadrant Euclidean space vector partitions. Citing the theory in Andrews [12, chapter 12]) these coefficients enumerate respectively, the number of partitions of:

(a) *unrestricted* two dimensional vectors summing to vector $\langle m,n \rangle$,

(b) *distinct* two dimensional vectors summing to vector $\langle m,n \rangle$,

where m and n components are non-negative integers.

In the present book, we envision each integer lattice point in the vector space to be a stepping stone, where each one or more jumps from stone to stone is allowed according to specified rules of size and order of different jumps. As one of many interpretation theories or models for these vector partitions, at least in 2D and 3D

Euclidean space, this stepping stone analogy may suffice in some situations we can present later in our book.

We can derive a 2D vector partition identity to generalize the well-known *overpartition* q-series identity as follows. Starting from (13.3.4), replace a by a/b and t by bz, hence for $|bz|<1$,

$$F_2(x,y;a/b,bz)=\prod_{j,k\geq 0}\frac{1-x^j y^k az}{1-x^j y^k bz} \tag{13.3.5}$$

$$=1+\frac{1-(a/b)}{(1-x)(1-y)}\frac{bz}{1!}+\begin{vmatrix}\frac{1-(a/b)}{(1-x)(1-y)} & -1\\ \frac{1-(a/b)^2}{(1-x^2)(1-y^2)} & \frac{1-(a/b)}{(1-x)(1-y)}\end{vmatrix}\frac{b^2z^2}{2!}$$

$$+\begin{vmatrix}\frac{1-(a/b)}{(1-x)(1-y)} & -1 & 0\\ \frac{1-(a/b)^2}{(1-x^2)(1-y^2)} & \frac{1-(a/b)}{(1-x)(1-y)} & -2\\ \frac{1-(a/b)^3}{(1-x^3)(1-y^3)} & \frac{1-(a/b)^2}{(1-x^2)(1-y^2)} & \frac{1-(a/b)}{(1-x)(1-y)}\end{vmatrix}\frac{b^3z^3}{3!}$$

$$+\begin{vmatrix}\frac{1-(a/b)}{(1-x)(1-y)} & -1 & 0 & 0\\ \frac{1-(a/b)^2}{(1-x^2)(1-y^2)} & \frac{1-(a/b)}{(1-x)(1-y)} & -2 & 0\\ \frac{1-(a/b)^3}{(1-x^3)(1-y^3)} & \frac{1-(a/b)^2}{(1-x^2)(1-y^2)} & \frac{1-(a/b)}{(1-x)(1-y)} & -3\\ \frac{1-(a/b)^4}{(1-x^4)(1-y^4)} & \frac{1-(a/b)^3}{(1-x^3)(1-y^3)} & \frac{1-(a/b)^2}{(1-x^2)(1-y^2)} & \frac{1-(a/b)}{(1-x)(1-y)}\end{vmatrix}\frac{b^4z^4}{4!}$$

$$+\ etc.$$

Next allow $a\to -1$ and $b\to 1$, giving us

$$\prod_{j,k\geq 0}\frac{1+x^j y^k z}{1-x^j y^k z} \tag{13.3.6}$$

$$=1+\frac{2}{(1-x)(1-y)}\frac{z}{1!}+\begin{vmatrix}\frac{2}{(1-x)(1-y)} & -1\\ 0 & \frac{2}{(1-x)(1-y)}\end{vmatrix}\frac{z^2}{2!}$$

$$+\begin{vmatrix}\frac{2}{(1-x)(1-y)} & -1 & 0\\ 0 & \frac{2}{(1-x)(1-y)} & -2\\ \frac{2}{(1-x^3)(1-y^3)} & 0 & \frac{2}{(1-x)(1-y)}\end{vmatrix}\frac{z^3}{3!}$$

$$+\begin{vmatrix}\frac{2}{(1-x)(1-y)} & -1 & 0 & 0\\ 0 & \frac{2}{(1-x)(1-y)} & -2 & 0\\ \frac{2}{(1-x^3)(1-y^3)} & 0 & \frac{2}{(1-x)(1-y)} & -3\\ 0 & \frac{2}{(1-x^3)(1-y^3)} & 0 & \frac{2}{(1-x)(1-y)}\end{vmatrix}\frac{z^4}{4!}$$

$$+\ etc.$$

This generalizes the overpartition identity. An overpartition of n is a non-increasing sequence of natural numbers whose sum is n in which the first occurrence of a number may be overlined, introduced by Corteel and Lovejoy in 2004 [115]. Let $\overline{p}(n)$ denote the number of overpartitions of an integer n. For example, $\overline{p}(3) = 8$ because there are 8 possible overpartitions of 3:

$$(3); (\overline{3}); (2;1); (2;\overline{1}); (\overline{2};1); (\overline{2};\overline{1}); (1;1;1); (\overline{1};1;1).$$

Since the overlined parts form a partition into distinct parts and the non-overlined parts form an ordinary partition, the generating function for overpartitions is given by

$$\sum_{n=0}^{\infty} \overline{p}(n) q^n = \frac{(-q;q)_\infty}{(q;q)_\infty} = \left(\frac{1+q}{1-q}\right)\left(\frac{1+q^2}{1-q^2}\right)\left(\frac{1+q^3}{1-q^3}\right)\left(\frac{1+q^4}{1-q^4}\right)\cdots.$$

See for example, Corteel and Lovejoy [115], and Bringmann and Lovejoy [77].

A further interesting consideration of the approach leading from (13.3.5) to (13.3.6) and overpartition generalization is that the reciprocal product of (13.3.6) can be obtained similarly. In (13.3.5), allow $a \to 1$ and $b \to -1$, or simply replace z with $-z$ in (13.3.6) to see that

$$\prod_{j,k \geq 0} \frac{1 - x^j y^k z}{1 + x^j y^k z} \tag{13.3.7}$$

$$= 1 - \frac{2}{(1-x)(1-y)} \frac{z}{1!} + \begin{vmatrix} \frac{2}{(1-x)(1-y)} & -1 \\ 0 & \frac{2}{(1-x)(1-y)} \end{vmatrix} \frac{z^2}{2!}$$

$$- \begin{vmatrix} \frac{2}{(1-x)(1-y)} & -1 & 0 \\ 0 & \frac{2}{(1-x)(1-y)} & -2 \\ \frac{2}{(1-x^3)(1-y^3)} & 0 & \frac{2}{(1-x)(1-y)} \end{vmatrix} \frac{z^3}{3!}$$

$$+ \begin{vmatrix} \frac{2}{(1-x)(1-y)} & -1 & 0 & 0 \\ 0 & \frac{2}{(1-x)(1-y)} & -2 & 0 \\ \frac{2}{(1-x^3)(1-y^3)} & 0 & \frac{2}{(1-x)(1-y)} & -3 \\ 0 & \frac{2}{(1-x^3)(1-y^3)} & 0 & \frac{2}{(1-x)(1-y)} \end{vmatrix} \frac{z^4}{4!}$$

$$- + \textit{etc.}$$

This then generalizes the classical identity due to Gauss, (see Chapter 3, theorem 3.6)

$$\sum_{n=-\infty}^{\infty} (-1)^n q^{n^2} = \frac{(q;q)_\infty}{(-q;q)_\infty} = \left(\frac{1-q}{1+q}\right)\left(\frac{1-q^2}{1+q^2}\right)\left(\frac{1-q^3}{1+q^3}\right)\left(\frac{1-q^4}{1+q^4}\right)\cdots, \tag{13.3.8}$$

prompting the question of whether (13.3.7) may have some 2D cases that someway extend or mimic (13.3.8).

Since we have dwelt here on some 2D ideas for the overpartitions generating function and for the Gauss identity, it's worth taking a look at the 3D version.

Corollary 13.3 *The 3-space example of (13.3.1) is*

$$F_3(x,y,z;a,t) = \prod_{j,k,h\geq 0} \frac{1-x^j y^k z^h at}{1-x^j y^k z^h t} \tag{13.3.9}$$

$$= 1 + \frac{1-a}{(1-x)(1-y)(1-z)} \frac{t}{1!} + \begin{vmatrix} \frac{1-a}{(1-x)(1-y)(1-z)} & -1 \\ \frac{1-a^2}{(1-x^2)(1-y^2)(1-z^2)} & \frac{1-a}{(1-x)(1-y)(1-z)} \end{vmatrix} \frac{t^2}{2!}$$

$$+ \begin{vmatrix} \frac{1-a}{(1-x)(1-y)(1-z)} & -1 & 0 \\ \frac{1-a^2}{(1-x^2)(1-y^2)(1-z^2)} & \frac{1-a}{(1-x)(1-y)(1-z)} & -2 \\ \frac{1-a^3}{(1-x^3)(1-y^3)(1-z^3)} & \frac{1-a^2}{(1-x^2)(1-y^2)(1-z^2)} & \frac{1-a}{(1-x)(1-y)(1-z)} \end{vmatrix} \frac{t^3}{3!}$$

$$+ \begin{vmatrix} \frac{1-a}{(1-x)(1-y)(1-z)} & -1 & 0 & 0 \\ \frac{1-a^2}{(1-x^2)(1-y^2)(1-z^2)} & \frac{1-a}{(1-x)(1-y)(1-z)} & -2 & 0 \\ \frac{1-a^3}{(1-x^3)(1-y^3)(1-z^3)} & \frac{1-a^2}{(1-x^2)(1-y^2)(1-z^2)} & \frac{1-a}{(1-x)(1-y)(1-z)} & -3 \\ \frac{1-a^4}{(1-x^4)(1-y^4)(1-z^4)} & \frac{1-a^3}{(1-x^3)(1-y^3)(1-z^3)} & \frac{1-a^2}{(1-x^2)(1-y^2)(1-z^2)} & \frac{1-a}{(1-x)(1-y)(1-z)} \end{vmatrix} \frac{t^4}{4!}$$

$$+ \; etc.$$

The identities in the present book broadly follow the approach taken by Andrews [12], who develops his theory of integer partitions starting from a historical viewpoint and uses the q-binomial theorem to derive the Euler identities for the number of partitions of a positive integer. He then also uses (13.1.1) to yield the Jacobi triple product, the Heine q-series summation and transform, and other q-series sum-product identities relevant to partitions. We have followed this approach in Chapter 3, including also the quintuple product identity.

13.4 FINITE PRODUCT 2D, 3D AND nD CASES OF q-BINOMIAL EXTENSIONS

The reader may at this point recall that the (13.1.1) form of the familiar q-binomial theorem includes cases of finite products where the term-by-term expansion of the infinite numerator product coincides eventually (and cancels with) the infinite denominator product. After all, the binomial theorem itself began by representing a finite product of terms in the expansion $(a+b)^n$. For example, some finite q-binomial cases are,

$$F_1(q;1,t) = 1,$$

$$F_1(q;q,t) = \frac{1}{1-t},$$

$$F_1(q;q^2,t) = \frac{1}{(1-t)(1-qt)},$$

$$F_1(q;q^3,t) = \frac{1}{(1-t)(1-qt)(1-q^2t)},$$

$$\vdots$$

It is from these finite product representations that we enter the realm of research into the function $p(n,m)$, the number of partitions of n into parts less than or equal to m, for $m = 1,2,3,4$. Chapter 6 of the book by Andrews and Eriksson [29] deals with this very interesting integer partitions topic, where we see firstly the generating function

$$\sum_{k=0}^{\infty} p(k,m)q^k = \frac{1}{(1-q)(1-q^2)\cdots(1-q^m)},$$

having the particular case

$$\sum_{k=0}^{\infty} p(k,3)q^k = \frac{1}{(1-q)(1-q^2)(1-q^3)},$$

from which, after some simplifying, we derive the result

$$p(n,3) = \left[\left[\frac{(n+3)^2}{12}\right]\right]$$

where $[[x]]$ is the nearest integer to x.

13.4.1 2D, 3D, and nD Unrestricted Partitions on a Cluster of Vectors

The approach taken by Andrews and Eriksson [29] for $p(n,m)$ can of course lend itself to the vector partition versions given here. We start by similarly considering the 2D cases of (13.3.4), explicitly

$$F_2(x,y;(xy)^2,xyt) = \prod_{\langle 1,1\rangle \le \langle j,k\rangle \le \langle 2,2\rangle} \frac{1}{1-x^j y^k t} \tag{13.4.1}$$

$$= \frac{1}{(1-xyt)(1-x^2yt)}$$

$$\times \frac{1}{(1-xy^2t)(1-x^2y^2t)}$$

$$= 1 + x(1+x)y(1+y)\frac{t}{1!} + \begin{vmatrix} x(1+x)y(1+y) & -1 \\ x^2(1+x^2)y^2(1+y^2) & x(1+x)y(1+y) \end{vmatrix} \frac{t^2}{2!}$$

$$+ \begin{vmatrix} x(1+x)y(1+y) & -1 & 0 \\ x^2(1+x^2)y^2(1+y^2) & x(1+x)y(1+y) & -2 \\ x^3(1+x^3)y^3(1+y^3) & x^2(1+x^2)y^2(1+y^2) & x(1+x)y(1+y) \end{vmatrix} \frac{t^3}{3!}$$

$$+ \ etc.$$

This identity enables the following corollaries to be asserted. Firstly, just from considering the coefficient of t^2, we get

Corollary 13.4 *The number of unrestricted partitions of the vector* $\langle m,n\rangle$ *from the set of vectors*

$$\{\langle 1,1\rangle,\langle 1,2\rangle,\langle 2,1\rangle,\langle 2,2\rangle\},$$

into exactly two parts is the coefficient of $x^m y^n$ *in the expansion of*

$$\begin{vmatrix} x(1+x)y(1+y) & -1 \\ x^2(1+x^2)y^2(1+y^2) & x(1+x)y(1+y) \end{vmatrix} \frac{1}{2!}.$$

This simplifies to

$$x^2(x+1)^2y^2(y+1)^2+x^2(x^2+1)y^2(y^2+1).$$

Next case is the product form

$$\begin{aligned} F_2(x,y;(xy)^3,xyt) &= \prod_{\langle 1,1\rangle\le\langle j,k\rangle\le\langle 3,3\rangle} \frac{1}{1-x^jy^kt} \qquad (13.4.2)\\ &= \frac{1}{(1-xyt)(1-x^2yt)(1-x^3yt)} \\ &\quad\times\frac{1}{(1-xy^2t)(1-x^2y^2t)(1-x^3y^2t)} \\ &\quad\times\frac{1}{(1-xy^3t)(1-x^2y^3t)(1-x^3y^3t)} \end{aligned}$$

$$= 1+xy\left(\frac{1-x^3}{1-x}\right)\left(\frac{1-y^3}{1-y}\right)\frac{t}{1!}+\begin{vmatrix} xy\left(\frac{1-x^3}{1-x}\right)\left(\frac{1-y^3}{1-y}\right) & -1 \\ x^2y^2\left(\frac{1-x^6}{1-x^2}\right)\left(\frac{1-y^6}{1-y^2}\right) & xy\left(\frac{1-x^3}{1-x}\right)\left(\frac{1-y^3}{1-y}\right) \end{vmatrix}\frac{t^2}{2!}$$

$$+\begin{vmatrix} xy\left(\frac{1-x^3}{1-x}\right)\left(\frac{1-y^3}{1-y}\right) & -1 & 0 \\ x^2y^2\left(\frac{1-x^6}{1-x^2}\right)\left(\frac{1-y^6}{1-y^2}\right) & xy\left(\frac{1-x^3}{1-x}\right)\left(\frac{1-y^3}{1-y}\right) & -2 \\ x^3y^3\left(\frac{1-x^9}{1-x^3}\right)\left(\frac{1-y^9}{1-y^3}\right) & x^2y^2\left(\frac{1-x^6}{1-x^2}\right)\left(\frac{1-y^6}{1-y^2}\right) & xy\left(\frac{1-x^3}{1-x}\right)\left(\frac{1-y^3}{1-y}\right) \end{vmatrix}\frac{t^3}{3!}$$

$$+ \textit{ etc.}$$

This identity admits the following implied corollaries. From coefficient of t^3, we have

Corollary 13.5 *The number of unrestricted partitions of the vector* $\langle m,n\rangle$ *from the vectors*

$$\begin{gathered} \langle 1,1\rangle,\langle 1,2\rangle,\langle 1,3\rangle, \\ \langle 2,1\rangle,\langle 2,2\rangle,\langle 2,3\rangle, \\ \langle 3,1\rangle,\langle 3,2\rangle,\langle 3,3\rangle, \end{gathered}$$

into exactly three parts is the coefficient of $x^m y^n$ *in the expansion of*

$$\begin{vmatrix} xy\left(\frac{1-x^3}{1-x}\right)\left(\frac{1-y^3}{1-y}\right) & -1 & 0 \\ x^2y^2\left(\frac{1-x^6}{1-x^2}\right)\left(\frac{1-y^6}{1-y^2}\right) & xy\left(\frac{1-x^3}{1-x}\right)\left(\frac{1-y^3}{1-y}\right) & -2 \\ x^3y^3\left(\frac{1-x^9}{1-x^3}\right)\left(\frac{1-y^9}{1-y^3}\right) & x^2y^2\left(\frac{1-x^6}{1-x^2}\right)\left(\frac{1-y^6}{1-y^2}\right) & xy\left(\frac{1-x^3}{1-x}\right)\left(\frac{1-y^3}{1-y}\right) \end{vmatrix}\frac{1}{3!}.$$

This simplifies to

$$\frac{x^3(1-x^3)^3(1-y^3)^3y^3}{(1-x)^3(1-y)^3}+\frac{2x^3(1-x^9)(1-y^9)y^3}{(1-x^3)(1-y^3)}+\frac{3x^3(1-x^3)(1-x^6)(1-y^3)(1-y^6)y^3}{(1-x)(1-x^2)(1-y)(1-y^2)}.$$

The reader can easily see an infinite set of similar corollaries are possible here. We give one more illustrative example where the domain set of vectors for the partitions are vectors $\langle a,b\rangle$ where the components a and b are each positive integers no greater than 4. First start with the identity,

$$F_2(x,y;(xy)^4,xyt)=\prod_{\langle 1,1\rangle\leq\langle j,k\rangle\leq\langle 4,4\rangle}\frac{1}{1-x^jy^kt} \tag{13.4.3}$$

$$=\frac{1}{(1-xyt)(1-x^2yt)(1-x^3yt)(1-x^4yt)}$$

$$\times\frac{1}{(1-xy^2t)(1-x^2y^2t)(1-x^3y^2t)(1-x^4y^2t)}$$

$$\times\frac{1}{(1-xy^3t)(1-x^2y^3t)(1-x^3y^3t)(1-x^4y^3t)}$$

$$\times\frac{1}{(1-xy^4t)(1-x^2y^4t)(1-x^3y^4t)(1-x^4y^4t)}$$

$$=1+\frac{xy(1-x^4)(1-y^4)}{(1-x)(1-y)}\frac{t}{1!}+\begin{vmatrix}\frac{xy(1-x^4)(1-y^4)}{(1-x)(1-y)} & -1\\ \frac{x^2y^2(1-x^8)(1-y^8)}{(1-x^2)(1-y^2)} & \frac{xy(1-x^4)(1-y^4)}{(1-x)(1-y)}\end{vmatrix}\frac{t^2}{2!}$$

$$+\begin{vmatrix}\frac{xy(1-x^4)(1-y^4)}{(1-x)(1-y)} & -1 & 0\\ \frac{x^2y^2(1-x^8)(1-y^8)}{(1-x^2)(1-y^2)} & \frac{xy(1-x^4)(1-y^4)}{(1-x)(1-y)} & -2\\ \frac{1-(xy)^12}{(1-x^3)(1-y^3)} & \frac{x^2y^2(1-x^8)(1-y^8)}{(1-x^2)(1-y^2)} & \frac{xy(1-x^4)(1-y^4)}{(1-x)(1-y)}\end{vmatrix}\frac{t^3}{3!}$$

$$+\begin{vmatrix}\frac{xy(1-x^4)(1-y^4)}{(1-x)(1-y)} & -1 & 0 & 0\\ \frac{x^2y^2(1-x^8)(1-y^8)}{(1-x^2)(1-y^2)} & \frac{xy(1-x^4)(1-y^4)}{(1-x)(1-y)} & -2 & 0\\ \frac{x^3y^3(1-x^{12})(1-y^{12})}{(1-x^3)(1-y^3)} & \frac{x^2y^2(1-x^8)(1-y^8)}{(1-x^2)(1-y^2)} & \frac{xy(1-x^4)(1-y^4)}{(1-x)(1-y)} & -3\\ \frac{x^4y^4(1-x^{16})(1-y^{16})}{(1-x^4)(1-y^4)} & \frac{x^3y^3(1-x^{12})(1-y^{12})}{(1-x^3)(1-y^3)} & \frac{x^2y^2(1-x^8)(1-y^8)}{(1-x^2)(1-y^2)} & \frac{xy(1-x^4)(1-y^4)}{(1-x)(1-y)}\end{vmatrix}\frac{t^4}{4!}$$

$$+\ etc.$$

We next state the partition theoretic interpretation of the coefficient of t^4 from this.

Corollary 13.6 *The number of unrestricted partitions of the vector $\langle m,n\rangle$ using any of the vectors*

$$\begin{gathered}\langle 1,1\rangle,\langle 1,2\rangle,\langle 1,3\rangle,\langle 1,4\rangle,\\ \langle 2,1\rangle,\langle 2,2\rangle,\langle 2,3\rangle,\langle 2,4\rangle,\\ \langle 3,1\rangle,\langle 3,2\rangle,\langle 3,3\rangle,\langle 3,4\rangle,\\ \langle 4,1\rangle,\langle 4,2\rangle,\langle 4,3\rangle,\langle 4,4\rangle,\end{gathered}$$

into exactly four parts is the coefficient of $x^m y^n$ in the expansion of

$$\begin{vmatrix} \frac{xy(1-x^4)(1-y^4)}{(1-x)(1-y)} & -1 & 0 & 0 \\ \frac{x^2y^2(1-x^8)(1-y^8)}{(1-x^2)(1-y^2)} & \frac{xy(1-x^4)(1-y^4)}{(1-x)(1-y)} & -2 & 0 \\ \frac{x^3y^3(1-x^{12})(1-y^{12})}{(1-x^3)(1-y^3)} & \frac{x^2y^2(1-x^8)(1-y^8)}{(1-x^2)(1-y^2)} & \frac{xy(1-x^4)(1-y^4)}{(1-x)(1-y)} & -3 \\ \frac{x^4y^4(1-x^{16})(1-y^{16})}{(1-x^4)(1-y^4)} & \frac{x^3y^3(1-x^{12})(1-y^{12})}{(1-x^3)(1-y^3)} & \frac{x^2y^2(1-x^8)(1-y^8)}{(1-x^2)(1-y^2)} & \frac{xy(1-x^4)(1-y^4)}{(1-x)(1-y)} \end{vmatrix} \frac{1}{4!}.$$

This simplifies to

$$a^4+6a^2b+8ac+3b^2+6d$$

where

$$\begin{aligned} a &= \frac{xy(1-x^4)(1-y^4)}{(1-x)(1-y)}, \\ b &= \frac{x^2y^2(1-x^8)(1-y^8)}{(1-x^2)(1-y^2)}, \\ c &= \frac{x^3y^3(1-x^{12})(1-y^{12})}{(1-x^3)(1-y^3)}, \\ d &= \frac{x^4y^4(1-x^{16})(1-y^{16})}{(1-x^4)(1-y^4)}. \end{aligned}$$

We therefore state a general form of this in the

Corollary 13.7 *A finite product 2D example of (13.2.3) is*

$$F_2(x,y;(xy)^m,xyt) = \prod_{\langle 1,1\rangle\le\langle j,k\rangle\le\langle m,m\rangle} \frac{1}{1-x^jy^kt} \tag{13.4.4}$$

$$= 1+\frac{xy(1-x^m)(1-y^m)}{(1-x)(1-y)}\frac{t}{1!}+\begin{vmatrix} \frac{xy(1-x^m)(1-y^m)}{(1-x)(1-y)} & -1 \\ \frac{x^2y^2(1-x^{2m})(1-y^{2m})}{(1-x^2)(1-y^2)} & \frac{xy(1-x^m)(1-y^m)}{(1-x)(1-y)} \end{vmatrix}\frac{t^2}{2!}$$

$$+\begin{vmatrix} \frac{xy(1-x^m)(1-y^m)}{(1-x)(1-y)} & -1 & 0 \\ \frac{x^2y^2(1-x^{2m})(1-y^{2m})}{(1-x^2)(1-y^2)} & \frac{xy(1-x^m)(1-y^m)}{(1-x)(1-y)} & -2 \\ \frac{x^3y^3(1-x^{3m})(1-y^{3m})}{(1-x^3)(1-y^3)} & \frac{x^2y^2(1-x^{2m})(1-y^{2m})}{(1-x^2)(1-y^2)} & \frac{xy(1-x^m)(1-y^m)}{(1-x)(1-y)} \end{vmatrix}\frac{t^3}{3!}$$

$$+\begin{vmatrix} \frac{xy(1-x^m)(1-y^m)}{(1-x)(1-y)} & -1 & 0 & 0 \\ \frac{x^2y^2(1-x^{2m})(1-y^{2m})}{(1-x^2)(1-y^2)} & \frac{xy(1-x^m)(1-y^m)}{(1-x)(1-y)} & -2 & 0 \\ \frac{x^3y^3(1-x^{3m})(1-y^{3m})}{(1-x^3)(1-y^3)} & \frac{x^2y^2(1-x^{2m})(1-y^{2m})}{(1-x^2)(1-y^2)} & \frac{xy(1-x^m)(1-y^m)}{(1-x)(1-y)} & -3 \\ \frac{x^4y^4(1-x^{4m})(1-y^{4m})}{(1-x^4)(1-y^4)} & \frac{x^3y^3(1-x^{3m})(1-y^{3m})}{(1-x^3)(1-y^3)} & \frac{x^2y^2(1-x^{2m})(1-y^{2m})}{(1-x^2)(1-y^2)} & \frac{xy(1-x^m)(1-y^m)}{(1-x)(1-y)} \end{vmatrix}\frac{t^4}{4!}$$

$$+\ etc.$$

Note the finite first product of (13.2.3) is a product of m^2 denominator terms. Furthermore, the finite first product in (13.2.1) given next has a denominator product of m^3 terms.

Corollary 13.8 *A finite product 3-space extension of (13.2.3) is*

$$F_3(x,y,z;0,xyzt) = \prod_{\langle 1,1,1\rangle \leq \langle j,k,l\rangle \leq \langle m,m,m\rangle} \frac{1}{1-x^jy^kz^lt} \tag{13.4.5}$$

$$= 1 + \frac{xy(1-x^m)(1-y^m)(1-z^m)}{(1-x)(1-y)(1-z)}\frac{t}{1!} + \begin{vmatrix} \frac{xy(1-x^m)(1-y^m)(1-z^m)}{(1-x)(1-y)(1-z)} & -1 \\ \frac{x^2y^2(1-x^{2m})(1-y^{2m})(1-z^{2m})}{(1-x^2)(1-y^2)(1-z^2)} & \frac{xy(1-x^m)(1-y^m)(1-z^m)}{(1-x)(1-y)(1-z)} \end{vmatrix}\frac{t^2}{2!}$$

$$+\begin{vmatrix} \frac{xy(1-x^m)(1-y^m)(1-z^m)}{(1-x)(1-y)(1-z)} & -1 & 0 \\ \frac{x^2y^2(1-x^{2m})(1-y^{2m})(1-z^{2m})}{(1-x^2)(1-y^2)(1-z^2)} & \frac{xy(1-x^m)(1-y^m)(1-z^m)}{(1-x)(1-y)(1-z)} & -2 \\ \frac{x^3y^3(1-x^{3m})(1-y^{3m})(1-z^{3m})}{(1-x^3)(1-y^3)(1-z^3)} & \frac{x^2y^2(1-x^{2m})(1-y^{2m})(1-z^{2m})}{(1-x^2)(1-y^2)} & \frac{xy(1-x^m)(1-y^m)(1-z^m)}{(1-x)(1-y)(1-z)} \end{vmatrix}\frac{t^3}{3!}$$

$$+\begin{vmatrix} \frac{xy(1-x^m)(1-y^m)(1-z^m)}{(1-x)(1-y)(1-z)} & -1 & 0 & 0 \\ \frac{x^2y^2(1-x^{2m})(1-y^{2m})(1-z^{2m})}{(1-x^2)(1-y^2)(1-z^2)} & \frac{xy(1-x^m)(1-y^m)(1-z^m)}{(1-x)(1-y)(1-z)} & -2 & 0 \\ \frac{x^3y^3(1-x^{3m})(1-y^{3m})(1-z^{3m})}{(1-x^3)(1-y^3)(1-z^3)} & \frac{x^2y^2(1-x^{2m})(1-y^{2m})(1-z^{2m})}{(1-x^2)(1-y^2)(1-z^2)} & \frac{xy(1-x^m)(1-y^m)(1-z^m)}{(1-x)(1-y)(1-z)} & -3 \\ \frac{x^4y^4(1-x^{4m})(1-y^{4m})(1-z^{4m})}{(1-x^4)(1-y^4)(1-z^4)} & \frac{x^3y^3(1-x^{3m})(1-y^{3m})(1-z^{3m})}{(1-x^3)(1-y^3)(1-z^3)} & \frac{x^2y^2(1-x^{2m})(1-y^{2m})(1-z^{2m})}{(1-x^2)(1-y^2)(1-z^2)} & \frac{xy(1-x^m)(1-y^m)(1-z^m)}{(1-x)(1-y)(1-z)} \end{vmatrix}\frac{t^4}{4!}$$

$$+\ etc.$$

Let us now take the case of corollary 13.2, namely

Corollary 13.9 *The 2D unrestricted partitions example of (13.1.1) is*

$$F_2(x,y;0,t) = \prod_{j,k\geq 0} \frac{1}{1-x^jy^kt} \tag{13.4.6}$$

$$= 1 + \frac{1}{(1-x)(1-y)}\frac{t}{1!} + \begin{vmatrix} \frac{1}{(1-x)(1-y)} & -1 \\ \frac{1}{(1-x^2)(1-y^2)} & \frac{1}{(1-x)(1-y)} \end{vmatrix}\frac{t^2}{2!}$$

$$+\begin{vmatrix} \frac{1}{(1-x)(1-y)} & -1 & 0 \\ \frac{1}{(1-x^2)(1-y^2)} & \frac{1}{(1-x)(1-y)} & -2 \\ \frac{1}{(1-x^3)(1-y^3)} & \frac{1}{(1-x^2)(1-y^2)} & \frac{1}{(1-x)(1-y)} \end{vmatrix}\frac{t^3}{3!}$$

$$+\begin{vmatrix} \frac{1}{(1-x)(1-y)} & -1 & 0 & 0 \\ \frac{1}{(1-x^2)(1-y^2)} & \frac{1}{(1-x)(1-y)} & -2 & 0 \\ \frac{1}{(1-x^3)(1-y^3)} & \frac{1}{(1-x^2)(1-y^2)} & \frac{1}{(1-x)(1-y)} & -3 \\ \frac{1}{(1-x^4)(1-y^4)} & \frac{1}{(1-x^3)(1-y^3)} & \frac{1}{(1-x^2)(1-y^2)} & \frac{1}{(1-x)(1-y)} \end{vmatrix} \frac{t^4}{4!}$$

$$+ \ etc.$$

We can easily work out the power series in x and y for each determinant given here by applying the formulas given in Chapter 31 of the current book. These are in fact equivalent to the Bell Polynomial summations as inferred in Andrews [12, chapter 12]. These determinants when worked out this way, give us closed forms for the *number of 2D unrestricted partitions* into exactly integer m parts for the coefficient of $x^j y^k$ in the mth order determinant coefficient of t^m. So, we see that (13.4.6) reads in *determinant expanded* form, as

$$F_2(x,y;0,t) = \prod_{j,k\geq 0} \frac{1}{1-x^j y^k t} \tag{13.4.7}$$

$$= 1 + a\frac{t}{1!} + \left(a^2+b\right)\frac{t^2}{2!}$$

$$+\left(a^3+3ab+2c\right)\frac{t^3}{3!}$$

$$+\left(a^4+6a^2b+8ac+3b^2+6d\right)\frac{t^4}{4!}$$

$$+\left(a^5+10a^3b+20a^2c+15ab^2+30ad+20bc+24e\right)\frac{t^5}{5!}$$

$$+\left(a^6+15a^4b+40a^3c+45a^2b^2+90a^2d+120abc+144ae+15b^3+90bd+40c^2+120f\right)\frac{t^6}{6!}$$

$$+ \ etc., \ where$$

$$a = \frac{1}{(1-x)(1-y)}, b = \frac{1}{(1-x^2)(1-y^2)}, c = \frac{1}{(1-x^3)(1-y^3)},$$

$$d = \frac{1}{(1-x^4)(1-y^4)}, e = \frac{1}{(1-x^5)(1-y^5)}, f = \frac{1}{(1-x^6)(1-y^6)}, etc.$$

The above result allows us to state the first few theorems on this enumeration of vector partitions. It suits us best to use (13.2.3) in the case $F_2(x,y;0,xyt)$. The following are the immediate consequences.

Theorem 13.6 *The number of unrestricted partitions of the vector $\langle m,n\rangle$ with $m \geq 1$, $n \geq 1$, into exactly two parts is the coefficient of $x^m y^n$ in the expansion of*

$$\left(a^2+b\right)\frac{1}{2!} \ where \ a = \frac{1}{(1-x)(1-y)}, \ b = \frac{1}{(1-x^2)(1-y^2)}.$$

This simplifies to

$$\frac{(1+xy)}{(1-x)(1-x^2)(1-y)(1-y^2)}.$$

Likewise we see that

Theorem 13.7 *The number of unrestricted partitions of the vector* $\langle m, n\rangle$ *with* $m \geq 1$, $n \geq 1$, *into exactly three parts is the coefficient of* $x^m y^n$ *in the expansion of*

$$\left(a^3 + 3ab + 2c\right)\frac{1}{3!} \text{ where}$$

$$a = \frac{1}{(1-x)(1-y)},\ b = \frac{1}{(1-x^2)(1-y^2)},\ c = \frac{1}{(1-x^3)(1-y^3)}.$$

This simplifies to

$$\frac{(1+xy+xy^2+x^2y+x^2y^2+x^3y^3)}{(1-x)(1-x^2)(1-x^3)(1-y)(1-y^2)(1-y^3)}.$$

Continuing this way we have next

Theorem 13.8 *The number of unrestricted partitions of the vector* $\langle m, n\rangle$ *with* $m \geq 1$, $n \geq 1$, *into exactly four parts is the coefficient of* $x^m y^n$ *in the expansion of*

$$\left(a^4 + 6a^2b + 8ac + 3b^2 + 6d\right)/24 \text{ where}$$

$$a = \frac{1}{(1-x)(1-y)},\ b = \frac{1}{(1-x^2)(1-y^2)},$$

$$c = \frac{1}{(1-x^3)(1-y^3)},\ d = \frac{1}{(1-x^4)(1-y^4)}.$$

This simplifies to

$$\frac{(1+xy+xy^2+xy^3+x^2y+2x^2y^2+x^2y^3+x^2y^4+x^3y+x^3y^2+2x^3y^3)}{(1-x)(1-x^2)(1-x^3)(1-x^4)(1-y)(1-y^2)(1-y^3)(1-y^4)}$$

$$+\frac{(x^3y^4+x^3y^5+x^4y^2+x^4y^3+2x^4y^4+x^4y^5+x^5y^3+x^5y^4+x^5y^5+x^6y^6)}{(1-x)(1-x^2)(1-x^3)(1-x^4)(1-y)(1-y^2)(1-y^3)(1-y^4)}.$$

13.4.2 2D, 3D, and nD Distinct Partitions on a Cluster of Vectors

The approach taken in the previous subsection for *unrestricted vector partitions* on a finite cluster of vectors applies equally to the *distinct vector partitions* on those same clusters. We now present some 2D cases of (13.3.4) related to vector partition clusters into distinct parts

$$F_2(x,y;(xy)^{-2}, -(xy)^3 t) = \prod_{\langle 1,1\rangle \leq \langle j,k\rangle \leq \langle 2,2\rangle} (1 + x^j y^k t)$$

$$= (1+xyt)(1+x^2yt)$$
$$\times (1+xy^2t)(1+x^2y^2t)$$

$$= 1 + x(1+x)y(1+y)\frac{t}{1!} + \begin{vmatrix} x(1+x)y(1+y) & -1 \\ -x^2(1+x^2)y^2(1+y^2) & x(1+x)y(1+y) \end{vmatrix}\frac{t^2}{2!}$$

$$+\begin{vmatrix} x(1+x)y(1+y) & -1 & 0 \\ -x^2(1+x^2)y^2(1+y^2) & x(1+x)y(1+y) & -2 \\ x^3(1+x^3)y^3(1+y^3) & -x^2(1+x^2)y^2(1+y^2) & x(1+x)y(1+y) \end{vmatrix} \frac{(t)^3}{3!}$$

$$+\begin{vmatrix} x(1+x)y(1+y) & -1 & 0 & 0 \\ -x^2(1+x^2)y^2(1+y^2) & x(1+x)y(1+y) & -2 & 0 \\ x^3(1+x^3)y^3(1+y^3) & -x^2(1+x^2)y^2(1+y^2) & x(1+x)y(1+y) & -3 \\ -x^4(1+x^4)y^4(1+y^4) & x^3(1+x^3)y^3(1+y^3) & -x^2(1+x^2)y^2(1+y^2) & x(1+x)y(1+y) \end{vmatrix} \frac{t^4}{4!}.$$

This identity expands as a fourth-degree polynomial in t. Hence the coefficient of t^4, which is the above 4×4 determinant reduces to x^6y^6. Remarkably, we see that this determinant is equivalent to

$$a^4+6a^2b+8ac+3b^2+6d$$

where

$$a=x(1+x)y(1+y),\ b=-x^2(1+x^2)y^2(1+y^2),$$
$$c=x^3(1+x^3)y^3(1+y^3),\ d=-x^4(1+x^4)y^4(1+y^4);$$

which reduces impressively to x^6y^6 in any calculating machine app.

Corollary 13.10 *The number of unrestricted partitions of the vector $\langle m,n\rangle$ from the set of vectors*

$$\{\langle 1,1\rangle,\langle 1,2\rangle,\langle 2,1\rangle,\langle 2,2\rangle\},$$

into exactly four distinct parts is 1, which is the coefficient of x^6y^6 in the expansion of

$$\prod_{\langle 1,1\rangle\le\langle j,k\rangle\le\langle 2,2\rangle}(1+x^jy^kt).$$

The single partition with four parts here is trivially $\langle 1,1\rangle+\langle 1,2\rangle+\langle 2,1\rangle+\langle 2,2\rangle=\langle 6,6\rangle$.

Next, we consider the finite product of nine terms of identity

$$F_2(x,y;(xy)^{-3},-(xy)^4t)=\prod_{\langle 1,1\rangle\le\langle j,k\rangle\le\langle 3,3\rangle}(1+x^jy^kt)$$

$$=(1+xyt)(1+x^2yt)(1+x^3yt)$$
$$\times(1+xy^2t)(1+x^2y^2t)(1+x^3y^2t)$$
$$\times(1+xy^3t)(1+x^2y^3t)(1+x^3y^3t)$$

$$=1+x(1+x+x^2)y(1+y+y^2)\frac{t}{1!}+\begin{vmatrix} x(1+x+x^2)y(1+y+y^2) & -1 \\ -x^2(1+x^2+x^4)y^2(1+y^2+y^4) & x(1+x+x^2)y(1+y+y^2) \end{vmatrix}\frac{t^2}{2!}$$

$$+\begin{vmatrix} x(1+x+x^2)y(1+y+y^2) & & -1 & 0 \\ -x^2(1+x^2+x^4)y^2(1+y^2+y^4) & x(1+x+x^2)y(1+y+y^2) & -2 & \\ x^3(1+x^3+x^6)y^3(1+y^3+y^6) & -x^2(1+x^2+x^4)y^2(1+y^2+y^4) & x(1+x+x^2)y(1+y+y^2) & \end{vmatrix}\frac{t^3}{3!}$$

$$+\begin{vmatrix} x(1+x+x^2)y(1+y+ & y^2) & -1 & 0 & 0 \\ -x^2(1+x^2+x^4)y^2(1+y^2+y^4) & x(1+x+x^2)y(1+y+y^2) & -2 & 0 & \\ x^3(1+x^3+x^6)y^3(1+y^3+y^6) & -x^2(1+x^2+x^4)y^2(1+y^2+y^4) & x(1+x+x^2)y(1+y+y^2) & -3 & \\ -x^4(1+x^4+x^8)y^4(1+y^4+y^8) & x^3(1+x^3+x^6)y^3(1+y^3+y^6) & -x^2(1+x^2+x^4)y^2(1+y^2+y^4) & x(1+x+x^2)y(1+y+y^2) & \end{vmatrix}\frac{t^4}{4!}$$

$$+\cdots+x^{18}y^{18}\frac{t^9}{9!}$$

This identity is a polynomial in t of degree 9. It admits the following implied statements, given as two corollaries. From coefficient of t^2 in (13.4.7, we have

Corollary 13.11 *The number of vector partitions of the vector $\langle m,n\rangle$ into exactly two distinct parts from the vectors*

$$\langle 1,1\rangle,\langle 1,2\rangle,\langle 1,3\rangle,$$
$$\langle 2,1\rangle,\langle 2,2\rangle,\langle 2,3\rangle,$$
$$\langle 3,1\rangle,\langle 3,2\rangle,\langle 3,3\rangle,$$

is the coefficient of $x^m y^n$ in the expansion of

$$\begin{vmatrix} x(1+x+x^2)y(1+y+y^2) & -1 \\ -x^2(1+x^2+x^4)y^2(1+y^2+y^4) & x(1+x+x^2)y(1+y+y^2) \end{vmatrix} \frac{1}{2!}.$$

This simplifies to

$$\frac{x^2(1-x^3)y^2(1-y^3)(x+y)(1+xy)}{(1-x)(1-y)}.$$

The tool for simplifying this was WolframAlpha, the input

$$a^2+b, \text{ where } a=x(1+x+x^2)y(1+y+y^2),\ b=-x^2(1+x^2+x^4)y^2(1+y^2+y^4).$$

Next, from the coefficient of t^3 we get

Corollary 13.12 *The number of partitions of the vector $\langle m,n\rangle$ using only the vectors*

$$\langle 1,1\rangle,\langle 1,2\rangle,\langle 1,3\rangle,$$
$$\langle 2,1\rangle,\langle 2,2\rangle,\langle 2,3\rangle,$$
$$\langle 3,1\rangle,\langle 3,2\rangle,\langle 3,3\rangle,$$

into exactly three distinct parts parts is the coefficient of $x^m y^n$ in the expansion of

$$-\begin{vmatrix} \frac{xy(1-x^3)(1-y^3)}{(1-x)(1-y)} & -1 & 0 \\ \frac{x^2y^2(1-x^6)(1-y^6)}{(1-x^2)(1-y^2)} & \frac{xy(1-x^3)(1-y^3)}{(1-x)(1-y)} & -2 \\ \frac{x^3y^3(1-x^9)(1-y^9)}{(1-x^3)(1-y^3)} & \frac{x^2y^2(1-x^6)(1-y^6)}{(1-x^2)(1-y^2)} & \frac{xy(1-x^3)(1-y^3)}{(1-x)(1-y)} \end{vmatrix} \frac{1}{3!}.$$

This simplifies to

$$\frac{x^6y^6(1-x^6)^3(1-y^6)^3}{(1-x^2)^3(1-y^2)^3}+\frac{3x^3y^3(1-x^3)(1-y^3)(1-x^6)(1-y^6)}{(1-x)(1-x^2)(1-y)(1-y^2)}+\frac{2(1-x^9)(1-y^9)}{(1-x^3)(1-y^3)}.$$

Here the simplification comes from

$$b^3+3ab+2c,$$

where

$$\begin{aligned} a &= \frac{xy(1-x^3)(1-y^3)}{(1-x)(1-y)}, \\ b &= \frac{x^2y^2(1-x^6)(1-y^6)}{(1-x^2)(1-y^2)}, \\ c &= \frac{x^3y^3(1-x^9)(1-y^9)}{(1-x^3)(1-y^3)}. \end{aligned}$$

submitted in an online query. The reader will note at this juncture, that the example corollaries given here are a few cases of an infinite set of similar corollaries that could be chosen instead. We give one more illustrative example where the domain set of vectors for the partitions are vectors $\langle a,b\rangle$ where the components a and b are each positive integers no greater than 4. First start with the identity,

$$F_2(x,y;(xy)^{-4},-(xy)^5t) = \prod_{\langle 1,1\rangle \le \langle j,k\rangle \le \langle 4,4\rangle} (1+x^jy^kt)$$

$$\begin{aligned} &= (1+xyt)(1+x^2yt)(1+x^3yt)(1+x^4yt) \\ &\times(1+xy^2t)(1+x^2y^2t)(1+x^3y^2t)(1+x^4y^2t) \\ &\times(1+xy^3t)(1+x^2y^3t)(1+x^3y^3t)(1+x^4y^3t) \\ &\times(1+xy^4t)(1+x^2y^4t)(1+x^3y^4t)(1+x^4y^4t), \end{aligned}$$

$$= 1 - \frac{xy(1-x^4)(1-y^4)}{(1-x)(1-y)}\frac{t}{1!} + \begin{vmatrix} \frac{xy(1-x^4)(1-y^4)}{(1-x)(1-y)} & -1 \\ \frac{x^2y^2(1-x^8)(1-y^8)}{(1-x^2)(1-y^2)} & \frac{xy(1-x^4)(1-y^4)}{(1-x)(1-y)} \end{vmatrix}\frac{t^2}{2!}$$

$$- \begin{vmatrix} \frac{xy(1-x^4)(1-y^4)}{(1-x)(1-y)} & -1 & 0 \\ \frac{x^2y^2(1-x^8)(1-y^8)}{(1-x^2)(1-y^2)} & \frac{xy(1-x^4)(1-y^4)}{(1-x)(1-y)} & -2 \\ \frac{x^3y^3(1-x^{12})(1-y^{12})}{(1-x^3)(1-y^3)} & \frac{x^2y^2(1-x^8)(1-y^8)}{(1-x^2)(1-y^2)} & \frac{xy(1-x^4)(1-y^4)}{(1-x)(1-y)} \end{vmatrix}\frac{t^3}{3!}$$

$$+ \begin{vmatrix} \frac{xy(1-x^4)(1-y^4)}{(1-x)(1-y)} & -1 & 0 & 0 \\ \frac{x^2y^2(1-x^8)(1-y^8)}{(1-x^2)(1-y^2)} & \frac{1-(xy)^{-4}}{(1-x)(1-y)} & -2 & 0 \\ \frac{x^3y^3(1-x^{12})(1-y^{12})}{(1-x^3)(1-y^3)} & \frac{x^2y^2(1-x^8)(1-y^8)}{(1-x^2)(1-y^2)} & \frac{1-(xy)^{-4}}{(1-x)(1-y)} & -3 \\ \frac{x^4y^4(1-x^{16})(1-y^{16})}{(1-x^4)(1-y^4)} & \frac{x^3y^3(1-x^{12})(1-y^{12})}{(1-x^3)(1-y^3)} & \frac{x^2y^2(1-x^8)(1-y^8)}{(1-x^2)(1-y^2)} & \frac{1-(xy)^{-4}}{(1-x)(1-y)} \end{vmatrix}\frac{t^4}{4!}$$

$$-+ etc.$$

We next state the partition theoretic interpretation of the coefficient of t^3 from this.

Corollary 13.13 *The number of partitions of the vector* $\langle m,n\rangle$ *using any of the vectors*

$$\langle 1,1\rangle, \langle 1,2\rangle, \langle 1,3\rangle, \langle 1,4\rangle,$$

$$\langle 2,1\rangle,\langle 2,2\rangle,\langle 2,3\rangle,\langle 2,4\rangle,$$
$$\langle 3,1\rangle,\langle 3,2\rangle,\langle 3,3\rangle,\langle 3,4\rangle,$$
$$\langle 4,1\rangle,\langle 4,2\rangle,\langle 4,3\rangle,\langle 4,4\rangle,$$

into exactly three distinct parts is the coefficient of $x^m y^n$ *in the expansion of*

$$-\begin{vmatrix} \frac{xy(1-x^4)(1-y^4)}{(1-x)(1-y)} & -1 & 0 \\ \frac{x^2y^2(1-x^8)(1-y^8)}{(1-x^2)(1-y^2)} & \frac{xy(1-x^4)(1-y^4)}{(1-x)(1-y)} & -2 \\ \frac{x^3y^3(1-x^{12})(1-y^{12})}{(1-x^3)(1-y^3)} & \frac{x^2y^2(1-x^8)(1-y^8)}{(1-x^2)(1-y^2)} & \frac{xy(1-x^4)(1-y^4)}{(1-x)(1-y)} \end{vmatrix}\frac{1}{3!}.$$

This simplifies to

$$\frac{2x^3y^3(1-x^{12})(1-y^{12})}{(1-x^3)(1-y^3)}+\frac{x^6y^6(1-x^8)^3(1-y^8)^3}{(1-x^2)^3(1-y^2)^3}+\frac{3x^3y^3(1-x^4)(1-x^8)(1-y^4)(1-y^8)}{(1-x)(1-x^2)(1-y)(1-y^2)}.$$

Here, the simplification comes from

$$b^3+3ab+2c$$

where

$$\begin{aligned} a &= \frac{xy(1-x^4)(1-y^4)}{(1-x)(1-y)}, \\ b &= \frac{x^2y^2(1-x^8)(1-y^8)}{(1-x^2)(1-y^2)}, \\ c &= \frac{x^3y^3(1-x^{12})(1-y^{12})}{(1-x^3)(1-y^3)}. \end{aligned}$$

We next state, out of interest, a 3D form of this where the lattice points summed upon are in the range of the $x-y-z$ extended $m\times m$ cube in the first hyperquadrant of the 3D space, enclosing the vectors

$$\langle 1,1,1\rangle \le \langle j,k,h\rangle \le \langle m,m,m\rangle.$$

We consider the finite product of m^3 terms derived from

$$H_3(x,y,z;m^3;t)=\frac{G_3(x,y,z;t)}{G_3(x,y,z;(xyz)^{m+1}t)}$$

Corollary 13.14 *A finite product 3D example of (13.3.2) determined from first principles on the logarithmic derivative is*

$$H_3(x,y,z;m^3;t)=\prod_{\langle 1,1,1\rangle\le\langle j,k,h\rangle\le\langle m,m,m\rangle}(1+x^jy^kz^ht) \tag{13.4.8}$$

$$=1+\frac{xyz(1-x^m)(1-y^m)(1-z^m)}{(1-x)(1-y)(1-z)}\frac{t}{1!}$$

$$+\begin{vmatrix} \frac{xyz(1-x^m)(1-y^m)(1-z^m)}{(1-x)(1-y)(1-z)} & -1 \\ \frac{-x^2y^2z^2(1-x^{2m})(1-y^{2m})(1-z^{2m})}{(1-x^2)(1-y^2)(1-z^2)} & \frac{xyz(1-x^m)(1-y^m)(1-z^m)}{(1-x)(1-y)(1-z)} \end{vmatrix} \frac{t^2}{2!}$$

$$+\begin{vmatrix} \frac{xyz(1-x^m)(1-y^m)(1-z^m)}{(1-x)(1-y)(1-z)} & -1 & 0 \\ \frac{-x^2y^2z^2(1-x^{2m})(1-y^{2m})(1-z^{2m})}{(1-x^2)(1-y^2)(1-z^2)} & \frac{xyz(1-x^m)(1-y^m)(1-z^m)}{(1-x)(1-y)(1-z)} & -2 \\ \frac{x^3y^3z^3(1-x^{3m})(1-y^{3m})(1-z^{3m})}{(1-x^3)(1-y^3)(1-z^3)} & \frac{-x^2y^2z^2(1-x^{2m})(1-y^{2m})(1-z^{2m})}{(1-x^2)(1-y^2)} & \frac{xyz(1-x^m)(1-y^m)(1-z^m)}{(1-x)(1-y)(1-z)} \end{vmatrix} \frac{t^3}{3!}$$

$$+ \textit{ and so on up to}$$

$$+x^{\frac{3m^2(m+1)}{2}} y^{\frac{3m^2(m+1)}{2}} z^{\frac{3m^2(m+1)}{2}} \frac{t^{(m^3)}}{(m^3)!}$$

Note the right side product of (13.4.8) is a product of m^3 terms. Also, taking $H_3(x,y,z;3^3;t)$ as our next example, the right side product (13.4.5) is a product of $3^3 = 27$ terms. Hence we have a polynomial in t of degree 27.

Corollary 13.15 *A finite product 3-space example of (13.2.6) is*

$$H_3(x,y,z;3^3;t) = \prod_{\langle 1,1,1\rangle \leq \langle j,k,h\rangle \leq \langle 3,3,3\rangle} (1+x^j y^k z^h t) \tag{13.4.9}$$

$$\begin{aligned} = &(1+x^1y^3z^3t) \; \times(1+x^2y^3z^3t) \; \times(1+x^3y^3z^3t) \\ &\times(1+x^1y^2z^3t) \; \times(1+x^2y^2z^3t) \; \times(1+x^3y^2z^3t) \\ &\times(1+x^1y^1z^3t) \; \times(1+x^2y^1z^3t) \; \times(1+x^3y^1z^3t) \\ \\ &\times(1+x^1y^3z^2t) \; \times(1+x^2y^3z^2t) \; \times(1+x^3y^3z^2t) \\ &\times(1+x^1y^2z^2t) \; \times(1+x^2y^2z^2t) \; \times(1+x^3y^2z^2t) \\ &\times(1+x^1y^1z^2t) \; \times(1+x^2y^1z^2t) \; \times(1+x^3y^1z^2t) \\ \\ &\times(1+x^1y^3z^1t) \; \times(1+x^2y^3z^1t) \; \times(1+x^3y^3z^1t) \\ &\times(1+x^1y^2z^1t) \; \times(1+x^2y^2z^1t) \; \times(1+x^3y^2z^1t) \\ &\times(1+x^1y^1z^1t) \; \times(1+x^2y^1z^1t) \; \times(1+x^3y^1z^1t) \end{aligned}$$

$$= 1 + \frac{xyz(1-x^3)(1-y^3)(1-z^3)}{(1-x)(1-y)(1-z)} \frac{t}{1!}$$

$$+\begin{vmatrix} \frac{xyz(1-x^3)(1-y^3)(1-z^3)}{(1-x)(1-y)(1-z)} & -1 \\ \frac{-x^2y^2z^2(1-x^6)(1-y^6)(1-z^6)}{(1-x^2)(1-y^2)(1-z^2)} & \frac{xyz(1-x^3)(1-y^3)(1-z^3)}{(1-x)(1-y)(1-z)} \end{vmatrix} \frac{t^2}{2!}$$

$$+\begin{vmatrix} \frac{xyz(1-x^3)(1-y^3)(1-z^3)}{(1-x)(1-y)(1-z)} & -1 & 0 \\ \frac{-x^2y^2z^2(1-x^6)(1-y^6)(1-z^6)}{(1-x^2)(1-y^2)(1-z^2)} & \frac{xyz(1-x^3)(1-y^3)(1-z^3)}{(1-x)(1-y)(1-z)} & -2 \\ \frac{x^3y^3z^3(1-x^9)(1-y^9)(1-z^9)}{(1-x^3)(1-y^3)(1-z^3)} & \frac{-x^2y^2z^2(1-x^6)(1-y^6)(1-z^6)}{(1-x^2)(1-y^2)} & \frac{xyz(1-x^3)(1-y^3)(1-z^3)}{(1-x)(1-y)(1-z)} \end{vmatrix} \frac{t^3}{3!}$$

$$+ \textit{ and so on up to}$$

$$+x^{54}y^{54}z^{54}\frac{t^{27}}{27!}$$

13.5 PARTITIONS IN 2D TRIANGLE, 3D PYRAMID OR nD HYPERPYRAMID LATTICES

The determinant method employed to derive the nD q-binomial extension is easy to apply to the lattice points in the sector of the 2D first quadrant above or below the line $y = x$ and/or on it. Hence, by applying the methodology from earlier sections, we see easily that

$$\begin{aligned}
&\frac{1}{(1-xyt)} \\
&\times \frac{1}{(1-xy^2t)(1-x^2y^2t)} \\
&\times \frac{1}{(1-xy^3t)(1-x^2y^3t)(1-x^3y^3t)} \\
&\times \frac{1}{(1-xy^4t)(1-x^2y^4t)(1-x^3y^4t)(1-x^4y^4t)} \\
&\times \frac{1}{(1-xy^5t)(1-x^2y^5t)(1-x^3y^5t)(1-x^4y^5t)(1-x^5y^5t)} \\
&\times \textit{etc.} \\
&= 1 + \frac{xy}{(1-y)(1-xy)}\frac{t}{1!} \\
&+ \begin{vmatrix} \frac{xy}{(1-y)(1-xy)} & -1 \\ \frac{x^2y^2}{(1-y^2)(1-x^2y^2)} & \frac{xy}{(1-y)(1-xy)} \end{vmatrix} \frac{t^2}{2!} \\
&+ \begin{vmatrix} \frac{xy}{(1-y)(1-xy)} & -1 & 0 \\ \frac{x^2y^2}{(1-y^2)(1-x^2y^2)} & \frac{xy}{(1-y)(1-xy)} & -2 \\ \frac{x^3y^3}{(1-y^3)(1-x^3y^3)} & \frac{x^2y^2}{(1-y^2)(1-x^2y^2)} & \frac{xy}{(1-y)(1-xy)} \end{vmatrix} \frac{t^3}{3!} \\
&+ \begin{vmatrix} \frac{xy}{(1-y)(1-xy)} & -1 & 0 & 0 \\ \frac{x^2y^2}{(1-y^2)(1-x^2y^2)} & \frac{xy}{(1-y)(1-xy)} & -2 & 0 \\ \frac{x^3y^3}{(1-y^3)(1-x^3y^3)} & \frac{x^2y^2}{(1-y^2)(1-x^2y^2)} & \frac{xy}{(1-y)(1-xy)} & -3 \\ \frac{x^4y^4}{(1-y^4)(1-x^4y^4)} & \frac{x^3y^3}{(1-y^3)(1-x^3y^3)} & \frac{x^2y^2}{(1-y^2)(1-x^2y^2)} & \frac{xy}{(1-y)(1-xy)} \end{vmatrix} \frac{t^4}{4!} \\
&+ \textit{etc.}
\end{aligned} \tag{13.5.1}$$

where we note the key to this is the summation

$$xy + (x+x^2)y^2 + (x+x^2+x^3)y^3 + (x+x^2+x^3+x^4)y^4 + \cdots = \frac{xy}{(1-y)(1-xy)}; \tag{13.5.2}$$

this being the sum taken over the 2D lattice points in the lower half quadrant beneath the line $y = x$. Clearly, this principle will apply to any radial region of lattice points whether it be in the 2D space or the nD space. Therefore we get a 2D partition identity for any exact number of parts for the vector partition. The partition into exactly n parts for the particular lattice generating (13.5.2) is the coefficient of t^n in the power series in t of identity (13.5.1).

A 3D pyramid lattice extension of the above 2D result is as follows.

$$\begin{aligned}
&\frac{1}{(1-xyzt)} \\
&\times\frac{1}{(1-xyz^2t)(1-xy^2z^2t)} \\
&\times\frac{1}{(1-x^2yz^2t)(1-x^2y^2z^2t)} \\
&\times\frac{1}{(1-xyz^3t)(1-xy^2z^3t)(1-xy^3z^3t)} \\
&\times\frac{1}{(1-x^2yz^3t)(1-x^2y^2z^3t)(1-x^2y^3z^3t)} \\
&\times\frac{1}{(1-x^3yz^3t)(1-x^3y^2z^3t)(1-x^3y^3z^3t)} \\
&\times \textit{ etc.} \\
&= 1+f(x,y,z)\frac{t}{1!}+\begin{vmatrix} f(x,y,z) & -1 \\ f(x^2,y^2,z^2) & f(x,y,z)\end{vmatrix}\frac{t^2}{2!} \\
&+\begin{vmatrix} f(x,y,z) & -1 & 0 \\ f(x^2,y^2,z^2) & f(x,y,z) & -2 \\ f(x^3,y^3,z^3) & f(x^2,y^2,z^2) & f(x,y,z)\end{vmatrix}\frac{t^3}{3!} \\
&+\begin{vmatrix} f(x,y,z) & -1 & 0 & 0 \\ f(x^2,y^2,z^2) & f(x,y,z) & -2 & 0 \\ f(x^3,y^3,z^3) & f(x^2,y^2,z^2) & f(x,y,z) & -3 \\ f(x^4,y^4,z^4) & f(x^3,y^3,z^3) & f(x^2,y^2,z^2) & f(x,y,z)\end{vmatrix}\frac{t^4}{4!}+\textit{etc.}
\end{aligned} \tag{13.5.3}$$

where

$$f(x,y,z) = xyz + \frac{xy}{(1-x)(1-y)}\left(\frac{z^2}{1-z} - \frac{xz^2}{1-xz} - \frac{yz^2}{1-yz} + \frac{x^2y^2z^2}{1-xyz}\right).$$

We note $f(x,y,z)$ is the generating function for the lattice point vectors comprising the 3D vector parts of the partitions enumerated by (13.5.3) and that

$$f(x,y,z) = xyz + \begin{pmatrix} xy & +xy^2 \\ +x^2y & +x^2y^2 \end{pmatrix} z^2 + \begin{pmatrix} xy & +xy^2 & +xy^3 \\ +x^2y & +x^2y^2 & +x^2y^3 \\ +x^3y & +x^3y^2 & +x^3y^3 \end{pmatrix} z^3 \tag{13.5.4}$$

$$+\begin{pmatrix} xy & +xy^2 & +xy^3 & +xy^4 \\ +x^2y & +x^2y^2 & +x^2y^3 & +x^2y^4 \\ +x^3y & +x^3y^2 & +x^3y^3 & +x^3y^4 \\ +x^4y & +x^4y^2 & +x^4y^3 & +x^4y^4 \end{pmatrix} z^4 + \cdots$$

$$= xyz + \frac{xy(1-x^2)(1-y^2)}{(1-x)(1-y)}z^2 + \frac{xy(1-x^3)(1-y^3)}{(1-x)(1-y)}z^3 + \cdots$$

$$= xyz + \frac{xy}{(1-x)(1-y)}\left(\frac{z^2}{1-z} - \frac{xz^2}{1-xz} - \frac{yz^2}{1-yz} + \frac{x^2y^2z^2}{1-xyz}\right);$$

this being the sum taken over the 3D lattice points in an upper inverted square pyramid. Clearly, this principle will apply to any radial region of lattice points whether it be in the 2D, 3D space or nD space. So we get a 3D partition identity for any exact number of parts for the vector partition. The number of partitions into exactly n parts for the particular lattice generated by (13.5.4) has its partition generating function as the coefficient of t^n in the power series in t of identity (13.5.3).

Think About It...

Pascal's triangle is named for 17th century French mathematician Blaise Pascal. However, the Chinese Jia Xian found the triangle in the 11th century. In China it is named the *Yang Hui triangle* after a Chinese mathematician in the 13th century.

The triangle was also studied by 11th century Persian mathematician, astronomer, and poet, Omar Khayyam.

Brittanica
www.britannica.com/science/Pascals-triangle

FURTHER READING

ABRAMOWITZ, M. and STEGUN, I. *Handbook of Mathematical Functions.* Dover Publications Inc., New York, 1972. 10th Printing, 1972, with corrections.

ANDREWS, G.E. *The Theory of Partitions*, volume 2 of *Encyclopedia of Mathematics and its Applications.* Addison-Wesley Publishing Company, Advanced Book Program, Reading, Massachusetts, USA, 1976.

ANDREWS, G.E. and ERIKSSON, K. *Integer Partitions.* Cambridge University Press, Cambridge, UK, New York, USA, Port Melbourne, Australia, Madrid, Spain, Cape Town, South Africa, 2004.

BAXTER, R.J. *Exactly Solved Models in Statistical Mechanics.* Academic Press, New York, 1982. Reprinted 1989 and 2007.

BRINGMANN, K. and LOVEJOY , J. Ranks and congruences for overpartition pairs. *Int. J. Number Theory*, 4:303–322, 2008.

CAMPBELL, G.B. Multiplicative functions over Riemann zeta function products. *J. Ramanujan Soc.*, 7(1):52–63, 1992.

CAMPBELL, G.B. Dirichlet summations and products over primes. *Int. J. Math. Math. Sci.*, 16(2):359–372, 1993.

CAMPBELL, G.B. A generalized formula of Hardy. *Int. J. Math. Math. Sci.*, 17(2):369–378, 1994.

CAMPBELL, G.B. A new class of infinite products, and Euler's totient. *Int. J. Math. and Math. Sci.*, 17(3):417–422, 1994.

CAMPBELL, G.B. *Combinatorial identities in number theory related to q-series and arithmetical functions.* Doctor of philosophy thesis, School of Mathematical Sciences, The Australian National University, October 1997.

CAMPBELL, G.B. A closer look at some new identities. *Int. J. Math. and Math. Sci.*, 21(3):581–586, 1998.

CAMPBELL, G.B. Infinite products over hyperpyramid lattices. *Int. J. Math. and Math. Sci.*, 23(4):271–277, 2000.

CAMPBELL, G.B. Some n-space q-binomial theorem extensions and similar identities. arxiv preprint, June 2019. arXiv:1906.07526v1 [math.NT]: https://arxiv.org/abs/1906.07526.

CORTEEL, S. and LOVEJOY, J. Overpartitions. *Trans. Amer. Math. Soc.*, 356:1623–1635, 2004.

GAUSS, C.F. Disquisitiones generales circa seriem infinitam . . .. In *Werke 3*, volume II, pages 123–162. Comm. soc. reg. sci. Gött. rec., 1876. article first printed 1813; reprinted in Werke 3 1876.

HEINE, E. Untersuchungen über die Reihe ... . *J. Reine angew. Math.*, 34:285–328, 1847.

HEINE, E. *Handbuch der Kugelfunctionen*, Theorie und Andwendungen, volume 1. Reimer, Berlin, 1878.

MACDONALD, I.G. *Symmetric Functions And Hall Polynomials.* Oxford:Clarendon Press; New York/London: Oxford University Press, 2nd edition, 1995.

CHAPTER 14

Weighted Vector Partitions as Hybrid n-Space Variations

14.1 EXAMPLES OF HYBRID VARIATIONS FOR IDENTITIES GENERATING WEIGHTED VECTOR PARTITIONS

THE METHODS of the book by Macdonald [191] would have determined these formulae earlier but in his examples he considers only a single q-variable and bivariate infinite products. Although Cramer's rule (see Mac Lane and Birkhoff [190]) is used in examples of his book, the method has not been specifically applied to products such as those in our text here.

In relation to some research of the past fifty-three years, the interesting papers on vector partition generating functions such as Cheema [108], Cheema and Motzkin [109], Gordon [141] and Wright [302] show that our current methods have not been used in many of the old problems in vector partition theory.

We continue with a vector partition generating function variation on the work so far, by stating

Definition 14.1 *We define the function $G_n(t)$ for all non-negative integers n, and for suitable complex numbers a and t, by*

$$G_0(-;a,t) = \frac{1-at}{1-t} \tag{14.1.1}$$

and for all of $|q_1|, |q_2|, ..., |q_n| < 1$, by

$$G_n(q_1, q_2, ..., q_n; a, t) = \prod_{\alpha_1,\alpha_2,...,\alpha_n \geq 0} \left(\frac{1 - q_1^{\alpha_1} q_2^{\alpha_2} ... q_n^{\alpha_n} at}{1 - q_1^{\alpha_1} q_2^{\alpha_2} ... q_n^{\alpha_n} t} \right)^{\frac{1}{\alpha_1!\alpha_2!...\alpha_n!}} \tag{14.1.2}$$

$$\equiv \sum_{k=0}^{\infty} {}_nB_k(q_1, q_2, ..., q_n; a) t^k$$

where $\alpha_1, \alpha_2, ..., \alpha_n$ may be all zero which in that particular case is $G_0(-;a,t)$.

DOI: 10.1201/9781003174158-14

By the way, note that by this definition the right side $G_1(q;a,t)$, is a variant on the standard q-binomial expression (13.1.1). Next, based on this definition we can assert the

Theorem 14.1 *An n-space hybrid of the extended q-binomial theorem.*

$$ {}_nB_k(q_1,q_2,...,q_n;a) = \det(a_{ij})/k! \tag{14.1.3}$$

where the determinant is of order k, and

$$a_{ij} = \begin{cases} (1-a^{i-j+1})\exp\left({q_1}^{i-j+1}+{q_2}^{i-j+1}+...+{q_n}^{i-j+1}\right), & i \geq j; \\ -i, & i = j-1; \\ 0, & otherwise. \end{cases}$$

14.2 PROOF OF THE n-SPACE HYBRID q-BINOMIAL THEOREM

As with the full n-space q-binomial theorem proof, we mention that the cases $n=2$ and $n=3$ of equation (14.1.3) are new to the literature. Also, for $n=1$, the q-binomial theorem variation, there is no proof in the literature. However, we first prove the n-space identity, before then giving a diverse range of examples.

Proof 2 *We see that, similar to the earlier proof,*

$$\begin{aligned} G_n(q_1,q_2,...,q_n;a,t) &= \exp\left(\sum_{\alpha_1,\alpha_2,...,\alpha_n \geq 0} \frac{1}{\alpha_1!\alpha_2!...\alpha_n!}\log\left(\frac{1-q_1^{\alpha_1}q_2^{\alpha_2}...q_n^{\alpha_n}at}{1-q_1^{\alpha_1}q_2^{\alpha_2}...q_n^{\alpha_n}t}\right)\right) \\ &= \exp\left(\sum_{k=1}^{\infty}\left(1-a^k\right)\exp\left({q_1}^k+{q_2}^k+...+{q_n}^k\right)\frac{t^k}{k}\right). \end{aligned}$$

Since $G_n(q_1,q_2,q_3,...,q_n;a,t) = G_n(t) = \sum_{k=0}^{\infty} {}_nB_k t^k$, differentiating both sides of the equation

$$\log\left(\sum_{k=0}^{\infty} {}_nB_k t^k\right) = \sum_{k=1}^{\infty}\left(1-a^k\right)\exp\left({q_1}^k+{q_2}^k+...+{q_n}^k\right)\frac{t^k}{k}$$

with respect to t, then equating like powers of t, leads to the set of simultaneous equations for any positive integer k

$$1 \times {}_nB_1 = (1-a)\exp(q_1+q_2+...+q_n),$$

$$2 \times {}_nB_2 = \left(1-a^2\right)\exp\left({q_1}^2+{q_2}^2+...+{q_n}^2\right) + {}_nB_1 \times (1-a)\exp(q_1+q_2+...+q_n),$$

$$\begin{aligned} 3 \times {}_nB_3 = &\left(1-a^3\right)\exp\left({q_1}^3+{q_2}^3+...+{q_n}^3\right) + {}_nB_1 \times \left(1-a^2\right)\exp\left({q_1}^2+{q_2}^2+...+{q_n}^2\right) \\ &+ {}_nB_2 \times (1-a)\exp(q_1+q_2+...+q_n), \end{aligned}$$

$$4 \times {}_nB_4 = \left(1-a^4\right)\exp\left({q_1}^4+{q_2}^4+...+{q_n}^4\right) + {}_nB_1 \times \left(1-a^3\right)\exp\left({q_1}^3+{q_2}^3+...+{q_n}^3\right)$$

$$+{}_nB_2 \times \left(1-a^2\right)\exp\left({q_1}^2+{q_2}^2+...+{q_n}^2\right)+{}_nB_3\times(1-a)\exp\left(q_1+q_2+...+q_n\right),$$

etc., down to the kth equation

$$k \times {}_nB_k$$
$$=\left(1-a^k\right)\exp\left({q_1}^k+{q_2}^k+...+{q_n}^k\right)$$
$$+{}_nB_1\times\left(1-a^{k-1}\right)\exp\left({q_1}^{k-1}+{q_2}^{k-1}+...+{q_n}^{k-1}\right)+...$$
$$...+{}_nB_{k-2}\times\left(1-a^2\right)\exp\left({q_1}^2+{q_2}^2+...+{q_n}^2\right)$$
$$+{}_nB_{k-1}\times(1-a)\exp\left(q_1+q_2+...+q_n\right).$$

We note that for the above, the first equation has one unknown, the first and second equations together are two equations in two unknowns the first, second and third equations are three equations in three unknowns, and so on. The set of k equations here is a solvable set of k equations in k unknowns, and is solved by Cramer's Rule involving determinants. Applying this rule yields the theorem.

The above proof is essentially the same as that given earlier here, simply again applying Faà de Bruno's theorem.

We write theorem 14.1 as a single equation identity as follows.

$$G_n(q_1,q_2,q_3,...,q_n;a,t)=\prod_{\alpha_1,\alpha_2,...,\alpha_n\geq 0}\left(\frac{1-q_1^{\alpha_1}q_2^{\alpha_2}...q_n^{\alpha_n}at}{1-q_1^{\alpha_1}q_2^{\alpha_2}...q_n^{\alpha_n}t}\right)^{\frac{1}{\alpha_1!\alpha_2!...\alpha_n!}} \tag{14.2.1}$$

$$=1+(1-a)\exp\left(q_1+q_2+...+q_n\right)\frac{t}{1!}$$

$$+\begin{vmatrix}(1-a)\exp\left(q_1+q_2+...+q_n\right) & -1\\ (1-a^2)\exp\left({q_1}^2+{q_2}^2+...+{q_n}^2\right) & (1-a)\exp\left(q_1+q_2+...+q_n\right)\end{vmatrix}\frac{t^2}{2!}$$

$$+\begin{vmatrix}(1-a)\exp\left(q_1+q_2+...+q_n\right) & -1 & 0\\ \left(1-a^2\right)\exp\left({q_1}^2+{q_2}^2+...+{q_n}^2\right) & (1-a)\exp\left(q_1+q_2+...+q_n\right) & -2\\ \left(1-a^3\right)\exp\left({q_1}^3+{q_2}^3+...+{q_n}^3\right) & \left(1-a^2\right)\exp\left({q_1}^2+{q_2}^2+...+{q_n}^2\right) & (1-a)\exp\left(q_1+q_2+...+q_n\right)\end{vmatrix}\frac{t^3}{3!}$$

$$+\begin{vmatrix}(1-a)\exp\left(\sum_{j=1}^n q_j\right) & -1 & 0 & 0\\ \left(1-a^2\right)\exp\left(\sum_{j=1}^n {q_j}^2\right) & (1-a)\exp\left(\sum_{j=1}^n q_j\right) & -2 & 0\\ \left(1-a^3\right)\exp\left(\sum_{j=1}^n {q_j}^3\right) & \left(1-a^2\right)\exp\left(\sum_{j=1}^n {q_j}^2\right) & (1-a)\exp\left(\sum_{j=1}^n q_j\right) & -3\\ \left(1-a^4\right)\exp\left(\sum_{j=1}^n {q_j}^4\right) & \left(1-a^3\right)\exp\left(\sum_{j=1}^n {q_j}^3\right) & \left(1-a^2\right)\exp\left(\sum_{j=1}^n {q_j}^2\right) & (1-a)\exp\left(\sum_{j=1}^n q_j\right)\end{vmatrix}\frac{t^4}{4!}$$

$$+etc.$$

The following 1-space example of equation (14.2.1) is a variant on the q-binomial theorem itself. This result is not in the literature and is equivalent to a statement about weighted integer partitions.

Corollary 14.1

$$G_1(q;a,t) = \prod_{k=0}^{\infty} \left(\frac{1-atq^k}{1-tq^k} \right)^{1/k!} = 1 + (1-a)e^q \frac{t}{1!} \tag{14.2.2}$$

$$+ \begin{vmatrix} (1-a)e^q & -1 \\ (1-a^2)e^{q^2} & (1-a)e^q \end{vmatrix} \frac{t^2}{2!} + \begin{vmatrix} (1-a)e^q & -1 & 0 \\ (1-a^2)e^{q^2} & (1-a)e^q & -2 \\ (1-a^3)e^{q^3} & (1-a^2)e^{q^2} & (1-a)e^q \end{vmatrix} \frac{t^3}{3!}$$

$$+ \begin{vmatrix} (1-a)e^q & -1 & 0 & 0 \\ (1-a^2)e^{q^2} & (1-a)e^q & -2 & 0 \\ (1-a^3)e^{q^3} & (1-a^2)e^{q^2} & (1-a)e^q & -3 \\ (1-a^4)e^{q^4} & (1-a^3)e^{q^3} & (1-a^2)e^{q^2} & (1-a)e^q \end{vmatrix} \frac{t^4}{4!} + etc.$$

An interesting case of this is with $a = 0$ where

Corollary 14.2

$$G_1(q;0,t) = \prod_{k=0}^{\infty} \left(\frac{1}{1-tq^k} \right)^{1/k!} \tag{14.2.3}$$

$$= 1 + e^q \frac{t}{1!} + \begin{vmatrix} e^q & -1 \\ e^{q^2} & e^q \end{vmatrix} \frac{t^2}{2!} + \begin{vmatrix} e^q & -1 & 0 \\ e^{q^2} & e^q & -2 \\ e^{q^3} & e^{q^2} & e^q \end{vmatrix} \frac{t^3}{3!} + \begin{vmatrix} e^q & -1 & 0 & 0 \\ e^{q^2} & e^q & -2 & 0 \\ e^{q^3} & e^{q^2} & e^q & -3 \\ e^{q^4} & e^{q^3} & e^{q^2} & e^q \end{vmatrix} \frac{t^4}{4!} + etc.$$

$$= 1 + e^q \frac{t}{1!} + \left(e^{q^2} + e^{2q} \right) \frac{t^2}{2!} + \left(2e^{q^3} + 3e^{q^2+q} + e^{3q} \right) \frac{t^3}{3!}$$

$$+ \left(6e^{q^4} + 8e^{q^3+q} + 3e^{2q^2} + 6e^{q^2+2q} + e^{4q} \right) \frac{t^4}{4!} + etc.$$

Inferences from limiting cases of the coefficients lead us to affirm as checks and balances for example, that:

$$\begin{vmatrix} a & -1 & 0 & 0 \\ a & a & -2 & 0 \\ a & a & a & -3 \\ a & a & a & a \end{vmatrix} = a(a+1)(a+2)(a+3), \tag{14.2.4}$$

and

$$\begin{vmatrix} a & -1 & 0 & 0 \\ a^2 & a & -2 & 0 \\ a^3 & a^2 & a & -3 \\ a^4 & a^3 & a^2 & a \end{vmatrix} = 4!a^4. \tag{14.2.5}$$

Both (14.2.4) and (14.2.5) are easy to prove, extend to generalized order determinants, and are consistent with the application of Faà De Bruno's formula. It might be a worthwhile thing to publish a comprehensive list of cases of the types (14.2.4) and (14.2.5) applicable to the generating functions in our book and other possible follow-up papers related to the analysis of classes of vector partitions.

Next, repeating a variation on the recurrence logic from the start of section 14.2, we see that

Corollary 14.3

$$G_1(q;0;t)/G_1(q^2;0;t^2) = \prod_{k=0}^{\infty} \left(1+tq^k\right)^{1/k!} \tag{14.2.6}$$

$$= 1 + e^q \frac{t}{1!} + \begin{vmatrix} -e^q & -1 \\ e^{q^2} & -e^q \end{vmatrix} \frac{t^2}{2!} + \begin{vmatrix} -e^q & -1 & 0 \\ e^{q^2} & -e^q & -2 \\ -e^{q^3} & e^{q^2} & -e^q \end{vmatrix} \frac{t^3}{3!} + \begin{vmatrix} -e^q & -1 & 0 & 0 \\ e^{q^2} & -e^q & -2 & 0 \\ -e^{q^3} & e^{q^2} & -e^q & -3 \\ e^{q^4} & -e^{q^3} & e^{q^2} & -e^q \end{vmatrix} \frac{t^4}{4!} + etc.$$

$$= 1 + e^q \frac{t}{1!} + \left(e^{2q} - e^{q^2}\right) \frac{t^2}{2!} + \left(2e^{q^3} - 3e^{q^2+q} + e^{3q}\right) \frac{t^3}{3!}$$
$$+ \left(e^{4q} + 3e^{2q^2} - 6e^{q^4} - 6e^{2q+q^2} + 8e^{q+q^3}\right) \frac{t^4}{4!} + etc.$$

The statement of theorem 14.1 and it's form in equation (14.1.1) reminds us that considerations associated with oscillating functions near the boundaries of convergence might apply or at least be worth covering off for exclusion for such results. For an account of this phenomenon see Hardy and Littlewood's paper [151] on their "high indices theorem". In particular, the reader should heed the early papers by Hardy (see [150] and [149]) where it is clear that the behavior of our functions near their poles and the radii of convergence should be examined and studied in order to eliminate the possibility that our formulas do not oscillate wildly in certain neighborhoods, or are only picking up major terms and omitting complicated oscillating minor terms.

Corollary 14.4 *The 2-space example of equation (14.2.1) is for* $|x| < 1$, $|y| < 1$, $|a| < 1$, $|t| < 1$,

$$G_2(x,y;a,t) = \prod_{j,k \geq 0} \left(\frac{1 - x^j y^k at}{1 - x^j y^k t}\right)^{\frac{1}{j!k!}} \tag{14.2.7}$$

$$= 1 + (1-a)e^{x+y} \frac{t}{1!} + \begin{vmatrix} (1-a)e^{x+y} & -1 \\ (1-a^2)e^{x^2+y^2} & (1-a)e^{x+y} \end{vmatrix} \frac{t^2}{2!}$$

$$+ \begin{vmatrix} (1-a)e^{x+y} & -1 & 0 \\ (1-a^2)e^{x^2+y^2} & (1-a)e^{x+y} & -2 \\ (1-a^3)e^{x^3+y^3} & (1-a^2)e^{x^2+y^2} & (1-a)e^{x+y} \end{vmatrix} \frac{t^3}{3!}$$

$$+ \begin{vmatrix} (1-a)e^{x+y} & -1 & 0 & 0 \\ (1-a^2)e^{x^2+y^2} & (1-a)e^{x+y} & -2 & 0 \\ (1-a^3)e^{x^3+y^3} & (1-a^2)e^{x^2+y^2} & (1-a)e^{x+y} & -3 \\ (1-a^3)e^{x^4+y^4} & (1-a^3)e^{x^3+y^3} & (1-a^2)e^{x^2+y^2} & (1-a)e^{x+y} \end{vmatrix} \frac{t^4}{4!} + etc.$$

The function $G_2(x,y;a,t)$ is interesting for a variety of reasons and can lead to new and interesting combinatorial analysis involving weighted partitions of vectors. In other words, we can examine stepping stone jumps between integer lattice points in 2-space, where each jump to the next stepping stone involves carrying a weight assigned to the coefficient, and our "vector partition science" will be based on the total weight carried in order to combine all possible stepping stone jumps assigned by rule. In 3-space this same weighted stepping stone combinatorial analysis will also apply, and the 3-space version of (14.2.1) given now is:

Corollary 14.5 *The 3-space example of equation (14.2.1) is for* $|x|<1,\ |y|<1,\ |z|<1,\ |a|<1,\ |t|<1,$

$$G_3(x,y,z;a,t) = \prod_{h,j,k\geq 0}\left(\frac{1-x^h y^j z^k at}{1-x^h y^j z^k t}\right)^{\frac{1}{h!j!k!}} \tag{14.2.8}$$

$$= 1+(1-a)e^{x+y+z}\frac{t}{1!}+\begin{vmatrix} (1-a)e^{x+y+z} & -1 \\ (1-a^2)e^{x^2+y^2+z^2} & (1-a)e^{x+y+z}\end{vmatrix}\frac{t^2}{2!}$$

$$+\begin{vmatrix} (1-a)e^{x+y+z} & -1 & 0 \\ (1-a^2)e^{x^2+y^2+z^2} & (1-a)e^{x+y+z} & -2 \\ (1-a^3)e^{x^3+y^3+z^3} & (1-a^2)e^{x^2+y^2+z^2} & (1-a)e^{x+y+z}\end{vmatrix}\frac{t^3}{3!}$$

$$+\begin{vmatrix} (1-a)e^{x+y+z} & -1 & 0 & 0 \\ (1-a^2)e^{x^2+y^2+z^2} & (1-a)e^{x+y+z} & -2 & 0 \\ (1-a^3)e^{x^3+y^3+z^3} & (1-a^2)e^{x^2+y^2+z^2} & (1-a)e^{x+y+z} & -3 \\ (1-a^3)e^{x^4+y^4+z^4} & (1-a^3)e^{x^3+y^3+z^3} & (1-a^2)e^{x^2+y^2+z^2} & (1-a)e^{x+y+z}\end{vmatrix}\frac{t^4}{4!}+etc.$$

We reiterate here that the identities in the present book follow the approach taken by Andrews [12, chapter 2], who develops his theory of integer partitions starting from a historical viewpoint and uses the q-binomial theorem to derive the Euler identities for the number of partitions of a positive integer. It would be interesting if (14.2.4) and (14.2.5) could yield higher Jacobi triple product analogies or higher 2-space and 3-space type Heine q-series summations and transforms, and other 2-space and 3-space q-series sum-product identities relevant to vector partitions.

The 4-space example of theorem 14.1 is next stated here.

Corollary 14.6 *The 4D version of equation (14.1.1) is for* $|w|<1,\ |x|<1,\ |y|<1,\ |z|<1,\ |a|<1,\ |t|<1,$

$$G_4(w,x,y,z;a,t) = \prod_{h,i,j,k\geq 0}\left(\frac{1-w^h x^i y^j z^k at}{1-w^h x^i y^j z^k t}\right)^{\frac{1}{h!i!j!k!}} \tag{14.2.9}$$

$$= 1+(1-a)e^{w+x+y+z}\frac{t}{1!}+\begin{vmatrix} (1-a)e^{w+x+y+z} & -1 \\ (1-a^2)e^{w^2+x^2+y^2+z^2} & (1-a)e^{w+x+y+z}\end{vmatrix}\frac{t^2}{2!}$$

$$+\begin{vmatrix} (1-a)e^{w+x+y+z} & -1 & 0 \\ (1-a^2)e^{w^2+x^2+y^2+z^2} & (1-a)e^{w+x+y+z} & -2 \\ (1-a^3)e^{w^3+x^3+y^3+z^3} & (1-a^2)e^{w^2+x^2+y^2+z^2} & (1-a)e^{w+x+y+z}\end{vmatrix}\frac{t^3}{3!}$$

$$+\begin{vmatrix} (1-a)e^{w+x+y+z} & -1 & 0 & 0 \\ (1-a^2)e^{w^2+x^2+y^2+z^2} & (1-a)e^{w+x+y+z} & -2 & 0 \\ (1-a^3)e^{w^3+x^3+y^3+z^3} & (1-a^2)e^{w^2+x^2+y^2+z^2} & (1-a)e^{w+x+y+z} & -3 \\ (1-a^3)e^{w^4+x^4+y^4+z^4} & (1-a^3)e^{w^3+x^3+y^3+z^3} & (1-a^2)e^{w^2+x^2+y^2+z^2} & (1-a)e^{w+x+y+z} \end{vmatrix} \frac{t^4}{4!} + etc.$$

14.3 EXERCISES

We consider the case of (14.2.7) where $a=0$.

$$G_2(x,y;0,t) = \prod_{j,k\geq 0} \left(\frac{1}{1-x^j y^k t} \right)^{\frac{1}{j!k!}} \tag{14.3.1}$$

$$= 1 + e^{x+y}\frac{t}{1!} + \begin{vmatrix} e^{x+y} & -1 \\ e^{x^2+y^2} & e^{x+y} \end{vmatrix} \frac{t^2}{2!} + \begin{vmatrix} e^{x+y} & -1 & 0 \\ e^{x^2+y^2} & e^{x+y} & -2 \\ e^{x^3+y^3} & e^{x^2+y^2} & e^{x+y} \end{vmatrix} \frac{t^3}{3!}$$

$$+ \begin{vmatrix} e^{x+y} & -1 & 0 & 0 \\ e^{x^2+y^2} & e^{x+y} & -2 & 0 \\ e^{x^3+y^3} & e^{x^2+y^2} & e^{x+y} & -3 \\ e^{x^4+y^4} & e^{x^3+y^3} & e^{x^2+y^2} & e^{x+y} \end{vmatrix} \frac{t^4}{4!} + etc..$$

$$= 1 + e^{x+y}\frac{t}{1!} + \left(e^{2x+2y} + e^{x^2+y^2}\right)\frac{t^2}{2!} + \left(e^{3x^2+3y^2} + 3e^{x^2+x+y+y^2} + 2e^{x^3+y^3}\right)\frac{t^3}{3!}$$

$$+ \left(e^{4x+4y} + 9e^{2x^2+2y^2} + 8e^{x^3+x+y+y^3} + 6e^{x^4+y^4}\right)\frac{t^4}{4!} + etc..$$

The product (14.3.1) for $G_2(x,y;0,t)$ can be written as

$$\prod_{j,k\geq 0} \left\{ 1 + \binom{\frac{-1}{j!k!}}{1} x^j y^k t + \binom{\frac{-1}{j!k!}}{2} x^{2j} y^{2k} t^2 + \binom{\frac{-1}{j!k!}}{3} x^{3j} y^{3k} t^3 + etc. \right\}. \tag{14.3.2}$$

This infinite product of infinite series is the clue to the basic building blocks for our *weighted vector partitions.* At this point in our analysis, we can view the combinatorial contribution of each term inside the curly brackets. This helps to theorize a physical model to show us what is going on in identity (14.3.1). We note that the following points:

(a) Each curly bracketed sum given in (14.3.1) contributes to the parts of the 2D vector partition all 2D vectors of gradient $\frac{k}{j}$. That is, the vectors $\langle j,k \rangle$, $\langle 2j,2k \rangle$, $\langle 3j,3k \rangle$, etc. the nth term multiplied by the associated scalar term $\binom{\frac{-1}{j!k!}}{n}$.

(b) Curly bracketed terms map to a contribution to a function of a vector partition as per the notation start of Chapter 12,

$$\binom{\frac{-1}{j!k!}}{1} x^j y^k \mapsto \binom{\frac{-1}{j!k!}}{1} (\langle j,k\rangle^1) \mapsto \binom{\frac{-1}{j!k!}}{1} \langle j,k\rangle,$$

$$\binom{\frac{-1}{j!k!}}{2} x^{2j} y^{2k} \mapsto \binom{\frac{-1}{j!k!}}{2} (\langle j,k\rangle^2) \mapsto \binom{\frac{-1}{j!k!}}{2} (\langle j,k\rangle + \langle j,k\rangle),$$

$$\binom{\frac{-1}{j!k!}}{3} x^{3j} y^{3k} \mapsto \binom{\frac{-1}{j!k!}}{3} (\langle j,k\rangle^3) \mapsto \binom{\frac{-1}{j!k!}}{3} (\langle j,k\rangle + \langle j,k\rangle) + \langle j,k\rangle),$$

and so on.

(c) We see now why VPVs, that is vectors $\langle j,k\rangle$ where $\gcd(j,k)=1$, assume a unique place in the 2D generating functions.

Continuing this way we see in the 3D situation based on (14.2.8) where $a=0$ that

$$G_3(x,y,z;0,t) = \prod_{h,j,k\geq 0} \left(\frac{1}{1-x^h y^j z^k t}\right)^{\frac{1}{h!j!k!}} \tag{14.3.3}$$

$$= 1 + e^{x+y+z}\frac{t}{1!} + \begin{vmatrix} e^{x+y+z} & -1 \\ e^{x^2+y^2+z^2} & e^{x+y+z} \end{vmatrix} \frac{t^2}{2!} + \begin{vmatrix} e^{x+y+z} & -1 & 0 \\ e^{x^2+y^2+z^2} & e^{x+y+z} & -2 \\ e^{x^3+y^3+z^3} & e^{x^2+y^2+z^2} & e^{x+y+z} \end{vmatrix} \frac{t^3}{3!}$$

$$+ \begin{vmatrix} e^{x+y+z} & -1 & 0 & 0 \\ e^{x^2+y^2+z^2} & e^{x+y+z} & -2 & 0 \\ e^{x^3+y^3+z^3} & e^{x^2+y^2+z^2} & e^{x+y+z} & -3 \\ e^{x^4+y^4+z^4} & e^{x^3+y^3+z^3} & e^{x^2+y^2+z^2} & e^{x+y+z} \end{vmatrix} \frac{t^4}{4!} + etc.$$

Next, as an exercise, it may be interesting to prove or disprove that an infinite variable case may exist, such that the following is true. (Excluding zero denominator terms.)

$$G_\infty(1^{-s},2^{-s},3^{-s},\ldots;0,t) = \prod_{a_1,a_2,a_3,\ldots\geq 1} \left(\frac{1}{1-(1^{-a_1 s}2^{-a_2 s}3^{-a_3 s}\ldots)t}\right)^{\frac{1}{a_1!\,a_2!\,a_3!}\cdots} \tag{14.3.4}$$

$$= 1 + e^{\zeta(s)}\frac{t}{1!} + \begin{vmatrix} e^{\zeta(s)} & -1 \\ e^{\zeta(2s)} & e^{\zeta(s)} \end{vmatrix} \frac{t^2}{2!} + \begin{vmatrix} e^{\zeta(s)} & -1 & 0 \\ e^{\zeta(2s)} & e^{\zeta(s)} & -2 \\ e^{\zeta(3s)} & e^{\zeta(2s)} & e^{\zeta(s)} \end{vmatrix} \frac{t^3}{3!}$$

$$+ \begin{vmatrix} e^{\zeta(s)} & -1 & 0 & 0 \\ e^{\zeta(2s)} & e^{\zeta(s)} & -2 & 0 \\ e^{\zeta(3s)} & e^{\zeta(2s)} & e^{\zeta(s)} & -3 \\ e^{\zeta(4s)} & e^{\zeta(3s)} & e^{\zeta(2s)} & e^{\zeta(s)} \end{vmatrix} \frac{t^4}{4!} + etc.$$

Think About It...

Without mathematics, there's nothing you can do.

Everything around you is mathematics. Everything around you is numbers.

Shakuntala Devi
Indian writer and mental calculator

FURTHER READING

ANDREWS, G.E. *The Theory of Partitions*, volume 2 of *Encyclopedia of Mathematics and its Applications*. Addison-Wesley Publishing Company, Advanced Book Program, Reading, Massachusetts, USA, 1976.

CHAVEZ, F. and DUHR, C. Three-mass triangle integrals and single-valued polylogarithms. *J. High Energy Phys.*, 2012. article id.114.

CHEEMA, M.S. and MOTZKIN, T.S. Multipartitions and multipermutations. *Proc. Symp. Pure Math.*, 19:37–39, 1971.

GORDON, B. Two theorems on multipartite partitions. *J. London Math. Soc.*, 38:459–464, 1963.

HARDY, G.H. *An extension of a theorem on oscillating series.* In *Collected Papers, volume VI*, pages 500–506. Clarendon Press, Oxford, 1974.

HARDY, G.H. *On certain oscillating series.* In *Collected Papers, volume VI*, pages 146–167. Clarendon Press, Oxford, 1974.

HARDY, G.H. and LITTLEWOOD, J.E. *A further note on the converse of Abel's theorem.* In *Collected Papers of Hardy, volume VI*, pages 699–716. Clarendon Press, Oxford, 1974.

MAC LANE, S. and BIRKHOFF, G. *A survey of modern algebra.* N.Y., Macmillan, fourth edition, 1977.

MACDONALD, I.G. *Symmetric Functions And Hall Polynomials.* Oxford: Clarendon Press; New York/London: Oxford University Press, 2nd edition, 1995.

WRIGHT, E.M. Partitions of multipartite numbers. *Proc. Amer. Math. Soc.*, 28:880–890, 1956.

CHAPTER 15

Functional Equations for n-Space Vector Partitions

15.1 AN n-SPACE q-BINOMIAL FUNCTIONAL EQUATION

WE RECALL our earlier chapter definition as follows,

Definition 15.1 *Define the function $F_n(t)$ for all complex numbers a, t with $|a|, |t| < 1$, and for $n \geq 1$ with $|q_1|, |q_2|, ..., |q_n| < 1$ by the sequence of functions $F_n(t)$ with*

$$F_0(-; a, t) = \frac{1 - at}{1 - t} \tag{15.1.1}$$

and for all of $|q_1|, |q_2|, |q_3|, ..., |q_n| < 1$, by

$$F_n(q_1, q_2, q_3, ..., q_n; a, t) = \prod_{\alpha_1, \alpha_2, \alpha_3, ..., \alpha_n \geq 0} \frac{1 - q_1^{\alpha_1} q_2^{\alpha_2} ... q_n^{\alpha_n} at}{1 - q_1^{\alpha_1} q_2^{\alpha_2} ... q_n^{\alpha_n} t} \equiv \sum_{k=1}^{\infty} {}_nA_k t^k \tag{15.1.2}$$

where $\alpha_1, \alpha_2, \alpha_3, ..., \alpha_n$ may be all zero.

We now state a functional equation theorem for $F_n(t)$, next highlighting the 2D, 3D, and 4D cases as examples in a corollary.

Theorem 15.1 *If the left side of (15.1.2) is for the moment considered only as a function, $F_n(t)$, of t, then the following finite product holds,*

$$\frac{1 - at}{1 - t} = \frac{F_n(t) \left(\prod_{S_2} F_n(q_{i_1} q_{i_2} t)\right) \left(\prod_{S_4} F_n(q_{i_1} q_{i_2} q_{i_3} q_{i_4} t)\right) ...}{\left(\prod_{S_1} F_n(q_{i_1} t)\right) \left(\prod_{S_3} F_n(q_{i_1} q_{i_2} q_{i_3} t)\right) \left(\prod_{S_5} F_n(q_{i_1} q_{i_2} q_{i_3} q_{i_4} q_{i_5} t)\right) ...} \tag{15.1.3}$$

the numerator products each being over the nth order even symmetric combinations whilst the denominator products are over the nth order odd symmetric combinations of the variables. We have used the notation $S_1, S_2, S_3, S_4, ...$ to denote the respective nth order symmetric function product of terms.

DOI: 10.1201/9781003174158-15

It is worth noting (naively) that we have used the abbreviation $F(t) = F_n(q_1, q_2, q_3, ..., q_n; a, t)$ and the right side of (15.1.1) is a finite product of such functions going up to the nth order symmetric function product of terms. It is also seen here that (15.1.1) is a functional equation for a *generalized right side of (15.1.2)* form of the q-binomial product. As an illustration, we next give the $n = 1$ up to $n = 4$ examples of the functional equation (15.1.1), with $q_1 \to x$, $q_2 \to y$, $q_3 \to z$. With these variables, we assert

Corollary 15.1 *If for* $|w| < 1, |x| < 1, |y| < 1, |z| < 1,$ *and all complex numbers* a *and* t,

$$\begin{aligned}
F_1(t) = F_1(z;a,t) &= \prod_{j\geq 0} \frac{1-z^j at}{1-z^j t},\\
F_2(t) = F_2(y,z;a,t) &= \prod_{j,k\geq 0} \frac{1-y^j z^k at}{1-y^j z^k t},\\
F_3(t) = F_3(x,y,z;a,t) &= \prod_{j,k,l\geq 0} \frac{1-x^j y^k z^l at}{1-x^j y^k z^l t},\\
F_4(t) = F_4(w,x,y,z;a,t) &= \prod_{j,k,l,m\geq 0} \frac{1-w^j x^k y^l z^m at}{1-w^j x^k y^l z^m t},\\
\text{etc.} & \qquad (15.1.4)
\end{aligned}$$

then

$$\begin{aligned}
\frac{1-at}{1-t} &= \frac{F_1(t)}{F_1(zt)},\\
\frac{1-at}{1-t} &= \frac{F_2(t)F_2(yzt)}{F_2(yt)F_2(zt)},\\
\frac{1-at}{1-t} &= \frac{F_3(t)F_3(yzt)F_3(xyt)F_3(xzt)}{F_3(xt)F_3(yt)F_3(zt)F_3(xyzt)},\\
\frac{1-at}{1-t} &= \frac{F_4(t)F_4(yzt)F_4(xyt)F_4(xzt)F_4(wxt)F_4(wyt)F_4(wzt)F_4(wxyzt)}{F_4(wt)F_4(xt)F_4(yt)F_4(zt)F_4(wxyt)F_4(wxzt)F_4(wyzt)F_4(xyzt)},\\
\text{etc.} & \qquad (15.1.5)
\end{aligned}$$

The above functional equations are a starting point for finding theorems concerning higher dimensional vector partitions. This may well apply to partition congruences for higher dimension vector partitions; it being plausible that the concept of Crank of a partition is applicable for 2D, 3D, 4D, and so on, scenarios.

15.2 AN n-SPACE BINARY VECTOR PARTITIONS FUNCTIONAL EQUATION

We state binary partition definitions and similarly define their generating functions as follows. Worth noting that the functions we are dealing with here are **vector partitions** in the sense that $b_n(q_1, q_2, \ldots, q_n)$ is the number of unrestricted partitions of the vector $\langle q_1, q_2, \ldots, q_n \rangle$ with non-negative integer components $q_1, q_2, \ldots, q_n$, including every possible binary partition of each of $q_1, q_2, \ldots, q_n$, under the operation of the traditional vector addition. The reader may well recognize the function $b(k)$ as the classical number of unrestricted binary partitions.

Definition 15.2 *Define the function $B_n(q_1, q_2, \ldots, q_n)$ for complex numbers $q_1, q_2, \ldots, q_n$, and for $|x| < 1$*

$$B_1(x) = \prod_{k=0}^{\infty} \frac{1}{1 - x^{2^k}} = \sum_{k=0}^{\infty} b(k) x^k, \tag{15.2.1}$$

where $b(k) = b_1(k)$ for convenience; and for all of $|q_1|, |q_2|, |q_3|, \ldots, |q_n| < 1$, by

$$B_n(q_1, q_2, q_3, \ldots, q_n) = \prod_{k_1, k_2, k_3, \ldots, k_n \geq 0} \frac{1}{1 - q_1^{2^{k_1}} q_2^{2^{k_2}} \cdots q_n^{2^{k_n}}} \tag{15.2.2}$$

$$= \sum_{k_1, k_2, \ldots, k_n \geq 0} b_n(q_1, q_2, \ldots, q_n) {q_1}^{k_1} {q_2}^{k_2} \cdots {q_n}^{k_n}.$$

We now state a functional equation theorem for $B_n(q_1, q_2, q_3, \ldots, q_n)$, next highlighting the 2D, 3D, and 4D cases as examples in a corollary.

Theorem 15.2 *We consider the left side of (15.2.2) function, $B_n(q_1, q_2, q_3, \ldots, q_n)$, of vector components $q_1, q_2, q_3, \ldots, q_n$. Then the following finite product holds,*

$$\frac{1}{1 - q_1 q_2 q_3 \cdots q_n} \tag{15.2.3}$$

$$= \frac{B_n(q_1, q_2, q_3, \ldots, q_n) \left(\prod_{S_2} B_n(q_1^2, q_2^2, q_3, \ldots, q_n)\right) \left(\prod_{S_4} B_n(q_1^2, q_2^2, q_3^2, q_4^2, q_5, \ldots, q_n)\right) \cdots}{\left(\prod_{S_1} B_n(q_1^2, q_2, \ldots, q_n)\right) \left(\prod_{S_3} B_n(q_1^2, q_2^2, q_3^2, q_4, \ldots, q_n)\right) \left(\prod_{S_5} B_n(q_1^2, q_2^2, q_3^2, q_4^2, q_5^2 \ldots, q_n)\right) \cdots}$$

the numerator products each being over the nth order even symmetric combinations whilst the denominator products are over the nth order odd symmetric combinations of the squares of the variables. We have used the notation $S_1, S_2, S_3, S_4, \ldots$ to denote the respective nth order symmetric function product of terms. For example $\left(\prod_{S_2} B_n(q_1^2, q_2^2, q_3, \ldots, q_n)\right)$ means symmetric combinations two at a time across all possible of the q_i terms where two of the $q_1, q_2, q_3, \ldots, q_n$ variables are squared. Similarly $\left(\prod_{S_3} B_n(q_1^2, q_2^2, q_3^2, \ldots, q_n)\right)$ means three variables q_i of the $q_1, q_2, q_3, \ldots, q_n$ variables are squared at a time, then we use $\left(\prod_{S_4} B_n(q_1^2, q_2^2, q_3^2, q_4^2, \ldots, q_n)\right)$ to mean four variables q_i of the $q_1, q_2, q_3, \ldots, q_n$ variables are squared at a time, etc.

As expected, and seen from the corollary below, which we rate as a theorem, the functional equation is simpler for two variables than for three variables, which is simpler than for four variables, and so on. It is worth noting that we have used the notation $B_n(q_1, q_2, q_3, \ldots, q_n)$ in (15.2.3) with a finite product of such functions going up to the nth order symmetric function product of terms. It is also seen here that (15.2.3) is a functional equation for a *generalized right side of (15.2.2)* form of the binary partition generating function product. As an illustration, we next give the $n = 1$ through to $n = 4$ examples of the functional equation (15.2.3), with $q_1 \to x$, $q_2 \to y$, $q_3 \to z$. With these variables, we assert

Theorem 15.3 *If for all complex numbers w, x, y, and z, such that $|w| < 1$, $|x| < 1, |y| < 1, |z| < 1$,*

$$\begin{aligned}
B_1(z) &= \prod_{k\geq 0} \frac{1}{1-z^{2^k}} = \sum_{\alpha\geq 0}^{\infty} b(\alpha) z^{\alpha}, \\
B_2(y,z) &= \prod_{j,k\geq 0} \frac{1}{1-y^{2^j} z^{2^k}} = \sum_{\alpha,\beta\geq 0}^{\infty} b_2(\alpha,\beta) y^{\alpha} z^{\beta}, \\
B_3(x,y,z) &= \prod_{j,k,l\geq 0} \frac{1}{1-x^{2^j} y^{2^k} z^{2^l}} = \sum_{\alpha,\beta,\gamma\geq 0}^{\infty} b_3(\alpha,\beta,\gamma) x^{\alpha} y^{\beta} z^{\gamma}, \\
B_4(w,x,y,z) &= \prod_{j,k,l,m\geq 0} \frac{1}{1-w^{2^j} x^{2^k} y^{2^l} z^{2^m}} = \sum_{\alpha,\beta,\gamma,\delta\geq 0}^{\infty} b_4(\alpha,\beta,\gamma,\delta) w^{\alpha} x^{\beta} y^{\gamma} z^{\delta}, \\
&\textit{etc.}
\end{aligned} \tag{15.2.4}$$

then

$$\begin{aligned}
\frac{1}{1-z} &= \frac{B_1(z)}{B_1(z^2)}, \\
\frac{1}{1-yz} &= \frac{B_2(y,z)B_2(y^2,z^2)}{B_2(y^2,z)B_2(y,z^2)}, \\
\frac{1}{1-xyz} &= \frac{B_3(x,y,z)B_3(x,y^2,z^2)B_3(x^2,y^2,z)B_3(x^2,y,z^2)}{B_3(x^2,y,z)B_3(x,y^2,z)B_3(x,y,z^2)B_3(x^2,y^2,z^2)}, \\
\frac{1}{1-wxyz} &= \frac{B_4(w,x,y,z)B_4(w,x,y^2,z^2)B_4(w,x^2,y^2,z)B_4(w,x^2,y,z^2)}{B_4(w^2,x,y,z)B_4(w,x^2,y,z)B_4(w,x,y^2,z)B_4(w,x,y,z^2)} \\
&\quad\times \frac{B_4(w^2,x^2,y,z)B_4(w^2,x,y^2,z)B_4(w^2,x,y,z^2)B_4(w^2,x^2,y^2,z^2)}{B_4(w^2,x^2,y^2,z)B_4(w^2,x^2,y,z^2)B_4(w^2,x,y^2,z^2,)B_4(w,x^2,y^2,z^2)}, \\
&\textit{etc.}
\end{aligned} \tag{15.2.5}$$

where the numerator and denominator combinations are derived from the even and the odd symmetric functions respectively.

Although the functional equations of this chapter are not evident in the literature, a good further reading exercise is to study several chapters of Andrews' 1976 book [12] and also the 2004 smaller book by Andrews and Eriksson [29] to cover off background material on multipartitite partitions and binary partitions. The papers by Alkauskas [6], Alkauskas and Mendel-Strasse [7] from 2003 and 2010 could also be examined by the reader.

Think About It...

Mathematics is the result of mysterious powers which no one understands, and which the unconscious recognition of beauty must play an important part.

Out of an infinity of designs a mathematician chooses one pattern for beauty's sake and pulls it down to earth.

Marston Morse
American Mathematician, 1892–1977

FURTHER READING

ALKAUSKAS, G. Generalization of the Rödseth-Gupta Theorem on Binary Partitions. *Lithuanian Mathematical Journal*, April 2003.

ALKAUSKAS, G. and MENDEL-STRASSE, G. Congruence Properties of the Function that Counts Compositions into Powers of 2. *J. Integer Seq.*, 13, 2010. Article 10.5.3.

ANDREWS, G.E. *The Theory of Partitions*, volume 2 of *Encyclopedia of Mathematics and its Applications*. Addison-Wesley Publishing Company, Advanced Book Program, Reading, Massachusetts, USA, 1976.

ANDREWS, G.E. and ERIKSSON, K. *Integer Partitions*. Cambridge University Press, Cambridge, UK, New York, USA, Port Melbourne, Australia, Madrid, Spain, Cape Town, South Africa, 2004.

CHAPTER 16

Binary Partitions and their Vector Generalizations

16.1 ELEMENTARY IDEAS FOR BINARY PARTITIONS

A BINARY PARTITION is an integer partition where each number in the sum is of the form 2^n where $n \in \mathbb{N}$. The binary integer partition function was first studied by Euler in the 18th century. The topic was first considered by Euler, who found some fundamental facts about binary partitions. That our computers are all based on a binary systemic approach makes the topic *a priori* interesting. In an earlier chapter on congruences of partitions, the Rödseth-Gupta Theorem on binary partitions was given, and suggested approaches to finding congruence relations for binary partition generalizations in 2D, 3D, etc suggested utilizing the functional equations already presented in other chapters.

16.2 A FEW FINITE AND INFINITE PRODUCTS FOR BINARY PARTITIONS

Consider the following finite products:-

$$\begin{aligned}
(1+x) &= \frac{1-x^2}{1-x}, \\
(1+x)(1+x^2) &= \frac{1-x^4}{1-x}, \\
(1+x)(1+x^2)(1+x^4) &= \frac{1-x^8}{1-x}, \\
(1+x)(1+x^2)(1+x^4)(1+x^8) &= \frac{1-x^{16}}{1-x},
\end{aligned}$$

and generally

$$(1+x)(1+x^2)(1+x^4)(1+x^8)\cdots(1+x^{2^n}) = \frac{1-x^{2^{(n+1)}}}{1-x}.$$

DOI: 10.1201/9781003174158-16

It is easy to see how this leads to the well-known results:-

$$(1+x)(1+x^2)(1+x^4)(1+x^8)\cdots(1+x^{2^n})\ldots = \frac{1}{1-x}, \quad for\,|x|<1; \tag{16.2.1}$$

and

$$\left(\frac{1+x^{1/2}}{2}\right)\left(\frac{1+x^{1/4}}{2}\right)\left(\frac{1+x^{1/8}}{2}\right)\left(\frac{1+x^{1/16}}{2}\right)\cdots\left(\frac{1+x^{2^{-n}}}{2}\right)\cdots = \frac{1-x}{\log(\frac{1}{x})},$$

$for\,|x| \neq 1\,or\,0.$

16.3 2D VERSION OF EVERY INTEGER IS A UNIQUE SUM OF DISTINCT BINARY POWERS

We are fortunate to have the identities from the previous section. They permit us easy access to some very simple 2D binary partition results. Let us start by considering the following identities, each easy to prove from the finite products above.

$$\begin{matrix}(1+xy) & \times(1+x^2y) \\ \times(1+xy^2) & (1+x^2y^2)\end{matrix} = \begin{matrix}\frac{1-x^3y^3}{(1-xy)} & \times\frac{1-x^4y^2}{(1-x^2y)} \\ \times\frac{(1-xy^2)}{1-x^2y^4} & \end{matrix} \tag{16.3.1}$$

$$= x^6y^6 + x^5y^5 + x^5y^4 + x^4y^5 + x^4y^4 + x^4y^3 + x^3y^4 + 2x^3y^3 + x^3y^2 + x^2y^3 + x^2y^2 + x^2y + xy^2 + xy + 1$$

$$= 1 + \begin{array}{llllll} & & & & & +x^6y^6 \\ & & & +x^4y^5 & +x^5y^5 & \\ & & +x^3y^4 & +x^4y^4 & +x^5y^4 & \\ & +x^2y^3 & +2x^3y^3 & +x^4y^3 & & \\ +xy^2 & +x^2y^2 & +x^3y^2 & & & \\ +xy & +x^2y & & & & \end{array},$$

where in this last equation we've taken poetic licence to mimick the partition grid, which would look like this

2D binary product grid – Order 2

6							1
5					1	1	
4				1	1	1	
3			1	2	1		
2		1	1	1			
1		1	1				
0	1						
	0	**1**	**2**	**3**	**4**	**5**	**6**

In a similar fashion we see easily that

$$\begin{matrix} (1+xy) & (1+x^2y) & (1+x^4y) \\ (1+xy^2) & (1+x^2y^2) & (1+x^4y^2) \\ (1+xy^4) & (1+x^2y^4) & (1+x^4y^4) \end{matrix} = \begin{matrix} \frac{(1-x^8y^8)}{(1-xy)} & \frac{(1-x^8y^4)}{(1-x^2y)} & \frac{(1-x^8y^2)}{(1-x^4y)} \\ \frac{(1-x^4y^8)}{(1-xy^2)} & & \\ \frac{(1-x^2y^8)}{(1-xy^4)} & & \end{matrix} \tag{16.3.2}$$

$$= x^{21}y^{21} + x^{20}y^{20} + x^{19}y^{20} + x^{17}y^{20} + x^{20}y^{19} + x^{19}y^{19} + x^{18}y^{19} + x^{17}y^{19} + x^{16}y^{19}$$
$$+x^{15}y^{19} + x^{19}y^{18} + 2x^{18}y^{18} + x^{17}y^{18} + 2x^{16}y^{18} + 2x^{15}y^{18} + x^{14}y^{18} + x^{13}y^{18} + x^{20}y^{17}$$
$$+x^{19}y^{17} + x^{18}y^{17} + 2x^{17}y^{17} + 2x^{16}y^{17} + 2x^{15}y^{17} + 3x^{14}y^{17} + x^{13}y^{17} + x^{12}y^{17} + x^{11}y^{17}$$
$$+x^{19}y^{16} + 2x^{18}y^{16} + 2x^{17}y^{16} + 3x^{16}y^{16} + 3x^{15}y^{16} + 3x^{14}y^{16} + 3x^{13}y^{16} + 2x^{12}y^{16} + x^{11}y^{16}$$
$$+x^{10}y^{16} + x^{19}y^{15} + 2x^{18}y^{15} + 2x^{17}y^{15} + 3x^{16}y^{15} + 4x^{15}y^{15} + 4x^{14}y^{15} + 4x^{13}y^{15} + 3x^{12}y^{15}$$
$$+2x^{11}y^{15} + 2x^{10}y^{15} + x^{9}y^{15} + x^{18}y^{14} + 3x^{17}y^{14} + 3x^{16}y^{14} + 4x^{15}y^{14}y^{14} + 6x^{14}y^{14} + 5x^{13}$$
$$+5x^{12}y^{14} + 4x^{11}y^{14} + 2x^{10}y^{14} + 2x^{9}y^{14} + x^{8}y^{14} + x^{18}y^{13} + x^{17}y^{13} + 3x^{16}y^{13} + 4x^{15}y^{13}$$
$$+5x^{14}y^{13} + 6x^{13}y^{13} + 6x^{12}y^{13} + 5x^{11}y^{13} + 5x^{10}y^{13} + 3x^{9}y^{13} + 2x^{8}y^{13} + x^{7}y^{13} + x^{1}7y^{12}$$
$$+2x^{16}y^{12} + 3x^{15}y^{12} + 5x^{14}y^{12} + 6x^{13}y^{12} + 7x^{12}y^{12} + 6x^{11}y^{12} + 6x^{10}y^{12} + 4x^{9}y^{12}$$
$$+3x^{8}y^{12} + 2x^{7}y^{12} + x^{6}y^{12} + x^{17}y^{11} + x^{16}y^{11} + 2x^{15}y^{11} + 4x^{14}y^{11} + 5x^{13}y^{11} + 6x^{12}y^{11}$$
$$+7x^{11}y^{11} + 6x^{10}y^{11} + 6x^{9}y^{11} + 5x^{8}y^{11} + 2x^{7}y^{11} + 2x^{6}y^{11} + x^{5}y^{11} + x^{16}y^{10} + 2x^{15}y^{10}$$
$$+2x^{14}y^{10} + 5x^{13}y^{10} + 6x^{12}y^{10} + 6x^{11}y^{10} + 7x^{10}y^{10} + 6x^{9}y^{10} + 5x^{8}y^{10} + 4x^{7}y^{10} + 2x^{6}y^{10}$$
$$+x^{5}y^{10} + x^{4}y^{10} + x^{15}y^{9} + 2x^{14}y^{9} + 3x^{13}y^{9} + 4x^{12}y^{9} + 6x^{11}y^{9} + 6x^{10}y^{9} + 7x^{9}y^{9}$$
$$+6x^{8}y^{9} + 5x^{7}y^{9} + 3x^{6}y^{9} + 2x^{5}y^{9} + x^{4}y^{9} + x^{14}y^{8} + 2x^{13}y^{8} + 3x^{12}y^{8} + 5x^{11}y^{8}$$
$$+5x^{10}y^{8} + 6x^{9}y^{8} + 6x^{8}y^{8} + 5x^{7}y^{8} + 4x^{6}y^{8} + 3x^{5}y^{8} + x^{4}y^{8} + x^{3}y^{8} + x^{1}3y^{7} + 2x^{12}y^{7}$$
$$+2x^{11}y^{7} + 4x^{10}y^{7} + 5x^{9}y^{7} + 5x^{8}y^{7} + 6x^{7}y^{7} + 4x^{6}y^{7} + 3x^{5}y^{7} + 3x^{4}y^{7} + x^{3}y^{7}$$
$$+x^{12}y^{6} + 2x^{11}y^{6} + 2x^{10}y^{6} + 3x^{9}y^{6} + 4x^{8}y^{6} + 4x^{7}y^{6} + 4x^{6}y^{6} + 3x^{5}y^{6} + 2x^{4}y^{6}$$
$$+2x^{3}y^{6} + x^{2}y^{6} + x^{11}y^{5} + x^{10}y^{5} + 2x^{9}y^{5} + 3x^{8}y^{5} + 3x^{7}y^{5} + 3x^{6}y^{5} + 3x^{5}y^{5}$$
$$+2x^{4}y^{5} + 2x^{3}y^{5} + x^{2}y^{5} + x^{10}y^{4} + x^{9}y^{4} + x^{8}y^{4} + 3x^{7}y^{4} + 2x^{6}y^{4} + 2x^{5}y^{4} + 2x^{4}y^{4}$$
$$+x^{3}y^{4} + x^{2}y^{4} + xy^{4} + x^{8}y^{3} + x^{7}y^{3} + 2x^{6}y^{3} + 2x^{5}y^{3} + x^{4}y^{3} + 2x^{3}y^{3} + x^{2}y^{3}$$
$$+x^{6}y^{2} + x^{5}y^{2} + x^{4}y^{2} + x^{3}y^{2} + x^{2}y^{2} + xy^{2} + x^{4}y + x^{2}y + xy + 1$$

It is clear that this identity can be extended to any finite case which can then be proved by induction. The infinite case is evident also. We see the coefficients in the above expansion are represented by the grid below.

2D binary distinct partition grid – Order 3

21																						1
20																		1		1	1	
19																1	1	1	1	1	1	
18														1	1	2	2	1	2	1		
17												1	1	1	3	2	2	2	1	1	1	
16											1	1	2	3	3	3	3	2	2	1		
15										1	2	3	3	4	4	4	3	2	2	1		
14									1	2	2	4	5	5	6	4	3	3	1			
13							1	1	2	3	5	5	6	6	5	4	3	1	1			
12							1	2	3	4	6	6	7	6	5	3	2	1				
11						1	2	2	5	6	6	7	6	5	4	2	1	1				
10					1	1	2	4	5	6	7	6	6	5	2	2	1					
9					1	2	3	5	6	7	6	6	4	3	2	1						
8				1	1	3	4	5	6	6	5	5	3	2	1							
7				1	3	3	4	6	5	5	4	2	2	1								
6			1	2	2	3	4	4	4	3	2	2	1									
5			1	2	2	3	3	3	3	2	1	1										
4		1	1	1	2	2	2	3	1	1	1											
3			1	2	1	2	2	1	1													
2		1	1	1	1	1	1															
1		1	1		1																	
0	1																					
	0	**1**	**2**	**3**	**4**	**5**	**6**	**7**	**8**	**9**	**10**	**11**	**12**	**13**	**14**	**15**	**16**	**17**	**18**	**19**	**20**	**21**

16.4 THE 2D BINARY, n-ARY, AND 10-ARY FORMULAS

A 2D generalization of our earlier formula for distinct binary partitions has the cases:

$$(p+q) = \frac{p^2 - q^2}{p-q}, \tag{16.4.1}$$

$$(p+q)(p^2+q^2) = \frac{p^4 - q^4}{p-q}, \tag{16.4.2}$$

$$(p+q)(p^2+q^2)(p^4+q^4) = \frac{p^8 - q^8}{p-q}, \tag{16.4.3}$$

$$\vdots$$

$$(p+q)(p^2+q^2)(p^4+q^4)\cdots(p^{2^{n-1}}+q^{2^{n-1}}) = \frac{p^{2^n} - q^{2^n}}{p-q}. \tag{16.4.4}$$

Hence, adding together both sides of (16.4.1) to (16.4.4) gives the identity,

$$\begin{aligned} 1+(p+q)+(p+q)(p^2+q^2)+(p+q)(p^2+q^2)(p^4+q^4)+\ldots \\ +(p+q)(p^2+q^2)(p^4+q^4)\cdots(p^{2^{n-1}}+q^{2^{n-1}}) \end{aligned} \tag{16.4.5}$$

$$= \frac{(p+p^2+p^4+\ldots+p^{2^n})-(q+q^2+q^4+\ldots+q^{2^n})}{p-q}.$$

This is an interesting fundamental result in distinct 2D binary partitions. Before going further on this we repeat these equations, this time including the expansions.

$$(p+q) = \frac{p^2-q^2}{p-q} \tag{16.4.6}$$

$$(p+q)(p^2+q^2) = \frac{p^4-q^4}{p-q}, \tag{16.4.7}$$

$$= p^3+p^2q+pq^2+q^3,$$

$$(p+q)(p^2+q^2)(p^4+q^4) = \frac{p^8-q^8}{p-q}, \tag{16.4.8}$$

$$= p^7+p^6q+p^5q^2+p^4q^3+p^3q^4+p^2q^5+pq^6+q^7,$$

$$(p+q)(p^2+q^2)(p^4+q^4)(p^8+q^8) = \frac{p^{16}-q^{16}}{p-q} \tag{16.4.9}$$

$$= p^{15}+p^{14}q+p^{13}q^2+p^{12}q^3+p^{11}q^4+p^{10}q^5+p^9q^6+p^8q^7$$
$$+p^7q^8+p^6q^9+p^5q^{10}+p^4q^{11}+p^3q^{12}+p^2q^{13}+pq^{14}+q^{15},$$

$$\vdots$$

$$(p+q)(p^2+q^2)(p^4+q^4)\cdots(p^{2^{n-1}}+q^{2^{n-1}}) = \frac{p^{2^n}-q^{2^n}}{p-q} \tag{16.4.10}$$

$$= p^{2^n-1}+p^{2^n-2}q+p^{2^n-3}q^2+p^{2^n-4}q^3+\ldots+p^3q^{2^n-4}+p^2q^{2^n-3}+pq^{2^n-2}+q^{2^n-1},$$

where each term p^aq^b has the restriction that a and b are positive integers and $a+b = 2^m-1$, for every $m \in N^+$.

The above identities combine to give us a theorem for 2D binary component partitions into distinct parts. Therefore we have the

Theorem 16.1 *The number $B_2(a,b)$ for a and b both non negative integers, of vector partitions with parts of the form $\langle 2^\alpha, 2^\beta \rangle$ with $\alpha \geq 0$ and $\beta \geq 0$, into distinct parts is*

$$B_2(a,b) = \begin{cases} 1, & \text{if condition } B_2 \text{ holds,} \\ 0, & \text{otherwise.} \end{cases}$$

where condition B_2 is : a ***and*** *b* ***are positive integers and*** *$a+b = 2^m-1$,* ***for every*** *$m \in N^+$. The generating function identity encoding this is*

$$1+\sum_{k=0}^{\infty}\prod_{j=0}^{k}(p^{2^k}+q^{2^k}) = 1+\sum_{a,b\geq 0} B_2(a,b)p^aq^b \tag{16.4.11}$$

$$= \frac{(p+p^2+p^4+\ldots+p^{2^m}+\ldots)-(q+q^2+q^4+\ldots+q^{2^m}+\ldots)}{p-q}.$$

$$= 1+(p+q)+(p+q)(p^2+q^2)+\ldots+(p+q)(p^2+q^2)(p^4+q^4)\cdots(p^{2^m}+q^{2^m})+\ldots$$

Part of the partition grid for the generating function identity

$$\frac{(p+p^2+p^4+\ldots+p^{2^m}+\ldots)-(q+q^2+q^4+\ldots+q^{2^m}+\ldots)}{p-q},$$

is as shown below.

2D binary partition grid for $\{(\sum \mathbf{p}^{2^m})-(\sum \mathbf{q}^{2^m})\}/(\mathbf{p}-\mathbf{q})$

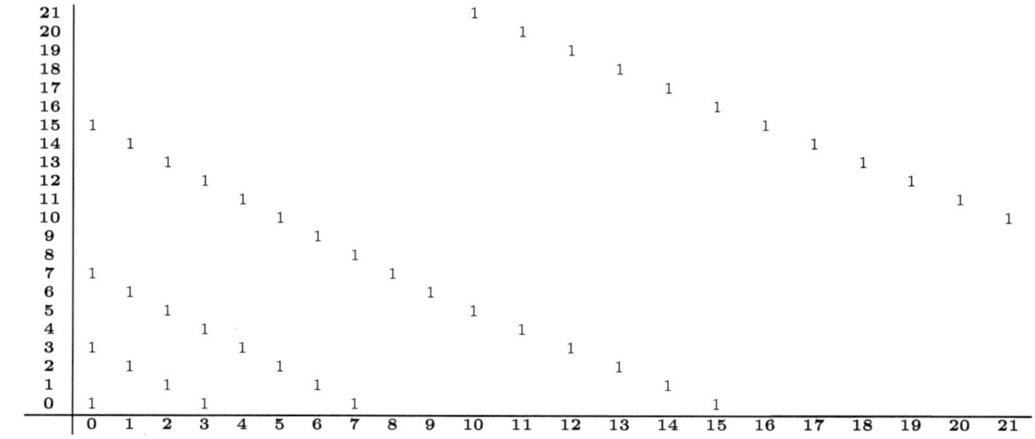

We note that in the above grid, each 1 occurs at coordinates that add to 2^m-1 for some positive integer m. Note also, that the above grid depicts the binary case of the n-ary generalization for $n=2,3,4,5,6,\ldots$. Considering examples using the $n=10$, 10-ary, or decimal, case allows us to bring in familiar base ten numbering, so providing a familiar presentation instead of asking the reader to think in number systems based n generally.

So next, our objective is to write down the base n, n-ary version of the above binary 2D generating function.

To start the rationale, let us rewrite equations (16.4.6) to (16.4.10), using binary numbers as the indices in the expanded polynomials. This will enable us to write the base n version of the binary cases given above.

$$(p+q)=\frac{p^2-q^2}{p-q}=p^{1_2}+q^{1_2}, \tag{16.4.12}$$

$$(p+q)(p^2+q^2)=\frac{p^4-q^4}{p-q}=p^{11_2}+p^{10_2}q+pq^{10_2}+q^{11_2}, \tag{16.4.13}$$

$$(p+q)(p^2+q^2)(p^4+q^4)=\frac{p^8-q^8}{p-q} \tag{16.4.14}$$

$$=p^{111_2}+p^{110_2}q+p^{101_2}q^{10_2}+p^{100_2}q^{11_2}+p^{11_2}q^{100_2}+p^{10_2}q^{101_2}+pq^{110_2}+q^{111_2},$$

$$(p+q)(p^2+q^2)(p^4+q^4)(p^8+q^8)=\frac{p^{16}-q^{16}}{p-q} \tag{16.4.15}$$

$$= p^{1111_2} + p^{1110_2}q + p^{1101_2}q^{10_2} + p^{1100_2}q^{11_2} + p^{1011_2}q^{100_2} + p^{1010_2}q^{101_2}$$
$$+ p^{1001_2}q^{110_2} + p^{1000_2}q^{111_2} + p^{111_2}q^{1000_2} + p^{110_2}q^{1001_2} + p^{101_2}q^{1010_2}$$
$$+ p^{100_2}q^{1011_2} + p^{11_2}q^{1100_2} + p^{10_2}q^{1101_2} + pq^{1110_2} + q^{1111_2},$$
$$\vdots$$

$$(p+q)(p^2+q^2)(p^4+q^4)\cdots(p^{2^{n-1}}+q^{2^{n-1}}) = \frac{p^{2^m}-q^{2^m}}{p-q} = 1 + \sum_{\substack{a,b\geq 0;\\ a+b\neq 0}} B_2(a,b)p^a q^b \tag{16.4.16}$$
$$= p^{\sum_{k=0}^{m-1}2^k} + p^{\sum_{k=0}^{m-2}2^k}q + p^{\sum_{k=0}^{m-3}2^k}q^2 + \ldots + p^2 q^{\sum_{k=0}^{m-3}2^k} + pq^{\sum_{k=0}^{m-2}2^k} + q^{\sum_{k=0}^{m-1}2^k},$$

where each term $p^a q^b$ has the restriction that a and b are positive integers and $a+b = 2^m - 1$, for every $m \in N^+$.

The above identities combine to give us an approach for 2D n-ary component partitions into distinct parts. All we need to do is replace a_2 in binary, with a_n in n-ary.

Therefore we see the **base n** or **n-ary** partition analogy case is:

$$(p+q) = p^{1_n} + q^{1_n}, \tag{16.4.17}$$

$$(p+q)(p^n+q^n) = p^{11_n} + p^{10_n}q + pq^{10_n} + q^{11_n}, \tag{16.4.18}$$

$$(p+q)(p^n+q^n)(p^{n^2}+q^{n^2}) \tag{16.4.19}$$
$$= p^{111_n} + p^{110_n}q + p^{101_n}q^{10_n} + p^{100_n}q^{11_n} + p^{11_n}q^{100_n} + p^{10_n}q^{101_n} + pq^{110_n} + q^{111_n},$$

$$(p+q)(p^n+q^n)(p^{n^2}+q^{n^2})(p^{n^3}+q^{n^3}) \tag{16.4.20}$$

$$= p^{1111_n} + p^{1110_n}q + p^{1101_n}q^{10_n} + p^{1100_n}q^{11_n} + p^{1011_n}q^{100_n} + p^{1010_n}q^{101_n}$$
$$+ p^{1001_n}q^{110_n} + p^{1000_n}q^{111_n} + p^{111_n}q^{1000_n} + p^{110_n}q^{1001_n} + p^{101_n}q^{1010_n}$$
$$+ p^{100_n}q^{1011_n} + p^{11_n}q^{1100_n} + p^{10_n}q^{1101_n} + pq^{1110_n} + q^{1111_n},$$
$$\vdots$$

$$(p+q)(p^{10}+q^{10})(p^{10^2}+q^{10^2})\cdots(p^{10^{n-1}}+q^{10^{n-1}}) \tag{16.4.21}$$

$$= p^{\sum_{k=0}^{m-1}n^k} + p^{\sum_{k=0}^{m-2}n^k}q^1 + p^{\sum_{k=0}^{m-3}n^k}q^{10_n} + \ldots + p^{10_n}q^{\sum_{k=0}^{m-3}n^k}$$
$$+ p^1 q^{\sum_{k=0}^{m-2}n^k} + q^{\sum_{k=0}^{m-1}n^k},$$

where each term $p^a q^b$ has the restriction a and b are positive integers base n with digits comprised of only 1s and 0s, and $a+b = \sum_{k=0}^{m-1} n^k$, for every $m \in N^+$.

The above identities combine to give us a theorem for 2D **base n** or **n-ary** component partitions into distinct parts. Therefore we have the

Theorem 16.2 *The number $B_n(a,b)$ for a and b both non negative integers, of vector partitions with parts of the form $\langle 2^\alpha, 2^\beta \rangle$ with $\alpha \geq 0$ and $\beta \geq 0$, into distinct parts is*

$$B_n(a,b) = \begin{cases} 1, & \text{if condition } B_n \text{ holds,} \\ 0, & \text{otherwise,} \end{cases}$$

where condition B_n is : a ***and b are positive integers base n with digits comprised of only 1s and 0s, and*** $a+b = \sum_{k=0}^{m-1} n^k$***, for every*** $m \in N^+$. *The generating function identity encoding this is*

$$1+\sum_{k=0}^{\infty}\prod_{j=0}^{k}(p^{n^k}+q^{n^k}) = 1+\sum_{\substack{a,b\geq 0;\\ a+b\neq 0}} B_n(a,b)p^a q^b \tag{16.4.22}$$

$$= 1+(p+q)+(p+q)(p^n+q^n)+\ldots+(p+q)(p^n+q^n)(p^{n^2}+q^{n^2})\cdots(p^{n^m}+q^{n^m})+\ldots$$

The base 10 or decimal case in which $n = 10$ is derived by the set of successive equations:

$$(p+q) = p^1+q^1 \tag{16.4.23}$$

$$(p+q)(p^{10}+q^{10}) = p^{11}+p^{10}q^1+p^1q^{10}+q^{11}, \tag{16.4.24}$$

$$(p+q)(p^{10}+q^{10})(p^{10^2}+q^{10^2}) \tag{16.4.25}$$

$$= p^{111}+p^{110}q^1+p^{101}q^{10}+p^{100}q^{11}+p^{11}q^{100}+p^{10}q^{101}+p^1q^{110}+q^{111},$$

$$(p+q)(p^{10}+q^{10})(p^{10^2}+q^{10^2})(p^{10^3}+q^{10^3}) \tag{16.4.26}$$

$$= p^{1111}+p^{1110}q^1+p^{1101}q^{10}+p^{1100}q^{11}+p^{1011}q^{100}+p^{1010}q^{101}+p^{1001}q^{110}+p^{1000}q^{111}$$
$$+p^{111}q^{1000}+p^{110}q^{1001}+p^{101}q^{1010}+p^{100}q^{1011}+p^{11}q^{1100}+p^{10}q^{1101}+p^1q^{1110}+q^{1111},$$

$$\vdots$$

$$(p+q)(p^{10}+q^{10})(p^{10^2}+q^{10^2})\cdots(p^{10^{m-1}}+q^{10^{m-1}}) \tag{16.4.27}$$

$$= p^{\sum_{k=0}^{m-1}10^k}+p^{\sum_{k=0}^{m-2}10^k}q^1+p^{\sum_{k=0}^{m-3}10^k}q^{10}+\ldots+p^{10}q^{\sum_{k=0}^{m-3}10^k}$$
$$+p^1q^{\sum_{k=0}^{m-2}10^k}+q^{\sum_{k=0}^{m-1}10^k},$$

where each term $p^a q^b$ has the restriction that a and b are positive integers in base 10 with digits comprised of only 1s and 0s, and $a+b = \sum_{k=0}^{m-1} 10^k$, for every $m \in N^+$.

The above identities combine to give us a theorem for 2D **base 10** component partitions into distinct parts. Therefore we have the

Theorem 16.3 *The number $B_{10}(a,b)$ for a and b both non negative integers, of vector partitions with parts of the form $\langle 2^\alpha, 2^\beta \rangle$ with $\alpha \geq 0$ and $\beta \geq 0$, with $\alpha+\beta \neq 0$, into distinct parts is*

$$B_{10}(a,b) = \begin{cases} 1, & \textit{if condition } B_{10} \textit{ holds,} \\ 0, & \textit{otherwise,} \end{cases}$$

where condition B_{10} is : a ***and*** *b* ***are positive integers base 10 with digits comprised of only 1s and 0s, and*** $a+b=\sum_{k=0}^{m-1} 10^k$, ***for every*** $m \in N^+$. *The generating function identity encoding this is*

$$1+\sum_{k=0}^{\infty}\prod_{j=0}^{k}(p^{10^k}+q^{10^k}) = 1+\sum_{\substack{a,b\geq 0;\\ a+b\neq 0}} B_{10}(a,b)p^a q^b \tag{16.4.28}$$

$$=1+(p+q)+(p+q)(p^{10}+q^{10})+\ldots+(p+q)(p^{10}+q^{10})(p^{10^2}+q^{10^2})\cdots(p^{10^m}+q^{10^m})+\ldots$$

16.5 SOME BINARY INTEGER PARTITION PRELIMINARY RESULTS

We start by revisiting the elementary identity,

$$(1+x)(1+x^2)(1+x^4)(1+x^8)\cdots(1+x^{2^n})\ldots = \frac{1}{1-x}, \quad \textit{for } |x|<1. \tag{16.5.1}$$

By considering both sides of (16.5.1) as power series it is clear that every positive integer is a sum of distinct binary powers in exactly one way. Hence, each positive integer has a unique binary representation. That is $1_{10}=1_2$, $2_{10}=10_2$, $3_{10}=11_2$, $4_{10}=100_2$, $5_{10}=101_2$, $6_{10}=110_2$, etc.

Next, we give an infinite product relationship that is easily derived from the following cases of (16.5.1).

$$\begin{aligned}
(1+x)(1+x^2)(1+x^4)(1+x^8)(1+x^{16})(1+x^{32})\cdots &= \frac{1}{1-x},\\
(1+x^2)(1+x^4)(1+x^8)(1+x^{16})(1+x^{32})\cdots &= \frac{1}{1-x^2},\\
(1+x^4)(1+x^8)(1+x^{16})(1+x^{32})\cdots &= \frac{1}{1-x^4},\\
(1+x^8)(1+x^{16})(1+x^{32})\cdots &= \frac{1}{1-x^8},\\
(1+x^{16})(1+x^{32})\cdots &= \frac{1}{1-x^{16}},\\
\textit{etc.} &
\end{aligned} \tag{16.5.2}$$

Now, the product of all left sides here is equal to the product of all right sides here so we arrive at for $|x|<1$,

$$(1+x)(1+x^2)^2(1+x^4)^3(1+x^8)^4\cdots(1+x^{2^n})^{n+1}\ldots \tag{16.5.3}$$

$$= \frac{1}{(1-x)(1-x^2)(1-x^4)(1-x^8)\cdots} = B(x).$$

This gives us an alternative generating function for the integer binary partition function. It is also suggestive of the theorem,

Theorem 16.4 *The number of unrestricted binary partitions $b(n)$ of a positive integer n is equal to the number of partitions of n such that*

$$n = a_1 2^0 + a_2 2^1 + a_3 2^2 + a_4 2^3 + \ldots + a_m 2^{m-1},$$

where $0 \le a_1 \le 1, 0 \le a_2 \le 2, 0 \le a_3 \le 3, \ldots, 0 \le a_m \le m$. Another way of putting this is partitions of n into powers of 2 where 2^0 is used at most once, 2^1 is used at most twice, 2^2 is used at most 3 times, etc and generally, 2^{m-1} is used at most m times.

Although this theorem and (16.5.3) is shallow in the sense of it being fundamental for integer binary partitions, it doesn't seem to be in the literature. Whether this result is known is immaterial, as we will apply it to vector partitions with binary components for $2D$ and $3D$ cases, and assert the nD generalization.

Next, let us consider the 2D infinite binary product given as $\mathbf{B}_2(y,z)$ below. This product enumerates the number of partitions of vector $\langle a,b\rangle$ into distinct parts of the form $\langle 2^c, 2^d\rangle$ for c and d non-negative integers under vector addition.

We see that each rightward, downward sloping diagonal of $\mathbf{B}_2(y,z)$ is a case of (16.5.1). Therefore, the following 2D equivalence is seen, leading to a new theorem on 2D binary partitions of vectors $\langle a,b\rangle$ for a and b positive integers.

16.6 SOME EASY 2D BINARY PARTITION TRANSFORM GENERATING FUNCTIONS

Theorem 16.5 ***Distinct 2D binary vector partition transform.*** *The following infinite 2D binary product relation holds for $|y| < 1$, $|z| < 1$. The first product here is the generating function for the number of partitions of 2D vector $\langle m,n\rangle$ into distinct vectors whose components are non-negative powers of 2. The relation is*

$$\prod_{m,n\ge 0}(1+y^{2^m}z^{2^n}) = \frac{1}{1-xy}\prod_{k\ge 1}\frac{1}{(1-y^{2^k}z)(1-yz^{2^k})} = \sum_{m,n\ge 0} \boldsymbol{b}_2(m,n)y^m z^n. \quad (16.6.1)$$

In longhand expansion, this equation is as per below. Each diagonal of the first tableaux here is a case of (16.5.1). The second tableaux here relates to the right side of the equation in the theorem, after successive application of (16.5.1). We arrive then at

$$\mathbf{B}_2(y,z) = \quad (16.6.2)$$

$$(1+y^{2^0}z^{2^0})(1+y^{2^0}z^{2^1})(1+y^{2^0}z^{2^2})(1+y^{2^0}z^{2^3})(1+y^{2^0}z^{2^4})(1+y^{2^0}z^{2^5})\cdots$$
$$(1+y^{2^1}z^{2^0})(1+y^{2^1}z^{2^1})(1+y^{2^1}z^{2^2})(1+y^{2^1}z^{2^3})(1+y^{2^1}z^{2^4})(1+y^{2^1}z^{2^5})\cdots$$
$$(1+y^{2^2}z^{2^0})(1+y^{2^2}z^{2^1})(1+y^{2^2}z^{2^2})(1+y^{2^2}z^{2^3})(1+y^{2^2}z^{2^4})(1+y^{2^2}z^{2^5})\cdots$$
$$(1+y^{2^3}z^{2^0})(1+y^{2^3}z^{2^1})(1+y^{2^3}z^{2^2})(1+y^{2^3}z^{2^3})(1+y^{2^3}z^{2^4})(1+y^{2^3}z^{2^5})\cdots$$

$$(1+y^{2^4}z^{2^0})(1+y^{2^4}z^{2^1})(1+y^{2^4}z^{2^2})(1+y^{2^4}z^{2^3})(1+y^{2^4}z^{2^4})(1+y^{2^4}z^{2^5})\cdots$$
$$(1+y^{2^5}z^{2^0})(1+y^{2^5}z^{2^1})(1+y^{2^5}z^{2^2})(1+y^{2^5}z^{2^3})(1+y^{2^5}z^{2^4})(1+y^{2^5}z^{2^5})\cdots$$
$$\vdots \qquad \vdots \qquad \vdots \qquad \vdots \qquad \vdots \qquad \vdots \qquad \ddots$$

$$= \frac{1}{(1-yz)}$$
$$\times \frac{1}{(1-yz^{2^1})(1-yz^{2^2})(1-yz^{2^3})(1-yz^{2^4})(1-yz^{2^5})\cdots}$$
$$\times \frac{1}{(1-zy^{2^1})(1-zy^{2^2})(1-zy^{2^3})(1-zy^{2^4})(1-zy^{2^5})\cdots}.$$

This equation has the following 10×10 patch of grid.

2D binary partition grid for $\prod_{m,n\geq 0}(1+y^{2^m}z^{2^n})$

⋮	⋮	⋮	⋮	⋮	⋮	⋮	⋮	⋮	⋮	⋮	
10		1	2	3	4	5	7	7	9	12	⋯
9		1	2	2	4	5	5	7	9	9	⋯
8	1	1	1	2	3	4	5	7	7	7	⋯
7			1	3	3	4	6	5	5	7	⋯
6		1	2	2	3	4	4	4	5	5	⋯
5		1	2	2	3	3	3	3	4	4	⋯
4	1	1	1	2	2	2	3	2	2	3	⋯
3		1	2	1	2	2	1	1	2	2	⋯
2	1	1	1	1	1	1		1	1	1	⋯
1	1	1		1				1			⋯
	1	2	3	4	5	6	7	8	9	10	⋯

(16.6.3)

So identity (16.6.2) is equivalent to stating that

Theorem 16.6 *The number of distinct vector 2D binary partitions of a 2D vector $\langle m,n\rangle$ in the first quadrant is equal to the number of unrestricted 2D binary partitions of 2D vector $\langle m,n\rangle$ into binary component vectors with parts having a component equal to 1.*

From observation of the first tableau in (16.6.2) we see that

$$\mathbf{B}_2(y,z) = (1+yz)\frac{\mathbf{B}_2(y^2,z)\mathbf{B}_2(y,z^2)}{\mathbf{B}_2(y^2,z^2)}.$$

Also, it is seen that our earlier discussed 2D binary vector unrestricted partition generating function $B_2(y,z)$, satisfies

$$\mathbf{B}_2(y,z) = \frac{B_2(y,z)}{B_2(y^2,z^2)}.$$

Then this has implications from the set of functional equations already given. Again consider the equation

$$(1+y^{2^0}z^{2^0})(1+y^{2^0}z^{2^1})(1+y^{2^0}z^{2^2})(1+y^{2^0}z^{2^3})(1+y^{2^0}z^{2^4})(1+y^{2^0}z^{2^5})\cdots \quad (16.6.4)$$
$$= 1+y^{\mathbf{b}(1)}z+y^{\mathbf{b}(2)}z^2+y^{\mathbf{b}(3)}z^3+\cdots+y^{\mathbf{b}(n)}z^n+\cdots := \sum_{n\geq 0} y^{\mathbf{b}(n)}z^n,$$

where $\mathbf{b}(n)$ is the number of 1s in the binary number representing n. Hence for example, $\mathbf{b}(2^5)=1$, $\mathbf{b}(2^5+2^3+2^1)=3$, $\mathbf{b}(2^{11}+2^5+2^3+2^1)=4$, $\mathbf{b}(7)=3$, $\mathbf{b}(2^{100}+2^6+2^5+2^3+1)=5$, and so on, with the convenient definition $\mathbf{b}(0)=0$.

Hence, substituting cases of (16.6.4) into (16.6.2) and comparing them to (16.6.1) we see that

$$\mathbf{B}_2(y,z) = (\sum_{n\geq 0} y^{\mathbf{b}(n)}z^n)(\sum_{n\geq 0} y^{2\mathbf{b}(n)}z^n)(\sum_{n\geq 0} y^{4\mathbf{b}(n)}z^n)(\sum_{n\geq 0} y^{8\mathbf{b}(n)}z^n)\cdots \quad (16.6.5)$$
$$= \sum_{m,n\geq 0} \mathbf{b}_2(m,n)y^m z^n.$$

In the next chapter, we revisit this further. We next give a different example, albeit a little contrived.

Next, we consider the 2D infinite binary product tableau given below. This product enumerates the number of partitions of vector $\langle a,b\rangle$ into at most $d+1$ parts of the form $\langle 2^c,2^d\rangle$ for c and d non-negative integers under vector addition. The tableau is easily transformed by applying (16.5.3) repeatedly to each diagonal from left down, and replacing each successive term of LHS of the case of (16.5.3) with the corresponding successive term of the RHS of (16.5.3).

Hence the following 2D binary vector partition identity result ensues.

Theorem 16.7 ***Unrestricted 2D binary vector partition transform.*** *The following infinite 2D binary product relation holds for* $|y|<1$, $|z|<1$. *The second product here is the generating function for the number of partitions of 2D vector* $\langle m,n\rangle$ *into unrestricted vectors whose components are non-negative powers of 2. The relation is*

$$\prod_{m,n\geq 0}(1+y^{2^m}z^{2^n})^{n+1} = \prod_{m,n\geq 0}\frac{1}{(1-y^{2^m}z^{2^n})} = \sum_{m,n\geq 0} b_2(m,n)y^m z^n.$$

Proof: The equation from the theorem in longhand is as follows. Each diagonal of the first tableaux here is a case of (16.5.1). The second tableaux here relates to the right side of the equation in the theorem, after successive application of (16.5.1). We arrive then at

$$(1+y^{2^0}z^{2^0})^1(1+y^{2^0}z^{2^1})^1(1+y^{2^0}z^{2^2})^1(1+y^{2^0}z^{2^3})^1(1+y^{2^0}z^{2^4})^1(1+y^{2^0}z^{2^5})^1\cdots$$

$$(1+y^{2^1}z^{2^0})^1(1+y^{2^1}z^{2^1})^2(1+y^{2^1}z^{2^2})^2(1+y^{2^1}z^{2^3})^2(1+y^{2^1}z^{2^4})^2(1+y^{2^1}z^{2^5})^2\cdots$$
$$(1+y^{2^2}z^{2^0})^1(1+y^{2^2}z^{2^1})^2(1+y^{2^2}z^{2^2})^3(1+y^{2^2}z^{2^3})^3(1+y^{2^2}z^{2^4})^3(1+y^{2^2}z^{2^5})^3\cdots$$
$$(1+y^{2^3}z^{2^0})^1(1+y^{2^3}z^{2^1})^2(1+y^{2^3}z^{2^2})^3(1+y^{2^3}z^{2^3})^4(1+y^{2^3}z^{2^4})^4(1+y^{2^3}z^{2^5})^4\cdots$$
$$(1+y^{2^4}z^{2^0})^1(1+y^{2^4}z^{2^1})^2(1+y^{2^4}z^{2^2})^3(1+y^{2^4}z^{2^3})^4(1+y^{2^4}z^{2^4})^5(1+y^{2^4}z^{2^5})^5\cdots$$
$$(1+y^{2^5}z^{2^0})^1(1+y^{2^5}z^{2^1})^2(1+y^{2^5}z^{2^2})^3(1+y^{2^5}z^{2^3})^4(1+y^{2^5}z^{2^4})^5(1+y^{2^5}z^{2^5})^6\cdots$$
$$\vdots \quad \vdots \quad \vdots \quad \vdots \quad \vdots \quad \vdots \quad \ddots$$

$$= \frac{1}{(1-y^{2^0}z^{2^0})(1-y^{2^0}z^{2^1})(1-y^{2^0}z^{2^2})(1-y^{2^0}z^{2^3})(1-y^{2^0}z^{2^4})(1-y^{2^0}z^{2^5})\cdots}$$
$$\times \frac{1}{(1-y^{2^1}z^{2^0})(1-y^{2^1}z^{2^1})(1-y^{2^1}z^{2^2})(1-y^{2^1}z^{2^3})(1-y^{2^1}z^{2^4})(1-y^{2^1}z^{2^5})\cdots}$$
$$\times \frac{1}{(1-y^{2^2}z^{2^0})(1-y^{2^2}z^{2^1})(1-y^{2^2}z^{2^2})(1-y^{2^2}z^{2^3})(1-y^{2^2}z^{2^4})(1-y^{2^2}z^{2^5})\cdots}$$
$$\times \frac{1}{(1-y^{2^3}z^{2^0})(1-y^{2^3}z^{2^1})(1-y^{2^3}z^{2^2})(1-y^{2^3}z^{2^3})(1-y^{2^3}z^{2^4})(1-y^{2^3}z^{2^5})\cdots}$$
$$\times \frac{1}{(1-y^{2^4}z^{2^0})(1-y^{2^4}z^{2^1})(1-y^{2^4}z^{2^2})(1-y^{2^4}z^{2^3})(1-y^{2^4}z^{2^4})(1-y^{2^4}z^{2^5})\cdots}$$
$$\times \frac{1}{(1-y^{2^5}z^{2^0})(1-y^{2^5}z^{2^1})(1-y^{2^5}z^{2^2})(1-y^{2^5}z^{2^3})(1-y^{2^5}z^{2^4})(1-y^{2^5}z^{2^5})\cdots}$$
$$\vdots$$
$$= B_2(y,z). \quad \blacksquare$$

So, the generating function $B_2(y,z)$ for 2D unrestricted binary vector partitions has an alternate generating function, as did $B(z)$ for binary integer unrestricted partitions. Hence we can state the following

Theorem 16.8 *The number of unrestricted binary vector partitions, $b_2(m,n)$ of a vector $\langle m,n\rangle$ is equal to the number of partitions of vector $\langle m,n\rangle$ into 2D vectors with each $\lambda_i = 0,1,2,3,\ldots$*

$$\langle 2^{\lambda_1},2^0\rangle,\ \langle 2^{\lambda_2},2^1\rangle,\ \langle 2^{\lambda_3},2^2\rangle,\ \langle 2^{\lambda_4},2^3\rangle,\ \ldots,\ \langle 2^{\lambda_m},2^{m-1}\rangle,\ \ldots,$$

where $2^{\lambda_1} = 2^0$ is used at most once, $2^{\lambda_2} = 2^1$ is used at most twice, $2^{\lambda_3} = 2^2$ is used at most 3 times, etc and generally, $2^{\lambda_m} = 2^{m-1}$ is used at most m times.

Theorem 16.9 ***2D binary vector partition transform with triangular numbers.*** *The following infinite 2D binary product relation holds for* $|y|<1,\ |z|<1$. *The second product here is the generating function for the number of partitions of 2D vector* $\langle m,n\rangle$ *into unrestricted vectors whose components are non-negative powers of 2.*

$$\prod_{m,n\geq 0}\left(1+y^{2^m}z^{2^n}\right)^{\frac{(m+1)(m+2)}{2}}=\prod_{m,n\geq 0}\frac{1}{(1-y^{2^m}z^{2^n})^{m+1}}.$$

Proof: The equation from the theorem in longhand is as follows. Each diagonal of the first tableaux here is a case of (16.5.3). The second tableaux here relates to the right side of the equation in the theorem, after successive application of (16.5.3). We arrive then at

$$\begin{array}{c}
(1+y^{2^0}z^{2^0})^1(1+y^{2^0}z^{2^1})^1(1+y^{2^0}z^{2^2})^1(1+y^{2^0}z^{2^3})^1(1+y^{2^0}z^{2^4})^1(1+y^{2^0}z^{2^5})^1\cdots\\
(1+y^{2^1}z^{2^0})^1(1+y^{2^1}z^{2^1})^3(1+y^{2^1}z^{2^2})^3(1+y^{2^1}z^{2^3})^3(1+y^{2^1}z^{2^4})^3(1+y^{2^1}z^{2^5})^3\cdots\\
(1+y^{2^2}z^{2^0})^1(1+y^{2^2}z^{2^1})^3(1+y^{2^2}z^{2^2})^6(1+y^{2^2}z^{2^3})^6(1+y^{2^2}z^{2^4})^6(1+y^{2^2}z^{2^5})^6\cdots\\
(1+y^{2^3}z^{2^0})^1(1+y^{2^3}z^{2^1})^3(1+y^{2^3}z^{2^2})^6(1+y^{2^3}z^{2^3})^{10}(1+y^{2^3}z^{2^4})^{10}(1+y^{2^3}z^{2^5})^{10}\cdots\\
(1+y^{2^4}z^{2^0})^1(1+y^{2^4}z^{2^1})^3(1+y^{2^4}z^{2^2})^6(1+y^{2^4}z^{2^3})^{10}(1+y^{2^4}z^{2^4})^{15}(1+y^{2^4}z^{2^5})^{15}\cdots\\
(1+y^{2^5}z^{2^0})^1(1+y^{2^5}z^{2^1})^3(1+y^{2^5}z^{2^2})^6(1+y^{2^5}z^{2^3})^{10}(1+y^{2^5}z^{2^4})^{15}(1+y^{2^5}z^{2^5})^{21}\cdots\\
\vdots\qquad\vdots\qquad\vdots\qquad\vdots\qquad\vdots\qquad\vdots\qquad\ddots
\end{array}$$

$$\begin{aligned}
&=\frac{1}{(1-y^{2^0}z^{2^0})(1-y^{2^0}z^{2^1})(1-y^{2^0}z^{2^2})(1-y^{2^0}z^{2^3})(1-y^{2^0}z^{2^4})(1-y^{2^0}z^{2^5})\cdots}\\
&\times\frac{1}{(1-y^{2^1}z^{2^0})(1-y^{2^1}z^{2^1})^2(1-y^{2^1}z^{2^2})^2(1-y^{2^1}z^{2^3})^2(1-y^{2^1}z^{2^4})^2(1-y^{2^1}z^{2^5})^2\cdots}\\
&\times\frac{1}{(1-y^{2^2}z^{2^0})(1-y^{2^2}z^{2^1})^2(1-y^{2^2}z^{2^2})^3(1-y^{2^2}z^{2^3})^3(1-y^{2^2}z^{2^4})^3(1-y^{2^2}z^{2^5})^3\cdots}\\
&\times\frac{1}{(1-y^{2^3}z^{2^0})(1-y^{2^3}z^{2^1})^2(1-y^{2^3}z^{2^2})^3(1-y^{2^3}z^{2^3})^4(1-y^{2^3}z^{2^4})^4(1-y^{2^3}z^{2^5})^4\cdots}\\
&\times\frac{1}{(1-y^{2^4}z^{2^0})(1-y^{2^4}z^{2^1})^2(1-y^{2^4}z^{2^2})^3(1-y^{2^4}z^{2^3})^4(1-y^{2^4}z^{2^4})^5(1-y^{2^4}z^{2^5})^5\cdots}\\
&\times\frac{1}{(1-y^{2^5}z^{2^0})(1-y^{2^5}z^{2^1})^2(1-y^{2^5}z^{2^2})^3(1-y^{2^5}z^{2^3})^4(1-y^{2^5}z^{2^4})^5(1-y^{2^5}z^{2^5})^6\cdots}\\
&\qquad\vdots
\end{aligned}$$

■

So, we have here stated that the generating functions for two differently derived classes of 2D partitions say $\mathbf{C1}_2(y,z)$ and $\mathbf{C2}_2(y,z)$ have equivalence. Hence we can state the following

Theorem 16.10 *Define* $\mathbf{C1}_2(y,z)$ *as the generating function for partitions of vector* $\langle m,n\rangle$ *for* m *and* n *positive integers such that each partition involves precisely for* $\lambda = 0,1,2,3,\ldots,$ *the parts,*

$$\langle 2^\lambda, 2^0\rangle, \langle 2^\lambda, 2^1\rangle, \langle 2^\lambda, 2^2\rangle, \langle 2^\lambda, 2^3\rangle, \ldots, \langle 2^\lambda, 2^{m-1}\rangle, \ldots,$$

where $2^\lambda = 2^0$ *is used at most once,* $2^\lambda = 2^1$ *is used at most three times,* $2^\lambda = 2^2$ *is used at most six times, etc and generally,* $2^\lambda = 2^{m-1}$ *is used at most* $\frac{m(m+1)}{2}$ *times.*

Define $\mathbf{C2}_2(y,z)$ *as the generating function for partitions of vector* $\langle m,n\rangle$ *for* m *and* n *positive integers such that each partition involves precisely for* $\lambda = 0,1,2,3,\ldots,$ *the parts,*

$$\langle 2^\lambda, 2^0\rangle, \langle 2^\lambda, 2^1\rangle, \langle 2^\lambda, 2^2\rangle, \langle 2^\lambda, 2^3\rangle, \ldots, \langle 2^\lambda, 2^{m-1}\rangle, \ldots,$$

where $2^\lambda = 2^0$ *is used at most once,* $2^\lambda = 2^1$ *is used at most three times,* $2^\lambda = 2^2$ *is used at most six times, etc and generally,* $2^\lambda = 2^{m-1}$ *is used at most* $\frac{m(m+1)}{2}$ *times.*

16.7 A BINARY PARTITION 2-SPACE SIMPLE IDENTITY

We can apply the method to other vector partition generating functions. An example is now given. The following theorem is based on the ideas associated with the elementary identity

$$(1+x)\left(1+x^2\right)\left(1+x^4\right)\left(1+x^8\right)\cdots = \frac{1}{1-x}.$$

In fact, the combinatorial interpretation of this is "*each positive integer is uniquely represented by a sum of distinct powers of 2*". So, we are here looking at an extension of this result in the

Theorem 16.11

$$\prod_{k\geq 0}\left(\frac{1}{1-qt^{2^k}}\right) = \prod_{\substack{j,k\geq 0\\ j\leq k}}\left(1+q^{2^j}t^{2^k}\right) = 1+\sum_{k=1}^{\infty} A_k t^k \tag{16.7.1}$$

where

$$A_k = \frac{1}{k!}\begin{vmatrix} q & -1 & 0 & 0 & \cdots & 0 \\ q^2+2q & q & -2 & 0 & \cdots & 0 \\ q^3 & q^2+q & q & -3 & \cdots & 0 \\ q^4+2q^2+4q & q^3 & q^2+q & q & \ddots & \vdots \\ \vdots & \vdots & \vdots & \vdots & \ddots & -(k-1) \\ \sum_{\substack{2^j\mid k\\ j\geq 0}} 2^j q^{k/2^j} & \sum_{\substack{2^j\mid(k-1)\\ j\geq 0}} 2^j q^{(k-1)/2^j} & \sum_{\substack{2^j\mid(k-2)\\ j\geq 0}} 2^j q^{(k-2)/2^j} & \sum_{\substack{2^j\mid(k-3)\\ j\geq 0}} 2^j q^{(k-3)/2^j} & \cdots & q \end{vmatrix}$$

The combinatorial interpretation of (16.7.1) is

Theorem 16.12 *If* $B(j,k)$ *is the number of vector partitions of* $\langle j,k\rangle$ *into distinct parts of kind* $\langle 2^a, 2^b\rangle$ *in which* $a \leq b$ *with non-negative integers* a *and* b*, then* $B(j,k)$ *equals also the number of partitions into "unrestricted" parts of kind* $\langle 1, 2^b\rangle$ *in which* b *is a non negative integer, and* $B(j,k)$ *is the coefficient of* $q^j t^k$ *in (16.7.1).*

Each side of (16.7.1) satisfies the equation $f(t)(1-qt) = f(t^2)$ and this equation also leads to a set of recurrences solvable using Cramer's rule.

In Mathematica, Maple or WolframAlpha online we can easily check that

$$\prod_{k\geq 0}\left(\frac{1}{1-qt^{2^k}}\right) = 1+qt+q(q+1)t^2+q^2(q+1)t^3+q(q^3+q^2+q+1)t^4 \qquad (16.7.2)$$
$$+q^2(q^3+q^2+q+1)t^5+q^2(q^4+q^3+q^2+2q+1)t^6$$
$$+q^3(q^4+q^3+q^2+2q+1)t^7$$
$$+q(q^7+q^6+q^5+2q^4+2q^3+q^2+q+1)t^8$$
$$+q^2(q^7+q^6+q^5+2q^4+2q^3+q^2+q+1)t^9$$
$$+q^2(q^8+q^7+q^6+2q^5+2q^4+2q^3+2q^2+2q+1)t^{10}+\ldots$$

Also, as a matter of interest, utilizing a form $\prod_{k=0}^{\infty}\prod_{j=0}^{k}(1+q^{2^j}t^{2^k})$, on calulating engines, the two product expansions in (16.7.1) can be easily verified; both of them yielding the series given in (16.7.2).

So in longhand we have that

$$\frac{1}{(1-qt)(1-qt^2)(1-qt^4)(1-qt^8)(1-qt^{16})\ldots}$$
$$=(1+qt)$$
$$(1+qt^2)(1+q^2t^2)$$
$$(1+qt^4)(1+q^2t^4)(1+q^4t^4)$$
$$(1+qt^8)(1+q^2t^8)(1+q^4t^8)(1+q^4t^8)$$
$$(1+qt^{16})(1+q^2t^{16})(1+q^4t^{16})(1+q^4t^{16})(1+q^8t^{16})$$
$$\vdots$$

So let us define $\alpha_k(q)$ and $\beta_2(j,k)$ from

$$\prod_{k\geq 0}\left(\frac{1}{1-qt^{2^k}}\right) = \prod_{\substack{j,k\geq 0\\ j\leq k}}\left(1+q^{2^j}t^{2^k}\right) = 1+\sum_{k=1}^{\infty}\alpha_k(q)t^k = \sum_{j=0,k=0}^{\infty}\beta_2(j,k)q^jt^k. \quad (16.7.3)$$

So we see from (16.7.2) that

$$\begin{aligned}
\alpha_1(q) &= q,\\
\alpha_2(q) &= q(q+1),\\
\alpha_3(q) &= q^2(q+1),\\
\alpha_4(q) &= q(q^3+q^2+q+1),\\
\alpha_5(q) &= q^2(q^3+q^2+q+1),\\
\alpha_6(q) &= q^2(q^4+q^3+q^2+2q+1),\\
\alpha_7(q) &= q^3(q^4+q^3+q^2+2q+1),\\
\alpha_8(q) &= q(q^7+q^6+q^5+2q^4+2q^3+q^2+q+1),\\
\alpha_9(q) &= q^2(q^7+q^6+q^5+2q^4+2q^3+q^2+q+1),\\
\alpha_{10}(q) &= q^2(q^8+q^7+q^6+2q^5+2q^4+2q^3+2q^2+2q+1),
\end{aligned}$$

and so on. Hence, from this we see that

$$\beta_2(1,1)=1, \beta_2(1,2)=1, \beta_2(2,2)=1, \beta_2(1,3)=0, \beta_2(2,3)=1, \beta_2(1,4)=1, \beta_2(2,4)=1, etc.$$

All of the $\beta_2(j,k)$ values are encoded in the following grid table.

Partition grid for $\mathbf{1/((1-qt)(1-qt^2)(1-qt^4)(1-qt^8)...)=\sum_{j,k\geq 0}^{\infty}\beta_2(j,k)q^jt^k}$

2D binary partition grid for $\prod_{n\geq 0} 1/(1-qt^{2^n})$

⋮	⋮	⋮	⋮	⋮	⋮	⋮	⋮	⋮	⋮	⋮	⋮	⋮	⋮
13			1	2	2	3	3	2	2	2	1	1	1
12		1	2	2	3	3	2	2	2	1	1	1	
11			1	2	2	2	2	2	1	1	1		
10		1	2	2	2	2	2	1	1	1			
9		1	1	1	2	2	1	1	1				
8	1	1	1	2	2	1	1	1					
7			1	2	1	1	1						
6		1	2	1	1	1							
5		1	1	1	1								
4	1	1	1	1									
3		1	1										
2	1	1											
1	1												
	1	2	3	4	5	6	7	8	9	10	11	12	13

To illustrate theorem 16.12, we give an arbitrary case for the 2D vector $\langle 3,6\rangle$:

Corollary 16.1 *$\beta_2(3,6)=2$ is the number of vector partitions of $\langle 3,6\rangle$ into distinct parts of kind $\langle 2^a,2^b\rangle$ in which $a\leq b$ with non-negative integers a and b. The two partitions are $\langle 2,4\rangle+\langle 1,2\rangle$ and $\langle 1,4\rangle+\langle 2,2\rangle$. Also $\beta_2(3,6)=2$ equals the number of partitions into "unrestricted" parts of kind $\langle 1,2^b\rangle$ in which b is a non negative integer. The two partitions are $\langle 1,2\rangle+\langle 1,2\rangle+\langle 1,2\rangle$ and $\langle 1,4\rangle+\langle 1,1\rangle+\langle 1,1\rangle$. Then also $\beta_2(3,6)=2$ is the coefficient of q^3t^6 in (16.7.1).*

The vector partitions $\beta_2(j,k)$ defined for (16.7.1) are easily visualized by number entries in the 2D grid above.

We see that our methods set up the platform for possible research for partitions into binary component vectors, and for other n-space infinite product generating functions.

16.8 FIRST QUADRANT 2D BINARY PARTITIONS

We employ the first quadrant expansion of binary powers in x and y.

$$(1+xy)(1+x^2y)(1+x^4y)(1+x^8y)(1+x^{16}y)(1+x^{32}y)(1+x^{64}y)(1+x^{128}y)(1+x^{256}y)\cdots$$

$$(1+xy^2)(1+x^2y^2)(1+x^4y^2)(1+x^8y^2)(1+x^{16}y^2)(1+x^{32}y^2)(1+x^{64}y^2)(1+x^{128}y^2)\cdots$$

$$(1+xy^4)(1+x^2y^4)(1+x^4y^4)(1+x^8y^4)(1+x^{16}y^4)(1+x^{32}y^4)(1+x^{64}y^4)(1+x^{128}y^4)\cdots$$
$$(1+xy^8)(1+x^2y^8)(1+x^4y^8)(1+x^8y^8)(1+x^{16}y^8)(1+x^{32}y^8)(1+x^{64}y^8)(1+x^{128}y^8)\cdots$$
$$(1+xy^{16})(1+x^2y^{16})(1+x^4y^{16})(1+x^8y^{16})(1+x^{16}y^{16})(1+x^{32}y^{16})(1+x^{64}y^{16})(1+x^{128}y^{16})\cdots$$
$$(1+xy^{32})(1+x^2y^{32})(1+x^4y^{32})(1+x^8y^{32})(1+x^{16}y^{32})(1+x^{32}y^{32})(1+x^{64}y^{32})(1+x^{128}y^{32})\cdots$$

etc.

We note that each diagonal infinite product with terms going one row down and one column right is a case of the product

$$(1+z)(1+z^2)(1+z^4)(1+z^8)(1+z^{16})\cdots = \frac{1}{1-z},$$

so the above tableau product is therefore equal to

$$\frac{1}{(1-xy)}$$
$$\times \frac{1}{(1-x^2y)(1-x^4y)(1-x^8y)(1-x^{16}y)(1-x^{32}y)(1-x^{64}y)(1-x^{128}y)(1-x^{256}y)\cdots}$$
$$\times \frac{1}{(1-xy^2)(1-xy^4)(1-xy^8)(1-xy^{16})(1-xy^{32})(1-xy^{64})(1-xy^{128})(1-xy^{256})\ldots}$$

Hence, we have described two equivalent generating functions for 2D first quadrant binary component partitions. We see that this approach works to give us a theorem on 2D vector partitions where the two vector components a and b of vector $\langle a,b\rangle$ are integers of the form $a=2^j$, $b=2^k$ for all non-negative integers j and k. In the usual notation, the above identity is the generating function version of the theorem just below it

$$\frac{1}{1-xy}\prod_{j\geq 1}\left(\frac{1}{(1-x^{2^j}y)(1-xy^{2^j})}\right) = \prod_{\substack{j,k\geq 0\\ j\leq k}}\left(1+x^{2^j}y^{2^k}\right) \tag{16.8.1}$$

So then we can formulate a

Theorem 16.13 *Consider the set V_1 of all first quadrant 2D vectors $\langle a,b\rangle$, where the components $a=2^j$ and $b=2^k$ are binary powers with $j,k\geq 0$. Next consider the set V_2 of all first quadrant 2D vectors $\langle c,d\rangle$, where the components are either:*

(a) $c=1$ and $d=1$; or

(b) $c=1$ and $d=2^j$ with $j\geq 1$; or

(c) $c=2^k$ and $d=1$ with $k\geq 1$.

Then the number of partitions of vector $\langle m,n\rangle$ with $m\geq 1$, $n\geq 1$, into distinct partitions from the set V_1 is equal to the number of unrestricted partitions from the set V_2.

16.9 FIRST QUADRANT UPPER DIAGONAL 2D BINARY PARTITIONS

Consider the first quadrant upper diagonal expansion of binary powers in x and y. Then for this, we state the

Theorem 16.14

$$\begin{gathered}(1+xy)\\(1+xy^2)(1+x^2y^2)\\(1+xy^4)(1+x^2y^4)(1+x^4y^4)\\(1+xy^8)(1+x^2y^8)(1+x^4y^8)(1+x^8y^8)\\(1+xy^{16})(1+x^2y^{16})(1+x^4y^{16})(1+x^8y^{16})(1+x^{16}y^{16})\\(1+xy^{32})(1+x^2y^{32})(1+x^4y^{32})(1+x^8y^{32})(1+x^{16}y^{32})(1+x^{32}y^{32})\\ \textit{etc.}\end{gathered}$$

$$=\frac{1}{(1-xy)(1-xy^2)(1-xy^4)(1-xy^8)(1-xy^{16})(1-xy^{32})(1-xy^{64})(1-xy^{128})\cdots}$$

Hence, we have again described two equivalent generating functions for 2D first quadrant binary component partitions. We then have another theorem on 2D vector partitions.

Theorem 16.15 *Consider the set U_1 of all first quadrant 2D vectors $\langle a,b\rangle$, where the components $a=2^j$ and $b=2^k$ are binary powers with $0\le j\le k$ or equivalently $a\le b$. Next consider the set U_2 of all first quadrant 2D vectors $\langle c,d\rangle$, where the components are either:*

(a) $c=1$ and $d=1$; or

(b) $c=1$ and $d=2^j$ with $j\ge 0$.

Then the number of partitions of vector $\langle m,n\rangle$ with $m\ge 1$, $n\ge 1$, into distinct partitions from the set U_1 is equal to the number of unrestricted partitions from the set U_2.

16.10 FIRST HYPERQUADRANT 3D BINARY PARTITIONS

Following similar logic for the 3D first hyperquadrant points for binary numbers 2^a with positive integers a components of vector $\langle m,n,p\rangle$ it is easy to see that (as illustrated in the pyramid of Figure 16.1)

$$\frac{1}{1-xyz}\prod_{a,b\ge 1}\left(\frac{1}{(1-xy^{2^a}z^{2^b})(1-x^{2^a}yz^{2^b})(1-x^{2^a}y^{2^b}z)}\right)=\prod_{a,b,c\ge 0}\left(1+x^{2^a}y^{2^b}z^{2^c}\right) \tag{16.10.1}$$

So let us next look at the infinite product of binary powers in the 3D pyramid.

$$\begin{gathered}
(\mathbf{1+xyz}) \\
(1+xy^2z)(1+x^2y^2z) \\
(1+xy^2z^2)(\mathbf{1+x^2y^2z^2}) \\
(1+xy^4z)(1+x^2y^4z)(1+x^4y^4z) \\
(1+xy^4z^2)(1+x^2y^4z^2)(1+x^4y^4z^2) \\
(1+xy^4z^4)(1+x^2y^4z^4)(\mathbf{1+x^4y^4z^4}) \\
(1+xy^8z)(1+x^2y^8z)(1+x^4y^8z)(1+x^8y^8z) \\
(1+xy^8z^2)(1+x^2y^8z^2)(1+x^4y^8z^2)(1+x^8y^8z^2) \\
(1+xy^8z^4)(1+x^2y^8z^4)(1+x^4y^8z^4)(1+x^8y^8z^4) \\
(1+xy^8z^8)(1+x^2y^8z^8)(1+x^4y^8z^8)(\mathbf{1+x^8y^8z^8}) \\
etc
\end{gathered}$$

where we see for example, that the **bold** terms form the infinite product for $\frac{\mathbf{1}}{\mathbf{1-xyz}}$. Hence, picking out every instance of the basic binary infinite product as we have done for the **bold** product, we see quite easily the following infinite product transform (the first product here depicted in Figure 16.1 in the context of the extended pyramid arranged form of binary product terms)

$$\begin{gathered}
(1+xyz) \\
(1+x^1y^2z^1)(1+x^2y^2z^1) \\
(1+x^1y^2z^2)(1+x^2y^2z^2) \\
(1+x^1y^4z)(1+x^2y^4z^1)(1+x^4y^4z^1) \\
(1+x^1y^4z^2)(1+x^2y^4z^2)(1+x^4y^4z^2) \\
(1+x^1y^4z^4)(1+x^2y^4z^4)(1+x^4y^4z^4) \\
(1+x^1y^8z^1)(1+x^2y^8z^1)(1+x^4y^8z^1)(1+x^8y^8z^1) \\
(1+x^1y^8z^2)(1+x^2y^8z^2)(1+x^4y^8z^2)(1+x^8y^8z^2) \\
(1+x^1y^8z^4)(1+x^2y^8z^4)(1+x^4y^8z^4)(1+x^8y^8z^4) \\
(1+x^1y^8z^8)(1+x^2y^8z^8)(1+x^4y^8z^8)(1+x^8y^8z^8) \\
etc
\end{gathered}$$

$$= \frac{1}{(1-x^1y^1z^1)}$$

$$\frac{1}{(1-x^1y^2z^1)(1-x^2y^2z^1)}$$

$$\frac{1}{(1-x^1y^2z^2)}$$
$$\frac{1}{(1-x^1y^4z^1)(1-x^2y^4z^1)(1-x^4y^4z^1)}$$
$$\frac{1}{(1-x^1y^4z^2)}$$
$$\frac{1}{(1-x^1y^4z^4)}$$
$$\frac{1}{(1-x^1y^8z^1)(1-x^2y^8z^1)(1-x^4y^8z^1)(1-x^8y^8z^1)}$$
$$\frac{1}{(1-x^1y^8z^2)}$$
$$\frac{1}{(1-x^1y^8z^4)}$$
$$\frac{1}{(1-x^1y^8z^8)}$$
$$etc$$

Both sides of this 3D equation enumerate vector partitions in 3-space. So the transformation identity above has an interpretation in vector partitions implying the

Theorem 16.16 *Consider the infinitely extended 3D pyramid defined in the x-y-z Euclidean space by the inequalities*

$$0 \le x \le y,\ 1 \le z \le y,\ y \ge 1.$$

Define all integer component lattice point vectors inside that pyramid to be of form $\langle a,b,c\rangle$ where:

$$0 \le a \le b,\ 1 \le c \le b,\ b \ge 1.$$

Then $b_3(a,b,c)$, the number of 3D binary partitions of $\langle a,b,c\rangle$ have the two equivalent forms:

Form A) All distinct 3D binary vector partitions of kind

$$\langle a,b,c\rangle = \sum_{a_i,a_j,a_k \ge 0} \langle 2^{a_i}, 2^{a_j}, 2^{a_k}\rangle, \tag{16.10.2}$$

Form B) All unrestricted 3D binary vector partitions of kind

$$\langle a,b,c\rangle = \sum_{a_i,a_j,a_k \ge 0} \left(\langle 1, 2^{a_j}, 2^{a_k}\rangle + \langle 2^{a_i}, 1, 2^{a_k}\rangle + \langle 2^{a_i}, 2^{a_j}, 1\rangle\right). \tag{16.10.3}$$

Another way of putting this is

"The number of distinct 3D binary vector partitions of $\langle a,b,c\rangle$ inside the infinite extended pyramid $0 \le x \le y$, $1 \le z \le y$, $y \ge 1$ equals the number of unrestricted 3D binary vector partitions of $\langle a,b,c\rangle$ with at least one component a, b, or c equal to 1, in the same pyramid."

This idea of vector partitions and their generating functions is greatly assisted by visualizations, the above theorem is depicted here yet again, but with the added help of a pyramidal structure beside the equations.

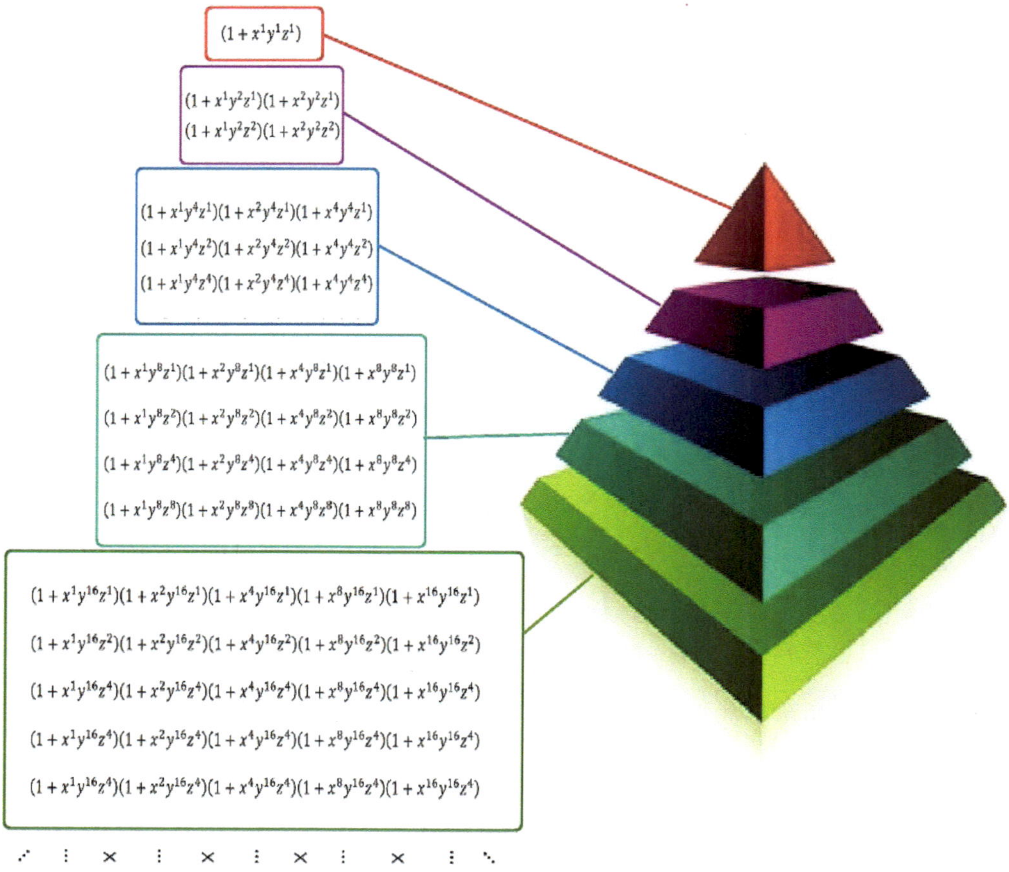

Figure 16.1 3D Pyramid Binary Product Form illustrating groupings of product factor terms arranging them in a pyramid shape.

$$= \frac{1}{(1-xyz)} \times \frac{1}{\left\{\begin{matrix}(1-xy^2z) & (1-x^2y^2z) \\ (1-xy^2z^2) & \end{matrix}\right\}} \times \frac{1}{\left\{\begin{matrix}(1-xy^4z) & (1-x^2y^4z) & (1-x^4y^4z) \\ (1-xy^4z^2) & & \\ (1-xy^4z^4) & & \end{matrix}\right\}} \tag{16.10.4}$$

$$\times \frac{1}{\left\{\begin{matrix}(1-xy^8z) & (1-x^2y^8z) & (1-x^4y^8z) & (1-x^4y^8z) \\ (1-xy^8z^2) & & & \\ (1-xy^8z^4) & & & \\ (1-xy^8z^8) & & & \end{matrix}\right\}} \times etc.$$

We remind the reader that the basic tenets of this chapter are laid out in Andrews' classical book [12]. Recommended further specialized reading related to this chapter is as follows:

- The two fascinating papers 40 years apart by Mahler [194, 195].
- Rodseth's interesting paper from 1970 on n-ary partitions [224].
- The paper by Campbell [86] and Campbell and Zujev [100] on the binary series that Ramanujan misunderstood.
- The year 2000 paper by Campbell [91] on Hyperpyramid lattice Visible Point Vector identities.

Think About It...

Mathematics is not a careful march down a well-cleared highway, but a journey into a strange wilderness, where the explorers often get lost.

Rigor should be a signal to the historians that the maps have been made, and the real explorers have gone elsewhere.

W.S. Anglin
Mathematics author

FURTHER READING

ANDREWS, G.E. *The Theory of Partitions*, volume 2 of *Encyclopedia of Mathematics and its Applications.* Addison-Wesley Publishing Company, Advanced Book Program, Reading, Massachusetts, USA, 1976.

CAMPBELL, G.B. A generalized formula of Hardy. *Int. J. Math. Math. Sci.*, 17(2):369–378, 1994.

CAMPBELL, G.B. Infinite products over hyperpyramid lattices. *Int. J. Math. and Math. Sci.*, 23(4):271–277, 2000.

CAMPBELL, G.B. and ZUJEV, A. The series that Ramanujan misunderstood. arxiv preprint, Oct 2016. arXiv:1610.03693v1 [math.NT], https://arxiv.org/abs/1610.03693v1.

MAHLER, K. On a Special Functional Equation. *J. London Math. Soc.*, 15:115– 123, 1940.

MAHLER, K. On a Special Function. *J. Number Theory*, 12:20–26, 1980.

RÖDSETH, Ö. Some arithmetical properties of m-ary partitions. *Proc. Cambridge Philos. Soc.*, 68:447–453, 1970.

CHAPTER 17

n-ary Partitions and their Vector Generalizations

17.1 INTEGER n-ARY PARTITIONS

In this section, we state some introductory results related to n-ary partitions, where n is a positive integer greater than 1. The study of these partitions is allied to basic concepts for number systems in base n, especially for distinct n-ary partitions.

Theorem 17.1 ***Distinct n-ary partitions and base n numbers:*** *The number systems base 2, base 3, base 4, base 5, ..., base n, and so on each represents every positive integer by use of:*

(a) 2 digits namely 0 and 1 for the binary system,

(b) 3 digits, namely 0,1,2 for the base 3 system,

(c) 4 digits, namely 0,1,2,3 for the base 4 system,

(d) 5 digits, namely 0,1,2,3,4 for the base 5 system,

etc

(e) n digits, namely $0, 1, 2, \ldots, 9, a_{10}, a_{11}, \ldots a_{n-1}$ *for a base n system.*

DOI: 10.1201/9781003174158-17

Each of these is equivalent successively to the infinite products:

$$\begin{aligned}
\frac{1}{1-q} &= \prod_{k=0}^{\infty}(1+q^{2^k}) && (17.1.1)\\
&= \prod_{k=0}^{\infty}(1+q^{3^k}+q^{2.3^k}) && (17.1.2)\\
&= \prod_{k=0}^{\infty}(1+q^{4^k}+q^{2.4^k}+q^{3.4^k}) && (17.1.3)\\
&= \prod_{k=0}^{\infty}(1+q^{5^k}+q^{2.5^k}+q^{3.5^k}+q^{4.5^k}) && (17.1.4)\\
&\vdots\\
&= \prod_{k=0}^{\infty}(1+q^{n^k}+q^{2.n^k}+q^{3.n^k}+\ldots+q^{a_{n-1}.n^k}) && (17.1.5)\\
&= 1+q+q^2+q^3+q^4+q^5+q^6+\ldots && (17.1.6)
\end{aligned}$$

17.2 A BASE 10 OR 10-ARY SET OF CASES

Because we live most of our lives using a decimal system base 10, let's take that example, namely,

$$\begin{aligned}
\frac{1}{1-q} &= \prod_{k=0}^{\infty}(1+q^{10^k}+q^{2.10^k}+q^{3.10^k}+\ldots+q^{9.10^k}) && (17.2.1)\\
&= 1+q+q^2+q^3+q^4+q^5+q^6+\ldots
\end{aligned}$$

The identity (17.2.1) when interpreted in plain language, as a 10-ary or decimal partition statement simply says

"Every positive integer can be represented using the digits $0,1,2,3,4,5,6,7,8,9$ in the base 10 (decimal) number system." This is a fact assumed everyday by people all over the world.

(17.2.1) when interpreted mathematically as

"The number of integer 10-ary distinct partitions of positive integer m in the form

$$m = a_1.10^0 + a_2.10^1 + a_3.10^2 + \ldots + a_k.10^k, \qquad (17.2.2)$$

where each a_i is a non-negative integer less than 10, is 1."

Theorem 17.1 is well known and appears often in the literature. However, an extensive search has not found the following interesting results:

$$\prod_{k=0}^{\infty}(1+q^{2.10^k}+q^{3.10^k}) = 1+q^2+q^3+q^{20}+q^{22}+q^{23}+q^{30}+q^{32}+q^{33}+O(q^{200}),$$

$$\prod_{k=0}^{\infty}(1+q^{1.10^k}+q^{5.10^k}) = 1+q+q^5+q^{10}+q^{11}+q^{15}+q^{50}+q^{51}+q^{55}+O(q^{99}),$$

$$\prod_{k=0}^{\infty}(1+q^{5.10^k}+q^{7.10^k}) = 1+q^5+q^7+q^{50}+q^{55}+q^{57}+q^{70}+q^{75}+q^{77}+O(q^{486}),$$

$$\prod_{k=0}^{\infty}(1+q^{8.10^k}+q^{9.10^k}) = 1+q^8+q^9+q^{80}+q^{88}+q^{89}+q^{90}+q^{98}+q^{99}+O(q^{800}),$$

$$\prod_{k=0}^{\infty}(1+q^{1.10^k}+q^{2.10^k}) = 1+q+q^2+q^{10}+q^{11}+q^{12}+q^{20}+q^{21}+q^{22}+q^{100}+O(q^{101}),$$

where the left side indices next to 10^k are the digits in the right side indices with zeroes are included in every possible combination for positive integers.

The above five equations are cases of the

Theorem 17.2 ***The general 10-ary distinct partition generating function.*** *If $a_1, a_2, ..., a_m$ are different integers with $1 \leq a_1 < a_2 < \ldots < a_m \leq 9$ then*

$$\prod_{k=0}^{\infty}(1+q^{a_1.10^k}+q^{a_2.10^k}+\ldots+q^{a_m.10^k}) = 1+\sum_{a_1,a_2,...,a_m}^{\infty} q^{D_{10}(a_1,a_2,...,a_m)} \tag{17.2.3}$$

where the right side indices $D_{10}(a_1, a_2, ..., a_m)$ are the base 10 positive integers whose digits are "$a_1, a_2, ..., a_m$ and zeroes" included exactly once per every possible combination.

This theorem gives us new results on 10-ary partitions quite easily. Consider the identities, each fairly easy to prove:

$$1+q+q^2+q^3+q^4+q^5+q^6+q^7+q^8+0q^9 = (1+q+q^2)(1+q^3+q^6), \tag{17.2.4}$$

$$1+q+q^2+q^3+q^4+q^5+0q^6+q^7+q^8+q^9 = (1+q+q^2)(1+q^3+q^7), \tag{17.2.5}$$

$$1+q+q^2+0q^3+q^4+q^5+q^6+q^7+q^8+q^9 = (1+q+q^2)(1+q^4+q^7), \tag{17.2.6}$$

$$1+0q+0q^2+q^3+q^4+q^5+q^6+q^7+q^8+q^9 \tag{17.2.7}$$
$$= (1+q)(1-q+q^2+q^4+q^6+q^8) = (1+q)(1-q+q^2)(1+q^4+q^5+q^6),$$

$$1+q+0q^2+0q^3+q^4+q^5+q^6+q^7+q^8+q^9 = (1+q)(1+q^4+q^6+q^8), \tag{17.2.8}$$

$$1+q+0q^2+q^3+q^4+0q^5+q^6+q^7+q^8+q^9 = (1+q)(1+q^3+q^6+q^8), \tag{17.2.9}$$

$$1+q+0q^2+q^3+q^4+q^5+q^6+0q^7+q^8+q^9 \tag{17.2.10}$$
$$= (1+q)^3(1-q+q^2)(1-q+q^2-q^3+q^4).$$

These identities are easy to verify on a calculating website or app. Note that each product term and each initial polynomial in each of the seven equations (17.2.4) to (17.2.10) are applicable to cases of theorem 17.2 above. We now follow through this process to obtain seven corollaries from each of these. So, step one to achieve our corollaries is applying the product operator to each factor in each of (17.2.4) to (17.2.10). We see easily that:

$$\prod_{k=0}^{\infty}(1+q^{1.10^k}+q^{2.10^k}+q^{3.10^k}+q^{4.10^k}+q^{5.10^k}+q^{6.10^k}+q^{7.10^k}+q^{8.10^k}+0q^{9.10^k}) \tag{17.2.11}$$

$$=\left(\prod_{k=0}^{\infty}(1+q^{1.10^k}+q^{2.10^k})\right)\left(\prod_{k=0}^{\infty}(1+q^{3.10^k}+q^{6.10^k})\right),$$

$$\prod_{k=0}^{\infty}(1+q^{1.10^k}+q^{2.10^k}+q^{3.10^k}+q^{4.10^k}+q^{5.10^k}+0q^{6.10^k}+q^{7.10^k}+q^{8.10^k}+q^{9.10^k}) \tag{17.2.12}$$

$$=\left(\prod_{k=0}^{\infty}(1+q^{1.10^k}+q^{2.10^k})\right)\left(\prod_{k=0}^{\infty}(1+q^{3.10^k}+q^{7.10^k})\right),$$

$$\prod_{k=0}^{\infty}(1+q^{1.10^k}+q^{2.10^k}+0q^{3.10^k}+q^{4.10^k}+q^{5.10^k}+q^{6.10^k}+q^{7.10^k}+q^{8.10^k}+q^{9.10^k}) \tag{17.2.13}$$

$$=\left(\prod_{k=0}^{\infty}(1+q^{1.10^k}+q^{2.10^k})\right)\left(\prod_{k=0}^{\infty}(1+q^{4.10^k}+q^{7.10^k})\right),$$

$$\prod_{k=0}^{\infty}(1+0q^{1.10^k}+0q^{2.10^k}+q^{3.10^k}+q^{4.10^k}+q^{5.10^k}+q^{6.10^k}+q^{7.10^k}+q^{8.10^k}+q^{9.10^k}) \tag{17.2.14}$$

$$=\left(\prod_{k=0}^{\infty}(1+q^{1.10^k})\right)\left(\prod_{k=0}^{\infty}(1-q^{1.10^k}+q^{2.10^k}+q^{4.10^k}+q^{6.10^k}+q^{8.10^k})\right)$$

$$=\left(\prod_{k=0}^{\infty}(1+q^{1.10^k})\right)\left(\prod_{k=0}^{\infty}(1-q^{1.10^k}+q^{2.10^k})\right)\left(\prod_{k=0}^{\infty}(1+q^{4.10^k}+q^{5.10^k}+q^{6.10^k})\right),$$

$$\prod_{k=0}^{\infty}(1+q^{1.10^k}+0q^{2.10^k}+0q^{3.10^k}+q^{4.10^k}+q^{5.10^k}+q^{6.10^k}+q^{7.10^k}+q^{8.10^k}+q^{9.10^k}) \tag{17.2.15}$$

$$=\left(\prod_{k=0}^{\infty}(1+q^{1.10^k})\right)\left(\prod_{k=0}^{\infty}(1+q^{4.10^k}+q^{6.10^k}+q^{8.10^k})\right),$$

$$\prod_{k=0}^{\infty}(1+q^{1.10^k}+0q^{2.10^k}+q^{3.10^k}+q^{4.10^k}+0q^{5.10^k}+q^{6.10^k}+q^{7.10^k}+q^{8.10^k}+q^{9.10^k}) \tag{17.2.16}$$

$$=\left(\prod_{k=0}^{\infty}(1+q^{1.10^k})\right)\left(\prod_{k=0}^{\infty}(1+q^{3.10^k}+q^{6.10^k}+q^{8.10^k})\right),$$

$$\prod_{k=0}^{\infty}(1+q^{1.10^k}+0q^{2.10^k}+q^{3.10^k}+q^{4.10^k}+q^{5.10^k}+q^{6.10^k}+0q^{7.10^k}+q^{8.10^k}+q^{9.10^k}) \tag{17.2.17}$$

$$=\left(\prod_{k=0}^{\infty}(1+q^{1.10^k})^3\right)\left(\prod_{k=0}^{\infty}(1-q^{1.10^k}+q^{2.10^k})\right)\left(\prod_{k=0}^{\infty}(1-q^{1.10^k}+q^{2.10^k}-q^{3.10^k}+q^{4.10^k})\right).$$

The above seven equations are a selection from many possible identities we could have chosen as examples to apply to our theorem 17.2. So, let us now apply this theorem to equations (17.2.11) to (17.2.17), thereby obtaining directly the following seven corollaries.

Firstly a required notation for our base 10 number sequences.

Definition 17.1 *For a base n number system having digits $0,1,2,...,n-1$,*

$$D_n(list\ of\ digits\ included)$$

is the sequence of non-negative integers in base n having only digits from a list of digits included as digits of numbers in the sequence.

We illustrate this definition by noting the resulting sequences used in the seven corollaries below.

Corollary 17.1 *Consider the following non-negative integer base 10 sequences:*

$$D_{10}(0,1,2,3,4,5,6,7,8) = \{non\text{-}negative\ integers\ without\ 9\ as\ a\ digit\}.$$

This sequence is $0,1,2,3,4,5,6,7,8,10,11,12,13,14,15,16,17,18,20,21,22,23,24,$ $25,26,27,28,30,31,32,33,34,35,36,37,38,40,41,42,43,44,45,46,7,48,50,51,52,$ $53,54,55,56,57,58,\ldots$.

$$D_{10}(0,1,2) = \{non\text{-}negative\ integers\ with\ only\ 0,1\ and\ 2\ as\ digits\}.$$

This sequence is $0,1,2,10,11,12,20,21,22,100,101,102,110,111,112,120,121,$ $122,200,201,202,210,211,212,220,221,222,1000,1001,1002,1010,1011,1012,$ $1020,1021,1022,\ldots$.

$$D_{10}(0,3,6) = \{non\text{-}negative\ integers\ with\ only\ 0,3\ and\ 6\ as\ digits\}.$$

This sequence is $0,3,6,30,33,36,60,63,66,300,303,306,330,333,336,360,363,$

366, 600, 603, 606, 630, 633, 636, 660, 663, 666, 3000, 3003, 3006, 3030, 3033, 3036, 3060, 3063, 3066, ...

Then the following product of series holds.

$$\sum_{k\in D_{10}(0,1,2,3,4,5,6,7,8)} q^k = \left(\sum_{k\in D_{10}(0,1,2)} q^k\right)\left(\sum_{k\in D_{10}(0,3,6)} q^k\right). \tag{17.2.18}$$

This when written in full as power series is:

$$\begin{aligned}&1+q^1+q^2+q^3+q^4+q^5+q^6+q^7+q^8+q^{10}+q^{11}+q^{12}\\&\quad+q^{13}+q^{14}+q^{15}+q^{16}+q^{17}+q^{18}+q^{20}+q^{21}+\dots\end{aligned} \tag{17.2.19}$$

$$= \left(1+q^1+q^2+q^{10}+q^{11}+q^{12}+q^{20}+q^{21}+q^{22}+q^{100}+q^{101}+q^{102}+q^{110}+q^{111}+\dots\right)$$

$$\times\left(1+q^3+q^6+q^{30}+q^{33}+q^{36}+q^{60}+q^{63}+q^{66}+q^{300}+q^{303}+q^{306}+q^{330}+q^{333}+\dots\right).$$

Corollary 17.2

$$\sum_{k\in D_{10}(0,1,2,3,4,5,7,8,9)} q^k = \left(\sum_{k\in D_{10}(0,1,2)} q^k\right)\left(\sum_{k\in D_{10}(0,3,7)} q^k\right), \tag{17.2.20}$$

This written in full as power series is:

$$\begin{aligned}&1+q^1+q^2+q^3+q^4+q^5+q^7+q^8+q^9+q^{10}+q^{11}\\&\quad+q^{12}+q^{13}+q^{14}+q^{15}+q^{17}+q^{18}+q^{19}+q^{20}+q^{21}+\dots\end{aligned} \tag{17.2.21}$$

$$= \left(1+q^1+q^2+q^{10}+q^{11}+q^{12}+q^{20}+q^{21}+q^{22}+q^{100}+q^{101}+q^{102}+q^{110}+q^{111}+\dots\right)$$

$$\times\left(1+q^3+q^7+q^{30}+q^{33}+q^{37}+q^{70}+q^{73}+q^{77}+q^{300}+q^{303}+q^{307}+q^{330}+q^{333}+\dots\right).$$

Corollary 17.3

$$\sum_{k\in D_{10}(0,1,2,4,5,6,7,8,9)} q^k = \left(\sum_{k\in D_{10}(0,1,2)} q^k\right)\left(\sum_{k\in D_{10}(0,4,7)} q^k\right), \tag{17.2.22}$$

This written in full as power series is:

$$\begin{aligned}&1+q^1+q^2+q^4+q^5+q^6+q^7+q^8+q^9+q^{10}+q^{11}+q^{12}\\&\quad+q^{14}+q^{15}+q^{16}+q^{17}+q^{18}+q^{19}+q^{20}+q^{21}+\dots\end{aligned} \tag{17.2.23}$$

$$= \left(1+q^1+q^2+q^{10}+q^{11}+q^{12}+q^{20}+q^{21}+q^{22}+q^{100}+q^{101}+q^{102}+q^{110}+q^{111}+\dots\right)$$

$$\times\left(1+q^4+q^7+q^{40}+q^{44}+q^{47}+q^{70}+q^{74}+q^{77}+q^{400}+q^{404}+q^{407}+q^{440}+q^{444}+\dots\right).$$

Corollary 17.4

$$\sum_{k\in D_{10}(0,3,4,5,6,7,8,9)} q^k = \left(\sum_{k\in D_{10}(0,1)} q^k\right)\left(\sum_{k\in D_{10}(0,1,2,4,6,8)} (-q)^k\right) \tag{17.2.24}$$

$$= \left(\sum_{k\in D_{10}(0,1)} q^k\right)\left(\sum_{k\in D_{10}(0,1,2)} (-q)^k\right)\left(\sum_{k\in D_{10}(0,4,5,6)} q^k\right),$$

This written in full as power series is:

$$\begin{aligned}1+q^3+q^4+q^5+q^6+q^7+q^8+q^9+q^{30}+q^{33}+q^{34}+q^{35}\\ +q^{36}+q^{37}+q^{38}+q^{39}+q^{40}+q^{43}+q^{44}+q^{45}+\dots\end{aligned} \tag{17.2.25}$$

$$=(1+q^1+q^{10}+q^{11}+q^{100}+q^{101}+q^{111}+q^{1000}+q^{1001}+q^{1011}+q^{1111}+q^{10000}+q^{10001}+\dots)$$

$$\times(1-q^1+q^2+q^4+q^6+q^8+q^{10}-q^{11}+q^{12}+q^{14}+q^{16}+q^{18}+q^{20}-q^{21}+q^{22}+q^{24}+\dots)$$

$$=(1+q^1+q^{10}+q^{11}+q^{100}+q^{101}+q^{111}+q^{1000}+q^{1001}+q^{1011}+q^{1111}+q^{10000}+q^{10001}+\dots)$$
$$\times(1-q^1+q^2+q^{10}-q^{11}+q^{12}+q^{20}-q^{21}+q^{22}+q^{100}-q^{101}+q^{102}+q^{110}-q^{111}+q^{112}+\dots)$$

$$\times(1+q^4+q^5+q^6+q^{40}+q^{44}+q^{45}+q^{46}+q^{50}+q^{54}+q^{55}+q^{56}+q^{60}+q^{64}+q^{65}+q^{66}+\dots).$$

Corollary 17.5

$$\sum_{k\in D_{10}(0,1,4,5,6,7,8,9)} q^k = \left(\sum_{k\in D_{10}(0,1)} q^k\right)\left(\sum_{k\in D_{10}(0,4,6,8)} q^k\right) \tag{17.2.26}$$

This written in full as power series is:

$$\begin{aligned}1+q^1+q^4+q^5+q^6+q^7+q^8+q^9+q^{10}+q^{11}+q^{14}\\ +q^{15}+q^{16}+q^{17}+q^{18}+q^{19}+q^{40}+q^{41}+q^{44}+q^{45}+\dots\end{aligned} \tag{17.2.27}$$

$$=\left(1+q^1+q^{10}+q^{11}+q^{100}+q^{101}+q^{111}+q^{1000}+q^{1001}+q^{1011}+q^{1111}+q^{10000}+\dots\right)$$
$$\times\left(1+q^4+q^6+q^8+q^{40}+q^{44}+q^{46}+q^{48}+q^{60}+q^{64}+q^{66}+q^{68}+q^{80}+q^{84}+q^{86}+\dots\right)$$

Corollary 17.6

$$\sum_{k\in D_{10}(0,1,3,4,6,7,8,9)} q^k = \left(\sum_{k\in D_{10}(0,1)} q^k\right)\left(\sum_{k\in D_{10}(0,3,6,8)} q^k\right) \tag{17.2.28}$$

This written in full as power series is:

$$1+q^1+q^3+q^4+q^6+q^7+q^8+q^9+q^{10}+q^{11}+q^{13}$$
$$+q^{14}+q^{16}+q^{17}+q^{18}+q^{19}+q^{30}+q^{31}+q^{33}+q^{34}+\dots \quad (17.2.29)$$

$$=\left(1+q^1+q^{10}+q^{11}+q^{100}+q^{101}+q^{111}+q^{1000}+q^{1001}+q^{1011}+q^{1111}+q^{10000}+\dots\right)$$

$$\times\left(1+q^3+q^6+q^8+q^{30}+q^{33}+q^{36}+q^{38}+q^{60}+q^{63}+q^{66}+q^{68}+q^{80}+q^{83}+q^{86}+\dots\right)$$

Corollary 17.7

$$\sum_{k\in D_{10}(0,1,3,4,5,6,8,9)} q^k \quad (17.2.30)$$

$$=\left(\sum_{k\in D_{10}(0,1)} q^k\right)^3\left(\sum_{k\in D_{10}(0,1,2)} (-q)^k\right)\left(\sum_{k\in D_{10}(0,1,2,3,4)} (-q)^k\right)$$

This written in full as power series is:

$$1+q^1+q^3+q^4+q^5+q^6+q^8+q^9+q^{10}+q^{11}+q^{13}$$
$$+q^{14}+q^{15}+q^{16}+q^{18}+q^{19}+q^{30}+q^{31}+q^{33}+q^{34}+\dots \quad (17.2.31)$$

$$=\left(1+q^1+q^{10}+q^{11}+q^{100}+q^{101}+q^{111}+q^{1000}+q^{1001}+q^{1011}+q^{1111}+q^{10000}+\dots\right)^3$$

$$\times\left(1-q^1+q^2+q^{10}-q^{11}+q^{12}+q^{20}-q^{21}+q^{22}+q^{100}-q^{101}+q^{102}+q^{110}-q^{111}+\dots\right)$$

$$\times\left(1-q^1+q^2-q^3+q^4+q^{10}-q^{11}+q^{12}-q^{13}+q^{14}+q^{20}-q^{21}+q^{22}-q^{23}+q^{24}+\dots\right)$$

17.3 THE n-ARY INTEGER PARTITION SET OF CASES

We follow the previous section's train of thinking to enable us to derive an n-ary system of integer partitions using base n numbers, in full generality.

Recall our n-ary example infinite product, which underpins the base n number system,

$$\frac{1}{1-q} = \prod_{k=0}^{\infty}(1+q^{n^k}+q^{2.n^k}+q^{3.n^k}+\dots+q^{(n-1).n^k}) \quad (17.3.1)$$
$$= 1+q+q^2+q^3+q^4+q^5+q^6+\dots$$

The identity (17.3.1) when interpreted in plain language, as a n-ary or base n partition statement simply says

"Every positive integer can be represented using the digits $0,1,2,3,\ldots,n-1$ in the base n (n-ary) number system." As mentioned earlier, the decimal base 10 version of this is a fact assumed everyday by people all over the world.

Alternatively (17.3.1) can be interpreted as

"The number of integer n-ary distinct partitions of positive integer k in the form

$$k = a_1.n^0 + a_2.n^1 + a_3.n^2 + \ldots + a_{n-1}.n^k, \tag{17.3.2}$$

where each a_i is a non-negative integer less than n, is 1."

As in the previous section, using j_n to mean integer j in base n, an extensive search has not found a general theorem, nor the following interesting cases:

$$\prod_{k=0}^{\infty}(1+q^{2.n^k}+q^{3.n^k}) \quad for\ selected\ positive\ integer\ n \geq 4 \tag{17.3.3}$$

$$= 1+q^{2_n}+q^{3_n}+q^{20_n}+q^{22_n}+q^{23_n}+q^{30_n}+q^{32_n}+q^{33_n}+O(q^{200_n}),$$

$$\prod_{k=0}^{\infty}(1+q^{1.n^k}+q^{5.n^k}) \quad for\ selected\ positive\ integer\ n \geq 6 \tag{17.3.4}$$

$$= 1+q+q^{5_n}+q^{10_n}+q^{11_n}+q^{15_n}+q^{50_n}+q^{51_n}+q^{55_n}+O(q^{99_n}),$$

$$\prod_{k=0}^{\infty}(1+q^{5.n^k}+q^{7.n^k}) \quad for\ selected\ positive\ integer\ n \geq 8 \tag{17.3.5}$$

$$= 1+q^{5_n}+q^{7_n}+q^{50_n}+q^{55_n}+q^{57_n}+q^{70_n}+q^{75_n}+q^{77_n}+O(q^{486_n}),$$

$$\prod_{k=0}^{\infty}(1+q^{8.n^k}+q^{9.n^k}) \quad for\ selected\ positive\ integer\ n \geq 10 \tag{17.3.6}$$

$$= 1+q^{8_n}+q^{9_n}+q^{80_n}+q^{88_n}+q^{89_n}+q^{90_n}+q^{98_n}+q^{99_n}+O(q^{800_n}),$$

$$\prod_{k=0}^{\infty}(1+q^{1.n^k}+q^{2.n^k}) \quad for\ selected\ positive\ integer\ n \geq 3 \tag{17.3.7}$$

$$= 1+q+q^{2_n}+q^{10_n}+q^{11_n}+q^{12_n}+q^{20_n}+q^{21_n}+q^{22_n}+q^{100_n}+O(q^{101_n}),$$

where, as stated for each equation, the infinite product indices next to n^k are all less than n, and on the equation summation sides the indices of q are the base n unique **digits along with zeroes** formed from the product term digits a in $q^{a.n^k}$ included in every possible distinct combination for positive integers. Of course, choosing $n = 10$ in any of (17.3.3) to (17.3.7) allows us to use familiar decimal system integers.

The above five equations are cases of the

Theorem 17.3 ***The general n-ary distinct partition generating function.*** *If $a_1, a_2, \ldots, a_m$ are different integers with $1 \leq a_1 < a_2 < \ldots < a_m \leq n$ then*

$$\prod_{k=0}^{\infty}(1+q^{a_1.n^k}+q^{a_2.n^k}+\ldots+q^{a_m.n^k}) = 1+\sum_{k\in D_n(a_1,a_2,\ldots,a_m)} q^k \tag{17.3.8}$$

where the right side indices k are all base n positive integers with digits "$a_1, a_2, \ldots, a_m$ and/or zeroes in base n numbers" a subset of $\{0,1,2,\ldots,n-1\}$.

This theorem gives us new results on n-ary partitions quite easily. So we use the terminology

$$\mathrm{D}_n(\textit{list of digits included})$$

from the previous section replacing 10-ary with n-ary. We could at this point re-use the seven identities from the previous section, but this time we take other simply proven identities. So, consider the five equations:

$$1+q+q^2+q^3+q^4+0q^5+q^6=(1+q^2)(1+q+q^4), \tag{17.3.9}$$

$$1+q+q^2+q^3+0q^4+q^5+q^6=(1+q)(1+q^2+q^5), \tag{17.3.10}$$

$$1+q+q^2+0q^3+q^4+q^5+q^6=(1+q^4)(1+q+q^2), \tag{17.3.11}$$

$$1+q+0q^2+q^3+q^4+q^5+q^6=(1+q)(1+q^3+q^5), \tag{17.3.12}$$

$$1+0q+q^2+q^3+q^4+q^5+q^6=(1+q^2)(1+q^3+q^4). \tag{17.3.13}$$

These identities (17.3.9) to (17.3.13) are easy to verify on a calculating website or app. Applying the principles described in theorem 17.3 we see that our five above equations yield the following five corollaries.

Corollary 17.8 *For n a positive integer selected from $\{7,8,9,10,\ldots\}$,*

$$\sum_{k\in D_n(0,1,2,3,4,6)} q^k = \left(\sum_{k\in D_n(0,2)} q^k\right)\left(\sum_{k\in D_n(0,1,4)} q^k\right) \tag{17.3.14}$$

This written in full as power series is:

$$\begin{aligned}1&+q^{1_n}+q^{2_n}+q^{3_n}+q^{4_n}+q^{6_n}+q^{10_n}+q^{11_n}+q^{12_n}+q^{13_n}+q^{14_n}+q^{16_n}+q^{20_n}+q^{21_n}+q^{22_n}\\&+q^{23_n}+q^{24_n}+q^{26_n}+q^{30_n}+q^{31_n}+q^{32_n}+q^{33_n}+q^{34_n}+q^{36_n}+q^{40_n}+\ldots\end{aligned} \tag{17.3.15}$$

$$\begin{aligned}&=\left(1+q^{2_n}+q^{20_n}+q^{22_n}+q^{200_n}+q^{202_n}+q^{222_n}+q^{2000_n}+q^{2002_n}+q^{2022_n}+q^{2222_n}+\ldots\right)\\&\times\left(1+q^{1_n}+q^{4_n}+q^{10_n}+q^{11_n}+q^{14_n}+q^{40_n}+q^{41_n}+q^{44_n}+q^{400_n}+q^{401_n}+q^{404_n}+\ldots\right).\end{aligned}$$

Corollary 17.9 *For n a positive integer selected from $\{7,8,9,10,\ldots\}$,*

$$\sum_{k\in D_n(0,1,2,3,5,6)} q^k = \left(\sum_{k\in D_n(0,1)} q^k\right)\left(\sum_{k\in D_n(0,2,5)} q^k\right) \tag{17.3.16}$$

This written in full as power series is:

$$\begin{aligned}1&+q^{1_n}+q^{2_n}+q^{3_n}+q^{5_n}+q^{6_n}+q^{10_n}+q^{11_n}+q^{12_n}+q^{13_n}+q^{15_n}+q^{16_n}+q^{20_n}+q^{21_n}+q^{22_n}\\&+q^{23_n}+q^{25_n}+q^{26_n}+q^{30_n}+q^{31_n}+q^{32_n}+q^{33_n}+q^{35_n}+q^{36_n}+q^{40_n}+\ldots\end{aligned} \tag{17.3.17}$$

$$\begin{aligned}&=\left(1+q^{12_n}+q^{10_n}+q^{11_n}+q^{100_n}+q^{101_n}+q^{111_n}+q^{1000_n}+q^{1001_n}+q^{1011_n}+q^{1111_n}+\ldots\right)\\&\times\left(1+q^{2_n}+q^{5_n}+q^{20_n}+q^{22_n}+q^{25_n}+q^{50_n}+q^{52_n}+q^{55_n}+q^{200_n}+q^{202_n}+q^{205_n}+\ldots\right).\end{aligned}$$

Corollary 17.10 *For n a positive integer selected from $\{7,8,9,10,\ldots\}$,*

$$\sum_{k\in D_n(0,1,2,4,5,6)} q^k = \left(\sum_{k\in D_n(0,4)} q^k\right)\left(\sum_{k\in D_n(0,1,2)} q^k\right) \tag{17.3.18}$$

This written in full as power series is:

$$\begin{aligned} &1+q^{1_n}+q^{2_n}+q^{4_n}+q^{5_n}+q^{6_n}+q^{10_n}+q^{11_n}+q^{12_n}+q^{14_n}+q^{15_n}+q^{16_n}+q^{20_n}+q^{21_n}+q^{22_n} \\ &\qquad +q^{24_n}+q^{25_n}+q^{26_n}+q^{30_n}+q^{31_n}+q^{32_n}+q^{34_n}+q^{35_n}+q^{36_n}+q^{40_n}+\ldots \end{aligned} \tag{17.3.19}$$

$$\begin{aligned} &=\left(1+q^{4_n}+q^{40}+q^{44_n}+q^{400_n}+q^{404_n}+q^{444_n}+q^{4000_n}+q^{4004_n}+q^{4044_n}+q^{4444_n}+\ldots\right) \\ &\times\left(1+q^{1_n}+q^{2_n}+q^{10_n}+q^{11_n}+q^{12_n}+q^{20_n}+q^{21_n}+q^{22_n}+q^{100_n}+q^{101_n}+q^{102_n}+\ldots\right). \end{aligned}$$

Corollary 17.11 *For n a positive integer selected from $\{7,8,9,10,\ldots\}$,*

$$\sum_{k\in D_n(0,1,3,4,5,6)} q^k = \left(\sum_{k\in D_n(0,1)} q^k\right)\left(\sum_{k\in D_n(0,1,3,5)} q^k\right) \tag{17.3.20}$$

This written in full as power series is:

$$\begin{aligned} &1+q^{1_n}+q^{3_n}+q^{4_n}+q^{5_n}+q^{6_n}+q^{10_n}+q^{11_n}+q^{13_n}+q^{14_n}+q^{15_n}+q^{16_n}+q^{20_n}+q^{21_n}+q^{23_n} \\ &\qquad +q^{24_n}+q^{25_n}+q^{26_n}+q^{30_n}+q^{31_n}+q^{33_n}+q^{34_n}+q^{35_n}+q^{36_n}+q^{40_n}+\ldots \end{aligned} \tag{17.3.21}$$

$$\begin{aligned} &=\left(1+q^{1_n}+q^{10_n}+q^{11_n}+q^{100_n}+q^{101_n}+q^{111_n}+q^{1000_n}+q^{1001_n}+q^{1011_n}+q^{1111_n}+\ldots\right) \\ &\times\left(1+q^{1_n}+q^{3_n}+q^{5_n}+q^{10_n}+q^{11_n}+q^{13_n}+q^{15_n}+q^{20_n}+q^{21_n}+q^{23_n}+q^{25_n}+\ldots\right). \end{aligned}$$

Corollary 17.12 *For n a positive integer selected from $\{7,8,9,10,\ldots\}$,*

$$\sum_{k\in D_n(0,2,3,4,5,6)} q^k = \left(\sum_{k\in D_n(0,1)} q^k\right)\left(\sum_{k\in D_n(0,3,4)} q^k\right) \tag{17.3.22}$$

This written in full as power series is:

$$\begin{aligned} &1+q^{2_n}+q^{3_n}+q^{4_n}+q^{5_n}+q^{6_n}+q^{10_n}+q^{12_n}+q^{13_n}+q^{14_n}+q^{15_n}+q^{16_n}+q^{20_n}+q^{22_n}+q^{23_n} \\ &\qquad +q^{24_n}+q^{25_n}+q^{26_n}+q^{30_n}+q^{32_n}+q^{33_n}+q^{34_n}+q^{35_n}+q^{36_n}+q^{40_n}+\ldots \end{aligned} \tag{17.3.23}$$

$$\begin{aligned} &=\left(1+q^{1_n}+q^{10_n}+q^{11_n}+q^{100_n}+q^{101_n}+q^{111_n}+q^{1000_n}+q^{1001_n}+q^{1011_n}+q^{1111_n}+\ldots\right) \\ &\times\left(1+q^{3_n}+q^{4_n}+q^{30_n}+q^{33_n}+q^{34_n}+q^{40_n}+q^{43_n}+q^{44_n}+q^{300_n}+q^{303_n}+q^{304_n}+\ldots\right). \end{aligned}$$

17.4 THE BINARY VECTOR PARTITION SET OF CASES

In the examples that follow here we use the term $\gcd(j,k) = n$ for *integer n is the greatest common divisor of integers j and k* abbreviated to $(j,k) = n$. Recalling our analysis from earlier chapters, it is evident that when applying generating function forms to radial-from-origin regions over vector partitions, the theory of $\langle j,k\rangle$ where $(j,k) = 1$ simplifies many considerations. That is due to each vector $\langle j,k\rangle$ where $(j,k) = 1$ representing a minimal length arrow of a unique gradient.

$$\prod_{\substack{j,k\geq 1;\\(j,k)=1}} \frac{1}{1-x^j y^k} = \prod_{\substack{j,k\geq 1;\\(j,k)=1}} \left\{1+x^j y^k + x^{2j}y^{2k} + x^{3j}y^{3k} + \ldots\right\} = \sum_{a,b\geq 0} \mathbf{v}_2(a,b)x^a y^b \tag{17.4.1}$$

$$= \prod_{\substack{j,k\geq 1;\\(j,k)=1}} \prod_{h=0}^{\infty} \left[1+x^{j2^h}y^{k2^h}\right] = \prod_{\substack{j,k\geq 1; h\geq 0\\ \gcd(j,k)=2^h}} \left[1+x^{j2^h}y^{k2^h}\right] = \sum_{a,b\geq 0} \mathbf{w}_2(a,b)x^a y^b. \tag{17.4.2}$$

Note that the curly-bracketed term in (17.4.1) contributes vector parts of rational gradient k/j with $(j,k) = 1$ in the first 2D quadrant exactly one part per partition. So $\mathbf{v}_2(a,b)$ is the number of partitions of 2D first quadrant $\langle a,b\rangle$ into distinct vectors of positive rational gradient. In traditional integer partition language, this would also mean $\mathbf{v}_2(a,b)$ is the number of partitions into unrestricted Visible Point Vector (VPV) parts. Hence we have the

Theorem 17.4 *For 2D first quadrant vectors, the number of partitions of $\langle a,b\rangle$ into distinct vector directions of positive rational gradient is equal to the number of partitions of $\langle a,b\rangle$ into unrestricted Visible Point Vectors.*

Note that the square-bracketed term in (17.4.2) contributes vector parts of rational gradient j/k with $(j,k) = 2^h$ for $h = 0,1,2...$ in the first 2D quadrant exactly one part per partition. So $\mathbf{w}_2(a,b)$ is the number of partitions of $\langle a,b\rangle$ into vectors each part of form $2^h \langle j,k\rangle$ positive rational gradient .

So, in the present section, we consider using this VPV concept and applying a binary number 2^m for $m = 0,1,2,3...$ multiple to each term of the curly bracketed radial-from-origin generating function terms. So, we consider first

$$\prod_{\substack{j,k\geq 1;\\(j,k)=1}} \left(1+\Upsilon(x^j y^k)\right) = \prod_{\substack{j,k\geq 1;\\(j,k)=1}} \left\{1+x^j y^k + x^{2j}y^{2k} + x^{4j}y^{4k} + \ldots\right\} \tag{17.4.3}$$

$$= \sum_{a,b\geq 0} \mathbf{b}_2(a,b)x^a y^b.$$

where $\Upsilon(z) = z + z^2 + z^4 + z^8 + \ldots$, and $\mathbf{b}_2(a,b)$ is the number of partitions into vectors $\langle a,b\rangle$ where $\gcd(a,b)$ is a non-negative power of 2.

In closing this chapter we remark that the 10-ary identities given here are probably new to the literature as integer (or 1D) partition identities. It is clear that the principles involved apply on any radial from the origin line through 2D, 3D or nD vector space. Hence we could easily write new identities for nD vector partitions into vector parts that are on 2 or 3 or 4 and so on, radial-from-origin straight lines. Using the methods outlined in Chapter 12, we give next an example of applying theory from section 12.6 from chapter 12.

17.5 nD VECTOR PARTITIONS INTO PARTS: UNRESTRICTED BINARY PARTITIONS ON A LINE AND UNRESTRICTED TERNARY PARTITIONS ON ANOTHER LINE

From applying theory in Chapter 12, section 12.6, we can state the following

Theorem 17.5 *Take an nD lattice point vector of fixed non-negative integers* $\langle a_1, a_2, ..., a_n\rangle$ *(not all zero) where* $\gcd(a_1, a_2, ..., a_n)$*, and this defines a straight line from the origin through the n-space. Suppose then, that all lattice point vectors on that line are given by the notation*

$$\mathbb{L}_n(a_1, a_2, ..., a_n) = \{\langle a_1, a_2, ..., a_n\rangle, \langle 2a_1, 2a_2, ..., 2a_n\rangle, \langle 3a_1, 3a_2, ..., 3a_n\rangle, \ldots\}.$$

Similarly define for "n-tuple coprime" the lattice point vector $\langle b_1, b_2, ..., b_n\rangle$*; integer lattice point vectors on the line through the n-space, as follows:*

$$\mathbb{L}_n(b_1, b_2, ..., b_n) = \{\langle b_1, b_2, ..., b_n\rangle, \langle 2b_1, 2b_2, ..., 2b_n\rangle, \langle 3b_1, 3b_2, ..., 3b_n\rangle, \ldots\}.$$

Let $\mathbb{V}_n(A_1, A_2, ..., A_n)$ *be the number of vector partitions of the n-space vector* $\langle A_1, A_2, ..., A_n\rangle$ *with* $A_1, A_2, ..., A_{n-1}$ *each a non-negative integer and* A_n *a positive integer, into unrestricted binary parts from the line* $\mathbb{L}_n(a_1, a_2, ..., a_n)$*, and into unrestricted ternary parts from the line* $\mathbb{L}_n(b_1, b_2, ..., b_n)$*.*

Then $\mathbb{V}_n(A_1, A_2, ..., A_n)$ *is the coefficient of* ${q_1}^{A_1}{q_2}^{A_2}...{q_n}^{A_n}$ *in either side of the expression,*

$$\prod_{m=1}^{\infty} \frac{1}{(1-({q_1}^{a_1}{q_2}^{a_2}...{q_n}^{a_n})^{2^m})(1-({q_1}^{b_1}{q_2}^{b_2}...{q_n}^{b_n})^{3^m})} \tag{17.5.1}$$

$$= \left(\sum_{m\geq 0} b(m)({q_1}^{a_1}{q_2}^{a_2}\cdots {q_n}^{a_n})^m\right) \times \left(\sum_{m\geq 0} t(m)({q_1}^{b_1}{q_2}^{b_2}\cdots {q_n}^{b_n})^m\right),$$

where $b(m)$ *is the number of unrestricted binary partitions of* m*; and* $t(m)$ *is the number of unrestricted ternary partitions of* m*. That is,* $\mathbb{V}_n(A_1, A_2, ..., A_n)$ *can be found exactly as a combination of ordinary* $b(m)$ *and* $t(m)$ *integer partition functions.*

This theorem is in fact variation of theorem 12.5 from Chapter 12. In that earlier version, we used the unrestricted integer partition function $p(n)$ applied to the parts of the vector partition contributed by each separate line. In the cases described above in the theorem, the principles employed are the same, but apply to two lines instead of three lines, and we have used binary partitions for one line and ternary partitions for the other line.

Think About It...

Why are numbers beautiful? It's like asking why is Beethoven's Ninth Symphony beautiful. If you don't see why, someone can't tell you. I know numbers are beautiful. If they aren't beautiful, nothing is.

Paul Erdös
1913–1996

FURTHER READING

ALKAUSKAS, G. Generalization of the Rödseth-Gupta Theorem on Binary Partitions. *Lithuanian Mathematical Journal*, April 2003.

ALKAUSKAS, G. and MENDEL-STRASSE, G. Congruence Properties of the Function that Counts Compositions into Powers of 2. *Journal of Integer Sequences*, 13, 2010. Article 10.5.3.

CAMPBELL, G.B. A generalized formula of Hardy. *Int. J. Math. Math. Sci.*, 17(2):369-378, 1994.

CAMPBELL, G.B. Infinite products over hyperpyramid lattices. *Int. J. Math. and Math. Sci.*, 23(4):271-277, 2000.

CAMPBELL, G.B. Some n-space q-binomial theorem extensions and similar identities. arxiv preprint, June 2019. arXiv:1906.07526v1 [math.NT]: https://arxiv.org/abs/1906.07526.

CAMPBELL, G.B. and ZUJEV, A. The series that Ramanujan misunderstood. arxiv preprint, Oct 2016. arXiv:1610.03693v1 [math.NT], https://arxiv.org/abs/1610.03693v1.

CHEEMA, M.S. and MOTZKIN, T.S. Multipartitions and multipermutations. *Proc. Symp. Pure Math.*, 19:37-39, 1971.

CHURCHHOUSE, R.F. Congruence properties of the binary partition function. *Proc. Cambridge Philos. Soc.*, 66:371-376, 1969.

HARDY, G.H. *On certain oscillating series.* In *Collected Papers, volume VI*, pages 146-167. Clarendon Press, Oxford, 1974.

MAHLER, K. On a Special Functional Equation. J. London Math. Soc., 15:115-123, 1940.

MAHLER, K. On a Special Function. *J. Number Theory*, 12:20-26, 1980.

RÖDSETH, Ö. Some arithmetical properties of m-ary partitions. *Proc. Cambridge Philos. Soc.*, 68:447-453, 1970.

CHAPTER 18

Some Binary and n-ary Partition Analytic Formulas

18.1 HISTORY OF THE BINARY PARTITION OSCILLATING SERIES

In 1907 G.H. Hardy ([148]) published a version of the formula, for $|z| < 1$

$$\sum_{k=1}^{\infty}(z^{2^{-k}}-1)+\sum_{k=0}^{\infty} z^{2^k} = \frac{\log\log(1/z)}{\log 2} - \frac{\gamma}{\log 2} + \frac{1}{2} - \sum{}' \Gamma\left(\frac{2ki\pi}{\log 2}\right)\left(\frac{1}{\log(1/z)}\right)^{\frac{2ki\pi}{\log 2}}, \tag{18.1.1}$$

with the sum $\sum'$ over all nonzero integers k.

The $\sum'$ term is a complex periodic function of $\log(1/z)$ that oscillates between finite limits. From examining the right side of (18.1.1) it's the series $\sum_{k=0}^{\infty} z^{2^k}$ contributing to the oscillations. Many authors have considered (18.1.1) and it is easily proved by the application of the residue theorem. Equation (18.1.2) below, is derived either from (18.1.1) or by first principles using the residue theorem, or a Mellin inversion. Most of the results in this chapter depend on summations and products worked out by Campbell [86], but for a more recent version depicting oscillations see Campbell and Zujev [100]. However, many of the inferences from these given in the present chapter may not be in the extant literature.

Earlier we remarked about the identity for $|z| \neq 1$ or 0,

$$\left(\frac{1+z^{1/2}}{2}\right)\left(\frac{1+z^{1/4}}{2}\right)\left(\frac{1+z^{1/16}}{2}\right)\cdots\left(\frac{1+z^{2^{-n}}}{2}\right)\cdots = \frac{1-z}{\log(1/z)}.$$

Clearly this result is equivalent to, for $|z| \neq 1$ or 0,

$$\cdots\left(\frac{1+z^{2^{-n}}}{2}\right)\cdots\left(\frac{1+z^{1/4}}{2}\right)\left(\frac{1+z^{1/2}}{2}\right) \times (1+z)(1+z^2)\cdots(1+z^{2^{n-1}})\cdots \tag{18.1.2}$$

DOI: 10.1201/9781003174158-18

$$= \frac{1}{\log(1/z)}.$$

Define for the moment that the left side of (18.1.1) is $f(z)$. Then

$$\frac{f(z)}{1} - \frac{f(z^2)}{2} + \frac{f(z^3)}{3} - etc \tag{18.1.3}$$

is equal to

$$\cdots\left(\frac{1+z^{2^{-n}}}{2}\right)\cdots\left(\frac{1+z^{1/4}}{2}\right)\left(\frac{1+z^{1/2}}{2}\right)\times(1+z)(1+z^2)\cdots(1+z^{2^{n-1}})\cdots,$$

which implies that (18.1.2) can be derived from the transformation using (18.1.3). Note that the oscillating $\sum'$ rightmost in (18.1.1) cancels in the transform (18.1.3). Performing this transform to both sides of (18.1.1), an interesting implied result that comes out is the well-known sum

$$\sum_{k=1}^{\infty}(-1)^{k-1}\frac{\log(k)}{k} = \frac{1}{2}(\log 2)^2 - \gamma\log 2 \simeq -0.159868903742430971 76.$$

18.2 SOME LACUNARY SERIES AND PRODUCTS FOR n-ARY PARTITIONS

The following formulas were given in Campbell [86], and will be applied subsequently in this chapter. They can be proven by applying a Mellin inversion, or from first principles using the residue theorem.

Firstly, we state the n-ary version of the first sum due to Hardy in the previous section,

Theorem 18.1 ***An n-ary Bilateral Oscillating Series.*** *If $\Re n > 1$, and $\Re t > 0$ with $z = e^{-t}$,*

$$\sum_{k=1}^{\infty}(z^{n^{-k}}-1) + \sum_{k=0}^{\infty} z^{n^k} = \frac{\log\log(1/z)}{\log n} - \frac{\gamma}{\log n} + \frac{1}{2} - \sum{}'\Gamma\left(\frac{2ki\pi}{\log n}\right)\left(\frac{1}{\log(1/z)}\right)^{\frac{2ki\pi}{\log n}}, \tag{18.2.1}$$

with the sum $\sum'$ over all nonzero integers k.

Next, we state the result

Theorem 18.2 ***Another n-ary Bilateral Oscillating Series.*** *If $\Re z < 1$, $\Re m > 0$, $\Re n > 1$,*

$$\sum_{k=-\infty}^{\infty} n^{km} z^{n^k} = \frac{1}{\log n}\left(\frac{1}{\log(1/z)}\right)^m \tag{18.2.2}$$

$$\times\left\{\Gamma(m) - \sum{}'\Gamma\left(m+\frac{2ki\pi}{\log n}\right)\left(\frac{1}{\log(1/z)}\right)^{\frac{2ki\pi}{\log n}}\right\},$$

with the sum $\sum'$ over all nonzero integers k.

As a third and more elaborate follow-up to the two theorems above we give the following Faulhaber-type summation, also due to Campbell [86].

Theorem 18.3 ***An n-ary Faulhaber type Oscillating Series.*** *For non-negative integers* m, $\Re z > 0$, $\Re n > 1$,

$$\sum_{k=1}^{\infty}(-k)^m(z^{n^{-k}}-1)+\sum_{k=0}^{\infty}k^m z^{n^k} \tag{18.2.3}$$

$$= \frac{1}{(m+1)(\log a)^{m+1}}\sum_{j=0}^{m+1}\binom{m+1}{j}\Gamma^{(m+1-j)}(1)\left(\log\left(\frac{1}{\log(1/z)}\right)\right)^j$$

$$-\frac{1}{m+1}B_{m+1}$$

$$-\frac{1}{(\log a)^{m+1}}\sum{}'\int_0^{\infty}\left(\log\left(\frac{t}{\log(1/z)}\right)\right)^m\left(\frac{t}{\log(1/z)}\right)^{\frac{2ki\pi}{\log n}}\frac{dt}{te^t},$$

where $\Gamma^{(j)}$ *is the* j*-th derivative of the gamma function,* B_m *the* m*-th Bernoulli number,* $\sum'$ *is the sum over all nonzero integers* k*; and* $k^m := 1$ *at* $k = m = 0$.

There are four features of this theorem worth noting at this point:

Feature 1) The derivatives at unity of the gamma function $\Gamma^{(m+1-j)}(1)$ are easily found from the expansion valid for $\Re z < 1$, (see Abramowitz and Stegun [1, pp259, 6.3.14]),

$$\frac{\Gamma'(1+z)}{\Gamma(1+z)} = -\gamma+\zeta(2)z-\zeta(3)z^2+\zeta(4)z^3-\ldots,$$

so that $\Gamma'(1) = -\gamma$, $\Gamma^{(2)}(1) = \gamma^2+\frac{1}{6}\pi^2$, $\Gamma^{(3)}(1) = -\gamma^2-\frac{1}{2}\gamma\pi^2-2\zeta(3)$, $\Gamma^{(4)}(1) = 8\gamma\zeta(3)+\gamma^4+\gamma^2\pi^2+\frac{3}{20}\pi^4$, and so on; all easily verified on a calculation device.

Feature 2) The relationship of the last integral in theorem 18.3 to the gamma function is obviated by realizing

$$\int_0^{\infty}\left(\log\left(\frac{t}{\log(1/z)}\right)\right)^m\left(\frac{t}{\log(1/z)}\right)^{\frac{2ki\pi}{\log n}}\frac{dt}{te^t}$$

$$=\left(\frac{t}{\log(1/z)}\right)^{\frac{2ki\pi}{\log n}}\sum_{j=0}^{m}\binom{m}{j}\left(\frac{1}{\log(1/z)}\right)^j\Gamma^{(m-j)}\left(\frac{2ki\pi}{\log n}\right).$$

Taking this into account, the summation

$$\sum{}'\int_0^{\infty}\left(\log\left(\frac{t}{\log(1/z)}\right)\right)^m\left(\frac{t}{\log(1/z)}\right)^{\frac{2ki\pi}{\log n}}\frac{dt}{te^t}, \tag{18.2.4}$$

is seen to converge bilaterally, owing to the fact [1, pp 257, 6.1.45] that for fixed m and $1+\delta < \Re t < \Re n\,(\delta > 0)$,

$$\left|\Gamma^{(m)}\left(1+\frac{2ki\pi}{\log t}\right)\right| \sim \exp\left(-\frac{1}{2}\pi\left|\frac{2k\pi}{\log t}\right|\right)\sum_{j=0}^{m}A(m,j)\left|\frac{2k\pi}{\log t}\right|^{\frac{1}{2}-j},$$

where

$$A(m,j) = \sqrt{2\pi}\binom{m}{j}\frac{\Gamma\left(\frac{1}{2}+j\right)}{\Gamma\left(\frac{1}{2}\right)}\left(-\frac{1}{2}\pi\right)^{m-j}.$$

So, then we have

$$A(1,0) = -\sqrt{\frac{\pi^3}{2}}, \quad A(1,1) = \sqrt{\frac{\pi}{2}};$$

$$A(2,0) = \frac{1}{2}\sqrt{\frac{\pi^5}{2}}, \quad A(2,1) = -\sqrt{\frac{\pi^3}{2}}, \quad A(2,2) = \frac{3}{2}\sqrt{\frac{\pi}{2}};$$

and so on.

Feature 3) Hardy's method of proving theorem 18.1 relied on a simple Mellin inversion applied to the left side series. Such a straightforward application designed to arrive at theorem 18.3 would involve the rather non-trivial evaluation of residues leading to

$$\frac{1}{2\pi i}\int_{-c-i\infty}^{-c+i\infty}\Gamma(-u)x^u\left\{\sum_{j=0}^{m}\mathbf{S}_m^{(j)}\frac{a^{uj}j!}{(1-a^u)^{j+1}}\right\}du = \sum_{k=1}^{\infty}(-k)^m\left(1-e^{-a^{-k}x}\right)$$

plus the additional terms in the last three equation terms lines of theorem 18.3, where also $\mathbf{S}_m^{(j)}$ are Stirling numbers of the second kind.

Feature 4) It can be seen from the above, that (18.2.4) is a sum of $m+1$ oscillating functions, each multiplied by a power of $\log x$. Each oscillating function is of order $o(1)$ and oscillates between finite limits. In Hardy's original paper [148, Vol. VI, ibid., 699-716] estimates were given for these limits in terms of n. In Mahler [195] an elementary method for obtaining such estimates was also given. However, if $n=2$ in (18.2.1) it is easy to use any online calculator to establish that the oscillatory term only enters in after about the fourth decimal place, so for $n=2$, $o(0.001)$ seems a conservative estimate for the oscillatory function under $\sum'$.

18.3 BINARY AND n-ARY VERSIONS OF PRODUCTS FOR DISTINCT PARTITIONS

Now, from the series given in the previous section, we can derive the following infinite products, each of which includes generating functions for integer n-ary partitions, and so include the integer binary partitions. From there we can "build" results on n-ary component vector partitions, such as the 2D and 3D relations we give in other chapters here.

Theorem 18.4 ***The n-ary Distinct Partitions Product.*** *If* $\Re n > 1$, *and* $\Re t > 0$ *with* $z = e^{-t}$,

$$\left(\prod_{k=1}^{\infty}\frac{1+z^{n^{-k}}}{2}\right)\times\left(\prod_{k=0}^{\infty}\left(1+z^{n^k}\right)\right) = \left(\frac{n}{2}\right)^{\frac{1}{2}\frac{\log 2}{\log n}}\left(\frac{1}{\log(1/z)}\right)^{\frac{\log 2}{\log n}} \qquad (18.3.1)$$

$$\times \exp\left\{\frac{-1}{\log n}\sum{}'\Gamma\left(\frac{2ki\pi}{\log n}\right)\zeta\left(1+\frac{2ki\pi}{\log n}\right)\left(1-2^{\frac{-2ki\pi}{\log n}}\right)\left(\frac{1}{\log(1/z)}\right)^{\frac{2ki\pi}{\log n}}\right\}$$

with the sum $\sum'$ *over all nonzero integers* k.

We note that if n is a positive integer greater than 1, $\prod_{k=0}^{\infty}(1+z^{n^k})$ is the generating function for distinct n-ary integer partitions. That is, we can define

$$\mathbf{D}_n(z) = \prod_{k=0}^{\infty}(1+z^{n^k}) = \sum_{k=0}^{\infty}\mathbf{d}_n(k)z^k.$$

Hence, in particular, $\mathbf{D}_2(z) = \frac{1}{1-z}$, and $\mathbf{d}_2(k) = 1$. Here is the n-ary version for $\mathbf{d}_n(k)$.

Theorem 18.5 ***n-ary Distinct Partitions formula.*** *For* $\{n \geq 2\} \subset \mathbb{N}$, *the number of partitions of positive integer* k *into distinct* n*-th powers is*

$$\mathbf{d}_n(k) = \begin{cases} 1, & \textit{for } k \textit{ base } n \textit{ only in digits } 0\textit{s and } 1\textit{s, with at least one } 1, \\ 0, & \textit{otherwise.} \end{cases}$$

Also, the particular case of (18.3.1) with $n = 2$ yields the familiar binary product

$$\left(\prod_{k=1}^{\infty}\frac{1+z^{2^{-k}}}{2}\right) \times \left(\prod_{k=0}^{\infty}\left(1+z^{2^k}\right)\right) = \frac{1}{\log(1/z)}.$$

Other cases of (18.3.1) are:

Corollary 18.1 ***The 3-ary (Ternary) Distinct Partitions Product.*** *If* $\Re t > 0$ *with* $z = e^{-t}$,

$$\left(\prod_{k=1}^{\infty}\frac{1+z^{3^{-k}}}{2}\right) \times \left(\prod_{k=0}^{\infty}\left(1+z^{3^k}\right)\right) = \left(\frac{3}{2}\right)^{\frac{1}{2}\frac{\log 2}{\log 3}}\left(\frac{1}{\log(1/z)}\right)^{\frac{\log 2}{\log 3}} \tag{18.3.2}$$

$$\times \exp\left\{\frac{-1}{\log 3}\sum{}'\Gamma\left(\frac{2ki\pi}{\log 3}\right)\zeta\left(1+\frac{2ki\pi}{\log 3}\right)\left(1-2^{\frac{-2ki\pi}{\log 3}}\right)\left(\frac{1}{\log(1/z)}\right)^{\frac{2ki\pi}{\log 3}}\right\}$$

with the sum $\sum'$ *over all nonzero integers* k.

Corollary 18.2 ***The 4-ary (Base 4) Distinct Partitions Product.*** *If* $\Re t > 0$ *with* $z = e^{-t}$,

$$\left(\prod_{k=1}^{\infty}\frac{1+z^{4^{-k}}}{2}\right) \times \left(\prod_{k=0}^{\infty}\left(1+z^{4^k}\right)\right) = \sqrt[4]{2} \times \sqrt{\frac{1}{\log(1/z)}} \tag{18.3.3}$$

$$\times \exp\left\{\frac{-1}{2\log 2}\sum{}'\Gamma\left(\frac{ki\pi}{\log 2}\right)\zeta\left(1+\frac{ki\pi}{\log 2}\right)\left(1-2^{\frac{-ki\pi}{\log 2}}\right)\left(\frac{1}{\log(1/z)}\right)^{\frac{ki\pi}{\log 2}}\right\}$$

with the sum $\sum'$ *over all nonzero integers* k.

For a proof of (18.3.1) see Campbell [86].

Theorem 18.6 ***n-ary partitions into no more than k terms of kind*** n^k**.** *If* $\Re n > 1$, *and* $\Re t > 0$ *with* $z = e^{-t}$,

$$\left(\prod_{k=1}^{\infty}\left(\frac{1+z^{n^{-k}}}{2}\right)^{(-k)^m}\right) \times \left(\prod_{k=0}^{\infty}\left(1+z^{n^k}\right)^{k^m}\right) = \exp\left(\mathbf{G}_1(z) + \sum{}' \mathbf{G}_2(z)\right) \quad (18.3.4)$$

where $\sum'$ *over all nonzero integers k, and* $k := 1$ *at* $k = m = 0$,

$$\mathbf{G}_1(z) = \frac{-1}{m+1}B_{m+1}\log 2 + \left(\frac{1}{\log n}\right)^{m+1}$$
$$\times \sum (m; m_1, m_2, m_3)\Gamma^{(m_1)}(1)\eta^{(m_2)}(1)\left(\log\frac{1}{\log(1/z)}\right)^{m_3} \quad (18.3.5)$$
$$\mathbf{G}_2(z) = \left(\frac{1}{\log n}\right)^{m+1}\int_0^{\infty}\left\{\sum_{r=1}^{\infty}\frac{(-1)^r(\log(\frac{t}{r\log(1/z)}))^m}{r^{1+\frac{2ki\pi}{\log n}}}\right\}\left(\frac{t}{\log(1/z)}\right)^{\frac{2ki\pi}{\log n}}\frac{dt}{te^t},$$

where the sum in $\mathbf{G}_1(z)$ *is over* $m = m_1 + m_2 + m_3$ *in non-negative integers and*

$$(m; m_1, m_2, m_3) = \frac{m!}{m_1!m_2!m_3!}, \quad (18.3.6)$$

and also

$$\eta^{(m)}(1) = \sum_{r=1}^{\infty}(-1)^{r+1}\frac{1}{r}\left(\log\left(\frac{1}{r}\right)\right). \quad (18.3.7)$$

Related to theorem 18.2 is the further result

Theorem 18.7 ***The n-ary Unrestricted Partitions Product.*** *If* $\Re n > 1$, *and* $\Re t > 0$ *with* $z = e^{-t}$,

$$\left(\prod_{k=1}^{\infty}\frac{\log(1/z)}{n^k(1-z^{n^{-k}})}\right) \times \left(\prod_{k=0}^{\infty}\left(\frac{1}{1-z^{n^k}}\right)\right) = \left(\frac{n}{2}\right)^{\frac{1}{2}\frac{\log 2}{\log n}}\left(\frac{1}{\log(1/z)}\right)^{\frac{\log 2}{\log n}} \quad (18.3.8)$$
$$\times \exp\left\{\frac{-1}{\log n}\sum{}'\Gamma\left(\frac{2ki\pi}{\log n}\right)\zeta\left(1+\frac{2ki\pi}{\log n}\right)\left(\frac{1}{\log(1/z)}\right)^{\frac{2ki\pi}{\log n}}\right\}$$

with the sum $\sum'$ *over all nonzero integers k.*

Think About It...

I believe that numbers and functions of Analysis are not the arbitrary result of our minds;

I think that they exist outside of us, with the same character of necessity as the things of objective reality, and we meet them or discover them, and study them, as do the physicists, the chemists and the zoologists.

Charles Hermite

French Mathematician, 1822–1901

FURTHER READING

ABRAMOWITZ, M. and STEGUN, I. *Handbook of Mathematical Functions.* Dover Publications Inc., New York, 1972. 10th Printing, 1972, with corrections.

CAMPBELL, G.B. A generalized formula of Hardy. *Int. J. Math. Math. Sci.*, 17(2):369–378, 1994.

CAMPBELL, G.B. and ZUJEV, A. The series that Ramanujan misunderstood. arxiv preprint, Oct 2016. arXiv:1610.03693v1 [math.NT], https://arxiv.org/abs/1610.03693v1.

HARDY, G.H. *Ramanujan.* Cambridge University Press, London and New York, 1940. (reprinted by Chelsea, New York).

MAHLER, K. On a Special Function. *J. Number Theory*, 12:20–26, 1980.

IV

Visible Point Vector Partition Theory

CHAPTER 19

Features of the Visible Lattice Points

19.1 VISIBLE POINTS AND THE gcd FUNCTION

BEFORE ENTERING further into theory involving visible lattice points and partitions concerning Visible Point Vectors (VPVs) it seems appropriate to present a few known facts about these concepts. The mathematical literature has a small but diverse range of references on this topic. The highly regarded 1976 Number Theory book by Tom Apostol [32] has an image of the 2D visible lattice points depicted on its front cover, and proofs of many of the facts presented in the present chapter. The 1992 paper by Mosseri et al. [203] and the work by Baake et al. [39, 40] contain proofs of many of the facts listed below. The more recent 2015 paper by Baake and Huck [38] covers important features of the Ergodic properties of visible lattice points. The papers by Campbell in 1994, [87] and [88], introduce the concept of VPV Identities. The reader may also follow a connection between the visible lattice points and the famous Extended Lindelöf Hypothesis that is related to the Riemann Hypothesis. (See the 2017 paper by Takeda [293])

In the 2D first quadrant, we have the familiar grid depicted below for the visible points $\bullet$ and invisible points $\circ$, from the viewpoint of the origin $\mathbf{O}$.

$\vdots$	$\vdots$	$\vdots$	$\vdots$	$\vdots$	$\vdots$	$\vdots$	$\vdots$	$\vdots$	$\vdots$	$\vdots$	$\vdots\cdots$
9	$\circ$	$\bullet$	$\bullet$	$\circ$	$\bullet$	$\bullet$	$\circ$	$\bullet$	$\bullet$	$\circ$	$\cdots$
8	$\circ$	$\bullet$	$\circ$	$\bullet$	$\circ$	$\bullet$	$\circ$	$\bullet$	$\circ$	$\bullet$	$\cdots$
7	$\circ$	$\bullet$	$\bullet$	$\bullet$	$\bullet$	$\bullet$	$\bullet$	$\circ$	$\bullet$	$\bullet$	$\cdots$
6	$\circ$	$\bullet$	$\circ$	$\circ$	$\circ$	$\bullet$	$\circ$	$\bullet$	$\circ$	$\circ$	$\cdots$
5	$\circ$	$\bullet$	$\bullet$	$\bullet$	$\bullet$	$\circ$	$\bullet$	$\bullet$	$\bullet$	$\bullet$	$\cdots$
4	$\circ$	$\bullet$	$\circ$	$\bullet$	$\circ$	$\bullet$	$\circ$	$\bullet$	$\circ$	$\bullet$	$\cdots$
3	$\circ$	$\bullet$	$\bullet$	$\circ$	$\bullet$	$\bullet$	$\circ$	$\bullet$	$\bullet$	$\circ$	$\cdots$
2	$\circ$	$\bullet$	$\circ$	$\bullet$	$\circ$	$\bullet$	$\circ$	$\bullet$	$\circ$	$\bullet$	$\cdots$
1	$\bullet$	$\bullet$	$\bullet$	$\bullet$	$\bullet$	$\bullet$	$\bullet$	$\bullet$	$\bullet$	$\bullet$	$\cdots$
0	$\mathbf{O}$	$\bullet$	$\circ$	$\circ$	$\circ$	$\circ$	$\circ$	$\circ$	$\circ$	$\circ$	$\cdots$
y/x	0	1	2	3	4	5	6	7	8	9	$\cdots$

(19.1.1)

DOI: 10.1201/9781003174158-19

While in the 2D grid above we show visible and invisible points in the first quadrant extended to each axis, on occasion we may define visible and invisible points in radial-from-origin extended sectors. In this grid, an arrow drawn from origin **O** to a bullet point corresponds to a Visible Point Vector (VPV).

Next, we depict the 3D cube infinitely extended lattice layers (first hyperquadrant)- visible and invisible points

layer $(x, y, 2)$; $z = 2 \quad \mapsto$

⋮	⋮	⋮	⋮	⋮	⋮	⋮	⋮	⋮	⋮	⋮	⋰⋯
9	●	●	●	●	●	●	●	●	●	●	⋯
8	○	●	○	●	○	●	○	●	○	●	⋯
7	●	●	●	●	●	●	●	●	●	●	⋯
6	○	●	○	○	○	●	○	●	○	○	⋯
5	●	●	●	●	●	●	●	●	●	●	⋯
4	○	●	○	●	○	●	○	●	○	●	⋯
3	●	●	●	●	●	●	●	●	●	●	⋯
2	○	●	○	●	○	●	○	●	○	●	⋯
1	●	●	●	●	●	●	●	●	●	●	⋯
0	○	●	○	●	○	●	○	●	○	●	⋯
y/x	0	1	2	3	4	5	6	7	8	9	⋯

layer $(x, y, 1)$; $z = 1 \quad \mapsto$

⋮	⋮	⋮	⋮	⋮	⋮	⋮	⋮	⋮	⋮	⋮	⋰⋯
9	●	●	●	●	●	●	●	●	●	●	⋯
8	●	●	●	●	●	●	●	●	●	●	⋯
7	●	●	●	●	●	●	●	●	●	●	⋯
6	●	●	●	●	●	●	●	●	●	●	⋯
5	●	●	●	●	●	●	●	●	●	●	⋯
4	●	●	●	●	●	●	●	●	●	●	⋯
3	●	●	●	●	●	●	●	●	●	●	⋯
2	●	●	●	●	●	●	●	●	●	●	⋯
1	●	●	●	●	●	●	●	●	●	●	⋯
0	●	●	●	●	●	●	●	●	●	●	⋯
y/x	0	1	2	3	4	5	6	7	8	9	⋯

layer $(x, y, 0)$; $z = 0 \quad \mapsto$

⋮	⋮	⋮	⋮	⋮	⋮	⋮	⋮	⋮	⋮	⋮	⋰⋯
9	○	●	●	○	●	●	○	●	●	○	⋯
8	○	●	○	●	○	●	○	●	○	●	⋯
7	○	●	●	●	●	●	●	○	●	●	⋯
6	○	●	○	○	○	●	○	●	○	○	⋯
5	○	●	●	●	●	○	●	●	●	●	⋯
4	○	●	○	●	○	●	○	●	○	●	⋯
3	○	●	●	○	●	●	○	●	●	○	⋯
2	○	●	○	●	○	●	○	●	○	●	⋯
1	●	●	●	●	●	●	●	●	●	●	⋯
0	**O**	●	○	○	○	○	○	○	○	○	⋯
y/x	0	1	2	3	4	5	6	7	8	9	⋯

In the above arrays, we depict 2D layers of the 3D grid, showing visible and invisible points for each layer of $z=0, z=1, z=2$ and so on in the first extended cube hyperquadrant including each axis x, y and z.

19.2 FACTS ABOUT VISIBLE POINTS

The following statements are known to be true.

Fact 1. A *Visible from the Origin Lattice Point* (a,b) in 2D with integers a and b is represented by the vector $\langle a,b \rangle$ where $\gcd(a,b)=1$. That is a and b are coprime. Any lattice point $\langle c,d \rangle$ in the 2D integer lattice having $\gcd(c,d)>1$ is said to be *Invisible from the Origin* point $\langle 0,0 \rangle$.

Fact 2. A Visible from the Origin Lattice Point (a,b,c) in 3D with integers a, b and c is represented by the vector $\langle a,b,c \rangle$ where $\gcd(a,b,c)=1$. That is a, b and c are coprime as a triple or 3-tuple. Any lattice point $\langle d,e,f \rangle$ in the 2D integer lattice having $\gcd(d,e,f)>1$ is said to be *Invisible from the Origin* point $\langle 0,0,0 \rangle$.

Fact 3. A Visible Lattice Point from the Origin in nD space is represented by the vector $< a_1, a_2, \ldots, a_n >$ where $\gcd(a_1, a_2, \ldots, a_n)=1$. That is $a_1, a_2, \ldots$, and a_n are coprime as an n-tuple.

Fact 4. The probability that a point in the first quadrant 2D integer lattice is a Visible from the Origin Point, is $1/\zeta(2)=6/\pi^2$.

Fact 5. The probability that a point in the first extended cube hyperquadrant 3D lattice is a Visible from the Origin Point is $1/\zeta(3)$; the reciprocal of Apery's constant.

Fact 6. The probability that a point in the nD lattice is a Visible Point from the Origin is $1/\zeta(n)$; the reciprocal of the Riemann zeta function of n. (First proved by Lehmer [182] in 1900.)

Fact 7. In the 2D lattice, the first invisible 2×2 square with closest corner to origin is $(14,20),(14,21),(15,20),(15,21)$ (See Gosper and Schroeppel 1972 [145] and Amer. Math. Monthly, May 1971, pp.487)

Fact 8. In the 2D lattice, the first invisible 3×3 filled square with closest corner to origin is $(104,6200)$, $(105,6200)$, $(105,6200)$, $(104,6201)$, $(105,6201)$, $(105,6201)$, $(104,6202)$, $(105,6202)$, $(105,6202)$, (See Gosper and Schroeppel 1972 [145] and Amer. Math. Monthly, May 1971, p487) Verification: $\{\gcd(104,6200)$, $\gcd(105,6200)$, $\gcd(105,6200)$, $\gcd(104,6201)$, $\gcd(105,6201)$, $\gcd(105,6201)$, $\gcd(104,6202)$, $\gcd(105,6202)$, $\gcd(105,6202)\}$ $=\{8,5,5,13,3,3,2,7,7\}$.

Fact 9. In the 2D integer point lattice, a straight line through the origin with gradient $\sqrt{2}$ with equation $y=\sqrt{2}x$, extends forever lengthwise from the origin; and does not ever intersect with an integer lattice point.

Fact 10. In the 3D integer point lattice, a straight line through the origin with component gradients $\sqrt{2}$ implying equation $z = \sqrt{2}y = \sqrt{2}x$, extends forever lengthwise from the origin; and does not ever intersect with an integer lattice point.

Fact 11. The facts 9 and 10 imply any line radial from the origin of irrational gradient has no visible point on it.

19.3 THE FAREY FRACTIONS RELATION TO THE 2D VISIBLE POINTS

The Farey fraction sequence of order n denoted F_n, is the sequence of completely reduced fractions, either between 0 and 1, or without this restriction, which when in lowest terms have denominators less than or equal to n, arranged in order of increasing size. Hence, there is a correspondence between the Farey sequence and the set of 2D Visible Points in the lower sector between, say, the line $y = x$ and the x axis.

With the restricted definition, each Farey fraction sequence starts at the value 0, given by the fraction $\frac{0}{1}$, and ends with the value 1, given by the fraction $\frac{1}{1}$.

A Farey fraction sequence is sometimes called a Farey series, which is not strictly correct, because it is not a sum.

The Farey fraction sequences of orders 1 to 8 are:

$$F_1 = \left\{\frac{0}{1}, \frac{1}{1}\right\}$$

$$F_2 = \left\{\frac{0}{1}, \frac{1}{2}, \frac{1}{1}\right\}$$

$$F_3 = \left\{\frac{0}{1}, \frac{1}{3}, \frac{1}{2}, \frac{2}{3}, \frac{1}{1}\right\}$$

$$F_4 = \left\{\frac{0}{1}, \frac{1}{4}, \frac{1}{3}, \frac{1}{2}, \frac{2}{3}, \frac{3}{4}, \frac{1}{1}\right\}$$

$$F_5 = \left\{\frac{0}{1}, \frac{1}{5}, \frac{1}{4}, \frac{1}{3}, \frac{2}{5}, \frac{1}{2}, \frac{3}{5}, \frac{2}{3}, \frac{3}{4}, \frac{4}{5}, \frac{1}{1}\right\}$$

$$F_6 = \left\{\frac{0}{1}, \frac{1}{6}, \frac{1}{5}, \frac{1}{4}, \frac{1}{3}, \frac{2}{5}, \frac{1}{2}, \frac{3}{5}, \frac{2}{3}, \frac{3}{4}, \frac{4}{5}, \frac{5}{6}, \frac{1}{1}\right\}$$

$$F_7 = \left\{\frac{0}{1}, \frac{1}{7}, \frac{1}{6}, \frac{1}{5}, \frac{1}{4}, \frac{2}{7}, \frac{1}{3}, \frac{2}{5}, \frac{3}{7}, \frac{1}{2}, \frac{4}{7}, \frac{3}{5}, \frac{2}{3}, \frac{5}{7}, \frac{3}{4}, \frac{4}{5}, \frac{5}{6}, \frac{6}{7}, \frac{1}{1}\right\}$$

$$F_8 = \left\{\frac{0}{1}, \frac{1}{8}, \frac{1}{7}, \frac{1}{6}, \frac{1}{5}, \frac{1}{4}, \frac{2}{7}, \frac{1}{3}, \frac{3}{8}, \frac{2}{5}, \frac{3}{7}, \frac{1}{2}, \frac{4}{7}, \frac{3}{5}, \frac{5}{8}, \frac{2}{3}, \frac{5}{7}, \frac{3}{4}, \frac{4}{5}, \frac{5}{6}, \frac{6}{7}, \frac{7}{8}, \frac{1}{1}\right\}$$

Clearly, the numerator plotted against the denominator for each term of F_n results in the visible points in the 2D first quadrant triangle patch defined by

$$\{(x, y) \,:\, 0 \leq x < y \leq n \,\text{and}\, \gcd(x.y) = 1\}.$$

19.3.1 Farey Sunburst

Plotting the numerators versus the denominators of a Farey sequence gives a shape like the one in Figure 19.1 for F_7.

Reflecting this shape around the diagonal and main axes generates the Farey sunburst, shown here. The Farey sunburst of order n connects the 2D visible integer grid points from the origin in the square of side $2n$, centered at the origin. Using Pick's theorem, the area of the sunburst is $4(|F_n|-1)$, where $|F_n|$ is the number of fractions in F_n.

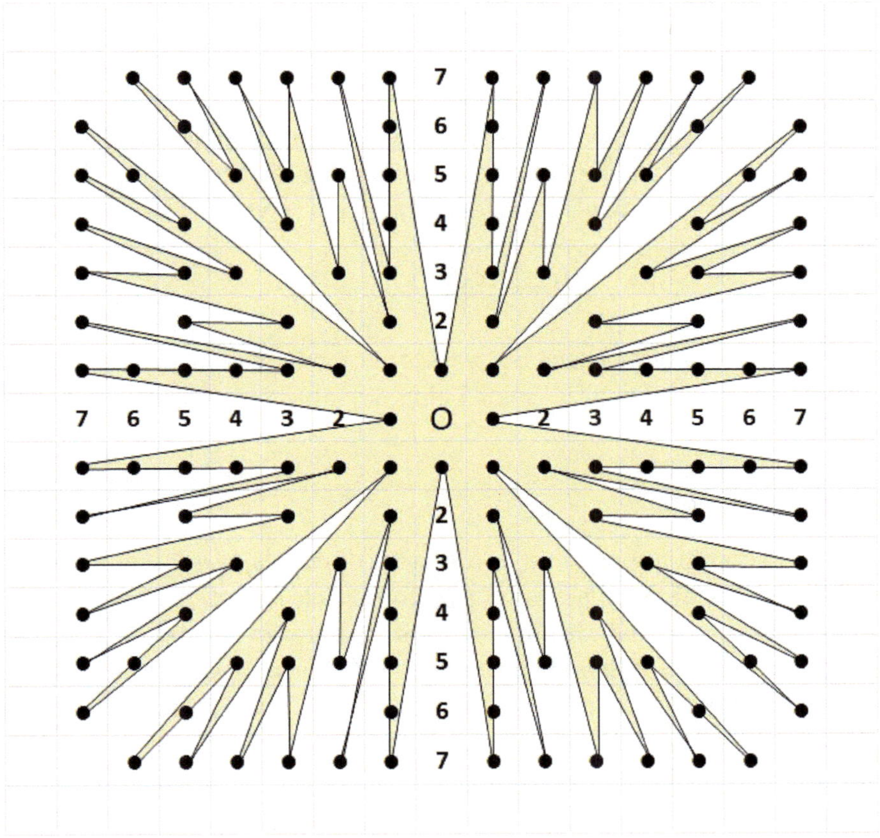

Figure 19.1 Farey Sunburst of Order 7

19.4 ORCHARD VISIBILITY AND THE 2D VISIBLE LATTICE POINTS

19.4.1 Polya's Orchard Visibility Problem

A tree is planted at each lattice point in a circular orchard which has center at the origin and radius r. If the radius of trees exceeds 1/r units, one is unable to see out of the orchard in any direction. However, if the radii of the trees are $\frac{1}{\sqrt{r^2+1}}$, one can see out at certain angles. The problem involves formulating different scenarios of radii of trees and radius of the orchard.

19.4.2 Euclid's Orchard

An array of "trees" of unit height located at integer-coordinate points in a point lattice. When viewed from the Origin corner along the line y=x in normal perspective, a quadrant of Euclid's orchard turns into the modified Dirichlet function. (See Gosper et al. [145])

Euclid's orchard is a two-dimensional array of one-dimensional "trees" of unit height planted at the lattice points in one quadrant of a square lattice. Euclid's orchard is the set of line segments from $(x,y,0)$ to $(x,y,1)$, where x and y are positive integers.

The trees visible from the origin are those at lattice points $(x,y,0)$, where x and y are coprime, i.e., where the fraction $\frac{x}{y}$ is in reduced form. The name Euclid's orchard is derived from the Euclidean algorithm.

If the orchard is projected relative to the origin onto the plane $x+y=1$ (or, equivalently, drawn in perspective from a viewpoint at the origin) the tops of the trees form a graph of Thomae's function. The point $(x,y,1)$ projects to

$$\left(\frac{x}{x+y},\frac{y}{x+y},\frac{1}{x+y}\right).$$

The proportion of points in the $n\times n$ grid that have trees on them is approximately $\frac{6}{\pi^2}$ based on the probability of a point being a visible point, with the error tending to zero as $n\to\infty$.

19.4.3 Opaque Forests

In discrete geometry, an opaque set is a system of curves or another set in the plane that blocks all lines of sight across a polygon, circle, or other shape. Opaque sets have also been called barriers, beam detectors, opaque covers, or (in cases where they have the form of a forest of line segments or other curves) opaque forests. Opaque sets were introduced by Stefan Mazurkiewicz in 1916 [197], and the problem of minimizing their total length was posed by Frederick Bagemihl in 1959 [41].

For instance, visibility through a unit square can be blocked by its four boundary edges, with length 4, but a shorter opaque forest blocks visibility across the square with length $\sqrt{2}+\frac{1}{2}\sqrt{6}\approx 2.63895843\ldots$. It is not known whether this is the shortest possible opaque set for the square, and for most other shapes this problem similarly remains unsolved. The shortest opaque set for any bounded convex set in the plane has length at most the perimeter of the set, and at least half the perimeter. For the square, a slightly stronger lower bound than half the perimeter is known. Another convex set whose opaque sets are commonly studied is the unit circle, for which the shortest connected opaque set has length $2+\pi$. Without the assumption of connectivity, the shortest opaque set for the circle has length at least π and at most 4.7998.

Several published algorithms claiming to find the shortest opaque set for a convex polygon were later shown to be incorrect. Nevertheless, it is possible to find an opaque set with a guaranteed approximation ratio in linear time, or to compute the subset of the plane whose visibility is blocked by a given system of line segments in polynomial time.

Figure 19.2 Greatest Common Divisors for Lattice Points $|x|$ and $|y| \leq 25$

EXERCISES

19.1 *Invisible cubes in 3D.* In this chapter, we showed that the 2D Visible Points hide the Invisible Points, and that the first invisible 2×2 square with the closest corner to origin is $(14, 20), (14, 21), (15, 20), (15, 21)$. In the 3D case the visible points are where the co-ordinates a, b, and c are co-prime, so $\gcd(a, b, c) = 1$. Find the co-ordinates of the closest $2 \times 2 \times 2$ invisible cube to the Origin in the first hyperquadrant. Show that the form of the solution is 8 adjacent ordered pairs each with gcd greater than 1.

19.2 *Invisible tesseracts in 4D.* In the 4D case, the visible points are where the co-ordinates a, b, c and d are co-prime, so $\gcd(a, b, c, d) = 1$. Find the co-ordinates of the closest $2 \times 2 \times 2 \times 2$ invisible tesseract (4D hypercube) to the Origin in

Figure 19.3 Greatest Common Divisors for Lattice Points $|x|$ and $|y| \leq 100$

the first hyperquadrant. Show that the form of the solution will be 16 adjacent ordered pairs in the 4-space each with gcd greater than 1.

19.3 *A Dirichlet series involving 2D Visible Points.* Show that if

$$f(s) = \sum_{\substack{a,b\geq 1 \\ (a,b)=1}} \frac{1}{(ab)^s} \quad \Re s > 1$$

then

$$f(s) + \frac{f(s)}{2^s} + \frac{f(s)}{3^s} + \frac{f(s)}{4^s} + \ldots = \zeta(s)^2,$$

so $f(s) = \zeta(s)$.

19.4 *A Dirichlet series involving nD Visible Points.* Show that if

$$f_n(s) = \sum_{\substack{a_1,a_2,\ldots a_n\geq 1 \\ (a_1,a_2,\ldots a_n)=1}} \frac{1}{(a_1 a_2 \cdots a_n)^s} \quad \Re s > 1$$

then

$$f_n(s)+\frac{f_n(s)}{2^s}+\frac{f_n(s)}{3^s}+\frac{f_n(s)}{4^s}+\ldots=\zeta(s)^n,$$

so $f_n(s)=\zeta(s)^{n-1}$.

Think About It...

When you come to the end of all the light you know,
determined by exactly repeatable, visible imprints
and it's time to step into the darkness of the unknown,
faith is knowing that one of two things shall happen:
either you will be given something solid to stand on
or you will be taught to fly.

Edward Teller
Theoretical physicist

GLOSSARY

Euclid's Orchard: Euclid's orchard is a two-dimensional array of one-dimensional "trees" of unit height planted at the lattice points in one quadrant of a square lattice.

Visible Lattice Point: A point in n-dimensional Euclidean space whose coordinates are $(a_1, a_2, \ldots, a_n)$ where $a_1, a_2, \ldots, a_n$ are co-prime integers as an n-tuple.

Farey sequence: The Farey fraction sequence of order n denoted F_n, is the sequence of completely reduced fractions, either between 0 and 1 or without this restriction, which when in lowest terms have denominators less than or equal to n, arranged in order of increasing size.

Polya's Orchard Visibility Problem: Each lattice point is a tree in a circular orchard of center at the origin and radius r. If the tree radius is more than 1/r units, one cannot see out of the orchard in any direction. But, if the trees' radii are $1/\sqrt(r^2+1)$, one can see out at certain angles. The problem formulates scenarios of tree radii and orchard radius.

Opaque forest: In discrete geometry, an opaque set is a system of curves or another set in the plane that block all lines of sight across a polygon, circle, or other shape. Opaque sets have also been called barriers, beam detectors, opaque covers, or (in cases where they have the form of a forest of line segments or other curves) opaque forests.

FURTHER READING

APOSTOL, T. *Introduction to Analytic Number Theory.* Springer-Verlag, New York, 1976.

BAAKE, M. and HUCK, C. Ergodic properties of visible lattice points. *Proc. Steklov Inst. Math.*, 288(1):165–188, 2015. (English summary) Reprint of Tr. Mat. Inst. Steklova Vol. 288, pages 184-208 (2015).

BAAKE, M., GRIMM, U. and WARRINGTON, D.H. Some remarks on the visible points of a lattice. *J. Phys. A: Math. Gen.*, 27:2669–2674, 1994.

BAAKE, M., MOODY, R.V. and PLEASANTS, P.A.B. Diffraction from Visible Lattice Points and kth Power Free Integers. arxiv preprint, June 1999. http://arxiv.org/abs/math.MG/9906132/.

BAGEMIHL, F. Some opaque subsets of a square. *Michigan Math. J.*, 6(2):99–103, 1959.

CAMPBELL, G.B. A new class of infinite products, and Euler's totient. *Int. J. Math. and Math. Sci.*, 17(3):417–422, 1994.

CAMPBELL, G.B. Infinite products over visible lattice points. *Int. J. Math. and Math. Sci.*, 17(4):637–654, 1994.

GOSPER, R.W. and SCHROEPPEL, R. 2 dimensional visible lattice point invisible polygons. In BEELER, M., GOSPER, R.W. and SCHROEPPEL, R., editor, HAKMEM: MIT Artificial Intelligence Laboratory, Memo AIM-239, page 19. Cambridge, MA, 1972. ITEM 48 (Gosper, Schroeppel), Feb 1972.

LEHMER, D.N. An asymptotic evaluation of certain totient sums. *Amer. J. Math.*, 22:293–355, 1900.

MAZURKIEWICZ, S. Sur un ensemble fermeé, punctiforme, qui rencontre toute droite passant par un certain domaine. *Prace Mat.-Fiz. (in Polish and French)*, 27:11–16, 1916.

MOSSERI, R. Visible Points in a Lattice. *J. Phys. A: Math. Gen.*, 25:L25–L29, 1992.

TAKEDA, W. Visible lattice points and the extended Lindelöf hypothesis. *J. Number Theory*, 180:297–309, 2017. MR3679799 Reviewed.

CHAPTER 20

Visible Point Vector Identities in the First Hyperquadrant

20.1 INTRODUCING THE VPV IDENTITIES

IN THE 1990S and up to 2000 the author published a series of papers introducing the so-called Visible Point Vector (VPV) identities. In these papers such as for example Campbell [87], [88], [90], the identities of the present chapter were given. They attracted some interest at the time, but most people ignored them as not relevant to their own research. Perhaps the proofs in the early papers were too obscure or cryptic. So, our approach here is to introduce the deeper n-dimensional identities by first trying to simply derive the easier 2-dimensional VPV identities. Some of this work was revisited in 2019 by the author in [93]. Other useful papers as background for this chapter are by the author circa 2000 [89, 90, 91].

20.2 DERIVING THE 2D FIRST QUADRANT VPV IDENTITY

We will derive the 2D VPV identity for the first quadrant integer lattice points. Let $(j,k)=1$ mean that j and k are coprime integers. We see that

$$\left(\sum_{m=1}^{\infty}\frac{y^m}{m^a}\right)\left(\sum_{n=1}^{\infty}\frac{z^n}{n^b}\right)$$

$$=\left(\frac{y^1}{1^a}+\frac{y^2}{2^a}+\frac{y^3}{3^a}+\frac{y^4}{4^a}+\frac{y^5}{5^a}+\frac{y^6}{6^a}+\frac{y^7}{7^a}+\cdots\right)$$

$$\times\left(\frac{z^1}{1^b}+\frac{z^2}{2^b}+\frac{z^3}{3^b}+\frac{z^4}{4^b}+\frac{z^5}{5^b}+\frac{z^6}{6^b}+\frac{z^7}{7^b}+\cdots\right)$$

$$=\frac{y^1z^1}{1^a1^b}+\frac{y^2z^1}{2^a1^b}+\frac{y^3z^1}{3^a1^b}+\frac{y^4z^1}{4^a1^b}+\frac{y^5z^1}{5^a1^b}+\frac{y^6z^1}{6^a1^b}+\frac{y^7z^1}{7^a1^b}+\cdots$$

DOI: 10.1201/9781003174158-20

$$\begin{aligned}
&+\frac{y^1z^2}{1^a2^b}+\frac{y^2z^2}{2^a2^b}+\frac{y^3z^2}{3^a2^b}+\frac{y^4z^2}{4^a2^b}+\frac{y^5z^2}{5^a2^b}+\frac{y^6z^2}{6^a2^b}+\frac{y^7z^2}{7^a2^b}+\cdots\\
&+\frac{y^1z^3}{1^a3^b}+\frac{y^2z^3}{2^a3^b}+\frac{y^3z^3}{3^a3^b}+\frac{y^4z^3}{4^a3^b}+\frac{y^5z^3}{5^a3^b}+\frac{y^6z^3}{6^a3^b}+\frac{y^7z^3}{7^a3^b}+\cdots\\
&+\frac{y^1z^4}{1^a4^b}+\frac{y^2z^4}{2^a4^b}+\frac{y^3z^4}{3^a4^b}+\frac{y^4z^4}{4^a4^b}+\frac{y^5z^4}{5^a4^b}+\frac{y^6z^4}{6^a4^b}+\frac{y^7z^4}{7^a4^b}+\cdots\\
&+\frac{y^1z^5}{1^a5^b}+\frac{y^2z^5}{2^a5^b}+\frac{y^3z^5}{3^a5^b}+\frac{y^4z^5}{4^a5^b}+\frac{y^5z^5}{5^a5^b}+\frac{y^6z^5}{6^a5^b}+\frac{y^7z^5}{7^a5^b}+\cdots\\
&+\frac{y^1z^6}{1^a6^b}+\frac{y^2z^6}{2^a6^b}+\frac{y^3z^6}{3^a6^b}+\frac{y^4z^6}{4^a6^b}+\frac{y^5z^6}{5^a6^b}+\frac{y^6z^6}{6^a6^b}+\frac{y^7z^6}{7^a6^b}+\cdots\\
&+\frac{y^1z^7}{1^a7^b}+\frac{y^2z^7}{2^a7^b}+\frac{y^3z^7}{3^a7^b}+\frac{y^4z^7}{4^a7^b}+\frac{y^5z^7}{5^a7^b}+\frac{y^6z^7}{6^a7^b}+\frac{y^7z^7}{7^a7^b}+\cdots\\
&+\ \vdots\ +\ \vdots\ +\ \vdots\ +\ \vdots\ +\ \vdots\ +\ \vdots\ +\ \vdots\ \ \ddots
\end{aligned}$$

$$= \sum_{m,n\geq 1}^{\infty} \frac{y^m z^n}{m^a n^b}$$

$$= \sum_{\substack{h\geq 1\\ (j,k)=1}}^{\infty} \frac{(y^j z^k)^h}{h^{a+b}(j^a k^b)}$$

$$= \sum_{(j,k)=1}^{\infty} \frac{1}{(j^a k^b)} \sum_{h=1}^{\infty} \frac{(y^j z^k)^h}{h^{a+b}}$$

$$= \sum_{(j,k)=1}^{\infty} \frac{1}{(j^a k^b)} \log\left(\frac{1}{1-y^j z^k}\right) \quad \textit{if} \quad a+b=1.$$

Therefore, we have shown that

$$\left(\sum_{m=1}^{\infty}\frac{y^m}{m^a}\right)\left(\sum_{n=1}^{\infty}\frac{z^n}{n^b}\right) = \sum_{(j,k)=1}^{\infty} \frac{1}{(j^a k^b)} \log\left(\frac{1}{1-y^j z^k}\right) \quad \textit{if} \quad a+b=1.$$

Exponentiating both sides gives us the 2D first quadrant VPV identity,

$$\prod_{\substack{(j,k)=1\\ j,k\geq 1}} \left(\frac{1}{1-y^j z^k}\right)^{\frac{1}{j^a k^b}} = \exp\left\{\left(\sum_{j=1}^{\infty}\frac{y^j}{j^a}\right)\left(\sum_{k=1}^{\infty}\frac{z^k}{k^b}\right)\right\} \quad \textit{if} \quad a+b=1. \qquad (20.2.1)$$

This approach will be useful to return to when considering the derivation of 2D identities from first principles. Clearly, we have summed on the 2D first quadrant lattice points, that is on points with positive integer coordinates. However, this was done by summing on all positive integer multiples of the Visible Points in the first 2D quadrant, the first 20 rows and columns of VPVs shown here.

2D first quadrant visible lattice points

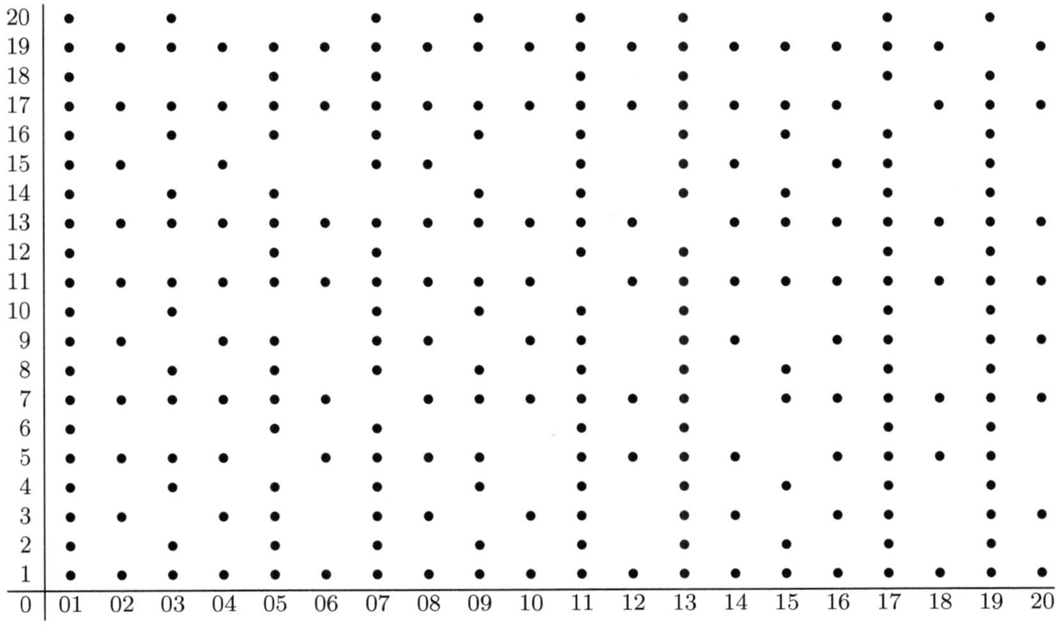

We give immediate interesting cases of (20.2.1) now.
Taking $a = 0$, $b = 1$ gives

$$\prod_{\substack{(j,k)=1\\ j,k\geq 1}} \left(\frac{1}{1-y^j z^k}\right)^{\frac{1}{k}} = \left(\frac{1}{1-z}\right)^{\frac{y}{1-y}} \tag{20.2.2}$$

$$= \sum_{n=0}^{\infty} z^n (-1)^n \binom{\frac{y}{y-1}}{n}$$

$$= 1 - \frac{yz}{y-1} + \frac{yz^2}{2!(y-1)^2} + \frac{(y-2)yz^3}{3!(y-1)^3} + \frac{(y-2)y(2y-3)z^4}{4!(y-1)^4}$$
$$+ \frac{(y-2)y(2y-3)(3y-4)z^5}{5!(y-1)^5} + O(z^6).$$

A finite partition grid area for $\left(\frac{1}{1-z}\right)^{\frac{y}{1-y}}$ listing coefficients of $y^a z^b$ with $0 \leq a \leq 9$, $0 \leq b \leq 10$ is given in the following table.

2D partitions from the exponential generating function $\left(\frac{1}{1-z}\right)^{\frac{y}{1-y}}$

b/a	0	1	2	3	4	5	6	7	8	9
10		$\frac{362880}{10!}$	$\frac{1389456}{10!}$	$\frac{3588732}{10!}$	$\frac{7684388}{10!}$	$\frac{14669429}{10!}$	$\frac{25869458}{10!}$	$\frac{43015399}{10!}$	$\frac{68326540}{10!}$	$\frac{104604811}{10!}$
9		$\frac{40320}{9!}$	$\frac{149904}{9!}$	$\frac{377612}{9!}$	$\frac{790728}{9!}$	$\frac{1478985}{9!}$	$\frac{2559101}{9!}$	$\frac{4179861}{9!}$	$\frac{6527781}{9!}$	$\frac{9833391}{9!}$
8		$\frac{5040}{8!}$	$\frac{18108}{8!}$	$\frac{44308}{8!}$	$\frac{90409}{8!}$	$\frac{165140}{8!}$	$\frac{279512}{8!}$	$\frac{447168}{8!}$	$\frac{684762}{8!}$	$\frac{1012368}{8!}$
7		$\frac{720}{7!}$	$\frac{2484}{7!}$	$\frac{5872}{7!}$	$\frac{11619}{7!}$	$\frac{20635}{7!}$	$\frac{34026}{7!}$	$\frac{53116}{7!}$	$\frac{79470}{7!}$	$\frac{114918}{7!}$
6		$\frac{120}{6!}$	$\frac{394}{6!}$	$\frac{893}{6!}$	$\frac{1702}{6!}$	$\frac{2921}{6!}$	$\frac{4666}{6!}$	$\frac{7070}{6!}$	$\frac{10284}{6!}$	$\frac{14478}{6!}$
5		$\frac{24}{5!}$	$\frac{74}{5!}$	$\frac{159}{5!}$	$\frac{289}{5!}$	$\frac{475}{5!}$	$\frac{729}{5!}$	$\frac{1064}{5!}$	$\frac{1494}{5!}$	$\frac{2034}{5!}$
4		$\frac{6}{4!}$	$\frac{17}{4!}$	$\frac{34}{4!}$	$\frac{58}{4!}$	$\frac{90}{4!}$	$\frac{131}{4!}$	$\frac{182}{4!}$	$\frac{244}{4!}$	$\frac{318}{4!}$
3		$\frac{2}{3!}$	$\frac{5}{3!}$	$\frac{9}{3!}$	$\frac{14}{3!}$	$\frac{20}{3!}$	$\frac{27}{3!}$	$\frac{35}{3!}$	$\frac{44}{3!}$	$\frac{54}{3!}$
2		$\frac{1}{2!}$	$\frac{2}{2!}$	$\frac{3}{2!}$	$\frac{4}{2!}$	$\frac{5}{2!}$	$\frac{6}{2!}$	$\frac{7}{2!}$	$\frac{8}{2!}$	$\frac{9}{2!}$
1		1	1	1	1	1	1	1	1	1
0	1									
b/a	0	1	2	3	4	5	6	7	8	9

Or equivalent to (20.2.2) is it's reciprocal of both sides,

$$\prod_{\substack{(j,k)=1 \\ j,k\geq 1}} \left(1-y^j z^k\right)^{\frac{1}{k}} = (1-z)^{\frac{y}{1-y}} \tag{20.2.3}$$

$$= \sum_{n=0}^{\infty} z^n (-1)^n \binom{\frac{y}{1-y}}{n}$$

$$= 1 + \frac{yz}{y-1} + \frac{y(2y-1)z^2}{2!(y-1)^2} + \frac{y(2y-1)(3y-2)z^3}{3!(y-1)^3} + \frac{y(2y-1)(3y-2)(4y-3)z^4}{4!(y-1)^4}$$
$$+ \frac{y(2y-1)(3y-2)(4y-3)(5y-4)z^5}{5!(y-1)^5} + O(z^6).$$

A finite partition grid portion for $(1-z)^{\frac{y}{1-y}}$ as an exponential generating function for coefficients of $y^a z^b$ with $0 \leq a \leq 9$, $0 \leq b \leq 10$ is given in the table grid below. It was recently found that the limiting case of this grid, given in Chapter 5 has interesting congruence properties applying to the coefficients. We see that if $\varphi(k)$ is the Euler totient function, $\prod_{k=1}^{\infty}(1-z^k)^{\frac{\varphi(k)}{k}} = e^{\frac{z}{z-1}} = \sum_{k=0}^{\infty} \frac{\alpha(k)}{k!}$, which is equation (5.7.3) in Chapter 5, exhibits that coefficients always end in digit 1 or 9, so, one easily conjectures the congruence properties there. (See OEIS sequence A293116 [237]) These remarks about congruence properties of exponential generating function coefficients are evidently applicable more generally, applying to the previous grid table above, but little seems to be known of this at present.

2D partitions from the exponential generating function $(1-z)^{\frac{y}{1-y}}$

b/a	0	1	2	3	4	5	6	7	8	9
10		$\frac{-362880}{10!}$	$\frac{663696}{10!}$	$\frac{517572}{10!}$	$\frac{-77572}{10!}$	$\frac{-667381}{10!}$	$\frac{-1003552}{10!}$	$\frac{-990011}{10!}$	$\frac{-637670}{10!}$	$\frac{-26939}{10!}$
9		$\frac{-40320}{9!}$	$\frac{69264}{9!}$	$\frac{60724}{9!}$	$\frac{1344}{9!}$	$\frac{-64041}{9!}$	$\frac{-108509}{9!}$	$\frac{-119061}{9!}$	$\frac{-93141}{9!}$	$\frac{-35631}{9!}$
8		$\frac{-5040}{8!}$	$\frac{8028}{8!}$	$\frac{7964}{8!}$	$\frac{1537}{8!}$	$\frac{-6444}{8!}$	$\frac{-12808}{8!}$	$\frac{-15728}{8!}$	$\frac{-14454}{8!}$	$\frac{-9072}{8!}$
7		$\frac{-720}{7!}$	$\frac{1044}{7!}$	$\frac{1184}{7!}$	$\frac{435}{7!}$	$\frac{-643}{7!}$	$\frac{-1644}{7!}$	$\frac{-2296}{7!}$	$\frac{-2442}{7!}$	$\frac{-2022}{7!}$
6		$\frac{-120}{6!}$	$\frac{154}{6!}$	$\frac{203}{6!}$	$\frac{112}{6!}$	$\frac{-49}{6!}$	$\frac{-224}{6!}$	$\frac{-370}{6!}$	$\frac{-456}{6!}$	$\frac{-462}{6!}$
5		$\frac{-24}{5!}$	$\frac{26}{5!}$	$\frac{42}{5!}$	$\frac{31}{5!}$	$\frac{5}{5!}$	$\frac{-29}{5!}$	$\frac{-64}{5!}$	$\frac{-94}{5!}$	$\frac{-114}{5!}$
4		$\frac{-6}{4!}$	$\frac{5}{4!}$	$\frac{10}{4!}$	$\frac{10}{4!}$	$\frac{6}{4!}$	$\frac{-1}{4!}$	$\frac{-10}{4!}$	$\frac{-20}{4!}$	$\frac{-30}{4!}$
3		$\frac{-2}{3!}$	$\frac{1}{3!}$	$\frac{3}{3!}$	$\frac{4}{3!}$	$\frac{4}{3!}$	$\frac{3}{3!}$	$\frac{1}{3!}$	$\frac{-2}{3!}$	$\frac{-6}{3!}$
2		$\frac{1}{2!}$	$\frac{0}{2!}$	$\frac{-1}{2!}$	$\frac{-2}{2!}$	$\frac{-3}{2!}$	$\frac{-4}{2!}$	$\frac{-5}{2!}$	$\frac{-6}{2!}$	$\frac{-7}{2!}$
1		1	1	1	1	1	1	1	1	1
0	1									

From the case of (20.2.3) with y^2 and z^2 replacing y and z, and dividing both sides of (20.2.3) into it we have,

$$\prod_{\substack{(j,k)=1\\ j,k\geq 1}} \left(1+y^j z^k\right)^{\frac{1}{k}} = \frac{(1-z^2)^{\frac{y^2}{1-y^2}}}{(1-z)^{\frac{y}{1-y}}}. \tag{20.2.4}$$

$$= \sum_{n=0}^{\infty} \frac{\Gamma\left(n+\frac{y}{1-y}\right)}{\Gamma\left(\frac{y}{1-y}\right)} {}_3F_2\left(\frac{1}{2}-\frac{n}{2}, -\frac{n}{2}, -\frac{y^2}{1-y^2}; \frac{1}{2}-\frac{n}{2}-\frac{y}{2(1-y)}, 1-\frac{n}{2}-\frac{y}{2(1-y)}; 1\right) \frac{z^n}{n!}.$$

So, in (20.2.1) we have derived the VPV 2D first quadrant identity summed upon the visible points in that quadrant as depicted in the shaded area of Figure 16.1.

We note that from the power series in z for (20.2.3), the series is finite when $y = \frac{1}{2}, \frac{2}{3}, \frac{3}{4}, \frac{4}{5}, \frac{5}{6}, \ldots$ leading us to the following identities,

$$\prod_{\substack{(j,k)=1\\ j,k\geq 1}} \left(1-\left(\frac{1}{2}\right)^j z^k\right)^{\frac{1}{k}} = 1-z,$$

$$\prod_{\substack{(j,k)=1\\ j,k\geq 1}} \left(1-\left(\frac{2}{3}\right)^j z^k\right)^{\frac{1}{k}} = (1-z)^2 = 1-2z+z^2,$$

$$\prod_{\substack{(j,k)=1\\ j,k\geq 1}} \left(1-\left(\frac{3}{4}\right)^j z^k\right)^{\frac{1}{k}} = (1-z)^3 = 1-3z+3z^2-z^3,$$

$$\prod_{\substack{(j,k)=1\\ j,k\geq 1}} \left(1-\left(\frac{4}{5}\right)^j z^k\right)^{\frac{1}{k}} = (1-z)^4 = 1-4z+6z^2-4z^3+z^4,$$

$$\prod_{\substack{(j,k)=1\\ j,k\geq 1}} \left(1-\left(\frac{5}{6}\right)^j z^k\right)^{\frac{1}{k}} = (1-z)^5 = 1-5z+10z^2-10z^3+5z^4-z^5,$$

and so on. These infinite products tell an interesting story as they are a product of infinite series that reduce to a finite polynomial.

Similarly, for (20.2.2), the series is finite when $y = 2, \frac{3}{2}, \frac{4}{3}, \frac{5}{4}, \frac{6}{5}, \ldots$ leading us to the following identities,

$$\prod_{\substack{(j,k)=1\\ j,k\geq 1}} \left(\frac{1}{1-2^j z^k}\right)^{\frac{1}{k}} = (1-z)^2 = 1-2z+z^2,$$

$$\prod_{\substack{(j,k)=1\\ j,k\geq 1}} \left(\frac{1}{1-\left(\frac{3}{2}\right)^j z^k}\right)^{\frac{1}{k}} = (1-z)^3 = 1-3z+3z^2-z^3,$$

$$\prod_{\substack{(j,k)=1\\ j,k\geq 1}} \left(\frac{1}{1-\left(\frac{4}{3}\right)^j z^k}\right)^{\frac{1}{k}} = (1-z)^4 = 1-4z+6z^2-4z^3+z^4,$$

$$\prod_{\substack{(j,k)=1\\ j,k\geq 1}} \left(\frac{1}{1-\left(\frac{5}{4}\right)^j z^k}\right)^{\frac{1}{k}} = (1-z)^5 = 1-5z+10z^2-10z^3+5z^4-z^5,$$

$$\prod_{\substack{(j,k)=1\\ j,k\geq 1}} \left(\frac{1}{1-\left(\frac{6}{5}\right)^j z^k}\right)^{\frac{1}{k}} = (1-z)^6 = 1-6z+15z^2-20z^3+15z^4-6z^5+z^6,$$

and so on.

Another natural case of (20.2.1) to consider is where $a = b = \frac{1}{2}$, so then

$$\prod_{\substack{(j,k)=1\\ j,k\geq 1}} \left(\frac{1}{1-y^j z^k}\right)^{\frac{1}{\sqrt{jk}}} = \exp\left\{\left(\sum_{j=1}^{\infty} \frac{y^j}{\sqrt{j}}\right)\left(\sum_{k=1}^{\infty} \frac{z^k}{\sqrt{k}}\right)\right\}, \tag{20.2.5}$$

and equivalently,

$$\prod_{\substack{(j,k)=1\\ j,k\geq 1}} \left(1-y^j z^k\right)^{\frac{1}{\sqrt{jk}}} = \exp\left\{-\left(\sum_{j=1}^{\infty} \frac{y^j}{\sqrt{j}}\right)\left(\sum_{k=1}^{\infty} \frac{z^k}{\sqrt{k}}\right)\right\}. \tag{20.2.6}$$

From the case of (20.2.5) with y^2 and z^2 replacing y and z, and dividing both sides of (20.2.5) into it we have

$$\prod_{\substack{(j,k)=1\\ j,k\geq 1}} \left(1+y^j z^k\right)^{\frac{1}{\sqrt{jk}}} = \exp\left\{\left(\sum_{j=1}^{\infty}\frac{y^j}{\sqrt{j}}\right)\left(\sum_{k=1}^{\infty}\frac{z^k}{\sqrt{k}}\right)-\left(\sum_{j=1}^{\infty}\frac{y^{2j}}{\sqrt{j}}\right)\left(\sum_{k=1}^{\infty}\frac{z^{2k}}{\sqrt{k}}\right)\right\}. \tag{20.2.7}$$

20.3 DERIVING THE n-DIMENSIONAL FIRST HYPERQUADRANT VPV IDENTITY

They are, however, generating functions for weighted vector partitions, and evidently contribute a new branch to the literature on partitions of vectors.

So, we establish the following conventions for the VPV identities.

Definition 20.1 *We use the notation* $(x_1,x_2,x_3,...,x_n)$, *to mean "the greatest common divisor of all of* $x_1,x_2,x_3,...,x_n$ *together; the same as* $\gcd(x_1,x_2,x_3,...,x_n)$*". It is important to distinguish between this and the ordered n-tuple utilized for the vector* $\langle x_1,x_2,x_3,...,x_n\rangle$. *In either case, we will be concerned with lattice points in the relevant Euclidean space, hence any vector or gcd will be over integer coordinates.*

Definition 20.2 *Any Euclidean vector* $\langle x_1,x_2,x_3,...,x_n\rangle$ *for which* $(x_1,x_2,x_3,...,x_n) = 1$ *we call a visible point vector, abbreviated VPV.*

Theorem 20.1 *The first hyperquadrant VPV identity. If* $i = 1,2,3,...,n$ *then for each* $x_i \in \mathbb{C}$ *such that* $|x_i| < 1$ *and* $b_i \in \mathbb{C}$ *such that* $\sum_{i=1}^{n} b_i = 1$,

$$\prod_{\substack{(a_1,a_2,...,a_n)=1\\ a_1,a_2,...,a_n\geq 1}} \left(\frac{1}{1-x_1^{a_1}x_2^{a_2}x_3^{a_3}\cdots x_n^{a_n}}\right)^{\frac{1}{a_1^{b_1}a_2^{b_2}a_3^{b_3}\cdots a_n^{b_n}}} \tag{20.3.1}$$

$$= \exp\left\{\prod_{i=1}^{n}\left(\sum_{j=1}^{\infty}\frac{x_i^{\,j}}{j^{b_i}}\right)\right\} = \exp\left\{\prod_{i=1}^{n} Li_{b_i}(x_i)\right\}.$$

There follow numerous example corollaries of this theorem, all of them susceptible to the combinatorial analysis of the previous sections. However, firstly we give the lemma and proof underpinning theorem 20.1.

Lemma 20.1 *Consider an infinite region raying out of the origin in any n-dimensional Euclidean vector space. The set of all lattice point vectors apart from the origin in that hyper-sector region is precisely the set of positive integer multiples of the VPVs in that region.*

Proof of Lemma 20.1.

Each VPV will have integer coordinates whose greatest common divisor is unity. Viewed from the origin, all other lattice points are obscured behind the VPV end points. If x is a VPV in the region then all vectors in that region from the origin with direction of x preserved are enumerated by a sequence $1x, 2x, 3x, ...,$ and the greatest common divisor of the components of nx is clearly n. This is because if the scalar n is non-integer at least one of the coordinates of nx would be a non-integer. Therefore, if the VPVs in the region are countably given by $x_1, x_2, x_3, ...,$ then all lattice point vectors from the origin in the region are

$$1x_1, 2x_1, 3x_1, ...; \quad 1x_2, 2x_2, 3x_2, ...; \quad 1x_3, 2x_3, 3x_3, ... \text{ etc.}$$

Completion of the proof comes with the recognition that the set of all VPVs in any *rayed from the origin* region in any Euclidean vector space is a countable set. Proof of this last assertion is by induction on the dimension, knowing that the lattice points are countable in any two-dimensional region. As we count each lattice point vector in the desired region we decide whether it is a VPV simply by observing whether its coordinates are relatively prime as a whole. ■

This then brings us to the

Proof of theorem 20.1.
We start with the multiple sum

$$\sum_{a_1, a_2, ..., a_n \in \mathbb{Z}^+} \frac{x_1^{a_1} x_2^{a_2} x_3^{a_3} \cdots x_n^{a_n}}{a_1^{b_1} a_2^{b_2} a_3^{b_3} \cdots a_n^{b_n}} = \prod_{i=1}^{n} \sum_{j=1}^{\infty} \frac{x_i^j}{j^{b_i}}$$

which, due to Lemma 20.3.1, also equals, letting $b = \sum_{i=1}^{n} b_i$,

$$\sum_{\substack{(a_1, a_2, ..., a_n)=1 \\ a_1, a_2, ..., a_n \geq 1}} \left(\frac{x_1^{a_1} x_2^{a_2} \cdots x_n^{a_n}}{1^b} + \frac{(x_1^{a_1} x_2^{a_2} \cdots x_n^{a_n})^2}{2^b} + \frac{(x_1^{a_1} x_2^{a_2} \cdots x_n^{a_n})^3}{3^b} + \cdots \right) \frac{1}{a_1^{b_1} a_2^{b_2} a_3^{b_3} \cdots a_n^{b_n}}$$

$$= \sum_{\substack{(a_1, a_2, ..., a_n)=1 \\ a_1, a_2, ..., a_n \geq 1}} \frac{-\log(1 - x_1^{a_1} x_2^{a_2} \cdots x_n^{a_n})}{a_1^{b_1} a_2^{b_2} a_3^{b_3} \cdots a_n^{b_n}}$$

Exponentiating both sides then yields the theorem. ■

The cases of theorem 20.1 with $n = 2, n = 3, n = 4, n = 5$, are stated easily in the forms,

Corollary 20.1 *If* $|y| < 1, |z| < 1,$ *and* $s + t = 1$, *then*

$$\prod_{\substack{(a,b)=1 \\ a,b \geq 1}} \left(\frac{1}{1 - y^a z^b} \right)^{\frac{1}{a^s b^t}} = \exp\left\{ \left(\sum_{j=1}^{\infty} \frac{y^j}{j^s} \right) \left(\sum_{k=1}^{\infty} \frac{z^k}{k^t} \right) \right\}. \tag{20.3.2}$$

Corollary 20.2 *If* $|x| < 1, |y| < 1, |z| < 1$ *and* $s + t + u = 1$, *then*

$$\prod_{\substack{(a,b,c)=1 \\ a,b,c \geq 1}} \left(\frac{1}{1 - x^a y^b z^c} \right)^{\frac{1}{a^s b^t c^u}} = \exp\left\{ \left(\sum_{i=1}^{\infty} \frac{x^i}{i^s} \right) \left(\sum_{j=1}^{\infty} \frac{y^j}{j^t} \right) \left(\sum_{k=1}^{\infty} \frac{z^k}{k^u} \right) \right\}. \tag{20.3.3}$$

Corollary 20.3 *If $|w|<1, |x|<1, |y|<1, |z|<1$ and $r+s+t+u=1$, then*

$$\prod_{\substack{(a,b,c,d)=1\\a,b,c,d\geq 1}}\left(\frac{1}{1-w^a x^b y^c z^d}\right)^{\frac{1}{a^r b^s c^t d^u}} = \exp\left\{\left(\sum_{h=1}^{\infty}\frac{w^h}{h^r}\right)\left(\sum_{i=1}^{\infty}\frac{x^i}{i^s}\right)\left(\sum_{j=1}^{\infty}\frac{y^j}{j^t}\right)\left(\sum_{k=1}^{\infty}\frac{z^k}{k^u}\right)\right\}. \tag{20.3.4}$$

Corollary 20.4 *If $|v|<1, |w|<1, |x|<1, |y|<1, |z|<1$ and $q+r+s+t+u=1$, then*

$$\prod_{\substack{(a,b,c,d,e)=1\\a,b,c,d,e\geq 1}}\left(\frac{1}{1-v^a w^b x^c y^d z^e}\right)^{\frac{1}{a^q b^r c^s d^t e^u}} \tag{20.3.5}$$

$$= \exp\left\{\left(\sum_{g=1}^{\infty}\frac{v^g}{g^q}\right)\left(\sum_{h=1}^{\infty}\frac{w^h}{h^r}\right)\left(\sum_{i=1}^{\infty}\frac{x^i}{i^s}\right)\left(\sum_{j=1}^{\infty}\frac{y^j}{j^t}\right)\left(\sum_{k=1}^{\infty}\frac{z^k}{k^u}\right)\right\}.$$

The reader will recognize the polylogarithm occurring on the right sides of (20.3.2) through to (20.3.5). The many known particular values of the polylogarithms as combinations of generalized Mordell-Tornheim-Witten (MTW) zeta-function values along with their derivatives, and recently found connections with multiple zeta values (MZVs) implies a lot of possible future research. For example, taking the substitution where $b_k = \frac{1}{n}$ for all $1 \leq k \leq n$, we have interesting new identities such as,

$$\prod_{\substack{(a,b)=1\\a,b\geq 1}}\left(\frac{1}{1-y^a z^b}\right)^{\frac{1}{\sqrt{(ab)}}} = \exp\left\{\left(\sum_{j=1}^{\infty}\frac{y^j}{\sqrt{j}}\right)\left(\sum_{k=1}^{\infty}\frac{z^k}{\sqrt{k}}\right)\right\}, \tag{20.3.6}$$

$$\prod_{\substack{(a,b,c)=1\\a,b,c\geq 1}}\left(\frac{1}{1-x^a y^b z^c}\right)^{\frac{1}{\sqrt[3]{abc}}} = \exp\left\{\left(\sum_{i=1}^{\infty}\frac{x^i}{\sqrt[3]{i}}\right)\left(\sum_{j=1}^{\infty}\frac{y^j}{\sqrt[3]{j}}\right)\left(\sum_{k=1}^{\infty}\frac{z^k}{\sqrt[3]{k}}\right)\right\}. \tag{20.3.7}$$

$$\prod_{\substack{(a,b,c,d)=1\\a,b,c,d\geq 1}}\left(\frac{1}{1-w^a x^b y^c z^d}\right)^{\frac{1}{\sqrt[4]{abcd}}} = \exp\left\{\left(\sum_{h=1}^{\infty}\frac{w^h}{\sqrt[4]{h}}\right)\left(\sum_{i=1}^{\infty}\frac{x^i}{\sqrt[4]{i}}\right)\left(\sum_{j=1}^{\infty}\frac{y^j}{\sqrt[4]{j}}\right)\left(\sum_{k=1}^{\infty}\frac{z^k}{\sqrt[4]{k}}\right)\right\}. \tag{20.3.8}$$

$$\prod_{\substack{(a,b,c,d,e)=1\\a,b,c,d,e\geq 1}}\left(\frac{1}{1-v^a w^b x^c y^d z^e}\right)^{\frac{1}{\sqrt[5]{abcde}}} \tag{20.3.9}$$

$$= \exp\left\{\left(\sum_{g=1}^{\infty}\frac{v^g}{\sqrt[5]{g}}\right)\left(\sum_{h=1}^{\infty}\frac{w^h}{\sqrt[5]{h}}\right)\left(\sum_{i=1}^{\infty}\frac{x^i}{\sqrt[5]{i}}\right)\left(\sum_{j=1}^{\infty}\frac{y^j}{\sqrt[5]{j}}\right)\left(\sum_{k=1}^{\infty}\frac{z^k}{\sqrt[5]{k}}\right)\right\}.$$

20.4 HYPERQUADRANT LATTICES AND THEIR HYPERDIAGONAL LINE FUNCTIONS

The above equations (20.3.6) to (20.3.9) when represented by their 2D, 3D, 4D, and 5D grids in their respective first quadrants and first hyperquadrants, obviously become unwieldy to visualize after the easily envisioned 2D extended rectangular lattice, and the challenging vision of the diagrammatic 3D extended cubic lattice. What human consciousness can say they see a 4D extended tesseract (4D cube) lattice, or a 5D extended hypercube lattice? However, we understand this dimensional concept, there is a logical higher space extension also of the 2D first quadrant diagonal of lattice points upon the line $y=x$. In the 3D cube first hyperquadrant in an x-y-z plane of lattice points there is the diagonal line represented by the equation $x=y=z$ and the coefficients of those lattice point vectors arise from setting the generating function $f(x,y,z)$ to say $f(z,z,z)$ which then is the generating function for the 3D diagonal lattice point *vectors whose three co-ordinates are equal* in that hyperquadrant. Similarly for the 4D case generating function $f(w,x,y,z)$ calculate $f(z,z,z,z)$ to give the 4D hyperdiagonal generating function for lattice points for the number of partitions into *vectors whose four co-ordinates are equal* along the line with equation $w=x=y=z$ through the 4D extended tesseract (hypecube). And so on for the 5D function generated by $f(v,w,x,y,z)$ giving us the function $f(z,z,z,z,z)$ generating the coefficients (vector partition sums) for the lattice points along the 5D line $v=w=x=y=z$. This concept can be applied to any of our 2D, 3D, 4D, etc. lattice point vector grids throughout our present volume. It applies to VPV identities as *hyperquadrant* lattice functions, *square hyperpyramid* lattice functions and *skewed hyperpyramid* lattice point vector identities. This includes the many possibilities of applying *polylogarithm* formulas and *Parametric Euler sum* identities; examples given in our present book being just the starting place of a large number of possibilities.

We are reminded by (20.3.2) of the functional equation due originally to Riemann in his famous paper [221] on the Riemann zeta function. Both have the $s+t=1$ caveat. Riemann's zeta function reflection formula is equivalent to

$$\Gamma\left(\frac{s}{2}\right)\pi^{-s/2}\zeta(s)=\Gamma\left(\frac{t}{2}\right)\pi^{-t/2}\zeta(t) \tag{20.4.1}$$

where $s+t=1$, but equation (20.3.2) is quite a different relationship in a context amenable to the critical line Riemann zeta function $\zeta\left(\frac{1}{2}+it\right)$ for nontrivial zeroes.

There are several further corollary cases that we can state here, that may be susceptible to the analysis of the earlier sections. There are natural and simple cases of theorem 20.1 to consider.

20.5 SOME HYPERDIAGONAL LINE GENERATING FUNCTIONS

We return to section 20.3 and the (20.3.6) to (20.3.9) equations to find particular cases that follow easily from setting $v=w=x=y=z$ into them. Incidentally, the lattice point vectors on the line with equations $v=w=x=y=z$ through $5D$first hyperquadrant are listed (in Chapter 12 notation) as

$$\mathbb{L}_5(1,1,1,1,1) = \{\langle 1,1,1,1,1\rangle, \langle 2,2,2,2,2\rangle, \langle 3,3,3,3,3\rangle, \ldots\}.$$

So we have 2D, 3D, 4D, and 5D examples of the *hyperdiagonal line* generating functions mentioned in our previous section.

If $|z| < 1$, then

$$\prod_{\substack{\gcd(a,b)=1\\ a,b\geq 1}} \left(\frac{1}{1-z^{a+b}}\right)^{\frac{1}{\sqrt{(ab)}}} = \exp\left\{\left(\sum_{k=1}^{\infty} \frac{z^k}{\sqrt{k}}\right)^2\right\} \tag{20.5.1}$$

$$= 1 + z^2 + \sqrt{2}z^3 + \left(1 + \frac{2}{\sqrt{3}}\right)z^4 + \left(1 + \sqrt{\frac{2}{3}} + \sqrt{2}\right)z^5 + \left(2 + \frac{1}{\sqrt{2}} + \frac{2}{\sqrt{3}} + \frac{2}{\sqrt{5}}\right)z^6$$

$$+ \left(1 + \sqrt{\frac{2}{5}} + 4\sqrt{\frac{2}{3}} + \sqrt{2} + \frac{1}{\sqrt{3}}\right)z^7 + \left(\frac{8}{3} + \frac{3}{\sqrt{2}} + \frac{5}{\sqrt{3}} + \frac{2}{\sqrt{5}} + \frac{2}{\sqrt{7}} + \frac{2}{\sqrt{15}}\right)z^8$$

$$+ \left(2 + \sqrt{\frac{2}{7}} + 3\sqrt{\frac{2}{5}} + 4\sqrt{\frac{2}{3}} + \frac{17}{3\sqrt{2}} + \sqrt{3} + \frac{1}{\sqrt{5}}\right)z^9$$

$$+ \left(\frac{14}{3} + 2\sqrt{\frac{3}{5}} + \frac{3}{\sqrt{2}} + 3\sqrt{3} + \frac{4}{\sqrt{5}} + \frac{7}{\sqrt{6}} + \frac{2}{\sqrt{7}} + \frac{2}{\sqrt{21}}\right)z^{10} + O(z^{11}).$$

The coefficients of z^n for positive integer n in (20.5.1) are the diagonal entries on the 2D y-z grid for identity (20.3.6), namely, lattice points along the line $y = z$ in the first quadrant.

$$\prod_{\substack{\gcd(a,b,c)=1\\ a,b,c\geq 1}} \left(\frac{1}{1-z^{a+b+c}}\right)^{\frac{1}{\sqrt[3]{abc}}} = \exp\left\{\left(\sum_{k=1}^{\infty} \frac{z^k}{\sqrt[3]{k}}\right)^3\right\} \tag{20.5.2}$$

$$= 1 + z^3 + \frac{3}{\sqrt[3]{2}}z^4 + \left(\frac{3}{\sqrt[3]{4}} + \sqrt[3]{9}\right)z^5 + \left(1 + \frac{3}{\sqrt[3]{4}} + \sqrt[3]{36}\right)z^6$$

$$+ \left(3 + \sqrt[3]{\frac{9}{4}} + \frac{3}{\sqrt[3]{2}} + \sqrt[3]{3} + \frac{3}{\sqrt[3]{5}}\right)z^7 + O(z^8).$$

The coefficients of z^n for positive integer n in (20.5.2) are the diagonal entries on the 3D x-y-z grid for identity (20.3.7), namely, lattice points along the 3D line with equation $x = y = z$ in the first hyperquadrant with x, y, and z all positive integers.

$$\prod_{\substack{\gcd(a,b,c,d)=1\\ a,b,c,d\geq 1}} \left(\frac{1}{1-z^{a+b+c+d}}\right)^{\frac{1}{\sqrt[4]{abcd}}} = \exp\left\{\left(\sum_{k=1}^{\infty} \frac{z^k}{\sqrt[4]{k}}\right)^4\right\}, \tag{20.5.3}$$

$$\prod_{\substack{\gcd(a,b,c,d,e)=1\\ a,b,c,d,e\geq 1}} \left(\frac{1}{1-z^{a+b+c+d+e}}\right)^{\frac{1}{\sqrt[5]{abcde}}} = \exp\left\{\left(\sum_{k=1}^{\infty} \frac{z^k}{\sqrt[5]{k}}\right)^5\right\}. \tag{20.5.4}$$

Each of the equations (20.5.1) to (20.5.4) can give us a combinatorial theorem on a relation between two arithmetical functions. For example, (20.5.1) can be rewritten as the equation

$$\prod_{k=2}^{\infty}\left(\frac{1}{1-z^k}\right)^{\sum \frac{1}{\sqrt{(ab)}}} = \exp\left\{\left(\sum_{k=1}^{\infty}\frac{z^k}{\sqrt{k}}\right)^2\right\}, \tag{20.5.5}$$

where $\sum \frac{1}{\sqrt{(ab)}}$ is the sum over all possible coprime a and b positive integers such that $a+b=k$.

Similarly, (20.5.2) can be rewritten as the equation

$$\prod_{k=3}^{\infty}\left(\frac{1}{1-z^k}\right)^{\sum \frac{1}{\sqrt[3]{abc}}} = \exp\left\{\left(\sum_{k=1}^{\infty}\frac{z^k}{\sqrt[3]{k}}\right)^3\right\}, \tag{20.5.6}$$

where $\sum \frac{1}{\sqrt[3]{abc}}$ is the sum over all possible triplewise coprime a, b, and c positive integers such that $a+b+c=k$.

The identities (20.5.5) and (20.5.6) seem *a priori* to have some useful aspect in relation to examining the *abc Conjecture*. This is due to the feature of the identities having coprime numbers in both a sum and also in a product within each term of the infinite product left sides.

We note also that both (20.5.5) and (20.5.6) may fit the conditions allowing us to apply the saddle point asymptotic Theorem of Meinardus, thereby determining the asymptotic behavior of coefficients of the infinite products that seem to relate to the *abc Conjecture*.

Let us first enlarge the theorem's positive coordinate hyperquadrant to include lattice points on each axis except for the highest or nth dimension. In other words, the product operator for variable z on each left side of (20.5.7) to (20.5.10) runs over each integer 1, 2, 3,... whereas for the non-z variables v,w,x,y, the product is over 0, 1, 2, 3,....

In our general term notation, with $b_1+b_2+b_3+...+b_n=1$, applied to slightly enlarge the number of lattice point vectors in the n-space hyperquadrant *radial from the origin* region, we sum on the particular lattice points with vectors defined by $\langle b_1,b_2,b_3,...,b_n\rangle$ such that

$$b_k = \begin{cases} 0, & 1 \le k \le n-1; \\ 1, & k = n. \end{cases}$$

Using this we can easily obtain the following infinite products in respectively 2D, 3D, 4D, and 5D space involving weighted VPV partitions in their combinatorial interpretations.

Corollary 20.5 *For each of* $|v|,|w|,|x|,|y|,|z|<1$,

$$\prod_{\substack{(a,b)=1\\ a\ge 0, b>0}} \left(1-y^a z^b\right)^{\frac{1}{b}} = (1-z)^{\frac{1}{1-y}}, \tag{20.5.7}$$

$$\prod_{\substack{(a,b,c)=1\\a,b\geq 0,c>0}} \left(1-x^a y^b z^c\right)^{\frac{1}{c}} = (1-z)^{\frac{1}{(1-x)(1-y)}}, \tag{20.5.8}$$

$$\prod_{\substack{(a,b,c,d)=1\\a,b,c\geq 0,d>0}} \left(1-w^a x^b y^c z^d\right)^{\frac{1}{d}} = (1-z)^{\frac{1}{(1-w)(1-x)(1-y)}}, \tag{20.5.9}$$

$$\prod_{\substack{(a,b,c,d,e)=1\\a,b,c,d\geq 0,e>0}} \left(1-v^a w^b x^c y^d z^e\right)^{\frac{1}{e}} = (1-z)^{\frac{1}{(1-v)(1-w)(1-x)(1-y)}}. \tag{20.5.10}$$

The above four infinite products and their reciprocals are worth deeper analysis as simple examples of weighted VPV partitions, giving us *exact results* reminiscent of the integer partition theorems. (20.5.7) to (20.5.10) are slight variations on particular cases of (20.5.1) to (20.5.2) and their 4D and 5D forms.

Let us for a moment examine the derivatives of the logarithms of both sides of (20.5.7) to (20.5.10). For each equation, we then are essentially performing the same operation, namely,

$$z\frac{\partial}{\partial z}\log(f(z)) = \frac{z}{f} \times \frac{\partial f}{\partial z}.$$

The results of that operation applied to each infinite product (20.5.7) to (20.5.10) are the multivariate summations,

Corollary 20.6 *For each of* $|v|, |w|, |x|, |y|, |z| < 1$,

$$\sum_{\substack{(a,b)=1\\a\geq 0,b>0}} \frac{y^a z^b}{1-y^a z^b} = \frac{z}{(1-y)(1-z)}, \tag{20.5.11}$$

$$\sum_{\substack{(a,b,c)=1\\a,b\geq 0,c>0}} \frac{x^a y^b z^c}{1-x^a y^b z^c} = \frac{z}{(1-x)(1-y)(1-z)}, \tag{20.5.12}$$

$$\sum_{\substack{(a,b,c,d)=1\\a,b,c\geq 0,d>0}} \frac{w^a x^b y^c z^d}{1-w^a x^b y^c z^d} = \frac{z}{(1-w)(1-x)(1-y)(1-z)}, \tag{20.5.13}$$

$$\sum_{\substack{(a,b,c,d,e)=1\\a,b,c,d\geq 0,e>0}} \frac{v^a w^b x^c y^d z^e}{1-v^a w^b x^c y^d z^e} = \frac{z}{(1-v)(1-w)(1-x)(1-y)(1-z)}. \tag{20.5.14}$$

Each of the above equations (20.5.11) to (20.5.14) are the integer multiple sums of the sum over VPVs in 2D, 3D, 4D, and 5D first hyperquadrants extended for z taken only over positive integer values, and all other variables y, x, w etc. summed over non-negative integers thereby including the y axis, x axis, w axis and so on, points.

They may be interesting to examine in the "near bijection" context that has been applied to the classical Euler pentagonal number theorem. This is a large topic probably beyond the scope of our present chapter. However, we can easily write the cases of corollary 20.5 above where $v = w = x = y = z$, so then

Corollary 20.7 *For* $|z| < 1$,

$$\prod_{\substack{(a,b)=1\\ a\geq 0,b>0}} \left(1-z^{a+b}\right)^{\frac{1}{b}} = \prod_{k=1}^{\infty}\left(1-z^k\right)^{\sum \frac{1}{b}} = (1-z)^{\frac{1}{1-z}} \tag{20.5.15}$$

$$= 1+z+2z^2+\frac{7z^3}{2}+\frac{35z^4}{6}+\frac{113z^5}{12}+\frac{593z^6}{40}+\frac{2743z^7}{120}+\frac{87251z^8}{2520}+O(z^9),$$

where $\sum \frac{1}{b}$ *is the sum over all possible coprime a and b integers* $a \geq 0, b > 0$ *such that* $a+b=k$.

$$\prod_{\substack{(a,b,c)=1\\ a,b\geq 0,c>0}} \left(1-z^{a+b+c}\right)^{\frac{1}{c}} = \prod_{k=1}^{\infty}\left(1-z^k\right)^{\sum \frac{1}{c}} = (1-z)^{\frac{1}{(1-z)^2}} \tag{20.5.16}$$

$$= 1+z+3z^2+7z^3+\frac{91z^4}{6}+\frac{95z^5}{3}+\frac{3859z^6}{60}+\frac{1531z^7}{12}+\frac{624707z^8}{2520}+O(z^9),$$

where $\sum \frac{1}{c}$ *is the sum over all possible coprime a, b, and c integers* $a,b \geq 0, c > 0$ *such that* $a+b+c=k$.

$$\prod_{\substack{(a,b,c,d)=1\\ a,b,c\geq 0,d>0}} \left(1-z^{a+b+c+d}\right)^{\frac{1}{d}} = \prod_{k=1}^{\infty}\left(1-z^k\right)^{\sum \frac{1}{d}} = (1-z)^{\frac{1}{(1-z)^3}} \tag{20.5.17}$$

$$= 1+z+4z^2+\frac{23z^3}{2}+30z^4+\frac{301z^5}{4}+\frac{22057z^6}{120}+\frac{52621z^7}{120}+\frac{215011z^8}{210}+O(z^9),$$

where $\sum \frac{1}{d}$ *is the sum over all possible coprime a, b, c, and d integers* $a,b,c \geq 0, d > 0$ *such that* $a+b+c+d=k$.

$$\prod_{\substack{(a,b,c,d,e)=1\\ a,b,c,d\geq 0,e>0}} \left(1-z^{a+b+c+d+e}\right)^{\frac{1}{e}} = \prod_{k=1}^{\infty}\left(1-z^k\right)^{\sum \frac{1}{e}} = (1-z)^{\frac{1}{(1-z)^4}} \tag{20.5.18}$$

$$= 1+z+5z^2+17z^3+\frac{154z^4}{3}+\frac{446z^5}{3}+\frac{6302z^6}{15}+\frac{5804z^7}{5}+\frac{790871z^8}{252}+O(z^9),$$

where $\sum \frac{1}{e}$ *is the sum over all possible coprime a, b, c, d, and e integers* $a,b,c,d \geq 0, e > 0$ *such that* $a+b+c+d+e=k$.

It would be an interesting study to examine the exponential generating function coefficients for (20.5.15) to (20.5.18) as weighted integer partitions that may exhibit congruence properties as was recently discovered for the exponential generating function for $\exp\left(\frac{z}{z-1}\right)$. (See Chapter 5.)

20.6 INTERPRETING A 3D VPV IDENTITY FOR WEIGHTED VECTOR PARTITIONS

The idea in this section is readily generalized to apply to nD VPV identities, but this we may do in a later chapter. Returning to the more general identities, take the example of equation (20.5.8). The right side product is a case of the binomial theorem, which when applied gives us,

$$\prod_{\substack{(a,b,c)=1\\a,b\geq 0,c>0}} \left(1-x^a y^b z^c\right)^{\frac{1}{c}} = (1-z)^{\frac{1}{(1-x)(1-y)}} \tag{20.6.1}$$

$$= 1 - \frac{z}{n1!} + \frac{(n-1)z^2}{n^2 2!} + \frac{(n-1)(2n-1)z^3}{n^3 3!} + \frac{(n-1)(2n-1)(3n-1)z^4}{n^4 4!}$$
$$+ \frac{(n-1)(2n-1)(3n-1)(4n-1)z^5}{n^5 5!} + \cdots, \quad where \quad n = (1-x)(1-y).$$

Looking closer at this, we see that (20.6.1) encodes a theorem about weighted 3-dimensional VPV partitions in the first hyperquadrant including the x and y axes but not the z axis.

It is also clear that the power series terminates to a polynomial in z whenever $(1-x)(1-y)$ equals any of $1, \frac{1}{2}, \frac{1}{3}, \frac{1}{4}, \ldots$. For partitions of these VPVs in the first hyperquadrant of Euclidean 3-space, each vector $\langle a,b,c \rangle$ has integer coordinates that satisfy $a,b \geq 0, c > 0$. Consider the left side of equation (20.6.1) written in the form

$$\prod_{\substack{(a,b,c)=1\\a,b\geq 0,c>0}} \left\{1 - \binom{1/c}{1} x^a y^b z^c + \binom{1/c}{2} x^{2a} y^{2b} z^{2c} - \binom{1/c}{3} x^{3a} y^{3b} z^{3c} + - \ldots \right\}. \tag{20.6.2}$$

This, when multiplied out in full will give a sum like

$$\sum_{a,b\geq 0,c>0} (-1)^{f_1(a,b,c)} \left(\prod_{f_2(a,b,c)} \left(\frac{1/c}{\gcd(a,b,c)} \right)^{f_3(a,b,c)} \right) x^a y^b z^c, \tag{20.6.3}$$

where $f_1(a,b,c)$, $f_2(a,b,c)$, and $f_3(a,b,c)$, are specified on qualities of the parts of the 3D vector partition.

The coefficient of $x^a y^b z^c$ in (20.6.3) is the **weight** of any partition generated by (20.6.1).

20.7 DERIVING SOME 2D TO 5D VPV IDENTITY CASES

We consider cases of (20.3.2) to (20.3.5) given here as the following identities in 2D, 3D, 4D, and 5D.

If $|y|<1, |z|<1$, then

$$\prod_{\substack{(a,b)=1\\a,b\geq 1}} \left(\frac{1}{1-y^a z^b}\right)^{\frac{1}{a^1 b^0}} = \exp\left\{\left(\sum_{i=1}^{\infty}\frac{y^i}{i^1}\right)\left(\sum_{j=1}^{\infty}\frac{z^j}{j^0}\right)\right\}. \tag{20.7.1}$$

If $|x|<1, |y|<1, |z|<1$ then

$$\prod_{\substack{(a,b,c)=1\\a,b,c\geq 1}} \left(\frac{1}{1-x^a y^b z^c}\right)^{\frac{1}{a^1 b^1 c^{-1}}} = \exp\left\{\left(\sum_{i=1}^{\infty}\frac{x^i}{i^1}\right)\left(\sum_{j=1}^{\infty}\frac{y^j}{j^1}\right)\left(\sum_{k=1}^{\infty}\frac{z^k}{k^{-1}}\right)\right\}. \tag{20.7.2}$$

If $|w|<1, |x|<1, |y|<1, |z|<1$ then

$$\prod_{\substack{(a,b,c,d)=1\\a,b,c,d\geq 1}} \left(\frac{1}{1-w^a x^b y^c z^d}\right)^{\frac{1}{a^1 b^1 c^1 d^{-2}}} = \exp\left\{\left(\sum_{h=1}^{\infty}\frac{w^h}{h^1}\right)\left(\sum_{i=1}^{\infty}\frac{x^i}{i^1}\right)\left(\sum_{j=1}^{\infty}\frac{y^j}{j^1}\right)\left(\sum_{k=1}^{\infty}\frac{z^k}{k^{-2}}\right)\right\}. \tag{20.7.3}$$

If $|v|<1, |w|<1, |x|<1, |y|<1, |z|<1$ then

$$\prod_{\substack{(a,b,c,d,e)=1\\a,b,c,d,e\geq 1}} \left(\frac{1}{1-v^a w^b x^c y^d z^e}\right)^{\frac{1}{a^1 b^1 c^1 d^1 e^{-3}}} \tag{20.7.4}$$

$$= \exp\left\{\left(\sum_{g=1}^{\infty}\frac{v^g}{g^1}\right)\left(\sum_{h=1}^{\infty}\frac{w^h}{h^1}\right)\left(\sum_{i=1}^{\infty}\frac{x^i}{i^1}\right)\left(\sum_{j=1}^{\infty}\frac{y^j}{j^1}\right)\left(\sum_{k=1}^{\infty}\frac{z^k}{k^{-3}}\right)\right\}.$$

It is not hard to see that these identities can be easily written up to 5D, 6D, 7D space etc. So, if we do this for 2D to 5D, and sum the elementary power series in the exponential term right hand sides, we arrive at

Corollary 20.8 *For each of* $|v|, |w|, |x|, |y|, |z| < 1$,

$$\prod_{\substack{(a,b)=1\\a,b\geq 1}} \left(1-y^a z^b\right)^{\frac{1}{a}} = (1-y)^{\frac{z}{1-z}}, \tag{20.7.5}$$

$$\prod_{\substack{(a,b,c)=1\\a,b,c\geq 1}} \left(1-x^a y^b z^c\right)^{\frac{c}{ab}} = ((1-x)(1-y))^{\frac{z}{(1-z)^2}}, \tag{20.7.6}$$

$$\prod_{\substack{(a,b,c,d)=1\\a,b,c,d\geq 1}} \left(1-w^a x^b y^c z^d\right)^{\frac{d^2}{abc}} = ((1-w)(1-x)(1-y))^{\frac{z(1+z)}{(1-z)^3}}, \tag{20.7.7}$$

$$\prod_{\substack{(a,b,c,d,e)=1\\a,b,c,d,e\geq 1}} \left(1-v^a w^b x^c y^d z^e\right)^{\frac{e^3}{abcd}} = ((1-v)(1-w)(1-x)(1-y))^{\frac{z(1+4z+z^2)}{(1-z)^4}}. \tag{20.7.8}$$

The identities (20.7.6) to (20.7.8) are probably new to the literature. The index functions on the right sides will be recognized by many as the well-known values of the non-positive integer polylogarithms, $Li_0(z), Li_{-1}(z), Li_{-2}(z), Li_{-3}(z)$.

So next we briefly give definitions and versions of the *abc Conjecture*, before in the section ensuing that, continuing discussion of equation (20.7.6) in the context of the conjecture. This shows plausible uses of VPV identities to examine known open problems from a new perspective.

20.8 DIVERSIONARY NOTE ON THE abc CONJECTURE

The *abc Conjecture* (also known as the Oesterlé-Masser conjecture) is a conjecture in number theory, first proposed by Joseph Oesterlé (1988) and David Masser (1985) (see [206] and [196]). It is stated in terms of three positive integers, a, b and c (hence the name) that are relatively prime and satisfy $a+b=c$. If d denotes the product of the distinct prime factors of abc, the conjecture essentially states that d is usually not much smaller than c. In other words: if a and b are composed from large powers of primes, then c is usually not divisible by large powers of primes. A number of famous conjectures and theorems in number theory would follow immediately from the *abc Conjecture* or its versions. Goldfeld (1996) (see [139]) described the *abc Conjecture* as *the most important unsolved problem in Diophantine analysis.*

The **abc Conjecture** originated as the outcome of attempts by Oesterlé and Masser to understand the Szpiro conjecture about elliptic curves, which involves more geometric structures in its statement than the abc Conjecture. The *abc Conjecture* was shown to be equivalent to the modified Szpiro's conjecture [291, 292].

Various attempts to prove the *abc Conjecture* have been made, but none are currently accepted by the mainstream mathematical community and as of 2022, the conjecture is still largely regarded as unproven.

Before we state the conjecture we introduce the notion of the radical of an integer: for a positive integer n, the radical of n, denoted $\mathrm{rad}(n)$, is the product of the distinct prime factors of n. For example

$\mathrm{rad}(16) = \mathrm{rad}(2^4) = \mathrm{rad}(2) = 2,$

$\mathrm{rad}(17) = 17,$

$\mathrm{rad}(18) = \mathrm{rad}(2 \times 3^2) = 2 \times 3 = 6,$

$\mathrm{rad}(1000000) = \mathrm{rad}(2^6 \times 5^6) = 2 \times 5 = 10.$

If a, b, and c are coprime positive integers such that $a+b=c$, it turns out that "usually" $c < \mathrm{rad}(abc)$. The *abc Conjecture* deals with the exceptions. Specifically, it states that:

Conjecture 20.1 *abc Conjecture (version I). For every positive real number ε, there exist only finitely many triples (a,b,c) of coprime positive integers, with $a+b=c$, such that $c > rad(abc)^{1+\varepsilon}$.*

An equivalent formulation:

abc Conjecture (version II). For every positive real number ε, there exists a constant K_ε such that for all triples (a,b,c) of coprime positive integers, with $a+b=c$: $c < K_\varepsilon rad(abc)^{1+\varepsilon}$.

Another equivalent formulation:

abc Conjecture (version III). For every positive real number ε, there exist only finitely many triples (a,b,c) of coprime positive integers with $a+b=c$ such that $q(a,b,c)>1+\varepsilon$.

This third equivalent formulation of the conjecture involves the quality $q(a,b,c)$ of the triple (a,b,c), defined as

$$q(a,b,c)=\frac{\log(c)}{\log(\operatorname{rad}(abc))}.$$

For example:

$q(4,127,131)=\frac{\log(131)}{\log(\operatorname{rad}(4\times 127\times 131))}=\frac{\log(131)}{\log(2\times 127\times 131)}=0.46820...$

$q(3,125,128)=\frac{\log(128)}{\log(\operatorname{rad}(3\times 125\times 128))}=\frac{\log(128)}{\log(30)}=1.426565...$

A typical triple (a,b,c) of coprime positive integers with $a+b=c$ will have $c<\operatorname{rad}(abc)$, i.e., $q(a,b,c)<1$. Triples with $q>1$ such as in the second example are rather special, they consist of numbers divisible by high powers of small prime numbers.

Whereas it is known that there are infinitely many triples (a,b,c) of coprime positive integers with $a+b=c$ such that $q(a,b,c)>1$, the conjecture predicts that only finitely many of those have $q>1.01$ or $q>1.001$ or even $q>1.0001$, etc. In particular, if the conjecture is true, then there must exist a triple (a,b,c) that achieves the maximal possible quality $q(a,b,c)$.

20.9 APPLYING A 3D VPV IDENTITY TO THE abc CONJECTURE

Three versions of the *abc Conjecture* were presented in the previous section. This enables us to relate the following identity and try to gain some insight into the conjecture. The feature in common between the *abc Conjecture* and a 3D VPV identity is the aspect that $a+b=c$ may be approached with $\gcd(a,b,c)=1$. Hence, there is scope to examine this potential crossover between theories.

So, we consider a form of equation (20.7.6) and state the identity as follows, applying a substitution parameter t.

We can choose complex number t so that for each of $|tx|,|ty|,|z/t|<1$, for $t\neq 0$,

$$\prod_{\substack{(a,b,c)=1\\ a,b,c\geq 1}}\left(1-t^{a+b-c}x^a y^b z^c\right)^{\frac{c}{ab}}=((1-tx)(1-ty))^{\frac{z}{(t-z)^2}}.$$

Consider the case where $x\to 1, y\to 1$ so that the identity becomes

$$\prod_{\substack{(a,b,c)=1\\ a,b,c\geq 1}}\left(1-t^{a+b-c}z^c\right)^{\frac{c}{ab}}=(1-t)^{\frac{2z}{(t-z)^2}},\tag{20.9.1}$$

convergent for $0<|z|<|t|<1$. We can expand the right side of (20.9.1) as a power series in z as follows

$$(1-t)^{\frac{2z}{(t-z)^2}}=1+\frac{2\log(1-t)}{t^2}z+\frac{2\log(1-t)(2t+\log(1-t))}{t^4}z^2\tag{20.9.2}$$

$$+\frac{2\log(1-t)(9t^2+2\log^2(1-t)+12t\log(1-t))}{3t^6}z^3$$

$$+\frac{2\log(1-t)(12t^3+30t^2\log(1-t)+\log^3(1-t)+12t\log^2(1-t))}{3t^8}z^4$$

$$+\frac{2\log(1-t)(75t^4+300t^3\log(1-t)+210t^2\log^2(1-t)+2\log^4(1-t)+40t\log^3(1-t)}{15t^{10}}z^5$$

$$+O(z^6).$$

This is equivalent to the the statement

$$(1-t)^{\frac{2z}{(t-z)^2}} = \sum_{n=0}^{\infty} c_n z^n; \text{ for } c_0 = 1;\ c_1 = \frac{2\log(1-t)}{t^2};\ c_2 = \frac{2\log(1-t)(2t+\log(1-t))}{t^4}$$

and the other coefficients are determined from the recurrence

$$nc_n - (3t(n+1) - 2\log(1-t))\,c_{n+1} + t\,(3t(n+2) + 2\log(1-t))\,c_{n+2} - (n+3)t^3 c_{n+3} = 0.$$

We can similarly expand the right side of (20.9.1) as a power series in t as follows

$$(1-t)^{\frac{2z}{(t-z)^2}} = 1 - \frac{2}{z}t - \frac{z+2}{z^2}t^2 - \frac{2(z^2-1)}{3z^3}t^3 + \frac{-3z^3+3z^2+18z+28}{6z^4}t^4 \qquad (20.9.3)$$

$$+\frac{-6z^4+10z^3+45z^2+80z+106}{15z^5}t^5 + O(t^6)$$

The *abc Conjecture* is concerned with cases where $a+b=c$, and also asserts:

"For every positive real number ε, there exist only finitely many triples (a,b,c) of coprime positive integers, with $a+b=c$, such that $c > \text{rad}(abc)^{1+\varepsilon}$."

In the infinite product (20.9.1) we have covered off all cases where $a+b=c$, and convergence of the product to a limit implies certain things about the comparison of both $c-a-b$ and $\frac{c}{ab}$ terms as c increases, perhaps suggesting features of $\text{rad}(abc)$.

Our analysis, while not settling the truth or otherwise of the conjecture, may yet provide new insights. The identities (20.7.6) to (20.7.8) can be used to apply to extended versions of the *abc Conjecture*, to say an *abcd Conjecture* or an *abcde Conjecture*, but this idea may find expression in further researches, utilizing the Visible Point Vector identities to yield estimates for an entirely new theory relating asymptotic properties of integer sum and product divisibilities.

20.10 HYPERDIAGONAL LINE GENERATING FUNCTIONS FOR DIFFERENT nD SLOPES

At the start of section 20.5 of this chapter we saw how to calculate the hyperdiagonal generating function for an nD generated vector partition grid. Basically, we have the

Statement 5 *If the nD generating function for the entire nD grid is the n variable* $f_n(z_1, z_2, z_3, ..., z_n)$*, the equation of the nD hyperdiagonal line from the origin is* $z_1 = z_2 = z_3 = ... = z_n$ *and so the generating function for the nD vector partitions along that line is given by the single variable function* $f_n(z, z, ..., z)$.

That is, even though an nD space grid is impossible for humans to envision, we can define a straight line through the hyperdiagonal from the nD origin point, and formulate an exact value of the vector partition function at any point along that hyperdiagonal line.

Furthermore, taking an arbitrary lattice point vector $\langle a_1, a_2, a_3, ..., a_n\rangle$ in the nD grid we can show that the vector partition function (or nD coefficient) for that lattice point is exactly evaluated as follows.

Statement 6 *If the nD generating function for the entire nD grid is the n variable* $f_n(z_1, z_2, z_3, ..., z_n)$*, the equation of the nD hyper-radial-from-origin line from the origin to the arbitrary point* $\langle a_1, a_2, a_3, ..., a_n\rangle$ *is based on knowing values of*

$$\langle a_1, a_2, a_3, ..., a_n\rangle = r\langle c_1, c_2, c_3, ..., c_n\rangle \tag{20.10.1}$$

where $\langle c_1, c_2, c_3, ..., c_n\rangle$ *is a VPV with* $\gcd(c_1, c_2, c_3, ..., c_n) = 1$ *and* r *is the unique positive integer that makes this true. So the straight line from the origin to the point* $\langle a_1, a_2, a_3, ..., a_n\rangle$ *has the defining equation* $c_1 z_1 = c_2 z_2 = c_3 z_3 = ... = c_n z_n$ *and so the generating function for the nD vector partitions along that line is given by the single variable function* $f_n(c_1 z, c_2 z, c_3 z, ..., c_n z)$*, which has the coefficient of* z^r *equal to the vector partition function (or nD coefficient) for that lattice point* $\langle a_1, a_2, a_3, ..., a_n\rangle$ *exactly evaluated.*

EXERCISES

20.1 *2D first quadrant sums using* gcd. For each of $|y|, |z| < 1$, show that

$$\sum_{\substack{\gcd(a,b)=1\\ a\geq 0, b\geq 1}} \frac{y^a z^b}{1 - y^a z^b} = \frac{z}{(1-y)(1-z)}. \tag{P20.1}$$

Similarly, show that for $\Re y > 1$, $\Re z > 1$,

$$\sum_{\substack{\gcd(a,b)=1\\ a,b\geq 1}} \frac{1}{a^y b^z} = \frac{\zeta(y)\zeta(z)}{\zeta(y+z)}. \tag{P20.2}$$

Infer from (P20.1) and (P20.2) that they encode two similar statements about partitions into lattice point vectors $\langle a, b\rangle$ with $gcd(a, b) = 1$ and for (P20.1) we have non-negative integers a and positive integers b; whereas for (P20.2) we have positive integers for both a and b.

20.2 *Find the particular cases of (P20.1).* Prove the following:

$$\sum_{\substack{\gcd(a,b)=1\\ a\geq 0,b\geq 1}} \frac{1}{3^n 2^n - 1} = \frac{2^{-n}}{(1-3^{-n})(1-2^{-n})} \quad (for\ \Re n > 1),$$

$$= \frac{1}{2^n}\left(1+\frac{1}{2^n}+\frac{1}{3^n}+\frac{1}{4^n}+\frac{1}{6^n}+\frac{1}{8^n}+\frac{1}{9^n}+\frac{1}{12^n}+\cdots\right);$$

$$\sum_{\substack{\gcd(a,b)=1\\ a\geq 0,b\geq 1}} \frac{z^{a+b}}{1-z^{a+b}} = \frac{z}{(1-z)^2}, \quad |z|<1;$$

$$\sum_{\substack{\gcd(a,b)=1\\ a\geq 0,b\geq 1}} tan^{2(a+b)}(\theta) = \frac{1}{cos^2(\theta)}.$$

20.3 *Find the particular cases of (P20.2).* Prove the following:

$$\sum_{\substack{\gcd(a,b)=1\\ a,b\geq 1}} \frac{1}{a^2 b^2} = \frac{5}{2},$$

$$\sum_{\substack{\gcd(a,b)=1\\ a,b\geq 1}} \frac{1}{a^2 b^3} = \frac{\pi^2\zeta(3)}{6\zeta(5)},$$

$$\sum_{\substack{\gcd(a,b)=1\\ a,b\geq 1}} \frac{1}{a^3 b^5} = \frac{9450\zeta(3)\zeta(5)}{\pi^8}.$$

20.4 *3D first hyperquadrant sums using* gcd. For each of $|x|,|y|,|z|<1$, show that

$$\sum_{\substack{\gcd(a,b,c)=1\\ a,b\geq 0,c>0}} \frac{x^a y^b z^c}{1-x^a y^b z^c} = \frac{z}{(1-x)(1-y)(1-z)}. \qquad \text{(P20.3)}$$

Similarly, show that for $\Re x>1$, $\Re y>1$, $\Re z>1$,

$$\sum_{\substack{\gcd(a,b,c)=1\\ a,b,c\geq 1}} \frac{1}{a^x b^y c^z} = \frac{\zeta(x)\zeta(y)\zeta(z)}{\zeta(x+y+z)}. \qquad \text{(P20.4)}$$

Infer from (P20.3) and (P20.4) that they encode two similar statements about partitions into lattice point vectors $\langle a,b,c\rangle$ with $gcd(a,b,c)=1$ and for (P20.3) we have non-negative integers a, b and positive integers c; whereas for (P20.4) we have positive integers for a, b, and c.

20.5 *4D hyperquadrant VPV sums using* gcd. For each of $|w|,|x|,|y|,|z|<1$, show that

$$\sum_{\substack{\gcd(a,b,c,d)=1\\ a,b,c\geq 0,d>0}} \frac{w^a x^b y^c z^d}{1-w^a x^b y^c z^d} = \frac{z}{(1-w)(1-x)(1-y)(1-z)}. \qquad \text{(P20.5)}$$

Similarly, show that for $\Re w$, $\Re x$, $\Re y$, and $\Re z$ all > 1,

$$\sum_{\substack{\gcd(a,b,c,d)=1 \\ a,b,c,d \geq 1}} \frac{1}{a^w b^x c^y d^z} = \frac{\zeta(w)\zeta(x)\zeta(y)\zeta(z)}{\zeta(w+x+y+z)}. \quad \text{(P20.6)}$$

Infer from (P20.5) and (P20.6) that they encode two similar statements about partitions into lattice point vectors $\langle a,b,c,d\rangle$ with $gcd(a,b,c,d) = 1$ and for (P20.5) we have non-negative integers a, b, c and positive integers d; whereas for (P20.6) we have positive integers for a, b, c, and d.

20.6 *5D hyperquadrant VPV sums using* gcd. For each of $|v|, |w|, |x|, |y|, |z| < 1$, show that

$$\sum_{\substack{\gcd(a,b,c,d,e)=1 \\ a,b,c,d \geq 0, e>0}} \frac{v^a w^b x^c y^d z^e}{1 - v^a w^b x^c y^d z^e} = \frac{z}{(1-v)(1-w)(1-x)(1-y)(1-z)}. \quad \text{(P20.7)}$$

Similarly, show that for $\Re v$, $\Re w$, $\Re x$, $\Re y$, and $\Re z$ all > 1,

$$\sum_{\substack{\gcd(a,b,c,d,e)=1 \\ a,b,c,d,e \geq 1}} \frac{1}{a^v b^w c^x d^y e^z} = \frac{\zeta(w)\zeta(w)\zeta(x)\zeta(y)\zeta(z)}{\zeta(v+w+x+y+z)}. \quad \text{(P20.8)}$$

Infer from (P20.7) and (P20.8) that they encode two similar statements about partitions into lattice point vectors $\langle a,b,c,d\rangle$ with $gcd(a,b,c,d,e) = 1$ and for (P20.7) we have non-negative integers a, b, c, d and positive integers e; whereas for (P20.8) we have positive integers for a, b, c, d and e.

20.7 *nD hyperquadrant sums using* gcd. For each of $|q_1|, |q_2|, |q_3|, \ldots, |q_n| < 1$, show that

$$\sum_{\substack{\gcd(a_1,a_2,a_3,\ldots,a_n)=1 \\ a_1,a_2,\ldots,a_{n-1} \geq 0, a_n \geq 1}} \frac{q_1^{a_1} q_2^{a_2} q_3^{a_3} \cdots q_n^{a_n}}{1 - q_1^{a_1} q_2^{a_2} q_3^{a_3} \cdots q_n^{a_n}} = \frac{q_n}{(1-q_1)(1-q_2)(1-q_3)\cdots(1-q_n)}. \quad \text{(P20.9)}$$

Similarly, show that for $\Re q_1$, $\Re q_2$, $\Re q_3$, $\ldots$, and $\Re q_n$ all > 1,

$$\sum_{\substack{\gcd(a_1,a_2,a_3,\ldots,a_n)=1 \\ a_1,a_2 \ldots a_n \geq 1}} \frac{1}{a_1^{q_1} a_2^{q_2} a_3^{q_3} \cdots a_n^{q_n}} = \frac{\zeta(q_1)\zeta(q_2)\zeta(q_3)\ldots\zeta(q_n)}{\zeta(q_1+q_2+q_3+\cdots+q_n)}. \quad \text{(P20.10)}$$

Infer from (P20.7) and (P20.10) that they encode two similar statements about partitions into lattice point vectors $\langle a_1, a_2, a_3, \ldots, a_n\rangle$ with $gcd(a_1, a_2, a_3, \ldots, a_n) = 1$ and for (P20.7) we have non-negative integers a_1, a_2, $\ldots$, a_{n-1} and positive integers a_n; whereas for (P20.10) we have positive integers for a_1, a_2, $\ldots$, up to a_n.

20.8 *Derive the following from corollary 20.5.* For each of $|y|, |z| < 1$,

$$\prod_{\substack{\gcd(a,b)=1\\ a\geq 0, b>0}} \left(1+y^a z^b\right)^{\frac{1}{b}} = \frac{(1-z^2)^{\frac{1}{1-y^2}}}{(1-z)^{\frac{1}{1-y}}} \tag{P20.11}$$

$$= \sum_{n=0}^{\infty} \frac{\Gamma\left(n+\frac{1}{1-y}\right)}{\Gamma\left(\frac{1}{1-y}\right)} \, {}_3F_2\left(\frac{1}{2}-\frac{n}{2}, -\frac{n}{2}, \frac{-1}{1-y^2}; \frac{1}{2}-\frac{n}{2}-\frac{1}{2(1-y)}, 1-\frac{n}{2}-\frac{1}{2(1-y)}; 1\right) \frac{z^n}{n!}.$$

20.9 *Verify the previous exercise.* The reader may choose to verify (P20.11), by entering the code

series $(\mathbf{1-z^2})$^$(\mathbf{1/((1-x^2)(1-y^2))})/(\mathbf{1-z})$^$(\mathbf{1/((1-x)(1-y))})$ at $\mathbf{z=0}$.

at an online calculating engine.

20.10 *Also using corollary 20.5, derive the following 3D result.*

For each of $|x|, |y|, |z| < 1$,

$$\prod_{\substack{\gcd(a,b,c)=1\\ a,b\geq 0, c>0}} \left(1+x^a y^b z^c\right)^{\frac{1}{c}} = \frac{(1-z^2)^{\frac{1}{(1-x^2)(1-y^2)}}}{(1-z)^{\frac{1}{(1-x)(1-y)}}} = \sum_{n=0}^{\infty} \frac{\Gamma(n+\frac{1}{(1-x)(1-y)})\,\mathbf{F(n)}z^n}{\Gamma(\frac{1}{(1-x)(1-y)})n!}, \tag{P20.12}$$

where $\mathbf{F(n)}$ is the hypergeometric series

$${}_3F_2\left(\frac{1}{2}-\frac{n}{2}, \frac{-n}{2}, \frac{-1}{(1-x^2)(1-y^2)}; \frac{1}{2}-\frac{n}{2}-\frac{1}{2(1-x)(1-y)}, 1-\frac{n}{2}-\frac{1}{2(1-x)(1-y)}; 1\right).$$

20.11 *Using corollary 20.5, derive the following 4D result.*

For each of $|w|, |x|, |y|, |z| < 1$,

$$\prod_{\substack{\gcd(a,b,c,d)=1\\ a,b,c\geq 0, d>0}} \left(1+w^a x^b y^c z^d\right)^{\frac{1}{d}} = \frac{(1-z^2)^{\frac{1}{(1-w^2)(1-x^2)(1-y^2)}}}{(1-z)^{\frac{1}{(1-w)(1-x)(1-y)}}}. \tag{P20.13}$$

20.12 *Using corollary 20.5, derive the following 5D result.*

For each of $|v|, |w|, |x|, |y|, |z| < 1$,

$$\prod_{\substack{\gcd(a,b,c,d,e)=1\\ a,b,c,d\geq 0, e>0}} \left(1+v^a w^b x^c y^d z^e\right)^{\frac{1}{e}} = \frac{(1-z^2)^{\frac{1}{(1-v^2)(1-w^2)(1-x^2)(1-y^2)}}}{(1-z)^{\frac{1}{(1-v)(1-w)(1-x)(1-y)}}}. \tag{P20.14}$$

20.13 *Derive the following from corollary 20.8.*

For each of $|y|, |z| < 1$,

$$\prod_{\substack{\gcd(a,b)=1 \\ a,b\geq 1}} \left(1+y^a z^b\right)^{\frac{1}{a}} = \frac{(1-y^2)^{\frac{z^2}{1-z^2}}}{(1-y)^{\frac{z}{1-z}}}. \tag{P20.15}$$

20.14 *Also using corollary 20.8, derive the following 3D result.*

For each of $|x|, |y|, |z| < 1$,

$$\prod_{\substack{\gcd(a,b,c)=1 \\ a,b,c\geq 1}} \left(1+x^a y^b z^c\right)^{\frac{c}{ab}} = \frac{((1-x^2)(1-y^2))^{\frac{z^2}{(1-z^2)^2}}}{((1-x)(1-y))^{\frac{z}{(1-z)^2}}}. \tag{P20.16}$$

20.15 *Using corollary 20.8, derive the following 4D result.*

For each of $|w|, |x|, |y|, |z| < 1$,

$$\prod_{\substack{\gcd(a,b,c,d)=1 \\ a,b,c,d\geq 1}} \left(1+w^a x^b y^c z^d\right)^{\frac{d^2}{abc}} = \frac{((1-w^2)(1-x^2)(1-y^2))^{\frac{z^2(1+z^2)}{(1-z^2)^3}}}{((1-w)(1-x)(1-y))^{\frac{z(1+z)}{(1-z)^3}}}. \tag{P20.17}$$

20.16 *Using corollary 20.8, derive the following 5D result.*

For each of $|v|, |w|, |x|, |y|, |z| < 1$,

$$\prod_{\substack{\gcd(a,b,c,d,e)=1 \\ a,b,c,d,e\geq 1}} \left(1+v^a w^b x^c y^d z^e\right)^{\frac{e^3}{abcd}} = \frac{((1-v^2)(1-w^2)(1-x^2)(1-y^2))^{\frac{z^2(1+4z^2+z^4)}{(1-z^2)^4}}}{((1-v)(1-w)(1-x)(1-y))^{\frac{z(1+4z+z^2)}{(1-z)^4}}}. \tag{P20.18}$$

Think About It...

It is our task, both in science and in society at large, to prove the conventional wisdom wrong and to make our unpredictable dreams come true.

Freeman Dyson (Physicist/Mathematician, 1923–2020).
From Eros to Gaia, p.94, Pantheon

FURTHER READING

CAMPBELL, G.B. A new class of infinite products, and Euler's totient. *Int. J. Math. and Math. Sci.*, 17(3):417–422, 1994.

CAMPBELL, G.B. Infinite products over visible lattice points. *Int. J. Math. and Math. Sci.*, 17(4):637–654, 1994.

CAMPBELL, G.B. *Combinatorial identities in number theory related to q-series and arithmetical functions.* Doctor of philosophy thesis, School of Mathematical Sciences, The Australian National University, October 1997.

CAMPBELL, G.B. A closer look at some new identities. *Int. J. Math. and Math. Sci.*, 21(3):581–586, 1998.

CAMPBELL, G.B. Infinite products over hyperpyramid lattices. *Int. J. Math. and Math. Sci.*, 23(4):271–277, 2000.

CAMPBELL, G.B. Some n-space q-binomial theorem extensions and similar identities. arxiv preprint, June 2019. arXiv:1906.07526v1 [math.NT]: https://arxiv.org/abs/1906.07526.

GOLDFELD, D. Beyond the last theorem. *Math Horizons*, 4:26–34, September 1996.

MASSER, D.W. *Open problems.* Proceedings of the Symposium on Analytic Number Theory. London: Imperial College, 1985. Chen, W. W. L. (ed.).

OESTERLÉ, J. Nouvelles approches du "théomème" de Fermat. *Asteérisque, Séminaire Bourbaki exp.*, 694(161):165–186, 1988. ISSN 0303-1179, MR 0992208.

RIEMANN, G.F.B. Über die Anzahl der Primzahlen unter einer gegebenen Grösse. *Monatsber. Königl. Preuss. Akad. Wiss. Berlin*, pages 671–680, Nov 1859.

SLOANE, N.J.A. Expansion of e.g.f. exp(x/(x-1)). The On-Line Encyclopedia of Integer Sequences (OEIS), 2023. sequence A293116. (www.oeis.org/A293116).

SZPIRO, L. Propriétés numériques du faisceau dualisant rélatif. *Astérisque.*, 86:44–78, 1981. Seminaire sur les pinceaux des courbes de genre au moins deux. Zbl 0517.14006.

SZPIRO, L. Présentation de la théorie d'Arakelov. *Contemporary Mathematics*, 67:279–293, 1987. Zbl 0634.14012. 27.

CHAPTER 21

Visible Point Vector Identities in Hyperpyramid Lattices

21.1 VPV IDENTITIES IN SQUARE HYPERPYRAMID REGIONS

In the decade up to 2000, the author published papers that culminated in the 2000 paper on hyperpyramid "Visible Point Vector" (VPV) identities. (See Campbell [84, 85, 87, 88, 89, 90] then [91].) These papers although breaking new ground, were not at the time taken any further than a statement of a general theorem and a few prominent examples. Since then, these identities have not been developed further in the literature, despite there being evidently a large number of ways the Parametric Euler Sum Identities of the 21st century along with experimental computation results are definitely applicable. Partitions and lattice sums have been developed "in their separate lanes" in mathematical analysis, statistical mechanics physics, and in chemistry, without any great attempt to relate these areas of contemporary research. Add to this the possibility that light diffusion lattice models, random walk regimes, and stepping stone weighted partitions seem fundamentally applicable in contexts of VPV identities, and it becomes clear that the transition from integer partitions to vector partitions may be a path for future researches. In order to bring some of the developments by the author since 2000 into the public domain, the research developments have been recently summarized in Campbell [93, 96, 97, 98, 99], which include content from several chapters of the present book.

So, to be clear, we state the basic 2D, 3D, and 4D identities as follows.

Statement 7 *For* $|y|, |z| < 1$,

$$\prod_{\substack{gcd(a,b)=1 \\ a<b \\ a\geq 0,\, b\geq 1}} \left(\frac{1}{1-y^a z^b}\right)^{\frac{1}{b}} = \left(\frac{1-yz}{1-z}\right)^{\frac{1}{1-y}} \tag{21.1.1}$$

DOI: 10.1201/9781003174158-21

$$= 1 + \frac{z}{1!} + \begin{vmatrix} 1 & -1 \\ \frac{1-y^2}{1-y} & 1 \end{vmatrix} \frac{z^2}{2!} + \begin{vmatrix} 1 & -1 & 0 \\ \frac{1-y^2}{1-y} & 1 & -2 \\ \frac{1-y^3}{1-y} & \frac{1-y^2}{1-y} & 1 \end{vmatrix} \frac{z^3}{3!} + \begin{vmatrix} 1 & -1 & 0 & 0 \\ \frac{1-y^2}{1-y} & 1 & -2 & 0 \\ \frac{1-y^3}{1-y} & \frac{1-y^2}{1-y} & 1 & -3 \\ \frac{1-y^4}{1-y} & \frac{1-y^3}{1-y} & \frac{1-y^2}{1-y} & 1 \end{vmatrix} \frac{z^4}{4!} + etc.$$

Statement 8 *For each of* $|x|, |y|, |z| < 1$,

$$\prod_{\substack{gcd(a,b,c)=1 \\ a,b<c \\ a,b\geq 0,\ c>0}} \left(\frac{1}{1 - x^a y^b z^c} \right)^{\frac{1}{c}} = \left(\frac{(1-xz)(1-yz)}{(1-z)(1-xyz)} \right)^{\frac{1}{(1-x)(1-y)}} \tag{21.1.2}$$

$$= 1 + \frac{z}{1!} + \begin{vmatrix} 1 & -1 \\ \frac{(1-x^2)(1-y^2)}{(1-x)(1-y)} & 1 \end{vmatrix} \frac{z^2}{2!} + \begin{vmatrix} 1 & -1 & 0 \\ \frac{(1-x^2)(1-y^2)}{(1-x)(1-y)} & 1 & -2 \\ \frac{(1-x^3)(1-y^3)}{(1-x)(1-y)} & \frac{(1-x^2)(1-y^2)}{(1-x)(1-y)} & 1 \end{vmatrix} \frac{z^3}{3!}$$

$$+ \begin{vmatrix} 1 & -1 & 0 & 0 \\ \frac{(1-x^2)(1-y^2)}{(1-x)(1-y)} & 1 & -2 & 0 \\ \frac{(1-x^3)(1-y^3)}{(1-x)(1-y)} & \frac{(1-x^2)(1-y^2)}{(1-x)(1-y)} & 1 & -3 \\ \frac{(1-x^4)(1-y^4)}{(1-x)(1-y)} & \frac{(1-x^3)(1-y^3)}{(1-x)(1-y)} & \frac{(1-x^2)(1-y^2)}{(1-x)(1-y)} & 1 \end{vmatrix} \frac{z^4}{4!} + etc.$$

Statement 9 *For each of* $|w|, |x|, |y|, |z| < 1$,

$$\prod_{\substack{gcd(a,b,c,d)=1 \\ a,b,c<d \\ a,b,c\geq 0,\ d>0}} \left(\frac{1}{1 - w^a x^b y^c z^d} \right)^{\frac{1}{d}} = \left(\frac{(1-wz)(1-xz)(1-yz)(1-wxyz)}{(1-z)(1-wxz)(1-wyz)(1-xyz)} \right)^{\frac{1}{(1-w)(1-x)(1-y)}}, \tag{21.1.3}$$

$$= 1 + \frac{z}{1!} + \begin{vmatrix} 1 & -1 \\ \frac{(1-w^2)(1-x^2)(1-y^2)}{(1-w)(1-x)(1-y)} & 1 \end{vmatrix} \frac{z^2}{2!}$$

$$+ \begin{vmatrix} 1 & -1 & 0 \\ \frac{(1-w^2)(1-x^2)(1-y^2)}{(1-w)(1-x)(1-y)} & 1 & -2 \\ \frac{(1-w^3)(1-x^3)(1-y^3)}{(1-w)(1-x)(1-y)} & \frac{(1-w^2)(1-x^2)(1-y^2)}{(1-w)(1-x)(1-y)} & 1 \end{vmatrix} \frac{z^3}{3!}$$

$$+ \begin{vmatrix} 1 & -1 & 0 & 0 \\ \frac{(1-w^2)(1-x^2)(1-y^2)}{(1-w)(1-x)(1-y)} & 1 & -2 & 0 \\ \frac{(1-w^3)(1-x^3)(1-y^3)}{(1-w)(1-x)(1-y)} & \frac{(1-w^2)(1-x^2)(1-y^2)}{(1-w)(1-x)(1-y)} & 1 & -3 \\ \frac{(1-w^4)(1-x^4)(1-y^4)}{(1-w)(1-x)(1-y)} & \frac{(1-w^3)(1-x^3)(1-y^3)}{(1-w)(1-x)(1-y)} & \frac{(1-w^2)(1-x^2)(1-y^2)}{(1-w)(1-x)(1-y)} & 1 \end{vmatrix} \frac{z^4}{4!} + etc.$$

Each of the above identities gives us exact enumerations of certain functions of vector partitions. Suppose we say the z variable is a "vertical axis" upon which to plot the 2D, 3D, and 4D graph for the type of partitions and partition functions under consideration. Then the power series in z give us exact determinant representations at each 1D, 2D or 3D layer corresponding to $z = 1, 2, 3, \ldots$.

The proofs of the power series determinant coefficient function in (21.1.1) to (21.1.3) rely only on applying Cramer's Rule to the coefficient recurrences, as well as differentiating the logarithms of both sides of the infinite products and their closed form evaluations.

So, in ensuing pages, we give the simplest n-space hyperpyramid VPV theorem due to the author in [91]. The so-called "Skewed Hyperpyramid n-space Identities" from [91] we shall cover in a later chapter. The application of the determinant coefficient technique of our earlier work on hyperquadrant lattices is applicable and bearing some semblance to the q-binomial from earlier chapters' variants. Note that for most identities in this chapter, the left side products are taken over a set of integer lattice points inside an inverted 2D triangle lattice, or 3D pyramid, or 3+ dimension hyperpyramid on the Euclidean cartesian space.

In the first 15 years of the 21st century the summations found by the Borwein brothers Peter and Jonathan, their father David with their colleagues, see [63] to [69] have renewed interest in the old Euler Sums. Their results give us particular values of polylogarithms and related functions popularized by Lewin [183, 184, 185] involving the generalized Harmonic numbers. This work has been developed some way over nearly two decades so now we speak of the Mordell-Tornheim-Witten sums, which are polylogarithm generalizations all seen to be applicable to the VPV identities, but that connection is not yet fully worked through in the present literature. These newer researches including experimentally calculated results can, many of them, be substituted into VPV identities to give us exact results for weighted vector partitions. To make sense of these new results, we need to go back to fundamental definitions and ideas for partitions of vectors as distinct from those well considered already for integer partitions.

We examine the following correspondences. From the elementary generating function for unrestricted integer partitions, we have,

$$\prod_{n=1}^{\infty} \left(1 + x^{1n} + x^{2n} + x^{3n} + \ldots\right) = 1 + p(1)x + p(2)x^2 + p(3)x^3 + \ldots. \tag{21.1.4}$$

So, equations (21.1.1) to (21.1.3) similarly imply

$$\prod_{\substack{gcd(a,b)=1;\, a<b \\ a\geq 0,\, b\geq 1}} \left(1 + \binom{1/b}{1}(y^a z^b) + \binom{1/b}{2}(y^a z^b)^2 + \binom{1/b}{3}(y^a z^b)^3 + \ldots\right) \tag{21.1.5}$$

$$= \sum_{\substack{a<b \\ a\geq 0,\, b\geq 1}} V_2(a,b) y^a z^b.$$

$$\prod_{\substack{gcd(a,b,c)=1; \\ a,b<c \\ a,b\geq 0,\, c\geq 1}} \left(1 + \binom{1/c}{1}(x^a y^b z^c) + \binom{1/c}{2}(x^a y^b z^c)^2 + \binom{1/c}{3}(x^a y^b z^c)^3 + \ldots\right)$$

(21.1.6)

$$= \sum_{\substack{a,b<c \\ a,b\geq 0,\ c\geq 1}} V_3(a,b,c)x^a y^b z^c.$$

$$\prod_{\substack{gcd(a,b,c,d)=1; \\ a,b,c<d \\ a,b,c\geq 0,\ d\geq 1}} \left(1+\binom{1/d}{1}(w^a x^b y^c z^d)+\binom{1/d}{2}(w^a x^b y^c z^d)^2+\binom{1/d}{3}(w^a x^b y^c z^d)^3+\ldots\right)$$

$$= \sum_{\substack{a,b,c<d \\ a,b,c\geq 0,\ d\geq 1}} V_4(a,b,c,d)w^a x^b y^c z^d. \tag{21.1.7}$$

Equations (21.1.5) to (21.1.7) ought to give us exact closed-form generating functions V_2, V_3, and V_4, in 2D, 3D, and 4D space respectively, mapping each integer partition across all partitions of each $p(n)$ onto the new V function. We put this notion aside for later researches.

21.2 DERIVING 2D VPV IDENTITIES IN EXTENDED TRIANGLE REGIONS

In this section, we derive the $2D$ Visible Point Vector identities from essentially creating them out of a simple summation transformation based on the simple idea that the integer lattice points in the first quadrant have co-ordinates that are either coprime integer pairs namely, "lattice points visible from the origin", or co-ordinates that are the integer multiples of the coprime pairs. As we did in the hyperquadrant chapter, we again start with a simple $2D$ summation. Consider

$$\sum_{n=1}^{\infty}\left(\sum_{m=1}^{n}\frac{y^m}{m^a}\right)\frac{z^n}{n^b}$$

$$= \left(\frac{y^1}{1^a}\right)\frac{z^1}{1^b}+\left(\frac{y^1}{1^a}+\frac{y^2}{2^a}\right)\frac{z^2}{2^b}+\left(\frac{y^1}{1^a}+\frac{y^2}{2^a}+\frac{y^3}{3^a}\right)\frac{z^3}{3^b}+\left(\frac{y^1}{1^a}+\frac{y^2}{2^a}+\frac{y^3}{3^a}+\frac{y^4}{4^a}\right)\frac{z^4}{4^b}$$

$$+\left(\frac{y^1}{1^a}+\frac{y^2}{2^a}+\frac{y^3}{3^a}+\frac{y^4}{4^a}+\frac{y^5}{5^a}\right)\frac{z^5}{5^b}+\left(\frac{y^1}{1^a}+\frac{y^2}{2^a}+\frac{y^3}{3^a}+\frac{y^4}{4^a}+\frac{y^5}{5^a}+\frac{y^6}{6^a}\right)\frac{z^6}{6^b}+\cdots$$

$$= \frac{y^1z^1}{1^a1^b}$$

$$+\frac{y^1z^2}{1^a2^b}+\frac{y^2z^2}{2^a2^b}$$

$$+\frac{y^1z^3}{1^a3^b}+\frac{y^2z^3}{2^a3^b}+\frac{y^3z^3}{3^a3^b}$$

$$+\frac{y^1z^4}{1^a4^b}+\frac{y^2z^4}{2^a4^b}+\frac{y^3z^4}{3^a4^b}+\frac{y^4z^4}{4^a4^b}$$

$$+\frac{y^1z^5}{1^a5^b}+\frac{y^2z^5}{2^a5^b}+\frac{y^3z^5}{3^a5^b}+\frac{y^4z^5}{4^a5^b}+\frac{y^5z^5}{5^a5^b}$$

$$+\frac{y^1z^6}{1^a6^b}+\frac{y^2z^6}{2^a6^b}+\frac{y^3z^6}{3^a6^b}+\frac{y^4z^6}{4^a6^b}+\frac{y^5z^6}{5^a6^b}+\frac{y^6z^6}{6^a6^b}$$

$$+\frac{y^1z^7}{1^a7^b}+\frac{y^2z^7}{2^a7^b}+\frac{y^3z^7}{3^a7^b}+\frac{y^4z^7}{4^a7^b}+\frac{y^5z^7}{5^a7^b}+\frac{y^6z^7}{6^a7^b}+\frac{y^7z^7}{7^a7^b}$$

$$+\ \vdots\ +\ \vdots\ +\ \vdots\ +\ \vdots\ +\ \vdots\ +\ \vdots\ +\ \vdots\ \ddots$$

$$=\sum_{m,n\geq 1; m\leq n}^{\infty}\frac{y^m z^n}{m^a n^b}$$

$$=\sum_{\substack{h,j,k\geq 1\\ j\leq k;\,(j,k)=1}}\frac{(y^j z^k)^h}{h^{a+b}(j^a k^b)}$$

$$=\sum_{\substack{j,k\geq 1\\ j\leq k;\,(j,k)=1}}\frac{1}{(j^a k^b)}\sum_{h=1}^{\infty}\frac{(y^j z^k)^h}{h^{a+b}}$$

$$=\sum_{\substack{j,k\geq 1\\ j\leq k;\,(j,k)=1}}\frac{1}{(j^a k^b)}\log\left(\frac{1}{1-y^j z^k}\right)\quad if\quad a+b=1.$$

Therefore, we have shown that

$$\sum_{n=1}^{\infty}\left(\sum_{m=1}^{n}\frac{y^m}{m^a}\right)\frac{z^n}{n^b}=\sum_{\substack{j,k\geq 1\\ j\leq k;\,(j,k)=1}}\frac{1}{(j^a k^b)}\log\left(\frac{1}{1-y^j z^k}\right)\quad if\quad a+b=1.$$

Exponentiating both sides gives us the $2D$ first extended triangle VPV identity, where in this $2D$ case the nD pyramid reduces to the form of a triangle shaped array of lattice point vectors, and so we can state the

Theorem 21.1 ***A* 2D *first quadrant triangle VPV identity.*** *For* $|y|<1$, $|z|<1$,

$$\prod_{\substack{j,k\geq 1\\ j\leq k;\,(j,k)=1}}\left(\frac{1}{1-y^j z^k}\right)^{\frac{1}{j^a k^b}}=\exp\left\{\sum_{n=1}^{\infty}\left(\sum_{m=1}^{n}\frac{y^m}{m^a}\right)\frac{z^n}{n^b}\right\}\quad if\quad a+b=1. \tag{21.2.1}$$

As with our earlier exploits into the $2D$ first quadrant case, for the present result, we take some simple example cases where new and interesting results arise.

So, let us take the case where $a=0, b=1$, giving us

$$\prod_{\substack{j,k\geq 1\\ j\leq k;\,(j,k)=1}}\left(\frac{1}{1-y^j z^k}\right)^{\frac{1}{k}}=\exp\left\{\sum_{n=1}^{\infty}\left(\sum_{m=1}^{n}y^m\right)\frac{z^n}{n}\right\}$$

$$=\exp\left\{\sum_{n=1}^{\infty}\left(y\frac{1-y^n}{1-y}\right)\frac{z^n}{n}\right\}=\exp\left\{\frac{y}{1-y}\log\left(\frac{1-yz}{1-z}\right)\right\}.$$

So, we arrive then at the following pair of equivalent results,

$$\prod_{\substack{j,k\geq 1\\ j\leq k;\,(j,k)=1}} \left(\frac{1}{1-y^j z^k}\right)^{\frac{1}{k}} = \left(\frac{1-yz}{1-z}\right)^{\frac{y}{1-y}}, \tag{21.2.2}$$

and

$$\prod_{\substack{j,k\geq 1\\ j\leq k;\,(j,k)=1}} \left(1-y^j z^k\right)^{\frac{1}{k}} = \left(\frac{1-z}{1-yz}\right)^{\frac{y}{1-y}}. \tag{21.2.3}$$

From here, multiply both sides of (21.2.2) and the case of (21.2.3) with $y \mapsto y^2$ and $z \mapsto z^2$ to get,

$$\prod_{\substack{j,k\geq 1;\, j\leq k\\ gcd(j,k)=1}} \left(1+y^j z^k\right)^{\frac{1}{k}} = \left(\frac{1-yz}{1-z}\right)^{\frac{y}{1-y}} \left(\frac{1-z^2}{1-y^2z^2}\right)^{\frac{y^2}{1-y^2}}. \tag{21.2.4}$$

Obviously multiplying both sides of (21.2.3) and (21.2.4) give a restated (21.2.4).

Particular cases: $y = \frac{1}{2}$ gives us from (21.2.3) and (21.2.4) the remarkable two results,

$$\prod_{\substack{j,k\geq 1;\, j\leq k\\ gcd(j,k)=1}} \left(1-\frac{z^k}{2^j}\right)^{\frac{1}{k}} = \frac{2-2z}{2-z} = 1-\frac{z}{2}-\frac{z^2}{4}-\frac{z^3}{8}-\frac{z^4}{16}-\frac{z^5}{32}-\dots$$

$$= \left(1-\frac{z}{2}\right)$$

$$\sqrt{\left(1-\frac{z^2}{2^1}\right)}$$

$$\sqrt[3]{\left(1-\frac{z^3}{2^1}\right)\left(1-\frac{z^3}{2^2}\right)}$$

$$\sqrt[4]{\left(1-\frac{z^4}{2^1}\right)\left(1-\frac{z^4}{2^3}\right)}$$

$$\sqrt[5]{\left(1-\frac{z^5}{2^1}\right)\left(1-\frac{z^5}{2^2}\right)\left(1-\frac{z^5}{2^3}\right)\left(1-\frac{z^5}{2^4}\right)}$$

$$\sqrt[6]{\left(1-\frac{z^6}{2^1}\right)\left(1-\frac{z^6}{2^5}\right)}$$

$$\vdots,$$

$$\prod_{\substack{j,k\geq 1; j\leq k \\ gcd(j,k)=1}} \left(1+\frac{z^k}{2^j}\right)^{\frac{1}{k}} = \frac{1-\frac{z}{2}}{1-z}\sqrt[3]{\frac{1-z^2}{1-\frac{z^2}{4}}} = 1+\frac{z}{2}+\frac{z^2}{4}+\frac{3z^3}{8}+\frac{z^4}{4}+\frac{5z^5}{16}+\dots$$

$$= \left(1+\frac{z}{2}\right)$$

$$\sqrt{\left(1+\frac{z^2}{2^1}\right)}$$

$$\sqrt[3]{\left(1+\frac{z^3}{2^1}\right)\left(1+\frac{z^3}{2^2}\right)}$$

$$\sqrt[4]{\left(1+\frac{z^4}{2^1}\right)\left(1+\frac{z^4}{2^3}\right)}$$

$$\sqrt[5]{\left(1+\frac{z^5}{2^1}\right)\left(1+\frac{z^5}{2^2}\right)\left(1+\frac{z^5}{2^3}\right)\left(1+\frac{z^5}{2^4}\right)}$$

$$\sqrt[6]{\left(1+\frac{z^6}{2^1}\right)\left(1+\frac{z^6}{2^5}\right)}$$

$$\vdots$$

These two equations can be easily verified on a calculating engine like Mathematica or WolframAlpha by expanding each side into it's Taylor series around $z=0$ and comparing coefficients of like powers of z. Next, take the cases of (21.2.3) and (21.2.4) with $y=2$, both of which converge if $|z|<\frac{1}{2}$, so then, after a slight adjustment to both sides we have,

$$\prod_{\substack{j,k\geq 1; j<k \\ gcd(j,k)=1}} \left(1-2^j z^k\right)^{\frac{1}{k}} = 1-\frac{z^2}{(1-z)^2} = 1-z^2-2z^3-3z^4-4z^5-\dots-nz^{n+1}-\dots$$

$$= \sqrt{(1-2^1z^2)}$$

$$\sqrt[3]{(1-2^1z^3)(1-2^2z^3)}$$

$$\sqrt[4]{(1-2^1z^4)(1-2^3z^4)}$$

$$\sqrt[5]{(1-2^1z^5)(1-2^2z^5)(1-2^3z^5)(1-2^4z^5)}$$

$$\sqrt[6]{(1-2^1z^6)(1-2^5z^6)}$$

$$\sqrt[7]{(1-2^1z^7)(1-2^2z^7)(1-2^3z^7)(1-2^4z^7)(1-2^5z^7)(1-2^6z^7)} \quad etc.$$

which is also easy to verify on a calculating engine term by term from the power series of each side. The notably simple coefficients make this result somewhat tantalizing, as there seems no obvious reason for such coefficients to come out of the products of binomial series roots.

We remark at this juncture that equations (21.2.3) and it's reciprocal equation (21.2.4) are amenable to applying the limit as $y \to 1$. In fact we have as follows that,

$$\lim_{y \to 1} \left(\frac{1-z}{1-yz} \right)^{\frac{y}{1-y}} = e^{\frac{z}{z-1}}$$

and also from considering equation (21.2.4) there is the limit, easily evaluated,

$$\lim_{y \to 1} \left(\frac{1-yz}{1-z} \right)^{\frac{y}{1-y}} \left(\frac{1-z^2}{1-y^2z^2} \right)^{\frac{y^2}{1-y^2}} = e^{\frac{z}{1-z^2}}.$$

Therefore, applying these two limits to equations (21.2.3) and (21.2.4) respectively we obtain the two interesting results (21.2.5) and (21.2.6) given here.

$$\prod_{k=1}^{\infty} \left(1 - z^k\right)^{\frac{\varphi(k)}{k}} = e^{\frac{z}{z-1}} = \sum_{k=0}^{\infty} \frac{\alpha(k) z^k}{k!} \tag{21.2.5}$$

$$= 1 - \frac{z}{1!} - \frac{z^2}{2!} - \frac{z^3}{3!} + \frac{z^4}{4!} + \frac{19z^5}{5!} + \frac{151z^6}{6!} + \frac{1091z^7}{7!}$$

$$+ \frac{7841z^8}{8!} + \frac{56519z^9}{9!} + \frac{396271z^{10}}{10!} + O(z^{11}),$$

where $\varphi(k)$ is the Euler totient function, the number of positive integers less than and coprime to k. (21.2.5) demonstrates that sequence $\alpha(k)$ has the exponential generating function $e^{\frac{z}{z-1}}$. The first 31 coefficients generated by the series are,

n	α(n) from (21.2.5)
0	1
1	−1
2	−1
3	−1
4	1
5	19
6	151
7	1091
8	7841
9	56519
10	396271
11	2442439
12	7701409
13	−145269541
14	−4833158329
15	−104056218421
16	−2002667085119
17	−37109187217649
18	−679877731030049
19	−12440309297451121
20	−227773259993414719
21	−4155839606711748061
22	−74724654677947488521
23	−1293162252850914402221
24	−20381626111249718908319
25	−244110863655032038665001
26	267543347653261450406351
27	172316772106087159102974551
28	8944973491570029894272392801
29	361702062324149751903132843499
30	13353699077321671584329389125031

Amazingly $\gcd(\alpha(k), k!) = 1$, for all values of k up to 34, and mostly beyond that, and $\alpha(k) \equiv 1 \; or \; 9 \; (mod \; 10)$, and also the recurrence relation

$$\alpha(n) + (n-1)(n-2)\,\alpha(n-2) = (2n-3)\,\alpha(n-1)$$

holds. (See OEIS sequence A293116 [237]) This recurrence relation allows us to write continued fractions for the ratios $\alpha(n+1)/\alpha(n)$.

Next we examine

$$\prod_{k=1}^{\infty} \left(1+z^k\right)^{\frac{\varphi(k) z^k}{k}} = e^{\frac{z}{1-z^2}} = \sum_{k=0}^{\infty} \frac{\beta(k) z^k}{k!} \tag{21.2.6}$$

$$= 1 + \frac{z}{1!} + \frac{z^2}{2!} + \frac{7z^3}{3!} + \frac{25z^4}{4!} + \frac{181z^5}{5!} + \frac{1201z^6}{6!}$$

$$+ \frac{10291z^7}{7!} + \frac{97777z^8}{8!} + \frac{202709z^9}{9!} + O(z^{10}),$$

where again, $\varphi(k)$ is the Euler totient function. The first 30 coefficients for this exponential generating function are given by

n	β(n) from (21.2.6)
0	1
1	1
2	1
3	7
4	25
5	181
6	1201
7	10291
8	97777
9	**202709** × **5**
10	**1743223** × **7**
11	151573951
12	2173233481
13	31758579997
14	**104811403133** × **5**
15	8838296029291
16	164416415570401
17	**449336774160103** × **7**
18	65057767274601217
19	**278248694109978827** × **5**
20	31671795881695430521
21	747996624368605997701
22	**1430133056253846489397** × **13**
23	480126276655495820797987
24	**369787228453820056451963** × **5** × **7**
25	362507506102249314587009401
26	10537197720733307853696168001
27	317940307036989922598772811951
28	9915242854364117895813434738857
29	**2563767360580175587315966478129** × **5**^{3}
30	10674537251310602152370412880372081

where the **bolded** numbers are shown here to be "not coprime" to their factorial denominator $n!$ in the exponential generating function. Again, we could investigate the difference equations arising and form theorems regarding the congruence relations of these e.g.f. coefficients. (See also the last section of chapter 5.)

Next we take (21.2.1) with the case that $a = 1$ and $b = 0$, so then

$$\prod_{\substack{j,k \geq 1 \\ j \leq k;\, (j,k)=1}} \left(\frac{1}{1-y^j z^k}\right)^{\frac{1}{j}} = \exp\left\{\sum_{n=1}^{\infty} \left(\sum_{m=1}^{n} \frac{y^m}{m}\right) z^n\right\}$$

$$= \exp\left\{\frac{1}{1-z}\sum_{n=1}^{\infty}\frac{y^n z^n}{n}\right\} = \exp\left\{\frac{1}{1-z}\log\left(\frac{1}{1-yz}\right)\right\}.$$

This leads us to establish that

$$\prod_{\substack{j,k\geq 1\\ j\leq k;\,(j,k)=1}} \left(\frac{1}{1-y^j z^k}\right)^{\frac{1}{j}} = \left(\frac{1}{1-yz}\right)^{\frac{1}{1-z}}, \tag{21.2.7}$$

which is equivalent to

$$\prod_{\substack{j,k\geq 1\\ j\leq k;\,(j,k)=1}} \left(1-y^j z^k\right)^{\frac{1}{j}} = (1-yz)^{\frac{1}{1-z}}. \tag{21.2.8}$$

From multiplying both sides of (21.2.7) in which $y \mapsto y^2$ and $z \mapsto z^2$ with both sides of (21.2.8), we obtain

$$\prod_{\substack{j,k\geq 1\\ j\leq k;\,(j,k)=1}} \left(1+y^j z^k\right)^{\frac{1}{j}} = \frac{(1-y^2z^2)^{\frac{1}{1-z^2}}}{(1-yz)^{\frac{1}{1-z}}}. \tag{21.2.9}$$

Particular cases:

$z = \frac{1}{2}$ gives us from (21.2.8) and (21.2.9) the remarkable result that

$$\prod_{\substack{j,k\geq 1;\, j\leq k\\ gcd(j,k)=1}} \left(1-\frac{y^j}{2^k}\right)^{\frac{1}{j}} = \left(1-\frac{y}{2}\right)^2 = 1-\frac{y}{4}+\frac{y^2}{4}$$

$$= \left(1-\frac{y^1}{2^1}\right)$$

$$\left(1-\frac{y^1}{2^2}\right)$$

$$\left(1-\frac{y^1}{2^3}\right)\sqrt{\left(1-\frac{y^2}{2^3}\right)}$$

$$\left(1-\frac{y^1}{2^4}\right)\sqrt[3]{\left(1-\frac{y^3}{2^4}\right)}$$

$$\left(1-\frac{y^1}{2^5}\right)\sqrt{\left(1-\frac{y^2}{2^5}\right)}\sqrt[3]{\left(1-\frac{y^3}{2^5}\right)}\sqrt[4]{\left(1-\frac{y^4}{2^5}\right)}$$

$$\left(1-\frac{y^1}{2^6}\right)\sqrt[5]{\left(1-\frac{y^5}{2^6}\right)}$$

$$\vdots,$$

and the result,

$$\prod_{\substack{j,k\geq 1; j\leq k \\ gcd(j,k)=1}} \left(1+\frac{y^j}{2^k}\right)^{\frac{1}{j}} = \frac{\sqrt[3]{(4-y^2)^4}}{\sqrt[3]{4}(2-y)^2} = 1+y+\frac{5y^2}{12}+\frac{y^3}{6}+\frac{11y^4}{144}+\frac{5y^5}{144}+O(y^6)$$

$$= \left(1+\frac{y^1}{2^1}\right)$$

$$\left(1+\frac{y^1}{2^2}\right)$$

$$\left(1+\frac{y^1}{2^3}\right)\sqrt{\left(1+\frac{y^2}{2^3}\right)}$$

$$\left(1+\frac{y^1}{2^4}\right)\sqrt[3]{\left(1+\frac{y^3}{2^4}\right)}$$

$$\left(1+\frac{y^1}{2^5}\right)\sqrt{\left(1+\frac{y^2}{2^5}\right)}\sqrt[3]{\left(1+\frac{y^3}{2^5}\right)}\sqrt[4]{\left(1+\frac{y^4}{2^5}\right)}$$

$$\left(1+\frac{y^1}{2^6}\right)\sqrt[5]{\left(1+\frac{y^5}{2^6}\right)}$$

$$\vdots$$

These two equations can be verified on a calculating engine like Mathematica or WolframAlpha by expanding each side into it's Taylor series around $y = 0$ and comparing coefficients of like powers of y. However, the calculation is an infinite series for each coefficient, unlike in the previous examples, where it is a finite sum.

21.3 DERIVING 3D VPV IDENTITIES IN SQUARE PYRAMID REGIONS

We start by considering the infinite inverted pyramid with square layered arrays of lattice point vectors as per the following diagram, with VPVs bolded.

$$
\begin{gathered}
\begin{array}{ccccccccccc}
 & & & & & \langle\mathbf{1,6,6}\rangle & & & & & \\
 & & & & \langle\mathbf{1,5,6}\rangle & & \langle 2,6,6\rangle & & & & \\
 & & & \langle\mathbf{1,4,6}\rangle & & \langle\mathbf{2,5,6}\rangle & & \langle 3,6,6\rangle & & & \\
 & & \langle\mathbf{1,3,6}\rangle & & \langle 2,4,6\rangle & & \langle\mathbf{3,5,6}\rangle & & \langle 4,6,6\rangle & & \\
 & \langle\mathbf{1,2,6}\rangle & & \langle\mathbf{2,3,6}\rangle & & \langle\mathbf{3,4,6}\rangle & & \langle\mathbf{4,5,6}\rangle & & \langle\mathbf{5,6,6}\rangle & \\
\langle\mathbf{1,1,6}\rangle & & \langle 2,2,6\rangle & & \langle 3,3,6\rangle & & \langle 4,4,6\rangle & & \langle\mathbf{5,5,6}\rangle & & \langle 6,6,6\rangle \\
 & \langle\mathbf{2,1,6}\rangle & & \langle\mathbf{3,2,6}\rangle & & \langle\mathbf{4,3,6}\rangle & & \langle\mathbf{5,4,6}\rangle & & \langle\mathbf{6,5,6}\rangle & \\
 & & \langle\mathbf{3,1,6}\rangle & & \langle 4,2,6\rangle & & \langle\mathbf{5,3,6}\rangle & & \langle 6,4,6\rangle & & \\
 & & & \langle\mathbf{4,1,6}\rangle & & \langle\mathbf{5,2,6}\rangle & & \langle 6,3,6\rangle & & & \\
 & & & & \langle\mathbf{5,1,6}\rangle & & \langle 6,2,6\rangle & & & & \\
 & & & & & \langle\mathbf{6,1,6}\rangle & & & & &
\end{array}
\\[1em]
\begin{array}{ccccccccc}
 & & & & \langle\mathbf{1,5,5}\rangle & & & & \\
 & & & \langle\mathbf{1,4,5}\rangle & & \langle\mathbf{2,5,5}\rangle & & & \\
 & & \langle\mathbf{1,3,5}\rangle & & \langle\mathbf{2,4,5}\rangle & & \langle\mathbf{3,5,5}\rangle & & \\
 & \langle\mathbf{1,2,5}\rangle & & \langle\mathbf{2,3,5}\rangle & & \langle\mathbf{3,4,5}\rangle & & \langle\mathbf{4,5,5}\rangle & \\
\langle\mathbf{1,1,5}\rangle & & \langle\mathbf{2,2,5}\rangle & & \langle\mathbf{3,3,5}\rangle & & \langle\mathbf{4,4,5}\rangle & & \langle 5,5,5\rangle \\
 & \langle\mathbf{2,1,5}\rangle & & \langle\mathbf{3,2,5}\rangle & & \langle\mathbf{4,3,5}\rangle & & \langle\mathbf{5,4,5}\rangle & \\
 & & \langle\mathbf{3,1,5}\rangle & & \langle\mathbf{4,2,5}\rangle & & \langle\mathbf{5,3,5}\rangle & & \\
 & & & \langle\mathbf{4,1,5}\rangle & & \langle\mathbf{1,2,5}\rangle & & & \\
 & & & & \langle\mathbf{5,1,5}\rangle & & & &
\end{array}
\\[1em]
\begin{array}{ccccccc}
 & & & \langle\mathbf{1,4,4}\rangle & & & \\
 & & \langle\mathbf{1,3,4}\rangle & & \langle 2,4,4\rangle & & \\
 & \langle\mathbf{1,2,4}\rangle & & \langle\mathbf{2,3,4}\rangle & & \langle\mathbf{3,4,4}\rangle & \\
\langle\mathbf{1,1,4}\rangle & & \langle 2,2,4\rangle & & \langle\mathbf{3,3,4}\rangle & & \langle 4,4,4\rangle \\
 & \langle\mathbf{2,1,4}\rangle & & \langle\mathbf{3,2,4}\rangle & & \langle\mathbf{4,3,4}\rangle & \\
 & & \langle\mathbf{3,1,4}\rangle & & \langle 4,2,4\rangle & & \\
 & & & \langle\mathbf{4,1,4}\rangle & & &
\end{array}
\\[1em]
\begin{array}{ccccc}
 & & \langle\mathbf{1,3,3}\rangle & & \\
 & \langle\mathbf{1,2,3}\rangle & & \langle\mathbf{2,3,3}\rangle & \\
\langle\mathbf{1,1,3}\rangle & & \langle\mathbf{2,2,3}\rangle & & \langle 3,3,3\rangle \\
 & \langle\mathbf{2,1,3}\rangle & & \langle\mathbf{3,2,3}\rangle & \\
 & & \langle\mathbf{3,1,3}\rangle & &
\end{array}
\\[1em]
\begin{array}{ccc}
 & \langle\mathbf{1,2,2}\rangle & \\
\langle\mathbf{1,1,2}\rangle & & \langle 2,2,2\rangle \\
 & \langle\mathbf{2,1,2}\rangle &
\end{array}
\\[1em]
\langle\mathbf{1,1,1}\rangle
\end{gathered}
\tag{21.3.1}
$$

From this we create a $3D$ summation over integer co-ordinates in the above lattice point vectors. We consider the sum,

$$
\sum_{n=1}^{\infty}\left(\sum_{l=1}^{n}\frac{x^l}{l^a}\right)\left(\sum_{m=1}^{n}\frac{y^m}{m^b}\right)\frac{z^n}{n^c}
$$

$$
=\left(\frac{x^1}{1^a}\right)\left(\frac{y^1}{1^b}\right)\frac{z^1}{1^c}
$$

$$
+\left(\frac{x^1}{1^a}+\frac{x^2}{2^a}\right)\left(\frac{y^1}{1^b}+\frac{y^2}{2^b}\right)\frac{z^2}{2^c}
$$

$$
+\left(\frac{x^1}{1^a}+\frac{x^2}{2^a}+\frac{x^3}{3^a}\right)\left(\frac{y^1}{1^b}+\frac{y^2}{2^b}+\frac{y^3}{3^b}\right)\frac{z^3}{3^c}
$$

$$
+\left(\frac{x^1}{1^a}+\frac{x^2}{2^a}+\frac{x^3}{3^a}+\frac{x^4}{4^a}\right)\left(\frac{y^1}{1^b}+\frac{y^2}{2^b}+\frac{y^3}{3^b}+\frac{y^4}{4^b}\right)\frac{z^4}{4^c}
$$

$$
+\left(\frac{x^1}{1^a}+\frac{x^2}{2^a}+\frac{x^3}{3^a}+\frac{x^4}{4^a}+\frac{x^5}{5^a}\right)\left(\frac{y^1}{1^b}+\frac{y^2}{2^b}+\frac{y^3}{3^b}+\frac{y^4}{4^b}+\frac{y^5}{5^b}\right)\frac{z^5}{5^c}+\cdots
$$

$$
=\frac{x^1y^1z^1}{1^a1^b1^c}
$$

$$
+\frac{x^1y^1z^2}{1^a1^b2^c}+\frac{x^1y^2z^2}{1^a2^b2^c}
$$

$$
\begin{aligned}
&+\frac{x^2y^1z^2}{2^a1^b2^c}+\frac{x^2y^2z^2}{2^a2^b2^c} \\[1em]
&+\frac{x^1y^1z^3}{1^a1^b3^c}+\frac{x^1y^2z^3}{1^a2^b3^c}+\frac{x^1y^3z^3}{1^a3^b3^c} \\
&+\frac{x^2y^1z^3}{2^a1^b3^c}+\frac{x^2y^2z^3}{2^a2^b3^c}+\frac{x^2y^3z^3}{2^a3^b3^c} \\
&+\frac{x^3y^1z^3}{3^a1^b3^c}+\frac{x^3y^2z^3}{3^a2^b3^c}+\frac{x^3y^3z^3}{3^a3^b3^c} \\[1em]
&+\frac{x^1y^1z^4}{1^a1^b4^c}+\frac{x^1y^2z^4}{1^a2^b4^c}+\frac{x^1y^3z^4}{1^a3^b4^c}+\frac{x^1y^4z^4}{1^a4^b4^c} \\
&+\frac{x^2y^1z^4}{2^a1^b4^c}+\frac{x^2y^2z^4}{2^a2^b4^c}+\frac{x^2y^3z^4}{2^a3^b4^c}+\frac{x^2y^4z^4}{2^a4^b4^c} \\
&+\frac{x^3y^1z^4}{3^a1^b4^c}+\frac{x^3y^2z^4}{3^a2^b4^c}+\frac{x^3y^3z^4}{3^a3^b4^c}+\frac{x^3y^4z^4}{3^a4^b4^c} \\
&+\frac{x^4y^1z^4}{4^a1^b4^c}+\frac{x^4y^2z^4}{4^a2^b4^c}+\frac{x^4y^3z^4}{4^a3^b4^c}+\frac{x^4y^4z^4}{4^a4^b4^c} \\[1em]
&+\frac{x^1y^1z^5}{1^a1^b5^c}+\frac{x^1y^2z^5}{1^a2^b5^c}+\frac{x^1y^3z^5}{1^a3^b5^c}+\frac{x^1y^4z^5}{1^a4^b5^c}+\frac{x^1y^5z^5}{1^a5^b5^c} \\
&+\frac{x^2y^1z^5}{2^a1^b5^c}+\frac{x^2y^2z^5}{2^a2^b5^c}+\frac{x^2y^3z^5}{2^a3^b5^c}+\frac{x^2y^4z^5}{2^a4^b5^c}+\frac{x^2y^5z^5}{2^a5^b5^c} \\
&+\frac{x^3y^1z^5}{3^a1^b5^c}+\frac{x^3y^2z^5}{3^a2^b5^c}+\frac{x^3y^3z^5}{3^a3^b5^c}+\frac{x^3y^4z^5}{3^a4^b5^c}+\frac{x^3y^5z^5}{3^a5^b5^c} \\
&+\frac{x^4y^1z^5}{4^a1^b5^c}+\frac{x^4y^2z^5}{4^a2^b5^c}+\frac{x^4y^3z^5}{4^a3^b5^c}+\frac{x^4y^4z^5}{4^a4^b5^c}+\frac{x^4y^5z^5}{4^a5^b5^c} \\
&+\frac{x^5y^1z^5}{5^a1^b5^c}+\frac{x^5y^2z^5}{5^a2^b5^c}+\frac{x^5y^3z^5}{5^a3^b5^c}+\frac{x^5y^4z^5}{5^a4^b5^c}+\frac{x^5y^5z^5}{5^a5^b5^c} \\[1em]
&+\ \vdots\ +\ \vdots\ +\ \vdots\ +\ \vdots\ +\ \vdots\ \ddots
\end{aligned}
$$

$$
\begin{aligned}
&= \sum_{l,m,n\geq 1;\, l,m\leq n}^{\infty} \frac{x^l y^m z^n}{l^a m^b n^c} \\
&= \sum_{\substack{h,l,m,n\geq 1\\ l,m\leq n;\, \gcd(l,m,n)=1}} \frac{(x^l y^m z^n)^h}{h^{a+b+c}(l^a m^b n^c)} \\
&= \sum_{\substack{l,m,n\geq 1\\ l,m\leq n;\, \gcd(l,m,n)=1}} \frac{1}{(l^a m^b n^c)} \sum_{h=1}^{\infty} \frac{(x^l y^n z^n)^h}{h^{a+b+c}}
\end{aligned}
$$

$$= \sum_{\substack{l,m,n\geq 1 \\ l,m\leq n;\, \gcd(l,m,n)=1}} \frac{1}{(l^a m^b n^c)} \log\left(\frac{1}{1-x^l y^b z^c}\right) \quad if \quad a+b+c=1.$$

Therefore, we have shown that if $a+b+c=1$ then

$$\sum_{n=1}^{\infty}\left(\sum_{l=1}^{n}\frac{x^l}{l^a}\right)\left(\sum_{m=1}^{n}\frac{y^m}{m^b}\right)\frac{z^n}{n^c} = \sum_{\substack{l,m,n\geq 1 \\ l,m\leq n;\, \gcd(l,m,n)=1}} \frac{1}{(l^a m^b n^c)} \log\left(\frac{1}{1-x^l y^b z^c}\right).$$

Exponentiating both sides gives us the $3D$ "pyramid VPV identity".

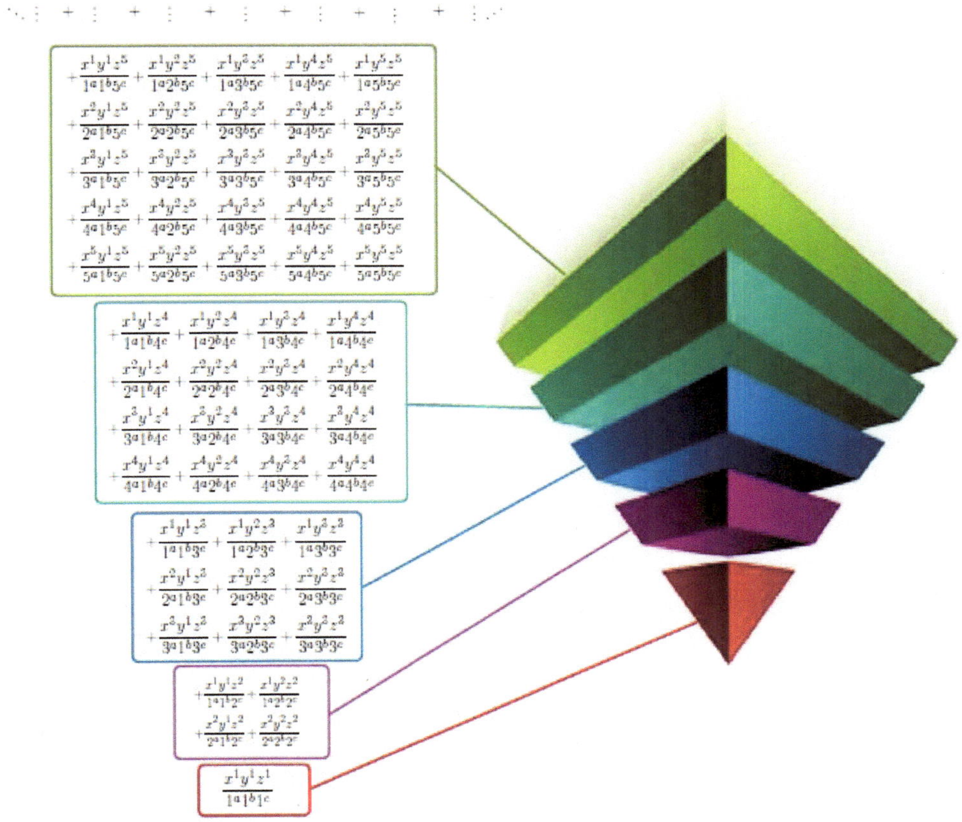

Figure 21.1 Inverted 3D hyperpyramid sum form (pyramid version)

The above image illustrates the 3D positioning of the summed lattice implying

Theorem 21.2 ***The*** *3D* ***first hyperquadrant pyramid VPV identity.*** *If* $|x|, |y|, |z| < 1$, *with* $a+b+c=1$,

$$\prod_{\substack{l,m,n\geq 1 \\ l,m\leq n;\, \gcd(l,m,n)=1}} \left(\frac{1}{1-x^l y^m z^n}\right)^{\frac{1}{l^a m^b n^c}} = \exp\left\{\sum_{n=1}^{\infty}\left(\sum_{l=1}^{n}\frac{x^l}{l^a}\right)\left(\sum_{m=1}^{n}\frac{y^m}{m^b}\right)\frac{z^n}{n^c}\right\}. \tag{21.3.2}$$

As we did for the $2D$ particular cases, we can examine some obvious example corollaries arising from this theorem. Firstly, take the case where $a=b=0, c=1$, so then,

$$\prod_{\substack{l,m,n\geq 1\\ l,m\leq n;\,\gcd(l,m,n)=1}} \left(\frac{1}{1-x^l y^m z^n}\right)^{\frac{1}{n}} = \exp\left\{\sum_{n=1}^{\infty}\left(\sum_{l=1}^{n} x^l\right)\left(\sum_{m=1}^{n} y^m\right)\frac{z^n}{n}\right\}$$

$$= \exp\left\{\sum_{n=1}^{\infty} xy\left(\frac{1-x^n}{1-x}\right)\left(\frac{1-y^n}{1-y}\right)\frac{z^n}{n}\right\}$$

$$= \exp\left\{\frac{xy}{(1-x)(1-y)}\log\left(\frac{(1-xz)(1-yz)}{(1-z)(1-xyz)}\right)\right\},$$

which brings us after exponentiating both sides to a set of $3D$ infinite products. So, we have

$$\prod_{\substack{l,m,n\geq 1\\ l,m\leq n;\,\gcd(l,m,n)=1}} \left(\frac{1}{1-x^l y^m z^n}\right)^{\frac{1}{n}} = \left(\frac{(1-xz)(1-yz)}{(1-z)(1-xyz)}\right)^{\frac{xy}{(1-x)(1-y)}}, \tag{21.3.3}$$

and the equivalent identity,

$$\prod_{\substack{l,m,n\geq 1\\ l,m\leq n;\,\gcd(l,m,n)=1}} \left(1-x^l y^m z^n\right)^{\frac{1}{n}} = \left(\frac{(1-z)(1-xyz)}{(1-xz)(1-yz)}\right)^{\frac{xy}{(1-x)(1-y)}}. \tag{21.3.4}$$

We see that (21.3.3) and (21.3.4) are generalizations of the 2D identities (21.2.2) and (21.2.3) from the previous section. Writing (21.3.4) in longhand, referencing the diagram (21.3.1), we see that

$$\left(\frac{(1-z)(1-xyz)}{(1-xz)(1-yz)}\right)^{\frac{xy}{(1-x)(1-y)}}$$
$$= (1-xyz)$$
$$\sqrt{(1-xyz^2)(1-xy^2z^2)(1-x^2yz^2)}$$
$$\sqrt[3]{(1-xyz^3)(1-x^2yz^3)(1-x^3yz^3)(1-x^2yz^3)}$$
$$\sqrt[3]{(1-x^2y^2z^3)(1-x^2y^3z^3)(1-x^3yz^3)(1-x^3y^2z^3)}$$
$$\sqrt[4]{(1-xyz^4)(1-x^2yz^4)(1-x^3yz^4)(1-x^4yz^4)}$$
$$\sqrt[4]{(1-xy^2z^4)(1-x^3y^2z^4)}$$
$$\sqrt[4]{(1-xy^3z^4)(1-x^2y^3z^4)(1-x^3y^3z^4)(1-x^4y^3z^4)}$$

$$\sqrt[4]{(1-xy^4z^4)(1-x^3y^4z^4)}$$

$$\sqrt[5]{(1-xyz^5)(1-x^2yz^5)(1-x^3yz^5)(1-x^4yz^5)(1-x^5yz^5)}$$

$$\sqrt[5]{(1-xy^2z^5)(1-x^2y^2z^5)(1-x^3y^2z^5)(1-x^4y^2z^5)(1-x^5y^2z^5)}$$

$$\sqrt[5]{(1-xy^3z^5)(1-x^2y^3z^5)(1-x^3y^3z^5)(1-x^4y^3z^5)(1-x^5y^3z^5)}$$

$$\sqrt[5]{(1-xy^4z^5)(1-x^2y^4z^5)(1-x^3y^4z^5)(1-x^4y^4z^5)(1-x^5y^4z^5)}$$

$$\sqrt[5]{(1-xy^5z^5)(1-x^2y^5z^5)(1-x^3y^5z^5)(1-x^4y^5z^5)}$$

etc.

This is easily verified on a calculating application if expanded on both sides as power series in z.

21.4 VPV IDENTITIES IN nD FIRST HYPERQUADRANT HYPERPYRAMID REGIONS

The n dimensional first hyperquadrant hyperpyramid VPV Identity is encoded in the following

Theorem 21.3 ***The nD first hyperquadrant hyperpyramid VPV identity.*** *If $i=1,2,3,\dots,n$ then for each $x_i \in \mathbb{C}$ such that $|x_i|<1$ and $b_i \in \mathbb{C}$ such that $\sum_{i=1}^{n} b_i = 1$,*

$$\prod_{\substack{\gcd(a_1,a_2,\dots,a_n)=1\\ a_1,a_2,\dots,a_{n-1}<a_n\\ a_1,a_2,\dots,a_n\geq 1}} \left(\frac{1}{1-x_1{}^{a_1}x_2{}^{a_2}x_3{}^{a_3}\cdots x_n{}^{a_n}}\right)^{\frac{1}{a_1{}^{b_1}a_2{}^{b_2}a_3{}^{b_3}\cdots a_n{}^{b_n}}} \tag{21.4.1}$$

$$= \exp\left\{\sum_{k=1}^{\infty}\prod_{i=1}^{n-1}\left(\sum_{j=1}^{k}\frac{x_i{}^j}{j^{b_i}}\right)\frac{x_n{}^k}{k^{b_n}}\right\}$$

$$= \exp\left\{\sum_{k=1}^{\infty}\left(\sum_{j=1}^{k}\frac{x_1{}^j}{j^{b_1}}\right)\left(\sum_{j=1}^{k}\frac{x_2{}^j}{j^{b_2}}\right)\left(\sum_{j=1}^{k}\frac{x_3{}^j}{j^{b_3}}\right)\cdots\left(\sum_{j=1}^{k}\frac{x_{n-1}{}^j}{j^{b_{n-1}}}\right)\frac{x_n{}^k}{k^{b_n}}\right\}.$$

This result is quite straight-forward to prove using the technique of our two previous sections. It was also given in Campbell [91] by summing on the VPV's in the n-space hyperpyramid, defined by the inequalities

$$x_1 < x_n, x_2 < x_n, x_3 < x_n, \dots, x_{n-1} < x_n \tag{21.4.2}$$

in the first n-space hyperquadrant, and applying the following

Lemma 21.1 *Consider an infinite region raying out of the origin in any Euclidean vector space. The set of all lattice point vectors apart from the origin in that region is precisely the set of positive integer multiples of the VPVs in that region.*

The corresponding theorem from Campbell [86] was summed very simply over all lattice point vectors in the first hyperquadrant.

Further consequences of the above theorem are given as follows.

The 2D case of theorem 21.3 is

Corollary 21.1 *If* $|yz|$ *and* $|z|<1$ *and* $s+t=1$ *then,*

$$\prod_{\substack{(a,b)=1\\ a<b\\ a\geq 0, b\geq 1}} \left(\frac{1}{1-y^a z^b}\right)^{\frac{1}{a^s b^t}} \tag{21.4.3}$$

$$= \exp\left\{\frac{z^1}{1^t} + \left(1+\frac{y^1}{1^s}\right)\frac{z^2}{2^t} + \left(1+\frac{y^1}{1^s}+\frac{y^2}{2^s}\right)\frac{z^3}{3^t} + \cdots\right\}$$

The 3D case of theorem 21.3 is

Corollary 21.2 *If* $|xyz|$, $|yz|$ *and* $|z|<1$ *and* $r+s+t=1$ *then,*

$$\prod_{\substack{(a,b,c)=1\\ a,b<c\\ a,b\geq 0, c\geq 1}} \left(\frac{1}{1-x^a y^b z^c}\right)^{\frac{1}{a^r b^s c^t}} \tag{21.4.4}$$

$$= \exp\left\{\frac{z^1}{1^t} + \left(1+\frac{x^1}{1^r}\right)\left(1+\frac{y^1}{1^s}\right)\frac{z^2}{2^t} + \left(1+\frac{x^1}{1^r}+\frac{x^2}{2^r}\right)\left(1+\frac{y^1}{1^s}+\frac{y^2}{2^s}\right)\frac{z^3}{3^t} + \cdots\right\}$$

The 4D case of theorem 21.3 is

Corollary 21.3 *If* $|wxyz|$, $|xyz|$, $|yz|$ *and* $|z|<1$ *and* $r+s+t+u=1$ *then,*

$$\prod_{\substack{(a,b,c,d)=1\\ a,b,c<d\\ a,b,c\geq 0, d\geq 1}} \left(\frac{1}{1-w^a x^b y^c z^d}\right)^{\frac{1}{a^r b^s c^t d^u}} = \exp\left\{\mathrm{P}_3(r,w;s,x;t,y;u,z)\right\} \tag{21.4.5}$$

where P_3, *is a 4D hyperpyramid function,*

$$\mathrm{P}_3(r,w;s,x;t,y;u,z) = \frac{z^1}{1^u} + \left(1+\frac{w^1}{1^r}\right)\left(1+\frac{x^1}{1^s}\right)\left(1+\frac{y^1}{1^t}\right)\frac{z^2}{2^u}$$
$$+ \left(1+\frac{w^1}{1^r}+\frac{w^2}{2^r}\right)\left(1+\frac{x^1}{1^s}+\frac{x^2}{2^s}\right)\left(1+\frac{y^1}{1^t}+\frac{y^2}{2^t}\right)\frac{z^3}{3^u} + \cdots$$

The approach we adopt to give the reader an intuitive sense for these identities is to state corollaries and then examples from them. The 2D case through to the 5D case of (21.4.1) are given in the following examples of the *square hyperpyramid identity.*

Corollary 21.4 *For* $|y|, |z| < 1$,

$$\prod_{\substack{(a,b)=1\\ a<b\\ a\geq 0, b\geq 1}} \left(\frac{1}{1-y^a z^b}\right)^{\frac{1}{b}} = \left(\frac{1-yz}{1-z}\right)^{\frac{1}{1-y}} \tag{21.4.6}$$

$$= 1 + \frac{z}{1!} + \begin{vmatrix} 1 & -1 \\ \frac{1-y^2}{1-y} & 1 \end{vmatrix} \frac{z^2}{2!} + \begin{vmatrix} 1 & -1 & 0 \\ \frac{1-y^2}{1-y} & 1 & -2 \\ \frac{1-y^3}{1-y} & \frac{1-y^2}{1-y} & 1 \end{vmatrix} \frac{z^3}{3!} + \begin{vmatrix} 1 & -1 & 0 & 0 \\ \frac{1-y^2}{1-y} & 1 & -2 & 0 \\ \frac{1-y^3}{1-y} & \frac{1-y^2}{1-y} & 1 & -3 \\ \frac{1-y^4}{1-y} & \frac{1-y^3}{1-y} & \frac{1-y^2}{1-y} & 1 \end{vmatrix} \frac{z^4}{4!} + etc.$$

In this case it is fairly easy to find the Taylor coefficients for the (21.4.6) right side function. Hence we get a closed form evaluation of the determinant coefficients. In Mathematica, and WolframAlpha one easily sees that the Taylor series is

$$\left(\frac{1-yz}{1-z}\right)^{\frac{1}{1-y}} = 1 + z + (y+2)\frac{z^2}{2!} + (2y^2+5y+6)\frac{z^3}{3!} + (6y^3+17y^2+26y+24)\frac{z^4}{4!}$$
$$+(24y^4+74y^3+129y^2+154y+120)\frac{z^5}{5!} + O(z^6)$$

and that the expansion is encapsulated by $\sum_{n=0}^{\infty} c_n z^n$ where $c_0 = 1$, $c_1 = 1$ with the recurrence

$$nyc_n + (n+2)c_{n+2} = (2+n+y+ny)c_{n+1}.$$

Incidentally, also in Mathematica, and WolframAlpha one easily sees, for example, that the code

$$Det[\{1,-1,0,0\},\{(1-y^2)/(1-y),1,-2,0\},\{(1-y^3)/(1-y),(1-y^2)/(1-y),1,-3\},$$
$$\{(1-y^4)/(1-y),(1-y^3)/(1-y),(1-y^2)/(1-y),1\}]$$

nicely verifies the coefficient given by

$$\begin{vmatrix} 1 & -1 & 0 & 0 \\ \frac{1-y^2}{1-y} & 1 & -2 & 0 \\ \frac{1-y^3}{1-y} & \frac{1-y^2}{1-y} & 1 & -3 \\ \frac{1-y^4}{1-y} & \frac{1-y^3}{1-y} & \frac{1-y^2}{1-y} & 1 \end{vmatrix} = 6y^3 + 17y^2 + 26y + 24.$$

It is interesting to compare our identity (21.3.3) given earlier in this chapter with the following result.

Corollary 21.5 *For each of* $|x|, |y|, |z| < 1$,

$$\prod_{\substack{(a,b,c)=1\\ a,b<c\\ a,b\geq 0, c>0}} \left(\frac{1}{1-x^a y^b z^c}\right)^{\frac{1}{c}} = \left(\frac{(1-xz)(1-yz)}{(1-z)(1-xyz)}\right)^{\frac{1}{(1-x)(1-y)}} \tag{21.4.7}$$

$$= 1 + \frac{z}{1!} + \begin{vmatrix} 1 & -1 \\ \frac{(1-x^2)(1-y^2)}{(1-x)(1-y)} & 1 \end{vmatrix} \frac{z^2}{2!} + \begin{vmatrix} 1 & -1 & 0 \\ \frac{(1-x^2)(1-y^2)}{(1-x)(1-y)} & 1 & -2 \\ \frac{(1-x^3)(1-y^3)}{(1-x)(1-y)} & \frac{(1-x^2)(1-y^2)}{(1-x)(1-y)} & 1 \end{vmatrix} \frac{z^3}{3!}$$

$$+ \begin{vmatrix} 1 & -1 & 0 & 0 \\ \frac{(1-x^2)(1-y^2)}{(1-x)(1-y)} & 1 & -2 & 0 \\ \frac{(1-x^3)(1-y^3)}{(1-x)(1-y)} & \frac{(1-x^2)(1-y^2)}{(1-x)(1-y)} & 1 & -3 \\ \frac{(1-x^4)(1-y^4)}{(1-x)(1-y)} & \frac{(1-x^3)(1-y^3)}{(1-x)(1-y)} & \frac{(1-x^2)(1-y^2)}{(1-x)(1-y)} & 1 \end{vmatrix} \frac{z^4}{4!} + etc.$$

Corollary 21.6 *For each of* $|w|, |x|, |y|, |z| < 1$,

$$\prod_{\substack{(a,b,c,d)=1 \\ a,b,c<d \\ a,b,c\geq 0, d>0}} \left(\frac{1}{1 - w^a x^b y^c z^d} \right)^{\frac{1}{d}} = \left(\frac{(1-wz)(1-xz)(1-yz)(1-wxyz)}{(1-z)(1-wxz)(1-wyz)(1-xyz)} \right)^{\frac{1}{(1-w)(1-x)(1-y)}}, \tag{21.4.8}$$

$$= 1 + \frac{z}{1!} + \begin{vmatrix} 1 & -1 \\ \frac{(1-w^2)(1-x^2)(1-y^2)}{(1-w)(1-x)(1-y)} & 1 \end{vmatrix} \frac{z^2}{2!}$$

$$+ \begin{vmatrix} 1 & -1 & 0 \\ \frac{(1-w^2)(1-x^2)(1-y^2)}{(1-w)(1-x)(1-y)} & 1 & -2 \\ \frac{(1-w^3)(1-x^3)(1-y^3)}{(1-w)(1-x)(1-y)} & \frac{(1-w^2)(1-x^2)(1-y^2)}{(1-w)(1-x)(1-y)} & 1 \end{vmatrix} \frac{z^3}{3!}$$

$$+ \begin{vmatrix} 1 & -1 & 0 & 0 \\ \frac{(1-w^2)(1-x^2)(1-y^2)}{(1-w)(1-x)(1-y)} & 1 & -2 & 0 \\ \frac{(1-w^3)(1-x^3)(1-y^3)}{(1-w)(1-x)(1-y)} & \frac{(1-w^2)(1-x^2)(1-y^2)}{(1-w)(1-x)(1-y)} & 1 & -3 \\ \frac{(1-w^4)(1-x^4)(1-y^4)}{(1-w)(1-x)(1-y)} & \frac{(1-w^3)(1-x^3)(1-y^3)}{(1-w)(1-x)(1-y)} & \frac{(1-w^2)(1-x^2)(1-y^2)}{(1-w)(1-x)(1-y)} & 1 \end{vmatrix} \frac{z^4}{4!} + etc.$$

Corollary 21.7 *For each of* $|v|, |w|, |x|, |y|, |z| < 1$,

$$\prod_{\substack{(a,b,c,d,e)=1 \\ a,b,c,d<e \\ a,b,c,d\geq 0, e>0}} \left(\frac{1}{1 - v^a w^b x^c y^d z^e} \right)^{\frac{1}{e}} \tag{21.4.9}$$

$$= \left(\frac{(1-vz)(1-wz)(1-xz)(1-yz)}{(1-z)(1-vwz)(1-vxz)(1-vyz)} \right)^{\frac{1}{(1-v)(1-w)(1-x)(1-y)}}$$

$$\times \left(\frac{(1-vwxz)(1-vwyz)(1-vxyz)(1-wxyz)}{(1-wxz)(1-wyz)(1-xyz)(1-vwxyz)} \right)^{\frac{1}{(1-v)(1-w)(1-x)(1-y)}}.$$

$$= 1 + \frac{z}{1!} + \begin{vmatrix} 1 & -1 \\ \frac{(1-v^2)(1-w^2)(1-x^2)(1-y^2)}{(1-v)(1-w)(1-x)(1-y)} & 1 \end{vmatrix} \frac{z^2}{2!}$$

$$+\begin{vmatrix} 1 & -1 & 0 \\ \dfrac{(1-v^2)(1-w^2)(1-x^2)(1-y^2)}{(1-v)(1-w)(1-x)(1-y)} & 1 & -2 \\ \dfrac{(1-v^3)(1-w^3)(1-x^3)(1-y^3)}{(1-v)(1-w)(1-x)(1-y)} & \dfrac{(1-v^2)(1-w^2)(1-x^2)(1-y^2)}{(1-v)(1-w)(1-x)(1-y)} & 1 \end{vmatrix}\frac{z^3}{3!}$$

$$+\begin{vmatrix} 1 & -1 & 0 & 0 \\ \dfrac{(1-v^2)(1-w^2)(1-x^2)(1-y^2)}{(1-v)(1-w)(1-x)(1-y)} & 1 & -2 & 0 \\ \dfrac{(1-v^3)(1-w^3)(1-x^3)(1-y^3)}{(1-v)(1-w)(1-x)(1-y)} & \dfrac{(1-v^2)(1-w^2)(1-x^2)(1-y^2)}{(1-v)(1-w)(1-x)(1-y)} & 1 & -3 \\ \dfrac{)1-v^4)(1-w^4)(1-x^4)(1-y^4)}{(1-v)(1-w)(1-x)(1-y)} & \dfrac{(1-v^3)(1-w^3)(1-x^3)(1-y^3)}{(1-v)(1-w)(1-x)(1-y)} & \dfrac{(1-v^2)(1-w^2)(1-x^2)(1-y^2)}{(1-v)(1-w)(1-x)(1-y)} & 1 \end{vmatrix}\frac{z^4}{4!}$$

$$+\ etc.$$

21.5 2D VPV IDENTITIES FOR A z-AXIS SYMMETRIC EXTENDED TRIANGLE LATTICE

As we did in the section 21.2 of this chapter, we again start with a simple $2D$ summation. Consider an infinite extension of the inverted triangle 2D lattice point vectors with the Visible Point Vectors bolded,

$$\begin{array}{ccccccccccc} \langle -5,5\rangle & \langle \mathbf{-4,5}\rangle & \langle \mathbf{-3,5}\rangle & \langle \mathbf{-2,5}\rangle & \langle -1,\mathbf{5}\rangle & \langle 0,5\rangle & \langle \mathbf{1,5}\rangle & \langle \mathbf{2,5}\rangle & \langle \mathbf{3,5}\rangle & \langle \mathbf{4,5}\rangle & \langle 5,5\rangle \\ & \langle -4,4\rangle & \langle \mathbf{-3,4}\rangle & \langle -2,4\rangle & \langle -1,\mathbf{4}\rangle & \langle 0,4\rangle & \langle \mathbf{1,4}\rangle & \langle 2,4\rangle & \langle \mathbf{3,4}\rangle & \langle 4,4\rangle & \\ & & \langle -3,3\rangle & \langle \mathbf{-2,3}\rangle & \langle -1,\mathbf{3}\rangle & \langle 0,3\rangle & \langle \mathbf{1,3}\rangle & \langle \mathbf{2,3}\rangle & \langle 3,3\rangle & & \\ & & & \langle -2,2\rangle & \langle -1,\mathbf{2}\rangle & \langle 0,2\rangle & \langle \mathbf{1,2}\rangle & \langle 2,2\rangle & & & \\ & & & & \langle -1,1\rangle & \langle \mathbf{0,1}\rangle & \langle \mathbf{1,1}\rangle & & & & \\ & & & & & \langle 0,0\rangle & & & & & \end{array} \tag{21.5.1}$$

Next we create the following summation with the sum covering the above co-ordinates

in infinite extended form.

$$\sum_{n=1}^{\infty}\left(\sum_{m=-n}^{n}\frac{y^m}{m^a}\right)\frac{z^n}{n^b}$$

$$=\left(\frac{y^{-1}}{(-1)^a}+1+\frac{y^1}{1^a}\right)\frac{z^1}{1^b}$$

$$+\left(\frac{y^{-2}}{(-2)^a}+\frac{y^{-1}}{(-1)^a}+1+\frac{y^1}{1^a}+\frac{y^2}{2^a}\right)\frac{z^2}{2^b}$$

$$+\left(\frac{y^{-3}}{(-3)^a}+\frac{y^{-2}}{(-2)^a}+\frac{y^{-1}}{(-1)^a}+1+\frac{y^1}{1^a}+\frac{y^2}{2^a}+\frac{y^3}{3^a}\right)\frac{z^3}{3^b}$$

$$+\left(\frac{y^{-4}}{(-4)^a}+\frac{y^{-3}}{(-3)^a}+\frac{y^{-2}}{(-2)^a}+\frac{y^{-1}}{(-1)^a}+1+\frac{y^1}{1^a}+\frac{y^2}{2^a}+\frac{y^3}{3^a}+\frac{y^4}{4^a}\right)\frac{z^4}{4^b}$$

$$+\ etc.$$

$$=\frac{y^{-1}z^1}{(-1)^a1^c}+\frac{y^0z^1}{1\times 1^c}+\frac{y^1z^1}{1^a1^c}$$

$$+\frac{y^{-2}z^2}{(-2)^a2^c}+\frac{y^{-1}z^2}{(-1)^a2^c}+\frac{y^0z^2}{1\times 2^c}+\frac{y^1z^2}{1^a2^c}+\frac{y^2z^2}{2^a2^c}$$

$$+\frac{y^{-3}z^3}{(-3)^a3^c}+\frac{y^{-2}z^3}{(-2)^a3^c}+\frac{y^{-1}z^3}{(-1)^a3^c}+\frac{y^0z^3}{1\times 3^c}+\frac{y^1z^3}{1^a3^c}+\frac{y^2z^3}{2^a3^c}+\frac{y^3z^3}{3^a3^c}$$

$$+\frac{y^{-4}z^4}{(-4)^a4^c}+\frac{y^{-3}z^4}{(-3)^a4^c}+\frac{y^{-2}z^4}{(-2)^a4^c}+\frac{y^{-1}z^4}{(-1)^a4^c}+\frac{y^0z^4}{1\times 4^c}+\frac{y^1z^4}{1^a4^c}+\frac{y^2z^4}{2^a4^c}+\frac{y^3z^4}{3^a4^c}+\frac{y^4z^4}{4^a4^c}$$

$$+\ etc.$$

$$=\sum_{|m|,n\geq 1;|m|\leq n}^{\infty}\frac{y^m z^n}{m^a n^b}$$

$$=\sum_{\substack{h,|j|,k\geq 1\\ |j|\leq k;(j,k)=1}}\frac{(y^j z^k)^h}{h^{a+b}(j^a k^b)}$$

$$=\sum_{\substack{|j|,k\geq 1\\ |j|\leq k;(j,k)=1}}\frac{1}{(j^a k^b)}\sum_{h=1}^{\infty}\frac{(y^j z^k)^h}{h^{a+b}}$$

$$=\sum_{\substack{|j|,k\geq 1\\ |j|\leq k;(j,k)=1}}\frac{1}{(j^a k^b)}\log\left(\frac{1}{1-y^j z^k}\right)\quad if\quad a+b=1.$$

Therefore, we have shown that

$$\sum_{n=1}^{\infty}\left(\sum_{m=-n}^{n}\frac{y^m}{m^a}\right)\frac{z^n}{n^b}=\sum_{\substack{j,k\geq 1\\ j\leq k;(j,k)=1}}\frac{1}{(j^a k^b)}\log\left(\frac{1}{1-y^j z^k}\right)\quad if\quad a+b=1.$$

Exponentiating both sides (and swapping sides) gives us the $2D$ first extended inverted symmetric triangle VPV identity, where in this $2D$ case the nD pyramid reduces to the form of a triangle-shaped array of lattice point vectors having the z-axis as the axis of symmetry, and so we can state the

Theorem 21.4 ***The* 2D *vertical symmetry extended triangle VPV identity.*** *For* $0<|yz|<1$, $0<|z/y|<1$, $0<|z|<1$, *with* $a+b=1$,

$$\prod_{\substack{|j|,k\geq 1\\ |j|\leq k;(j,k)=1}}\left(\frac{1}{1-y^j z^k}\right)^{\frac{1}{j^a k^b}}=\exp\left\{\sum_{n=1}^{\infty}\left(\sum_{m=-n}^{n}\frac{y^m}{m^a}\right)\frac{z^n}{n^b}\right\}\quad if\quad a+b=1.\qquad(21.5.2)$$

As with our earlier exploits into the $2D$ first quadrant case, for the present result we take some simple example cases where new and interesting results arise.

So, let us take the case where $a=0, b=1$, giving us for $0<|yz|<1$, $0<|z/y|<1$, $0<|z|<1$,

$$\prod_{\substack{|j|,k\geq 1\\ |j|\leq k;(j,k)=1}}\left(\frac{1}{1-y^j z^k}\right)^{\frac{1}{k}}=\exp\left\{\sum_{n=1}^{\infty}\left(\sum_{m=-n}^{n}y^m\right)\frac{z^n}{n}\right\}$$

$$= \exp\left\{\sum_{n=1}^{\infty}\left(\frac{y^{2n+1}-1}{y^n(y-1)}\right)\frac{z^n}{n}\right\} = \exp\left\{\frac{1}{1-y}\log\left(\frac{(1-yz)^y}{1-z/y}\right)\right\}.$$

So, we arrive then at the following pair of equivalent results, for $0<|yz|<1$, $0<|z/y|<1$, $0<|z|<1$,

$$\prod_{\substack{|j|,k\geq 1\\ |j|\leq k;\,(j,k)=1}}\left(\frac{1}{1-y^j z^k}\right)^{\frac{1}{k}} = \left(\frac{(1-yz)^y}{1-z/y}\right)^{\frac{1}{1-y}}, \tag{21.5.3}$$

and

$$\prod_{\substack{|j|,k\geq 1\\ |j|\leq k;\,(j,k)=1}}\left(1-y^j z^k\right)^{\frac{1}{k}} = \left(\frac{1-z/y}{(1-yz)^y}\right)^{\frac{1}{1-y}}. \tag{21.5.4}$$

From here, multiply both sides of (21.5.3) and the case of (21.5.4) with $y\mapsto y^2$ and $z\mapsto z^2$ to get,

$$\prod_{\substack{|j|,k\geq 1\\ |j|\leq k;\,(j,k)=1}}\left(1+y^j z^k\right)^{\frac{1}{k}} = \left(\frac{(1-yz)^y}{1-z/y}\right)^{\frac{1}{1-y}}\left(\frac{1-(z/y)^2}{(1-(yz)^2)^{y^2}}\right)^{\frac{1}{1-y^2}}. \tag{21.5.5}$$

Particular cases:
$y=\frac{1}{2}$ gives us from (21.5.4) and (21.5.5) the two results that

$$\prod_{\substack{|j|,k\geq 1\\ |j|\leq k;\,(j,k)=1}}\left(1-\frac{z^k}{2^j}\right)^{\frac{-1}{k}} = \frac{1-z/2}{(1-2z)^2}\sqrt[4]{\left(\frac{1-4z^2}{\sqrt[4]{1-z^2/4}}\right)^3}$$

$$= 1+\frac{7z}{2}+\frac{19z^2}{4}+\frac{61z^3}{8}+\frac{117z^4}{8}+\frac{423z^5}{16}+\frac{4861z^6}{96}+\frac{18259z^7}{192}$$
$$+\frac{140867z^8}{768}+\frac{538373z^9}{1536}+\frac{696379z^{10}}{1024}+O(z^{11})$$

$$= \frac{1}{(1-\frac{z}{2})}$$

$$\frac{1}{\sqrt{(1-2^1z^2)\left(1-\frac{z^2}{2^1}\right)}}$$

$$\frac{1}{\sqrt[3]{(1-2^2z^3)(1-2^1z^3)\left(1-\frac{z^3}{2^1}\right)\left(1-\frac{z^3}{2^2}\right)}}$$

$$\frac{1}{\sqrt[4]{(1-2^3z^4)(1-2^1z^4)\left(1-\frac{z^4}{2^1}\right)\left(1-\frac{z^4}{2^3}\right)}}$$

$$\frac{1}{\sqrt[5]{(1-2^4z^5)(1-2^3z^5)(1-2^2z^5)(1-2^1z^5)\left(1-\frac{z^5}{2^1}\right)\left(1-\frac{z^5}{2^2}\right)\left(1-\frac{z^5}{2^3}\right)\left(1-\frac{z^5}{2^4}\right)}}$$

$$\frac{1}{\sqrt[6]{(1-2^5z^6)(1-2^1z^6)\left(1-\frac{z^6}{2^1}\right)\left(1-\frac{z^6}{2^5}\right)}}$$

$$\vdots,$$

$$\prod_{\substack{|j|,k\geq 1\\ |j|\leq k;\,(j,k)=1}}\left(1+\frac{z^k}{2^j}\right)^{\frac{1}{k}} = \frac{2-z}{2-2z}\sqrt[3]{\frac{4-z^2}{4-4z^2}}$$

$$= 1+\frac{z}{2}+\frac{3z^2}{4}+\frac{5z^3}{8}+\frac{13z^4}{16}+\frac{23z^5}{32}+\frac{167z^6}{192}$$

$$+\frac{305z^7}{384}+\frac{59z^8}{64}+\frac{659z^9}{768}+O(z^{10})$$

$$= (1+2z)\left(1+\frac{z}{2}\right)$$

$$\sqrt{(1+2^1z^2)\left(1+\frac{z^2}{2^1}\right)}$$

$$\sqrt[3]{(1+2^2z^3)(1+2^1z^3)\left(1+\frac{z^3}{2^1}\right)\left(1+\frac{z^3}{2^2}\right)}$$

$$\sqrt[4]{(1+2^3z^4)(1+2^1z^4)\left(1+\frac{z^4}{2^1}\right)\left(1+\frac{z^4}{2^3}\right)}$$

$$\sqrt[5]{(1+2^4z^5)(1+2^3z^5)(1+2^2z^5)(1+2^1z^5)\left(1+\frac{z^5}{2^1}\right)\left(1+\frac{z^5}{2^2}\right)\left(1+\frac{z^5}{2^3}\right)\left(1+\frac{z^5}{2^4}\right)}$$

$$\sqrt[6]{(1+2^5z^6)(1+2^1z^6)\left(1+\frac{z^6}{2^1}\right)\left(1+\frac{z^6}{2^5}\right)}$$

$$\vdots.$$

These two equations can be easily verified on a calculating engine like Mathematica or WolframAlpha by expanding each side into it's Taylor series around $z = 0$ and comparing coefficients of like powers of z. Next, take the cases of (21.5.4) and (21.5.5) with $y = 2$, both of which converge if $|z| < 2$, so then, after a slight adjustment to both sides by a factor of $1-2z$,

We remark at this juncture that equations (21.5.4) and it's reciprocal equation (21.5.5) are amenable to applying the limit as y approaches 1. In fact we have as follows that,

$$\lim_{y\to 1}\left(\frac{1-z}{1-yz}\right)^{\frac{y}{1-y}}=e^{\frac{z}{z-1}}$$

and also from considering equation (21.5.5) there is the limit, easily evaluated,

$$\lim_{y\to 1}\left(\frac{1-yz}{1-z}\right)^{\frac{y}{1-y}}\left(\frac{1-z^2}{1-y^2z^2}\right)^{\frac{y^2}{1-y^2}}=e^{\frac{z}{1-z^2}}.$$

Therefore, applying these two limits to equations (21.5.4) and (21.5.5) respectively we obtain the two interesting results (21.5.6) and (21.5.7) given here.

$$\prod_{k=1}^{\infty}\left(1-z^k\right)^{\frac{\varphi(k)}{k}}=e^{\frac{z}{z-1}}=\sum_{k=0}^{\infty}\frac{\alpha(k)z^k}{k!} \tag{21.5.6}$$

$$=1-\frac{z}{1!}-\frac{z^2}{2!}-\frac{z^3}{3!}+\frac{z^4}{4!}+\frac{19z^5}{5!}+\frac{151z^6}{6!}+\frac{1091z^7}{7!}$$

$$+\frac{7841z^8}{8!}+\frac{56519z^9}{9!}+\frac{396271z^{10}}{10!}+O(z^{11}),$$

demonstrating sequence $\alpha(k)$ has the exponential generating function $e^{\frac{z}{z-1}}$. Amazingly $\gcd(\alpha(k),k!)=1$, for all values of k up to 34, and mostly beyond that, and $\alpha(k)\equiv 1\ or\ 9\ (mod\ 10)$, and also the recurrence relation

$$\alpha(n)+(n-1)(n-2)\,\alpha(n-2)=(2n-3)\,\alpha(n-1)$$

holds. (See OEIS sequence A293116 [237]) This recurrence relation allows us to write continued fractions for the ratios $\alpha(n+1)/\alpha(n)$.

$$\prod_{k=1}^{\infty}\left(1+z^k\right)^{\frac{\varphi(k)z^k}{k}}=e^{\frac{z}{1-z^2}}=\sum_{k=0}^{\infty}\frac{\beta(k)z^k}{k!} \tag{21.5.7}$$

$$=1+\frac{z}{1!}+\frac{z^2}{2!}+\frac{7z^3}{3!}+\frac{25z^4}{4!}+\frac{181z^5}{5!}+\frac{1201z^6}{6!}$$

$$+\frac{10291z^7}{7!}+\frac{97777z^8}{8!}+\frac{202709z^9}{9!}+O(z^{10}),$$

where $\varphi(k)$ is the Euler totient function, the number of positive integers less than and coprime to k.

Next we take (21.5.2) with the case that $a=1$ and $b=0$, so then

$$\prod_{\substack{j,k\geq 1\\ j\leq k;\,(j,k)=1}}\left(\frac{1}{1-y^jz^k}\right)^{\frac{1}{j}}=\exp\left\{\sum_{n=1}^{\infty}\left(\sum_{m=1}^{n}\frac{y^m}{m}\right)z^n\right\}$$

$$=\exp\left\{\frac{1}{1-z}\sum_{n=1}^{\infty}\frac{y^nz^n}{n}\right\}=\exp\left\{\frac{1}{1-z}\log\left(\frac{1}{1-yz}\right)\right\}.$$

This leads us to establish that

$$\prod_{\substack{j,k\geq 1\\ j\leq k;\,(j,k)=1}} \left(\frac{1}{1-y^j z^k}\right)^{\frac{1}{j}} = \left(\frac{1}{1-yz}\right)^{\frac{1}{1-z}}, \tag{21.5.8}$$

which is equivalent to

$$\prod_{\substack{j,k\geq 1\\ j\leq k;\,(j,k)=1}} \left(1-y^j z^k\right)^{\frac{1}{j}} = (1-yz)^{\frac{1}{1-z}}. \tag{21.5.9}$$

From multiplying both sides of (21.2.7) in which $y \mapsto y^2$ and $z \mapsto z^2$ with both sides of (21.2.8), we obtain

$$\prod_{\substack{j,k\geq 1\\ j\leq k;\,(j,k)=1}} \left(1+y^j z^k\right)^{\frac{1}{j}} = \frac{(1-y^2z^2)^{\frac{1}{1-z^2}}}{(1-yz)^{\frac{1}{1-z}}}. \tag{21.5.10}$$

Particular cases:
$z=\frac{1}{2}$ gives us from (21.5.9) and (21.5.10) the remarkable result that

$$\prod_{\substack{j,k\geq 1;\,j\leq k\\ gcd(j,k)=1}} \left(1-\frac{y^j}{2^k}\right)^{\frac{1}{j}} = \left(1-\frac{y}{2}\right)^2 = 1-\frac{y}{4}+\frac{y^2}{4}$$

$$= \left(1-\frac{y^1}{2^1}\right)$$

$$\left(1-\frac{y^1}{2^2}\right)$$

$$\left(1-\frac{y^1}{2^3}\right)\sqrt{\left(1-\frac{y^2}{2^3}\right)}$$

$$\left(1-\frac{y^1}{2^4}\right)\sqrt[3]{\left(1-\frac{y^3}{2^4}\right)}$$

$$\left(1-\frac{y^1}{2^5}\right)\sqrt{\left(1-\frac{y^2}{2^5}\right)}\sqrt[3]{\left(1-\frac{y^3}{2^5}\right)}\sqrt[4]{\left(1-\frac{y^4}{2^5}\right)}$$

$$\left(1-\frac{y^1}{2^6}\right)\sqrt[5]{\left(1-\frac{y^5}{2^6}\right)}$$

$$\vdots,$$

and the result,

$$\prod_{\substack{j,k\geq1;j\leq k\\ gcd(j,k)=1}}\left(1+\frac{y^j}{2^k}\right)^{\frac{1}{j}}=\frac{\sqrt[3]{(4-y^2)^4}}{\sqrt[3]{4}(2-y)^2}=1+y+\frac{5y^2}{12}+\frac{y^3}{6}+\frac{11y^4}{144}+\frac{5y^5}{144}+O(y^6)$$

$$=\left(1+\frac{y^1}{2^1}\right)$$

$$\left(1+\frac{y^1}{2^2}\right)$$

$$\left(1+\frac{y^1}{2^3}\right)\sqrt{\left(1+\frac{y^2}{2^3}\right)}$$

$$\left(1+\frac{y^1}{2^4}\right)\sqrt[3]{\left(1+\frac{y^3}{2^4}\right)}$$

$$\left(1+\frac{y^1}{2^5}\right)\sqrt{\left(1+\frac{y^2}{2^5}\right)}\sqrt[3]{\left(1+\frac{y^3}{2^5}\right)}\sqrt[4]{\left(1+\frac{y^4}{2^5}\right)}$$

$$\left(1+\frac{y^1}{2^6}\right)\sqrt[5]{\left(1+\frac{y^5}{2^6}\right)}$$

$$\vdots$$

These two equations can be verified on a calculating engine like Mathematica or WolframAlpha by expanding each side into it's Taylor series around $y=0$ and comparing coefficients of like powers of y. However, the calculation is an infinite series for each coefficient, unlike in the previous examples, where it is a finite sum.

21.6 3D VPV IDENTITIES FOR A RIGHT SQUARE PYRAMID LATTICE

As we did in section 21.3, we start with a simple $3D$ summation. We derive on a $3D$ inverted pyramid-shaped lattice that extends infinitely and occupies four adjacent hyperquadrants of the eight hyperquadrants that comprise the $X-Y-Z$ 3-space.

We depict this infinite inverted pyramid with square layered arrays of lattice point vectors as per the following diagram, with VPVs bolded.

$$
\begin{array}{ccccccccc}
\vdots & \vdots & \vdots & \vdots & \vdots & \vdots & \vdots & \vdots & \vdots \\
\langle \mathbf{-1,-3,3} \rangle & & \langle \mathbf{0,-2,3} \rangle & & \langle \mathbf{1,-1,3} \rangle & & \langle \mathbf{2,0,3} \rangle & & \langle \mathbf{3,1,3} \rangle \\
& \langle 0,-3,3 \rangle & & \langle \mathbf{1,-2,3} \rangle & & \langle \mathbf{2,-1,3} \rangle & & \langle 3,0,3 \rangle & \\
& & \langle \mathbf{1,-3,3} \rangle & & \langle \mathbf{2,-2,3} \rangle & & \langle \mathbf{3,-1,3} \rangle & & \\
& & & \langle \mathbf{2,-3,3} \rangle & & \langle \mathbf{3,-2,3} \rangle & & & \\
& & & & \langle 3,-3,3 \rangle & & & & \\
\\
& & & & \langle -2,2,2 \rangle & & & & \\
& & & \langle \mathbf{-2,1,2} \rangle & & \langle \mathbf{-1,2,2} \rangle & & & \\
& & \langle -2,0,2 \rangle & & \langle \mathbf{-1,1,2} \rangle & & \langle 0,2,2 \rangle & & \\
& \langle \mathbf{-2,-1,2} \rangle & & \langle \mathbf{-1,0,2} \rangle & & \langle \mathbf{0,1,2} \rangle & & \langle \mathbf{1,2,2} \rangle & \\
\langle -2,-2,2 \rangle & & \langle \mathbf{-1,-1,2} \rangle & & \langle 0,0,2 \rangle & & \langle \mathbf{1,1,2} \rangle & & \langle 2,2,2 \rangle \\
& \langle \mathbf{-1,-2,2} \rangle & & \langle \mathbf{0,-1,2} \rangle & & \langle \mathbf{1,3,2} \rangle & & \langle 2,0,2 \rangle & \\
& & \langle 0,-2,2 \rangle & & \langle \mathbf{1,-1,2} \rangle & & \langle 2,0,2 \rangle & & \\
& & & \langle \mathbf{1,-2,2} \rangle & & \langle \mathbf{2,-1,2} \rangle & & & \\
& & & & \langle 2,-2,2 \rangle & & & & \\
\\
& & & & \langle \mathbf{-1,1,1} \rangle & & & & \\
& & & \langle \mathbf{-1,0,1} \rangle & & \langle \mathbf{0,1,1} \rangle & & & \\
& & \langle \mathbf{-1,-1,1} \rangle & & \langle \mathbf{0,0,1} \rangle & & \langle \mathbf{1,1,1} \rangle & & \\
& & & \langle \mathbf{0,-1,1} \rangle & & \langle \mathbf{1,0,1} \rangle & & & \\
& & & & \langle \mathbf{1,-1,1} \rangle & & & & \\
& & & & \langle 0,0,0 \rangle & & & &
\end{array}
\tag{21.6.1}
$$

So now, we consider the sum, whose shape is an inverted $3D$ right pyramid whose apex is at the origin $\langle 0,0,0 \rangle$, given by (for $0 < each\ of\ |xz|, |z/x|, |yz|, |z/y|, |z| < 1$)

$$
\sum_{n=1}^{\infty} \left(\sum_{l=-n}^{n} \frac{x^l}{l^a} \right) \left(\sum_{m=-n}^{n} \frac{y^m}{m^b} \right) \frac{z^n}{n^c}
$$

$$
= \left(\frac{x^{-1}}{(-1)^a} + 1 + \frac{x^1}{1^a} \right) \left(\frac{y^{-1}}{(-1)^a} + 1 + \frac{y^1}{1^b} \right) \frac{z^1}{1^c}
$$

$$
+ \left(\frac{x^{-2}}{(-2)^a} + \frac{x^{-1}}{(-1)^a} + 1 + \frac{x^1}{1^a} + \frac{x^2}{2^a} \right) \left(\frac{y^{-2}}{(-2)^b} + \frac{y^{-1}}{(-1)^b} + 1 + \frac{y^1}{1^b} + \frac{y^2}{2^b} \right) \frac{z^2}{2^c}
$$

$$
+ \left(\frac{x^{-3}}{(-3)^a} + \frac{x^{-2}}{(-2)^a} + \frac{x^{-1}}{(-1)^a} + 1 + \frac{x^1}{1^a} + \frac{x^2}{2^a} + \frac{x^3}{3^a} \right)
$$

$$
\left(\frac{y^{-3}}{(-3)^a} + \frac{y^{-2}}{(-2)^b} + \frac{y^{-1}}{(-1)^b} + 1 + \frac{y^1}{1^b} + \frac{y^2}{2^b} + \frac{y^3}{3^b} \right) \frac{z^3}{3^c}
$$

$$
+ \left(\frac{x^{-4}}{(-4)^a} + \frac{x^{-3}}{(-3)^a} + \frac{x^{-2}}{(-2)^a} + \frac{x^{-1}}{(-1)^a} + 1 + \frac{x^1}{1^a} + \frac{x^2}{2^a} + \frac{x^3}{3^a} + \frac{x^4}{4^a} \right)
$$

$$
+ \left(\frac{y^{-4}}{(-4)^a} + \frac{y^{-3}}{(-3)^a} + \frac{y^{-2}}{(-2)^a} + \frac{y^{-1}}{(-1)^a} + 1 + \frac{y^1}{1^a} + \frac{y^2}{2^a} + \frac{y^3}{3^a} + \frac{y^4}{4^a} \right) \frac{z^4}{4^c}
$$

$$
+
$$

$$
\vdots
$$

$$
= \frac{x^{-1}y^1z^1}{(-1)^a\, 1^b\, 1^c} + \frac{x^0y^1z^1}{1 \times 1^b\, 1^c} + \frac{x^1y^1z^1}{1^a\, 1^b\, 1^c}
$$

$$
+ \frac{x^{-1}y^0z^1}{(-1)^a \times 1 \times 1^c} + \frac{x^0y^0z^1}{1 \times 1 \times 1^c} + \frac{x^1y^0z^1}{1^a \times 1 \times 1^c}
$$

$$+\frac{x^{-1}y^{-1}z^{1}}{(-1)^{a}(-1)^{b}1^{c}}+\frac{x^{0}y^{-1}z^{1}}{1\times(-1)^{b}1^{c}}+\frac{x^{1}y^{-1}z^{1}}{1^{a}(-1)^{b}1^{c}}$$

$$+\frac{x^{-2}y^{2}z^{2}}{(-2)^{a}2^{b}2^{c}}+\frac{x^{-1}y^{2}z^{2}}{(-1)^{a}2^{b}2^{c}}+\frac{x^{0}y^{2}z^{2}}{1\times 2^{b}2^{c}}+\frac{x^{1}y^{2}z^{2}}{1^{a}2^{b}2^{c}}+\frac{x^{2}y^{2}z^{2}}{2^{a}2^{b}2^{c}}$$

$$+\frac{x^{-2}y^{1}z^{2}}{(-2)^{a}1^{b}2^{c}}+\frac{x^{-1}y^{1}z^{2}}{(-1)^{a}1^{b}2^{c}}+\frac{x^{0}y^{1}z^{2}}{1\times 1^{b}2^{c}}+\frac{x^{1}y^{1}z^{2}}{1^{a}1^{b}2^{c}}+\frac{x^{2}y^{1}z^{2}}{2^{a}1^{b}2^{c}}$$

$$+\frac{x^{-2}y^{0}z^{2}}{(-2)^{a}\times 1\times 2^{c}}+\frac{x^{-1}y^{0}z^{2}}{(-1)^{a}\times 1\times 2^{c}}+\frac{x^{0}y^{0}z^{2}}{1\times 1\times 2^{c}}+\frac{x^{1}y^{0}z^{2}}{1^{a}\times 1\times 2^{c}}+\frac{x^{2}y^{0}z^{2}}{2^{a}\times 1\times 2^{c}}$$

$$+\frac{x^{-2}y^{-1}z^{2}}{(-2)^{a}(-1)^{b}2^{c}}+\frac{x^{-1}y^{-1}z^{2}}{(-1)^{a}(-1)^{b}2^{c}}+\frac{x^{0}y^{-1}z^{2}}{1\times(-1)^{b}2^{c}}+\frac{x^{1}y^{-1}z^{2}}{1^{a}(-1)^{b}2^{c}}+\frac{x^{2}y^{-1}z^{2}}{2^{a}(-1)^{b}2^{c}}$$

$$+\frac{x^{-2}y^{-2}z^{2}}{(-2)^{a}(-2)^{b}2^{c}}+\frac{x^{-1}y^{-2}z^{2}}{(-1)^{a}(-2)^{b}2^{c}}+\frac{x^{0}y^{-2}z^{2}}{1\times(-2)^{b}2^{c}}+\frac{x^{1}y^{-2}z^{2}}{1^{a}(-2)^{b}2^{c}}+\frac{x^{2}y^{-2}z^{2}}{2^{a}(-2)^{b}2^{c}}$$

$$+$$

$$\vdots$$

$$=\sum_{|l|,|m|,n\geq 1;\ |l|,|m|\leq n}^{\infty}\frac{x^{l}y^{m}z^{n}}{l^{a}m^{b}n^{c}}$$

$$=\sum_{\substack{h,|l|,|m|,n\geq 1\\ |l|,|m|\leq n;\gcd(l,m,n)=1}}\frac{(x^{l}y^{m}z^{n})^{h}}{h^{a+b+c}(l^{a}m^{b}n^{c})}$$

$$=\sum_{\substack{|l|,|m|,n\geq 1\\ |l|,|m|\leq n;\gcd(l,m,n)=1}}\frac{1}{(l^{a}m^{b}n^{c})}\sum_{h=1}^{\infty}\frac{(x^{l}y^{n}z^{n})^{h}}{h^{a+b+c}}$$

$$=\sum_{\substack{|l|,|m|,n\geq 1\\ |l|,|m|\leq n;\gcd(l,m,n)=1}}\frac{1}{(l^{a}m^{b}n^{c})}\log\left(\frac{1}{1-x^{l}y^{b}z^{c}}\right)\quad if\quad a+b+c=1.$$

Therefore, we have shown that if $a+b+c=1$ and for $0<|x|,|1/x|,|y|,|1/y|$ all $<|z|<1$, then

$$\sum_{n=1}^{\infty}\left(\sum_{l=-n}^{n}\frac{x^{l}}{l^{a}}\right)\left(\sum_{m=-n}^{n}\frac{y^{m}}{m^{b}}\right)\frac{z^{n}}{n^{c}}=\sum_{\substack{|l|,|m|,n\geq 1\\ |l|,|m|\leq n;\gcd(l,m,n)=1}}\frac{1}{(l^{a}m^{b}n^{c})}\log\left(\frac{1}{1-x^{l}y^{b}z^{c}}\right).$$

Simplifying both sides gives us the $3D$ "right square pyramid VPV identity", where in this $3D$ case the pyramid takes the form of layered square-shaped arrays of lattice point vectors as shown in the above workings, with the central axis rising vertically from the apex point $\langle 0,0,0\rangle$.

The identity is summarized in the

Theorem 21.5 ***The*** *3D* ***right square pyramid VPV identity.*** *For* $0 <$ *eachof* $|xz|, |z/x|, |yz|, |z/y|, |z| < 1$, *with* $a+b+c=1$,

$$\prod_{\substack{|l|,|m|,n\geq 1 \\ |l|,|m|\leq n;\, \gcd(l,m,n)=1}} \left(\frac{1}{1-x^l y^m z^n}\right)^{\frac{1}{l^a m^b n^c}} = \exp\left\{\sum_{n=1}^{\infty}\left(\sum_{l=-n}^{n}\frac{x^l}{l^a}\right)\left(\sum_{m=-n}^{n}\frac{y^m}{m^b}\right)\frac{z^n}{n^c}\right\}. \tag{21.6.2}$$

As we did for the $2D$ particular cases, we can examine some obvious example corollaries arising from this theorem. Firstly, take the case where $a=b=0, c=1$, so then,

$$\prod_{\substack{|l|,|m|,n\geq 1 \\ |l|,|m|\leq n;\, \gcd(l,m,n)=1}} \left(\frac{1}{1-x^l y^m z^n}\right)^{\frac{1}{n}} = \exp\left\{\sum_{n=1}^{\infty}\left(\sum_{l=-n}^{n}x^l\right)\left(\sum_{m=-n}^{n}y^m\right)\frac{z^n}{n}\right\}$$

$$= \exp\left\{\sum_{n=1}^{\infty}\left(\frac{x^{2n+1}-1}{x^n(x-1)}\right)\left(\frac{y^{2n+1}-1}{y^n(y-1)}\right)\frac{z^n}{n}\right\}$$

$$= \exp\left\{\frac{1}{(1-x)(1-y)}\sum_{n=1}^{\infty}\left((xy)^{n+1}+\left(\frac{1}{xy}\right)^n - x\left(\frac{x}{y}\right)^n - y\left(\frac{y}{x}\right)^n\right)\frac{z^n}{n}\right\}$$

$$= \exp\left\{\frac{1}{(1-x)(1-y)}\log\left(\frac{(1-xz/y)^x(1-yz/x)^y}{(1-xyz)^{xy}(1-z/(xy))}\right)\right\},$$

$$= \left\{\frac{(1-xz/y)^x(1-yz/x)^y}{(1-xyz)^{xy}(1-z/(xy))}\right\}^{\frac{1}{(1-x)(1-y)}}.$$

This then implies

Corollary 21.8 *For* $0 <$ *each of* $|xz|, |z/x|, |yz|, |z/y|, |z| < 1$, *with* $a+b+c=1$,

$$\prod_{\substack{|l|,|m|,n\geq 1 \\ |l|,|m|\leq n;\, \gcd(l,m,n)=1}} \left(\frac{1}{1-x^l y^m z^n}\right)^{\frac{1}{n}} = \left\{\frac{(1-xz/y)^x(1-yz/x)^y}{(1-xyz)^{xy}(1-z/(xy))}\right\}^{\frac{1}{(1-x)(1-y)}}, \tag{21.6.3}$$

and its reciprocal identity,

$$\prod_{\substack{|l|,|m|,n\geq 1 \\ |l|,|m|\leq n;\, \gcd(l,m,n)=1}} \left(1-x^l y^m z^n\right)^{\frac{1}{n}} = \left\{\frac{(1-xyz)^{xy}(1-z/(xy))}{(1-xz/y)^x(1-yz/x)^y}\right\}^{\frac{1}{(1-x)(1-y)}}. \tag{21.6.4}$$

We see that (21.6.3) and (21.6.4) are generalizations of the 2D identities (21.2.2) and (21.2.3) from an earlier section.

21.7 VPV IDENTITIES IN nD RIGHT-SQUARE HYPERPYRAMID REGIONS

The n-dimensional square hyperpyramid VPV Identity is encoded in the following

Theorem 21.6 ***The nD right-square hyperpyramid VPV identity.*** *If $i = 1,2,3,...,n$ then for each $x_i \in \mathbb{C}$ such that $|x_i| < 1$ and $b_i \in \mathbb{C}$ such that $\sum_{i=1}^{n} b_i = 1$,*

$$\prod_{\substack{\gcd(|a_1|,|a_2|,...,|a_{n-1}|,a_n)=1 \\ |a_1|,|a_2|,...,|a_{n-1}|\leq a_n \\ |a_1|,|a_2|,...,|a_{n-1}|,a_n\geq 1}} \left(\frac{1}{1-{x_1}^{a_1}{x_2}^{a_2}{x_3}^{a_3}\cdots {x_n}^{a_n}}\right)^{\frac{1}{{a_1}^{b_1}{a_2}^{b_2}{a_3}^{b_3}\cdots {a_n}^{b_n}}} \tag{21.7.1}$$

$$= \exp\left\{\sum_{k=1}^{\infty}\prod_{i=1}^{n-1}\left(\sum_{j=-k}^{k}\frac{{x_i}^j}{j^{b_i}}\right)\frac{{x_n}^k}{k^{b_n}}\right\}$$

$$= \exp\left\{\sum_{k=1}^{\infty}\left(\sum_{j=-k}^{k}\frac{{x_1}^j}{j^{b_1}}\right)\left(\sum_{j=-k}^{k}\frac{{x_2}^j}{j^{b_2}}\right)\left(\sum_{j=-k}^{k}\frac{{x_3}^j}{j^{b_3}}\right)\cdots\left(\sum_{j=-k}^{k}\frac{{x_{n-1}}^j}{j^{b_{n-1}}}\right)\frac{{x_n}^k}{k^{b_n}}\right\}.$$

This result is quite straight-forward to prove using the technique of our two previous sections. This methodology was also given in Campbell [91] but has not been worked through for corollaries of theorem 21.6 over the past 24 years.

Since in the previous section, we gave the 3D example of theorem 21.6, we state the 4D case of it now.

Corollary 21.9 *If $|wxyz|$, $|xyz|$, $|yz|$, and $|z| < 1$ and $r+s+t+u=1$ then,*

$$\prod_{\substack{(a,b,c,d)=1 \\ |a|,|b|,|c|\leq d \\ a,b,c\geq 0, d\geq 1}} \left(\frac{1}{1-w^a x^b y^c z^d}\right)^{\frac{1}{a^r b^s c^t d^u}} = \exp\{\mathrm{P}(r,w;s,x;t,y;u,z)\} \tag{21.7.2}$$

where

$$\mathrm{P}(r,w;s,x;t,y;u,z) = \sum_{n=1}^{\infty}\left(\sum_{k=-n}^{n}\frac{w^k}{k^r}\right)\left(\sum_{k=-n}^{n}\frac{x^k}{k^s}\right)\left(\sum_{k=-n}^{n}\frac{y^k}{k^t}\right)\frac{z^n}{n^u}.$$

Take the case where $r = s = t = 0, u = 1$, so then,

$$\prod_{\substack{|h|,|i|,|j|,k\geq 1 \\ |h|,|i|,|j|\leq k;\, \gcd(h,i,j,k)=1}} \left(\frac{1}{1-w^h x^i y^j z^k}\right)^{\frac{1}{k}}$$

$$= \exp\left\{\sum_{n=1}^{\infty}\left(\sum_{k=-n}^{n} w^k\right)\left(\sum_{k=-n}^{n} x^k\right)\left(\sum_{k=-n}^{n} y^k\right)\frac{z^n}{n}\right\}$$

$$= \exp\left\{\sum_{n=1}^{\infty}\left(\frac{w^{2n+1}-1}{w^n(w-1)}\right)\left(\frac{x^{2n+1}-1}{x^n(x-1)}\right)\left(\frac{y^{2n+1}-1}{y^n(y-1)}\right)\frac{z^n}{n}\right\}$$

$$= \left\{ \frac{(1-wxyz)^{wxy}(1-wz/(xy))^{w}(1-xz/(wy))^{x}(1-yz/(wx))^{y}}{(1-wxz/y)^{wx}(1-wyz/x)^{wy}(1-xyz/w)^{xy}(1-z/(wxy))} \right\}^{\frac{1}{(1-w)(1-x)(1-y)}}.$$

This then implies

Corollary 21.10 *For* $0 <$ *each of* $|wz|, |z/w|, |xz|, |z/x|, |yz|, |z/y|, |z| < 1$,

$$\prod_{\substack{|h|,|i|,|j|,k \geq 1 \\ |h|,|i|,|j| \leq k;\, \gcd(h,i,j,k)=1}} \left(\frac{1}{1-w^h x^i y^j z^k} \right)^{\frac{1}{n}} \tag{21.7.3}$$

$$= \left\{ \frac{(1-wxyz)^{wxy}(1-wz/(xy))^{w}(1-xz/(wy))^{x}(1-yz/(wx))^{y}}{(1-wxz/y)^{wx}(1-wyz/x)^{wy}(1-xyz/w)^{xy}(1-z/(wxy))} \right\}^{\frac{1}{(1-w)(1-x)(1-y)}},$$

and its reciprocal identity,

$$\prod_{\substack{|h|,|i|,|j|,k \geq 1 \\ |h|,|i|,|j| \leq k;\, \gcd(h,i,j,k)=1}} \left(1-w^h x^i y^j z^k \right)^{\frac{1}{n}} \tag{21.7.4}$$

$$= \left\{ \frac{(1-wxz/y)^{wx}(1-wyz/x)^{wy}(1-xyz/w)^{xy}(1-z/(wxy))}{(1-wxyz)^{wxy}(1-wz/(xy))^{w}(1-xz/(wy))^{x}(1-yz/(wx))^{y}} \right\}^{\frac{1}{(1-w)(1-x)(1-y)}}.$$

Note that both sides of (21.7.3) are formally equivalent to the power series

$$1 + \frac{z}{1!} + \begin{vmatrix} \frac{(1-w^3)(1-x^3)(1-y^3)}{w^1x^1y^1(1-w)(1-x)(1-y)} & -1 \\ \frac{(1-w^5)(1-x^5)(1-y^5)}{w^2x^2y^2(1-w)(1-x)(1-y)} & \frac{(1-w^3)(1-x^3)(1-y^3)}{w^1x^1y^1(1-w)(1-x)(1-y)} \end{vmatrix} \frac{z^2}{2!}$$

$$+ \begin{vmatrix} \frac{(1-w^3)(1-x^3)(1-y^3)}{w^1x^1y^1(1-w)(1-x)(1-y)} & -1 & 0 \\ \frac{(1-w^5)(1-x^5)(1-y^5)}{w^2x^2y^2(1-w)(1-x)(1-y)} & \frac{(1-w^3)(1-x^3)(1-y^3)}{w^1x^1y^1(1-w)(1-x)(1-y)} & -2 \\ \frac{(1-w^7)(1-x^7)(1-y^7)}{w^3x^3y^3(1-w)(1-x)(1-y)} & \frac{(1-w^5)(1-x^5)(1-y^5)}{w^2x^2y^2(1-w)(1-x)(1-y)} & \frac{(1-w^3)(1-x^3)(1-y^3)}{w^1x^1y^1(1-w)(1-x)(1-y)} \end{vmatrix} \frac{z^3}{3!} + etc.$$

21.8 ENVOI: RESEARCH EXERCISES

Study 1. **2D first quadrant Upper VPV combinatorial sum.**

For each of $|y|, |z| < 1$, show that

$$\sum_{\substack{\gcd(a,b)=1 \\ 0 \leq a < b \leq 1}} \frac{y^a z^b}{1-y^a z^b} = \frac{z}{(1-z)(1-yz)}. \tag{21.8.1}$$

For this example we work it through in more detail.

Proof: Starting with the left side of (21.8.1)

$$\begin{aligned}
LHS &= \frac{y^0z^1}{1-y^0z^1} \\
&+ \frac{y^0z^2}{1-y^0z^2}+\frac{y^1z^2}{1-y^1z^2} \\
&+ \frac{y^0z^3}{1-y^0z^3}+\frac{y^1z^3}{1-y^1z^3}+\frac{y^2z^3}{1-y^2z^3} \\
&+ \frac{y^0z^4}{1-y^0z^4}+\frac{y^1z^4}{1-y^1z^4}+\frac{y^2z^4}{1-y^3z^4} \\
&+ \frac{y^0z^5}{1-y^0z^5}+\frac{y^1z^5}{1-y^1z^5}+\frac{y^2z^5}{1-y^3z^5}+\frac{y^3z^5}{1-y^0z^5}+\frac{y^4z^5}{1-y^4z^5} \\
&+ \frac{y^0z^6}{1-y^0z^6}+\frac{y^1z^6}{1-y^1z^6}+\frac{y^5z^6}{1-y^5z^6} \\
&+ \; etc.
\end{aligned}$$

$$= \sum_{\substack{\gcd(j,k)=1\\0\le j<k\le 1}} \sum_{a=0}^{\infty}(y^{aj}z^{ak}) = \sum_{a=0}^{\infty} \sum_{\substack{\gcd(j,k)=1\\0\le j<k\le 1}} (y^{aj}z^{ak})$$

$$\begin{aligned}
&= y^0z^1 \\
&+ (y^0+y^1)z^2 \\
&+ (y^0+y^1+y^2)z^3 \\
&+ (y^0+y^1+y^2+y^3)z^4 \\
&+ (y^0+y^1+y^2+y^3+y^4)z^5 \\
&+ (y^0+y^1+y^2+y^3+y^4+y^5)z^6 \\
&+ \; etc.
\end{aligned}$$

$$= \frac{1-y}{1-y}z+\frac{1-y^2}{1-y}z^2+\frac{1-y^3}{1-y}z^3+\frac{1-y^4}{1-y}z^4+\cdots$$

$$= \frac{1}{1-y}\left(\frac{z}{1-z}-\frac{yz}{1-yz}\right)$$

$$= \frac{z}{(1-z)(1-yz)}. \qquad ■$$

Study 2. **2D first quadrant Upper VPV zeta function analogy for the previous exercise.**

Show that for $\Re y > 1$, $\Re z > 1$,

$$\sum_{\substack{\gcd(a,b)=1\\1\le a<b\le 2}} \frac{1}{a^y b^z} = \frac{\zeta(y)\zeta(z)}{\zeta(y+z)}. \tag{21.8.2}$$

Proof: Starting with the left side of (21.8.2)

$$\begin{aligned}
LHS &= \left(\frac{1}{1^y}\right)\frac{1}{2^z} \\
&+ \left(\frac{1}{1^y}+\frac{1}{2^y}\right)\frac{1}{3^z} \\
&+ \left(\frac{1}{1^y}+\frac{1}{3^y}\right)\frac{1}{4^z} \\
&+ \left(\frac{1}{1^y}+\frac{1}{2^y}+\frac{1}{3^y}+\frac{1}{4^y}\right)\frac{1}{5^z} \\
&+ \left(\frac{1}{1^y}+\frac{1}{5^y}\right)\frac{1}{6^z} \\
&+ \left(\frac{1}{1^y}+\frac{1}{2^y}+\frac{1}{3^y}+\frac{1}{4^y}+\frac{1}{5^y}+\frac{1}{6^y}\right)\frac{1}{7^z} \\
&+ \ etc.
\end{aligned}$$

$$= \sum_{\substack{\gcd(j,k)=1 \\ 0\le j<k\le 1}} \sum_{a=0}^{\infty} \frac{1}{j^{ay}k^{az}} = \sum_{a=0}^{\infty} \sum_{\substack{\gcd(j,k)=1 \\ 0\le j<k\le 1}} \frac{1}{j^{ay}k^{az}}$$

$$\begin{aligned}
&= y^0z^1 \\
&+ (y^0+y^1)z^2 \\
&+ (y^0+y^1+y^2)z^3 \\
&+ (y^0+y^1+y^2+y^3)z^4 \\
&+ (y^0+y^1+y^2+y^3+y^4)z^5 \\
&+ (y^0+y^1+y^2+y^3+y^4+y^5)z^6 \\
&+ \ etc.
\end{aligned}$$

$$\begin{aligned}
&= \frac{1-y}{1-y}z+\frac{1-y^2}{1-y}z^2+\frac{1-y^3}{1-y}z^3+\frac{1-y^4}{1-y}z^4+\cdots \\
&= \frac{1}{1-y}\left(\frac{z}{1-z}-\frac{yz}{1-yz}\right) \\
&= \frac{z}{(1-z)(1-yz)}. \qquad \blacksquare
\end{aligned}$$

Infer from (21.8.1) and (21.8.2) that they encode two similar statements about partitions into lattice point vectors $\langle a,b\rangle$ with $gcd(a,b)=1$ and for (21.8.1) we have non-negative integers a and positive integers b; whereas for (21.8.2) we have positive integers for both a and b.

Study 3. **Find the particular cases of (21.8.1).**

Prove the following:

$$\sum_{\substack{\gcd(a,b)=1\\0\le a<b\le 1}} \frac{1}{3^n\,2^n-1} = \frac{2^{-n}}{(1-3^{-n})(1-2^{-n})} \quad (for\ \Re n>1),$$

$$= \frac{1}{2^n}\left(1+\frac{1}{2^n}+\frac{1}{3^n}+\frac{1}{4^n}+\frac{1}{6^n}+\frac{1}{8^n}+\frac{1}{9^n}+\frac{1}{12^n}+\cdots\right);$$

$$\sum_{\substack{\gcd(a,b)=1\\0\le a<b\le 1}} \frac{z^{a+b}}{1-z^{a+b}} = \frac{z}{(1-z)^2}, \quad |z|<1;$$

$$\sum_{\substack{\gcd(a,b)=1\\0\le a<b\le 1}} tan^{2(a+b)}(\theta) = \frac{1}{cos^2(\theta)}.$$

Study 4. **Find the particular cases of (21.8.2).**

Prove the following:

$$\sum_{\substack{\gcd(a,b)=1\\a,b\ge 1}} \frac{1}{a^2b^2} = \frac{5}{2},$$

$$\sum_{\substack{\gcd(a,b)=1\\a,b\ge 1}} \frac{1}{a^2b^3} = \frac{\pi^2\zeta(3)}{6\zeta(5)},$$

$$\sum_{\substack{\gcd(a,b)=1\\a,b\ge 1}} \frac{1}{a^3b^5} = \frac{9450\zeta(3)\zeta(5)}{\pi^8}.$$

Study 5. **3D first hyperquadrant sums using gcd.**

For each of $|x|, |y|, |z| < 1$, show that

$$\sum_{\substack{\gcd(a,b,c)=1\\a,b\ge 0,c>0}} \frac{x^a\,y^b\,z^c}{1-x^a\,y^b\,z^c} = \frac{z}{(1-x)(1-y)(1-z)}. \tag{21.8.3}$$

Similarly, show that for $\Re x > 1$, $\Re y > 1$, $\Re z > 1$,

$$\sum_{\substack{\gcd(a,b,c)=1\\a,b,c\ge 1}} \frac{1}{a^x b^y c^z} = \frac{\zeta(x)\zeta(y)\zeta(z)}{\zeta(x+y+z)}. \tag{21.8.4}$$

Infer from (21.8.3) and (21.8.4) that they encode two similar statements about partitions into lattice point vectors $\langle a,b,c\rangle$ with $gcd(a,b,c)=1$ and for (21.8.3) we have non-negative integers a, b and positive integers c; whereas for (21.8.4) we have positive integers for a, b, and c.

Study 6. **4D hyperquadrant VPV sums using gcd.**

For each of $|w|, |x|, |y|, |z| < 1$, show that

$$\sum_{\substack{\gcd(a,b,c,d)=1 \\ a,b,c\geq 0, d>0}} \frac{w^a x^b y^c z^d}{1 - w^a x^b y^c z^d} = \frac{z}{(1-w)(1-x)(1-y)(1-z)}. \tag{21.8.5}$$

Similarly, show that for $\Re w$, $\Re x$, $\Re y$, and $\Re z$ all > 1,

$$\sum_{\substack{\gcd(a,b,c,d)=1 \\ a,b,c,d\geq 1}} \frac{1}{a^w b^x c^y d^z} = \frac{\zeta(w)\zeta(x)\zeta(y)\zeta(z)}{\zeta(w+x+y+z)}. \tag{21.8.6}$$

Infer from (21.8.5) and (21.8.6) that they encode two similar statements about partitions into lattice point vectors $\langle a, b, c, d\rangle$ with $gcd(a, b, c, d) = 1$ and for (21.8.5) we have non-negative integers a, b, c, and positive integers d; whereas for (21.8.6) we have positive integers for a, b, c, and d.

Study 7. **5D hyperquadrant VPV sums using gcd.**

For each of $|v|, |w|, |x|, |y|, |z| < 1$, show that

$$\sum_{\substack{\gcd(a,b,c,d,e)=1 \\ a,b,c,d\geq 0, e>0}} \frac{v^a w^b x^c y^d z^e}{1 - v^a w^b x^c y^d z^e} = \frac{z}{(1-v)(1-w)(1-x)(1-y)(1-z)}. \tag{21.8.7}$$

Similarly, show that for $\Re v$, $\Re w$, $\Re x$, $\Re y$, and $\Re z$ all > 1,

$$\sum_{\substack{\gcd(a,b,c,d,e)=1 \\ a,b,c,d,e\geq 1}} \frac{1}{a^v b^w c^x d^y e^z} = \frac{\zeta(v)\zeta(w)\zeta(x)\zeta(y)\zeta(z)}{\zeta(v+w+x+y+z)}. \tag{21.8.8}$$

Infer from (21.8.7) and (21.8.8) that they encode two similar statements about partitions into lattice point vectors $\langle a, b, c, d\rangle$ with $gcd(a, b, c, d, e) = 1$ and for (21.8.7) we have non-negative integers a, b, c, d, and positive integers e; whereas for (21.8.8) we have positive integers for a, b, c, d, and e.

Study 8. **nD hyperquadrant sums using gcd.**

For each of $|q_1|, |q_2|, |q_3|, \ldots, |q_n| < 1$, show that

$$\sum_{\substack{\gcd(a_1,a_2,a_3,\ldots,a_n)=1 \\ a_1,a_2,\ldots,a_{n-1}\geq 0, a_n\geq 1}} \frac{q_1^{a_1} q_2^{a_2} q_3^{a_3} \cdots q_n^{a_n}}{1 - q_1^{a_1} q_2^{a_2} q_3^{a_3} \cdots q_n^{a_n}} = \frac{q_n}{(1-q_1)(1-q_2)(1-q_3)\cdots(1-q_n)}. \tag{21.8.9}$$

Similarly, show that for $\Re q_1$, $\Re q_2$, $\Re q_3$, …, and $\Re q_n$ all > 1,

$$\sum_{\substack{\gcd(a_1,a_2,a_3,\ldots,a_n)=1 \\ a_1,a_2\ldots a_n\geq 1}} \frac{1}{a_1^{q_1} a_2^{q_2} a_3^{q_3} \cdots a_n^{q_n}} = \frac{\zeta(q_1)\zeta(q_2)\zeta(q_3)\ldots\zeta(q_n)}{\zeta(q_1+q_2+q_3+\cdots+q_n)}. \tag{21.8.10}$$

Infer from (21.8.7) and (21.8.8) that they encode two similar statements about partitions into lattice point vectors $\langle a_1, a_2, a_3, \ldots, a_n \rangle$ with $gcd(a_1, a_2, a_3, \ldots, a_n) = 1$ and for (21.8.7) we have non-negative integers a_1, a_2, ..., a_{n-1} and positive integers a_n; whereas for (21.8.10) we have positive integers for a_1, a_2, ..., up to a_n.

Think About It...

The merit of painting is in the exactness of reproduction. Painting is a science and all sciences are based on mathematics. No human inquiry can be a science unless it pursues its path through mathematical exposition and demonstration.

Leonardo da Vinci
Polymath, 1452–1519

FURTHER READING

BORWEIN, D., BORWEIN, J.M. and BRADLEY, D.M. Parametric Euler sum identities. *J. Math. Anal. Appl.*, 316:328–338, 2006.

BORWEIN, J.M., GLASSER, M.L., MCPHEDRAN, R.C., WAN, J.G. and ZUCKER, I.J. *Lattice Sums then and now*, volume 150 of *Encyclopedia of Mathematics and Its Applications.* Cambridge University Press, 2013.

CAMPBELL, G.B. Multiplicative functions over Riemann zeta function products. *J. Ramanujan Soc.*, 7(1):52–63, 1992.

CAMPBELL, G.B. Dirichlet summations and products over primes. *Int. J. Math. Math. Sci.*, 16(2):359–372, 1993.

CAMPBELL, G.B. A generalized formula of Hardy. *Int. J. Math. Math. Sci.*, 17(2):369–378, 1994.

CAMPBELL, G.B. A new class of infinite products, and Euler's totient. *Int. J. Math. and Math. Sci.*, 17(3):417–422, 1994.

CAMPBELL, G.B. Infinite products over visible lattice points. *Int. J. Math. and Math. Sci.*, 17(4):637–654, 1994.

CAMPBELL, G.B. *Combinatorial identities in number theory related to q-series and arithmetical functions.* Doctor of philosophy thesis, School of Mathematical Sciences, The Australian National University, October 1997.

CAMPBELL, G.B. A closer look at some new identities. *Int. J. Math. and Math. Sci.*, 21(3):581–586, 1998.

CAMPBELL, G.B. Infinite products over hyperpyramid lattices. *Int. J. Math. and Math. Sci.*, 23(4):271–277, 2000.

CAMPBELL, G.B. Some n-space q-binomial theorem extensions and similar identities. arxiv preprint, June 2019. arXiv:1906.07526v1 [math.NT]: https://arxiv.org/abs/1906.07526.

CAMPBELL, G.B. Continued Fractions for partition generating functions. arxiv preprint, Jan 2023. arXiv:2301.12945 [math.CO], https://arxiv.org/abs/2301.12945.

CAMPBELL, G.B. Vector Partition Identities for 2D, 3D and nD Lattices. arxiv preprint, Feb 2023. arXiv:10.48550/ARXIV.2302.01091 [math.NT], https://arxiv.org/abs/2302.01091.

CAMPBELL, G.B. Visible Point Partition Identities for Polylogarithms, and Parametric Euler Sums. arxiv preprint, Jun 2023. arXiv:2306.02241v1 [math.CO], https://arxiv.org/abs/2306.02241.

CAMPBELL, G.B. Visible Point Vector Partition Identities for Hyperpyramid Lattices. arxiv preprint, Sep 2023. arXiv:2309.16094v1 [math.CO], https://arxiv.org/abs/2309.16094.

LEWIN, L. *Polylogarithms and associated functions.* Elsevier Science Ltd.: North-Holland, New York, 1981. ISBN 0-444-00550-1,(First published London: Macdonald, 1958.).

LEWIN, L. The dilogarithm in algebraic fields. *J. Austral. Math. Soc. Ser. A*, 33:302–330, 1982.

LEWIN, L. (Ed.). *Structural Properties of Polylogarithms.* Amer. Math. Soc. - Providence, RI., 1991. ISBN 0-8218-1634-9.

SLOANE, N.J.A. Expansion of e.g.f. exp(x/(x-1)). The On-Line Encyclopedia of Integer Sequences (OEIS), 2023. sequence A293116. (www.oeis.org/A293116).

CHAPTER 22

Polylogarithms, and Parametric Euler Sum Identities

22.1 PURPOSE OF THIS CHAPTER

WE SET THE SCENE with known values and functional relations for dilogarithms, trilogarithms and polylogarithms of various orders, along with more recent Euler sum values and multidimensional computations paying homage to the three late Professors Borwein *et al.* We then apply many of these sum values to tabulate some sixty new combinatorial identities for weighted partitions into Visible Point Vectors in 2D, 3D, 4D and 5D cases suggesting new nD first hyperquadrant and hyperpyramid lattice point identities. Recent research developments are summarized in Campbell [93, 96, 97, 98, 99], which include content from this chapter.

22.2 EARLY HISTORY OF THE DILOGARITHM

The dilogarithm function was first studied by Euler [124] and Landen [181] in the second half of the eighteenth century; and in the second half 20th century in the books from Lewin [183, 184, 185]. So, the function known as the Euler dilogarithm, is defined by

$$Li_2(z) = \sum_{r=1}^{\infty} \frac{z^r}{r^2}, \ |z| \le 1. \tag{22.2.1}$$

A version extended beyond the unit circle is given by the integral

$$Li_2(z) = -\int_0^z \frac{\log(1-t)}{t} dt. \tag{22.2.2}$$

For z real and greater than unity the dilogarithm is complex. The function is transcendental, and is known to be irrational when z is rational.

DOI: 10.1201/9781003174158-22

Euler easily found functional equations and a few special values. The major functional equation results known today are:

$$Li_2(z)+Li_2(-z)=\frac{1}{2}Li_2{}^2(z), \tag{22.2.3}$$

$$Li_2(1-z)+Li_2(1-\frac{1}{z})=-\frac{1}{2}Li_2(z^2), \tag{22.2.4}$$

$$Li_2(z)+Li_2(1-z)=\frac{\pi^2}{6}-\log(z)\log(1-z), \tag{22.2.5}$$

$$Li_2(-z)-Li_2(1-z)+\frac{1}{2}Li_2(1-z^2)=-\frac{\pi^2}{12}-\log(z)\log(1+z). \tag{22.2.6}$$

In the literature, probably the best known bivariate functional equation is **Abel's equation**, easily verified by differentiation,

$$Li_2\left(\frac{x}{1-x}.\frac{y}{1-y}\right)=Li_2\left(\frac{x}{1-y}\right)+Li_2\left(\frac{y}{1-x}\right)-Li_2(x)-Li_2(y)-\log(1-x)\log(1-y). \tag{22.2.7}$$

This formula famously given by Abel was 20 years earlier given by Spence in the equivalent form,

$$Li_2\left(\frac{x}{1-x}.\frac{y}{1-y}\right)=Li_2\left(\frac{x}{1-y}\right)+Li_2\left(\frac{y}{1-x}\right)+Li_2\left(\frac{-x}{1-x}\right) \tag{22.2.8}$$

$$+Li_2\left(\frac{-y}{1-y}\right)+\frac{1}{2}\log^2\left(\frac{1-x}{1-y}\right).$$

Let $L(x)$ denote the Rogers L-function defined in terms of the usual dilogarithm by

$$L(x)=\frac{6}{\pi^2}\left(Li_2(x)+\frac{1}{2}\log(x)\log(1-x)\right)=\frac{6}{\pi^2}\left(\sum_{k=1}^{\infty}\frac{x^k}{k^2}+\frac{1}{2}\log(x)\log(1-x)\right). \tag{22.2.9}$$

Then $L(x)$ satisfies the functional equation

$$L(x)+L(y)=L(xy)+L\left(\frac{x(1-y)}{1-xy}\right)+L\left(\frac{y(1-x)}{1-xy}\right). \tag{22.2.10}$$

Related to this, are the *Watson identities* [297] defined as follows. Let α, $-\beta$, and $-\gamma^{-1}$ be the roots of the cubic equation

$$t^3+2t^2-t-1=0, \tag{22.2.11}$$

then the Rogers L-function satisfies

$$L(\alpha)-L(\alpha^2)=\frac{1}{7}, \tag{22.2.12}$$

$$L(\beta)+\frac{1}{2}L(\beta^2)=\frac{5}{7}, \tag{22.2.13}$$

$$L(\gamma)+\frac{1}{2}L(\gamma^2)=\frac{4}{7}. \tag{22.2.14}$$

A complete list of $Li_2(x)$ which can be evaluated in closed form is given by

$$Li_2(-1)=-\frac{\pi^2}{12}, \tag{22.2.15}$$

$$Li_2(0)=0, \tag{22.2.16}$$

$$Li_2(\frac{1}{2})=\frac{\pi^2}{12}-\frac{1}{2}(\log 2)^2, \tag{22.2.17}$$

$$Li_2(1)=-\frac{\pi^2}{6}, \tag{22.2.18}$$

$$Li_2(2)=-\frac{\pi^2}{4}-i\pi\log 2, \tag{22.2.19}$$

$$Li_2(-\phi)=-\frac{\pi^2}{10}-(\log\phi)^2, \tag{22.2.20}$$

$$Li_2(-\phi^{-1})=-\frac{\pi^2}{15}+\frac{1}{2}(\log\phi)^2, \tag{22.2.21}$$

$$Li_2(\phi^{-2})=\frac{\pi^2}{15}-(\log\phi)^2, \tag{22.2.22}$$

$$Li_2(\phi^{-1})=\frac{\pi^2}{10}-(\log\phi)^2, \tag{22.2.23}$$

where ϕ is the golden ratio $\frac{1+\sqrt{5}}{2}$. For proof of these see Bailey et al. [46] and Borwein et al. [68]. It may seem strange that the above nine values of $Li_2(x)$ is the full known list. Also, there are many interesting identities involving the dilogarithm function. Ramanujan gave the following:

$$Li_2\left(\frac{1}{3}\right)-\frac{1}{6}Li_2\left(\frac{1}{9}\right)=\frac{\pi^2}{18}-\frac{1}{6}(\log 3)^2, \tag{22.2.24}$$

$$Li_2\left(-\frac{1}{2}\right)+\frac{1}{6}Li_2\left(\frac{1}{9}\right)=-\frac{\pi^2}{18}+\log 2\log 3-\frac{1}{2}(\log 2)^2-\frac{1}{3}(\log 3)^2 \tag{22.2.25}$$

$$Li_2\left(\frac{1}{4}\right)+\frac{1}{3}Li_2\left(\frac{1}{9}\right)=\frac{\pi^2}{18}+2\log 2\log 3-2(\log 2)^2-\frac{2}{3}(\log 3)^2, \tag{22.2.26}$$

$$Li_2\left(-\frac{1}{3}\right)-\frac{1}{3}Li_2\left(\frac{1}{9}\right)=-\frac{\pi^2}{18}+\frac{1}{6}(\log 3)^2, \tag{22.2.27}$$

$$Li_2\left(-\frac{1}{8}\right)+Li_2\left(\frac{1}{9}\right)=-\frac{1}{2}\left(\log\left(\frac{9}{8}\right)\right)^2. \tag{22.2.28}$$

Proofs of these can be found in the Ramanujan Notebooks Vol IV by Berndt [62]. In 1997, Bailey et al. [46] showed that

$$\pi^2 = 36Li_2\left(\frac{1}{2}\right) - 36Li_2\left(\frac{1}{4}\right) - 12Li_2\left(\frac{1}{8}\right) + 6Li_2\left(\frac{1}{64}\right). \tag{22.2.29}$$

Lewin [185] gives 67 dilogarithm identities known as *ladders*, and Bailey and Broadhurst [43] found the impressive additional *17 dilogarithm* identity

$$0 = Li_2(\alpha_1^{-630}) - 2Li_2(\alpha_1^{-315}) - 3Li_2(\alpha_1^{-210}) - 10Li_2(\alpha_1^{-126}) \tag{22.2.30}$$

$$-7Li_2(\alpha_1^{-90}) + 18Li_2(\alpha_1^{-35}) + 84Li_2(\alpha_1^{-15}) + 90Li_2(\alpha_1^{-14}) - 4Li_2(\alpha_1^{-9})$$
$$+339Li_2(\alpha_1^{-8}) + 45Li_2(\alpha_1^{-7}) + 265Li_2(\alpha_1^{-6}) - 273Li_2(\alpha_1^{-5}) - 678Li_2(\alpha_1^{-4})$$

$$-1016Li_2(\alpha_1^{-3}) - 744Li_2(\alpha_1^{-2}) - 804Li_2(\alpha_1^{-1}) - 22050(\log\alpha_1)^2 + 2003\zeta(2),$$

where $\alpha_1 = (x^{10} + x^9 - x^7 - x^6 - x^5 - x^4 - x^3 + x + 1)_2 \approx 1.17628081825991750654$4 is the largest positive root of the polynomial in *Lehmer's Mahler measure problem* and $\zeta(z)$ is the Riemann zeta function.

The closed-form particular cases are all arrived at from elementary algebraic manipulation of the functional equations, and therefore will be given also as exercises. However, in these exercises clues and prompts may be useful expedients for the student.

Functional equations for the dilogarithm are most easily proved by differentiating both sides. As such, the proof of the functional equations will be part of the exercises at the end of this chapter.

In the literature, the Rogers L-function form used in (22.2.9) turns out to be convenient in simplifying the trilogarithm and 4th and 5th-degree polylogarithm functional equations.

The polylogarithm occurs naturally in the Visible Point Vector identities, and the functional equations applied to these will bring relationships for weighted vector partitions.

22.3 THE TRILOGARITHM FUNCTION

The trilogarithm $Li_3(z)$, sometimes also denoted L_3, is a special case of the polylogarithm $Li_n(z)$ for $n = 3$. Note that the notation $Li_3(x)$ for the trilogarithm is unfortunately similar to that for the logarithmic integral $Li(x)$.

Functional equations for the trilogarithm include

$$Li_3(z) + Li_3(-z) = \frac{1}{4}Li_3(z^2) \tag{22.3.1}$$

$$Li_3(-z) - Li_3(-z^{-1}) = -\frac{1}{6}(\log z)^3 - \frac{1}{6\pi^2}\log z \tag{22.3.2}$$

$$Li_3(z)+Li_3(1-z)+Li_3(1-z^{-1})=\zeta(3)+\frac{1}{6}(\log z)^3+\frac{\pi^2}{6}\log z-\frac{1}{2}(\log z)^2\log(1-z). \tag{22.3.3}$$

Analytic values for $Li_3(x)$ include

$$Li_3(-1)=-\frac{3}{4}\zeta(3), \tag{22.3.4}$$

$$Li_3(0)=0, \tag{22.3.5}$$

$$Li_3\left(\frac{1}{2}\right)=\frac{7}{8}\zeta(3)-\frac{1}{12}\pi^2\log 2+\frac{1}{6}(\log 2)^3, \tag{22.3.6}$$

$$Li_3(1)=\zeta(3), \tag{22.3.7}$$

$$Li_3(\phi^{-2})=\frac{4}{5}\zeta(3)-\frac{2}{15}\pi^2\log\phi+\frac{2}{3}(\log\phi)^3, \tag{22.3.8}$$

where $\zeta(3)$ is Apéry's constant and ϕ is the golden ratio $\frac{1+\sqrt{5}}{2}$.

Immediate *ad hoc* examples substituting the above trilogarithm values from (22.3.4) to (22.3.8) are available from our known derived hyperquadrant identities. The 2D, 3D, 4D, and 5D hyperquadrant identities are (22.3.9), (22.3.10), (22.3.11), (22.3.12) respectively.

If $|y|<1, |z|<1$, and $s+t=1$, then

$$\prod_{\substack{(a,b)=1\\ a,b\geq 1}}\left(\frac{1}{1-y^a z^b}\right)^{\frac{1}{a^s b^t}}=\exp\left\{\left(\sum_{j=1}^{\infty}\frac{y^j}{j^s}\right)\left(\sum_{k=1}^{\infty}\frac{z^k}{k^t}\right)\right\}. \tag{22.3.9}$$

If $|x|<1, |y|<1, |z|<1$ and $s+t+u=1$, then

$$\prod_{\substack{(a,b,c)=1\\ a,b,c\geq 1}}\left(\frac{1}{1-x^a y^b z^c}\right)^{\frac{1}{a^s b^t c^u}}=\exp\left\{\left(\sum_{i=1}^{\infty}\frac{x^i}{i^s}\right)\left(\sum_{j=1}^{\infty}\frac{y^j}{j^t}\right)\left(\sum_{k=1}^{\infty}\frac{z^k}{k^u}\right)\right\}. \tag{22.3.10}$$

If $|w|<1, |x|<1, |y|<1, |z|<1$ and $r+s+t+u=1$, then

$$\prod_{\substack{(a,b,c,d)=1\\ a,b,c,d\geq 1}}\left(\frac{1}{1-w^a x^b y^c z^d}\right)^{\frac{1}{a^r b^s c^t d^u}}=\exp\left\{\left(\sum_{h=1}^{\infty}\frac{w^h}{h^r}\right)\left(\sum_{i=1}^{\infty}\frac{x^i}{i^s}\right)\left(\sum_{j=1}^{\infty}\frac{y^j}{j^t}\right)\left(\sum_{k=1}^{\infty}\frac{z^k}{k^u}\right)\right\}. \tag{22.3.11}$$

If $|v|<1, |w|<1, |x|<1, |y|<1, |z|<1$ and $q+r+s+t+u=1$, then

$$\prod_{\substack{(a,b,c,d,e)=1\\ a,b,c,d,e\geq 1}}\left(\frac{1}{1-v^a w^b x^c y^d z^e}\right)^{\frac{1}{a^q b^r c^s d^t e^u}} \tag{22.3.12}$$

$$= \exp\left\{\left(\sum_{g=1}^{\infty}\frac{v^g}{g^q}\right)\left(\sum_{h=1}^{\infty}\frac{w^h}{h^r}\right)\left(\sum_{i=1}^{\infty}\frac{x^i}{i^s}\right)\left(\sum_{j=1}^{\infty}\frac{y^j}{j^t}\right)\left(\sum_{k=1}^{\infty}\frac{z^k}{k^u}\right)\right\}.$$

22.3.1 2D Hyperquadrant VPV Cases

Equation (22.3.9) has the following three example cases using our known sums for the trilogarithm and $Li_2(z)$. If $|z| < 1$ then

$$\prod_{\substack{(a,b)=1 \\ a,b\geq 1}} \left(1-(-1)^a z^b\right)^{\frac{b^2}{a^3}} = \exp\left\{\frac{3z(1+z)\zeta(3)}{4(1-z)^3}\right\}, \tag{22.3.13}$$

by (22.3.4) and $Li_{-2}(z) = \sum_{k=1}^{\infty} k^2 z^k = (z(1+z))/(1-z)^3$ when $|z| < 1$.

If $|z| < 1$ then

$$\prod_{\substack{(a,b)=1 \\ a,b\geq 1}} \left(1-\frac{z^b}{2^a}\right)^{\frac{b^2}{a^3}} = \exp\left\{\left(\frac{7}{8}\zeta(3)-\frac{1}{12}\pi^2\log 2+\frac{1}{6}(\log 2)^3\right)\left(\frac{z(1+z)}{(1-z)^3}\right)\right\}, \tag{22.3.14}$$

by (22.3.6) and our value for $Li_{-2}(z)$.

If $|z| < 1$ then

$$\prod_{\substack{(a,b)=1 \\ a,b\geq 1}} \left(1-\frac{z^b}{\phi^{2a}}\right)^{\frac{b^2}{a^3}} = \exp\left\{\left(\frac{4}{5}\zeta(3)-\frac{2}{15}\pi^2\log\phi+\frac{2}{3}(\log\phi)^3\right)\left(\frac{z(1+z)}{(1-z)^3}\right)\right\}, \tag{22.3.15}$$

with ϕ the golden ratio; substituting (22.3.8) and the value of $Li_{-2}(z)$.

A good way to summarize what can be deduced from the 2D VPV first quadrant identity from known dilogarithm and trilogarithm sums is to use the two identities: If $|y| < 1, |z| < 1$, then

$$\prod_{\substack{(a,b)=1 \\ a,b\geq 1}} \left(\frac{1}{1-y^a z^b}\right)^{\frac{b}{a^2}} = \exp\left\{\frac{zLi_2(y)}{(1-z)^2}\right\} \tag{22.3.16}$$

and

$$\prod_{\substack{(a,b)=1 \\ a,b\geq 1}} \left(\frac{1}{1-y^a z^b}\right)^{\frac{b^2}{a^3}} = \exp\left\{\frac{z(1+z)Li_3(y)}{(1-z)^3}\right\}. \tag{22.3.17}$$

22.3.2 3D Hyperquadrant VPV Cases

Next, the 3D first hyperquadrant VPV identity is: If $|x| < 1, |y| < 1, |z| < 1$ and $s+t+u=1$, then

$$\prod_{\substack{(a,b,c)=1 \\ a,b,c\geq 1}} \left(\frac{1}{1-x^a y^b z^c}\right)^{\frac{1}{a^s b^t c^u}} = \exp\left\{\left(\sum_{i=1}^{\infty}\frac{x^i}{i^s}\right)\left(\sum_{j=1}^{\infty}\frac{y^j}{j^t}\right)\left(\sum_{k=1}^{\infty}\frac{z^k}{k^u}\right)\right\}. \tag{22.3.18}$$

Using the nine known values of $L_2(z)$, and the five known values of $L_3(z)$, applied to (22.3.18) we get cases of the kinds given from (22.3.19) to (22.3.24) here:

If $|x|<1, |y|<1, |z|<1$ then

$$\prod_{\substack{(a,b,c)=1\\a,b,c\geq 1}} \left(\frac{1}{1-x^a y^b z^c}\right)^{\frac{c}{ab}} = \exp\{Li_1(x)Li_1(y)Li_{-1}(z)\} = (1-x)^{\frac{z\log(1-y)}{(1-z)^2}}, \quad (22.3.19)$$

$$\prod_{\substack{(a,b,c)=1\\a,b,c\geq 1}} \left(\frac{1}{1-x^a y^b z^c}\right)^{\frac{c^2}{ab^2}} = \exp\{Li_1(x)Li_2(y)Li_{-2}(z)\} \quad (22.3.20)$$

$$= (1-x)^{\frac{z(1+z)Li_2(y)}{(1-z)^3}},$$

$$\prod_{\substack{(a,b,c)=1\\a,b,c\geq 1}} \left(\frac{1}{1-x^a y^b z^c}\right)^{\frac{c^3}{ab^3}} = \exp\{Li_1(x)Li_3(y)Li_{-3}(z)\} \quad (22.3.21)$$

$$= \left(\frac{1}{1-x}\right)^{\frac{z(1+4z+z^2)Li_3(y)}{(1-z)^4}},$$

$$\prod_{\substack{(a,b,c)=1\\a,b,c\geq 1}} \left(\frac{1}{1-x^a y^b z^c}\right)^{\frac{c^3}{a^2b^2}} = \exp\{Li_2(x)Li_2(y)Li_{-3}(z)\} \quad (22.3.22)$$

$$= exp\left\{\frac{z(1+4z+z^2)Li_2(x)Li_2(y)}{(1-z)^4}\right\},$$

$$\prod_{\substack{(a,b,c)=1\\a,b,c\geq 1}} \left(\frac{1}{1-x^a y^b z^c}\right)^{\frac{c^4}{a^2b^3}} = \exp\{Li_2(x)Li_3(y)Li_{-4}(z)\} \quad (22.3.23)$$

$$= exp\left\{\frac{z(1+11z+11z^2+z^3)Li_2(x)Li_3(y)}{(1-z)^5}\right\},$$

$$\prod_{\substack{(a,b,c)=1\\a,b,c\geq 1}} \left(\frac{1}{1-x^a y^b z^c}\right)^{\frac{c^5}{a^3b^3}} = \exp\{Li_3(x)Li_3(y)Li_{-5}(z)\} \quad (22.3.24)$$

$$= exp\left\{\frac{z(1+26z+66z^2+26z^3+z^4)Li_3(x)Li_3(y)}{(1-z)^6}\right\}.$$

22.3.3 4D Hyperquadrant VPV Cases

If $|w|<1, |x|<1, |y|<1, |z|<1$ and $r+s+t+u=1$, then

$$\prod_{\substack{(a,b,c,d)=1 \\ a,b,c,d\geq 1}} \left(\frac{1}{1-w^a x^b y^c z^d}\right)^{\frac{1}{a^r b^s c^t d^u}} = \exp\{Li_r(w)Li_s(x)Li_t(y)Li_u(z)\}. \qquad (22.3.25)$$

We can apply all of our known dilogarithm and trilogarithm sums as well as use the facts $Li_1(z) = -\log(1-z)$, $Li_0(z) = z/(1-z)$, with also the known values

$$Li_{-n}(z) = \sum_{k=0}^{n} k!S(n+1,k+1)\left(\frac{z}{1-z}\right)^{k+1} \qquad (n=1,2,3,\ldots),$$

where $S(n,k)$ are Stirling numbers of the second kind. We see that for (22.3.25), most of the possible cases with closed form RHS's of the resulting identity case are given by the table:

Cases of r + s + t + u = 1 related to the 4D Hyperquadrant VPV

r	s	t	u	log (RHS of 4D identity (22.3.25))
3	−2	−3	3	$Li_3(w)Li_{-2}(x)Li_{-3}(y)Li_3(z)$
2	−2	−2	3	$Li_2(w)Li_{-2}(x)Li_{-2}(y)Li_3(z)$
1	−2	−1	3	$Li_1(w)Li_{-2}(x)Li_{-1}(y)Li_3(z)$
3	−1	−3	2	$Li_3(w)Li_{-1}(x)Li_{-3}(y)Li_2(z)$
2	−1	−2	2	$Li_2(w)Li_{-1}(x)Li_{-2}(y)Li_2(z)$
1	−1	−1	2	$Li_1(w)Li_{-1}(x)Li_{-1}(y)Li_2(z)$
0	−1	−1	3	$Li_0(w)Li_{-1}(x)Li_{-1}(y)Li_3(z)$
0	0	−1	2	$Li_0(w)Li_0(x)Li_{-1}(y)Li_2(z)$
0	0	0	1	$Li_0(w)Li_0(x)Li_0(y)Li_1(z)$
0	1	1	−1	$Li_0(w)Li_1(x)Li_1(y)Li_{-1}(z)$
1	1	1	−2	$Li_1(w)Li_1(x)Li_1(y)Li_{-2}(z)$
1	1	2	−3	$Li_1(w)Li_1(x)Li_2(y)Li_{-3}(z)$
1	2	2	−4	$Li_1(w)Li_2(x)Li_2(y)Li_{-4}(z)$
2	2	2	−5	$Li_2(w)Li_2(x)Li_2(y)Li_{-5}(z)$
1	1	3	−4	$Li_1(w)Li_1(x)Li_3(y)Li_{-4}(z)$
1	2	3	−5	$Li_1(w)Li_2(x)Li_3(y)Li_{-5}(z)$
2	2	3	−6	$Li_2(w)Li_2(x)Li_3(y)Li_{-6}(z)$
2	3	3	−7	$Li_2(w)Li_3(x)Li_3(y)Li_{-7}(z)$
3	3	3	−8	$Li_3(w)Li_3(x)Li_3(y)Li_{-8}(z)$

The above table has values of r, s, t, and u each less than or equal to 3, where $r+s+t+u=1$; and the resulting cases of (22.3.25).

22.3.4 5D Hyperquadrant VPV Cases

Cases of q + r + s + t + u = 1 related to the 5D Hyperquadrant VPV

q	r	s	t	u	log(RHS of 5D identity (22.3.26))
0	3	−2	−3	3	$Li_0(v)Li_3(w)Li_{-2}(x)Li_{-3}(y)Li_3(z)$
0	2	−2	−2	3	$Li_0(v)Li_2(w)Li_{-2}(x)Li_{-2}(y)Li_3(z)$
0	1	−2	−1	3	$Li_0(v)Li_1(w)Li_{-2}(x)Li_{-1}(y)Li_3(z)$
0	0	−1	−1	3	$Li_0(v)Li_0(w)Li_{-1}(x)Li_{-1}(y)Li_3(z)$
0	3	−1	−3	2	$Li_0(v)Li_3(w)Li_{-1}(x)Li_{-3}(y)Li_2(z)$
0	2	−1	−2	2	$Li_0(v)Li_2(w)Li_{-1}(x)Li_{-2}(y)Li_2(z)$
0	1	−1	−1	2	$Li_0(v)Li_1(w)Li_{-1}(x)Li_{-1}(y)Li_2(z)$
0	0	0	−1	2	$Li_0(v)Li_0(w)Li_0(x)Li_{-1}(y)Li_2(z)$
0	0	0	0	1	$Li_0(v)Li_0(w)Li_0(x)Li_0(y)Li_1(z)$
0	0	1	1	−1	$Li_0(v)Li_0(w)Li_1(x)Li_1(y)Li_{-1}(z)$
0	1	1	1	−2	$Li_0(v)Li_1(w)Li_1(x)Li_1(y)Li_{-2}(z)$
0	1	1	2	−3	$Li_0(v)Li_1(w)Li_1(x)Li_2(y)Li_{-3}(z)$
0	1	2	2	−4	$Li_0(v)Li_1(w)Li_2(x)Li_2(y)Li_{-4}(z)$
0	2	2	2	−5	$Li_0(v)Li_2(w)Li_2(x)Li_2(y)Li_{-5}(z)$
0	1	1	3	−4	$Li_0(v)Li_1(w)Li_1(x)Li_3(y)Li_{-4}(z)$
0	1	2	3	−5	$Li_0(v)Li_1(w)Li_2(x)Li_3(y)Li_{-5}(z)$
0	2	2	3	−6	$Li_0(v)Li_2(w)Li_2(x)Li_3(y)Li_{-6}(z)$
0	2	3	3	−7	$Li_0(v)Li_2(w)Li_3(x)Li_3(y)Li_{-7}(z)$
0	3	3	3	−8	$Li_0(v)Li_3(w)Li_3(x)Li_3(y)Li_{-8}(z)$
1	0	1	1	−2	$Li_0(v)Li_0(w)Li_1(x)Li_1(y)Li_{-1}(z)$
1	1	1	1	−3	$Li_0(v)Li_1(w)Li_1(x)Li_1(y)Li_{-2}(z)$
1	1	1	2	−4	$Li_0(v)Li_1(w)Li_1(x)Li_2(y)Li_{-3}(z)$
1	1	2	2	−5	$Li_0(v)Li_1(w)Li_2(x)Li_2(y)Li_{-4}(z)$
1	2	2	2	−6	$Li_0(v)Li_2(w)Li_2(x)Li_2(y)Li_{-5}(z)$
1	1	1	3	−5	$Li_0(v)Li_1(w)Li_1(x)Li_3(y)Li_{-4}(z)$
1	1	2	3	−6	$Li_0(v)Li_1(w)Li_2(x)Li_3(y)Li_{-5}(z)$
1	2	2	3	−7	$Li_0(v)Li_2(w)Li_2(x)Li_3(y)Li_{-6}(z)$
1	2	3	3	−8	$Li_0(v)Li_2(w)Li_3(x)Li_3(y)Li_{-7}(z)$
1	3	3	3	−9	$Li_0(v)Li_3(w)Li_3(x)Li_3(y)Li_{-8}(z)$

Next, we consider the range of closed-form RHS versions of the 5D hyperquadrant identity (22.3.26). The above table of finitely evaluable cases would include all integers q,r,s,t,u less than or equal to 3 such that $\mathbf{q+r+s+t+u=1}$. The identity related to this is as follows. If $|v|<1, |w|<1, |x|<1, |y|<1, |z|<1$ and $q+r+s+t+u=1$, then

$$\prod_{\substack{(a,b,c,d,e)=1\\ a,b,c,d,e\geq 1}} \left(\frac{1}{1-v^a w^b x^c y^d z^e}\right)^{\frac{1}{a^q b^r c^s d^t e^u}} \tag{22.3.26}$$

$$= \exp\{Li_q(v)Li_r(w)Li_s(x)Li_t(y)Li_u(z)\}.$$

Refer to the above table.

22.3.5 More Trilogarithm Equations

In 1997 Bailey et al., [46] showed that

$$\frac{35}{2}\zeta(3) - \pi^2 \log 2 = 36Li_3\left(\frac{1}{2}\right) - 18Li_3\left(\frac{1}{4}\right) - 4Li_3\left(\frac{1}{8}\right) + Li_3\left(\frac{1}{64}\right), \quad (22.3.27)$$

$$2(\log 2)^3 - 7\zeta(3) = -24Li_3\left(\frac{1}{2}\right) + 18Li_3\left(\frac{1}{4}\right) + 4Li_3\left(\frac{1}{8}\right) - Li_3\left(\frac{1}{64}\right), \quad (22.3.28)$$

$$10(\log 2)^3 - 2\pi^2 \log 2 = -48Li_3\left(\frac{1}{2}\right) + 54Li_3\left(\frac{1}{4}\right) + 12Li_3\left(\frac{1}{8}\right) - 3Li_3\left(\frac{1}{64}\right). \quad (22.3.29)$$

22.4 THE POLYLOGARITHM FUNCTION AND JONATHAN BORWEIN

The polylogarithm is the obvious extension of the dilogarithm and trilogarithm. The nth order polylogarithm is defined by

$$Li_n(z) = \sum_{r=1}^{\infty} \frac{z^r}{r^n}, \quad |z| \leq 1. \quad (22.4.1)$$

A version extended beyond the unit circle is given by the integral

$$Li_n(z) = \int_0^z \frac{Li_{n-1}(t)}{t} dt. \quad (22.4.2)$$

For z real and greater than unity the polylogarithm is complex.

Research on this function extends back over 200 years, but in the past 60 years, the main literature for it was written by Lewin (see [183] to [185]). However, with the rise of experimental computer science calculations notably by Jonathan Borwein ([63] to [65]) with colleagues including his father David Borwein and brother Peter. Euler sums and new polylogarithm sums have been discovered by experimental computation in the past 25 years in part from the work of Jonathan, his family and most notably David H. Bailey. All of this is critically important for VPV identities due to the fact that the new algorithmic results for polylogarithms find direct application to n-space VPV identities in the present volume.

Jonathan Borwein was a pioneer of present-day computer-assisted calculation and discovery methods, shown today in the spirit of the *Ramanujan Machine Project* [214]. Notwithstanding, we cannot unsee him as the bright Rhodes Scholar brimming with exuberant confidence and creative potential. That promise was borne out in a stellar career spanning 42 years. Many would know him as a leader of impeccable standards, a caring family man and a sought-after mentor who went the extra mile.

He cannot be replaced but his genius still inspires others to follow his golden path of understanding the obscure and beautiful that is research mathematics.

Figure 22.1 Jonathan M. Borwein, Rhodes Scholar, Oxford, 1974. (With permission from Judith Borwein.)

The Borwein trio and their colleagues such as David H. Bailey produced a wealth of Euler series sums and polylogarithm cases applicable to our present studies. Related summations and integrals were derived and published by Anthony Sofo [275] to [238]; which we cite in our next chapters. Recommended works by the three Borweins, Bailey and colleagues are for example [42] to [46] and [63] to [68].

So prolific was Prof Jonathan Borwein's output, *Microsoft Academic* listed him at number 21 on their list of top mathematics authors of all time. He also wrote over a dozen books, including a monograph on convex functions which was a *Choice 2011* Outstanding Academic Book, and over 450 refereed articles and proceedings.

22.5 MORDELL-TORNHEIM-WITTEN ENSEMBLES

The multidimensional Mordell-Tornheim-Witten (MTW) zeta function

$$\omega(s_1,\dots,s_{K+1}) = \sum_{m_1,\dots,m_K>0} \frac{1}{{m_1}^{s_1}{m_2}^{s_2}\dots {m_K}^{s_K}(m_1+m_2+\dots+m_K)^{s_{K+1}}} \tag{22.5.1}$$

has known relations going back 70 years, but remains mysterious with respect to many combinatorial phenomena, especially when we contemplate derivatives with respect to the s_i parameters. We refer to $K+1$ as the *depth* and $\sum_{j=1}^{k+1} s_j$ as the *weight* of ω.

We can get a more intuitive grasp of MTWs by writing the simplest ones to see what is going on with them. Hence we have

$$\omega(s_1,s_2) = \sum_{m_1>0} \frac{1}{{m_1}^{s_1}(m_1)^{s_2}}, \tag{22.5.2}$$

$$\omega(s_1,s_2,s_3) = \sum_{m_1,m_2>0} \frac{1}{{m_1}^{s_1}{m_2}^{s_2}(m_1+m_2)^{s_3}}, \tag{22.5.3}$$

$$\omega(s_1,s_2,s_3,s_4) = \sum_{m_1,m_2,m_3>0} \frac{1}{{m_1}^{s_1}{m_2}^{s_2}{m_3}^{s_3}(m_1+m_2+m_3)^{s_4}}, \tag{22.5.4}$$

There is a hyperpyramid lattice VPV identity first given in Campbell [87], which is evidently including MTW zeta functions. In fact we have

$$\prod_{\substack{(a_1,a_2,\dots,a_n)=1\\ a_1,\dots,a_{n-1}<a_n\\ a_1,\dots,a_{n-1}\geq 0\\ a_n\geq 1}} \left(\frac{1}{1-x_1^{a_1}x_2^{a_2}\dots x_n^{a_n}}\right)^{\frac{1}{a_1^{b_1}a_2^{b_2}\dots a_n^{b_n}}} = \exp\left\{\sum_{k=1}^{\infty}\prod_{i=1}^{n-1}\left(\sum_{j=1}^{k-1}\frac{x_i^j}{j^{b_i}}\right)\frac{x_n^k}{k^{b_n}}\right\} \tag{22.5.5}$$

where for $i=1,2,\dots,n$; $|x_i|<1$, b_i is a complex number with $\sum_{i=1}^{n} b_i = 1$.

After two decades of *Generalized Euler Sum* and *Polylogarithm Sum* research by Bailey, Bradley and the three Borweins setting the scenes (see Bailey-Borwein (David, Peter and Jon) or Bradley-Broadhurst-Plouffe [64], [46], [43], [67], [68] together in varied combinations), the 2016 work Bailey et al. [42] introduced and discussed a novel generalized MTW ensemble zeta function for positive integers M, N, and nonnegative integers s_i, t_j with constraints $M \geq N \geq 1$ together with a polylogarithm-integral representation:

$$\omega(s_1,\dots,s_M \mid t_1,\dots,t_N) = \sum_{\substack{m_1,\dots,m_M,n_1,\dots,n_N>0\\ \sum_{j=1}^M m_j=\sum_{k=1}^N n_k}} \prod_{j=1}^{M}\frac{1}{{m_j}^{s_j}}\prod_{k=1}^{N}\frac{1}{{n_k}^{t_k}} \tag{22.5.6}$$

$$= \frac{1}{2\pi}\int_0^{2\pi}\prod_{j=1}^{M} Li_{s_j}\left(e^{i\theta}\right)\prod_{k=1}^{N} Li_{t_k}\left(e^{-i\theta}\right)d\theta. \tag{22.5.7}$$

Here the *polylogarithm of order s* denotes $Li_s(z) := \sum_{n\geq 0} \frac{z^n}{n^s}$ and its analytic extensions [26] and the (complex) number s is its order. Note that if parameters are zero, there are convergence issues with this integral representation. One may use principal-value calculus, or an alternative representation such as (22.5.9) below. When $N = 1$ the representation (22.5.8) devolves to the classic MTW form, in that

$$\omega(s_1,...,s_{M+1}) = \omega(s_1,...,s_M \mid s_{M+1}). \tag{22.5.8}$$

These Euler Sum/Polylogarithm Sum/MTW Sum combinations are all applicable to the VPV identities, and can be adapted to the theorems on nD hyperquadrant, hyperpyramid and skewed hyperpyramid summations. Collecting these Summations with the cases arising from the more miscellaneous Euler Sum evaluations give in the 4o or more applicable papers by Sofo from 2008 to circa 2020, we see a vista opening wherein the numerous cases and theories can be applied to the VPV identities, and it seems easily hundreds of examples could be easily written from them. So, returning to our summations, we need a wider *MTW ensemble* involving outer derivatives, according to

$$\omega\begin{pmatrix} s_1,...,s_M, | t_1,...,t_N \\ d_1,...,d_M, | e_1,...,e_N \end{pmatrix} = \sum_{\substack{m_1,...,m_M,n_1,...,n_N>0 \\ \sum_{j=1}^{M} m_j = \sum_{k=1}^{N} n_k}} \prod_{j=1}^{M} \frac{(-\log m_j)^{d^j}}{{m_j}^{s_j}} \prod_{k=1}^{N} \frac{(-\log n_k)^{e^k}}{{n_k}^{t_k}} \tag{22.5.9}$$

$$= \frac{1}{2\pi} \int_0^{2\pi} \prod_{j=1}^{M} {Li_{s_j}}^{(d_j)}\left(e^{i\theta}\right) \prod_{k=1}^{N} {Li_{t_k}}^{(e_k)}\left(e^{-i\theta}\right) d\theta, \tag{22.5.10}$$

where the s-th outer derivative of a polylogarithm is denoted ${Li_s}^{(d)}(z) = \left(\frac{\partial}{\partial z}\right)^{(d)} Li_s(z)$.

22.6 FINITE EULER SUMS

Consider the following well known summations:

$$\sum_{k=1}^{n} k^1 = \frac{n}{2} + \frac{n^2}{2}, \tag{22.6.1}$$

$$\sum_{k=1}^{n} k^2 = \frac{n}{6} + \frac{n^2}{2} + \frac{n^3}{3}, \tag{22.6.2}$$

$$\sum_{k=1}^{n} k^3 = \frac{n^2}{4} + \frac{n^3}{3} + \frac{n^4}{4}, \tag{22.6.3}$$

$$\sum_{k=1}^{n} k^4 = \frac{-n}{30} + \frac{n^3}{3} + \frac{n^4}{2} + \frac{n^5}{5}, \tag{22.6.4}$$

and their generalizations:

$$\sum_{k=1}^{n} k^1 z^k = \frac{z - (1+n)z^{n+1} + nz^{n+2}}{(1-z)^2}, \tag{22.6.5}$$

$$\sum_{k=1}^{n} k^2 z^k = \frac{-n^2 z^{n+3} + (2n^2+2n-1)z^{n+2} - (n^2+2n+1)z^{n+1} + z^2 + z}{(1-z)^3}, \tag{22.6.6}$$

$$\sum_{k=1}^{n} k^3 z^k \tag{22.6.7}$$

$$= \frac{n^3 z^{n+4} + (3n^3+6n^2-4)z^{n+2} - (3n^3+3n^2-3n+1)z^{n+3} - (n+1)^3 z^{n+1} + z^3 + 4z^2 + z}{(1-z)^4},$$

$$\sum_{k=1}^{n} k^4 z^k \tag{22.6.8}$$

$$= -\frac{n^4 z^{n+5} + (-4n^4 - 4n^3 + 6n^2 - 4n + 1)z^{n+4} + (6n^4 + 12n^3 - 6n^2 - 12n + 11)z^{n+3}}{(1-z)^5}$$

$$-\frac{(-4n^4 - 12n^3 - 6n^2 + 12n + 11)z^{n+2} + (n+1)^4 z^{n+1} - z^4 - 11z^3 - 11z^2 - z}{(1-z)^5}.$$

These kind of finite sums lend themselves to substitution in the hyperpyramid identities; in fact for both the square and skewed hyperpyramid versions. It is illustrative here to give a few examples in 2D and 3D of the uses for the Euler Sums in the context of VPV identities here.

22.6.1 The $2D$ Square Hyperpyramid VPV Identity

Recall from an earlier chapter that:

If $|y|, |z| < 1$, with $a + b = 1$,

$$\prod_{\substack{m,n \geq 1 \\ m \leq n;\, \gcd(m,n)=1}} \left(\frac{1}{1 - y^m z^n} \right)^{\frac{1}{m^a n^b}} = \exp\left\{ \sum_{n=1}^{\infty} \left(\sum_{m=1}^{n} \frac{y^m}{m^a} \right) \frac{z^n}{n^b} \right\}. \tag{22.6.9}$$

Let us set by definition $\gcd(m,n) := (m,n)$ and apply equations (22.6.1) through to (22.6.4) substituted respectively into (22.6.9). This corresponds to the cases of (22.6.9) with:

$y = 1$, $a = -1$, implying from condition $a + b = 1$ that $b = 2$;
$y = 1$, $a = -2$, implying from condition $a + b = 1$ that $b = 3$;
$y = 1$, $a = -3$, implying from condition $a + b = 1$ that $b = 4$;
$y = 1$, $a = -4$, implying from condition $a + b = 1$ that $b = 5$;

so the resulting substitutions give us respectively the four equations

$$\prod_{\substack{m,n \geq 1 \\ m \leq n;\, (m,n)=1}} \left(\frac{1}{1 - z^n} \right)^{\frac{m^1}{n^2}} = \exp\left\{ \sum_{n=1}^{\infty} \left(\frac{n}{2} + \frac{n^2}{2} \right) \frac{z^n}{n^2} \right\}$$

$$= \exp\left\{\frac{1}{2}\log\left(\frac{1}{1-z}\right)+\frac{1}{2}\frac{z}{1-z}\right\},$$

$$\prod_{\substack{m,n\geq 1\\ m\leq n;\,(m,n)=1}} \left(\frac{1}{1-z^n}\right)^{\frac{m^2}{n^3}} = \exp\left\{\sum_{n=1}^{\infty}\left(\frac{n}{6}+\frac{n^2}{2}+\frac{n^3}{3}\right)\frac{z^n}{n^3}\right\}$$

$$= \exp\left\{\frac{1}{6}Li_2(z)+\frac{1}{2}\log\left(\frac{1}{1-z}\right)+\frac{1}{3}\frac{z}{1-z}\right\},$$

$$\prod_{\substack{m,n\geq 1\\ m\leq n;\,(m,n)=1}} \left(\frac{1}{1-z^n}\right)^{\frac{m^3}{n^4}} = \exp\left\{\sum_{n=1}^{\infty}\left(\frac{n^2}{4}+\frac{n^3}{3}+\frac{n^4}{4}\right)\frac{z^n}{n^4}\right\}$$

$$= \exp\left\{\frac{1}{4}Li_2(z)+\frac{1}{3}\log\left(\frac{1}{1-z}\right)+\frac{1}{4}\frac{z}{1-z}\right\},$$

$$\prod_{\substack{m,n\geq 1\\ m\leq n;\,(m,n)=1}} \left(\frac{1}{1-z^n}\right)^{\frac{m^4}{n^5}} = \exp\left\{\sum_{n=1}^{\infty}\left(\frac{-n}{30}+\frac{n^3}{3}+\frac{n^4}{2}+\frac{n^5}{5}\right)\frac{z^n}{n^5}\right\}$$

$$= \exp\left\{\frac{-1}{30}Li_3(z)+\frac{1}{3}\log\left(\frac{1}{1-z}\right)+\frac{1}{2}\frac{z}{(1-z)^2}+\frac{1}{5}\frac{z}{1-z}\right\}.$$

Based on the above equations and their reciprocal equations, we may assert the following four theorems.

Theorem 22.1 *For* $|z|<1$,

$$\prod_{\substack{m,n\geq 1\\ m\leq n;\,(m,n)=1}} \left(\frac{1}{1-z^n}\right)^{\frac{m^1}{n^2}} = \sqrt{\frac{1}{1-z}}\exp\left\{\frac{1}{2}\frac{z}{1-z}\right\},$$

$$\prod_{\substack{m,n\geq 1\\ m\leq n;\,(m,n)=1}} (1-z^n)^{\frac{m^1}{n^2}} = \sqrt{1-z}\exp\left\{\frac{-1}{2}\frac{z}{1-z}\right\}.$$

Theorem 22.2 *For* $|z|<1$,

$$\prod_{\substack{m,n\geq 1\\ m\leq n;\,(m,n)=1}} \left(\frac{1}{1-z^n}\right)^{\frac{m^2}{n^3}} = \sqrt{\frac{1}{1-z}}\exp\left\{\frac{1}{6}Li_2(z)+\frac{1}{3}\frac{z}{1-z}\right\},$$

$$\prod_{\substack{m,n\geq 1\\ m\leq n;\,(m,n)=1}} (1-z^n)^{\frac{m^2}{n^3}} = \sqrt{1-z}\exp\left\{\frac{-1}{6}Li_2(z)-\frac{1}{3}\frac{z}{1-z}\right\}.$$

Theorem 22.3 *For* $|z| < 1$,

$$\prod_{\substack{m,n\geq 1\\ m\leq n;\,(m,n)=1}} \left(\frac{1}{1-z^n}\right)^{\frac{m^3}{n^4}} = \sqrt[3]{\frac{1}{1-z}}\exp\left\{\frac{1}{4}Li_2(z)+\frac{1}{4}\frac{z}{1-z}\right\},$$

$$\prod_{\substack{m,n\geq 1\\ m\leq n;\,(m,n)=1}} (1-z^n)^{\frac{m^3}{n^4}} = \sqrt[3]{1-z}\exp\left\{\frac{-1}{4}Li_2(z)-\frac{1}{4}\frac{z}{1-z}\right\}.$$

Theorem 22.4 *For* $|z| < 1$,

$$\prod_{\substack{m,n\geq 1\\ m\leq n;\,(m,n)=1}} \left(\frac{1}{1-z^n}\right)^{\frac{m^4}{n^5}} = \sqrt[3]{\frac{1}{1-z}}\exp\left\{\frac{z(7-2z)}{10(1-z)^2}-\frac{1}{30}Li_3(z)\right\},$$

$$\prod_{\substack{m,n\geq 1\\ m\leq n;\,(m,n)=1}} (1-z^n)^{\frac{m^4}{n^5}} = \sqrt[3]{1-z}\exp\left\{\frac{z(2z-7)}{10(1-z)^2}+\frac{1}{30}Li_3(z)\right\}.$$

The above four theorems give us a single variable equation in each case that encodes a statement about "weighted integer partitions" of a kind not normally discussed in the partition literature. We could construct a step by step table for the function of the partition, much in the way this was done for the partition congruence function called the Crank, in Chapter 5. However, equations (22.6.5) through to (22.6.8) substituted respectively into (22.6.9) can supply us with two-variable 2D generalizations of theorem 22.1 to theorem 22.4. This corresponds to the cases of (22.6.9) with:

$a=-1$, implying from condition $a+b=1$ that $b=2$;
$a=-2$, implying from condition $a+b=1$ that $b=3$;
$a=-3$, implying from condition $a+b=1$ that $b=4$;
$a=-4$, implying from condition $a+b=1$ that $b=5$;

so the resulting substitutions (putting y for z in each case) give us respectively the four 2D equation workings,

$$\prod_{\substack{m,n\geq 1\\ m\leq n;\,(m,n)=1}} \left(\frac{1}{1-y^m z^n}\right)^{\frac{m^1}{n^2}} = \exp\left\{\sum_{n=1}^{\infty}\left(\frac{y-(1+n)y^{n+1}+ny^{n+2}}{(1-y)^2}\right)\frac{z^n}{n^2}\right\}$$

$$= \exp\left\{\frac{yLi_2(z)}{(1-y)^2}-\frac{yLi_2(yz)}{(1-y)^2}-\frac{y}{(1-y)^2}\log\left(\frac{1}{1-yz}\right)+\frac{y^2}{(1-y)^2}\log\left(\frac{1}{1-yz}\right)\right\}$$

$$= (1-yz)^{\frac{y}{1-y}}\exp\left\{\frac{y}{(1-y)^2}\left(Li_2(z)-Li_2(yz)\right)\right\};$$

$$\prod_{\substack{m,n\geq 1\\ m\leq n;\,(m,n)=1}} \left(\frac{1}{1-y^m z^n}\right)^{\frac{m^2}{n^3}}$$

$$= \exp\left\{\sum_{n=1}^{\infty}\left(\frac{-n^2y^{n+3}+(2n^2+2n-1)y^{n+2}-(n^2+2n+1)y^{n+1}+y^2+y}{(1-y)^3}\right)\frac{z^n}{n^3}\right\}$$

$$= \exp\left\{\sum_{n=1}^{\infty}\left(\frac{-y(y+1)(y^n-1)+2n(y-1)y^{n+1}-n^2(y-1)^2y^{n+1}}{(1-y)^3}\right)\frac{z^n}{n^3}\right\}$$

$$= \left(\frac{1}{1-yz}\right)^{\frac{y}{1-y}} \exp\left\{\frac{y(y+1)}{(1-y)^3}(Li_3(z)-Li_3(yz)) - \frac{2y}{(1-y)^2}Li_2(yz)\right\};$$

$$\prod_{\substack{m,n\geq 1\\ m\leq n;\,(m,n)=1}} \left(\frac{1}{1-y^mz^n}\right)^{\frac{m^3}{n^4}} = \exp\left\{\sum_{n=1}^{\infty}\left(\frac{n^3y^{n+4}+(3n^3+6n^2-4)y^{n+2}}{(1-y)^4}\right)\frac{z^n}{n^4}\right\}$$

$$\times \exp\left\{\sum_{n=1}^{\infty}\left(\frac{-(n+1)^3y^{n+1}+y^3+4y^2+y}{(1-y)^4}\right)\frac{z^n}{n^4}\right\}$$

$$= \exp\left\{\sum_{n=1}^{\infty}\left(\frac{-y(y^2+4y+1)(y^n-1)+3n(y^2-1)y^{n+1}-3n^2((y-1)^2y^{n+1})+n^3(y-1)^3y^{n+1}}{(1-y)^4}\right)\frac{z^n}{n^4}\right\}$$

$$= (1-yz)^{\frac{y}{1-y}} \exp\left\{\frac{y(y^2+4y+1)}{(1-y)^4}(Li_4(z)-Li_4(yz)) + \frac{3y(1+y)}{(1-y)^3}Li_3(yz) + \frac{3y}{(1-y)^2}Li_2(yz)\right\};$$

$$\prod_{\substack{m,n\geq 1\\ m\leq n;\,(m,n)=1}} \left(\frac{1}{1-y^mz^n}\right)^{\frac{m^4}{n^5}}$$

$$= \exp\left\{\sum_{n=1}^{\infty}\left(-\frac{n^4y^{n+5}+(-4n^4-4n^3+6n^2-4n+1)y^{n+4}}{(1-y)^5}\right)\frac{z^n}{n^5}\right\}$$

$$\times \exp\left\{\sum_{n=1}^{\infty}\left(-\frac{(6n^4+12n^3-6n^2-12n+11)y^{n+3}}{(1-y)^5}\right)\frac{z^n}{n^5}\right\}$$

$$\times \exp\left\{\sum_{n=1}^{\infty}\left(-\frac{(-4n^4-12n^3-6n^2+12n+11)y^{n+2}+(n+1)^4y^{n+1}}{(1-y)^5}\right)\frac{z^n}{n^5}\right\}$$

$$\times \exp\left\{\sum_{n=1}^{\infty}\left(+\frac{y^4+11y^3+11y^2+z}{(1-y)^5}\right)\frac{z^n}{n^5}\right\}$$

$$= \exp\left\{\frac{y(y+1)(y^2+10y+1)}{(1-y)^5}(Li_5(yz)-Li_5(z))\right\}$$

$$\times \exp\left\{\frac{4}{(1-y)^5}((-y^4+3y^3+y)Li_4(yz)+3Li_4(z))\right\}$$

$$\times \exp\left\{\frac{6}{(1-y)^5}((y^4-y^3+y)Li_3(yz)-Li_3(z))\right\}$$

$$\times \exp\left\{\frac{4}{(1-y)^5}((y-3y^3-y^4)Li_2(yz)-3Li_2(z))\right\}$$

$$\times \exp\left\{\frac{1}{(1-y)^5}((y^5-4y^4+6y^3+y)Li_1(yz)-4Li_1(z))\right\}$$

$$= \frac{(1-z)^{\frac{4}{(1-y)^5}}}{(1-yz)^{\frac{y^5-4y^4+6y^3+y}{(1-y)^5}}}$$

$$\times \exp\left\{\frac{y(y+1)(y^2+10y+1)}{(1-y)^5}(Li_5(yz)-Li_5(z))\right\}$$

$$\times \exp\left\{\frac{4}{(1-y)^5}((-y^4+3y^3+y)Li_4(yz)+3Li_4(z))\right\}$$

$$\times \exp\left\{\frac{6}{(1-y)^5}((y^4-y^3+y)Li_3(yz)-Li_3(z))\right\}$$

$$\times \exp\left\{\frac{4}{(1-y)^5}((y-3y^3-y^4)Li_2(yz)-3Li_2(z))\right\}.$$

Based on the above equations and their reciprocal equations, we may assert the following four 2D theorems.

Theorem 22.5 *For* $|z|<1$,

$$\prod_{\substack{m,n\geq 1\\ m\leq n;\,(m,n)=1}} \left(\frac{1}{1-y^m z^n}\right)^{\frac{m^1}{n^2}} = (1-yz)^{\frac{y}{1-y}} \exp\left\{\frac{y}{(1-y)^2}(Li_2(z)-Li_2(yz))\right\},$$

$$\prod_{\substack{m,n\geq 1\\ m\leq n;\,(m,n)=1}} (1-y^m z^n)^{\frac{m^1}{n^2}} = \left(\frac{1}{1-yz}\right)^{\frac{y}{1-y}} \exp\left\{-\frac{y}{(1-y)^2}(Li_2(z)-Li_2(yz))\right\}.$$

Theorem 22.6 *For* $|z|<1$,

$$\prod_{\substack{m,n\geq 1\\ m\leq n;\,(m,n)=1}} \left(\frac{1}{1-y^m z^n}\right)^{\frac{m^2}{n^3}} = \left(\frac{1}{1-yz}\right)^{\frac{y}{1-y}}$$

$$\times \exp\left\{\frac{y(y+1)}{(1-y)^3}(Li_3(z)-Li_3(yz))-\frac{2y}{(1-y)^2}Li_2(yz)\right\},$$

$$\prod_{\substack{m,n\geq 1\\ m\leq n;\,(m,n)=1}} (1-y^m z^n)^{\frac{m^2}{n^3}} = (1-yz)^{\frac{y}{1-y}}$$

$$\times \exp\left\{-\frac{y(y+1)}{(1-y)^3}(Li_3(z)-Li_3(yz))+\frac{2y}{(1-y)^2}Li_2(yz)\right\},$$

Theorem 22.7 *For* $|z|<1$,

$$\prod_{\substack{m,n\geq 1\\ m\leq n;\,(m,n)=1}} \left(\frac{1}{1-y^m z^n}\right)^{\frac{m^3}{n^4}} = (1-yz)^{\frac{y}{1-y}}$$

$$\times \exp\left\{\frac{y(y^2+4y+1)}{(1-y)^4}(Li_4(z)-Li_4(yz))+\frac{3y(1+y)}{(1-y)^3}Li_3(yz)+\frac{3y}{(1-y)^2}Li_2(yz)\right\},$$

$$\prod_{\substack{m,n\geq 1\\ m\leq n;\,(m,n)=1}} (1-y^m z^n)^{\frac{m^3}{n^4}} = \left(\frac{1}{1-yz}\right)^{\frac{y}{1-y}}$$

$$\times \exp\left\{-\frac{y(y^2+4y+1)}{(1-y)^4}(Li_4(z)-Li_4(yz)) - \frac{3y(1+y)}{(1-y)^3}Li_3(yz) - \frac{3y}{(1-y)^2}Li_2(yz)\right\}.$$

Theorem 22.8 *For* $|z|<1$,

$$\prod_{\substack{m,n\geq 1\\ m\leq n;\,(m,n)=1}} \left(\frac{1}{1-y^m z^n}\right)^{\frac{m^4}{n^5}} = \left\{\frac{(1-z)^4}{(1-yz)^{y^5-4y^4+6y^3+y}}\right\}^{\frac{1}{(1-y)^5}}$$

$$\times \exp\left\{\frac{y(y+1)(y^2+10y+1)}{(1-y)^5}(Li_5(yz)-Li_5(z))\right\}$$

$$\times \exp\left\{\frac{4}{(1-y)^5}((-y^4+3y^3+y)Li_4(yz)+3Li_4(z))\right\}$$

$$\times \exp\left\{\frac{6}{(1-y)^5}((y^4-y^3+y)Li_3(yz)-Li_3(z))\right\}$$

$$\times \exp\left\{\frac{4}{(1-y)^5}((y-3y^3-y^4)Li_2(yz)-3Li_2(z))\right\},$$

$$\prod_{\substack{m,n\geq 1\\ m\leq n;\,(m,n)=1}} (1-y^m z^n)^{\frac{m^4}{n^5}} = \left\{\frac{(1-yz)^{y^5-4y^4+6y^3+y}}{(1-z)^4}\right\}^{\frac{1}{(1-y)^5}}$$

$$\times \exp\left\{\frac{-y(y+1)(y^2+10y+1)}{(1-y)^5}(Li_5(yz)-Li_5(z))\right\}$$

$$\times \exp\left\{\frac{-4}{(1-y)^5}((-y^4+3y^3+y)Li_4(yz)+3Li_4(z))\right\}$$

$$\times \exp\left\{\frac{-6}{(1-y)^5}((y^4-y^3+y)Li_3(yz)-Li_3(z))\right\}$$

$$\times \exp\left\{\frac{-4}{(1-y)^5}((y-3y^3-y^4)Li_2(yz)-3Li_2(z))\right\}.$$

22.6.2 The $3D$ Square Hyperpyramid VPV Identity

Recall from an earlier chapter that:

If $|x|,|y|,|z|<1$, with $a+b+c=1$,

$$\prod_{\substack{l,m,n\geq 1\\ l,m\leq n;\,\gcd(l,m,n)=1}} \left(\frac{1}{1-x^l y^m z^n}\right)^{\frac{1}{l^a m^b n^c}} = \exp\left\{\sum_{n=1}^{\infty}\left(\sum_{l=1}^{n}\frac{x^l}{l^a}\right)\left(\sum_{m=1}^{n}\frac{y^m}{m^b}\right)\frac{z^n}{n^c}\right\}. \tag{22.6.10}$$

In one of the simpler examples we can give, let us apply equations (22.6.1) and (22.6.2) to (22.6.10). This corresponds to the case of (22.6.10) with $x=y=1$, $a=-1$, $b=-2$, implying from condition $a+b+c=1$ that $c=4$. The resulting substitution is

$$\prod_{\substack{l,m,n\geq 1\\ l,m\leq n;\gcd(l,m,n)=1}} \left(\frac{1}{1-z^n}\right)^{\frac{lm^2}{n^4}} = \exp\left\{\sum_{n=1}^{\infty}\left(\frac{n}{2}+\frac{n^2}{2}\right)\left(\frac{n}{6}+\frac{n^2}{2}+\frac{n^3}{3}\right)\frac{z^n}{n^4}\right\}$$

$$= \exp\left\{\sum_{n=1}^{\infty}\left(\frac{n^2}{12}+\frac{n^3}{3}+\frac{5n^4}{12}+\frac{n^5}{6}\right)\frac{z^n}{n^4}\right\}$$

$$= \exp\left\{\frac{1}{12}Li_2(z)+\frac{1}{3}Li_1(z)+\frac{5}{12}Li_0(z)+\frac{1}{6}Li_{-1}(z)\right\}$$

$$= \exp\left\{\frac{1}{12}Li_2(z)+\frac{1}{3}\log\left(\frac{1}{1-z}\right)+\frac{5}{12}\frac{z}{1-z}+\frac{1}{6}\frac{z}{(1-z)^2}\right\}$$

$$= \exp\left\{\frac{1}{12}Li_2(z)+\frac{z(7-5z))}{12(1-z)^2}+\frac{1}{3}\log\left(\frac{1}{1-z}\right)\right\}$$

$$= \sqrt[3]{\left(\frac{1}{1-z}\right)}\exp\left\{\frac{1}{12}Li_2(z)+\frac{z(7-5z))}{12(1-z)^2}\right\}.$$

Therefore, we have proven that if $|z|<1$,

$$\prod_{\substack{l,m,n\geq 1\\ l,m\leq n;\gcd(l,m,n)=1}} \left(\frac{1}{1-z^n}\right)^{\frac{lm^2}{n^4}} = \sqrt[3]{\left(\frac{1}{1-z}\right)}\exp\left\{\frac{1}{12}\left(Li_2(z)+\frac{z(7-5z)}{(1-z)^2}\right)\right\}, \tag{22.6.11}$$

and the equivalent result,

$$\prod_{\substack{l,m,n\geq 1\\ l,m\leq n;\gcd(l,m,n)=1}} (1-z^n)^{\frac{lm^2}{n^4}} = \sqrt[3]{(1-z)}\exp\left\{-\frac{1}{12}\left(Li_2(z)+\frac{z(7-5z)}{(1-z)^2}\right)\right\}. \tag{22.6.12}$$

There is a corresponding example we can give, if we apply equations (22.6.5) and (22.6.6) to (22.6.10). This corresponds to the case of (22.6.10) with $a=-1$, $b=-2$, implying from condition $a+b+c=1$ that $c=4$. The resulting substitution gives us

$$\prod_{\substack{l,m,n\geq 1\\ l,m\leq n;\gcd(l,m,n)=1}} \left(\frac{1}{1-x^l y^m z^n}\right)^{\frac{lm^2}{n^4}} \tag{22.6.13}$$

$$= \exp\left\{\sum_{n=1}^{\infty} A(n,x)B(n,y)\frac{z^n}{n^4}\right\}$$

where

$$A(n,x) = \frac{x-(1+n)x^{n+1}+nx^{n+2}}{(1-x)^2}$$

and

$$B(n,y) = \frac{-n^2y^{n+3}+(2n^2+2n-1)y^{n+2}-(n^2+2n+1)y^{n+1}+y^2+y}{(1-y)^3},$$

which, after expanding and simplifying is equivalent to the

Theorem 22.9

$$\prod_{\substack{l,m,n\geq 1\\ l,m\leq n;\,\gcd(l,m,n)=1}} \left(\frac{1}{1-x^l y^m z^n}\right)^{\frac{lm^2}{n^4}} \tag{22.6.14}$$

$$= \left(\frac{1}{1-xyz}\right)^{\frac{xy}{(1-x)^2(1-y)^3}} \times \exp\left\{\frac{-xy(2x+y-3)}{(1-x)^3(1-y)^4} Li_2(xyz)\right\}$$

$$\times \exp\left\{\frac{xy}{(1-x)^3(1-y)^3} Li_2(yz) + \frac{xy(y+1)}{(1-x)^3(1-y)^5} Li_4(xyz)\right\}$$

$$\times \exp\left\{\frac{-xy(xy+x+y-3)}{(1-x)^3(1-y)^5} Li_3(xyz) - \frac{xy(y+1)}{(1-x)^2(1-y)^5} Li_3(xz)\right\}$$

$$\times \exp\left\{-\frac{xy(y+1)}{(1-x)^3(1-y)^5} Li_4(xz) - \frac{2xy}{(1-x)^3(1-y)^4} Li_3(yz)\right\}.$$

$$\times \exp\left\{\frac{-xy(1+y)}{(1-x)^3(1-y)^5} Li_4(yz) + \frac{xy(1+y)}{(1-x)^3(1-y)^5} Li_4(z)\right\}.$$

Proof of theorem 22.9: The necessary workings and simplification in going from (22.6.13) to arrive at (22.6.14) are given in the following page and a half of analysis.

$$\prod_{\substack{l,m,n\geq 1\\ l,m\leq n;\,\gcd(l,m,n)=1}} \left(\frac{1}{1-x^l y^m z^n}\right)^{\frac{lm^2}{n^4}}$$

$$= \exp\left\{\frac{1}{(1-x)^3(1-y)^5} \sum_{n=1}^{\infty} (n^3x^{n+1}y^{n+1} - 2n^3x^{n+1}y^{n+2} + n^3x^{n+1}y^{n+3}) \frac{z^n}{n^4}\right\}$$

$$\times \exp\left\{\frac{1}{(1-x)^3(1-y)^5} \sum_{n=1}^{\infty} (-n^3x^{n+2}y^{n+1} + 2n^3x^{n+2}y^{n+2} - n^3x^{n+2}y^{n+3}) \frac{z^n}{n^4}\right\}$$

$$\times \exp\left\{\frac{1}{(1-x)^3(1-y)^5} \sum_{n=1}^{\infty} (+3n^2x^{n+1}y^{n+1} - 4n^2x^{n+1}y^{n+2} + n^2x^{n+1}y^{n+3}) \frac{z^n}{n^4}\right\}$$

$$\times \exp\left\{\frac{1}{(1-x)^3(1-y)^5}\sum_{n=1}^{\infty}(-2n^2x^{n+2}y^{n+1}+2n^2x^{n+2}y^{n+2}-n^2xy^{n+1})\frac{z^n}{n^4}\right\}$$

$$\times \exp\left\{\frac{1}{(1-x)^3(1-y)^5}\sum_{n=1}^{\infty}(+2n^2xy^{n+2}-n^2xy^{n+3}+3nx^{n+1}y^{n+1})\frac{z^n}{n^4}\right\}$$

$$\times \exp\left\{\frac{1}{(1-x)^3(1-y)^5}\sum_{n=1}^{\infty}(+x^{n+1}y^{n+1}-nx^{n+1}y^{n+2}+x^{n+1}y^{n+2})\frac{z^n}{n^4}\right\}$$

$$\times \exp\left\{\frac{1}{(1-x)^3(1-y)^5}\sum_{n=1}^{\infty}(-nx^{n+2}y^{n+1}-nx^{n+2}y^{n+2}-ny^2x^{n+1}-y^2x^{n+1})\frac{z^n}{n^4}\right\}$$

$$\times \exp\left\{\frac{1}{(1-x)^3(1-y)^5}\sum_{n=1}^{\infty}(+ny^2x^{n+2}-nyx^{n+1}-yx^{n+1}+nyx^{n+2}-2nxy^{n+1})\frac{z^n}{n^4}\right\}$$

$$\times \exp\left\{\frac{1}{(1-x)^3(1-y)^5}\sum_{n=1}^{\infty}(-xy^{n+1}+2nxy^{n+2}-xy^{n+2}+xy^2+xy)\frac{z^n}{n^4}\right\}$$

which works out through the following analysis to

$$\prod_{\substack{l,m,n\geq 1\\ l,m\leq n;\, \gcd(l,m,n)=1}}\left(\frac{1}{1-x^ly^mz^n}\right)^{\frac{lm^2}{n^4}}$$

$$= \exp\left\{\frac{1}{(1-x)^3(1-y)^5}(xyLi_1(xyz)-2xy^2Li_1(xyz)+xy^3Li_1(xyz))\right\}$$

$$\times \exp\left\{\frac{1}{(1-x)^3(1-y)^5}(-x^2yLi_1(xyz)+2x^2y^2Li_1(xyz)-x^2y^3Li_1(xyz))\right\}$$

$$\times \exp\left\{\frac{1}{(1-x)^3(1-y)^5}(+3xyLi_2(xyz)-4xy^2Li_2(xyz)+xy^3Li_2(xyz))\right\}$$

$$\times \exp\left\{\frac{1}{(1-x)^3(1-y)^5}(-2x^2yLi_2(xyz)+2x^2y^2Li_2(xyz)-xyLi_2(yz))\right\}$$

$$\times \exp\left\{\frac{1}{(1-x)^3(1-y)^5}(+2xy^2Li_2(yz)-xy^3Li_2(yz)+3xyLi_3(xyz))\right\}$$

$$\times \exp\left\{\frac{1}{(1-x)^3(1-y)^5}(+xyLi_4(xyz)-xy^2Li_3(xyz)+xy^2Li_4(xyz))\right\}$$

$$\times \exp\left\{\frac{1}{(1-x)^3(1-y)^5}(-x^2yLi_3(xyz)-x^2y^2Li_3(xyz)-xy^2Li_3(xz)-xy^2Li_4(xz))\right\}$$

$$\times \exp\left\{\frac{1}{(1-x)^3(1-y)^5}(+x^2y^2Li_3(xz)-xyLi_3(xz)-xyLi_4(xz)+x^2yLi_3(xz))\right\}$$

$$\times \exp\left\{\frac{1}{(1-x)^3(1-y)^5}(-2xyLi_3(yz)-xyLi_4(yz)+2xy^2Li_3(yz)-xy^2Li_4(yz))\right\}$$

$$\times \exp\left\{\frac{1}{(1-x)^3(1-y)^5}(+xy^2Li_4(z)+xyLi_4(z))\right\}$$

$$= \exp\left\{\frac{1}{(1-x)^3(1-y)^5}((x-1)x(y-1)^2y\log(1-xyz))\right\}$$
$$\times \exp\left\{\frac{1}{(1-x)^3(1-y)^5}(x(y-1)y(2x+y-3)Li_2(xyz))\right\}$$
$$\times \exp\left\{\frac{1}{(1-x)^3(1-y)^5}(-x(y-1)^2yLi_2(yz))\right\}$$
$$\times \exp\left\{\frac{1}{(1-x)^3(1-y)^5}(xy(y+1)Li_4(xyz))\right\}$$
$$\times \exp\left\{\frac{1}{(1-x)^3(1-y)^5}(-xy(xy+x+y-3)Li_3(xyz))\right\}$$
$$\times \exp\left\{\frac{1}{(1-x)^3(1-y)^5}((x-1)xy(y+1)Li_3(xz)-xy(y+1)Li_4(xz))\right\}$$
$$\times \exp\left\{\frac{1}{(1-x)^3(1-y)^5}(+2x(y-1)yLi_3(yz)-xy(1+y)Li_4(yz)+xy(1+y)Li_4(z))\right\}$$
$$= \left(\frac{1}{1-xyz}\right)^{\frac{xy}{(1-x)^2(1-y)^3}} \times \exp\left\{\frac{-xy(2x+y-3)}{(1-x)^3(1-y)^4}Li_2(xyz)\right\}$$
$$\times \exp\left\{\frac{xy}{(1-x)^3(1-y)^3}Li_2(yz)+\frac{xy(y+1)}{(1-x)^3(1-y)^5}Li_4(xyz)\right\}$$
$$\times \exp\left\{\frac{-xy(xy+x+y-3)}{(1-x)^3(1-y)^5}Li_3(xyz)-\frac{xy(y+1)}{(1-x)^2(1-y)^5}Li_3(xz)\right\}$$
$$\times \exp\left\{-\frac{xy(y+1)}{(1-x)^3(1-y)^5}Li_4(xz)-\frac{2xy}{(1-x)^3(1-y)^4}Li_3(yz)\right\}$$
$$\times \exp\left\{\frac{-xy(1+y)}{(1-x)^3(1-y)^5}Li_4(yz)+\frac{xy(1+y)}{(1-x)^3(1-y)^5}Li_4(z)\right\}.$$

■

The casual reader will see that the earlier simple summations arising from substituted equations (22.6.1) through (22.6.4) yield much more immediate and simple results than the more elaborate substituted equations (22.6.5) through (22.6.8).

22.7 PARAMETRIC EULER SUM IDENTITIES

We begin by defining various special functions which we will later use. For $\Re(x) > 0$, the gamma function and its logarithmic derivative (digamma function) are defined by

$$\Gamma(x) := \int_0^\infty e^{-t}t^{x-1}dt, \quad \Psi(x) := \frac{\Gamma'(x)}{\Gamma(x)}.$$

We note that

$$\sum_{n=1}^{\infty}\frac{x}{n(n-x)} := -\Psi(1-x)-\gamma, \quad \text{where} \quad \gamma := \lim_{N\to\infty}\left(\sum_{n=1}^{N}\frac{1}{n}-\log N\right)$$

is Euler's constant. In addition to these, we use the classical Riemann zeta function

$$\zeta(s) := \sum_{n=1}^{\infty} \frac{1}{n^s} \quad \Re(s) > 1,$$

and extend this to

$$\zeta(s,t) := \sum_{n>m>0} \frac{1}{n^s m^t} = \sum_{n=1}^{\infty} \frac{1}{n^s} \sum_{m=1}^{n-1} \frac{1}{m^t} \quad \Re(s) > 1,\ \Re(t) > 1. \tag{22.7.1}$$

defining a double Euler sum. The two-place function (22.7.1) was first introduced by Euler, who noted the reflection formula

$$\zeta(s,t) + \zeta(t,s) = \zeta(s)\zeta(t) - \zeta(s+t) \quad \Re(s) > 1,\ \Re(t) > 1, \tag{22.7.2}$$

and the reduction formula

$$\zeta(s,1) = \frac{1}{2} s\zeta(s+1) - \frac{1}{2} \sum_{k=1}^{s-2} \zeta(k+1)\zeta(s-k), \qquad 1 < s \in \mathbb{Z}, \tag{22.7.3}$$

which reduces the double Euler sum $\zeta(s,1)$ to a sum of products of classical Riemann ζ-values. From setting $s = 2$ in (22.7.3) yields the important sum,

$$\zeta(2,1) = \zeta(3). \tag{22.7.4}$$

The next cases are,

$$\zeta(3,1) = \frac{3}{2}\zeta(4) - \frac{1}{2}\zeta(2), \tag{22.7.5}$$

$$\zeta(4,1) = 2\zeta(5) - \frac{1}{2}\zeta(2)^2, \tag{22.7.6}$$

$$\zeta(5,1) = \frac{5}{2}\zeta(6) - \zeta(2)\zeta(3). \tag{22.7.7}$$

In 1994 Bailey et al. [45] formulated the following definitions for Euler sums:

$$s_h(m,n) = \sum_{k=1}^{\infty} \left(\frac{1}{1} + \frac{1}{2} + \cdots + \frac{1}{k}\right)^m \frac{1}{(k+1)^n}, \tag{22.7.8}$$

$$s_a(m,n) = \sum_{k=1}^{\infty} \left(\frac{1}{1} - \frac{1}{2} + \cdots - \frac{(-1)^{k+1}}{k}\right)^m \frac{1}{(k+1)^n}, \tag{22.7.9}$$

$$a_h(m,n) = \sum_{k=1}^{\infty} \left(\frac{1}{1} + \frac{1}{2} + \cdots + \frac{1}{k}\right)^m \frac{(-1)^{k+1}}{(k+1)^n}, \tag{22.7.10}$$

$$a_a(m,n) = \sum_{k=1}^{\infty} \left(\frac{1}{1} - \frac{1}{2} + \cdots + \frac{(-1)^{k+1}}{k}\right)^m \frac{(-1)^{k+1}}{(k+1)^n}, \tag{22.7.11}$$

$$\sigma_h(m,n) = \sum_{k=1}^{\infty} \left(\frac{1}{1^m} + \frac{1}{2^m} + \cdots + \frac{1}{k^m}\right) \frac{1}{(k+1)^n}, \tag{22.7.12}$$

$$\sigma_a(m,n) = \sum_{k=1}^{\infty}\left(\frac{1}{1^m}-\frac{1}{2^m}+\cdots-\frac{(-1)^{k+1}}{k^m}\right)\frac{1}{(k+1)^n}, \tag{22.7.13}$$

$$\alpha_h(m,n) = \sum_{k=1}^{\infty}\left(\frac{1}{1^m}+\frac{1}{2^m}+\cdots+\frac{1}{k^m}\right)\frac{(-1)^{k+1}}{(k+1)^n}, \tag{22.7.14}$$

$$\alpha_a(m,n) = \sum_{k=1}^{\infty}\left(\frac{1}{1^m}-\frac{1}{2^m}+\cdots+\frac{(-1)^{k+1}}{k^m}\right)\frac{(-1)^{k+1}}{(k+1)^n}, \tag{22.7.15}$$

and from these come three notable special forms,

$$s_h(m,n) = \sum_{k=1}^{\infty}(\gamma+\Psi(k+1))^m\frac{(-1)^{k+1}}{(k+1)^n}, \tag{22.7.16}$$

$$a_a(m,n) = \sum_{k=1}^{\infty}\left(\log 2+\frac{1}{2}(-1)^{k+1}\left[\Psi(\frac{k}{2}+\frac{1}{2})-\Psi(\frac{k}{2}+1)\right]\right)^m\frac{(-1)^{k+1}}{(k+1)^n}, \tag{22.7.17}$$

$$\alpha_h(m,n) = \left(1-\frac{2}{2^{m+n}}\right)\sum_{k=1}^{\infty}\frac{(-1)^{k+1}H_{k,m}}{k^n}, \tag{22.7.18}$$

where $H_{k,m} = \sum_{j=1}^{k} j^{-m}$ is a generalized harmonic number and $\Psi(n) = \Gamma'(z)/\Gamma(z)$ is the digamma function.

At this point, we note that each Euler sum defined by equations (22.7.8) to (22.7.15) and therefore also for (22.7.16) to (22.7.18) when viewed as lattice point multi-dimensional sums have the following characteristics. The first four sums (22.7.8) to (22.7.11) are $m+1$-dimensional lattice sums with m dimensions identical copies of each other, and the other four sums (22.7.12) to (22.7.15) are each 2D lattice sums. Of these, all eight sums are summations taken over radial from origin lattice point regions applicable to our higher dimensional VPV identities. In fact, the above definitions can be extended to power series that also conform to the radial from the origin region requirement for applying more general hyperquadrant and hyperpyramid VPV theorems.

There are many well known particular cases of the Euler sums as given in (22.7.8) to (22.7.15). A few of them for $s_h(m,n)$ were found by experimental means by Bailey et al. [45] and are:

$$s_h(3,2) = \frac{15}{2}\zeta(5)+\zeta(2)\zeta(3), \tag{22.7.19}$$

$$s_h(4,3) = -\frac{109}{8}\zeta(7)+\frac{37}{2}\zeta(3)\zeta(4)-5\zeta(2)\zeta(5), \tag{22.7.20}$$

$$s_h(5,4) = \frac{890}{9}\zeta(9)+66\zeta(4)\zeta(5)-\frac{4295}{24}\zeta(2)\zeta(5)-5\zeta(3)^3+\frac{265}{8}\zeta(2)\zeta(7), \tag{22.7.21}$$

each of which are cases of

$$s_h(m+1,m) = \sum_{k=1}^{\infty}\left(\frac{1}{1}+\frac{1}{2}+\cdots+\frac{1}{k}\right)^{m+1}\frac{1}{(k+1)^m}. \tag{22.7.22}$$

The sums (22.7.19) to (22.7.21) are likely to be true as they were verified to several hundred decimal places. All that we need with respect to them is an actual proof.

Equation (22.7.22) is directly applicable to the VPV hyperpyramid identity case, which may be established from methods in Chapter 21, and is here stated as,

$$\prod_{\substack{\gcd(j_1,j_2,\ldots,j_{m+1},k)=1\\ j_1,j_2,\ldots,j_{m+1}<k\\ j_1,j_2,\ldots,j_{m+1}\geq 0;k\geq 1}} \left(\frac{1}{1-y^{j_1+j_2+\ldots+j_{m+1}}z^k}\right)^{\frac{1}{k}} \tag{22.7.23}$$

$$= \exp\left[\sum_{k=1}^{\infty}\left(\frac{y}{1}+\frac{y^2}{2}+\cdots+\frac{y^k}{k}\right)^{m+1}\frac{z^{k+1}}{(k+1)^m}\right].$$

Applying our three experimentally obtained sums (22.7.19) to (22.7.21) successively, setting $y \to 1$ and $z \to 1$ in the three identities,

$$\prod_{\substack{\gcd(j_1,j_2,j_3,k)=1\\ j_1,j_2,j_3<k\\ j_1,j_2,j_3\geq 1;k\geq 2}} \left(\frac{1}{1-y^{j_1+j_2+j_3}z^k}\right)^{\frac{1}{k}} = \exp\left[\sum_{k=1}^{\infty}\left(\frac{y}{1}+\frac{y^2}{2}+\cdots+\frac{y^k}{k}\right)^{3}\frac{z^{k+1}}{(k+1)^2}\right]. \tag{22.7.24}$$

$$\prod_{\substack{\gcd(j_1,j_2,j_3,j_4,k)=1\\ j_1,j_2,j_3,j_4<k\\ j_1,j_2,j_3,j_4\geq 1;k\geq 2}} \left(\frac{1}{1-y^{j_1+j_2+j_3+j_4}z^k}\right)^{\frac{1}{k}} = \exp\left[\sum_{k=1}^{\infty}\left(\frac{y}{1}+\frac{y^2}{2}+\cdots+\frac{y^k}{k}\right)^{4}\frac{z^{k+1}}{(k+1)^3}\right]. \tag{22.7.25}$$

$$\prod_{\substack{\gcd(j_1,j_2,j_3,j_4,j_5,k)=1\\ j_1,j_2,j_3,j_4,j_5<k\\ j_1,j_2,j_3,j_4,j_5\geq 1;k\geq 2}} \left(\frac{1}{1-y^{j_1+j_2+j_3+j_4+j_5}z^k}\right)^{\frac{1}{k}} \tag{22.7.26}$$

$$= \exp\left[\sum_{k=1}^{\infty}\left(\frac{y}{1}+\frac{y^2}{2}+\cdots+\frac{y^k}{k}\right)^{5}\frac{z^{k+1}}{(k+1)^4}\right].$$

EXERCISES

22.1 *Research project on Euler sums.* Find a proof of the three experimentally obtained sums:

$$s_h(3,2) = \frac{15}{2}\zeta(5)+\zeta(2)\zeta(3), \tag{P22.1}$$

$$s_h(4,3) = -\frac{109}{8}\zeta(7)+\frac{37}{2}\zeta(3)\zeta(4)-5\zeta(2)\zeta(5), \tag{P22.2}$$

$$s_h(5,4) = \frac{890}{9}\zeta(9)+66\zeta(4)\zeta(5)-\frac{4295}{24}\zeta(2)\zeta(5)-5\zeta(3)^3+\frac{265}{8}\zeta(2)\zeta(7). \tag{P22.3}$$

22.2 *Research project for limiting cases.* We know already that (P22.1) to (P22.3) are cases of

$$s_h(m+1,m) = \sum_{k=1}^{\infty} \left(\frac{1}{1} + \frac{1}{2} + \cdots + \frac{1}{k}\right)^{m+1} \frac{1}{(k+1)^m}. \tag{P22.4}$$

Find the resulting infinite products as $y \to 1$ and $z \to 1$ for,

$$\prod_{\substack{\gcd(j_1,j_2,j_3,k)=1 \\ j_1,j_2,j_3<k \\ j_1,j_2,j_3\geq 1; k\geq 2}} \left(\frac{1}{1 - y^{j_1+j_2+j_3} z^k}\right)^{\frac{1}{k}}, \tag{P22.5}$$

$$\prod_{\substack{\gcd(j_1,j_2,j_3,j_4,k)=1 \\ j_1,j_2,j_3,j_4<k \\ j_1,j_2,j_3,j_4\geq 1; k\geq 2}} \left(\frac{1}{1 - y^{j_1+j_2+j_3+j_4} z^k}\right)^{\frac{1}{k}}, \tag{P22.6}$$

$$\prod_{\substack{\gcd(j_1,j_2,j_3,j_4,j_5,k)=1 \\ j_1,j_2,j_3,j_4,j_5<k \\ j_1,j_2,j_3,j_4,j_5\geq 1; k\geq 2}} \left(\frac{1}{1 - y^{j_1+j_2+j_3+j_4+j_5} z^k}\right)^{\frac{1}{k}}, \tag{P22.7}$$

and apply the three sums from the previous exercise to obtain three new closed form infinite products.

Think About It...

What is the universal?

The single case.

What is the particular?

Millions of cases.

Johann Wolfgang von Goethe (1749–1832)

Goethe On Science: An Anthology of Goethe's Scientific Writings (1996)

FURTHER READING

BAILEY, D.H. and BORWEIN, J.M. Computation and structure of character polylogarithms with applications to character Mordell-Tornheim-Witten sums. *Math. Comput.*, 85:295–324, 2016.

BAILEY, D.H. and BROADHURST, D.J. A Seventeenth-Order Polylogarithm Ladder: http : //arxiv.org/abs/math.CA/9906134/. arxiv preprint, June 1999.

BAILEY, D.H., BORWEIN, J.M. and GIRGENSOHN, R. Experimental Evaluation of Euler Sums. *Exper. Math.*, 3:17–30, 1994.

BAILEY, D.H., BORWEIN, P.B. and PLOUFFE, S. On the Rapid Computation of Various Polylogarithmic Constants. *Math. Comput.*, 66(218):903–913, April 1997.

BERNDT, B.C. *Ramanujan's Notebooks*, Part IV. New York: Springer-Verlag, 1994. ISBN 0-387-94109-6.

BORWEIN, D., BORWEIN, J.M. and BRADLEY, D.M. Parametric Euler sum identities. *J. Math. Anal. Appl.*, 316:328–338, 2006.

BORWEIN, D., BORWEIN, J.M. and GIRGENSOHN, R. Explicit evaluation of Euler sums. *Proc. Edinburgh Math. Soc.*, 38:277–294, 1995.

BORWEIN, J.M. and BAILEY, D.M. *Mathematics by Experiment: Plausible Reasoning in the 21st Century.* Wellesley, MA, 2003. Ed. A K Peters.

BORWEIN, J.M., BRADLEY, D.M., BROADHURST, D.J. and LISONEK, P. Special Values of Multidimensional Polylogarithms. *CECM*, 98(106), 1998.

BORWEIN, J.M., BRADLEY, D.M., BROADHURST, D.J. and LISONEK, P. Special values of multiple polylogarithms. *Trans. Am. Math. Soc.*, 353:907–941, 2001.

CAMPBELL, G.B. A new class of infinite products, and Euler's totient. *Int. J. Math. and Math. Sci.*, 17(3):417–422, 1994.

CAMPBELL, G.B. Some n-space q-binomial theorem extensions and similar identities. arxiv preprint, June 2019. arXiv:1906.07526v1 [math.NT]: https://arxiv.org/abs/1906.07526.

CAMPBELL, G.B. Continued Fractions for partition generating functions. arxiv preprint, Jan 2023. arXiv:2301.12945 [math.CO], https://arxiv.org/abs/2301.12945.

CAMPBELL, G.B. Vector Partition Identities for 2D, 3D and nD Lattices. arxiv preprint, Feb 2023. arXiv:10.48550/ARXIV.2302.01091 [math.NT], https://arxiv.org/abs/2302.01091.

CAMPBELL, G.B. Visible Point Partition Identities for Polylogarithms, and Parametric Euler Sums. arxiv preprint, Jun 2023. arXiv:2306.02241v1 [math.CO], https://arxiv.org/abs/2306.02241.

CAMPBELL, G.B. Visible Point Vector Partition Identities for Hyperpyramid Lattices. arxiv preprint, Sep 2023. arXiv:2309.16094v1 [math.CO], https://arxiv.org/abs/2309.16094.

EULER, L. *Institutiones calculi integralis*, volume 1. Marcum-Michaelum, Brousquet, Lausannae, 1768. http://doi.org/10.3931/e-rara-3788.

LANDEN, J. *Mathematical memoirs*, volume 1. Book on Demand Ltd. 112, 1780. Volume 1 Paperback - 11 November 2013.

LEWIN, L. *Polylogarithms and associated functions.* Elsevier Science Ltd.: North-Holland, New York, 1981. ISBN 0-444-00550-1,(First published London: Macdonald, 1958.).

LEWIN, L. The dilogarithm in algebraic fields. *J. Austral. Math. Soc. Ser. A*, 33:302–330, 1982.

LEWIN, L. (Ed.). *Structural Properties of Polylogarithms.* Amer. Math. Soc. - Providence, RI., 1991. ISBN 0-8218-1634-9.

The RAMANUJAN MACHINE. The Ramanujan Machine parallel computing project. (web page), 2023. [Online; accessed 22 October 2023]. www.ramanujanmachine.com.

SOFO, A. Integral identities for sums. *Math. Commun.*, 13:303–309, 2008.

SOFO, A. Families of Integrals of Polylogarithmic Functions. *Researchgate*, 7, 2019. Mathematics 10.3390/math7020143.

WATSON, G.N. A Note on Spence's Logarithmic Transcendent. *Quart. J. Math. Oxford Ser.*, 8:39–42, 1937.

CHAPTER 23

Visible Point Vector Identities from Particular Euler Sum Values

23.1 POLYLOGARITHMS NEAR TRIVIAL ZEROES OF THE RIEMANN ZETA FUNCTION

NEXT WE EXAMINE the limit as we approach the zeroes of the Riemann zeta function from the polylogarithm function whose behavior mimics the Riemann zeta function as the power series variable approaches unity. Although we will not hope to prove anything as profound as the Riemann Hypothesis itself, we can examine the behavior of the polylogarithm in the vicinity of the trivial and non-trivial zeroes of the Riemann zeta function, and along the way prove some new infinite product identities. The two examples of Visible Point Vector identities cited in WolframMathWorld (see Weisstein [298]) are cases of the author's identities, (see Campbell [88, 93, 97, 98, 99]) which include the

Theorem 23.1 *(Campbell [88]) If $|x|,|y|,|z|<1$ and $s,t,u,$ are complex numbers and $Li_s(z)=\sum_{k=1}^{\infty}\frac{z^k}{k^s}$,*

$$\prod_{\substack{a,b\geq 1\\(a,b)=1}}\left(1-y^a z^b\right)^{-a^{-s}b^{-t}}=\exp\left[Li_s(y)Li_t(z)\right], \tag{23.1.1}$$

for which $s+t=1$;and

$$\prod_{\substack{a,b,c\geq 1\\(a,b,c)=1}}\left(1-x^a y^b z^c\right)^{-a^{-s}b^{-t}c^{-u}}=\exp\left[Li_s(x)Li_t(y)Li_u(z)\right], \tag{23.1.2}$$

for which $s+t+u=1$.

 DOI: 10.1201/9781003174158-23

The n-dimensional extension of these is for all $|x_i| < 1$, $(0 \le i \le n)$

$$\prod_{\substack{a_1,a_2,\ldots,a_n \ge 1 \\ (a_1,a_2,\ldots,a_n)=1}} (1 - {x_1}^{a_1} {x_2}^{a_2} \cdots {x_n}^{a_n})^{-{a_1}^{-s_1} {a_2}^{-s_2} \cdots {a_n}^{-s_n}} \tag{23.1.3}$$

$$= \exp\left[Li_{s_1}(x_1) Li_{s_2}(x_2) \cdots Li_{s_n}(x_n)\right],$$

for which $s_1 + s_2 + \cdots + s_n = 1$.

Identities (23.1.1), (23.1.2), and (23.1.3) are easy to prove by differentiating the logarithm of both sides, and applying the well known

Lemma 23.1 *Consider an infinite region raying out of the origin in any n-dimensional Euclidean vector space. The set of all lattice point vectors apart from the origin in that region is precisely the set of positive integer multiples of the visible point vector (VPV's) in that region. (Campbell [88])*

We can make use of some well known summations and substitute them into (23.1.1) and (23.1.2) above. The following cases of the polylogarithm are well known. (see Abramowitz and Stegun [1])

Statement 10

$$Li_1(z) = -\ln(1-z), \tag{23.1.4}$$

$$Li_0(z) = \frac{z}{1-z}, \tag{23.1.5}$$

$$Li_{-1}(z) = \frac{z}{(1-z)^2}, \tag{23.1.6}$$

$$Li_{-2}(z) = \frac{z(1+z)}{(1-z)^3}, \tag{23.1.7}$$

$$Li_{-3}(z) = \frac{z(1+4z+z^2)}{(1-z)^4}, \tag{23.1.8}$$

$$Li_{-4}(z) = \frac{z(1+z)(1+10z+z^2)}{(1-z)^5}. \tag{23.1.9}$$

Some particular expressions for half-integer values of the argument z are:

Statement 11 *If ζ is the Riemann zeta function then*

$$Li_1(1/2) = \ln 2, \tag{23.1.10}$$

$$Li_2(1/2) = \frac{\pi^2}{12} - \frac{1}{2}(\ln 2)^2, \tag{23.1.11}$$

$$Li_3(1/2) = \frac{1}{6}(\ln 2)^3 - \frac{\pi^2}{12}(\ln 2)^2 - \frac{7}{8}\zeta(3). \tag{23.1.12}$$

No formulas of this type are known for higher integer orders (see Lewin [183, page 2]). However, one has for instance, the result from Borwein, Borwein and Girgensohn [64] that

Statement 12 *If the alternating double sum* $\zeta(\overline{3},\overline{1}) = \sum_{m>n>0}(-1)^{m+n}m^{-3}n^{-1}$, *then*

$$Li_4(1/2) = \frac{\pi^4}{360} - \frac{1}{24}(\ln 2)^4 - \frac{\pi^2}{24}(\ln 2)^4 - \frac{1}{2}\zeta(\overline{3},\overline{1}). \tag{23.1.13}$$

We can substitute from the statements 11 and 12 into 23.1.1 quite easily, to obtain the expressions for $\exp[Li_s(x)Li_{1-s}(y)]$. For the first few cases, namely $s = 1,2,3,4,5$, we have

Theorem 23.2 *If* $|x|,|y| < 1$,

$$\prod_{\substack{a,b\geq 1\\(a,b)=1}} \left(1-x^a y^b\right)^{-a^{-1}} = \left(\frac{1}{1-x}\right)^{\frac{y}{1-y}}, \tag{23.1.14}$$

$$\prod_{\substack{a,b\geq 1\\(a,b)=1}} \left(1-x^a y^b\right)^{-a^{-2}b^1} = \exp\left(Li_2(x)\frac{y}{(1-y)^2}\right), \tag{23.1.15}$$

$$\prod_{\substack{a,b\geq 1\\(a,b)=1}} \left(1-x^a y^b\right)^{-a^{-3}b^2} = \exp\left(Li_3(x)\frac{y(1+y)}{(1-y)^3}\right), \tag{23.1.16}$$

$$\prod_{\substack{a,b\geq 1\\(a,b)=1}} \left(1-x^a y^b\right)^{-a^{-4}b^3} = \exp\left(Li_4(x)\frac{y(1+4y+y^2)}{(1-y)^4}\right), \tag{23.1.17}$$

$$\prod_{\substack{a,b\geq 1\\(a,b)=1}} \left(1-x^a y^b\right)^{-a^{-5}b^4} = \exp\left(Li_5(x)\frac{y(1+y)(1+10y+y^2)}{(1-y)^5}\right), \tag{23.1.18}$$

These results explicitly stated are all new, and will allow us to examine the behavior of the Riemann zeta function in the neighborhoods of the trivial zeroes occurring at $-2,-4,-6,...$ by observing the right sides of identities (23.1.16), (23.1.18), etc. as $y \to 1$.

As an example limiting $Li_{-2}(y)$ as $y \to 1$ to the trivial zero of $\zeta(-2) = 0$, we see that (23.1.15) is equivalent to

$$\prod_{\substack{a,b\geq 1\\(a,b)=1}} \left(1-x^a y^b\right)^{-a^{-3}b^2} = \exp\left(Li_3(x)\left(y+2^2y^2+3^2y^3+4^2y^4+5^2y^5+\cdots\right)\right),$$

and we can examine this for the value $y=1-\delta$, for small positive real δ which means then we have, by (23.1.15), that

$$\prod_{\substack{a,b\geq 1\\(a,b)=1}} \left(1-x^a(1-\delta)^b\right)^{-a^{-3}b^2} = \exp\left(Li_3(x)\frac{(1-\delta)(2-\delta)}{(\delta)^3}\right)$$

where both sides approach 1 as $\delta\to 0$.

However, this is not very illuminating, as the limit of the polylogarithm approaches infinity. A better approach to take if we are to learn something new about the zeroes of the zeta function, may be to limit along the $s+t=1$ restriction on (23.1.15). The following particular cases are new, and incidental to theorem 23.2 above.

Corollary 23.1 *If* $|y|<1$,

$$\prod_{\substack{a,b\geq 1\\(a,b)=1}} \left(1-y^b\right)^{-a^{-2}b^1} = \exp\left(\frac{\pi^2}{6}\frac{y}{(1-y)^2}\right), \tag{23.1.19}$$

$$\prod_{\substack{a,b\geq 1\\(a,b)=1}} \left(1-y^b\right)^{-a^{-3}b^2} = \exp\left(\zeta(3)\frac{y(1+y)}{(1-y)^3}\right), \tag{23.1.20}$$

$$\prod_{\substack{a,b\geq 1\\(a,b)=1}} \left(1-y^b\right)^{-a^{-4}b^3} = \exp\left(\frac{\pi^4}{90}\frac{y(1+4y+y^2)}{(1-y)^4}\right), \tag{23.1.21}$$

$$\prod_{\substack{a,b\geq 1\\(a,b)=1}} \left(1-y^b\right)^{-a^{-5}b^4} = \exp\left(\zeta(5)\frac{y(1+y)(1+10y+y^2)}{(1-y)^5}\right), \tag{23.1.22}$$

In a similar way, the results from statements 11 and 12 yield incidentally, the

Corollary 23.2 *If* $|y|<1$,

$$\prod_{\substack{a,b\geq 1\\(a,b)=1}} \left(1-\frac{y^b}{2^a}\right)^{-\frac{1}{a}} = 2^{\frac{y}{1-y}}, \tag{23.1.23}$$

$$\prod_{\substack{a,b\geq 1\\(a,b)=1}} \left(1-\frac{y^b}{2^a}\right)^{-\frac{b}{a^2}} = \exp\left[\left(\frac{\pi^2}{12}-\frac{1}{2}(\ln 2)^2\right)\frac{y}{(1-y)^2}\right], \tag{23.1.24}$$

$$\prod_{\substack{a,b\geq 1\\(a,b)=1}} \left(1-\frac{y^b}{2^a}\right)^{-\frac{b^2}{a^3}} = \exp\left[\left(\frac{1}{6}(\ln 2)^3-\frac{\pi^2}{12}(\ln 2)^2-\frac{7}{8}\zeta(3)\right)\frac{y(1+y)}{(1-y)^3}\right], \tag{23.1.25}$$

$$\prod_{\substack{a,b\geq 1\\(a,b)=1}} \left(1-\frac{y^b}{2^a}\right)^{-\frac{b^3}{a^4}} = \exp\left[\left(\frac{\pi^4}{360}-\frac{1}{24}(\ln 2)^4-\frac{\pi^2}{24}(\ln 2)^4-\frac{1}{2}\zeta(\overline{3},\overline{1})\right)\frac{y(1+4y+y^2)}{(1-y)^4}\right], \tag{23.1.26}$$

with $\zeta(\overline{3},\overline{1})$ *as in (23.1.13).*

We note here that many of the Euler-Zagier sums from Borwein et al. [64], and Bailey et al. [44], will provide us with more new resulting infinite products. In the next section, we write the corresponding identities for the more interesting neighborhoods for the nontrivial zeroes of the Riemann zeta function where $y=\frac{1}{2}+iT$ for real numbers T.

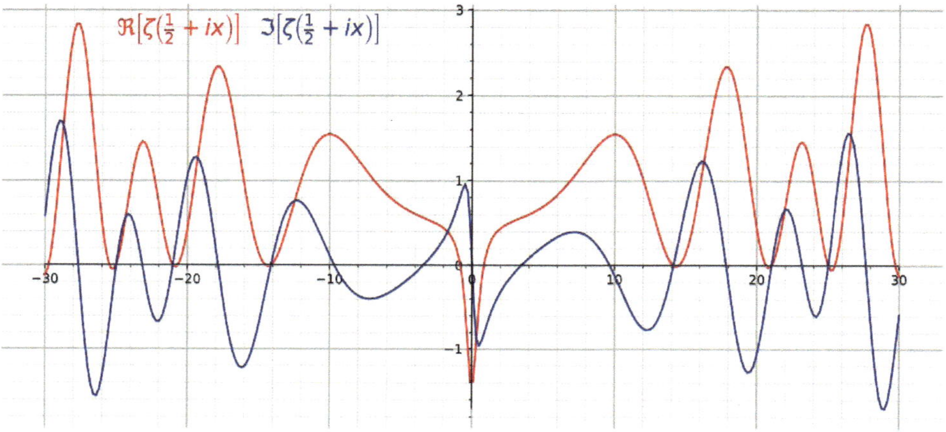

Figure 23.1 Riemann zeta function critical line plot

23.2 IDENTITIES NEAR NON-TRIVIAL ZEROES OF THE RIEMANN ZETA FUNCTION

We next consider cases of (23.1.1) to enable observation of the polylogarithm function as the Dirichlet series aspect of the argument approaches the non-trivial zeroes of the Riemann zeta function. In a similar manner to section 23.1, it is clear that in theorem 23.1 we can substitute $s=\frac{1}{2}+iT$ and $t=\frac{1}{2}-iT$, so $s+t=1$ as required and this gives:

Theorem 23.3 *If* $|x|,|y|<1$*; and* T *is a real number,*

$$\prod_{\substack{a,b\geq 1\\ \gcd(a,b)=1}} \left(\frac{1}{1-x^a y^b}\right)^{a^{\left(\frac{1}{2}+iT\right)}b^{\left(\frac{1}{2}-iT\right)}} = \exp\left[Li_{\left(\frac{1}{2}+iT\right)}(x)Li_{\left(\frac{1}{2}-iT\right)}(y)\right]. \tag{23.2.1}$$

Now, the sequence of known non-trivial zeroes $\frac{1}{2}+iT$ along the critical line of the Riemann zeta function to 31 decimal places starts out as (see LMFDB [188]):

Non-trivial Riemann zeta function zeroes imaginary part

ζ zero number	Imaginary partT
1	14.1347251417346937904572519835625
2	21.0220396387715549926284795938969
3	25.0108575801456887632137909925628
4	30.4248761258595132103118975305840
5	32.9350615877391896906623689640747
6	37.5861781588256712572177634807053
7	40.9187190121474951873981269146334
8	43.3270732809149995194961221654068
9	48.0051508811671597279424727494277
10	49.7738324776723021819167846785638

We can set T of (23.2.1) as any of the above first ten non-trivial zeroes. Then as x or y approaches unity the polylogarithm approaches a nontrivial zero of the zeta function. The resulting identities for each T enables study of the zeta function near any non-trivial zero.

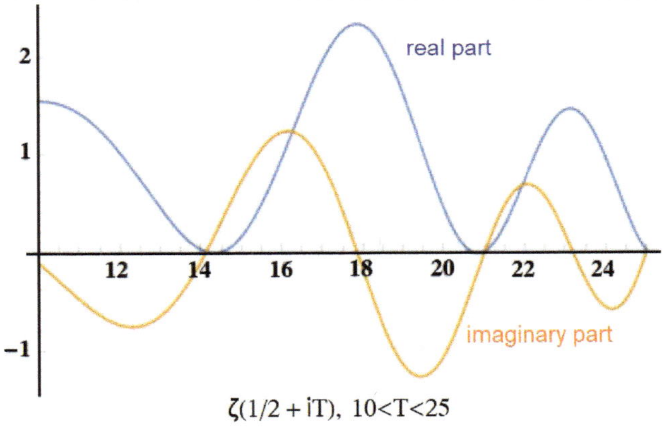

Figure 23.2 Riemann zeta function Re(1/2) line plot

23.3 NON-TRIVIAL ZEROES OF ZETA FUNCTION Re AND Im PARTS

We recall cases of the hyperquadrant theorem 20.1 of Chapter 20 with $n=2, n=3, n=4, n=5$, and restate them in the forms,

Corollary 23.3 *If* $|y|<1, |z|<1$, *and* $t_1+t_2=0$, *then*

$$\prod_{\substack{(a,b)=1 \\ a,b\geq 1}} \left(\frac{1}{1-y^a z^b}\right)^{1/(a^{\frac{1}{2}+it_1} b^{\frac{1}{2}+it_2})} = \exp\left\{\left(\sum_{j=1}^{\infty} \frac{y^j}{j^{\frac{1}{2}+it_1}}\right)\left(\sum_{k=1}^{\infty} \frac{z^k}{k^{\frac{1}{2}+it_2}}\right)\right\}. \quad (23.3.1)$$

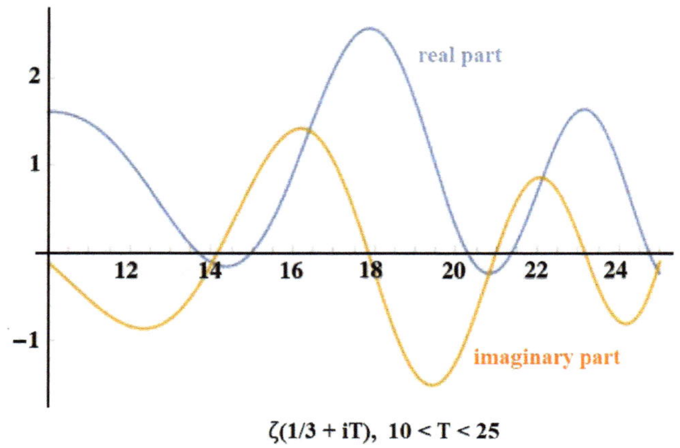

Figure 23.3 Riemann zeta function Re(1/3) line plot

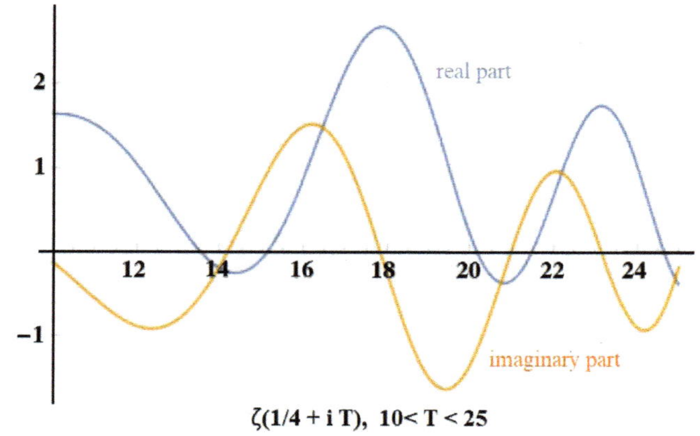

Figure 23.4 Riemann zeta function Re(1/4) line plot

Corollary 23.4 *If* $|x| < 1, |y| < 1, |z| < 1,$ *and* $t_1 + t_2 + t_3 = 0,$ *then*

$$\prod_{\substack{(a,b,c)=1\\ a,b,c\geq 1}} \left(\frac{1}{1 - x^a y^b z^c}\right)^{1/(a^{\frac{1}{3}+it_1} b^{\frac{1}{3}+it_2} c^{\frac{1}{3}+it_3})}$$

$$= \exp\left\{\left(\sum_{i=1}^{\infty} \frac{x^i}{i^{\frac{1}{3}+it_1}}\right)\left(\sum_{j=1}^{\infty} \frac{y^j}{j^{\frac{1}{3}+it_2}}\right)\left(\sum_{k=1}^{\infty} \frac{z^k}{k^{\frac{1}{3}+it_1}}\right)\right\}. \tag{23.3.2}$$

Corollary 23.5 *If* $|w| < 1, |x| < 1, |y| < 1, |z| < 1,$ *and* $t_1 + t_2 + t_3 + t_4 = 0,$ *then*

$$\prod_{\substack{(a,b,c,d)=1\\ a,b,c,d\geq 1}} \left(\frac{1}{1 - w^a x^b y^c z^d}\right)^{1/(a^{\frac{1}{4}+it_1} b^{\frac{1}{4}+it_2} c^{\frac{1}{4}+it_1} d^{\frac{1}{4}+it_4})} \tag{23.3.3}$$

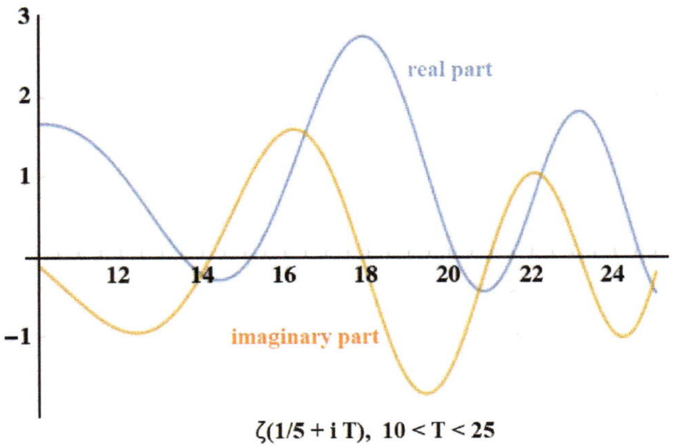

Figure 23.5 Riemann zeta function Re(1/5) line plot

$$= \exp\left\{\left(\sum_{h=1}^{\infty}\frac{w^h}{h^{\frac{1}{4}+it_1}}\right)\left(\sum_{i=1}^{\infty}\frac{x^i}{i^{\frac{1}{4}+it_2}}\right)\left(\sum_{j=1}^{\infty}\frac{y^j}{j^{\frac{1}{4}+it_3}}\right)\left(\sum_{k=1}^{\infty}\frac{z^k}{k^{\frac{1}{4}+it_4}}\right)\right\}.$$

Corollary 23.6 *If* $|v|<1, |w|<1, |x|<1, |y|<1, |z|<1,$ *and* $t_1+t_2+t_3+t_4+t_5=0,$ *then*

$$\prod_{\substack{(a,b,c,d,e)=1\\ a,b,c,d,e\geq 1}}\left(\frac{1}{1-v^aw^bx^cy^dz^e}\right)^{1/(a^{\frac{1}{5}+it_1}b^{\frac{1}{5}+it_2}c^{\frac{1}{5}+it_3}d^{\frac{1}{5}+it_4}e^{\frac{1}{5}+it_5})} \tag{23.3.4}$$

$$= \exp\left\{\left(\sum_{g=1}^{\infty}\frac{v^g}{g^{\frac{1}{5}+it_5}}\right)\left(\sum_{h=1}^{\infty}\frac{w^h}{h^{\frac{1}{5}+it_2}}\right)\left(\sum_{i=1}^{\infty}\frac{x^i}{i^{\frac{1}{5}+it_3}}\right)\left(\sum_{j=1}^{\infty}\frac{y^j}{j^{\frac{1}{5}+it_4}}\right)\left(\sum_{k=1}^{\infty}\frac{z^k}{k^{\frac{1}{5}+it_5}}\right)\right\}.$$

The reader may recall the polylogarithm occurring in the right sides of (23.3.1) through to (23.3.4).

Of course each polylogarithm is limiting to a case which is a Riemann zeta function, wherein the expression

$$L_{\sigma+iT}(z)=\sum_{k=1}^{\infty}\frac{z^k}{k^{(\sigma+iT)}}\mapsto\sum_{k=1}^{\infty}\frac{1}{k^{(\sigma+iT)}}=\zeta(\sigma+iT), \tag{23.3.5}$$

as $z\to 1$ is considered.

Over the past 165 years, the Riemann Hypothesis has become a perennial hot topic. A lot is known about the non-trivial zeta function zeroes whose real part is $\frac{1}{2}$, and millions of them have been identified along the critical line $\zeta\left(\frac{1}{2}+iT\right)$. (See Brent [70] and Brent et al. [71]) At the time of writing this, the RH is still an Open Question with respect to existence of nontrivial zeroes of the form $\zeta(\sigma+iT)$ where $0<\sigma<\frac{1}{2}$. So, if anyone were able to find or prove the existence of another nontrivial zero in $0<\sigma<\frac{1}{2}$ that would make the RH false; thereby causing major rethink of a large swathe of mathematics that depends on the RH being true. So then, what do our identities contribute to analysis with respect to the Riemann Hypothesis? Using contemporary apps and devices it is easily found for example that:

Riemann zeta function $\Re(\zeta(1/2+ix))=0$ and $\Im(\zeta(1/2+ix))=0$

$\Re(\zeta(1/2+ix))=0$	$\Im(\zeta(1/2+ix))=0$
$x \approx \pm 20.6540449693679\ldots$	$x \approx \pm \mathbf{14.1347251417347}\ldots$
$x \approx \pm 14.5179196282622\ldots$	$x \approx \pm 9.66690805613019\ldots$
$x \approx \pm \mathbf{14.1347251417347}\ldots$	$x \approx \pm 3.43621822608696\ldots$
$x \approx \pm 0.819545329425848\ldots$	$x=0$
	$x \approx 17.8455995404109\ldots$

Note that the first zero along $\zeta\left(\frac{1}{2}+ix\right)$ where the real value equals the imaginary value is the bold **14.1347251417347**.... The plots of the $\Re(\zeta(1/2+ix))$ and $\Im(\zeta(1/2+ix))$ are given in Figures 23.1 and 23.2; each showing the periodic, oscillating form of the curves for $\Re$ part and $\Im$ part. A seemingly miraculous feature of these two graphed curves for $\Re$ part and $\Im$ part is that, for all their irregularities and infinite number of periodic oscillations, they coincide at zero infinitely often, without revealing any obvious reason for this feature. If we examine the corresponding *real* cf *imaginary* oscillations for the curves for $\Re(\zeta(1/3+ix))$ and $\Im(\zeta(1/3+ix))$ in Figure 23.3, we see that the two curves never seem to quite intersect at zero, but often are close to zero. In other words, we get numerous *near misses* of *would-be* RH counterexamples. Then this feature of many near misses of counterexamples is observable in the $\zeta\left(\frac{1}{4}+iT\right)$, and $\zeta\left(\frac{1}{5}+iT\right)$ graphs given in Figures 23.4 and 23.5. Of course, tantalizingly, we don't have a proof either way that these two curves intersecting at *near misses* don't exceptionally nor occasionally intersect at zero some long, long distance along the line, to give thereby legitimate RH counterexamples.

So, why present the curve plots for Figures 23.3, 23.4, and 23.5?

Riemann zeta function $\Re(\zeta(1/3+ix))=0$ and $\Im(\zeta(1/3+ix))=0$

$\Re(\zeta(1/3+ix))=0$	$\Im(\zeta(1/3+ix))=0$
$x \approx \pm 20.2878228308336\ldots$	$x \approx \pm 14.1617633554208\ldots$
$x \approx \pm 14.9841684717239\ldots$	$x \approx \pm 9.70463607919240\ldots$
$x \approx \pm 13.6850707865916\ldots$	$x \approx \pm 3.36373749507605\ldots$
$x \approx \pm 0.902058175306342\ldots$	$x=0$
	$x \approx 17.8488748021877\ldots$

Riemann zeta function $\Re(\zeta(1/4+ix))=0$ and $\Im(\zeta(1/4+ix))=0$

$\Re(\zeta(1/4+ix))=0$	$\Im(\zeta(1/4+ix))=0$
$x \approx \pm 20.1848884083559\ldots$	$x \approx \pm 14.1756642345115\ldots$
$x \approx \pm 15.1142782132762\ldots$	$x \approx \pm 9.72422382047584\ldots$
$x \approx \pm 13.5642011378313\ldots$	$x \approx \pm 3.32579423316711\ldots$
$x \approx \pm 0.932995672808082\ldots$	$x=0$
	$x \approx 17.8509332328041\ldots$

Riemann zeta function $\Re(\zeta(1/5+ix))=0$ and $\Im(\zeta(1/5+ix))=0$

$\Re(\zeta(1/5+ix))=0$	$\Im(\zeta(1/5+ix))=0$
$x \approx \pm 20.1327630004499\ldots$	$x \approx \pm 14.1841280267195\ldots$
$x \approx \pm 15.1810158244925\ldots$	$x \approx \pm 9.73621176655371\ldots$
$x \approx \pm 13.5033266188502\ldots$	$x \approx \pm 3.30246316491894\ldots$
$x \approx \pm 0.948755232373065\ldots$	$x = 0$
	$x \approx 17.8523054476864\ldots$

The behavior of the real part and the imaginary part of $\zeta(1/2+ix))=0$ is oscillatory in each case. There seems no regularity nor predictable periodicity to the oscillations. And yet, there are known to be an infinite number of zeroes on the critical line for $\zeta(1/2+ix))$, implying that remarkably the x real number solution of $\Re(\zeta(1/2+ix))=0$ and the imaginary number solution $\Im(\zeta(1/2+ix))=0$ coincide and are identical for each of those famously infinite number of non-trivial zeroes whose real part is $\frac{1}{2}$. In the table for real and imaginary zeroes above we see the first such instance where we have

$$\Re(\zeta(1/2+i14.1347251417347\ldots))=\Im(\zeta(1/2+i14.1347251417347\ldots))=0. \quad (23.3.6)$$

The 2D, 3D, 4D, and 5D VPV identities shown above as equations (23.3.1), (23.3.2), (23.3.3), (23.3.4), respectively suggest that if limiting the polylogarithm functions then interesting results may be examined in the context of looking for further existing non-trivial ζ function zeroes in the strip $0<\Re\sigma<\frac{1}{2}$. That is, along the lines with real parts: $\sigma=\frac{1}{3}$, $\sigma=\frac{1}{4}$, $\sigma=\frac{1}{5}$, and so on. If we were able to find similar coincident equalities between the real and imaginary zeroes in the tales below that would disprove the Riemann Hypothesis. However, this is not looking likely from the small sample of results here; thus leaving RH still unresolved.

Think About It...

Although this may seem a paradox, all exact science is dominated by the idea of approximation.

If a man tells you he knows a thing exactly, then you can be safe in inferring that you are speaking to an inexact man.

Bertrand Russell
Mathematician/Philosopher, 1872–1970

FURTHER READING

ABRAMOWITZ, M. and STEGUN, I. *Handbook of Mathematical Functions.* Dover Publications Inc., New York, 1972. 10th Printing, 1972, with corrections.

BAILEY, D.H., BORWEIN, J.M. and CRANDALL, R.E. Computation and theory of extended Mordell Tornheim Witten sums. In *8th Asia-pacific Complex Systems Conference*, International Workshop on Operator Theory and Its Applications. Australian Mathematical Sciences Institute (AMSI), 2007.

BORWEIN, D., BORWEIN, J.M. and GIRGENSOHN, R. Explicit evaluation of Euler sums. Proc. *Edinburgh Math. Soc.*, 38:277–294, 1995.

BRENT, R.P. On the Zeros of the Riemann Zeta Function in the Critical Strip. *Math. Comput.*, 33:1361–1372, 1979.

BRENT, R.P., van de LUNE, J., te RIELE, J.J. and WINTER, D.T. On the Zeros of the Riemann Zeta Function in the Critical Strip. II. *Math. Comput.*, 39(160):681–688, October 1982.

CAMPBELL, G.B. Infinite products over visible lattice points. *Int. J. Math. and Math. Sci.*, 17(4):637–654, 1994.

CAMPBELL, G.B. Some n-space q-binomial theorem extensions and similar identities. arxiv preprint, June 2019. arXiv:1906.07526v1 [math.NT]: https://arxiv.org/abs/1906.07526.

CAMPBELL, G.B. Vector Partition Identities for 2D, 3D and nD Lattices. arxiv preprint, Feb 2023. arXiv:10.48550/ARXIV.2302.01091 [math.NT], https://arxiv.org/abs/2302.01091.

CAMPBELL, G.B. Visible Point Partition Identities for Polylogarithms, and Parametric Euler Sums. arxiv preprint, Jun 2023. arXiv:2306.02241v1 [math.CO], https://arxiv.org/abs/2306.02241.

CAMPBELL, G.B. Visible Point Vector Partition Identities for Hyperpyramid Lattices. arxiv preprint, Sep 2023. arXiv:2309.16094v1 [math.CO], https://arxiv.org/abs/2309.16094.

LEWIN, L. *Polylogarithms and associated functions.* Elsevier Science Ltd.: North-Holland, New York, 1981. ISBN 0-444-00550-1,(First published London: Macdonald, 1958.).

The LMFDB Collaboration. The L-functions and modular forms database. web page, 2023. [Online; accessed 18 February 2023]. www.lmfdb.org.

WEISSTEIN, Eric W. Visible Point Vector Identity. From MathWorld– A Wolfram Web Resource., July 2012. Article with 2D and 3D identities (https://mathworld.wolfram.com/VisiblePointVectorIdentity.html).

CHAPTER 24

Visible Point Vector Identities in Skewed Hyperpyramid Lattices

24.1 VISIBLE POINT VECTOR IDENTITIES IN ASYMMETRIC HYPERPYRAMID REGIONS

IN THE AUTHOR'S 2000 paper on hyperpyramid VPV identities, the square hyperpyramid theorem was published, but also there appeared the so-named "Skewed hyperpyramid theorem". Essentially, for the square hyperpyramid in 3D, the vector grid of lattice points has the horizontal cross section at height n as an $n \times n$ square of lattice points. In 4D the z axis of the hyperpyramid vector grid is "sort of considered" to be hypervertical, with a $n \times n \times n$ cube at height n. Recent research developments on hyperquadrant lattice sums and inverted hyperpyramid lattice sums are summarized in Campbell [93, 96, 97, 98, 99], which includes content from this chapter.

24.1.1 A more General Hyperpyramid Theorem

In Campbell [88], the main theorem of the paper came from summing on all lattice points in the first hyperquadrant. In the previous chapter, this is restricted to hyperpyramid lattices with symmetry in $n-1$ out of the n variables. This is evident by viewing Corollaries 23.1, 23.2, 24.1, and 24.2. If we choose to vary the shape of our hyperpyramid, the process is most easily illustrated as follows. Consider the variant of the 4D case of identity (21.4.1), from Chapter 21 where the x variable in the lattice is confined as follows:

$$\exp\left\{\sum_{k=1}^{\infty} \frac{z^k}{k^t}\right\} \prod_{\substack{(a,b,c,d)=1 \\ a,\frac{b}{2},c<d \\ a,b,c\geq 0,d>0}} \left(\frac{1}{1-w^a x^b y^c z^d}\right)^{\frac{1}{a^q b^r c^s d^t}} \tag{24.1.1}$$

$$= \exp\left\{\left(1+\frac{x^1}{1^r}\right)\frac{z^1}{1^t} + \left(1+\frac{w^1}{1^q}\right)\left(1+\frac{x^1}{1^r}+\frac{x^2}{2^r}+\frac{x^3}{3^r}\right)\left(1+\frac{y^1}{1^s}\right)\frac{z^2}{2^t}\right\}$$

DOI: 10.1201/9781003174158-24

$$\times \exp\left\{ + \left(1+\frac{w^1}{1^s}+\frac{w^2}{2^r}\right)\left(1+\frac{x^1}{1^s}+\frac{x^2}{2^s}+\frac{x^3}{3^s}+\frac{x^4}{4^s}+\frac{x^5}{5^s}\right)\left(1+\frac{y^1}{1^s}+\frac{y^2}{2^s}\right)\frac{z^3}{3^t}+\cdots\right\}.$$

It becomes clear from this that the general hyperpyramid sums can be all included in the following theorem. Also, the ensuing chapter on polylogarithms, the Sofo generalized Euler Sums and the related theory of the Mordell-Wittam-Tornheim functions lends itself neatly to presenting particular cases of skewed hyperpyramid VPV identities that encode light diffusion models. So then, we have the

Theorem 24.1 ***The skewed hyperpyramid VPV identity.*** *If $i = 1,2,3,...,n$ then for each $x_i \in \mathbb{C}$ such that $|x_i| < 1$ and $b_i \in \mathbb{C}$ such that $\sum_{i=1}^{n} b_i = 1$, and in addition for $\alpha_i \in \Re^+$, we have*

$$\exp\left\{\left(\sum_{k=1}^{\infty}\frac{{x_n}^k}{k^{b_n}}\right)\right\} \prod_{\substack{(a_1,a_2,\ldots,a_n)=1 \\ a_1\alpha_1,a_2\alpha_2,\ldots,a_{n-1}\alpha_{n-1}<a_n \\ a_1,a_2,\ldots,a_{n-1}>0 \\ a_n>1}} \left(\frac{1}{1-{x_1}^{a_1}{x_2}^{a_2}{x_3}^{a_3}\cdots{x_n}^{a_n}}\right)^{\frac{1}{{a_1}^{b_1}{a_2}^{b_2}{a_3}^{b_3}\cdots{a_n}^{b_n}}}$$

$$= \exp\left\{\sum_{k=1}^{\infty}\prod_{i=1}^{n-1}\left(\sum_{j=1}^{F(k)}\frac{{x_i}^j}{j^{b_i}}\right)\frac{{x_n}^k}{k^{b_n}}\right\}, \tag{24.1.2}$$

where

$$F(k) = \begin{cases} \left[\frac{k-1}{\alpha_i}\right], & \alpha_i > 1; \\ k-1, & \alpha_i = 1; \\ \left[\frac{k}{\alpha_i}-1\right], & \alpha_i < 1. \end{cases}$$

We see that this theorem counts terms on any possible asymmetric hyperpyramid lattice in the first hyperquadrant, where the hyperpyramid apex is at the origin. We may allow interesting limits to apply such as $\alpha_i \to 0$ for a particular α_i. This contributes a factor to the infinite series on the right side exponentiated sum (24.1.2) since $F(k)$ is arbitrarily large. If each α_i in the theorem is less than 1, the resulting identity is simpler than the case when some or all α_i's are greater than 1. In fact, applying the above theorem in a similar fashion to that of the square hyperpyramid cases, we easily arrive at examples such as the 2D case through to the 5D case of (24.1.2) are given in the following cases of the *skewed hyperpyramid identity.*

Corollary 24.1 *For $|y|,|z| < 1$,*

$$\prod_{\substack{(a,b)=1 \\ \frac{a}{11}<b \\ a\geq 0,b>0}} \left(\frac{1}{1-y^a z^b}\right)^{\frac{1}{b}} = \left(\frac{1-y^{11}z}{1-z}\right)^{\frac{1}{1-y}} \tag{24.1.3}$$

$$= 1+\frac{1-y^{11}}{1-y}\frac{z}{1!}+\begin{vmatrix} \frac{1-y^{11}}{1-y} & -1 \\ \frac{1-y^{22}}{1-y} & \frac{1-y^{11}}{1-y} \end{vmatrix}\frac{z^2}{2!}+\begin{vmatrix} \frac{1-y^{11}}{1-y} & -1 & 0 \\ \frac{1-y^{22}}{1-y} & \frac{1-y^{11}}{1-y} & -2 \\ \frac{1-y^{33}}{1-y} & \frac{1-y^{22}}{1-y} & \frac{1-y^{11}}{1-y} \end{vmatrix}\frac{z^3}{3!}$$

$$+\begin{vmatrix} \frac{1-y^{11}}{1-y} & -1 & 0 & 0 \\ \frac{1-y^{22}}{1-y} & \frac{1-y^{11}}{1-y} & -2 & 0 \\ \frac{1-y^{33}}{1-y} & \frac{1-y^{22}}{1-y} & \frac{1-y^{11}}{1-y} & -3 \\ \frac{1-y^{44}}{1-y} & \frac{1-y^{33}}{1-y} & \frac{1-y^{22}}{1-y} & \frac{1-y^{11}}{1-y} \end{vmatrix} \frac{z^4}{4!} + etc.$$

In this case, it is fairly easy to find the Taylor coefficients for the (24.1.3) right side function. Hence we get a closed form evaluation of the determinant coefficients. In Mathematica, and WolframAlpha one easily sees that the Taylor series is given by the expansion $\sum_{n=0}^{\infty} c_n z^n$ where $c_0 = 1$, $c_1 = \frac{1-y^{11}z}{1-z}$ with the recurrence

$$ny^{11}c_n + (2+n)c_{n+2} = \left(2 + y\frac{1-y^{11}z}{1-z} + n(1+y^{11})\right)c_{n+1}.$$

Incidentally, also in Mathematica, and WolframAlpha one easily sees, for example, that the code

$$Det[\{(1-y^{11})/(1-y), -1, 0, 0\}, \{(1-y^{22})/(1-y), (1-y^{11})/(1-y), -2, 0\},$$
$$\{(1-y^{33})/(1-y), (1-y^{22})/(1-y), (1-y^{11})/(1-y), -3\}, \{(1-y^{44})/(1-y),$$
$$(1-y^{33})/(1-y), (1-y^{22})/(1-y), (1-y^{11})/(1-y)\}]$$

nicely tells us a polynomial coefficient given by

$$\begin{vmatrix} \frac{1-y^{11}}{1-y} & -1 & 0 & 0 \\ \frac{1-y^{22}}{1-y} & \frac{1-y^{11}}{1-y} & -2 & 0 \\ \frac{1-y^{33}}{1-y} & \frac{1-y^{22}}{1-y} & \frac{1-y^{11}}{1-y} & -3 \\ \frac{1-y^{44}}{1-y} & \frac{1-y^{33}}{1-y} & \frac{1-y^{22}}{1-y} & \frac{1-y^{11}}{1-y} \end{vmatrix}$$

$$\begin{aligned} &= 6y^{43} + 17y^{42} + 34y^{41} + 58y^{40} + 90y^{39} + 131y^{38} + 182y^{37} + 244y^{36} + 318y^{35} + 405y^{34} \\ &+ 506y^{33} + 622y^{32} + 746y^{31} + 875y^{30} + 1006y^{29} + 1136y^{28} + 1262y^{27} + 1381y^{26} \\ &+ 1490y^{25} + 1586y^{24} + 1666y^{23} + 1727y^{22} + 1766y^{21} + 1774y^{20} + 1754y^{19} + 1709y^{18} \\ &+ 1642y^{17} + 1556y^{16} + 1454y^{15} + 1339y^{14} + 1214y^{13} + 1082y^{12} + 946y^{11} + 809y^{10} \\ &+ 666y^{9} + 540y^{8} + 430y^{7} + 335y^{6} + 254y^{5} + 186y^{4} + 130y^{3} + 85y^{2} + 50y + 24. \end{aligned}$$

Following on from this, a 3D version example identity is given by

Corollary 24.2 *For each of* $|x|, |y|, |z| < 1$,

$$\prod_{\substack{(a,b,c)=1 \\ \frac{a}{5}, \frac{b}{7} < c \\ a,b \geq 0, c > 0}} \left(\frac{1}{1 - x^a y^b z^c}\right)^{\frac{1}{c}} = \left(\frac{(1-x^5 z)(1-y^7 z)}{(1-z)(1-x^5 y^7 z)}\right)^{\frac{1}{(1-x)(1-y)}} \tag{24.1.4}$$

$$= 1 + \frac{(1-x^5)(1-y^7)}{(1-x)(1-y)} \frac{z}{1!} + \begin{vmatrix} \frac{(1-x^5)(1-y^7)}{(1-x)(1-y)} & -1 \\ \frac{(1-x^{10})(1-y^{14})}{(1-x)(1-y)} & \frac{(1-x^5)(1-y^7)}{(1-x)(1-y)} \end{vmatrix} \frac{z^2}{2!}$$

$$+ \begin{vmatrix} \frac{(1-x^5)(1-y^7)}{(1-x)(1-y)} & -1 & 0 \\ \frac{(1-x^{10})(1-y^{14})}{(1-x)(1-y)} & \frac{(1-x^5)(1-y^7)}{(1-x)(1-y)} & -2 \\ \frac{(1-x^{30})(1-y^{21})}{(1-x)(1-y)} & \frac{(1-x^{10})(1-y^{14})}{(1-x)(1-y)} & \frac{(1-x^5)(1-y^7)}{(1-x)(1-y)} \end{vmatrix} \frac{z^3}{3!}$$

$$+ \begin{vmatrix} \frac{(1-x^5)(1-y^7)}{(1-x)(1-y)} & -1 & 0 & 0 \\ \frac{(1-x^{10})(1-y^{14})}{(1-x)(1-y)} & \frac{(1-x^5)(1-y^7)}{(1-x)(1-y)} & -2 & 0 \\ \frac{(1-x^{30})(1-y^{21})}{(1-x)(1-y)} & \frac{(1-x^{10})(1-y^{14})}{(1-x)(1-y)} & \frac{(1-x^5)(1-y^7)}{(1-x)(1-y)} & -3 \\ \frac{(1-x^{40})(1-y^{28})}{(1-x)(1-y)} & \frac{(1-x^{30})(1-y^{21})}{(1-x)(1-y)} & \frac{(1-x^{10})(1-y^{14})}{(1-x)(1-y)} & \frac{(1-x^5)(1-y^7)}{(1-x)(1-y)} \end{vmatrix} \frac{z^4}{4!} + etc.$$

Continuing on, a 4D version example identity is given by

Corollary 24.3 *For each of* $|w|, |x|, |y|, |z| < 1$,

$$\prod_{\substack{(a,b,c,d)=1 \\ \frac{a}{13}, \frac{b}{6}, \frac{c}{17} < d \\ a,b,c \geq 0, d>0}} \left(\frac{1}{1 - w^a x^b y^c z^d} \right)^{\frac{1}{d}} = \left(\frac{(1-w^{13}z)(1-x^6 z)(1-y^{17}z)(1-w^{13}x^6 y^{17}z)}{(1-z)(1-w^{13}x^6 z)(1-w^{13}y^{17}z)(1-x^6 y^{17}z)} \right)^{\frac{1}{(1-w)(1-x)(1-y)}}, \tag{24.1.5}$$

$$= 1 + \frac{(1-w^{13})(1-x^6)(1-y^{17})}{(1-w)(1-x)(1-y)} \frac{z}{1!} + \begin{vmatrix} \frac{(1-w^{13})(1-x^6)(1-y^{17})}{(1-w)(1-x)(1-y)} & -1 \\ \frac{(1-w^{26})(1-x^{12})(1-y^{34})}{(1-w)(1-x)(1-y)} & \frac{(1-w^{13})(1-x^6)(1-y^{17})}{(1-w)(1-x)(1-y)} \end{vmatrix} \frac{z^2}{2!}$$

$$+ \begin{vmatrix} \frac{(1-w^{13})(1-x^6)(1-y^{17})}{(1-w)(1-x)(1-y)} & -1 & 0 \\ \frac{(1-w^{26})(1-x^{12})(1-y^{34})}{(1-w)(1-x)(1-y)} & \frac{(1-w^{13})(1-x^6)(1-y^{17})}{(1-w)(1-x)(1-y)} & -2 \\ \frac{(1-w^{39})(1-x^{18})(1-y^{51})}{(1-w)(1-x)(1-y)} & \frac{(1-w^{26})(1-x^{12})(1-y^{34})}{(1-w)(1-x)(1-y)} & \frac{(1-w^{13})(1-x^6)(1-y^{17})}{(1-w)(1-x)(1-y)} \end{vmatrix} \frac{z^3}{3!}$$

$$+ \begin{vmatrix} \frac{(1-w^{13})(1-x^6)(1-y^{17})}{(1-w)(1-x)(1-y)} & -1 & 0 & 0 \\ \frac{(1-w^{26})(1-x^{12})(1-y^{34})}{(1-w)(1-x)(1-y)} & \frac{(1-w^{13})(1-x^6)(1-y^{17})}{(1-w)(1-x)(1-y)} & -2 & 0 \\ \frac{(1-w^{39})(1-x^{18})(1-y^{51})}{(1-w)(1-x)(1-y)} & \frac{(1-w^{26})(1-x^{12})(1-y^{34})}{(1-w)(1-x)(1-y)} & \frac{(1-w^{13})(1-x^6)(1-y^{17})}{(1-w)(1-x)(1-y)} & -3 \\ \frac{(1-w^{52})(1-x^{24})(1-y^{68})}{(1-w)(1-x)(1-y)} & \frac{(1-w^{39})(1-x^{18})(1-y^{51})}{(1-w)(1-x)(1-y)} & \frac{(1-w^{26})(1-x^{12})(1-y^{34})}{(1-w)(1-x)(1-y)} & \frac{(1-w^{13})(1-x^6)(1-y^{17})}{(1-w)(1-x)(1-y)} \end{vmatrix} \frac{z^4}{4!} + etc.$$

And furthermore, a 5D version example identity is given by

Corollary 24.4 *For each of* $|v|, |w|, |x|, |y|, |z| < 1$,

$$\prod_{\substack{(a,b,c,d,e)=1 \\ a, \frac{b}{2}, \frac{c}{3}, \frac{d}{4} < e \\ a,b,c,d \geq 0, e>0}} \left(\frac{1}{1 - v^a w^b x^c y^d z^e} \right)^{\frac{1}{e}} \tag{24.1.6}$$

$$= \left(\frac{(1-vz)(1-w^2 z)(1-x^3 z)(1-y^4 z)(1-vw^2x^3 z)(1-vw^2y^3 z)(1-vx^3y^4 z)(1-w^2x^3y^4 z)}{(1-z)(1-vw^2 z)(1-vx^3 z)(1-vy^4 z)(1-w^2x^3 z)(1-w^2y^3 z)(1-x^3y^4 z)(1-vw^2x^3y^4 z)} \right)^A$$

$$where\ A = \frac{1}{(1-v)(1-w)(1-x)(1-y)}$$

$$= 1 + \frac{(1-v^1)(1-w^2)(1-x^3)(1-y^4)}{(1-v)(1-w)(1-x)(1-y)} \frac{z}{1!}$$

$$+\begin{vmatrix} \frac{(1-v^1)(1-w^2)(1-x^3)(1-y^4)}{(1-v)(1-w)(1-x)(1-y)} & -1 \\ \frac{(1-v^2)(1-w^4)(1-x^6)(1-y^8)}{(1-v)(1-w)(1-x)(1-y)} & \frac{(1-v^1)(1-w^2)(1-x^3)(1-y^4)}{(1-v)(1-w)(1-x)(1-y)} \end{vmatrix} \frac{z^2}{2!}$$

$$+\begin{vmatrix} \frac{(1-v^1)(1-w^2)(1-x^3)(1-y^4)}{(1-v)(1-w)(1-x)(1-y)} & -1 & 0 \\ \frac{(1-v^2)(1-w^4)(1-x^6)(1-y^8)}{(1-v)(1-w)(1-x)(1-y)} & \frac{(1-v^1)(1-w^2)(1-x^3)(1-y^4)}{(1-v)(1-w)(1-x)(1-y)} & -2 \\ \frac{(1-v^3)(1-w^6)(1-x^9)(1-y^{12})}{(1-v)(1-w)(1-x)(1-y)} & \frac{(1-v^2)(1-w^4)(1-x^6)(1-y^8)}{(1-v)(1-w)(1-x)(1-y)} & \frac{(1-v^1)(1-w^2)(1-x^3)(1-y^4)}{(1-v)(1-w)(1-x)(1-y)} \end{vmatrix} \frac{z^3}{3!}$$

$$+\begin{vmatrix} \frac{(1-v^1)(1-w^2)(1-x^3)(1-y^4)}{(1-v)(1-w)(1-x)(1-y)} & -1 & 0 & 0 \\ \frac{(1-v^2)(1-w^4)(1-x^6)(1-y^8)}{(1-v)(1-w)(1-x)(1-y)} & \frac{(1-v^1)(1-w^2)(1-x^3)(1-y^4)}{(1-v)(1-w)(1-x)(1-y)} & -2 & 0 \\ \frac{(1-v^3)(1-w^6)(1-x^9)(1-y^{12})}{(1-v)(1-w)(1-x)(1-y)} & \frac{(1-v^2)(1-w^4)(1-x^6)(1-y^8)}{(1-v)(1-w)(1-x)(1-y)} & \frac{(1-v^1)(1-w^2)(1-x^3)(1-y^4)}{(1-v)(1-w)(1-x)(1-y)} & -3 \\ \frac{(1-v^4)(1-w^8)(1-x^{12})(1-y^{16})}{(1-v)(1-w)(1-x)(1-y)} & \frac{(1-v^3)(1-w^6)(1-x^9)(1-y^{12})}{(1-v)(1-w)(1-x)(1-y)} & \frac{(1-v^2)(1-w^4)(1-x^6)(1-y^8)}{(1-v)(1-w)(1-x)(1-y)} & \frac{(1-v^1)(1-w^2)(1-x^3)(1-y^4)}{(1-v)(1-w)(1-x)(1-y)} \end{vmatrix} \frac{z^4}{4!}$$

$$+ \text{ etc.}$$

The identity featuring binomial coefficients

$$\prod_{\substack{(a,b,c,d,e)=1 \\ a,b,c,d<e \\ a,b,c,d\geq 0,e>0}} \left(\frac{1}{1-z^{a+b+c+d+e}}\right)^{\frac{1}{e}} = \left(\frac{(1-z^2)^4(1-z^4)^4}{(1-z)(1-z^3)^6(1-z^5)}\right)^{\frac{1}{(1-z)^4}} \tag{24.1.7}$$

comes from setting $v=w=x=y=z$ in (21.4.9) from Chapter 21. A similar effect can be contrived by setting $v=w=x=y$ in (24.1.7) to obtain

$$\prod_{\substack{(a,b,c,d,e)=1 \\ a,b,c,d<e \\ a,b,c,d\geq 0,e>0}} \left(\frac{1}{1-y^{a+b+c+d}z^e}\right)^{\frac{1}{e}} = \left(\frac{(1-yz)^4(1-y^3z)^4}{(1-z)(1-y^2z)^6(1-y^4z)}\right)^{\frac{1}{(1-y)^4}}. \tag{24.1.8}$$

The identity (24.1.7) comes from setting $v=w=x=y=z$ in (21.4.9). A *skewed version* arisen from theory of our present chapter is

$$\prod_{\substack{(a,b,c,d,e)=1 \\ a,\frac{b}{2},\frac{c}{3},\frac{d}{4}<e \\ a,b,c,d\geq 0,e>0}} \left(\frac{1}{1-z^{a+b+c+d+e}}\right)^{\frac{1}{e}} = \left(\frac{(1-z^2)(1-z^3)(1-z^9)(1-z^{10})}{(1-z)(1-z^6)^2(1-z^{11})}\right)^{\frac{1}{(1-z)^4}}. \tag{24.1.9}$$

Think About It...

One of the most amazing things about mathematics is the people who do math aren't interested in application, because mathematics itself is truly a beautiful art form.

It's structures and patterns, and that's what we love, and that's what we get off on.

Danica McKellar
American actress, mathematics writer, and education advocate

FURTHER READING

ABRAMOWITZ, M. and STEGUN, I. *Handbook of Mathematical Functions.* Dover Publications Inc., New York, 1972. 10th Printing, 1972, with corrections.

BAILEY, D.H., BORWEIN, J.M. and CRANDALL, R.E. Computation and theory of extended Mordell Tornheim Witten sums. In *8th Asia-pacific Complex Systems Conference, International Workshop on Operator Theory and Its Applications.* Australian Mathematical Sciences Institute (AMSI), 2007.

BORWEIN, D.; BORWEIN, J.M. and GIRGENSOHN, R. Explicit evaluation of Euler sums. *Proc. Edinburgh Math. Soc.*, 38:277–294, 1995.

CAMPBELL, G.B. Infinite products over visible lattice points. *Int. J. Math. and Math. Sci.*, 17(4):637–654, 1994.

CAMPBELL, G.B. Some n-space q-binomial theorem extensions and similar identities. arxiv preprint, June 2019. arXiv:1906.07526v1 [math.NT]: https://arxiv.org/abs/1906.07526.

CAMPBELL, G.B. Continued Fractions for partition generating functions. arxiv preprint, Jan 2023. arXiv:2301.12945 [math.CO], https://arxiv.org/abs/2301.12945.

CAMPBELL, G.B. Vector Partition Identities for 2D, 3D and nD Lattices. arxiv preprint, Feb 2023. arXiv:10.48550/ARXIV.2302.01091 [math.NT], https://arxiv.org/abs/2302.01091.

CAMPBELL, G.B. Visible Point Partition Identities for Polylogarithms, and Parametric Euler Sums. arxiv preprint, Jun 2023. arXiv:2306.02241v1 [math.CO], https://arxiv.org/abs/2306.02241.

CAMPBELL, G.B. Visible Point Vector Partition Identities for Hyperpyramid Lattices. arxiv preprint, Sep 2023. arXiv:2309.16094v1 [math.CO], https://arxiv.org/abs/2309.16094.

LEWIN, L. *Polylogarithms and associated functions.* Elsevier Science Ltd.: North-Holland, New York, 1981. ISBN 0-444-00550-1 (First published London: Macdonald, 1958).

The LMFDB Collaboration. The L-functions and modular forms database. web page, 2023. [Online; accessed 18 February 2023]. www.lmfdb.org.

WEISSTEIN, Eric W. Visible Point Vector Identity. From MathWorld- A Wolfram Web Resource., July 2012. Article with 2D and 3D identities (https://mathworld.wolfram.com/VisiblePointVectorIdentity.html).

CHAPTER 25

Harmonic Sums Applied to VPV Identities

25.1 THE SOFO PAPERS ON HARMONIC SUMS

IN THE YEARS from 2008 to 2021 about 44 papers were written by Anthony Sofo et al. related to *parametric Euler sums*, various integrals, and *hypergeometric series related Euler Sums*, in many ways similar to work done by Jonathan Borwein and his colleagues on polylogarithms and computational discovery. (See for example Sofo [238] to [279], and significantly Jon and/or David Borwein et al. [63] and [69].)

In this chapter we cover a few of the numerous sums and identities by Sofo, the Borweins and their colleagues, where the sums, integrals and theory created relating to Harmonic Sums and similar entities are clearly applicable to the Vector Partitions works in our current volume. There are many hundreds of particular result examples that could be taken from these researches and applied to the VPV identities. However, we shall content ourselves with some representative examples that demonstrate there are striking cases of substituted summations leading to VPV identities that are new to the literature, and applicable to lattice modeling situations.

A selection of numerous papers by Sofo from the past 15 years is given, each with applicable identities that will substitute into VPV identities from papers by Campbell [84, 85, 87, 88, 89, 90, 91, 93]. These are presented in the **Further Reading** references at the end of this chapter. In the book Bibliography, these are listed as Sofo [238, 239, 240, 241, 242, 246, 244, 245, 248, 247, 243, 251, 253, 280, 249, 250, 252, 254, 259, 256, 258, 276, 277, 255, 278, 257, 260, 281, 261, 262, 263, 267, 266, 264, 265, 269, 268, 270, 274, 273, 271, 272, 275, 279].

Following the notations of Sofo and Nimbran [279], we specify the

Definition 25.1 *We use the following notations defined by*

$$H_n = \sum_{k=1}^{n} \frac{1}{k}; \qquad H'_n = \sum_{k=1}^{n} \frac{(-1)^{k-1}}{k};$$

$$H_n^p = \left(\sum_{k=1}^{n} \frac{1}{k} \right)^p; \qquad H_n^{(p)} = \sum_{k=1}^{n} \frac{1}{k^p};$$

DOI: 10.1201/9781003174158-25

$$h_n = \sum_{k=1}^{n} \frac{1}{2k-1} = H_{2n} - \frac{1}{2}H_n; \qquad h'_n = \sum_{k=1}^{n} \frac{(-1)^{k-1}}{2k-1}.$$

We also note the well-known generating functions for $|z| < 1$

$$-\frac{\log(1-z)}{1-z} = \sum_{n=1}^{n} H_n z^n; \quad \text{and} \quad \frac{\log(1+z)}{1-z} = \sum_{n=1}^{n} H'_n z^n.$$

So then, Sofo and Nimbran [279] give numerous sums including

$$\sum_{n=1}^{\infty} \frac{H_n^{(2)}}{n^4} = \zeta^2(3) - \frac{1}{3}\zeta(6); \tag{25.1.1}$$

$$\sum_{n=1}^{\infty} \frac{H_n^{(4)}}{n^4} = \frac{13}{12}\zeta(8). \tag{25.1.2}$$

These are representative of the two general forms of identities

$$2\sum_{n=1}^{\infty} \frac{H_n^{(2p+1)}}{n^{2p+1}} = \zeta^2(2p+1) - \zeta(4p+2); \tag{25.1.3}$$

$$2\sum_{n=1}^{\infty} \frac{H_n^{(2p)}}{n^{2p}} = \left(1 - \frac{(4p)!B_{2p}^2}{2((2p)!)^2 B_{4p}}\right)\zeta(4p); \tag{25.1.4}$$

for positive integers p and where B_p are the signed Bernoulli numbers of the first kind.

Now, while the above cited papers by Sofo may be prospected to locate beautiful summations applicable to the VPV identities from earlier chapters, it is evident that the following generalizations will also be applicable.

Definition 25.2 *We use and extend the above notations from definition 25.1, so that for* $|z| < 1$

$$H_n(z) = \sum_{k=1}^{n} \frac{z^k}{k}; \qquad H'_n(z) = \sum_{k=1}^{n} \frac{(-z)^{k-1}}{k};$$

$$H_n^p(z) = \left(\sum_{k=1}^{n} \frac{z^k}{k}\right)^p; \qquad H_n^{(p)}(z) = \sum_{k=1}^{n} \frac{z^k}{k^p};$$

$$h_n(z) = \sum_{k=1}^{n} \frac{z^{2k-1}}{2k-1}; \qquad h'_n(z) = \sum_{k=1}^{n} \frac{(-z^{2k-1})^{k-1}}{2k-1}.$$

25.2 CREATING HARMONIC nD VPV IDENTITIES

The above definitions help us to state versions of the hyperpyramid theorem, and a general combinations of them. We create these new theorems as we please, from definitions 25.1 and 25.2, utilizing also polylogarithm functions

$$Li_s(z) = \sum_{k=1}^{\infty} \frac{z^k}{k^s} \tag{25.2.1}$$

wherever convenient. We recall from Chapter 21 in Section 21.4, the n dimensional square hyperpyramid VPV identity.

Theorem 25.1 ***The*** nD ***square hyperpyramid VPV identity.*** *If* $i = 1, 2, 3, ..., n$ *then for each* $x_i \in \mathbb{C}$ *such that* $|x_i| < 1$ *and* $b_i \in \mathbb{C}$ *such that* $\sum_{i=1}^{n} b_i = 1$,

$$\prod_{\substack{\gcd(a_1,a_2,...,a_n)=1 \\ a_1,a_2,...,a_{n-1}<a_n \\ a_1,a_2,...,a_n \geq 1}} \left(\frac{1}{1 - {x_1}^{a_1}{x_2}^{a_2}{x_3}^{a_3} \cdots {x_n}^{a_n}} \right)^{\frac{1}{{a_1}^{b_1}{a_2}^{b_2}{a_3}^{b_3} \cdots {a_n}^{b_n}}} \tag{25.2.2}$$

$$= \exp\left\{ \sum_{k=1}^{\infty} \prod_{i=1}^{n-1} \left(\sum_{j=1}^{k} \frac{{x_i}^j}{j^{b_i}} \right) \frac{{x_n}^k}{k^{b_n}} \right\}$$

$$= \exp\left\{ \sum_{k=1}^{\infty} \left(\sum_{j=1}^{k} \frac{{x_1}^j}{j^{b_1}} \right) \left(\sum_{j=1}^{k} \frac{{x_2}^j}{j^{b_2}} \right) \left(\sum_{j=1}^{k} \frac{{x_3}^j}{j^{b_3}} \right) \cdots \left(\sum_{j=1}^{k} \frac{{x_{n-1}}^j}{j^{b_{n-1}}} \right) \frac{{x_n}^k}{k^{b_n}} \right\}.$$

The 2D case of the theorem is: If $|yz|$ and $|z| < 1$ and $s + t = 1$ then,

$$\exp\left\{ \sum_{k=1}^{\infty} \frac{z^k}{k^t} \right\} \prod_{\substack{(a,b)=1 \\ a<b \\ a>0,b>1}} \left(\frac{1}{1 - y^a z^b} \right)^{\frac{1}{a^s b^t}} \tag{25.2.3}$$

$$= \exp\left\{ \frac{z^1}{1^t} + \left(1 + \frac{y^1}{1^s}\right) \frac{z^2}{2^t} + \left(1 + \frac{y^1}{1^s} + \frac{y^2}{2^s}\right) \frac{z^3}{3^t} + \cdots \right\}$$

The 3D case of theorem 25.1 is: if $|xyz|$, $|yz|$ and $|z| < 1$ and $r + s + t = 1$ then,

$$\exp\left\{ \sum_{k=1}^{\infty} \frac{z^k}{k^t} \right\} \prod_{\substack{(a,b,c)=1 \\ a,b<c \\ a,b>0,c>1}} \left(\frac{1}{1 - x^a y^b z^c} \right)^{\frac{1}{a^r b^s c^t}} \tag{25.2.4}$$

$$= \exp\left\{ \frac{z^1}{1^t} + \left(1 + \frac{x^1}{1^r}\right)\left(1 + \frac{y^1}{1^s}\right) \frac{z^2}{2^t} + \left(1 + \frac{x^1}{1^r} + \frac{x^2}{2^r}\right)\left(1 + \frac{y^1}{1^s} + \frac{y^2}{2^s}\right) \frac{z^3}{3^t} + \cdots \right\}$$

The 4D case of theorem 25.1 is: if $|wxyz|$, $|xyz|$, $|yz|$ and $|z|<1$ and $r+s+t+u=1$ then,

$$\exp\left\{\sum_{k=1}^{\infty}\frac{z^k}{k^u}\right\}\prod_{\substack{(a,b,c,d)=1\\ a,b,c<d\\ a,b,c>0,d>1}}\left(\frac{1}{1-w^a x^b y^c z^d}\right)^{\frac{1}{a^r b^s c^t d^u}}=\exp\{\mathrm{P}_3(r,w;s,x;t,y;u,z)\} \tag{25.2.5}$$

where P_3, is a 4D hyperpyramid function,

$$\mathrm{P}_3(r,w;s,x;t,y;u,z)=\frac{z^1}{1^u}+\left(1+\frac{w^1}{1^r}\right)\left(1+\frac{x^1}{1^s}\right)\left(1+\frac{y^1}{1^t}\right)\frac{z^2}{2^u}$$
$$+\left(1+\frac{w^1}{1^r}+\frac{w^2}{2^r}\right)\left(1+\frac{x^1}{1^s}+\frac{x^2}{2^s}\right)\left(1+\frac{y^1}{1^t}+\frac{y^2}{2^t}\right)\frac{z^3}{3^u}+\cdots$$

Using our Harmonic series notations, we may rewrite equations (25.2.3), (25.2.4), and (25.2.5) respectively as

If $|yz|$ and $|z|<1$ and $s+t=1$ then,

$$\prod_{\substack{(a,b)=1\\ a\leq b\\ a\geq 1,b\geq 1}}\left(\frac{1}{1-y^a z^b}\right)^{\frac{1}{a^s b^t}} \tag{25.2.6}$$

$$=\exp\left\{H_1^{(s)}(y)\frac{z^1}{1^t}+H_2^{(s)}(y)\frac{z^2}{2^t}+H_3^{(s)}(y)\frac{z^3}{3^t}+H_4^{(s)}(y)\frac{z^4}{4^t}+\cdots\right\}$$

If $|xyz|$, $|yz|$, and $|z|<1$ and $r+s+t=1$ then,

$$\prod_{\substack{(a,b,c)=1\\ a,b\leq c\\ a,b,c\geq 1}}\left(\frac{1}{1-x^a y^b z^c}\right)^{\frac{1}{a^r b^s c^t}} \tag{25.2.7}$$

$$=\exp\left\{H_1^{(r)}(x)H_1^{(s)}(y)\frac{z^1}{1^t}+H_2^{(r)}(x)H_2^{(s)}(y)\frac{z^2}{2^t}+H_3^{(r)}(x)H_3^{(s)}(y)\frac{z^3}{3^t}+\cdots\right\}$$

If $|wxyz|$, $|xyz|$, $|yz|$, and $|z|<1$ and $r+s+t+u=1$ then,

$$\prod_{\substack{(a,b,c,d)=1\\ a,b,c\leq d\\ a,b,c,d\geq 1}}\left(\frac{1}{1-w^a x^b y^c z^d}\right)^{\frac{1}{a^r b^s c^t d^u}}=\exp\{\mathrm{P}_3(r,w;s,x;t,y;u,z)\} \tag{25.2.8}$$

where P_3, is a 4D hyperpyramid function,

$$\mathrm{P}_3(r,w;s,x;t,y;u,z)=H_1^{(r)}(w)H_1^{(s)}(x)H_1^{(t)}(y)\frac{z^1}{1^u}+H_2^{(r)}(w)H_2^{(s)}(x)H_2^{(t)}(y)\frac{z^2}{2^u}$$
$$+H_3^{(r)}(w)H_3^{(s)}(x)H_3^{(t)}(y)\frac{z^3}{3^u}+H_4^{(r)}(w)H_4^{(s)}(x)H_4^{(t)}(y)\frac{z^4}{4^u}+\cdots$$

25.3 HARMONIC SUMS EQUAL TO RIEMANN ZETA COMBINATIONS

The following sums are in the literature and easily verified on-line or in Mathematica or Maple. These sums are applicable to the VPV identities, and they often represent boundary conditions for the variables and particular cases of Parametric Euler sums and Polylogarithm sums.

$$\sum_{n=1}^{\infty} \frac{H_n}{n2^n} = \frac{\pi^2}{12}, \tag{25.3.1}$$

$$\sum_{n=1}^{\infty} \frac{H_n^2}{(n+1)^2} = \frac{11\pi^4}{360}, \tag{25.3.2}$$

$$\sum_{n=1}^{\infty} \frac{H_n^2}{n^2} = \frac{17\pi^4}{360}, \tag{25.3.3}$$

$$\sum_{n=1}^{\infty} \frac{H_n}{n^3} = \frac{\pi^4}{72}, \tag{25.3.4}$$

$$\sum_{n=1}^{\infty} \frac{H_n}{n^4} = 3\zeta(5) - \frac{\pi^2}{6}\zeta(3), \tag{25.3.5}$$

$$\sum_{n=1}^{\infty} \frac{H_n}{n^5} = \frac{\pi^6}{540} - \frac{1}{2}\zeta(3)^2, \tag{25.3.6}$$

$$\sum_{n=1}^{\infty} \frac{H_n}{n^2} = 2\zeta(3), \tag{25.3.7}$$

$$\sum_{n=1}^{\infty} \frac{H_n^2}{n^3} = \frac{7}{2}\zeta(5) - \frac{\pi^2}{6}\zeta(3). \tag{25.3.8}$$

Many such summations like (25.3.1) to (25.3.8) are now known thanks to the work of the Borweins and their colleagues in papers between 1995 and 2016 like [64], [46], [67], [68], [63], and [42], in addition to the many papers by Sofo as referred to above. See also, our earlier Chapter 22 where Parametric Euler sums are introduced.

A useful transform is given by observing that sums involving k^{-n} can be re-expressed in terms of sums the form $(k+1)^{-n}$ via

$$\begin{aligned}\sum_{k=1}^{\infty}\left(\frac{1}{1^m}+\frac{1}{2^m}+\cdots+\frac{1}{k^m}\right)\frac{1}{k^n} &= \sum_{k=0}^{\infty}\left(\frac{1}{1^m}+\frac{1}{2^m}+\cdots+\frac{1}{(k+1)^m}\right)\frac{1}{(k+1)^n} \\ &= \sum_{k=1}^{\infty}\left(\frac{1}{1^m}+\frac{1}{2^m}+\cdots+\frac{1}{k^m}\right)\frac{1}{(k+1)^n}+\sum_{k=1}^{\infty}\frac{1}{(k)^{m+n}} \\ &\equiv \sigma_h(m,n)+\zeta(m+n)\end{aligned} \tag{25.3.9}$$

and

$$\sum_{k=1}^{\infty}\left(\frac{1}{1}+\frac{1}{2}+\cdots+\frac{1}{k}\right)^2\frac{1}{k^n} = s_h(2,n)+2s_h(1,n+1)\zeta(n+2); \tag{25.3.10}$$

both (25.3.9) and (25.3.10) utilizing the functions (see Bailey et al. [45])

$$s_h(m,n) = \sum_{k=1}^{\infty}\left(\frac{1}{1}+\frac{1}{2}+\cdots+\frac{1}{k}\right)^m \frac{1}{(k+1)^n}, \tag{25.3.11}$$

$$\sigma_h(m,n) = \sum_{k=1}^{\infty}\left(\frac{1}{1^m}+\frac{1}{2^m}+\cdots+\frac{1}{k^m}\right)\frac{1}{(k+1)^n}. \tag{25.3.12}$$

Note: The terminology and results pertaining to Harmonic Functions in this chapter can be applied easily to identities from Chapters 22 and 24. There are so many combinations of results obtainable this way, that we leave the reader to work through the connections between these two chapters and our present chapter.

25.4 HARMONIC SUMS AND POLYLOGARITHM INTEGRALS FOR USE IN VPV IDENTITIES

In this section, we simply state summations from the Sofo papers that would be applicable to the VPV hypequadrant and hyperpyramid variations. Often a numerical summation can be used on one or two (or three or four) of the variables in the identities given thus far in the book chapters to date. So, without further ado, we state some interesting summations due to Sofo. The results are not categorized nor presented according to any systematic principle, other than that they are adaptable to substitute into our known VPV identities. These are all easily verified in online calculating apps or Maple or Mathematica, but it's worth looking over derivations from evaluated integrals by Sofo. (see Sofo [274])

$$\sum_{n=1}^{\infty}\frac{H_n}{(2n+1)^2} = \frac{7}{4}\zeta(3) - \frac{1}{4}\pi^2\log 2;$$

$$\sum_{n=1}^{\infty}\frac{H_n}{(2n+1)^3} = \frac{1}{64}\pi^4 - \frac{7}{4}\zeta(3)\log 2;$$

$$\sum_{n=1}^{\infty}\frac{H_n}{(2n+1)^4} = -\frac{7}{32}\pi^2\zeta(3) + \frac{31}{8}\zeta(5) - \frac{1}{48}\pi^4\log 2;$$

$$\sum_{n=1}^{\infty}\frac{H_n}{(2n+1)^5} = -\frac{49}{64}\zeta(3)^2 - \frac{31}{16}\zeta(5)\log 2 + \frac{1}{384}\pi^6;$$

$$\sum_{n=1}^{\infty}\frac{H_n}{(2n+1)^6} = \frac{381}{64}\zeta(7) - \frac{93}{64}\zeta(5)\zeta(2) - \frac{105}{64}\zeta(4)\zeta(3) - \frac{63}{32}\zeta(6)\log 2;$$

$$\sum_{n=1}^{\infty}\frac{H_n}{(2n+1)^7} = -\frac{217}{128}\zeta(3)\zeta(5) - \frac{127\zeta(7)\log 2}{64} + \frac{17}{46080}\pi^8;$$

$$\sum_{n=1}^{\infty}\frac{H_n}{(2n+1)^8} = -\frac{7}{3840}\pi^6\zeta(3) - \frac{31}{1536}\pi^4\zeta(5) - \frac{127}{512}\pi^2\zeta(7) + \frac{511}{64}\zeta(9) - \frac{17}{80640}\pi^8\log 2;$$

$$\sum_{n=1}^{\infty} \frac{H_n}{(2n+1)^9} = -\frac{961}{1024}\zeta(5)^2 - \frac{889}{512}\zeta(3)\zeta(7) - \frac{511}{256}\zeta(9)\log 2 + \frac{31}{645120}\pi^{10}.$$

$$\sum_{n=1}^{\infty} \frac{H_n}{(2n+1)^{10}} = -\frac{17}{92160}\pi^8\zeta(3) - \frac{31}{1536}\pi^6\zeta(5) - \frac{127}{6144}\pi^4\zeta(7)$$

$$-\frac{511}{2048}\pi^2\zeta(9) + \frac{10235}{1024}\zeta(11) - \frac{31}{1451520}\pi^{10}\log 2.$$

The above nine summations will give us boundary cases when substituted in the VPV identities. If these nine sums were generalized to power series, according to the example generalization in going from definitions 25.1 to 25.2, we would have interesting multivariate identities from our substitutions. Worth mentioning is that also in the same Sofo paper of 2018 [274] for these summations, the following polylogarithm definite integrals, example cases are given now.

$$\int_0^1 (Li_3(-x))^2 dx = \frac{9}{16}\zeta(3)^2 + \frac{5}{8}\zeta(4) - \frac{3}{4}\zeta(2)\zeta(3)$$

$$+ (3\zeta(3) - 6\zeta(2) - 40)\log 2 + 4\zeta(2) + 12(\log 2)^2 + 20;$$

$$\int_0^1 Li_3(-x)Li_4(-x)dx = \frac{3}{4}\eta(4) + 2\zeta(3)\zeta(5) - \frac{49}{64}\zeta(6) - \frac{9}{16}\zeta(3)^2$$

$$+\frac{5}{4}\zeta(3) - \frac{15}{2}\zeta(2) + \left(10\zeta(2) - 6\zeta(3) + \frac{7}{4}\zeta(4) + 70\right)\log 2$$

$$-\frac{3}{2}\zeta(4) - \frac{17}{66}\zeta(5) + \frac{3}{2}\zeta(2)\zeta(3) - 20(\log 2)^2 - 35;$$

$$\int_0^1 \frac{Li_{2m}(-x)Li_{2m+1}(-x)}{x}dx = \frac{1}{2}\eta(2m+1)^2;$$

for non-negative integers m.

$$\int_0^1 \frac{Li_4(-x)Li_7(-x)}{x}dx = \eta(5)\eta(7) - \frac{1}{2}\eta(6)^2;$$

$$\int_0^1 \frac{Li_3(-x)Li_4(-x)}{x^2}dx = 2\zeta(5) - \frac{49}{64}\zeta(6) + \frac{23}{8}\zeta(4) - \frac{9}{16}\zeta(3)^2$$

$$+10\zeta(3) - \left(10\zeta(2) + 6\zeta(3) + \frac{7}{4}\zeta(4)\right)\log 2 + 10\zeta(2)$$

$$-\frac{3}{2}\zeta(2)\zeta(3) - 20(\log 2)^2 - \frac{21}{32}\zeta(3)\zeta(4);$$

$$\int_0^1 \left(\frac{Li_3(-x)}{x}\right)^2 dx = \frac{9}{8}\zeta(4) - \frac{9}{16}\zeta(3)^2 + 6\zeta(3) + 6\zeta(2)$$
$$-\frac{3}{4}\zeta(2)\zeta(3) - 6\zeta(2)\log 2 - 3\zeta(3)\log 2 - 12(\log 2)^2.$$

The above Harmonic summations and integrals of Polylogarithm functions can be adjusted or varied to substitute into VPV identities giving new particular exact cases. The reader of the present book will see that the above polylogarithm integral will apply to our VPV known identities too.

EXERCISES

25.1 *Harmonic Sum values.* If $n > 1$ show that

$$H_n = \sum_{k=1}^{n} \frac{1}{k} \qquad \text{(P25.1)}$$

is never an integer.

25.2 *Simple generating functions.* Show that the Cauchy product of $1/(1-z)$ and $-log(1-z)$ yields

$$-\frac{\log(1-z)}{1-z} = \sum_{n=1}^{\infty} H_n z^n, \qquad \text{(P25.2)}$$

while the Cauchy product of $1/(1-z)$ and $log(1+z)$ yields

$$\frac{\log(1+z)}{1-z} = \sum_{n=1}^{\infty} H'_n z^n, \qquad \text{(P25.3)}$$

25.3 *Euler's reflection formula.* The two-place function $\zeta(s,t)$ was first introduced by Euler. Prove the formula first given by Euler as the reflection formula

$$\zeta(s,t) + \zeta(t,s) = \zeta(s)\zeta(t) - \zeta(s+t), \quad \Re(s) > 1, \Re(t) > 1. \qquad \text{(P25.4)}$$

25.4 *Applying Harmonic Sum generating functions to VPV identities.* Apply the Harmonic Sums (25.3.3) to (25.3.8) to the formulas (25.2.6) and (25.2.7) to obtain new VPV infinite product identities in closed form.

Think About It...

Mathematics takes us still further from what is human, into the region of absolute necessity, to which not only the actual world, but every possible world, must conform.

Bertrand Russell
Philosopher, Mathematician, 1872–1970

GLOSSARY

Harmonic Number: The nth harmonic number is the sum of the reciprocals of the first n natural numbers:

$$H_n = 1 + \frac{1}{2} + \frac{1}{2} + \frac{1}{3} + \cdots + \frac{1}{n} = \sum_{k=1}^{n} \frac{1}{k}$$

Starting from $n = 1$, the sequence of harmonic numbers begins:

$$1, \frac{3}{2}, \frac{11}{6}, \frac{25}{12}, \frac{137}{60}, \frac{49}{20}, \frac{363}{140}, \frac{761}{280}, \frac{7129}{2520}, \frac{7381}{2520}, \cdots$$

Euler's constant γ: The constant γ is defined by

$$\gamma = \lim_{n\to\infty} \left(\sum_{k=1}^{n} \frac{1}{k} \right) - \log n = \lim_{n\to\infty} (H_n - \log n).$$

Riemann zeta function: The classical Riemann zeta function is defined by

$$\zeta(s) = \sum_{k=1}^{n} \frac{1}{k^s}, \quad \Re s > 1.$$

Euler sum: The double Euler sum is

$$\zeta(s,t) = \sum_{n>m>0} \frac{1}{n^s m^t} = \sum_{n=1}^{\infty} \frac{1}{n^s} \sum_{m=1}^{n-1} \frac{1}{m^t}.$$

Euler sum of depth N: The two-place function $\zeta(s,t)$ has a natural generalization

$$\zeta(s_1, s_2, s_3, \ldots, s_N) = \sum_{n_1>n_2>n_3>\cdots>n_N} \frac{1}{n_1^{s_1} n_2^{s_2} n_3^{s_3} \cdots n_N^{s_N}}, \tag{25.4.5}$$

which defines an Euler sum of depth N and weight $\sum_i s_i$. Some authors reverse the order of the variables.

FURTHER READING

BAILEY, D.H., BORWEIN, J.M. and GIRGENSOHN, R. Experimental Evaluation of Euler Sums. *Exper. Math.*, 3:17–30, 1994.

BORWEIN, D., BORWEIN, J.M. and BRADLEY, D.M. Parametric Euler sum identities. *J. Math. Anal. Appl.*, 316:328–338, 2006.

BORWEIN, J.M., GLASSER, M.L., MCPHEDRAN, R.C., WAN, J.G. and ZUCKER, I.J. *Lattice Sums then and now*, volume 150 of *Encyclopedia of Mathematics and Its Applications*. Cambridge University Press, 2013.

CAMPBELL, G.B. Multiplicative functions over Riemann zeta function products. *J. Ramanujan Soc.*, 7(1):52–63, 1992.

CAMPBELL, G.B. Dirichlet summations and products over primes. *Int. J. Math. Math. Sci.*, 16(2):359–372, 1993.

CAMPBELL, G.B. A new class of infinite products, and Euler's totient. *Int. J. Math. and Math. Sci.*, 17(3):417–422, 1994.

CAMPBELL, G.B. Infinite products over visible lattice points. *Int. J. Math. and Math. Sci.*, 17(4):637–654, 1994.

CAMPBELL, G.B. *Combinatorial identities in number theory related to q-series and arithmetical functions.* Doctor of philosophy thesis, School of Mathematical Sciences, The Australian National University, October 1997.

CAMPBELL, G.B. A closer look at some new identities. *Int. J. Math. and Math. Sci.*, 21(3):581–586, 1998.

CAMPBELL, G.B. Infinite products over hyperpyramid lattices. *Int. J. Math. and Math. Sci.*, 23(4):271–277, 2000.

CAMPBELL, G.B. Some n-space q-binomial theorem extensions and similar identities. arxiv preprint, June 2019. arXiv:1906.07526v1 [math.NT]: https://arxiv.org/abs/1906.07526.

CAMPBELL, G.B. Continued Fractions for partition generating functions. arxiv preprint, Jan 2023. arXiv:2301.12945 [math.CO], https://arxiv.org/abs/2301.12945.

CAMPBELL, G.B. Vector Partition Identities for 2D, 3D and nD Lattices. arxiv preprint, Feb 2023. arXiv:10.48550/ARXIV.2302.01091 [math.NT], https://arxiv.org/abs/2302.01091.

CAMPBELL, G.B. Visible Point Partition Identities for Polylogarithms, and Parametric Euler Sums. arxiv preprint, Jun 2023. arXiv:2306.02241v1 [math.CO], https://arxiv.org/abs/2306.02241.

CAMPBELL, G.B. Visible Point Vector Partition Identities for Hyperpyramid Lattices. arxiv preprint, Sep 2023. arXiv:2309.16094v1 [math.CO], https://arxiv.org/abs/2309.16094.

SOFO, A. Integral identities for sums. *Math. Commun.*, 13:303–309, 2008.

SOFO, A. Double Integral Representation of Sums. *Int. J. Math. Anal.*, 3(10):483–489, 2009.

SOFO, A. Harmonic numbers and double binomial coefficients. *Integral Transforms Spec. Funct.*, 20(11):847–857, 2009.

SOFO, A. Integral forms of sums associated with harmonic numbers. *Appl. Math. Comput.*, 207(2):365–372, 2009.

SOFO, A. Series associated with Polygamma functions. *Scientia Series A, Mathematical Sciences*, pages 95–104, 2009.

SOFO, A. Closed Form Representations of Harmonic Sums. *Hacettepe J. Math. Stat.*, 39(2):255–263, 2010.

SOFO, A. Double Integral Identities and Zeta Functions. *Int. J. Contemp. Math. Sci.*, 5(17-20):869–876, 2010.

SOFO, A. Harmonic sums and integral representations. *J. Appl. Anal.*, 16(2):265–277, 2010. pp1869–6082 (online).

SOFO, A. Integrals and Polygamma Representations for Binomial Sums. *Journal of Integer Sequences*, 13:1–12, 2010.

SOFO, A. Sums in Terms of Polygamma Functions. *Bull. Math. Anal. Appl.*, 2(3):40–52, 2010.

SOFO, A. Triple integral identities and Zeta functions. *Appl. Anal. Discret. Math.*, 4(2):347–360, 2010.

SOFO, A. Harmonic Number Sums in Closed Form. *Math. Commun.* 16(2):335–345, 2011.

SOFO, A. Harmonic Number Sums in Higher Powers. *J. Math. Anal.*, 2(2):15–22, 2011.

SOFO, A. Harmonic Numbers and Cubed Binominal Coefficients. *Int. J. Comb*, 2011.

SOFO, A. Integral Identities for Rational Series Involving Binomial Coefficients. *Bull. Malays. Math. Sci. Soc.*, 34(3):631–637, 2011.

SOFO, A. Summation Formula Involving Harmonic Numbers. *Anal. Math.*, 37(1):51–64, 2011.

SOFO, A. Euler-related sums. *Math. Sci.*, 6(10), 2012.

SOFO, A. Finite Sums in Pascal's Triangle. *Fibonacci Q.*, 50(4):337–345, 2012.

SOFO, A. Harmonic numbers of order two. *Miscolc Math. Notes*, 13(2):499–514, 2012.

SOFO, A. Mixed binomial sum identities. *Integral Transforms Spec. Funct.*, 24(3):187–200, 2012.

SOFO, A. New classes of harmonic number identities. *J. Integer Seq.*, 15, 2012.

SOFO, A. Reciprocal power sums. *Integers*, 12, 2012.

SOFO, A. Generalized hypergeometric function identities at argument ±1. *Integral Transforms Spec. Funct.*, 25(11):909–921, 2014.

SOFO, A. Shifted harmonic sums of order two. *Commun. Korean Math. Soc.*, 29(2):239–255, 2014.

SOFO, A. New families of alternating harmonic number sums. *Tbilisi Math. J.*, 8(2):195–209, 2015.

SOFO, A. Quadratic alternating harmonic number sums. *J. Number Theory*, 154:144–159, 2015.

SOFO, A. Harmonic numbers at half integer and binomial squared sums. *Honam Math. J.*, 38(2):279–294, 2016.

SOFO, A. Harmonic numbers at half integer values. *Integral Transforms Spec. Funct.*, 27(6):430–442, 2016.

SOFO, A. Identities for Alternating Inverse Squared Binomial and Harmonic Number Sums. *Mediterr. J. Math.*, 13(4):1407–1418, 2016.

SOFO, A. Integrals of logarithmic and hypergeometric functions. *Commun. Math.*, 24:7–22, 2016.

SOFO, A. Polylogarithmic connections with Euler sums. *Sarajevo J. Math.*, 12:17–32, 2016.

SOFO, A. Sums of quadratic half integer harmonic numbers of alternating type. *J. Classical Anal.*, 8(2):99–111, 2016.

SOFO, A. A master integral in four parameters. *J. Math. Anal. Appl.*, 448:81–92, 2017.

SOFO, A. Alternating cubic Euler sums with binomial squared terms. *Int. J. Number Theory,* 14(05):1357–1374, 2018.

SOFO, A. Evaluation of integrals with hypergeometric and logarithmic functions. *Open Math.*, 16(1):63–74, 2018.

SOFO, A. General order Euler sums with multiple argument. *J. Number Theory,* 189:255–271, 2018.

SOFO, A. Integrals of polylogarithmic functions with negative argument. *Acta Univ. Sapientiae Math.*, 10:347–367, 2018.

SOFO, A. Families of Integrals of Polylogarithmic Functions. *Researchgate*, 7, 2019. Mathematics 10.3390/math7020143.

SOFO, A. and CIMADEVILLA VILLACORTA, J. New identities for the Arctan function. *J. Math. Anal.*, 3(3):1–10, 2012.

SOFO, A. and CVIJOVIC, D. Extensions of Euler harmonic sums. *Appl. Anal. Discret. Math.*, 6:317–328, 2012. 2.

SOFO, A. and HASSANI, M. Quadratic harmonic number sums. *Appl. Anal. Discret. Math.*, 12, 2012.

SOFO, A. and NIMBRAN, A. Euler sums and integral connections. *Mathematics*, 7:833, 09 2019.

SOFO, A. and SRIVASTAVA, H.M. Identities for the Harmonic Numbers and Binomial Coefficients. *The Ramanujan J.*, 25(1):93–113, 2011.

SOFO, A. and SRIVASTAVA, H.M. A family of shifted harmonic sums. *The Ramanujan J.*, 37(1):89–108, 2014.

CHAPTER 26

The Ramanujan Trigonometric Function and Visible Point Identities

26.1 EXTENDING THE RAMANUJAN TRIGONOMETRIC FUNCTION

WE DEFINE AN EXTENSION of Ramanujan's trigonometric function to arbitrary dimensions, and give the Dirichlet series generating function. This links directly to visible point vector (VPV) identities in the author's papers, and possibly to *lattice sums* in Physics and Chemistry presented by Borwein et al. [69] or those given by Baake et. al. [37, 39, 40, 38] leading to aperiodic tilings. New generating functions and summations are given here, generalizing the Ramanujan function, Euler totient, and the Jordan totient functions, based on visible lattice point ideas. The literature on these Ramanujan Sums goes from Ramanujan's 1918 paper [217, pp. 179–199 of his Collected Papers] up to the 2023 work by Toth [295].

The function made well known by Ramanujan [217] for positive integers k, and n,

$$c_k(n) = \sum_{\substack{(j,k)=1 \\ 0<j<k}} \cos\left(\frac{2\pi nj}{k}\right) = \sum_{\substack{(j,k)=1 \\ 0<j<k}} e^{\frac{2\pi inj}{k}} \tag{26.1.1}$$

has been widely studied and referenced. See for example Hardy [148], Hardy and Wright [153], Sivaramakrishnan [232], and Apostol [32]. In this chapter we extend the function as follows

DOI: 10.1201/9781003174158-26

$$c_k(n_1,n_2) = \sum_{\substack{(j_1,j_2,k)=1\\ 0\le j_1<k\\ 0\le j_2<k\\ j_1+j_2\neq 0}} \cos\left(\frac{2\pi(j_1n_1+j_2n_2)}{k}\right) = \sum_{\substack{(j_1,j_2,k)=1\\ 0\le j_1<k\\ 0\le j_2<k\\ j_1+j_2\neq 0}} \exp\left(\frac{2\pi i(j_1n_1+j_2n_2}{k}\right), \tag{26.1.2}$$

$$\begin{aligned} c_k(n_1,n_2,n_3) &= \sum_{\substack{(j_1,j_2,j_3,k)=1\\ 0\le j_1<k\\ 0\le j_2<k\\ 0\le j_3<k\\ j_1+j_2+j_3\neq 0}} \cos\left(\frac{2\pi(j_1n_1+j_2n_2+j_3n_3)}{k}\right) \\ &= \sum_{\substack{(j_1,j_2,j_3,k)=1\\ 0\le j_1<k\\ 0\le j_2<k\\ 0\le j_3<k\\ j_1+j_2+j_3\neq 0}} \exp\left(\frac{2\pi i(j_1n_1+j_2n_2+j_3n_3}{k}\right), \end{aligned} \tag{26.1.3}$$

and generally

$$\begin{aligned} c_k(n_1,n_2,\ldots,n_m) &= \sum_{\substack{(j_1,j_2,\ldots,j_m,k)=1\\ 0\le j_1<k\\ \vdots\\ 0\le j_m<k\\ j_1+j_2+\ldots+j_m\neq 0}} \cos\left(\frac{2\pi(j_1n_1+j_2n_2+\ldots+j_mn_m)}{k}\right) \\ &= \sum_{\substack{(j_1,j_2,\ldots,j_m,k)=1\\ 0\le j_1<k\\ \vdots\\ 0\le j_m<k\\ j_1+j_2+\ldots+j_m\neq 0}} \exp\left(\frac{2\pi i(j_1n_1+j_2n_2+\ldots+j_mn_m}{k}\right) \end{aligned} \tag{26.1.4}$$

These functions are very natural to consider as their generating Dirichlet series exhibit a simplicity unexpected of a function requiring so many parameters. On the other hand, there seems an inherently natural aspect to considering the set of visible lattice points in any radial region of multidimensional Euclidean space. We shall see in the following sections that $c_k(n_1,n_2,\ldots,n_m)$ always reduces easily to a one dimensional $c_k(n)$ type function.

26.2 DIRICHLET SERIES GENERATING FUNCTIONS

In this section, we start with a definition then give the generating function for the new Ramanujan function, and some consequences.

Definition 26.1 *Let us define the operator*

$$\sum_{\substack{(j_1 j_2,\ldots,j_m,k)=1 \\ 0\leq j_1<k \\ \vdots \\ 0\leq j_m<k \\ j_1+j_2+\ldots+j_m\neq 0}} := \sum_{m\ k}, \tag{26.2.1}$$

so that

$$c_k(n_1,n_2,\ldots,n_m) = \sum_{m\ k} \cos\left(\frac{2\pi(j_1n_1+j_2n_2+\ldots+j_mn_m)}{k}\right). \tag{26.2.2}$$

It is well known that [153, pp. 139-143]

$$\frac{\sigma_{-s}(n)}{\zeta(s+1)} = \sum_{k=1}^{\infty} \frac{c_k(n)}{k^{s+1}}, \qquad \mathcal{R}s > 0, \tag{26.2.3}$$

where the left side contains the sum of $-s$th powers of the divisors of n, and the Riemann zeta function. If the Riemann hypothesis is true then (26.2.3) is valid for $\mathcal{R}s > -\frac{1}{2}$. We show that

Theorem 26.1 *If m is a positive integer then,*

$$\frac{\sigma_{-s-1+m}((n_1,n_2,\ldots,n_m))}{\zeta(s+1)} = \sum_{k=1}^{\infty} \frac{c_k(n_1,n_2,\ldots,n_m)}{k^{s+1}}, \qquad \mathcal{R}s > 0, \tag{26.2.4}$$

$$\exp\left(\sum_{k|(n_1,\ldots,n_m)} k^{m-1} z^k\right) = \prod_{k=1}^{\infty} (1-z^k)^{-c_k(n_1,\ldots,n_m)/k}, \qquad |z| < 1. \tag{26.2.5}$$

Like (26.2.3), (26.2.4) is probably also true for $\mathcal{R}s > -\frac{1}{2}$. It is easy to prove many results similar to (26.2.4) and (26.2.5). We shall prove theorem 26.1 in the next section. Features of the Ramanujan function such as multiplicativity generalize in the new versions.

Corollary 26.1 $c_k(n_1,n_2,\ldots,n_m)$ *is multiplicative in* k.

The paper by Apostol and Zuckermann [33] deals with the functional equation $F(mn)F((m,n)) = F(m)F(n)f((m,n))$ which many of the common arithmetical functions satisfy, and indeed so do our new functions $c_k(n_1,n_2,\ldots,n_m)$, in the form

$$c_{mn}(n_1,n_2,\ldots,n_\lambda)c_{(m,n)}(n_1,n_2,\ldots,n_\lambda)$$
$$= c_m(n_1,n_2,\ldots,n_\lambda)\sum_{k=1}^{\infty} \frac{c_k(n_1,n_2,\ldots,n_m)}{k} = 0f((m,n)). \tag{26.2.6}$$

Corollary 26.2 *If m is a positive integer,*

$$\sum_{k=1}^{\infty} \frac{c_k(n_1, n_2, \ldots, n_m)}{k} = 0. \tag{26.2.7}$$

Following the original paper by Ramanujan [217] on $c_n(k)$ this comes from the fact that the *sum of powers of divisors* function on left of (26.2.4) is a finite Dirichlet series, and at $s = 0$ in that equation depends on the well known convergence of

$$\sum_{k=1}^{\infty} \frac{\mu(k)}{k} = 0. \tag{26.2.8}$$

26.3 PROOF OF THEOREM 26.1

Theorem 26.1 depends on lemmas 26.1 and 26.2. The latter is an analytic restatement of the former.

Lemma 26.1 *(see Campbell [87]) Consider an infinite region raying out of the origin in any Euclidean vector n-dimensional space. The set of all lattice point vectors from the origin in that region is precisely the set of positive integer multiples of the visible point vectors (VPV's) in that region.*

This lemma underlies all of the author's cited papers [90] to [91] and is fundamental in obtaining VPV identities. A clearer picture of lemma 26.1 is gained from grid (26.3.1). It shows part of the 2D first quadrant visible-from-origin lattice points as the dots. All other lattice points in the part quadrant are shown as crosses. By lemma 26.1, the coordinates of the crosses are the positive integer (scalar) multiples of the coordinates of the dots; this being true in fact for the dots and crosses enclosed in any radial region from the origin. In the study of quasicrystals such as in Baake et al. (see Baake et al. [39], [40], [37]) and in other papers from Baake's research group, diffraction patterns and sectioning give rise to Fourier inversions of lattice patterns such as the visible points depicted here. There are some remarkable tiling patterns resulting from Dirichlet series generating functions hinting that the topics of our current chapter may well find expression in the new research areas associated with quasicrystals. (see also Campbell [92])

2D first quadrant visible point *bullets* and invisible from origin *circles*

Y									
8	●	○	●	○	●	○	●	○	
7	●	●	●	●	●	●	○	●	
6	●	○	○	○	●	○	●	○	
5	●	●	●	●	○	●	●	●	
4	●	○	●	○	●	○	●	○	
3	●	●	○	●	●	○	●	●	
2	●	○	●	○	●	○	●	○	
1	●	●	●	●	●	●	●	●	
0	1	2	3	4	5	6	7	8	X

(26.3.1)

An analytic restatement of lemma 26.1 is given in the

Lemma 26.2 *If* (a_k) *is an arbitrary sequence and* q_i *are variables chosen so that the following functions are all defined,*

$$\sum_{k=1}^{\infty}\left(a_k\prod_{h=1}^{m}\frac{1-q_h}{1-q_h^{1/k}}\right)=S_1+\sum_{k=2}^{\infty}\left(S_k\sum_{m\ k}(q_1^{j_1}\ldots q_m^{j_m})^{1/k}\right)\ \textit{where}\ S_k=\sum_{j=1}^{\infty}a_{jk}. \tag{26.3.2}$$

Proof of Lemma 26.2. The case with $m=1$ was given in Campbell [90] to [91], leading in [84] to proofs of (26.2.3) and the case $m=1$ of (26.2.5). The cases of (26.3.2) with $m=1,2,3$, are as follows, and correspond to lemma 26.1 in 1D, 2D, and 3D space:-

$$\sum_{k=1}^{\infty}a_k\frac{1-q_1}{1-q_1^{1/k}}=S_1+\sum_{k=2}^{\infty}\left(S_k\sum_{1\ k}q_1^{j_1/k}\right), \tag{26.3.3}$$

$$\sum_{k=1}^{\infty}a_k\frac{1-q_1}{1-q_1^{1/k}}\frac{1-q_2}{1-q_2^{1/k}}=S_1+\sum_{k=2}^{\infty}\left(S_k\sum_{2\ k}(q_1^{j_1}q_2^{j_2})^{1/k}\right), \tag{26.3.4}$$

$$\sum_{k=1}^{\infty}a_k\frac{1-q_1}{1-q_1^{1/k}}\frac{1-q_2}{1-q_2^{1/k}}\frac{1-q_3}{1-q_3^{1/k}}=S_1+\sum_{k=2}^{\infty}\left(S_k\sum_{3\ k}(q_1^{j_1}q_2^{j_2}q_3^{j_3})^{1/k}\right). \tag{26.3.5}$$

The proof of each of (26.3.3) to (26.3.5) is almost *a priori* when seen as interpreting lemma 26.1 if it were depicted in a generalized 3D or 4D version of the grid (26.3.1). To spell this out in the 2D case, we see that the left side of (26.3.2) is

$$\sum_{k=1}^{\infty}\left(a_k\prod_{h=1}^{m}\frac{1-q_h}{1-q_h^{1/k}}\right)$$

$$=\sum_{k=1}^{\infty}\left(a_k\prod_{h=1}^{m}\sum_{j=0}^{k-1}q_h^{j/k}\right) \tag{26.3.6}$$

$$=\sum_{k=1}^{\infty}\left(a_k\sum_{j=1}^{k-1}(q_1^{j_1}\cdots q_m^{1/k})\right) \tag{26.3.7}$$

$$=S_1+\sum_{k=2}^{\infty}\left(S_k\sum_{m\ k}(q_1^{j_1}\ldots q_m^{1/k})\right)\ \text{where}\ S_k=\sum_{j=1}^{\infty}a_{jk}. \tag{26.3.8}$$

The last step here depends on application of lemma 26.1 in m-space. ■

The VPV identities (Visible Point Vector) published by the author, such as for example in Campbell [87] or [91] are all particular cases of lemma 26.2.

Proof of Theorem 26.1. In lemma 26.2 set $q_k=exp(2\pi in_k)$, then set a_k equal to respectively k^{-s}, z^k. The resulting identities are (26.2.4) and (26.2.5). ■

26.4 A NEW JORDAN TOTIENT GENERATING FUNCTION, AND SOME RELATED RESULTS

The consequent identities from (26.3.3) were explored in some detail in the author's papers [90] to [91]. Borwein and Borwein [66, ex 10, pp.327] had some infinite products resembling those of the author, and some of the Dirichlet series implied from (26.3.3) were given in Lossers [189], Sivaramakrishnan [232], and Chandrasekharan [106]. Most books on arithmetical functions since Ramanujan's original paper [217] have included a treatment of the function (26.1.1) and give at least the result (26.2.3) cited above. However, (26.2.4) and (26.2.5) are not in the literature, and the identities (26.3.4) and (26.3.5) for example, lead us to a class of Dirichlet series and infinite products not hitherto considered. These results are near the surface and seem worth closer study. Lemma 26.2 is based on a summation over lattice points in an $m+1$ dimensional hyperpyramid. (see Campbell [91]) The methods for *lattice sums* given in Ninham et. al. [204] and Glasser and Zucker [138] using Möbius and Mellin inversions are also applicable to results from this lemma. This has been given a more systematic treatment by Baake et. al. (see [39] to [38]) and his colleagues in their investigations of aperiodic quasicrystal lattice structures and associated Fourier inversions.

Since some examples of (26.3.3) have been examined and led to new results in several of the author's papers, (see for example Campbell [84, 85] and [87] to [93]) we now consider the next simplest equation (26.3.4) with q_1, q_2 replaced respectively by e^{xz}, e^{yz}; and it too should lead to new identities. We have therefore the analysis

$$\sum_{k=1}^{\infty} a_k \left(\sum_{A=0}^{k-1} \sum_{B=0}^{k-1} e^{(Ax+By)z/k} \right) = S_1 + \sum_{k=2}^{\infty} \left(S_k \sum_{2\ k} e^{(j_1x+j_2y)z/k} \right), \tag{26.4.1}$$

$$\Leftrightarrow \sum_{k=1}^{\infty} a_k \left(\sum_{A=0}^{k-1} \sum_{B=0}^{k-1} \sum_{C=0}^{k-1} \frac{(Ax+By)^c z^c}{C! k^c} \right) = S_1 + \sum_{k=2}^{\infty} \left(S_k \sum_{2\ k} \sum_{C=0}^{\infty} \frac{(j_1x+j_2y)^c z^c}{C! k^c} \right), \tag{26.4.2}$$

$$\Leftrightarrow \left(\sum_{k=1}^{\infty} k^2 a_k \right) + \sum_{C=0}^{\infty} \sum_{k=1}^{\infty} \sum_{\substack{0 \le A < k \\ 0 \le B < k \\ A+B \neq 0}} \frac{(Ax+By)^c a_k z^c}{k^c C!} \tag{26.4.3}$$

$$= S_1 + \sum_{C=0}^{\infty} \sum_{k=2}^{\infty} \left(S_k \sum_{2\ k} \frac{(j_1x+j_2y)^c}{k^c} \right) \frac{z^c}{C!},$$

where, equating coefficients of like powers of z gives us the summation formulas

$$\sum_{k=1}^{\infty} k^2 a_k = S_1 + \sum_{k=2}^{\infty} S_k \sum_{2\ k} 1, \tag{26.4.4}$$

$$\sum_{k=1}^{\infty} \sum_{\substack{0 \le A < k \\ 0 \le B < k \\ A+B \neq 0}} \frac{(Ax+By)a_k}{k} = \sum_{k=2}^{\infty} \left(\frac{S_k}{k} \sum_{2\ k} (j_1x+j_2y) \right), \tag{26.4.5}$$

$$\sum_{k=1}^{\infty}\sum_{\substack{0\le A<k\\0\le B<k\\A+B\ne 0}}\frac{(Ax+By)^2 a_k}{k^2}=\sum_{k=2}^{\infty}\left(\frac{S_k}{k^2}\sum_{2\ k}(j_1x+j_2y)^2\right),\tag{26.4.6}$$

and so on. If this process from (26.4.1) to (26.4.6) is followed in an analogous manner for the general series (26.3.2), we have

Corollary 26.3 *For positive integers n,*

$$\sum_{k=1}^{\infty}\sum_{\substack{0\le A<k\\0\le B<k\\A+B\ne 0}}\frac{(Ax+By)^n a_k}{k^n}=\sum_{k=2}^{\infty}\left(\frac{S_k}{k^n}\sum_{2\ k}(j_1x+j_2y)^n\right).\tag{26.4.7}$$

In (26.4.4) the term

$$\sum_{2\ k}1=\sum_{2\ k}\left(\frac{j_1+j_2}{k}\right)^0=\varphi_0(2;k)$$

is the number of non-negative integer solutions of $(a,b,k)=1$ for a and b both less than k with $a+b\neq 0$. Suggestive from this, the following is a multidimensional totient function quite distinct from those of section 26.1.

Definition 26.2 *For all positive integers k, m, and $t+1$, let*

$$\varphi_t(m;k)=\sum_{m\ k}\left(\frac{j_1+\ldots+j_m}{k}\right)^t.\tag{26.4.8}$$

If t is zero and $m=1$ we would have the Euler totient function. If t is zero and $m=2$ we have again the function $\varphi_0(2;k)$. We shall see soon that $\varphi_0(m;k)$ is the Jordan totient function $J_m(k)$. The Dirichlet series generating functions for $\varphi_t(m;k)$ are found in a similar fashion to our derivation of corollary 26.3, and setting $x=y=1$. If this analogous process is followed we arrive at the

Theorem 26.2 *For positive integers t and suitably chosen functions a_i,*

$$\sum_{k=1}^{\infty}k^m a_k=S_1+\sum_{k=2}^{\infty}S_k\,\varphi_t(m;k),\tag{26.4.9}$$

whilst for $t=0$ we have

$$\sum_{k=1}^{\infty}k^m a_k=S_1+\sum_{k=2}^{\infty}S_k\,\varphi_0(m;k).\tag{26.4.10}$$

Therefore, if $a^k=k^{-z}$ we have

Corollary 26.4

$$\frac{\zeta(s-m)}{\zeta(s)}=1+\sum_{k=2}^{\infty}\frac{\varphi_0(m;k)}{k^s},\quad \mathcal{R}s>1.\tag{26.4.11}$$

So it is clear that $\varphi_0(m;k) = J_m(k)$, the Jordan totient function. This function is well known as the number of ordered sets of k elements chosen from a complete residue system (mod r) such that the greatest common divisor of each set is prime to r. It is also well known (see Sivaramakrishnan [232]), and easily deduced from writing left side of (26.4.11) as a product over primes, that

$$J_m(k) = k^m \prod_{p|k} (1-p^{-m}). \tag{26.4.12}$$

This function is therefore perhaps more simply defined as the number of non-negative integer solutions of $(a_1, a_2, \ldots, a_n, k) = 1$ with all of the a's less than k, and $a_1 + a_2 + \cdots + a_n \neq 0$. If in (26.4.9) we let $a^k = z^k k^{-1}$ then

Corollary 26.5

$$\frac{1}{1-z} \prod_{k=2} \left(\frac{1}{1-z^k} \right)^{J_m(k)/k} = \exp\left\{ \sum_{k=1}^{\infty} k^{m-1} z^k \right\}, \quad |z| < 1. \tag{26.4.13}$$

This is new, and related to many results in the author's papers cited, and also to a product in Borwein and Borwein [66, exercise 10, pp.327]. The right side of (26.4.13) is easily reduced by using the well known

$$\sum_{k=0}^{\infty} k^{m-1} z^k = \sum_{j=0}^{m-1} S_{m-1}^{(j)} z^j \frac{d^j}{dz^j} \left\{ \frac{1-z^n}{1-z} \right\}, \quad z \neq 1, \tag{26.4.14}$$

where $S_{m-1}^{(j)}$ are the Stirling numbers of the second kind. Hence we have the

Theorem 26.3 *If m is a positive integer*

$$\frac{1}{1-z} \prod_{k=2}^{\infty} \left(\frac{1}{1-z^k} \right)^{J_m(k)/k} = \exp\left\{ \sum_{j=0}^{m-1} S_{m-1}^{(j)} \frac{j! z^j}{(1-z)^{j+1}} \right\}, \quad |z| < 1. \tag{26.4.15}$$

The Stirling numbers of the second kind are of importance in combinatorial number theory. The Stirling number $S_m^{(n)}$ is the number of ways of partitioning a set of m elements into n non-empty sets. Formula (26.4.15) corresponds to a limiting case of the hyperpyramid lattice VPV identity first given in Campbell [87],

$$\prod_{\substack{(a_1,a_2,\ldots,a_n)=1 \\ a_1,\ldots,a_{n-1}<a_n \\ a_1,\ldots,a_{n-1}\geq 0 \\ a_n \geq 1}} \left(\frac{1}{1 - x_1^{a_1} x_2^{a_2} \ldots x_n^{a_n}} \right)^{\frac{1}{a_1^{b_1} a_2^{b_2} \ldots a_n^{b_n}}} = \exp\left\{ \sum_{k=1}^{\infty} \prod_{i=1}^{n-1} \left(\sum_{j=1}^{k-1} \frac{x_i^j}{j^{b_i}} \right) \frac{x_n^k}{k^{b_n}} \right\} \tag{26.4.16}$$

where for $i = 1, 2, \ldots, n$; $|x_i| < 1$, b_i is a complex number with $\sum_{i=1}^{n} b_i = 1$.

26.5 FURTHER MULTIDIMENSIONAL FORMULAS

In Campbell [85] it was shown that

Theorem 26.4 *If* (a_i), (b_i) *are arbitrary sequences chosen so that, together with choice of* x, *the following functions are defined then*

$$\sum_{k=1}^{n} a_k \frac{1-\exp(b_k x)}{1-\exp(b_k x/k)} = \left(\sum_{k=1}^{n} a_k\right) + \sum_{m=2}^{n} \sum_{k=1}^{[n/m]} \sum_{\substack{(j,m)=1 \\ 0<j<k}} a_{mk} \exp(b_{mk} j x/k), \qquad (26.5.1)$$

where $[n]$ *denotes the greatest integer in* n.

This theorem leads to new results on VPV identities, forming sums, or products over the lattice point values namely:

a) "*companion identities*", where interrelated identities divide up space radially in "*raying from origin*" infinitely extended sector regions. These identities have properties that map onto each other, such as for example, the *half first quadrant* above line $y=x$, mapping to *half first quadrant* below $y=x$. These are often infinite products, but they may be summations taken over the lattice points.

b) new Dirichlet series. These include generating functions for Ramanujan type trigonometric functions $c_k(n)$.

c) new Jacobi theta function types or extended types of identities.

d) new Diophantine equations.

In this section we give further results like (26.5.1) and some corresponding results as a consequence. Theorem 26.4 is a 2D VPV summation formula, and it was seen in Campbell [90] to [91] and can be inferred from earlier chapters of our current book that VPV identities can arise in any dimension. This is true in any particular radial region of space with apex at the origin. We now give the 3D version of theorem 26.4.

Theorem 26.5 *If* (a_i), (b_i), (c_i), *are each arbitrary sequences chosen so that, together with choice of* x, *the following functions are defined then*

$$\begin{aligned}
&\sum_{k=1}^{n} a_k \frac{1-\exp(b_k x)}{1-\exp(b_k x/k)} \frac{1-\exp(c_k x)}{1-\exp(c_k x/k)} \\
&= \left(\sum_{k=1}^{n} a_k\right) + \sum_{k=1}^{[n/2]} \sum_{2\ 2} a_{2k} \exp\left(b_{2k} j_1 + c_{2k} j_2)\frac{x}{2}\right) \\
&\quad + \sum_{k=1}^{[n/3]} \sum_{2\ 3} a_{3k} \exp\left((b_{3k} j_1 + c_{3k} j_2)\frac{x}{3}\right) \qquad (26.5.2) \\
&\quad + \sum_{k=1}^{[n/4]} \sum_{2\ 4} a_{4k} \exp\left((b_{4k} j_1 + c_{4k} j_2)\frac{x}{4}\right) \\
&\quad \vdots
\end{aligned}$$

$$+\sum_{k=1}^{[n/n]}\sum_{2\ n} a_{nk}\exp\left((b_{nk}j_1+c_{nk}j_2)\frac{x}{n}\right)$$

where $[n]$ *denotes the greatest integer in* n.

Most of the 3D VPV infinite products found in Campbell [84, 85, 87] are also corollaries of theorem 26.5. For example, if $n\to\infty, a_k = z^k/k, b_k = (k/x)logx, c_k = (k/x)logy$ we have the

Corollary 26.6 *If* a *and* b *are non-negative integers and* c *positive integers such that* $(a,b,c)=1$ *then*

$$\prod_{\substack{(a,b,c)=1\\ a,b<c;a,b\geq 0;c\geq 1}} (1-x^a y^b z^c)^{-1/c} = \left(\frac{(1-xz)(1-yz)}{(1-z)(1-xyz)}\right)^{\frac{1}{(1-x)(1-y)}} \tag{26.5.3}$$

valid for each of $|x|,|xz|,|yz|,|xyz|<1$.

This identity was given in [88] as one of four "companion identities" dividing up the first hyperquadrant in 3D space into three radial regions and the union region of these. Of course the simplicity of statement of (26.5.3) is slightly disguised in the notation of theorem 26.5. Also, the cases of theorem 26.5 with $m=2$ are trivially obtained. Equation (26.3.3) is also an easy consequence of (26.5.2). If in theorem 26.5 we let both b_k and c_k approach unity, we get the new result,

Corollary 26.7 *If* (a_k) *is an arbitrary sequence,*

$$\sum_{k=1}^{n} a_k k^2 = \left(\sum_{k=1}^{k} a_k\right) + \left(\sum_{k=1}^{[n/2]} a_{2k}J_2(2)\right) + \left(\sum_{k=1}^{[n/3]} a_{3k}J_2(3)\right) + \ldots + \left(\sum_{k=1}^{[n/n]} a_{nk}J_2(n)\right). \tag{26.5.4}$$

An obvious example is

$$\frac{n(n+1)(2n+1)}{6} = n + [n/2]J_2(2) + [n/3]J_2(3) + \ldots + [n/n]J_2(n). \tag{26.5.5}$$

It is easy to follow an analogous line of reasoning to that deriving theorem 5.2, then to obtain the analogue to corollary 5.4, which we rate here as a theorem,

Theorem 26.6 *If* (a_k) *is an arbitrary sequence and* m *any positive integer,*

$$\sum_{k=1}^{n} a_k k^m = \left(\sum_{k=1}^{n} a_k\right) + J_m(2)\left(\sum_{k=1}^{[n/2]} a_{2k}\right) + J_m(3)\left(\sum_{k=1}^{[n/3]} a_{3k}\right) + \ldots + J_m(n)\left(\sum_{k=1}^{[n/n]} a_{nk}\right). \tag{26.5.6}$$

We therefore see that (26.5.5) is a case of this. The known Dirichlet series generating function for the Jordan totient function is seen to be a limiting case of (26.5.6) if $a_k = k^{-(s+m)}$ and the appropriate convergence restrictions are in place. Partial sums of this generating function are seen in

Theorem 26.7 *If m and n are positive integers*

$$n = \left(\sum_{k=1}^{n} \frac{1}{k^m}\right) + \frac{J_m(2)}{2^m}\left(\sum_{k=1}^{[n/2]} \frac{1}{k^m}\right) + \frac{J_m(3)}{3^m}\left(\sum_{k=1}^{[n/3]} \frac{1}{k^m}\right) + \ldots + \frac{J_m(n)}{n^m}\left(\sum_{k=1}^{[n/n]} \frac{1}{k^m}\right), \tag{26.5.7}$$

$$\sum_{k=1}^{n} k^a = \left(\sum_{k=1}^{n} \frac{1}{k^{m-a}}\right) + \frac{J_m(2)}{2^{m-a}}\left(\sum_{k=1}^{[n/2]} \frac{1}{k^{m-a}}\right) + \frac{J_m(3)}{3^{m-a}}\left(\sum_{k=1}^{[n/3]} \frac{1}{k^{m-a}}\right) + \ldots + \frac{J_m(n)}{n^{m-a}}\left(\sum_{k=1}^{[n/n]} \frac{1}{k^{m-a}}\right), \tag{26.5.8}$$

$$\sum_{k=1}^{n} k^m = n + [n/2]J_m(2) + [n/3]J_m(3) + \ldots + [n/n]J_m(n). \tag{26.5.9}$$

Corollary 26.8 *If m and n are positive integers with $|z| < 1$*

$$\sum_{k=1}^{n} z^k k^m = \frac{1-z^n}{1-z} + J_m(2)\frac{1-z^{2[n/2]}}{1-z^2} + J_m(3)\frac{1-z^{3[n/3]}}{1-z^3} + \ldots + J_m(n)\frac{1-z^{n[n/n]}}{1-z^n}. \tag{26.5.10}$$

Clearly as z approaches unity in this we have (26.5.10), or if n is increased indefinitely (with $|z| < 1$) we have essentially the logarithmic derivative of (26.4.13). A very interesting corollary of theorem 26.5 is found from letting c_k approach zero. This results in many simple new identities. It is also interesting to compare (26.5.13) with (26.4.13) to (26.4.15) inferring a relation between the Jordan totient function and the Stirling numbers of the second kind. Another related yet distinct summation formula derived in the vein of the above analysis is

Theorem 26.8 *If $(a_k), (b_k)$, are each arbitrary sequences chosen so that, together with choice of x, the following functions are defined then*

$$\sum_{k=1}^{n} k\, a_k \frac{1-\exp(b_k x)}{1-\exp(b_k x/k)}$$

$$= \left(\sum_{k=1}^{n} a_k\right) + \left(\sum_{k=1}^{[n/2]} \sum_{2} a_{2k} \exp((b_{2k} j_1)\frac{x}{2})\right) + \left(\sum_{k=1}^{[n/3]} \sum_{3} a_{3k} \exp((b_{3k} j_1)\frac{x}{3})\right) \tag{26.5.11}$$

$$+ \left(\sum_{k=1}^{[n/4]} \sum_{4} a_{4k} \exp((b_{4k} j_1)\frac{x}{4})\right) + \ldots + \left(\sum_{k=1}^{[n/n]} \sum_{n} a_{nk} \exp((b_{nk} j_1)\frac{x}{n})\right). \tag{26.5.12}$$

This result with $a_k = z^k/k$, $b_k = k(logx)/x$, $n \to \infty$ is

Corollary 26.9 *If* $|z|, |xz|$ *are both* < 1 *and* ${}_m\varphi_k$ *is the number of solutions in integers of the conditions*

$$(a,m,k) = 1,\ 0 \le a \le k,\ 0 \le m \le k,\ a+m \neq 0, \tag{26.5.13}$$

for fixed m *and* k, *then*

$$\prod_{k=1}^{\infty} \left(\frac{1}{1-x^m z^k} \right)^{m\varphi_k/k} = \exp\left\{ \frac{1}{1-x} \left(\frac{z}{1-z} - \frac{xz}{1-xz} \right) \right\}. \tag{26.5.14}$$

This identity and similarly derived products in higher dimensions are non-trivial limiting cases of some of the results already given in Campbell [90]. That is, for example, (26.5.15) is a limiting case of an identity like (26.5.3). We may write down the generalized version of theorems 26.4 and 26.5 as follows:

Theorem 26.9 *If* $(a_k), ({}_1b_k), ({}_2b_k), \ldots ({}_hb_k)$ *are each arbitrary sequences of functions chosen so that, together with choice of* x, *the following series of functions are defined then*

$$\sum_{k=1}^{n} \left(a_k \prod_{\lambda=1}^{h} \frac{1-\exp({}_\lambda b_k x)}{1-\exp({}_\lambda b_k x/k)} \right)$$

$$\begin{aligned}
&= \left(\sum_{k=1}^{n} a_k \right) + \sum_{k=1}^{[n/2]} \sum_{h\ 2} a_{2k} \exp(({}_1b_{2k}j_1 + \ldots +_h b_{2k}j_h)\frac{x}{2}) \\
&\quad + \sum_{k=1}^{[n/3]} \sum_{h\ 3} a_{3k} \exp(({}_1b_{3k}j_1 + \ldots +_h b_{3k}j_h)\frac{x}{3}) \\
&\quad + \sum_{k=1}^{[n/4]} \sum_{h\ 4} a_{4k} \exp(({}_1b_{4k}j_1 + \ldots +_h b_{4k}j_h)\frac{x}{4}) \qquad (26.5.15) \\
&\quad \vdots \\
&\quad + \sum_{k=1}^{[n/n]} \sum_{h\ n} a_{nk} \exp(({}_1b_{nk}j_1 + \ldots +_h b_{nk}j_h)\frac{x}{n}).
\end{aligned}$$

We may now write down the general derivative with respect to x of both sides of this theorem as follows

Corollary 26.10 *If* m *is a positive integer; and* $(a_k), ({}_1b_k), ({}_2b_k), \ldots ({}_hb_k)$, *are each arbitrary sequences of functions chosen so that, together with choice of* x, *the following series of functions are all defined then*

$$\sum_{k=1}^{n} (a_k\, {}_hP_m(k))$$

$$= \sum_{k=1}^{[n/2]} a_{2k} \sum_{h\ 2} ((_1b_{2k}j_1 + \ldots +_h b_{2k}j_h)/2)^m$$

$$+ \sum_{k=1}^{[n/3]} a_{3k} \sum_{h\ 3} ((_1b_{3k}J_1 + \ldots +_h b_{3k}j_h)/3)^m$$

$$+ \sum_{k=1}^{[n/4]} a_{4k} \sum_{h\ 4} ((_1b_{4k}J_1 + \ldots +_h b_{4k}j_h)/4)^m \tag{26.5.16}$$

$$\vdots$$

$$+ \sum_{k=1}^{[n/n]} a_{nk} \sum_{h\ n} ((_1b_{nk}J_1 + \ldots +_h b_{nk}j_h)/n)^m,$$

where $_hP_m(k)$ is the coefficient of x^m in

$$\prod_{\lambda=1}^{h} \left\{ \sum_{\mu=1}^{\infty} T_\mu(_\lambda b_k x)^{\mu-1} \right\} \text{ where } T_\mu = -\sum_{\alpha=1}^{\mu} \binom{\mu}{\alpha} \frac{B_\alpha}{k^{\alpha-1}} \tag{26.5.17}$$

with B_α the Bernoulli numbers.

Theorem 26.8 results from theorem 26.9 if x approaches zero, whilst corollary 26.10 is the result of equating coefficients of powers of x from expanding the two sides of theorem 26.9 as power series in x. There is no problem about this as both sides of (26.5.17) are finite sums. The formally complex nature of $_hP_m(k)$ means that the simplest cases of corollary 26.10 are those in which $h = 2$ and $m = 1, 2, 3, \ldots$ successively. Clearly,

$$_2P_1(k) = 2T_1T_2\ {}_1b_k\ {}_2b_k, \tag{26.5.18}$$

$$_2P_2(k) = T_1T_3(_1b_k)(_2b_k)^3 + T_2T_2(_1b_k)^2(_2b_k)^2 + T_3T_1(_1b_k)^3(_2b_k)$$
$$= T_1T_3(_1b_k)(_2b_k)((_1b_k)^2 + (_2b_k)^2) + (T_2)^2(_1b_k)^2(_2b_k)^2, \tag{26.5.19}$$

so applying (26.5.27) and then (26.5.28) to corollary 5.11 gives us successively the two following corollaries.

Corollary 26.11 *If $(a_k), (_1b_k), (_2b_k)$ are each arbitrary sequences of functions chosen so that the following series of functions are all defined then for n positive integers greater than 1,*

$$\frac{1}{3} \sum_{k=1}^{n} \left(\frac{1}{k} a_{k1} b_k \right) = \sum_{k=1}^{[n/2]} \frac{1}{2} a_{2k} \sum_{2\ 2} (_1b_{2k}j_1 +_2 b_{2k}j_2)$$

$$+ \sum_{k=1}^{[n/3]} \frac{1}{3} a_{3k} \sum_{2\ 3} (_1b_{3k}j_1 +_2 b_{3k}j_2)$$

$$+\sum_{k=1}^{[n/4]} \frac{1}{4} a_{4k} \sum_{2\ 4} ({}_1b_{4k} j_1 +_2 b_{4k} j_2) \tag{26.5.20}$$

$$\vdots$$

$$+\sum_{k=1}^{[n/n]} \frac{1}{n} a_{nk} \sum_{2\ n} ({}_1b_{nk} j_1 +_2 b_{nk} j_2).$$

Corollary 26.12 *For the same conditions as corollary 26.11,*

$$\frac{1}{2}\sum_{k=1}^{n} a_k \left\{ \left(\frac{1}{6}k^2 - \frac{1}{4}k + \frac{1}{12} \right) ({}_1b_k^2 +_2 b_k^2) \frac{1}{4}(k-1)^2\, {}_1b_k\, {}_2b_k \right\}$$

$$= \sum_{k=1}^{[n/2]} \frac{1}{2} a_{2k} \sum_{2\ 2} ({}_1b_{2k} j_1 +_2 b_{2k} j_2) \tag{26.5.21}$$

$$+\sum_{k=1}^{[n/3]} \frac{1}{3} a_{3k} \sum_{2\ 3} ({}_1b_{3k} j_1 +_2 b_{3k} j_2) \tag{26.5.22}$$

$$+\sum_{k=1}^{[n/4]} \frac{1}{4} a_{4k} \sum_{2\ 4} ({}_1b_{4k} j_1 +_2 b_{4k} j_2) \tag{26.5.23}$$

$$\vdots \tag{26.5.24}$$

$$+\sum_{k=1}^{[n/n]} \frac{1}{n} a_{nk} \sum_{2\ n} ({}_1b_{nk} j_1 +_2 b_{nk} j_2). \tag{26.5.25}$$

It is clear that cases of our function $\phi_t(m;k)$ from definition 27.4.2 are a natural occurrence in the right sides of (26.5.15) and (26.5.16) if ${}_1b_k = 1$ and ${}_2b_k = 1$. When this is the case we have the two corollaries.

Corollary 26.13 *If (a_k) is an arbitrary sequence of functions chosen so that the following series of functions are all defined, and $\phi_m(n)$ is the mth power of the sum of the non-negative integers a, b, less than n such that $(a,b,n) = 1$ and $a+b \neq 0$, then for positive integers $n > 1$,*

$$\sum_{k=1}^{n} a_k k(k-1) = \frac{1}{2}\phi_1(2)\sum_{k=1}^{[n/2]} a_{2k} + \frac{1}{3}\phi_1(3)\sum_{k=1}^{[n/3]} a_{3k} + \ldots + \frac{1}{n}\phi_1(n)\sum_{k=1}^{[n/n]} a_{nk}, \tag{26.5.26}$$

$$\sum_{k=1}^{n} a_k \left(\frac{7}{12}k^2 - k + \frac{5}{12}\right) = \frac{1}{2}\phi_2(2)\sum_{k=1}^{[n/2]} a_{2k} + \frac{1}{3}\phi_2(3)\sum_{k=1}^{[n/3]} a_{3k} + \ldots + \frac{1}{n}\phi_2(n)\sum_{k=1}^{[n/n]} a_{nk}, \tag{26.5.27}$$

We next state some examples of this corollary. The cases are fairly obvious given the previous analysis,

Corollary 26.14

$$\sum_{k=1}^{n} k(k-1) = \frac{1}{2}\phi_1(2)[n/2] + \frac{1}{3}\phi_1(3)[n/3] + \ldots + \frac{1}{n}\phi_1(n)[n/n], \tag{26.5.28}$$

$$\sum_{k=1}^{n} (\frac{7}{12}k^2 - k + \frac{5}{12}) = \frac{1}{2}\phi_2(2)[n/2] + \frac{1}{3}\phi_2(3)[n/3] + \ldots + \frac{1}{n}\phi_2(n)[n/n], \tag{26.5.29}$$

$$\sum_{k=1}^{n} k(k-1)$$
$$= \frac{1}{2}\phi_1(2)[n/2](1+[n/2]) + \frac{1}{3}\phi_1(3)[n/3](1+[n/3]) + \ldots + \frac{1}{n}\phi_1(n)[n/n](1+[n/n]), \tag{26.5.30}$$

$$\sum_{k=1}^{n} k\left(\frac{7}{12}k^2 - k + \frac{5}{12}\right)$$
$$= \frac{1}{2}\phi_2(2)[n/2](1+[n/2]) + \frac{1}{3}\phi_2(3)[n/3](1+[n/3]) + \ldots + \frac{1}{n}\phi_2(n)[n/n](1+[n/n]), \tag{26.5.31}$$

These results are akin to those in Campbell [87] and Ramanujan [217]. Furthermore, if we permit n to increase indefinitely and let $a_k = k^{-(s+2)}$ in corollaries 26.13 and 26.14, we have the Dirichlet generating functions given by

Corollary 26.15 *If* $\Re s > 1$ *then*

$$\frac{\zeta(s) - \zeta(s+1)}{\zeta(s+2)} = \sum_{k=2}^{\infty} \frac{\phi_1(k)}{k^{(s+3)}}, \tag{26.5.32}$$

$$\frac{7\zeta(s) - 12\zeta(s+1) + 5\zeta(s+2)}{\zeta(s+3)} = \sum_{k=2}^{\infty} \frac{\phi_2(k)}{k^{(s+3)}}, \tag{26.5.33}$$

Likewise, if n too increases indefinitely and $a_k = z^k/k$ in corollaries 26.14 and 26.15 we have the infinite products,

Corollary 26.16 *If* $|z| < 1$ *then*

$$\prod_{k=2}^{\infty} (1-z^k)^{-\phi_1(k)/k^2} = \exp\left\{\frac{z}{(1-z)^2}\right\}, \tag{26.5.34}$$

$$\prod_{k=2}^{\infty} (1-z^k)^{-\phi_2(k)/k^2} = \left(\frac{1}{1-z}\right)^{\frac{5}{12}} \exp\left\{\frac{z(12z-5)}{12(1-z)^2}\right\}. \tag{26.5.35}$$

Recalling (26.4.11) and its ensuing paragraph, it was clear that

$$\frac{\zeta(s-m)}{\zeta(s)} = 1+\sum_{k=2}^{\infty}\frac{J_m(k)}{k^s}, \qquad \mathcal{R}s > 1. \tag{26.5.36}$$

This sum when compared to corollary 5.6 shows that

Corollary 26.17 *For positive integers k,*

$$\phi_1(k) = J_2(k) - J_1(k), \tag{26.5.37}$$

$$\phi_2(k) = \frac{7}{12}J_3(k) - J_2(k) + \frac{5}{12}J_1(k). \tag{26.5.38}$$

26.6 APPLICATION OF JACOBI THETA SERIES TO THE GENERALIZED SUMMATIONS

In Campbell [85], the Jacobi theta function was applied to the finite left hand side form of the summation formula (3.2). We use now the well known terminology for the theta function

$$\Theta_1(z,q) = \Theta_1(z) = 2q^{\frac{1}{2}}\sum_{k=1}^{\infty}(-1)^k q^{k(k+1)}\sin(2k+1)z, \tag{26.6.1}$$

$$\sum_{k=1}^{\infty}\frac{1}{k}\frac{q^{2k}}{1-q^{2k}}\sin 2k\alpha \sin 2k\alpha = \frac{1}{4}\log\left(\frac{\Theta_1((\alpha+\beta),q)\sin(\alpha-\beta)}{\Theta_1((\alpha-\beta),q)\sin(\alpha+\beta)}\right). \tag{26.6.2}$$

Application of (26.6.2) to (26.3.3) gives us the result (see Campbell [85, section 4])

Theorem 26.10 *If* $|q| < 1$ *and* $x^{1/k} \neq 1$ *then*

$$\exp\left(4\sum_{k=1}^{\infty}\frac{1}{k}\frac{q^{2k}}{1-q^{2k}}\frac{1-x}{1-x^{1/k}}\sin 2k\alpha \sin 2k\alpha\right)$$
$$= \frac{\Theta_1(\alpha+\beta)\sin(\alpha-\beta)}{\Theta_1(\alpha-\beta)\sin(\alpha+\beta)}\prod_{k=1}^{\infty}\left(\frac{\Theta_1((\alpha+\beta)k,q^k)\sin(\alpha-\beta)}{\Theta_1((\alpha-\beta)k,q^k)\sin(\alpha+\beta)}\right)^{f_k(x)/k} \tag{26.6.3}$$

where for positive integers j less than k,

$$f_k(x) = \sum_{(j,k)=1} x^{j/k} = \sum_{1\ k} x^{j_1/k}.$$

Particular cases of this statement include some of the known arithmetical functions, as highlighted in the following examples. To begin with, for positive integers n with $x = exp(2\pi i n)$ we have

Corollary 26.18 *(see Campbell [85])*

$$\exp\left(4\sum_{k|n}^{\infty}\frac{q^{2k}}{1-q^{2k}}\sin 2k\alpha\sin 2k\alpha\right)$$
$$=\frac{\Theta_1(\alpha+\beta)\sin(\alpha-\beta)}{\Theta_1(\alpha-\beta)\sin(\alpha+\beta)}\prod_{k=1}^{\infty}\left(\frac{\Theta_1((\alpha+\beta)k,q^k)\sin(\alpha-\beta)}{\Theta_1((\alpha-\beta)k,q^k)\sin(\alpha+\beta)}\right)^{c_k(n)/k}, \quad (26.6.4)$$

where $c_k(n)$ is Ramanujan's trigonometrical function as in (26.1.1).

If in (26.6.3) we allow x to approach unity, we have the result,

Corollary 26.19

$$\exp\left(4\sum_{k=1}^{\infty}\frac{q^{2k}}{1-q^{2k}}\sin 2k\alpha\sin 2k\alpha\right)$$
$$=\frac{\Theta_1(\alpha+\beta)\sin(\alpha-\beta)}{\Theta_1(\alpha-\beta)\sin(\alpha+\beta)}\prod_{k=1}^{\infty}\left(\frac{\Theta_1((\alpha+\beta)k,q^k)\sin(\alpha-\beta)}{\Theta_1((\alpha-\beta)k,q^k)\sin(\alpha+\beta)}\right)^{\varphi(n)/k}, \quad (26.6.5)$$

where $\varphi(n)$ is the Euler totient function.

If we follow an analogous line of reasoning to that which led us to (26.6.3) but from starting with lemma 26.1, we obtain the new result,

Theorem 26.11 *If $|q|<1$ and all of $x_1^{1/k}, x_2^{1/k}, \ldots, x_m^{1/k} \neq 1$ then*

$$\exp\left(4\sum_{k=1}^{\infty}\frac{1}{k}\frac{q^{2k}}{1-q^{2k}}\left(\frac{1-x_1}{1-x_1^{1/k}}\frac{1-x_2}{1-x_2^{1/k}}\cdots\frac{1-x_m}{1-x_m^{1/k}}\right)\sin 2k\alpha\sin 2k\alpha\right)$$
$$=\frac{\Theta_1(\alpha+\beta)\sin(\alpha-\beta)}{\Theta_1(\alpha-\beta)\sin(\alpha+\beta)}\prod_{k=1}^{\infty}\left(\frac{\Theta_1((\alpha+\beta)k,q^k)\sin(\alpha-\beta)}{\Theta_1((\alpha-\beta)k,q^k)\sin(\alpha+\beta)}\right)^{{}_mf_k(x)/k} \quad (26.6.6)$$

where

$$_mf_k(x)=\sum_{m\ k}\left({x_1}^{j_1}{x_2}^{j_2}\cdots{x_m}^{j_m}\right)^{1/k}. \quad (26.6.7)$$

We state two relevant corollaries with respect to the substance of the current work. Firstly, to obtain the corresponding result for the Jordan totient function, we need only allow all of the x's to approach unity, bearing in mind our work in section 26.4. We therefore have

Corollary 26.20

$$\exp\left(4\sum_{k=1}^{\infty}k^{m-1}\frac{q^{2k}}{1-q^{2k}}\sin 2k\alpha\sin 2k\alpha\right)$$
$$=\frac{\Theta_1(\alpha+\beta)\sin(\alpha-\beta)}{\Theta_1(\alpha-\beta)\sin(\alpha+\beta)}\prod_{k=1}^{\infty}\left(\frac{\Theta_1((\alpha+\beta)k,q^k)\sin(\alpha-\beta)}{\Theta_1((\alpha-\beta)k,q^k)\sin(\alpha+\beta)}\right)^{J_m(k)/k}, \quad (26.6.8)$$

where $J_m(k)$ is the Jordan totient function.

Secondly, to obtain the corresponding result for our new function, $c_k(n_1,n_2,\ldots,n_m)$ we need only allow all of the x_j's to approach $exp(2\pi i n_j)$, bearing in mind our work in section 26.1. We therefore have

Corollary 26.21

$$\exp\left(4\sum_{k|(n_1,n_2,\ldots,n_m)}\frac{q^{2k}}{1-q^{2k}}\sin 2k\alpha\sin 2k\alpha\right)$$
$$=\frac{\Theta_1(\alpha+\beta)\sin(\alpha-\beta)}{\Theta_1(\alpha-\beta)\sin(\alpha+\beta)}\prod_{k=1}^{\infty}\left(\frac{\Theta_1((\alpha+\beta)k,q^k)\sin(\alpha-\beta)}{\Theta_1((\alpha-\beta)k,q^k)\sin(\alpha+\beta)}\right)^{c_k(n_1,n_2,\ldots,n_m)/k}. \quad (26.6.9)$$

where $c_k(n_1,n_2,\ldots,n_m)$ is our new generalized Ramanujan totient function.

We have not attempted to use the full generality of corollary 26.3, and it is clear that we may return to this topic in future papers. The return to such investigations will no doubt be hastened if it turns out that the identities given in this paper can be shown to enumerate tilings in the sense described by Baake et. al. and his research team pertaining to quasicrystals. (see Baake et. al. [39], [40], [37], or Campbell [92])

EXERCISES

26.1 *Rewrite the 2D, 3D, and 4D finite summations.* We reduce equations (26.3.3), (26.3.4) and (26.3.5) to the simpler forms:

$$\sum_{k=1}^{\infty} a_k\frac{1-q}{1-q^{1/k}} = S_1+\sum_{k=2}^{\infty} S_k \sum_{\substack{(j,k)=1\\0<j<k}} q^{j/k}, \quad (P26.1)$$

$$\sum_{k=1}^{\infty} a_k\frac{1-x}{1-x^{1/k}}\frac{1-y}{1-y^{1/k}} = S_1+\sum_{k=2}^{\infty} S_k \sum_{\substack{(a,b,k)=1\\0\le a<k\\0\le b<k\\a+b\neq 0}} (x^a y^b)^{1/k}, \quad (P26.2)$$

$$\sum_{k=1}^{\infty} a_k\frac{1-x}{1-x^{1/k}}\frac{1-y}{1-y^{1/k}}\frac{1-z}{1-z^{1/k}} = S_1+\sum_{k=2}^{\infty} S_k \sum_{\substack{(a,b,c,k)=1\\0\le a<k\\0\le b<k\\0\le c<k\\a+b+c\neq 0}} (x^a y^b z^c)^{1/k}, \quad (P26.3)$$

where q, x, y and z are chosen for non-zero denominators, and we have arbitrary a_i functions defined for

$$\begin{aligned} S_1 &= a_1+a_2+a_3+\cdots,\\ S_2 &= a_2+a_4+a_6+\cdots,\\ S_3 &= a_3+a_6+a_9+\cdots,\\ S_4 &= a_4+a_8+a_{12}+\cdots, \end{aligned}$$

and so on.

26.2 *Cases of the Sum Transforms:* $a_k = 1/k^s$. Examples of P26.1, P26.2, and P26.3 setting $a_k = 1/k^s$ for $\Re(s) > 1$. With this substitution we get $S_1 = \zeta(s)$ and $S_n = \zeta(s)/n^s$ for any positive integer n. So then $S_1 = \zeta(s)$, $S_2 = \zeta(s)/2^s$, $S_1 = \zeta(s)/3^s$, etc. and our resulting cases of the three transforms are:

$$\sum_{k=1}^{\infty} k^{-s} \frac{1-q}{1-q^{1/k}} = \zeta(s)(1+\sum_{k=2}^{\infty} k^{-s} \sum_{\substack{(j,k)=1 \\ 0<j<k}} q^{j/k}), \tag{P26.4}$$

$$\sum_{k=1}^{\infty} k^{-s} \frac{1-x}{1-x^{1/k}} \frac{1-y}{1-y^{1/k}} = \zeta(s)(1+\sum_{k=2}^{\infty} k^{-s} \sum_{\substack{(a,b,k)=1 \\ 0\leq a<k \\ 0\leq b<k \\ a+b\neq 0}} (x^a y^b)^{1/k}), \tag{P26.5}$$

$$\sum_{k=1}^{\infty} k^{-s} \frac{1-x}{1-x^{1/k}} \frac{1-y}{1-y^{1/k}} \frac{1-z}{1-z^{1/k}} = \zeta(s)(1+\sum_{k=2}^{\infty} k^{-s} \sum_{\substack{(a,b,c,k)=1 \\ 0\leq a<k \\ 0\leq b<k \\ 0\leq c<k \\ a+b+c\neq 0}} (x^a y^b z^c)^{1/k}). \tag{P26.6}$$

26.3 *Cases of the Sum Transforms:* $a_k = w^k$. Examples of P26.1, P26.2, and P26.3 setting $a_k = w^k$ for $|w| < 1$. With this substitution we get $S_1 = w/(1-w)$ and $S_n = w^n/(1-w^n)$ for any positive integer n. The transforms become:

$$\sum_{k=1}^{\infty} w^k \frac{1-q}{1-q^{1/k}} = \frac{w}{1-w} + \sum_{k=2}^{\infty} \frac{w^k}{1-w^k} \sum_{\substack{(j,k)=1 \\ 0<j<k}} q^{j/k}, \tag{P26.7}$$

$$\sum_{k=1}^{\infty} w^k \frac{1-x}{1-x^{1/k}} \frac{1-y}{1-y^{1/k}} = \frac{w}{1-w} + \sum_{k=2}^{\infty} \frac{w^k}{1-w^k} \sum_{\substack{(a,b,k)=1 \\ 0\leq a<k \\ 0\leq b<k \\ a+b\neq 0}} (x^a y^b)^{1/k}, \tag{P26.8}$$

$$\sum_{k=1}^{\infty} w^k \frac{1-x}{1-x^{1/k}} \frac{1-y}{1-y^{1/k}} \frac{1-z}{1-z^{1/k}} = \frac{w}{1-w}(1+\sum_{k=2}^{\infty} \frac{w^k}{1-w^k} \sum_{\substack{(a,b,c,k)=1 \\ 0\leq a<k \\ 0\leq b<k \\ 0\leq c<k \\ a+b+c\neq 0}} (x^a y^b z^c)^{1/k}). \tag{P26.9}$$

26.4 *Famous limiting cases of (P26.4) and (P26.7).* Show that taking the limit as $q \to 1$ in (P26.4) and (P26.7) result respectively in the Dirichlet series and Lambert series generating functions:

$$\sum_{k=1}^{\infty} \frac{\varphi(k)}{k^s} = \frac{\zeta(s-1)}{\zeta(s)}, \tag{P26.10}$$

and

$$\sum_{k=1}^{\infty} \frac{\varphi(k)w^k}{1-w^k} = \frac{q}{(1-q)^2}, \tag{P26.11}$$

where $\varphi(k)$ is the Euler totient function.

Think About It...

The seeds from Ramanujan's garden have been blowing on the wind and have been sprouting all over the landscape.

[On the stimulating effects of Ramanujan's mathematical legacy.]

Freeman Dyson, (1923–2020)
Selected Papers of Freeman Dyson with Commentary (ed. 1996)

FURTHER READING

APOSTOL, T. *Introduction to Analytic Number Theory.* Springer-Verlag, New York, 1976.

APOSTOL, T.M. and ZUCKERMANN, H.S. On the functional equation $F(mn)F((m.n)) = F(m)F(n)f((m,n))$. *Pac. J. Math.*, 14:371–384, 1964.

BAAKE, M. and GRIMM, U. Combinatorial problems of quasi-crystallography. preprint *Institut fur mathematik*, Universitat Greifswald, December 2002.

BAAKE, M and HUCK, C. Ergodic properties of visible lattice points. *Proc. Steklov Inst. Math.*, 288(1):165–188, 2015. (English summary) Reprint of Tr. Mat. Inst. Steklova Vol. 288, pages 184–208 (2015).

BAAKE, M., GRIMM, U., and WARRINGTON, D.H. Some remarks on the visible points of a lattice. *J. Phys. A: Math. Gen.*, 27:2669–2674, 1994.

BAAKE, M.; MOODY, R.V.; and PLEASANTS, P.A.B. Diffraction from Visible Lattice Points and kth Power Free Integers. arxiv preprint, June 1999. http://arxiv.org/abs/math.MG/9906132/.

BORWEIN, J.M. and BORWEIN, P. *Pi and the AGM.* John Wiley and Sons, New York, 1986.

BORWEIN, J.M.; GLASSER, M.L.; MCPHEDRAN, R.C.; WAN, J.G.; and ZUCKER, I.J. *Lattice Sums then and now*, volume 150 of *Encyclopedia of Mathematics and Its Applications.* Cambridge University Press, 2013.

CAMPBELL, G.B. Multiplicative functions over Riemann zeta function products. *J. Ramanujan Soc.*, 7(1):52–63, 1992.

CAMPBELL, G.B. Dirichlet summations and products over primes. *Int. J. Math. Math. Sci.*, 16(2):359–372, 1993.

CAMPBELL, G.B. A new class of infinite products, and Euler's totient. *Int. J. Math. and Math. Sci.*, 17(3):417–422, 1994.

CAMPBELL, G.B. A closer look at some new identities. *Int. J. Math. and Math. Sci.*, 21(3):581–586, 1998.

CAMPBELL, G.B. Infinite products over hyperpyramid lattices. *Int. J. Math. and Math. Sci.*, 23(4):271–277, 2000.

CAMPBELL, G.B. Some n-space q-binomial theorem extensions and similar identities. arxiv preprint, June 2019. arXiv:1906.07526v1 [math.NT]: https://arxiv.org/abs/1906.07526.

CAMPBELL, G.B. Continued Fractions for partition generating functions. arxiv preprint, Jan 2023. arXiv:2301.12945 [math.CO], https://arxiv.org/abs/2301.12945.

CAMPBELL, G.B. Vector Partition Identities for 2D, 3D and nD Lattices. arxiv preprint, Feb 2023. arXiv:10.48550/ARXIV.2302.01091 [math.NT], https://arxiv.org/abs/2302.01091.

CAMPBELL, G.B. Visible Point Partition Identities for Polylogarithms, and Parametric Euler Sums. arxiv preprint, Jun 2023. arXiv:2306.02241v1 [math.CO], https://arxiv.org/abs/2306.02241.

CAMPBELL, G.B. Visible Point Vector Partition Identities for Hyperpyramid Lattices. arxiv preprint, Sep 2023. arXiv:2309.16094v1 [math.CO], https://arxiv.org/abs/2309.16094.

CHANDRASEKHARAN, K. *Arithmetical Functions*, volume 67. Springer- Verlag, New York, 1970.

GLASSER, M.L., and ZUCKER, I.J. Lattice Sums. *Th. Chem.: Adv. Persp.*, 5:67–139, 1980.

HARDY, G.H. and WRIGHT, E.M. *An Introduction to the Theory of Numbers.* Oxford University Press, Clarendon, London, 1971.

HARDY, G.H. *Ramanujan.* Cambridge University Press, London and New York, 1940. (reprinted by Chelsea, New York).

LOSSERS, O.P. The generalized totient function, solution to problem No. 6446. *Amer. Math. Monthly*, pages 434–434, 1985.

NINHAM, B.W., GLASSER, M.L., HUGHES, B.D., and FRANKEL, N.E. Möbius, Mellin, and Mathematical Physics. *Physica A*, 186:441–481, 1992.

RAMANUJAN, S. On certain trigonometrical sums and their application to the theory of numbers. In *Collected Papers of S. Ramanujan*, pages 179–199. Cambridge University Press, Cambridge (1927), 1962. reprinted by Chelsea, New York, 1962.

SIVARAMAKRISHNAN, R. *Classical Theory of Arithmetic Functions.* Marcel Dekker, Inc., (New York and Basel), 1989.

TÓTH, L. Sums of products of Ramanujan sums. *Annali dell'universita' di Ferrara*, 58:183–197, 2011.

CHAPTER 27

Other Non-weighted n-Space Vector Partition Theorems

27.1 SIMPLE BINARY PARTITION 2-SPACE IDENTITIES

It is clear that we may apply the first principles VPV method used in earlier chapters to other vector partition generating functions. Some good background material for these concepts are given in Alkauskas et al. [6, 7] for congruence properties of binary partition related functions. Also recommended is the paper by Rödseth [224] which over 50 years ago covered some ideas on m-ary partitions. However, to cover off the basics on binary partitions refer to Andrews' classical book [12] as well as earlier Chapters 12, 16, 17, and 18. The author's recent updates in Campbell [93, 96, 97, 98, 99], may also provide useful background and foreground material for the current chapter.

An example is now given. The following theorem is based around the ideas associated with the elementary identity

$$(1+x)\left(1+x^2\right)\left(1+x^4\right)\left(1+x^8\right)\cdots = \frac{1}{1-x}. \tag{27.1.1}$$

The easy combinatorial interpretation of this is "*each positive integer is uniquely represented by a sum of distinct powers of 2*". So, a minor variation on (27.1.1) is

$$\prod_{k=0}^{\infty}(1+yz^{2^k}) = \sum_{k=0}^{\infty} y^{\beta(k)} z^k; \tag{27.1.2}$$

in other words,

$$(1+yz)\left(1+yz^2\right)\left(1+yz^4\right)\left(1+yz^8\right)\cdots$$
$$= 1+yz+yz^2+y^2z^3+yz^4+y^2z^5+y^2z^6+y^3z^7+yz^8+\cdots+y^{\beta(m)}z^m+\cdots,$$

where $\beta(m)$ is the number of binary digits in m.

DOI: 10.1201/9781003174158-27

The partition grid for this is the sparsely populated (27.1.4). Incidentally, there is an interesting infinite product

$$\prod_{\substack{k\geq 1\\ p_k=k^{th}prime}} \left(1+\frac{z^{2^{k-1}}}{{p_k}^s}\right) \tag{27.1.3}$$

where the expansion of terms can be collected easily either as a power series in z or a Dirichlet series in s. This may be a suitable exercise for the interested researcher.

2D first quadrant partition grid for $\prod_{k=0}^{\infty}(1+yz^{2^k})$

⋮	⋮	⋮	⋮	⋮	⋮	⋮
19			1			
18		1				
17		1				
16	1					
15				1		
14			1			
13			1			
12		1				
11			1			
10		1				
9		1				
8	1					
7			1			
6		1				
5		1				
4	1					
3		1				
2	1					
1	1					
	1	2	3	4	5	6

(27.1.4)

The above result (27.1.2) may seem naïve. However, it is a simple lead-up to a class of identities in n-space of this kind.

A similar extension of this type of result is the *2D Upper Half* first quadrant lattice identity

Theorem 27.1

$$\prod_{k\geq 0}\left(\frac{1}{1-yz^{2^k}}\right)=\prod_{\substack{j,k\geq 0\\ j\leq k}}\left(1+y^{2^j}z^{2^k}\right)=1+\sum_{k=1}^{\infty}A_k z^k \tag{27.1.5}$$

where

$$A_k = \frac{1}{k!}\begin{vmatrix} y & -1 & 0 & 0 & \cdots & 0 \\ y^2+2y & y & -2 & 0 & \cdots & 0 \\ y^3 & y^2+y & y & -3 & \cdots & 0 \\ y^4+2y^2+4y & y^3 & y^2+y & y & \ddots & \vdots \\ \vdots & \vdots & \vdots & \vdots & \ddots & -(k-1) \\ \sum_{\substack{2^j \mid k \\ j\geq 0}} 2^j y^{k/2^j} & \sum_{\substack{2^j \mid (k-1) \\ j\geq 0}} 2^j y^{(k-1)/2^j} & \sum_{\substack{2^j \mid (k-2) \\ j\geq 0}} 2^j y^{(k-2)/2^j} & \sum_{\substack{2^j \mid (k-3) \\ j\geq 0}} 2^j y^{(k-3)/2^j} & \cdots & y \end{vmatrix}$$

The combinatorial interpretation of (27.1.5) is

Theorem 27.2 *If $B(j,k)$ is the number of vector partitions of $\langle j,k\rangle$ into distinct parts of kind $\langle 2^a,2^b\rangle$ in which $a \leq b$ with non-negative integers a and b, then $B(j,k)$ equals also the number of partitions into "unrestricted" parts of kind $\langle 1,2^b\rangle$ in which b is a non negative integer, and $B(j,k)$ is the coefficient of $y^j z^k$ in (27.1.5).*

Each side of (27.1.5) satisfies the equation $f(z)(1-yz) = f(z^2)$ and this equation also leads to a set of recurrences solvable using Cramer's rule.

In Mathematica or Wolframalpha online we can easily check that

$$\prod_{k\geq 0}\left(\frac{1}{1-yz^{2^k}}\right) = 1+yz+y(y+1)z^2+y^2(y+1)z^3+y(y^3+y^2+y+1)z^4 \quad (27.1.6)$$

$$+y^2(y^3+y^2+y+1)z^5+y^2(y^4+y^3+y^2+2y+1)z^6$$
$$+y^3(y^4+y^3+y^2+2y+1)z^7$$
$$+y(y^7+y^6+y^5+2y^4+2y^3+y^2+y+1)z^8$$
$$+y^2(y^7+y^6+y^5+2y^4+2y^3+y^2+y+1)z^9$$
$$+y^2(y^8+y^7+y^6+2y^5+2y^4+2y^3+2y^2+2y+1)z^{10}+\ldots$$

Also, for calculating machine verifying, copy a form $\prod_{k=0}^{\infty}\prod_{j=0}^{k}(1+y^{2^j}z^{2^k})$, in Mathematica, Maple or Wolframalpha. The two product expansions of (27.1.5) easily verify; both of them producing the series given in (27.1.6).

2D first quadrant partition grid for $\prod_{k=0}^{\infty}(1+yz^{2^k})$

⋮	⋮	⋮	⋮	⋮	⋮	⋮	⋮	⋮	⋮	⋮	⋮	⋮	⋮
13			1	2	2	3	3	2	2	2	1	1	1
12		1	2	2	3	3	2	2	2	1	1	1	
11			1	2	2	2	2	2	1	1	1		
10		1	2	2	2	2	2	1	1	1			
9		1	1	1	2	2	1	1	1				
8	1	1	1	2	2	1	1	1					
7			1	2	1	1	1						
6		1	2	1	1	1							
5		1	1	1	1								
4	1	1	1	1									
3		1	1										
2	1	1											
1	1												
	1	2	3	4	5	6	7	8	9	10	11	12	13

(27.1.7)

27.2 2D PARTITIONS INTO DISTINCT BINARY PARTS ON TWO STRAIGHT LINES

We apply the theory from Chapter 12, section 12.6 pertaining to 2D (or nD) vector partitions into parts on two (or positive integer m) straight radial-from-origin lines. Here, we adapt the rationale from that section as follows. It is instructive to start with the simplest example, so the reader will understand how this can be generalized.

2D Vector Partitions whose parts distinct binary powers on two straight lines.

The above analysis is with respect to vector partitions in the upper half first quadrant above and including the line $y=z$. The simplest version of this theory departing from the single radial from the origin line of lattice points, would be to state the result if dealing with 2D partitions from two such radial lines of lattice points. For example, consider the two lines with equations $y=z/2$ and $y=z/3$. The lattice point vectors along these lines in the first quadrant are listed as

$$\begin{aligned} S_1 &= \{\langle 1,2\rangle, \langle 2,4\rangle, \langle 3,6\rangle, \langle 4,8\rangle, \langle 5,10\rangle, \ldots\}; \\ S_2 &= \{\langle 1,3\rangle, \langle 2,6\rangle, \langle 3,9\rangle, \langle 4,12\rangle, \langle 5,15\rangle, \ldots\}. \end{aligned}$$

Following the above rationale, we see that the generating function for 2D vector partitions into distinct powers of 2 contained on the two lines S_1 and S_2 is

$$\left(\prod_{j\geq 0}(1+(yz^2)^{2^j})\right)\left(\prod_{k\geq 0}(1+(yz^3)^{2^k})\right)$$

$$= \left(1+y^1z^2+y^2z^4+y^3z^6+\ldots\right)\left(1+y^1z^3+y^2z^6+y^3z^9+\ldots\right) = \frac{1}{(1-yz^2)(1-yz^3)}$$

$$\begin{aligned}
= \; & 1 \\
& + yz^2(1+z) \\
& + y^2z^4(1+z+z^2) \\
& + y^3z^6(1+z+z^2+z^3) \\
& + y^4z^8(1+z+z^2+z^3+z^4) \\
& + y^5z^{10}(1+z+z^2+z^3+z^4+z^5) \\
& + y^6z^{12}(1+z+z^2+z^3+z^4+z^5+z^6) \\
& + y^7z^{14}(1+z+z^2+z^3+z^4+z^5+z^6+z^7) + O(y^8).
\end{aligned}$$

So, the number of 2D vector partitions into distinct powers of 2 number of parts on line S_1 and distinct powers of 2 number of parts on line S_2 is generated by

$$\frac{1}{(1-yz^2)(1-yz^3)}$$

which is seen to also generate: the number of vector partitions into up to two parts, using at most one part a vector from S_1, and at most one part a vector from S_2.

This leads us to state

2D first quadrant partition grid for $\frac{1}{(1-yz^2)(1-yz^3)}$

⋮	⋮	⋮	⋮	⋮	⋮	⋮	⋮	⋮	⋮
21								1	⋯
20								1	⋯
19								1	⋯
18							1	1	⋯
17							1	1	⋯
16							1	1	⋯
15						1	1	1	⋯
14						1	1	1	⋯
13						1	1		⋯
12					1	1	1		⋯
11					1	1			⋯
10					1	1			⋯
9				1	1				⋯
8				1	1				⋯
7				1					⋯
6			1	1					⋯
5			1						⋯
4			1						⋯
3		1							⋯
2		1							⋯
1									⋯
0	1								⋯
z/y	0	1	2	3	4	5	6	7	⋯

27.3 SIMPLE TERNARY PARTITION 2-SPACE IDENTITIES

There is a ternary version of our first section's binary identity worth examining also. It can be constructed as follows. For each diagonal shown in (27.3.1), apply the naïve identity, for $|y| < 1$, $|z| < 1$,

$$(1+yz+y^2z^2)(1+y^3z^3+y^6z^6)(1+y^9z^9+y^{18}z^{18})\cdots = \frac{1}{1-yz}.$$

Then clearly, for the same restrictions on y and z, we have

Theorem 27.3 *For* $|y| < 1$, $|z| < 1$,

$$\frac{1}{(1-yz)(1-yz^3)(1-yz^9)(1-yz^{27})\cdots} = \prod_{j=0}^{\infty}\prod_{j=0}^{k}(1+y^{3^j}z^{3^k}+y^{2\times 3^j}z^{2\times 3^k}) \quad (27.3.1)$$

$$= (1+yz+y^2z^2)$$
$$(1+yz^3+y^2z^6)(1+y^3z^3+y^6z^6)$$
$$(1+yz^9+y^2z^{18})(1+y^3z^9+y^6z^{18})(1+y^9z^9+y^{18}z^{18})$$
$$(1+yz^{27}+y^2z^{54})(1+y^3z^{27}+y^6z^{54})(1+y^9z^{27}+y^9z^{54})(1+y^{27}z^{27}+y^{54}z^{54})$$
$$\vdots \qquad \vdots \qquad \vdots \qquad \vdots$$
$$=1+yz+y^2z^2+(y^3+y)z^3+(y^4+y^2)z^4+(y^5+y^3)z^5+(y^6+y^4+y^2)z^6+(y^7+y^5+y^3)z^7$$
$$+(y^8+y^6+y^4)z^8+(y^9+y^7+y^5+y^3+y)z^9+(y^{10}+y^8+y^6+y^4+y^2)z^{10}$$
$$+(y^{11}+y^9+y^7+y^5+y^3)z^{11}+(y^{12}+y^{10}+y^8+y^6+2y^4+y^2)z^{12}$$
$$+(y^{13}+y^{11}+y^9+y^7+2y^5+y^3)z^{13}+y^4(y^{10}+y^8+y^6+y^4+2y^2+1)z^{14}+\ldots$$

And so we see that if $\tau_2(j,k)$ is the number of partitions of $\langle j,k\rangle$ into unrestricted parts of the form $\langle 1,3^n\rangle$ for non-negative integers n the coefficients are generated by

Partition grid for $1/((1-yz)(1-yz^3)(1-yz^9)(1-yz^{27})...)$

⋮	⋮	⋮	⋮	⋮	⋮	⋮	⋮	⋮	⋮	⋮	⋮	⋮	⋮	⋮
14				1		2		1		1		1		1
13			1		2		1		1		1		1	
12		1		2		1		1		1		1		
11			1		1		1		1		1			
10		1		1		1		1		1				
9	1		1		1		1		1					
8				1		1		1						
7			1		1		1							
6		1		1		1								
5			1		1									
4		1		1										
3	1		1											
2		1												
1	1													
	1	2	3	4	5	6	7	8	9	10	11	12	13	14

(27.3.2)

Also, the identity (27.3.1) implies the

Theorem 27.4 *If $\tau_2(j,k)$ is the number of partitions of $\langle j,k\rangle$ into unrestricted parts of the form $\langle 1,3^n\rangle$ for non-negative integers n, then $\tau_2(j,k)$ is also the number of partitions of $\langle j,k\rangle$ into distinct parts with $1 \le m \le n$ in the form, one of but not both of $\langle 3^m,3^n\rangle$, or $\langle 2\times 3^m, 2\times 3^n\rangle$.*

EXERCISES

27.1 *Number of vector binary partitions for a single vector.* Using theorem 27.1, verify the case for the 2D vector $\langle 3,6\rangle$, that $B(3,6)=2$ is the number of vector partitions of $\langle 3,6\rangle$ into distinct parts of kind $\langle 2^a,2^b\rangle$ in which $a \le b$ with non-negative integers a and b. The two partitions are $\langle 2,4\rangle+\langle 1,2\rangle$ and $\langle 1,4\rangle+\langle 2,2\rangle$. Also $B(3,6)=2$ equals the number of partitions into "unrestricted" parts of kind $\langle 1,2^b\rangle$ in which b is a non negative integer. The two partitions are $\langle 1,2\rangle+\langle 1,2\rangle+\langle 1,2\rangle$ and $\langle 1,4\rangle+\langle 1,1\rangle+\langle 1,1\rangle$. Then also $B(3,6)=2$ is the coefficient of y^3z^6 in (27.1.5). The vector partitions $B(j,k)$ defined for (27.1.5) are easily visualized by the 2-space grid (27.1.7) given above.

27.2 *Ternary partition $\tau_2(j,k)$.* Reading from the grid (27.3.2) we see that $\tau_2(5,7)=1$ and $\tau_2(4,12)=2$. The single partition that makes $\tau_2(5,7)=1$ contributed by the left side product of (27.3.1) is

$$\langle 5,7\rangle = \langle 1,3^1\rangle+\langle 1,3^0\rangle+\langle 1,3^0\rangle+\langle 1,3^0\rangle+\langle 1,3^0\rangle.$$

Think About It...

The mathematician's best work is art, a high perfect art, as daring as the most secret dreams of imagination, clear and limpid. Mathematical genius and artistic genius touch one another.

Gösta Mittag-Leffler
Swedish Mathematician, 1846–1927

FURTHER READING

ALKAUSKAS, G. Generalization of the Rödseth-Gupta Theorem on Binary Partitions. *Lith. Math. J.*, April 2003.

ALKAUSKAS, G. and MENDEL-STRASSE, G. Congruence Properties of the Function that Counts Compositions into Powers of 2. *J. Integer Seq.*, 13, 2010. Article 10.5.3.

ANDREWS, G.E. *The Theory of Partitions*, volume 2 of *Encyclopedia of Mathematics and its Applications*. Addison-Wesley Publishing Company, Advanced Book Program, Reading, Massachusetts, USA, 1976.

CAMPBELL, G.B. A generalized formula of Hardy. *Int. J. Math. Math. Sci.*, 17(2):369–378, 1994.

CAMPBELL, G.B. Infinite products over hyperpyramid lattices. *Int. J. Math. and Math. Sci.*, 23(4):271–277, 2000.

CAMPBELL, G.B. Some n-space q-binomial theorem extensions and similar identities. arxiv preprint, June 2019. arXiv:1906.07526v1 [math.NT]: https://arxiv.org/abs/1906.07526.

CAMPBELL, G.B. and ZUJEV, A. The series that Ramanujan misunderstood. arxiv preprint, Oct 2016. arXiv:1610.03693v1 [math.NT], https://arxiv.org/abs/1610.03693v1.

CAMPBELL, G.B. Some n-space q-binomial theorem extensions and similar identities. arxiv preprint, June 2019. arXiv:1906.07526v1 [math.NT]: https://arxiv.org/abs/1906.07526.

CAMPBELL, G.B. Continued Fractions for partition generating functions. arxiv preprint, Jan 2023. arXiv:2301.12945 [math.CO], https://arxiv.org/abs/2301.12945.

CAMPBELL, G.B. Vector Partition Identities for 2D, 3D and nD Lattices. arxiv preprint, Feb 2023. arXiv:10.48550/ARXIV.2302.01091 [math.NT], https://arxiv.org/abs/2302.01091.

CAMPBELL, G.B. Visible Point Partition Identities for Polylogarithms, and Parametric Euler Sums. arxiv preprint, Jun 2023. arXiv:2306.02241v1 [math.CO], https://arxiv.org/abs/2306.02241.

CAMPBELL, G.B. Visible Point Vector Partition Identities for Hyperpyramid Lattices. arxiv preprint, Sep 2023. arXiv:2309.16094v1 [math.CO], https://arxiv.org/abs/2309.16094.

CHURCHHOUSE, R.F. Congruence properties of the binary partition function. *Proc. Cambridge Philos. Soc.*, 66:371–376, 1969.

HARDY, G.H. *On certain oscillating series.* In *Collected Papers, volume VI*, pages 146–167. Clarendon Press, Oxford, 1974.

MAHLER, K. On a Special Functional Equation. *J. London Math. Soc.*, 15:115–123, 1940.

MAHLER, K. On a Special Function. *J. Number Theory*, 12:20–26, 1980.

RÖDSETH, Ö. Some arithmetical properties of m-ary partitions. *Proc. Cambridge Philos. Soc.*, 68:447–453, 1970.

CHAPTER 28

VPV Identity Cases Related to Some Exponential Relations

28.1 RATIONAL SOLUTIONS OF $x^y = y^x$

IN THIS CHAPTER we give some surprising Visible Point Vector identities related to rational and integer solutions of $x^y = y^x$ and also for the equation $x^y y^x = v^w w^v$. In the 18th century Christian Goldbach and later Leonhard Euler [123] famously solved the equation

$$x^y = y^x \tag{28.1.1}$$

in rational numbers. All possible solutions are well known to be

$$x = \left(1 + \frac{1}{n}\right)^n, \quad y = \left(1 + \frac{1}{n}\right)^{n+1}, \tag{28.1.2}$$

for all positive integers n. The best reference for this may be the 1990 paper by Márta Svéd [288].

28.2 SOLUTIONS TO $x^y y^x = v^w w^v$ IN RATIONALS AND INTEGERS

In this section, we show there is an infinite set of nontrivial solutions to

$$x^y \, y^x = v^w \, w^v \tag{28.2.1}$$

in rational numbers x, y, v, w. This was recently given in Campbell [95]. However, defining what is meant by a trivial solution may be open to discussion. For a clearly nontrivial example, we see that $x = \frac{1}{3}$, $y = \frac{1}{6}$, $v = \frac{1}{2}$, $w = \frac{4}{3}$, gives us

$$\left(\frac{1}{3}\right)^{\frac{1}{6}} \left(\frac{1}{6}\right)^{\frac{1}{3}} = \left(\frac{1}{2}\right)^{\frac{4}{3}} \left(\frac{4}{3}\right)^{\frac{1}{2}}.$$

DOI: 10.1201/9781003174158-28

Another evidently nontrivial example $x=\frac{1}{2}$, $y=\frac{1}{3}$, $v=\frac{1}{2}$, $w=\frac{4}{3}$, gives the solution

$$\left(\frac{1}{2}\right)^{\frac{1}{3}}\left(\frac{1}{3}\right)^{\frac{1}{2}}=\left(\frac{1}{2}\right)^{\frac{4}{3}}\left(\frac{4}{3}\right)^{\frac{1}{2}}.$$

Noting the right sides of the above two examples are the same, we get a third non-trivial case $x=\frac{1}{2}$, $y=\frac{1}{3}$, $v=\frac{1}{3}$, $w=\frac{1}{6}$, with the solution

$$\left(\frac{1}{2}\right)^{\frac{1}{3}}\left(\frac{1}{3}\right)^{\frac{1}{2}}=\left(\frac{1}{3}\right)^{\frac{1}{6}}\left(\frac{1}{6}\right)^{\frac{1}{3}}.$$

Note that trivial solutions exist if $x=v=1$. Another solution that may be trivial (or not) is $x=\frac{1}{2}$, $y=\frac{1}{2}$, $v=\frac{1}{2}$, $w=1$. Another possibly trivial solution is $x=4$, $y=\frac{3}{2}$, $v=1$, $w=\frac{81}{2}$. A definitely trivial set of solutions is $x=a^b\,b^a$, $y=1$, $v=a$, $w=b$, for specific fixed positive integers a, b, and c.

For rationals a, b, and c, an infinite set of rational solutions can be found setting $y=ax$, $v=bx$, $w=cx$ for any rational x and substituting into $x^y y^x=v^w w^v$. This gives the algebraic solution: For $a+1\neq b+c$ with $a\neq 0$,

$$x=\left(\frac{b^c c^b}{a}\right)^{\frac{1}{a-b-c+1}},\ y=a\left(\frac{b^c c^b}{a}\right)^{\frac{1}{a-b-c+1}},\ v=b\left(\frac{b^c c^b}{a}\right)^{\frac{1}{a-b-c+1}},\ w=c\left(\frac{b^c c^b}{a}\right)^{\frac{1}{a-b-c+1}}. \tag{28.2.2}$$

We seek rational solutions, so it's clear that choosing $a=b+c$ will solve our quest. Hence, an infinite number of rational solutions to $x^y y^x=v^w w^v$ are given by integer values of b and c with $b+c\neq 0$, such that

$$x=\left(\frac{b^c c^b}{b+c}\right),\ y=b^c c^b,\ v=b\left(\frac{b^c c^b}{b+c}\right),\ w=c\left(\frac{b^c c^b}{b+c}\right). \tag{28.2.3}$$

Some verifiable random cases of this are shown here.

(b,c;x,y,v,w): $(1, 1; \frac{1}{2}, 1, \frac{1}{2}, \frac{1}{2})$, $(2, 1; \frac{2}{3}, 2, \frac{4}{3}, \frac{2}{3})$, $(3, 1; \frac{3}{4}, 3, \frac{9}{4}, \frac{3}{4})$, $(4, 1; \frac{4}{5}, 4, \frac{16}{5}, \frac{4}{5})$, $(1, 2; \frac{2}{3}, 2, \frac{2}{3}, \frac{4}{3})$, $(2, 2; 4, 16, 8, 8)$, $(3, 2; \frac{72}{5}, 72, \frac{216}{5}, \frac{144}{5})$, $(6, 2; 288, 2304, 1728, 576)$, $(1, 3; \frac{3}{4}, 3, \frac{3}{4}, \frac{9}{4})$, $(3, 3; \frac{3^5}{2}, 3^6, \frac{3^6}{2}, \frac{3^6}{2})$, $(6, 3; 17496, 157464, 104976, 52488)$, $(2, 4; \frac{2^7}{3}, 2^8, \frac{2^8}{3}, \frac{2^9}{3})$, $(5, 3; \frac{30375}{8}, 30375, \frac{151875}{8}, \frac{91125}{8})$, etc.

WolframAlpha tells us that each side of

$$288^{2304}\times 2304^{288}=1728^{576}\times 576^{1728}$$

is approximately 6.843×10^{6634} which is a 6635 digit number in base 10.

It also tells us that each side of

$$17496^{157464}\times 157464^{17496}=104976^{52488}\times 52488^{104976}$$

is approximately 9.8661×10^{759039} which is a 759040 digit number in base 10.

Next, we remark that the results for $x^y = y^x$ are seen as a classical curiosity that never went beyond recreational mathematics. However, $x^y y^x = v^w w^v$ seems to have not been studied nor considered in the extended literature and the set of rational solutions given here do not seem to have been published anywhere, other than by Campbell in 2022 [95]. Also, an infinite number of positive integer solutions to $x^y y^x = v^w w^v$ evidently exist, but this is not established yet. There are probably an entire class of functional equations similar to those considered in this chapter, susceptible to rational or integer solutions, such as perhaps $v^x x^z z^v = u^v w^x y^z$, each having their own possible transforms as cases of VPV identities in combination.

28.3 VPV IDENTITY TRANSFORMS USING $\mathbf{x^y = y^x}$

The Visible Point Vector 2D identities for the first quadrant lattice points we recall are for $|X| < 1$ and $|Y| < 1$ and nonzero denominators,

$$\prod_{\substack{(j,k)=1 \\ j,k \geq 1}} \left(\frac{1}{1 - X^j Y^k} \right)^{\frac{1}{k}} = \left(\frac{1}{1-Y} \right)^{\frac{1}{1-X}}, \tag{28.3.1}$$

$$\prod_{\substack{(j,k)=1 \\ j,k \geq 1}} \left(1 - X^j Y^k \right)^{\frac{1}{k}} = (1-Y)^{\frac{1}{1-X}} \tag{28.3.2}$$

These can be applied to the previous section's rational and integer solutions, by transforming each solution of $x^y = y^x$ and $x^y y^x = v^w w^v$.

We start with the first six solutions of $x^y = y^x$ in the form of $x = \left(1 + \frac{1}{n}\right)^n$, $y = \left(1 + \frac{1}{n}\right)^{n+1}$, listed as follows,

$$\begin{aligned}
n = 1 &\Rightarrow x = 2^1,\ y = 2^2; \\
n = 2 &\Rightarrow x = \left(\frac{3}{2}\right)^2,\ y = \left(\frac{3}{2}\right)^3; \\
n = 3 &\Rightarrow x = \left(\frac{4}{3}\right)^3,\ y = \left(\frac{4}{3}\right)^4; \\
n = 4 &\Rightarrow x = \left(\frac{5}{4}\right)^4,\ y = \left(\frac{5}{4}\right)^5; \\
n = 5 &\Rightarrow x = \left(\frac{6}{5}\right)^5,\ y = \left(\frac{6}{5}\right)^6; \\
n = 6 &\Rightarrow x = \left(\frac{7}{6}\right)^6,\ y = \left(\frac{7}{6}\right)^7.
\end{aligned}$$

Hence, for example, the $n = 1$ only integer nontrivial solution,

$$(2)^{(4)} = (4)^{(2)} \tag{28.3.3}$$

can be written in the right side of (28.3.1) form

$$\left(\frac{1}{1-\frac{1}{2}}\right)^{\left(\frac{1}{1-\frac{3}{4}}\right)} = \left(\frac{1}{1-\frac{3}{4}}\right)^{\left(\frac{1}{1-\frac{1}{2}}\right)}, \tag{28.3.4}$$

where $X = \frac{1}{2}$, $Y = \frac{3}{4}$.

Applying (28.3.2) to (28.3.4) gives us the transform

$$\prod_{\substack{(j,k)=1 \\ j,k\geq 1}} \left\{1-\left(\frac{1}{2}\right)^j\left(\frac{1}{4}\right)^k\right\}^{\frac{1}{k}} = \prod_{\substack{(j,k)=1 \\ j,k\geq 1}} \left\{1-\left(\frac{1}{4}\right)^j\left(\frac{1}{2}\right)^k\right\}^{\frac{1}{k}}. \tag{28.3.5}$$

Likewise, the $n = 2$ case leads to the solution

$$\left(\left(\frac{3}{2}\right)^2\right)^{\left(\frac{3}{2}\right)^3} = \left(\left(\frac{3}{2}\right)^3\right)^{\left(\frac{3}{2}\right)^2}. \tag{28.3.6}$$

This solution can then be written in the right side of (28.3.1) form

$$\left(\frac{1}{1-Y}\right)^{\left(\frac{1}{1-X}\right)} = \left(\frac{1}{1-X}\right)^{\left(\frac{1}{1-Y}\right)}, \tag{28.3.7}$$

where $X = 1-\left(\frac{2}{3}\right)^2$, $Y = 1-\left(\frac{2}{3}\right)^3$.

Applying (28.3.2) to (28.3.6) gives us the transform

$$\prod_{\substack{(j,k)=1 \\ j,k\geq 1}} \left\{1-\left(1-\left(\frac{2}{3}\right)^2\right)^j\left(1-\left(\frac{2}{3}\right)^3\right)^k\right\}^{\frac{1}{k}} \tag{28.3.8}$$

$$= \prod_{\substack{(j,k)=1 \\ j,k\geq 1}} \left\{1-\left(1-\left(\frac{2}{3}\right)^3\right)^j\left(1-\left(\frac{2}{3}\right)^2\right)^k\right\}^{\frac{1}{k}}$$

which is the same as

$$\left(\left(\frac{2}{3}\right)^2\right)^{\left(\frac{3}{2}\right)^3} = \left(\left(\frac{2}{3}\right)^3\right)^{\left(\frac{3}{2}\right)^2}.$$

Again, we apply this same approach to the general rational solutions $x = \left(1+\frac{1}{n}\right)^n$, $y = \left(1+\frac{1}{n}\right)^{n+1}$

$$\left(\left(1+\frac{1}{n}\right)^n\right)^{\left(1+\frac{1}{n}\right)^{n+1}} = \left(\left(1+\frac{1}{n}\right)^{n+1}\right)^{\left(1+\frac{1}{n}\right)^n}. \tag{28.3.9}$$

This solution can then be written in the right side of (28.3.1) form

$$\left(\frac{1}{1-Y}\right)^{\left(\frac{1}{1-X}\right)}=\left(\frac{1}{1-X}\right)^{\left(\frac{1}{1-Y}\right)}, \tag{28.3.10}$$

where $X=1-\left(\frac{n}{n+1}\right)^{n}$, $Y=1-\left(\frac{n}{n+1}\right)^{(n+1)}$.

Applying (28.3.2) to (28.3.6) gives us the transform valid for all positive integers n,

$$\prod_{\substack{(j,k)=1\\ j,k\geq 1}}\left\{1-\left(1-\left(\frac{n}{n+1}\right)^{n}\right)^{j}\left(1-\left(\frac{n}{n+1}\right)^{(n+1)}\right)^{k}\right\}^{\frac{1}{k}} \tag{28.3.11}$$

$$=\prod_{\substack{(j,k)=1\\ j,k\geq 1}}\left\{1-\left(1-\left(\frac{n}{n+1}\right)^{(n+1)}\right)^{j}\left(1-\left(\frac{n}{n+1}\right)^{n}\right)^{k}\right\}^{\frac{1}{k}}$$

which is the same as

$$\left(\left(\frac{n}{n+1}\right)^{n}\right)^{\left(\frac{n+1}{n}\right)^{(n+1)}}=\left(\left(\frac{n}{n+1}\right)^{(n+1)}\right)^{\left(\frac{n+1}{n}\right)^{n}}.$$

28.4 VPV IDENTITY TRANSFORMS USING $\mathbf{x^y y^x = v^w w^v}$

Recall earlier we stated an algebraic solution of $x^y y^x = v^w w^v$:

For $a+1\neq b+c$ with $a\neq 0$,

$$x=\left(\frac{b^c c^b}{a}\right)^{\frac{1}{a-b-c+1}},\ y=a\left(\frac{b^c c^b}{a}\right)^{\frac{1}{a-b-c+1}},\ v=b\left(\frac{b^c c^b}{a}\right)^{\frac{1}{a-b-c+1}},\ w=c\left(\frac{b^c c^b}{a}\right)^{\frac{1}{a-b-c+1}}. \tag{28.4.1}$$

We then selected $a=b+c$ to find an infinite rational set of solutions x, y, v, and w, which appeared incidentally to yield an infinite set of positive integer solutions too.

Following a similar process to the previous section, we can deduce that

$$\left(\frac{1}{1-X}\right)^{\left(\frac{1}{1-Y}\right)}\left(\frac{1}{1-Y}\right)^{\left(\frac{1}{1-X}\right)}=\left(\frac{1}{1-V}\right)^{\left(\frac{1}{1-W}\right)}\left(\frac{1}{1-W}\right)^{\left(\frac{1}{1-V}\right)}, \tag{28.4.2}$$

where for non-zero denominators

$$X=\frac{x-1}{x}, Y=\frac{ax-1}{ax}, V=\frac{bx-1}{bx}, W=\frac{cx-1}{cx}, \quad with \quad x=\left(\frac{b^c c^b}{a}\right)^{\frac{1}{a-b-c+1}}. \tag{28.4.3}$$

Taking the reciprocal of both sides of (28.4.2) we then apply (28.3.2) to arrive for non-zero denominators at

$$\prod_{\substack{(j,k)=1\\ j,k\geq 1}} \left\{\left[1-\left(\frac{x-1}{x}\right)^j\left(\frac{ax-1}{ax}\right)^k\right]\left[1-\left(\frac{ax-1}{ax}\right)^j\left(\frac{x-1}{x}\right)^k\right]\right\}^{\frac{1}{k}} \tag{28.4.4}$$

$$= \prod_{\substack{(j,k)=1\\ j,k\geq 1}} \left\{\left[1-\left(\frac{bx-1}{bx}\right)^j\left(\frac{cx-1}{cx}\right)^k\right]\left[1-\left(\frac{cx-1}{cx}\right)^j\left(\frac{bx-1}{bx}\right)^k\right]\right\}^{\frac{1}{k}},$$

in which $x = \left(\frac{b^c c^b}{a}\right)^{\frac{1}{a-b-c+1}}$, provided that $a+1 \neq b+c$ and $a \neq 0$.

As an example, consider

$$288^{2304} \times 2304^{288} = 1728^{576} \times 576^{1728},$$

which is the same as

$$\left(\frac{1}{1-\frac{287}{288}}\right)^{\left(\frac{1}{1-\frac{2303}{2304}}\right)} \times \left(\frac{1}{1-\frac{2303}{2304}}\right)^{\left(\frac{1}{1-\frac{287}{288}}\right)}$$

$$= \left(\frac{1}{1-\frac{1727}{1728}}\right)^{\left(\frac{1}{1-\frac{575}{576}}\right)} \times \left(\frac{1}{1-\frac{575}{576}}\right)^{\left(\frac{1}{1-\frac{1727}{1728}}\right)};$$

and then consider the example

$$17496^{157464} \times 157464^{17496} = 104976^{52488} \times 52488^{104976},$$

which is the same as

$$\left(\frac{1}{1-\frac{17495}{17496}}\right)^{\left(\frac{1}{1-\frac{157463}{157464}}\right)} \times \left(\frac{1}{1-\frac{157463}{157464}}\right)^{\left(\frac{1}{1-\frac{17495}{17496}}\right)}$$

$$= \left(\frac{1}{1-\frac{104975}{104976}}\right)^{\left(\frac{1}{1-\frac{52487}{52488}}\right)} \times \left(\frac{1}{1-\frac{52487}{52488}}\right)^{\left(\frac{1}{1-\frac{104975}{104976}}\right)}.$$

Applying (28.4.4) then, we get the two VPV transform identities as follows.

$$\prod_{\substack{(j,k)=1\\ j,k\geq 1}} \left\{\left[1-\left(\frac{287}{288}\right)^j\left(\frac{2303}{2304}\right)^k\right]\left[1-\left(\frac{2303}{2304}\right)^j\left(\frac{287}{288}\right)^k\right]\right\}^{\frac{1}{k}} \tag{28.4.5}$$

$$= \prod_{\substack{(j,k)=1\\ j,k\geq 1}} \left\{ \left[1 - \left(\frac{1727}{1728}\right)^j \left(\frac{575}{576}\right)^k \right] \left[1 - \left(\frac{575}{576}\right)^j \left(\frac{1727}{1728}\right)^k \right] \right\}^{\frac{1}{k}};$$

and

$$\prod_{\substack{(j,k)=1\\ j,k\geq 1}} \left\{ \left[1 - \left(\frac{17495}{17496}\right)^j \left(\frac{157463}{157464}\right)^k \right] \left[1 - \left(\frac{157463}{157464}\right)^j \left(\frac{17495}{17496}\right)^k \right] \right\}^{\frac{1}{k}} \tag{28.4.6}$$

$$= \prod_{\substack{(j,k)=1\\ j,k\geq 1}} \left\{ \left[1 - \left(\frac{104975}{104976}\right)^j \left(\frac{52487}{52488}\right)^k \right] \left[1 - \left(\frac{52487}{52488}\right)^j \left(\frac{104975}{104976}\right)^k \right] \right\}^{\frac{1}{k}}.$$

Think About It...

Euler's work averages 800 pages a year from 1725 to 1783. He also wrote over 4500 letters and hundreds of manuscripts.

It has been estimated that Leonhard Euler was the author of a quarter of the combined output in mathematics, physics, mechanics, astronomy, and navigation in the 18th century.

Arjang A. Assad, (2007)
Leonhard Euler: A brief appreciation. Networks. 49 (3): 190–198

FURTHER READING

CAMPBELL, G.B. Fun with numbers: Rational solutions to $x^y y^x = v^w w^v$. *Aust. Math. Soc. Gazette*, 49(5):210–211, November 2022.

EULER, L. Introductio in analysin infinitorum. In *Opera Omnia*, volume 1. Marcum-Michaelum, Brousquet, Lausannae, 1748. Chapter 16.

SVED, M. On the Rational Solutions of $x^y = y^x$. *Mathematics Magazine*, 63(1):30–33, 1990. JSTOR, https://maa.org/programs/faculty-and-departments/classroom-capsules-and-notes/on-the-rational-solutions-of-xy-yx. Accessed 14 Sep. 2022.

V

Models, Interpretations, and Some Useful Tools

CHAPTER 29

2D and 3D Stepping Stones, Forests, Orchards and Light Diffusions

29.1 MODELS FOR OUR VECTOR PARTITION THEORIES AND ANALYSES

VECTOR PARTITION MODELS arise from the combinatory analysis in this book. Among the models we consider are:

(a) Counting the number of Stepping Stone Jumps.

(b) Spent Energies from carrying weights on Stepping Stone Jumps.

(c) Polya's Orchard Visibility Problem.

(d) Euclid's Orchard.

(e) Light Diffusion Models.

(f) Charge Flow Network Models.

Among the ideas that can apply to Vector Partition identities is the notion of stepping stone jumps between integer coordinate points, assuming an integer lattice point is a stepping stone in space. Imagine then jumping from stone to stone with integer coordinates. In the simplest cases, the jump is to another stone that is visible to you from your present position. An example of this would be, say jumping from point $\langle 1,1\rangle$ to point $\langle 2,3\rangle$, a *distance and direction* given by the vector $\langle 1,2\rangle$, since $\langle 1,1\rangle + \langle 1,2\rangle = \langle 2,3\rangle$ under vector addition.

Random walks: A random walk is a mathematical concept that describes a path that consists of a succession of random steps on some mathematical space. The simplest random walk is a one-dimensional walk, where a particle starts at a point and then takes a step to the left or right with equal probability. After taking many steps, the particle will likely be far from its starting point, and the distribution of its possible positions will be bell-shaped.

DOI: 10.1201/9781003174158-29

Random walks can be used to model a variety of phenomena, such as:

- A drunkard's walk
- The movement of a molecule in a gas
- The price of a stock over time
- The spread of a disease through a population

Random walks can be classified into two types:

1) Independent random walks: In an independent random walk, the direction of each step is independent of the direction of previous steps. This means that the probability of taking a step to the left is the same as the probability of taking a step to the right, regardless of what happened in previous steps.

2) Correlated random walks: In a correlated random walk, the direction of each step is correlated with the direction of previous steps. This means that the probability of taking a step to the left is more likely if the previous step was also to the left, and vice versa. Random walks are a powerful tool for modeling a variety of phenomena. They can be used to understand the behavior of complex systems and to make predictions about future events.

The above information and features of random walks have a corresponding aspect when looking at Vector Partitions. Basically a Vector Partition is a quiver of arrows; each arrow being a vector so that the entire quiver of arrows can be positioned head to tail going from the origin to a particular lattice point vector. Each arrow is a part of the vector partition which is the end to end path made from the arrows connecting the origin to the lattice point in space.

Vector Partition as a stepping stone jump path: A vector partition can be interpreted as a stepping stone set of jumps across a designated set of points. An integer partition is a 1D straight line set of jumps. In 2D space the jumps can be between integer lattice points as stepping stones.

Unweighted jumps: The jump may be all we care about; so counting the number of jumps may be sufficient. In that case, the number of vector partitions is actually the number of jumps, or the number of arrows in a quiver. If our vector is one-dimensional, along a line, then we have the theory of integer partitions as we know it, fitting into this category. Note that plane partition theory seems to require more than one-dimension generally.

Weighted jumps: However, as is common, each jump may "carry a weight" or have a value determined from a formula for that weight arising from an assigned rule for the jump. So we are not merely counting jumps for each partition, but assigning a *function of the partition*, wherein a vector partition maps to a value based on properties of each vector part of the partition. This concept is similar to the *crank* function applied to a partition to separate an integer partition into congruence classes as was described in chapter 5. Or we could say that the quiver of arrows contains arrows of different lengths and different weights assigned by a set of rules or restrictions.

29.2 2D STEPPING STONE JUMPS TO VISIBLE POINTS

Remember always that the coordinates of the first quadrant visible points correlate to the countable set of fractions between 0 and 1. We correlate the Order n VPVs with the Farey Fractions of Order n for positive integer n.

The options for vectors in the ensuing Upper VPV cases come from the Upper first quadrant in the diagram,

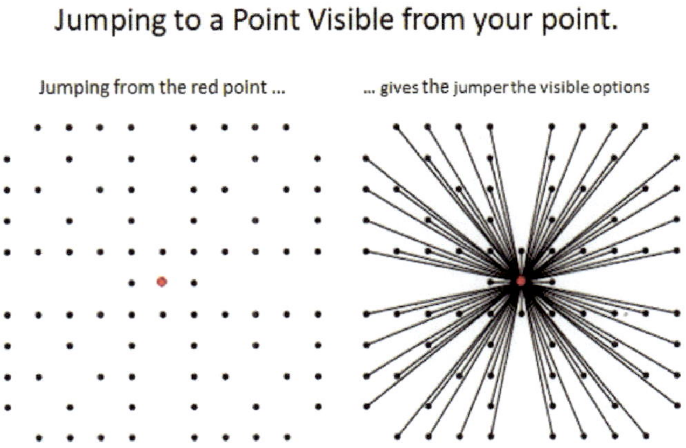

Figure 29.1 2D Visible Points Jump Options for Order 5

In this case we show the VPV jump options for Order 5, but the principle holds for any Order.

Partitions into Upper VPV Order 2, Distinct Parts

For the Upper Visible Point Vectors of Order 2 the co-ordinates correspond to the Farey Fractions of Order 2, excluding $\frac{1}{1}$,

$$F_2 = \left\{ \frac{0}{1}, \frac{1}{2}, \frac{1}{1} \right\}$$

This leads us to the following two diagrams. Figure 29.2 is the number of partitions into *distinct upper VPV parts* of Order 2. Figure 29.3 is the number of partitions into *distinct upper VPV even number of parts minus the number of partitions into distinct upper VPV even number of parts* of Order 2.

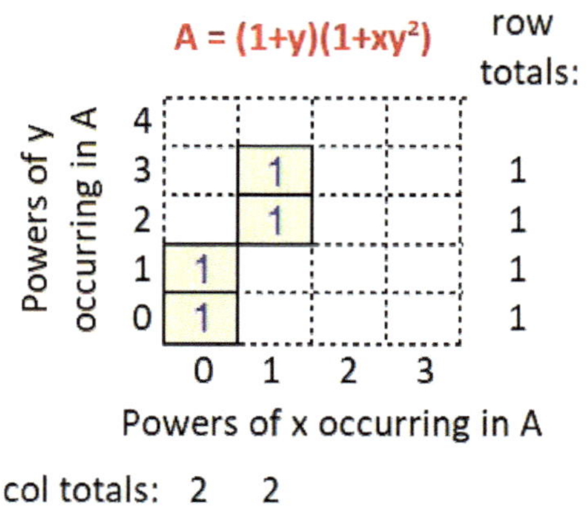

Figure 29.2 Upper VPV Order 2 Partitions into Distinct Parts

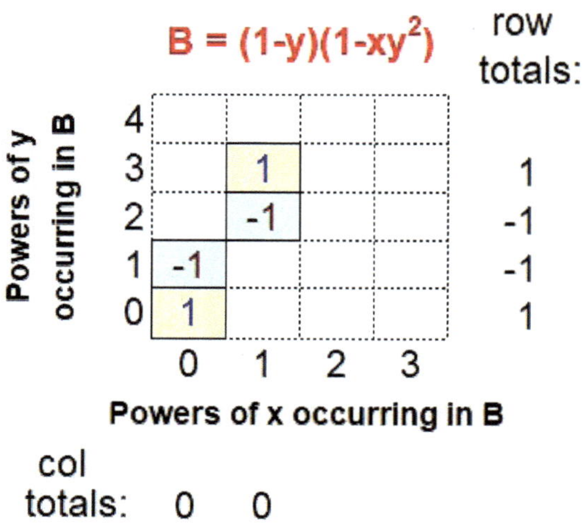

Figure 29.3 Upper VPV Order 2 Partitions into Distinct *Even♯ parts − Odd♯ parts*

Partitions into Upper VPV Order 3, Distinct Parts

For the Upper Visible Point Vectors of Order 3 the co-ordinates correspond to the Farey Fractions of Order 3, excluding $\frac{1}{1}$,

$$F_3 = \left\{ \frac{0}{1}, \frac{1}{3}, \frac{1}{2}, \frac{2}{3}, \frac{1}{1} \right\}$$

This leads us to the following two diagrams. Figure 29.4 is the number of partitions into *distinct upper VPV parts* of Order 3. Figure 29.5 is the number of partitions into *distinct upper VPV even number of parts minus the number of partitions into distinct upper VPV even number of parts* of Order 3.

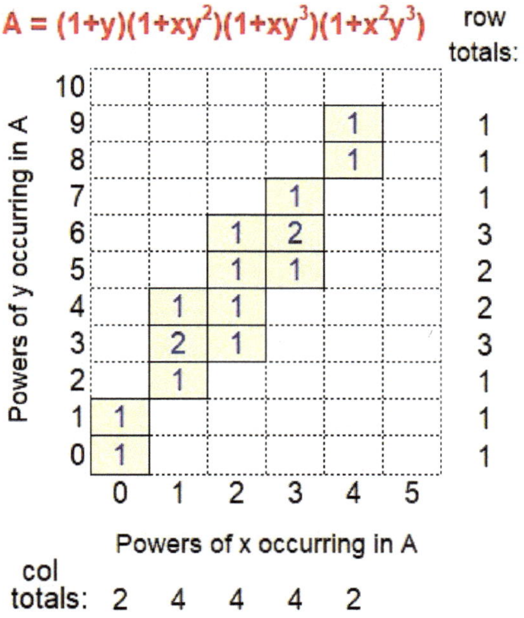

Figure 29.4 Upper VPV Order 3 Partitions into Distinct Parts

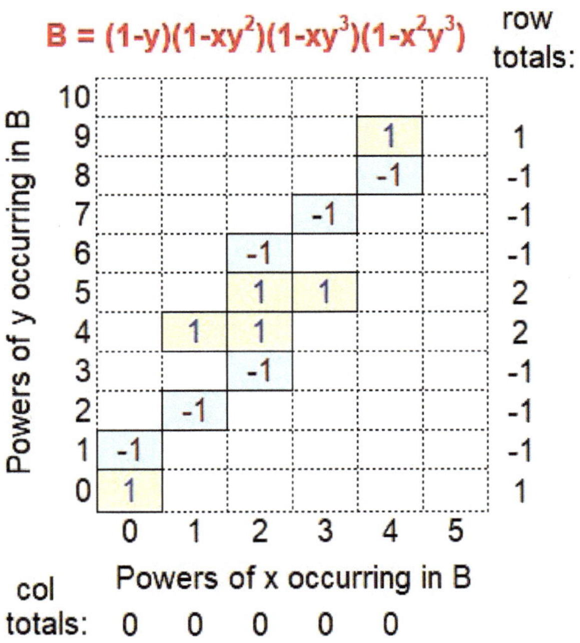

Figure 29.5 Upper VPV Order 3 Partitions into Distinct $Even\sharp\, parts - Odd\sharp\, parts$

Partitions into Upper VPV Order 4, Distinct Parts

For the Upper Visible Point Vectors of Order 4 the co-ordinates correspond to the Farey Fractions of Order 4, excluding $\frac{1}{1}$,

$$F_4 = \left\{ \frac{0}{1}, \frac{1}{4}, \frac{1}{3}, \frac{1}{2}, \frac{2}{3}, \frac{3}{4}, \frac{1}{1} \right\}$$

This leads us to the following two diagrams. Figure 29.6 is the number of partitions into *distinct upper VPV parts* of Order 4. Figure 29.7 is the number of partitions into *distinct upper VPV even number of parts minus the number of partitions into distinct upper VPV even number of parts* of Order 4.

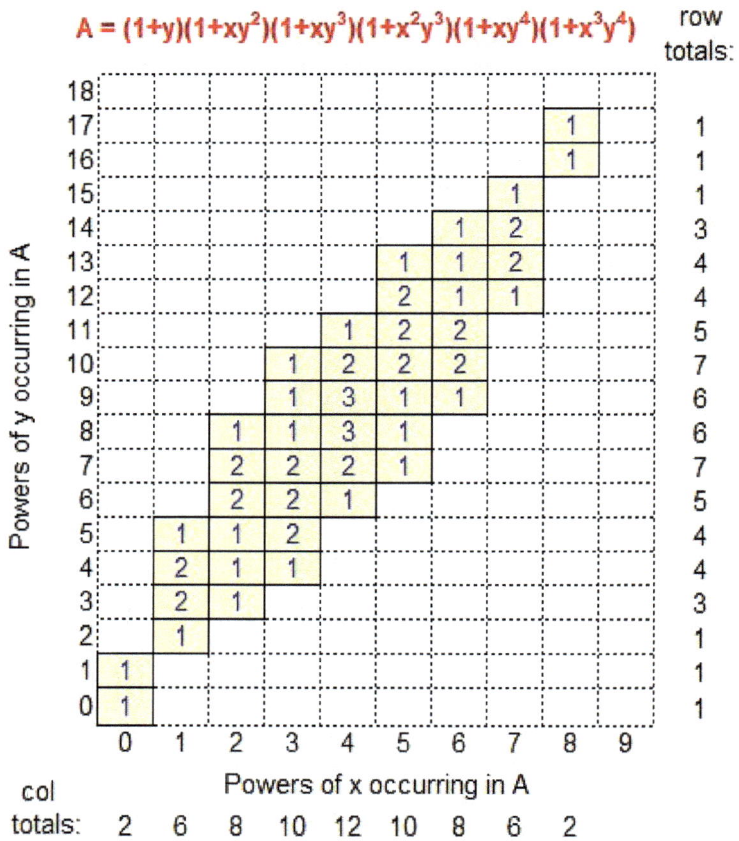

Figure 29.6 Upper VPV Order 4 Partitions into Distinct Parts

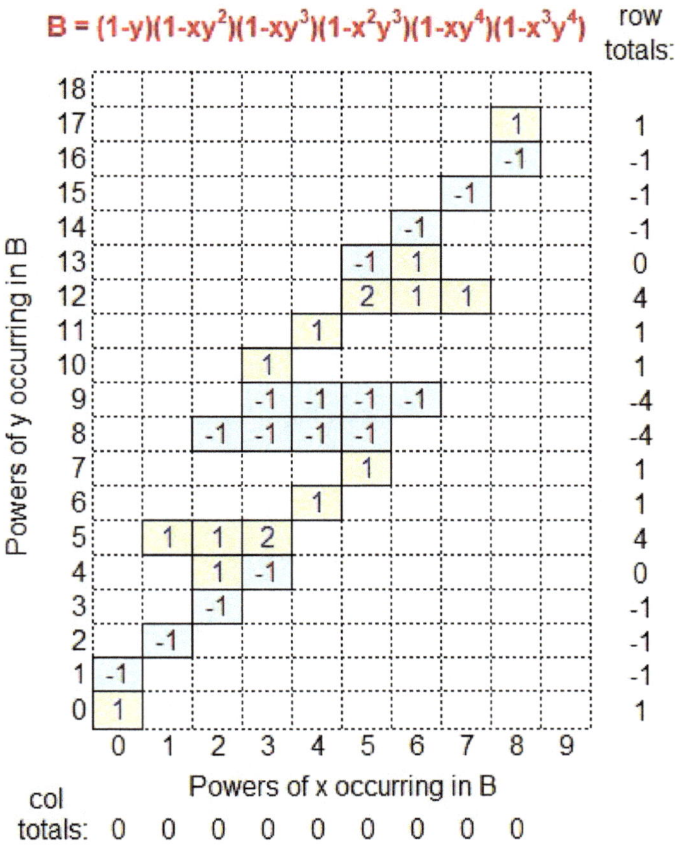

Figure 29.7 Upper VPV Order 4 Partitions into Distinct $Even\sharp\ parts - Odd\sharp\ parts$

Partitions into Upper VPV Order 5, Distinct Parts

For the Upper Visible Point Vectors of Order 5 the co-ordinates correspond to the Farey Fractions of Order 5, excluding $\frac{1}{1}$,

$$F_5 = \left\{ \frac{0}{1}, \frac{1}{5}, \frac{1}{4}, \frac{1}{3}, \frac{2}{5}, \frac{1}{2}, \frac{3}{5}, \frac{2}{3}, \frac{3}{4}, \frac{4}{5}, \frac{1}{1} \right\}$$

This leads us to the following two diagrams. Figure 29.8 is the number of partitions into *distinct upper VPV parts* of Order 5. Figure 29.9 is the number of partitions into *distinct upper VPV even number of parts minus the number of partitions into distinct upper VPV even number of parts* of Order 5.

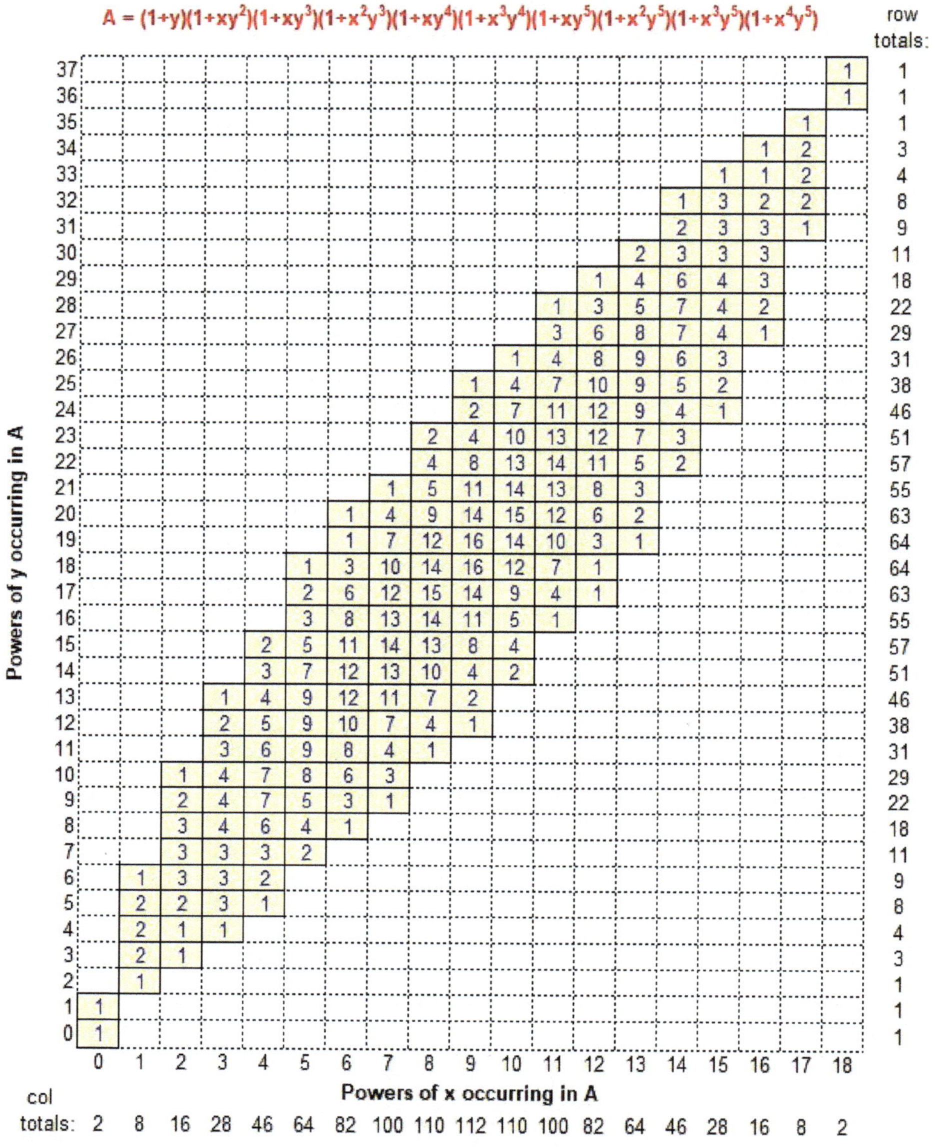

Figure 29.8 Upper VPV Order 5 Partitions into Distinct Parts

Partitions into Upper VPV Order 6, Distinct Parts

For the Upper Visible Point Vectors of Order 6 the co-ordinates correspond to the Farey Fractions of Order 6, excluding $\frac{1}{1}$,

$$F_6 = \left\{ \frac{0}{1}, \frac{1}{6}, \frac{1}{5}, \frac{1}{4}, \frac{1}{3}, \frac{2}{5}, \frac{1}{2}, \frac{3}{5}, \frac{2}{3}, \frac{3}{4}, \frac{4}{5}, \frac{5}{6}, \frac{1}{1} \right\}$$

This leads us to the following two diagrams. Figure 29.10 is the number of partitions into *distinct upper VPV parts* of Order 6. Figure 29.11 is the number of partitions into *distinct upper VPV even number of parts minus the number of partitions into distinct upper VPV even number of parts* of Order 6.

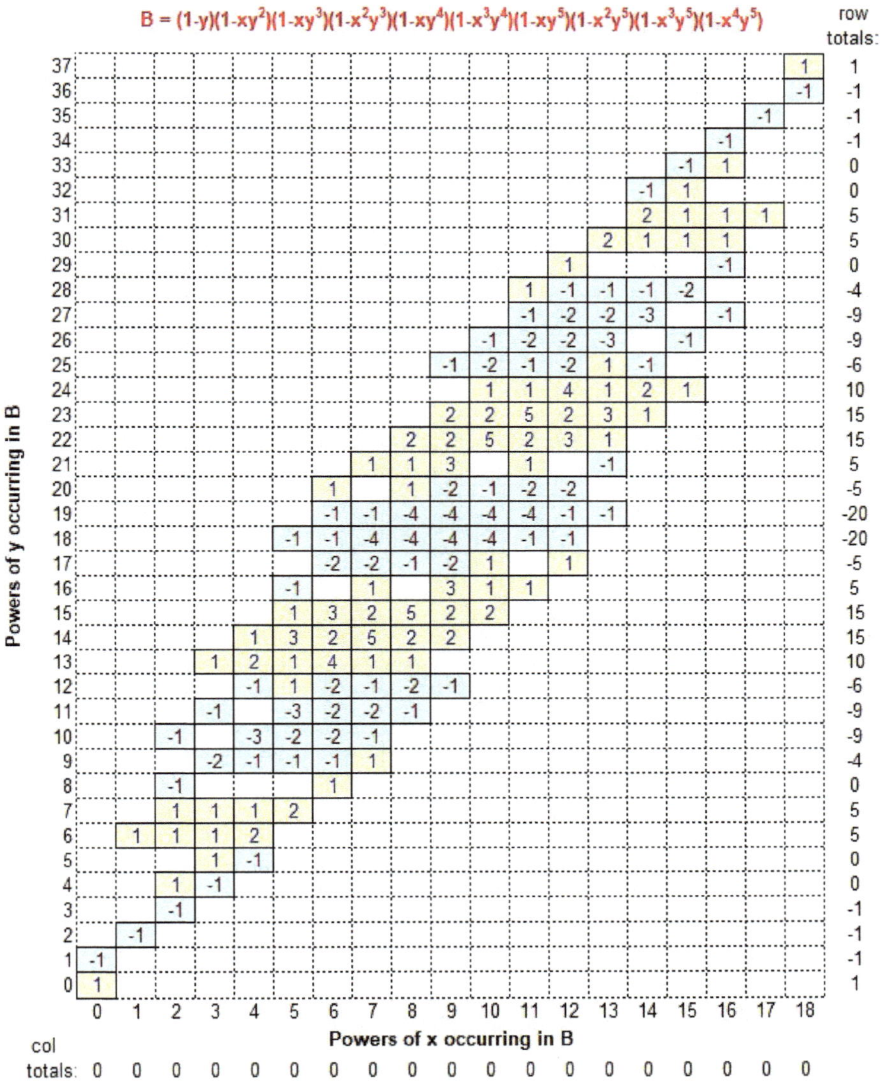

Figure 29.9 Upper VPV Order 5 Partitions into Distinct $Even\sharp\, parts - Odd\sharp\, parts$

Note that the **Upper VPV Order n Partition Grids** for positive integers n, with our examples in Figures 29.2 to 29.11, each have row totals and column totals. Each such total encodes a theorem on the vector partitions that comprise that row or column. Note that in Figures 29.3, 29.5, 29.7, 29.9, and 29.11, the sums of each vertical in the grid are zero, due to the factor $(1-y)$ in the generating function. So we are able to state the

Theorem 29.1 *The sum of all vector partitions along the vertical lines is zero for "number of partitions into even number of parts" minus "number of partitions into odd number of parts" for dissimilar visible point vector partitions $\langle a,b \rangle$ with $gcd(a,b)=1$ and $0<a<b$, with $0<b\leq 6$.*

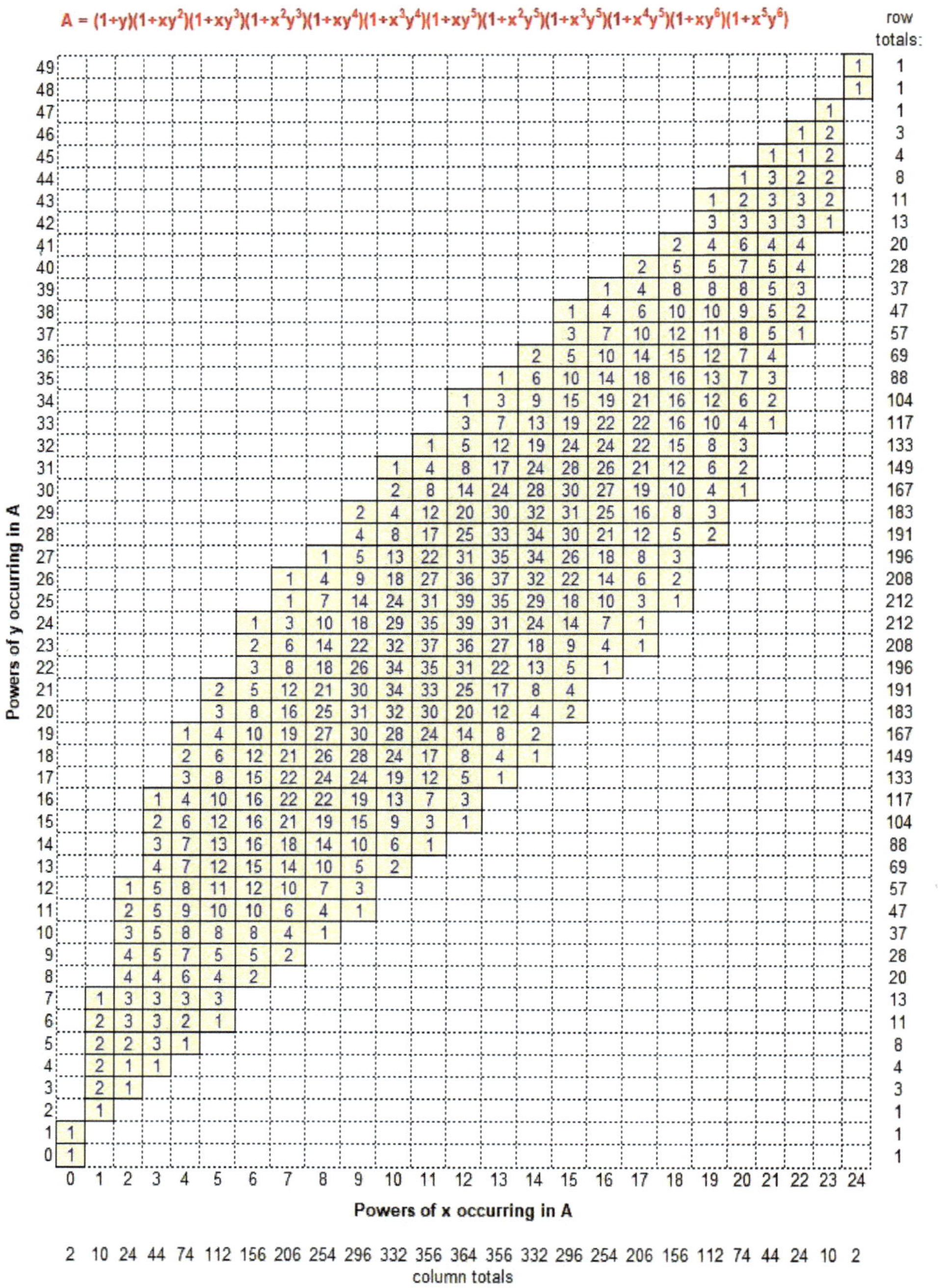

Figure 29.10 Upper VPV Order 6 Partitions into Distinct Parts

The proof of theorem 29.1 is trivial as the generating function has a factor of $(1-y)$.

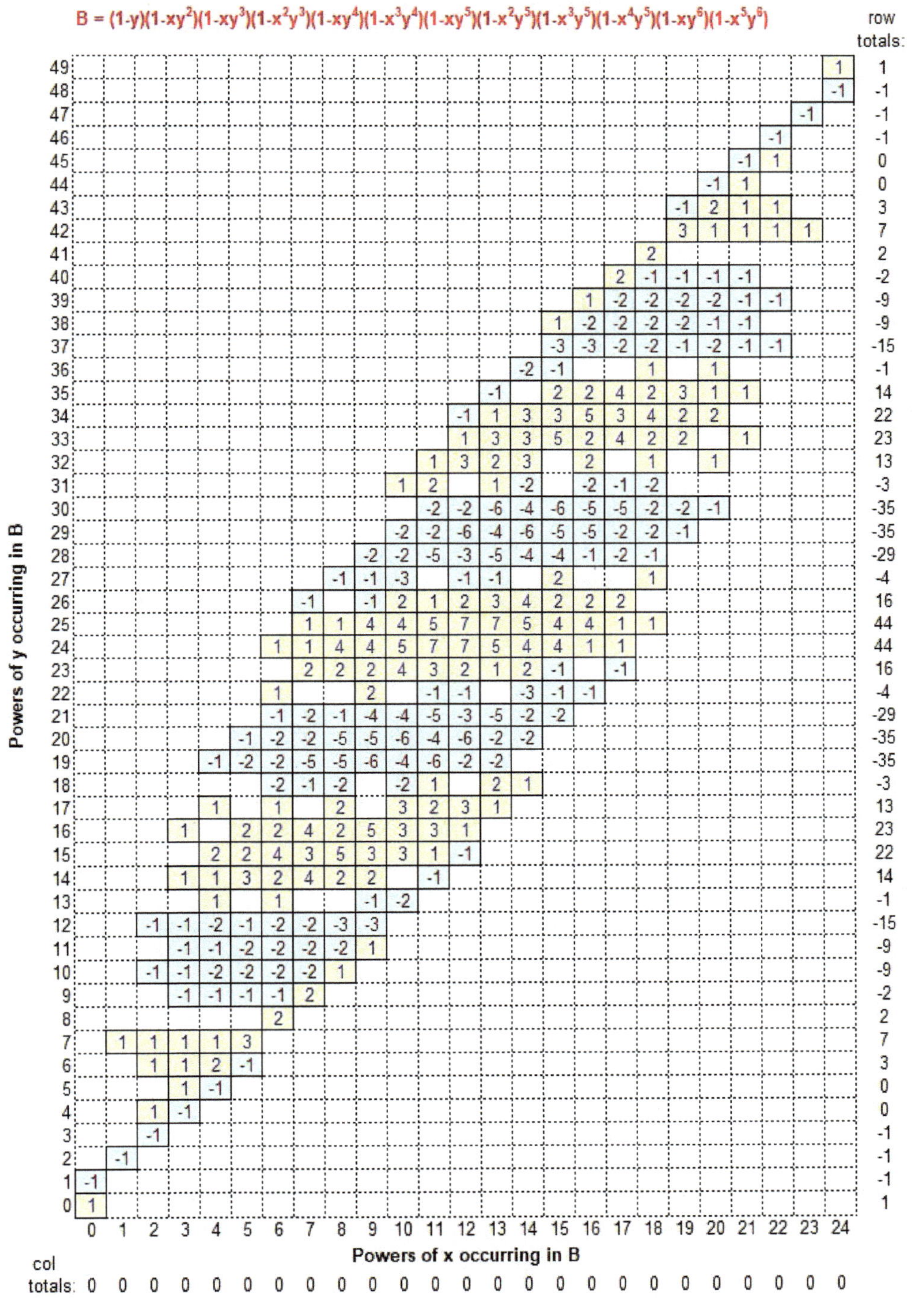

Figure 29.11 Upper VPV Order 6 Partitions into Distinct *Even♯ parts − Odd♯ parts*

29.3 STEPPING STONE WEIGHT VALUES FOR JUMPS

The author's *VPV identity papers* published in the mid to late 1990s (see Campbell [84, 85, 86, 89]) have many inferred combinatorial interpretations that were not fully examined.

29.4 ANY RADIAL-FROM-ORIGIN LINE HOLDS ALL PARALLEL VECTORS

We shall discuss some combinatory inferences here as well as stating some further results. We consider the reciprocals of the VPV identities already given in Chapter 20, corollary 20.5 as follows. Each one is easily proved by differentiating the logarithm of both sides.

Theorem 29.2 *If absolute values of x, y, z, and t are all less than unity,*

$$\prod_{\substack{(a,b)=1\\ a\geq 0;\ b\geq 1}} \left(\frac{1}{1-y^a z^b}\right)^{1/b} = \left(\frac{1}{1-z}\right)^{\frac{1}{1-y}}, \tag{29.4.1}$$

$$\prod_{\substack{(a,b,c)=1\\ a,b\geq 0;\ c\geq 1}} \left(\frac{1}{1-x^a y^b z^c}\right)^{1/c} = \left(\frac{1}{1-z}\right)^{\frac{1}{(1-x)(1-y)}}, \tag{29.4.2}$$

$$\prod_{\substack{(a,b,c,d)=1\\ a,b,c\geq 0;\ d\geq 1}} \left(\frac{1}{1-w^a x^b y^c z^d}\right)^{1/d} = \left(\frac{1}{1-z}\right)^{\frac{1}{(1-w)(1-x)(1-y)}}, \tag{29.4.3}$$

$$\prod_{\substack{(a,b,c,d,e)=1\\ a,b,c,d\geq 0;\ e\geq 1}} \left(\frac{1}{1-v^a w^b x^c y^d z^e}\right)^{1/e} = \left(\frac{1}{1-z}\right)^{\frac{1}{(1-v)(1-w)(1-x)(1-y)}}. \tag{29.4.4}$$

These identities are simplest cases of the general n-space infinite product theorem in Campbell [86], which is also given in Chapter 20 earlier in this book. Both the left side general term and the right sides of (29.4.1) to (29.4.4) are easy to expand as power series generating functions in z. To begin with, write the the left side of (29.4.1) as

$$\prod_{\substack{(a,b)=1\\ a\geq 0;\ b\geq 1}} \left[1+\binom{1/b}{\gcd(a,b)} y^a z^b + \binom{1/b}{\gcd(2a,2b)} y^{2a} z^{2b} + \binom{1/b}{\gcd(3a,3b)} y^{3a} z^{3b} + \dots\right] \tag{29.4.5}$$

Now, consider the identity (29.4.1). The square bracketed term under the product in (29.4.5) contributes "parts of the vector partition" that are integer coordinate vectors parallel to the line $z = \frac{b}{a}y$ in the first 2D quadrant of lattice point vectors. The square bracketed sum contributes exactly one part in the form

$$\binom{1/b}{\gcd(na,nb)} y^{na} z^{nb} = \binom{1/b}{n} y^{na} z^{nb}$$

to a partition in the appropriate 2D region.

So here we are looking at a "function of a partition" where, each partition of form

$$\sum_{j=1}^{m}\langle a_j,b_j\rangle=\langle a,b\rangle \quad \text{maps onto} \quad \prod_{j=1}^{m}\binom{1/b}{\gcd(ja,jb)}y^{ja}z^{jb}=y^a z^b.$$

So, we have examined the left side combinatory issues of (29.4.1). Next, the right side of this gives us

$$\left(\frac{1}{1-z}\right)^{\frac{1}{1-y}} \tag{29.4.6}$$

$$=1-\left(1-\frac{y}{1}\right)z+\left(1-\frac{y}{1}\right)\left(1-\frac{y}{2}\right)z^2-\left(1-\frac{y}{1}\right)\left(1-\frac{y}{2}\right)\left(1-\frac{y}{3}\right)z^3+\ldots$$

$$=1-\left(1-\frac{y}{1}\right)z$$

$$+\left(1-\frac{3y}{2}+\frac{y^2}{2}\right)z^2$$

$$-\left(1-\frac{11y}{6}+\frac{y^2}{1}-\frac{y^3}{6}\right)z^3$$

$$+\left(1-\frac{25y}{12}+\frac{35y^2}{24}-\frac{5y^3}{12}+\frac{y^4}{24}\right)z^4$$

$$-\left(1-\frac{137y}{60}+\frac{15y^2}{8}-\frac{17y^3}{24}+\frac{y^4}{8}-\frac{y^5}{120}\right)z^5$$

$$+\left(1-\frac{49y}{20}+\frac{203y^2}{90}-\frac{49y^3}{48}+\frac{35y^4}{144}-\frac{7y^5}{240}+\frac{y^6}{720}\right)z^6$$

$$-\left(1-\frac{363y}{140}+\frac{469y^2}{180}-\frac{967y^3}{720}+\frac{7y^4}{18}-\frac{23y^5}{360}+\frac{y^6}{180}-\frac{y^7}{5040}\right)z^7$$

$$+\left(1-\frac{761y}{280}+\frac{29531y^2}{10080}-\frac{267y^3}{160}+\frac{1069y^4}{1920}-\frac{9y^5}{80}+\frac{13y^6}{960}-\frac{y^7}{1120}+\frac{y^8}{40320}\right)z^8$$

$$-\left(1-\frac{7129y}{2520}+\frac{6515y^2}{2016}-\frac{4523y^3}{2268}+\frac{95y^4}{128}-\frac{3013y^5}{17280}+\frac{5y^6}{192}-\frac{29y^7}{12096}+\frac{y^8}{8064}-\frac{y^9}{362880}\right)z^9+etc.$$

which contains the well-known generating functions for *signed Stirling numbers of the first kind*, $s(m,n)$. These are defined such that the number of permutations of n elements which contain exactly m permutation cycles is the nonnegative number $|s(m,n)|=(-1)^{n-m}s(m,n)$. There are several notations in current use, but for now we employ the formula from Abramowitz and Stegun [1, pp. 822, 824] which states

$$\sum_{k=1}^{n+1}(-1)^{n+1-k}s(n+1,k)x^{n+1-k}=\prod_{k=1}^{n}(1+kx). \tag{29.4.7}$$

Let $x=\frac{-1}{y}$ so (29.4.7) becomes

$$\sum_{k=1}^{n+1}s(n+1,k)y^{k-1}=\prod_{k=1}^{n}(y-k)=\frac{(-1)^n}{k!}\prod_{k=1}^{n}(1-\frac{y}{k}). \tag{29.4.8}$$

Applying this to the coefficient functions of powers of z in (29.4.1) gives the equivalent expression

$$1+\left(s(2,1)y^0+s(2,2)y^1\right)z+\left(s(3,1)y^0+s(3,2)y^1+s(3,3)y^2\right)\frac{z^2}{2!}$$
$$+\left(s(4,1)y^0+s(4,2)y^1+s(4,3)y^2+s(4,4)y^3\right)\frac{z^3}{3!}$$
$$+\left(s(5,1)y^0+s(5,2)y^1+s(5,3)y^2+s(5,4)y^3+s(5,5)y^4\right)\frac{z^4}{4!}+etc.$$

This can apply to the left side of (29.4.1) so we can rewrite it as

$$\prod_{\substack{(a,b)=1\\ a\geq 0;\ b\geq 1}}\left(1+\frac{b}{1!}y^a z^b+\frac{b(b+1)}{2!}y^{2a}z^{2b}+\frac{b(b+1)(b+2)}{3!}y^{3a}z^{3b}+\ldots\right). \tag{29.4.9}$$

The above approach applied to the 2D identity (29.4.1) may benefit from a more rigorous approach, but it can analogously be applied to identities (29.4.2) to (29.4.4), without much change to the rationale but applied to 3D, 4D, and 5D Visible Point Vector scenarios.

Actually this way of expanding the product of sums in (29.4.9) is equivalent to conceptualizing a stepping-stone problem, where each jump carries a weight determined by a formula. Suppose we take the set of integer lattice points $\langle a,b\rangle$ such that $(a,b)=1;\ a\geq 0;\ b\geq 1$. (29.4.8) when expanded allows us to state a set of rules of jumping from $\langle 0,0\rangle$ to any point in the lattice $\langle A,B\rangle$ based on the type of weighted steps enumerated by the identity (29.4.1). One can leap to visible points if a weight of $\frac{b}{1!}$ is carried, and if a leap is jumping over one visible point the weight is $\frac{b(b+1)}{2!}$, over two visible points the weight is $\frac{b(b+1)(b+2)}{3!}$ and so on. Each directional jump may not be repeated. That is, each step must be in a different direction to go from $\langle 0,0\rangle$ to the point in the lattice $\langle A,B\rangle$. The sumtotal of all weights after all possible lattice paths have been traversed from the origin to the designated point is given by (29.4.1).

Equating like powers of z in (29.4.6) and (29.4.8) yields the identity,

$$\left(\begin{array}{c}\frac{1}{(1-w)(1-x)(1-y)}\\ k\end{array}\right)=\prod_{\substack{(a,b,c,d)=1\\ a,b,c,e|k\geq 0;\ d\geq 1}}\left(\begin{array}{c}1/d\\ e\end{array}\right)w^a x^b y^c. \tag{29.4.10}$$

The identities (29.4.4) to (29.4.8) can be used to derive (29.4.11) to (29.4.14). The method is simply to denote the right side of (29.4.1) as $f(x,y,z)$ and then form the quotient $f(x^2,y^2,z^2)/f(x,y,z)$. Hence we obtain

Theorem 29.3

$$\prod_{\substack{(a,b)=1\\ a\geq 0;\ b\geq 1}}\left(1+y^a z^b\right)^{b-1}=\frac{(1-z^2)^{\frac{1}{1-y^2}}}{(1-z)^{\frac{1}{1-y}}}, \tag{29.4.11}$$

$$\prod_{\substack{(a,b,c)=1\\ a,b\geq 0;\ c\geq 1}} \left(1+x^a y^b z^c\right)^{c^{-1}} = \frac{\left(1-z^2\right)^{\frac{1}{(1-x^2)(1-y^2)}}}{\left(1-z\right)^{\frac{1}{(1-x)(1-y)}}}, \tag{29.4.12}$$

$$\prod_{\substack{(a,b,c,d)=1\\ a,b,c\geq 0;\ d\geq 1}} \left(1+w^a x^b y^c z^d\right)^{d^{-1}} = \frac{\left(1-z^2\right)^{\frac{1}{(1-w^2)(1-x^2)(1-y^2)}}}{\left(1-z\right)^{\frac{1}{(1-w)(1-x)(1-y)}}}, \tag{29.4.13}$$

$$\prod_{\substack{(a,b,c,d,e)=1\\ a,b,c,d\geq 0;\ e\geq 1}} \left(1+v^a w^b x^c y^d z^e\right)^{e^{-1}} = \frac{\left(1-z^2\right)^{\frac{1}{(1-v^2)(1-w^2)(1-x^2)(1-y^2)}}}{\left(1-z\right)^{\frac{1}{(1-w)(1-w)(1-x)(1-y)}}}, \tag{29.4.14}$$

Some further examples of the vpv identities are exhibited in the so called *hyper-pyramid identities* as follows. Again, each one is easily proved by differentiating the logarithm of both sides.

Theorem 29.4

$$\prod_{\substack{(a,b)=1\\ a\geq 0;\ b\geq 1\\ a<b}} \left(1-y^a z^b\right)^{1/b} = \left(\frac{1-z}{1-yz}\right)^{\frac{1}{1-y}}, \tag{29.4.15}$$

$$\prod_{\substack{(a,b,c)=1\\ a,b\geq 0;\ c\geq 1\\ a,b<c}} \left(1-x^a y^b z^c\right)^{1/c} = \left(\frac{(1-z)(1-xyz)}{(1-xz)(1-yz)}\right)^{\frac{1}{(1-x)(1-y)}}, \tag{29.4.16}$$

$$\prod_{\substack{(a,b,c,d)=1\\ a,b,c\geq 0;\ d\geq 1\\ a,b,c<d}} \left(1-w^a x^b y^c z^d\right)^{1/d} = \left(\frac{(1-z)(1-wxz)(1-wyz)(1-xyz)}{(1-wz)(1-xz)(1-yz)(1-wxyz)}\right)^{\frac{1}{(1-w)(1-x)(1-y)}}, \tag{29.4.17}$$

These identities are simplest cases of the general n-space infinite product theorem in Campbell [86]. The right sides of (29.4.15) to (29.4.17) expand as power series generating functions in z. For example the right side of (29.4.16) is equivalent to

$$\sum_{k_1,k_2,k_3,k_4\geq 0} \binom{\frac{1}{(1-x)(1-y)}}{k_1}\binom{\frac{1}{(1-x)(1-y)}}{k_2}\binom{\frac{-1}{(1-x)(1-y)}}{k_3} \tag{29.4.18}$$
$$\times \binom{\frac{-1}{(1-x)(1-y)}}{k_4} x^{k_2+k_3}\, y^{k_2+k_4}\, (-z)^{k_1+k_2+k_3+k_4};$$

recalling the left side of (29.4.10) as

$$\prod_{\substack{(a,b,c,d)=1 \\ a,b,c,e|k\geq 0;\ d\geq 1}} \binom{1/d}{e} w^a x^b y^c (-1)^e. \tag{29.4.19}$$

Now, the identities (29.4.11) to (29.4.14) and (29.4.15) to (29.4.17) are susceptible also to the approach using Cramer's rule applied to obtain theorem 29.3. The cases corresponding to theorem 29.4 are:

Corollary 29.1 *If absolute values of x, y, z, and t are all less than unity,*

$$\prod_{\substack{(a,b)=1 \\ a\geq 0;\ b\geq 1 \\ a<b}} \left(1-y^a z^b\right)^{-b^{-1}} = 1+yz+\begin{vmatrix} 1 & -1 \\ \frac{1-y^2}{1-y} & 1 \end{vmatrix} \frac{z^2}{2!} \tag{29.4.20}$$

$$+\begin{vmatrix} 1 & -1 & 0 \\ \frac{1-y^2}{1-y} & 1 & -2 \\ \frac{1-y^3}{1-y} & \frac{1-y^2}{1-y} & 1 \end{vmatrix} \frac{z^3}{3!} + \ldots,$$

$$\prod_{\substack{(a,b,c)=1 \\ a,b\geq 0;\ c\geq 1 \\ a,b<c}} \left(1-x^a y^b z^c\right)^{-c^{-1}} = 1+xyz+\begin{vmatrix} 1 & -1 \\ \frac{(1-x^2)(1-y^2)}{(1-x)(1-y)} & 1 \end{vmatrix} \frac{z^2}{2!} \tag{29.4.21}$$

$$+\begin{vmatrix} 1 & -1 & 0 \\ \frac{(1-x^2)(1-y^2)}{(1-x)(1-y)} & 1 & -2 \\ \frac{(1-x^3)(1-y^3)}{(1-x)(1-y)} & \frac{(1-x^2)(1-y^2)}{(1-x)(1-y)} & 1 \end{vmatrix} \frac{z^3}{3!} + \ldots$$

29.5 ORCHARD VISIBILITY AND THE 2D VISIBLE LATTICE POINTS

29.5.1 Polya's Orchard Visibility Problem

A tree is planted at each lattice point in a circular orchard which has center at the origin and radius r. If the radius of trees exceeds $1/r$ units, one is unable to see out of the orchard in any direction. However, if the radii of the trees are $\frac{1}{\sqrt{r^2+1}}$, one can see out at certain angles. The problem involves formulating different scenarios of radii of trees and radius of the orchard. It may well be that some of the scenarios we have met in our book so far related to Visible Points in the lattice find the application here. For example, in Chapter 19 we mentioned the situation where a radial from origin line of gradient $\sqrt{2}$ through the 2D space never intersects any other integer lattice point. The rational approximation continued fraction convergents for $\sqrt{2}$, however, imply we can go as near to a point as we please without ever touching an integer lattice point. Therefore, it is clear, that the Orchard Visibility Problem may be examined from a continued fraction perspective.

29.5.2 Euclid's Orchard

An array of "trees" of unit height located at integer-coordinate points in a point lattice. When viewed from the Origin corner along the line y=x in normal perspective, a quadrant of Euclid's orchard turns into the modified Dirichlet function. (See Gosper et al. [145])

Euclid's orchard is a two-dimensional array of one-dimensional "trees" of unit height planted at the lattice points in one quadrant of a square lattice. Euclid's orchard is the set of line segments from $(x, y, 0)$ to $(x, y, 1)$, where x and y are positive integers.

The trees visible from the origin are those at lattice points $(x, y, 0)$, where x and y are coprime, i.e., where the fraction $\frac{x}{y}$ is in reduced form. The name Euclid's orchard is derived from the Euclidean algorithm.

If the orchard is projected relative to the origin onto the plane $x + y = 1$ (or, equivalently, drawn in perspective from a viewpoint at the origin) the tops of the trees form a graph of Thomae's function. The point $(x, y, 1)$ projects to

$$\left(\frac{x}{x+y}, \frac{y}{x+y}, \frac{1}{x+y}\right).$$

The proportion of points in the $n \times n$ grid that have trees on them is approximately $\frac{6}{\pi^2}$ based on the probability of a point being a visible point, with the error tending to zero as $n \to \infty$.

29.5.3 Opaque Forests

In discrete geometry, an opaque set is a system of curves or other set in the plane that blocks all lines of sight across a polygon, circle, or other shape. Opaque sets have also been called barriers, beam detectors, opaque covers, or (in cases where they have the form of a forest of line segments or other curves) opaque forests. Opaque sets were introduced by Stefan Mazurkiewicz in 1916, [197] and the problem of minimizing their total length was posed by Frederick Bagemihl in 1959. [41]

For instance, visibility through a unit square can be blocked by its four boundary edges, with length 4, but a shorter opaque forest blocks visibility across the square with length $\sqrt{2} + \frac{1}{2}\sqrt{6} \approx 2.63895843\ldots$. It is not known whether this is the shortest possible opaque set for the square, and for most other shapes this problem similarly remains unsolved. The shortest opaque set for any bounded convex set in the plane has length at most the perimeter of the set, and at least half the perimeter. For the square, a slightly stronger lower bound than half the perimeter is known. Another convex set whose opaque sets are commonly studied is the unit circle, for which the shortest connected opaque set has length $2 + \pi$. Without the assumption of connectivity, the shortest opaque set for the circle has length at least π and at most 4.7998.

Several published algorithms claiming to find the shortest opaque set for a convex polygon were later shown to be incorrect. Nevertheless, it is possible to find an opaque set with a guaranteed approximation ratio in linear time, or to compute the subset of the plane whose visibility is blocked by a given system of line segments in polynomial time.

29.6 LIGHT DIFFUSION MODELS

Suppose each visible point from the origin receives a light rayed from the origin, and in turn sends a relayed-on further light ray onto the points visible from there, and so on. Obviously, this scenario branches uniformly and infinitesimally and relates to counting the number of unrestricted VPV partitions taken to get from the origin to a designated point. This could be true in 2D space or 3D space and physically be represented in a mathematical model. The theory underlying 2D and 3D generalizes no problem to 4D, 5D etc. but we don't have a physical model anymore as it is humanly impossible to visualize in 4D, 5D and higher Euclidean space.

Consider the following mapping of the (29.6.1) vectors plus the origin point $\langle 0,0\rangle$, where each $\circ$ is considered as an infinitesimal small lens that receives light intensity L_n, and radiates it out at intensity L_{n+1} with each $L_{n+1} = f(L_n)$. i.e. L_{n+1} is a function of L_n.

$$
\begin{array}{c|c|c|c|c|c}
5 & & \langle 1,5\rangle & \langle 2,5\rangle & \langle 3,5\rangle & \langle 4,5\rangle \\
\hline
4 & & \langle 1,4\rangle & & \langle 3,4\rangle & \\
\hline
3 & & \langle 1,3\rangle & \langle 2,3\rangle & & \\
\hline
2 & & \langle 1,2\rangle & & & \\
\hline
1 & & & & & \\
\hline
0 & \langle 0,0\rangle & & & & \\
\hline
y/x & 0 & 1 & 2 & 3 & 4
\end{array}
\quad \mapsto \quad
\begin{array}{c|c|c|c|c|c}
5 & \uparrow & \circ & \circ & \circ & \circ \\
\hline
4 & \uparrow & \circ & & \circ & \nearrow \\
\hline
3 & \uparrow & \circ & \circ & \nearrow & \\
\hline
2 & \uparrow & \circ & \nearrow & & \\
\hline
1 & \uparrow & \nearrow & & & \\
\hline
0 & \circ & & & & \\
\hline
y/x & 0 & 1 & 2 & 3 & 4
\end{array}
\tag{29.6.1}
$$

Suppose the origin lens $\circ$ radiates with intensity Υ upward along $x=0$ and across to the diagonal $y=x$ in the radial region between lines. The arrows in the right side grid here depict this phenomenon. Each visible point designated by vectors $\langle a,b\rangle$ (with $\gcd(a,b)=1$ and $1 \le a < b \le 5$) will receive the light ray at intensity Υ, then in turn radiate that light onward in the same shaped radial upward sector around to the diagonal with intensity $f(\Upsilon)$, where f is some defined function. Each lens receiving light transmits that light further to a point visible to it with a new intensity of $f(f(\Upsilon))$.

Any product generating function for our present exercise would give us a means to examine light intensity radiating from an initial point, the origin, and extending via visible points in a radial region from the origin. The Visible Point Vector (VPV) identities can be construed as encoding this type of light diffusion.

Consider the following scenarios.

Scenario 1: In the first 2D quadrant of lattice points, shining a uniformly radiant beam of light from the origin will illuminate all of the *visible lattice points* and will not illuminate any *invisible lattice point.* If the visible lattice points have the property that they absorb 100% of the light received, but opaquely do not refract or further transmit any of the received light, then we say the light has zero diffusion at each visible point. Each visible lattice point is assigned radiance $r=1$, and each invisible lattice point is assigned radiance $r=0$. Then we are seeing "the number of partitions into exactly one part from the set of first quadrant 2D Visible Point Vectors",

Partition grid radiance r = 1 illumined points

$$
\begin{array}{c|cccccccccc}
\vdots & \vdots & \vdots & \vdots & \vdots & \vdots & \vdots & \vdots & \vdots & \cdot^{\cdot^{\cdot}} \\
8 & 1 & & 1 & & 1 & & 1 & & \cdots \\
7 & 1 & 1 & 1 & 1 & 1 & 1 & & 1 & \cdots \\
6 & 1 & & & & 1 & & & & \cdots \\
5 & 1 & 1 & 1 & 1 & & 1 & 1 & 1 & \cdots \\
4 & 1 & & 1 & & 1 & & 1 & & \cdots \\
3 & 1 & 1 & & 1 & 1 & & 1 & 1 & \cdots \\
2 & 1 & & 1 & & 1 & & 1 & & \cdots \\
1 & 1 & 1 & 1 & 1 & 1 & 1 & 1 & 1 & \cdots \\
\hline
0 & 1 & 2 & 3 & 4 & 5 & 6 & 7 & 8 & \cdots
\end{array}
\tag{29.6.2}
$$

Scenario 2: Shine a uniformly radiant light from the origin to the visible lattice points in the first 2D quadrant. Let the visible lattice points have the property that they absorb 100% of the light received, and have radiance $r = 1$, and then the light passes through each visible point, shining a uniformly radiant beam that can reach only to the next point that is visible from from the first visible point. So the first radial beam from origin to visible points sends $r = 1$ to each VPV, and the transmitted second radial beam sends $r = 1$ from the first VPVs transmitting/refracting to points visible from there. The sumtotal r of a 2D first quadrant point (a, b) is "the number of partitions into exactly two parts; each part being a member of the set of first quadrant 2D Visible Point Vectors",

Vectors for radiant from origin light model

$$
\begin{array}{cccccccc}
\vdots & \vdots & \vdots & \vdots & \vdots & \vdots & \vdots & \\
\langle 1,8\rangle & & \langle 3,8\rangle & & \langle 5,8\rangle & & \langle 7,8\rangle & \cdots \\
\langle 1,7\rangle & \langle 2,7\rangle & \langle 3,7\rangle & \langle 4,7\rangle & \langle 5,7\rangle & \langle 6,7\rangle & & \cdots \\
\langle 1,6\rangle & & & & \langle 5,6\rangle & & & \cdots \\
\langle 1,5\rangle & \langle 2,5\rangle & \langle 3,5\rangle & \langle 4,5\rangle & & \langle 6,5 & \langle 7,5\rangle & \cdots \\
\langle 1,4\rangle & & \langle 3,4\rangle & & \langle 5,4\rangle & & \langle 7,4\rangle & \cdots \\
\langle 1,3\rangle & \langle 2,3\rangle & & \langle 4,3\rangle & \langle 5,3\rangle & & \langle 7,3\rangle & \cdots \\
\langle 1,2\rangle & & \langle 3,2\rangle & & \langle 5,2\rangle & & \langle 7,2\rangle & \cdots \\
\langle 1,1\rangle & \langle 2,1\rangle & \langle 3,1\rangle & \langle 4,1\rangle & \langle 5,1\rangle & & \langle 7,1\rangle & \cdots
\end{array}
\tag{29.6.3}
$$

Theorem 29.5 *The radiance r at any 2D first quadrant lattice point $\langle a, b\rangle$ from Scenario 1 is the coefficient of $x^a y^b$ in*

$$
V(x,y) = \sum_{\substack{\gcd(a,b)=1 \\ a,b \geq 1}} x^a y^b, \quad |x| < 1 \text{ and } |y| < 1. \tag{29.6.4}
$$

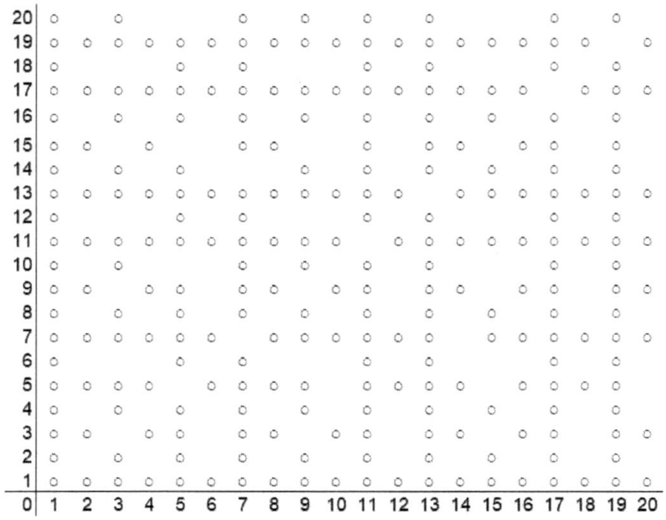

Figure 29.12 2D First Quadrant Visible Point Lenses, Order 20

The generating function for the number of partitions of $\langle a,b \rangle$ (for a and b both positive integers) into exactly two parts each part being a visible point vector is the coefficient of $x^a y^b$ in

$$V(x,y)^2 + V(x^2,y^2) \quad \text{where} \quad V(x,y) = \sum_{\substack{\gcd(a,b)=1 \\ a,b \geq 1}} x^a y^b, \tag{29.6.5}$$

valid for both $|x| < 1$ and $|y| < 1$. Clearly $V(x,y)$ is the infinite 2D sum taken on the visible lattice point vectors in (29.6.3). Since Scenario 2 includes Scenario 1, we need to add an instance of $V(x,y)$ to obtain the following

Theorem 29.6 *The radiance r at any 2D first quadrant lattice point $\langle a,b \rangle$ from Scenario 2 is the coefficient of $x^a y^b$ in*

$$V(x,y)^2 + V(x^2,y^2) + V(x,y) \quad \textit{where} \quad V(x,y) = \sum_{\substack{\gcd(a,b)=1 \\ a,b \geq 1}} x^a y^b. \tag{29.6.6}$$

29.6.1 Light Diffusion VPV n-Space Identities

We write an analogy to the n-space q-binomial theorem 13.2 of Chapter 13 as a single equation identity as follows. This will enable us to quantify the radiance r as defined for the lattice points in the first nD hyperquadrant with positive integer coordinates. We call this generating function $\mathbb{V}_n(q_1,q_2,q_3,\ldots,q_n;t)$, and quantify it here.

Theorem 29.7 *The n-space VPV generating function analogue of the nD q-binomial product is as follows. If each $|q_i| < 1$ with $i = 1,2,3,\ldots,n$; then*

$$\mathbb{V}_n(q_1,q_2,q_3,\ldots,q_n;t) = \prod_{\substack{\gcd(\alpha_1,\alpha_2,\alpha_3,\ldots,\alpha_n)=1 \\ \alpha_1,\alpha_2,\alpha_3,\ldots,\alpha_n \geq 1}} \frac{1}{1 - q_1^{\alpha_1} q_2^{\alpha_2} \cdots q_n^{\alpha_n} t} \tag{29.6.7}$$

$$= 1 + V_n(q_1,q_2,\cdots,q_n)\frac{t}{1!}$$

$$+\begin{vmatrix} V_n(q_1,q_2,\cdots,q_n) & -1 \\ V_n(q_1^2,q_2^2,\cdots,q_n^2) & V_n(q_1,q_2,\cdots,q_n) \end{vmatrix}\frac{t^2}{2!}$$

$$+\begin{vmatrix} V_n(q_1,q_2,\cdots,q_n) & -1 & 0 \\ V_n(q_1^2,q_2^2,\cdots,q_n^2) & V_n(q_1,q_2,\cdots,q_n) & -2 \\ V_n(q_1^3,q_2^3,\cdots,q_n^3) & V_n(q_1^2,q_2^2,\cdots,q_n^2) & V_n(q_1,q_2,\cdots,q_n) \end{vmatrix}\frac{t^3}{3!}$$

$$+\begin{vmatrix} V_n(q_1,q_2,\cdots,q_n) & -1 & 0 & 0 \\ V_n(q_1^2,q_2^2,\cdots,q_n^2) & V_n(q_1,q_2,\cdots,q_n) & -2 & 0 \\ V_n(q_1^3,q_2^3,\cdots,q_n^3) & V_n(q_1^2,q_2^2,\cdots,q_n^2) & V_n(q_1,q_2,\cdots,q_n) & -3 \\ V_n(q_1^4,q_2^4,\cdots,q_n^4) & V_n(q_1^3,q_2^3,\cdots,q_n^3) & V_n(q_1^2,q_2^2,\cdots,q_n^2) & V_n(q_1,q_2,\cdots,q_n) \end{vmatrix}\frac{t^4}{4!}$$

$$+etc.$$

where by definition

$$\begin{aligned} V_2(q_1,q_2) &= \sum_{\substack{\gcd(a,b)=1\\ a,b\geq 1}} q_1^a q_2^b \\ V_3(q_1,q_2,q_3) &= \sum_{\substack{\gcd(a,b,c)=1\\ a,b,c\geq 1}} q_1^a q_2^b q_3^c \\ \vdots \quad &\vdots \quad \vdots \\ V_n(q_1,q_2,\cdots,q_n) &= \sum_{\substack{\gcd(\alpha_1,\alpha_2,\alpha_3,\ldots,\alpha_n)=1\\ \alpha_1,\alpha_2,\alpha_3,\ldots,\alpha_n\geq 1}} q_1^{\alpha_1} q_2^{\alpha_2}\cdots q_n^{\alpha_n}. \end{aligned}$$

Theorem 29.7 is the generating function for nD partitions of the n-space lattice point vector

$$\langle \alpha_1,\alpha_2,\alpha_3,\ldots,\alpha_n\rangle, \quad \text{(for each positive integer } \alpha_i) \tag{29.6.8}$$

into unrestricted first hyperquadrant VPV parts.

Next we state the corresponding nD generating function $\mathbb{D}_n(q_1,q_2,q_3,\ldots,q_n;t)$ for distinct VPV partitions.

Theorem 29.8 *The n-space VPV generating function analogue of the nD product formula (13.3.2) from chapter 13 is as follows. If each* $|q_i| < 1$ *with* $i = 1,2,3,\ldots,n$*; then*

$$\mathbb{D}_n(q_1,q_2,q_3,\ldots,q_n;t) = \prod_{\substack{\gcd(\alpha_1,\alpha_2,\alpha_3,\ldots,\alpha_n)=1\\ \alpha_1,\alpha_2,\alpha_3,\ldots,\alpha_n\geq 1}} (1+q_1^{\alpha_1} q_2^{\alpha_2}\cdots q_n^{\alpha_n} t) \tag{29.6.9}$$

$$= 1 + V_n(q_1,q_2,\cdots,q_n)\frac{t}{1!}$$

$$+\begin{vmatrix} V_n(q_1,q_2,\cdots,q_n) & -1 \\ -V_n(q_1^2,q_2^2,\cdots,q_n^2) & V_n(q_1,q_2,\cdots,q_n) \end{vmatrix}\frac{t^2}{2!}$$

$$+\begin{vmatrix} V_n(q_1,q_2,\cdots,q_n) & -1 & 0 \\ -V_n(q_1^2,q_2^2,\cdots,q_n^2) & V_n(q_1,q_2,\cdots,q_n) & -2 \\ V_n(q_1^3,q_2^3,\cdots,q_n^3) & -V_n(q_1^2,q_2^2,\cdots,q_n^2) & V_n(q_1,q_2,\cdots,q_n) \end{vmatrix}\frac{t^3}{3!}$$

$$+\begin{vmatrix} V_n(q_1,q_2,\cdots,q_n) & -1 & 0 & 0 \\ -V_n(q_1^2,q_2^2,\cdots,q_n^2) & V_n(q_1,q_2,\cdots,q_n) & -2 & 0 \\ V_n(q_1^3,q_2^3,\cdots,q_n^3) & -V_n(q_1^2,q_2^2,\cdots,q_n^2) & V_n(q_1,q_2,\cdots,q_n) & -3 \\ -V_n(q_1^4,q_2^4,\cdots,q_n^4) & V_n(q_1^3,q_2^3,\cdots,q_n^3) & -V_n(q_1^2,q_2^2,\cdots,q_n^2) & V_n(q_1,q_2,\cdots,q_n) \end{vmatrix}\frac{t^4}{4!}$$

$$+etc.$$

Theorem 29.8 is the generating function for nD partitions of the n-space lattice point vector

$$\langle \alpha_1, \alpha_2, \alpha_3, \ldots, \alpha_n \rangle, \quad \text{(for each positive integer } \alpha_i) \tag{29.6.10}$$

into distinct first hyperquadrant VPV parts.

Think About It...

Jazz and rock are no longer to be despised as passing fads. Neither are chaos and complexity theory. But still, classical music and classical mathematics are not dead. Mozart lives, and so does Euler.

When the wheel of fashion turns once more, quantum mechanics and hard analysis will once again be in style.

Freeman Dyson
Mathematician, Physicist; 1923–2020

GLOSSARY

Euclid's Orchard: Euclid's orchard is a two-dimensional array of one-dimensional "trees" of unit height planted at the lattice points in one quadrant of a square lattice.

Visible Lattice Point: A point in n-dimensional Euclidean space whose coordinates are $(a_1, a_2, \ldots, a_n)$ where $a_1, a_2, \ldots, a_n$ are co-prime integers as an n-tuple.

Farey Sequence: The Farey fraction sequence of order n denoted F_n, is the sequence of completely reduced fractions, either between 0 and 1, or without this restriction, which when in lowest terms have denominators less than or equal to n, arranged in order of increasing size.

Polya's Orchard Visibility Problem: Each lattice point is a tree in a circular orchard of centre at the origin and radius r. If the tree radius is more than 1/r units, one cannot see out of the orchard in any direction. But, if the trees radii are $1/\sqrt{(r^2+1)}$, one can see out at certain angles. The problem formulates scenarios of trees radii and orchard radius.

Opaque Forest: In discrete geometry, an opaque set is a system of curves or other set in the plane that blocks all lines of sight across a polygon, circle, or other shape. Opaque sets have also been called barriers, beam detectors, opaque covers, or (in cases where they have the form of a forest of line segments or other curves) opaque forests.

FURTHER READING

BORWEIN, D.; BORWEIN, J.M.; and BRADLEY, D.M. Parametric Euler sum identities. *J. Math. Anal. Appl.*, 316:328–338, 2006.

BORWEIN, J.M.; GLASSER, M.L.; MCPHEDRAN, R.C.; WAN, J.G.; and ZUCKER, I.J. *Lattice Sums then and now*, volume 150 of Encyclopedia of Mathematics and Its Applications. Cambridge University Press, 2013.

CAMPBELL, G.B. Multiplicative functions over Riemann zeta function products. *J. Ramanujan Soc.*, 7(1):52–63, 1992.

CAMPBELL, G.B. Dirichlet summations and products over primes. *Int. J. Math. Math. Sci.*, 16(2):359–372, 1993.

CAMPBELL, G.B. A generalized formula of Hardy. *Int. J. Math. Math. Sci.*, 17(2):369–378, 1994.

CAMPBELL, G.B. A new class of infinite products, and Euler's totient. *Int. J. Math. and Math. Sci.*, 17(3):417–422, 1994.

CAMPBELL, G.B. Infinite products over visible lattice points. *Int. J. Math. and Math. Sci.*, 17(4):637–654, 1994.

CAMPBELL, G.B. *Combinatorial identities in number theory related to q-series and arithmetical functions.* Doctor of philosophy thesis, School of Mathematical Sciences, The Australian National University, October 1997.

CAMPBELL, G.B. A closer look at some new identities. *Int. J. Math. and Math. Sci.*, 21(3):581–586, 1998.

CAMPBELL, G.B. Infinite products over hyperpyramid lattices. *Int. J. Math. and Math. Sci.*, 23(4):271–277, 2000.

CAMPBELL, G.B. Some n-space q-binomial theorem extensions and similar identities. arxiv preprint, June 2019. arXiv:1906.07526v1 [math.NT]: https://arxiv.org/abs/1906.07526.

CAMPBELL, G.B. Continued Fractions for partition generating functions. arxiv preprint, Jan 2023. arXiv:2301.12945 [math.CO], https://arxiv.org/abs/2301.12945.

CAMPBELL, G.B. Vector Partition Identities for 2D, 3D and nD Lattices. arxiv preprint, Feb 2023. arXiv:10.48550/ARXIV.2302.01091 [math.NT], https://arxiv.org/abs/2302.01091.

CAMPBELL, G.B. Visible Point Partition Identities for Polylogarithms, and Parametric Euler Sums. arxiv preprint, Jun 2023. arXiv:2306.02241v1 [math.CO], https://arxiv.org/abs/2306.02241.

CAMPBELL, G.B. Visible Point Vector Partition Identities for Hyperpyramid Lattices. arxiv preprint, Sep 2023. arXiv:2309.16094v1 [math.CO], https://arxiv.org/abs/2309.16094.

LEWIN, L. *Polylogarithms and associated functions.* Elsevier Science Ltd.: North-Holland, New York, 1981. ISBN 0-444-00550-1,(First published London: Macdonald, 1958.).

LEWIN, L. The dilogarithm in algebraic fields. *J. Austral. Math. Soc. Ser. A*, 33:302–330, 1982.

LEWIN, L. (Ed.). *Structural Properties of Polylogarithms.* Amer. Math. Soc. - Providence, RI., 1991. ISBN 0-8218-1634-9.

CHAPTER 30

Euler Products Over Primes and New VPV Formulas

WE SAW IN EARLIER CHAPTERS, for example in Chapters 20 and 26, there are numerous fairly simple cases of VPV identities that may be new or unfamiliar to the reader. In the present chapter, we apply some basic formulas to transform some of these cases already given, so as to illustrate the possibilities for new and fairly simple identity expressions obtainable from transforming the VPV identities using Euler products over primes, and other well known infinite products.

30.1 EULER PRODUCTS OVER THE PRIMES

We derive classes of infinite products taken over the primes, for example, (see Campbell [85])

Theorem 30.1 *If* $\Re m \geq 1$ *and* $\Re n \geq 1$,

$$\prod_p \left(\frac{1}{1-p^{-n}}\right)^{(1-p^{-m})^{-1}} = \zeta(n) \prod_{\substack{\alpha,\beta \geq 1 \\ gcd(\alpha,\beta)=1}} \zeta(\alpha m + \beta n)^{1/\beta}. \tag{30.1.1}$$

and

$$\prod_p \left(\frac{1-p^{-(m+n)}}{1-p^{-n}}\right)^{(1-p^{-m})^{-1}} = \zeta(n) \prod_{\substack{1 \leq \alpha \leq \beta \\ \beta \geq 1 \\ gcd(\alpha,\beta)=1}} \zeta(\alpha m + \beta n)^{1/\beta}. \tag{30.1.2}$$

These infinite products were first given by Campbell [85], but have gone virtually unnoticed for over thirty years.

To illustrate what is in these products, each product term RHS of (30.1.1) is inserted here into the first quadrant 2D grid.

DOI: 10.1201/9781003174158-30

Riemann zeta functions inserted to the 2D All VPV first quadrant grid

$\vdots$	$\vdots$	$\vdots$	$\vdots$	$\vdots$	$\vdots$	$\vdots$	$\vdots$	$\vdots$	$\ddots$
7		$\zeta(m+7n)^{\frac{1}{7}}$	$\zeta(2m+7n)^{\frac{1}{7}}$	$\zeta(3m+7n)^{\frac{1}{7}}$	$\zeta(4m+7n)^{\frac{1}{7}}$	$\zeta(5m+7n)^{\frac{1}{7}}$	$\zeta(6m+7n)^{\frac{1}{7}}$		$\cdots$
6		$\zeta(m+6n)^{\frac{1}{6}}$				$\zeta(5m+6n)^{\frac{1}{6}}$		$\zeta(7m+6n)^{\frac{1}{6}}$	$\cdots$
5		$\zeta(m+5n)^{\frac{1}{5}}$	$\zeta(2m+5n)^{\frac{1}{5}}$	$\zeta(3m+5n)^{\frac{1}{5}}$	$\zeta(4m+5n)^{\frac{1}{5}}$		$\zeta(6m+5n)^{\frac{1}{5}}$	$\zeta(7m+5n)^{\frac{1}{5}}$	$\cdots$
4		$\zeta(m+4n)^{\frac{1}{4}}$		$\zeta(3m+4n)^{\frac{1}{4}}$		$\zeta(5m+4n)^{\frac{1}{4}}$		$\zeta(7m+4n)^{\frac{1}{4}}$	$\cdots$
3		$\zeta(m+3n)^{\frac{1}{3}}$	$\zeta(2m+3n)^{\frac{1}{3}}$		$\zeta(4m+3n)^{\frac{1}{3}}$	$\zeta(5m+3n)^{\frac{1}{3}}$		$\zeta(7m+3n)^{\frac{1}{3}}$	$\cdots$
2		$\zeta(m+2n)^{\frac{1}{2}}$		$\zeta(3m+2n)^{\frac{1}{2}}$		$\zeta(5m+2n)^{\frac{1}{2}}$		$\zeta(7m+2n)^{\frac{1}{2}}$	$\cdots$
1	$\zeta(n)$	$\zeta(m+n)$	$\zeta(2m+n)$	$\zeta(3m+n)$	$\zeta(4m+n)$	$\zeta(5m+n)$	$\zeta(6m+n)$	$\zeta(7m+n)$	
β/α	0	1	2	3	4	5	6	7	$\cdots$

Similarly, each product term RHS of (30.1.2) is inserted here into the upper first quadrant 2D grid.

Riemann zeta functions inserted to the 2D Upper VPV first quadrant grid

$\vdots$	$\vdots$	$\vdots$	$\vdots$	$\vdots$	$\vdots$	$\vdots$	$\vdots$	$\ddots$
7		$\zeta(m+7n)^{\frac{1}{7}}$	$\zeta(2m+7n)^{\frac{1}{7}}$	$\zeta(3m+7n)^{\frac{1}{7}}$	$\zeta(4m+7n)^{\frac{1}{7}}$	$\zeta(5m+7n)^{\frac{1}{7}}$	$\zeta(6m+7n)^{\frac{1}{7}}$	$\cdots$
6		$\zeta(m+6n)^{\frac{1}{6}}$				$\zeta(5m+6n)^{\frac{1}{6}}$		$\cdots$
5		$\zeta(m+5n)^{\frac{1}{5}}$	$\zeta(2m+5n)^{\frac{1}{5}}$	$\zeta(3m+5n)^{\frac{1}{5}}$	$\zeta(4m+5n)^{\frac{1}{5}}$			$\cdots$
4		$\zeta(m+4n)^{\frac{1}{4}}$		$\zeta(3m+4n)^{\frac{1}{4}}$				$\cdots$
3		$\zeta(m+3n)^{\frac{1}{3}}$	$\zeta(2m+3n)^{\frac{1}{3}}$					$\cdots$
2		$\zeta(m+2n)^{\frac{1}{2}}$						$\cdots$
1	$\zeta(n)$							
β/α	0	1	2	3	4	5	6	$\cdots$

Proof of theorem 30.1: We require the three known results:

$$\prod_{primes\, p} \frac{1}{1-\frac{1}{p^s}} = \zeta(s), \qquad \Re s > 1, \tag{30.1.3}$$

$$(1-y)^{(1-x)^{-1}} = \prod_{\substack{j\geq 0,k\geq 1 \\ gcd(j,k)=1}} (1-x^j y^k)^{1/k}, \quad |x|<1, |y|<1, \tag{30.1.4}$$

$$\left(\frac{1-y}{1-xy}\right)^{(1-x)^{-1}} = \prod_{\substack{j\geq 0,k\geq 1 \\ 0\leq j<k \\ gcd(j,k)=1}} (1-x^j y^k)^{1/k}, \quad |x|<1, |y|<1. \tag{30.1.5}$$

Equation (30.1.3) is the classical prime product representation of the Riemann zeta function, dated back to Euler. Equations (30.1.4) and (30.1.5) are familiar from workings of Chapters 20 and 24. Next, replace x with p^{-m} and y with p^{-n} in each of (30.1.4) and (30.1.5). Then in these two identities apply a product to both sides over all primes, using (30.1.3) on the right sides, so then we easily arrive at the theorem. ■

30.2 FURTHER EULER PRODUCTS FROM VPV CASES

We continue with examples of Euler products over primes by restating the simple VPV identities from Chapter 20.

$$\prod_{\substack{(j,k)=1\\ j,k\geq 1}} \left(1-\left(\frac{1}{2}\right)^j z^k\right)^{\frac{1}{k}} = 1-z,$$

$$\prod_{\substack{(j,k)=1\\ j,k\geq 1}} \left(1-\left(\frac{2}{3}\right)^j z^k\right)^{\frac{1}{k}} = (1-z)^2,$$

$$\prod_{\substack{(j,k)=1\\ j,k\geq 1}} \left(1-\left(\frac{3}{4}\right)^j z^k\right)^{\frac{1}{k}} = (1-z)^3,$$

$$\prod_{\substack{(j,k)=1\\ j,k\geq 1}} \left(1-\left(\frac{4}{5}\right)^j z^k\right)^{\frac{1}{k}} = (1-z)^4,$$

$$\prod_{\substack{(j,k)=1\\ j,k\geq 1}} \left(1-\left(\frac{5}{6}\right)^j z^k\right)^{\frac{1}{k}} = (1-z)^5,$$

and so on. In each of these substitute $z = p^{-s}$ and apply the product over primes operator $\prod_{primes\, p}$ to both sides of each equation. Therefore

$$\prod_{\substack{(j,k)=1;j,k\geq 1\\ primes\, p}} \left(1-\left(\frac{1}{2}\right)^j p^{-ks}\right)^{\frac{-1}{k}} = \zeta(s),$$

$$\prod_{\substack{(j,k)=1;j,k\geq 1\\ primes\, p}} \left(1-\left(\frac{2}{3}\right)^j p^{-ks}\right)^{\frac{-1}{k}} = (\zeta(s))^2,$$

$$\prod_{\substack{(j,k)=1;j,k\geq 1\\ primes\, p}} \left(1-\left(\frac{3}{4}\right)^j p^{-ks}\right)^{\frac{-1}{k}} = (\zeta(s))^3,$$

$$\prod_{\substack{(j,k)=1;j,k\geq 1\\ primes\, p}} \left(1-\left(\frac{4}{5}\right)^j p^{-ks}\right)^{\frac{-1}{k}} = (\zeta(s))^4,$$

$$\prod_{\substack{(j,k)=1;j,k\geq 1\\ primes\, p}} \left(1-\left(\frac{5}{6}\right)^j p^{-ks}\right)^{\frac{-1}{k}} = (\zeta(s))^5.$$

These products encode a theorem on weighted sums of divisor functions. Consider the infinite VPV product proven in Chapter 20

$$\prod_{\substack{(j,k)=1\\ j,k\geq 1}} \left(\frac{1}{1-y^j z^k}\right)^{\frac{1}{j^a k^b}} = \exp\left\{\left(\sum_{j=1}^{\infty} \frac{y^j}{j^a}\right)\left(\sum_{k=1}^{\infty} \frac{z^k}{k^b}\right)\right\} \quad if \quad a+b=1. \tag{30.2.1}$$

There is an obvious set of immediate interesting cases of (30.2.1) for example: Taking $a=0$, $b=1$ gives

$$\prod_{\substack{(j,k)=1\\ j,k\geq 1}} \left(\frac{1}{1-y^j z^k}\right)^{\frac{1}{k}} = \left(\frac{1}{1-z}\right)^{\frac{y}{1-y}} \tag{30.2.2}$$

$$= \sum_{n=0}^{\infty} z^n (-1)^n \binom{\frac{y}{y-1}}{n}.$$

30.3 SIMPLE BUT NEW PRODUCTS OVER PRIMES

In (30.2.2) substitute $y=p^{-s}$, $z=p^{-t}$ and apply the product over primes operator $\prod_{primes\ p}$ so we have

$$\prod_{\substack{(j,k)=1\\ j,k\geq 1}} (\zeta(js+kt))^{\frac{1}{k}} = \prod_{primes\ p} \left(\frac{p^t}{p^t-1}\right)^{\frac{1}{p^s-1}} \tag{30.3.1}$$

$$= \prod_{primes\ p} \left\{\sum_{n=0}^{\infty} \frac{(-1)^n}{p^{nt}} \binom{\frac{1}{1-p^s}}{n}\right\}.$$

We can adjust the products we form over hyperquadrant lattices to include the bordering axis thus including components of zero for such a bordering axis. So, we now restate the Chapter 20, corollary 20.5 set of four equations (20.5.7) to (20.5.10) covering the 2D to 5D VPV first hyperquadrant formulas as follows:

Corollary 30.1 *For each of* $|v|,|w|,|x|,|y|,|z|<1$,

$$\prod_{\substack{(a,b)=1\\ a\geq 0,b>0}} \left(1-y^a z^b\right)^{\frac{1}{b}} = (1-z)^{\frac{1}{1-y}}, \tag{30.3.2}$$

$$\prod_{\substack{(a,b,c)=1\\ a,b\geq 0,c>0}} \left(1-x^a y^b z^c\right)^{\frac{1}{c}} = (1-z)^{\frac{1}{(1-x)(1-y)}}, \tag{30.3.3}$$

$$\prod_{\substack{(a,b,c,d)=1\\ a,b,c\geq 0,d>0}} \left(1-w^a x^b y^c z^d\right)^{\frac{1}{d}} = (1-z)^{\frac{1}{(1-w)(1-x)(1-y)}}, \tag{30.3.4}$$

$$\prod_{\substack{(a,b,c,d,e)=1\\ a,b,c,d\geq 0,e>0}} \left(1-v^a w^b x^c y^d z^e\right)^{\frac{1}{e}} = (1-z)^{\frac{1}{(1-v)(1-w)(1-x)(1-y)}}. \tag{30.3.5}$$

So, we apply the Euler product over all primes to both sides of (30.3.2) to (30.3.5), to obtain

Theorem 30.2 *For each of* $|v|,|w|,|x|,|y|,|z|>1$,

$$\prod_{\substack{(a,b)=1\\ a\geq 0,b>0}} \zeta(ay+bz)^{\frac{1}{b}} = \prod_{\textit{all primes } p} \left(\frac{1}{1-\frac{1}{p^z}}\right)^{\frac{1}{1-\frac{1}{p^y}}}, \tag{30.3.6}$$

$$\prod_{\substack{(a,b,c)=1\\ a,b\geq 0,c>0}} \zeta(ax+by+cz)^{\frac{1}{c}} = \prod_{\textit{all primes } p} \left(\frac{1}{1-\frac{1}{p^z}}\right)^{\frac{1}{\left(1-\frac{1}{p^x}\right)\left(1-\frac{1}{p^y}\right)}}, \tag{30.3.7}$$

$$\prod_{\substack{(a,b,c,d)=1\\ a,b,c\geq 0,d>0}} \zeta(aw+bx+cy+dz)^{\frac{1}{d}} = \prod_{\textit{all primes } p} \left(\frac{1}{1-\frac{1}{p^z}}\right)^{\frac{1}{\left(1-\frac{1}{p^w}\right)\left(1-\frac{1}{p^x}\right)\left(1-\frac{1}{p^y}\right)}}, \tag{30.3.8}$$

$$\prod_{\substack{(a,b,c,d,e)=1\\ a,b,c,d\geq 0,e>0}} \zeta(av+bw+cx+dy+ez)^{\frac{1}{e}} = \prod_{\textit{all primes } p} \left(\frac{1}{1-\frac{1}{p^z}}\right)^{\frac{1}{\left(1-\frac{1}{p^v}\right)\left(1-\frac{1}{p^w}\right)\left(1-\frac{1}{p^x}\right)\left(1-\frac{1}{p^y}\right)}}. \tag{30.3.9}$$

These equations (30.3.6) to (30.3.9) are related to the multiplicative functions:

Definition 30.1 *If the prime decomposition of positive integer is* $k=\prod_{i=1}^{s} p_i^{a_i}$ *then define*

$$\begin{aligned}
\omega_2(m;k) &= \prod_{i=1}^{s} \binom{\frac{1}{(1-p_i^{-m})}}{a_i}, \\
\omega_3(m_1,m_2;k) &= \prod_{i=1}^{s} \binom{\frac{1}{(1-p_i^{-m_1})(1-p_i^{-m_2})}}{a_i}, \\
\omega_4(m_1,m_2,m_3;k) &= \prod_{i=1}^{s} \binom{\frac{1}{(1-p_i^{-m_1})(1-p_i^{-m_2})(1-p_i^{-m_3})}}{a_i}, \\
\omega_4(m_1,m_2,m_3,m_4;k) &= \prod_{i=1}^{s} \binom{\frac{1}{(1-p_i^{-m_1})(1-p_i^{-m_2})(1-p_i^{-m_3})(1-p_i^{-m_4})}}{a_i}.
\end{aligned}$$

Think About It...

This could well apply to intractable geniuses such as Évariste Galois, Srinivasa Ramanujan, or John von Neumann.

"The world is always ready to receive talent with open arms. Very often it does not know what to do with genius.

Talent is a docile creature. It bows its head meekly while the world slips the collar over it. It backs into the shafts like a lamb. It draws its load cheerfully, and is patient of the bit and of the whip. But genius is always impatient of its harness; its wild blood makes it hard to train."

Oliver Wendell Holmes, Sr.
1809–1894

FURTHER READING

CAMPBELL, G.B. A new class of infinite products, and Euler's totient. *Int. J. Math. and Math. Sci.*, 17(3):417–422, 1994.

CAMPBELL, G.B. Infinite products over visible lattice points. *Int. J. Math. and Math. Sci.*, 17(4):637–654, 1994.

CAMPBELL, G.B. *Combinatorial identities in number theory related to q-series and arithmetical functions.* Doctor of philosophy thesis, School of Mathematical Sciences, The Australian National University, October 1997.

CAMPBELL, G.B. A closer look at some new identities. *Int. J. Math. and Math. Sci.*, 21(3):581–586, 1998.

CAMPBELL, G.B. Infinite products over hyperpyramid lattices. *Int. J. Math. and Math. Sci.*, 23(4):271–277, 2000.

CAMPBELL, G.B. Some n-space q-binomial theorem extensions and similar identities. arxiv preprint, June 2019. arXiv:1906.07526v1 [math.NT]: https://arxiv.org/abs/1906.07526.

GOLDFELD, D. Beyond the last theorem. *Math Horizons*, 4:26–34, September 1996.

MASSER, D. W. Open problems. *Proceedings of the Symposium on Analytic Number Theory.* London: Imperial College, 1985. Chen, W. W. L. (ed.).

OESTERLÉ, J. Nouvelles approches du "théoréme" de Fermat. *Asteérisque*, Séminaire Bourbaki exp., 694(161):165–186, 1988. ISSN 0303-1179, MR 0992208.

RIEMANN, G.F.B. Über die Anzahl der Primzahlen unter einer gegebenen Grösse. *Monatsber. Königl. Preuss. Akad. Wiss. Berlin*, pages 671–680, Nov 1859.

SZPIRO, L. Propriétés numériques du faisceau dualisant rélatif. *Astérisque.*, 86:44–78, 1981. Seminaire sur les pinceaux des courbes de genre au moins deux. Zbl 0517.14006.

SZPIRO, L. Présentation de la théorie d'Arakelov. *Contemporary Mathematics*, 67:279–293, 1987. Zbl 0634.14012.

CHAPTER 31

Determinants, Bell Polynomial Expansions for Vector Partitions

31.1 REFERENCE CASES OF THE DETERMINANTS IN THIS BOOK

THE FOLLOWING cases of determinants are applicable to the results in this book. They are easy to verify in Mathematica or WolframAlpha online, by inputting the *Det* statements here. It would be nice to have the general $n \times n$ determinants evaluated in closed forms, preferably as product forms involving q-series functions, but these do not seem to be solved anywhere in the literature. So, for now, we content ourselves with the following esoteric cases. Some related references are Andrews and Burge [28], Carlitz [103], Krattenthaler [172, 173], Kuperberg [177, 179], or Abramowitz and Stegun [1].

$$Det[\{a_1,-1\},\{a_2,a_1\}] = \begin{vmatrix} a_1 & -1 \\ a_2 & a_1 \end{vmatrix} = {a_1}^2 + a_2. \tag{31.1.1}$$

$$Det[\{a_1,-1,0\},\{a_2,a_1,-2\},\{a_3,a_2,a_1\}] = \begin{vmatrix} a_1 & -1 & 0 \\ a_2 & a_1 & -2 \\ a_3 & a_2 & a_1 \end{vmatrix} = {a_2}^3 + 3a_1a_2 + 2a_3. \tag{31.1.2}$$

$$Det[\{a_1,-1,0,0\},\{a_2,a_1,-2,0\},\{a_3,a_2,a_1,-3\},\{a_4,a_3,a_2,a_1\}] \tag{31.1.3}$$

$$= \begin{vmatrix} a_1 & -1 & 0 & 0 \\ a_2 & a_1 & -2 & 0 \\ a_3 & a_2 & a_1 & -3 \\ a_4 & a_3 & a_2 & a_1 \end{vmatrix} = {a_1}^4 + 6{a_1}^2a_2 + 8a_1a_3 + 3{a_2}^2 + 6a_4.$$

$$Det[\{a_1,-1,0,0,0\},\{a_2,a_1,-2,0,0\},\{a_3,a_2,a_1,-3,0\}, \tag{31.1.4}$$

DOI: 10.1201/9781003174158-31

$$\{a_4,a_3,a_2,a_1,-4\},\{a_5,a_4,a_3,a_2,a_1\}]$$

$$= \begin{vmatrix} a_1 & -1 & 0 & 0 & 0 \\ a_2 & a_1 & -2 & 0 & 0 \\ a_3 & a_2 & a_1 & -3 & 0 \\ a_4 & a_3 & a_2 & a_1 & -4 \\ a_5 & a_4 & a_3 & a_2 & a_1 \end{vmatrix} = {a_1}^5+10{a_1}^3a_2+20{a_1}^2a_3+15a_1{a_2}^2+30a_1a_4+20a_2a_3+24a_5.$$

$$Det[\{a_1,-1,0,0,0,0\},\{a_2,a_1,-2,0,0,0\},\{a_3,a_2,a_1,-3,0,0\}, \tag{31.1.5}$$

$$\{a_4,a_3,a_2,a_1,-4,0\},\{a_5,a_4,a_3,a_2,a_1,-5\},\{a_6,a_5,a_4,a_3,a_2,a_1\}]$$

$$= \begin{vmatrix} a_1 & -1 & 0 & 0 & 0 & 0 \\ a_2 & a_1 & -2 & 0 & 0 & 0 \\ a_3 & a_2 & a_1 & -3 & 0 & 0 \\ a_4 & a_3 & a_2 & a_1 & -4 & 0 \\ a_5 & a_4 & a_3 & a_2 & a_1 & -5 \\ a_5 & a_5 & a_4 & a_3 & a_2 & a_1 \end{vmatrix}$$

$$= {a_1}^6+15{a_1}^4a_2+40{a_1}^3a_3+45{a_1}^2{a_2}^2+90{a_1}^2a_4+120a_1a_2a_3$$

$$+144a_1a_5+15{a_2}^3+90a_2a_4+40{a_3}^2+120a_5.$$

$$Det[\{a_1,-1,0,0,0,0,0\},\{a_2,a_1,-2,0,0,0,0\},\{a_3,a_2,a_1,-3,0,0,0\}, \tag{31.1.6}$$

$$\{a_4,a_3,a_2,a_1,-4,0,0\},\{a_5,a_4,a_3,a_2,a_1,-5,0\},\{a_6,a_5,a_4,a_3,a_2,a_1,-6\},$$

$$\{a_7,a_6,a_5,a_4,a_3,a_2,a_1\}]$$

$$= \begin{vmatrix} a_1 & -1 & 0 & 0 & 0 & 0 & 0 \\ a_2 & a_1 & -2 & 0 & 0 & 0 & 0 \\ a_3 & a_2 & a_1 & -3 & 0 & 0 & 0 \\ a_4 & a_3 & a_2 & a_1 & -4 & 0 & 0 \\ a_5 & a_4 & a_3 & a_2 & a_1 & -5 & 0 \\ a_6 & a_5 & a_4 & a_3 & a_2 & a_1 & -6 \\ a_7 & a_6 & a_5 & a_4 & a_3 & a_2 & a_1 \end{vmatrix}$$

$$= {a_1}^7+21{a_1}^5a_2+70{a_1}^4a_3+105{a_1}^3{a_2}^2+210{a_1}^3a_4+420{a_1}^2a_2a_3+504{a_1}^2a_5+105a_1{a_2}^3$$

$$+630a_1a_2a_4+280a_1{a_3}^2+840a_1a_6+210{a_2}^2a_3+504a_2a_5+420a_3a_4+720a_7.$$

These determinants can all derive from classical recurrences for Bell Polynomials. However, I have stated them all as the results herein are mostly cases from these formulas.

31.2 CALCULUS OF DETERMINANT EVALUATION

In the online arxiv e-print by Krattenthaler [173] there is a wealth of material on evaluations of $n \times n$ determinants. For example, we see that

Theorem 31.1 *Let n be a nonnegative integer, and let $L_1, L_2, ..., L_n$ and A, B be indeterminates. Then the following are true.*

$$\det_{1\le i,j\le n}\left(\begin{bmatrix} L_i+A+j \\ L_i+j \end{bmatrix}_q\right) = q^{\left(\sum_{i=1}^{n}(i-1)(L_i+i)\right)} \frac{\prod_{1\le i<j<n}[L_i-L_j]_q \prod_{i=1}^{n}[L_i+A+1]_q!}{\prod_{i=1}^{n}[L_i+n]_q! \prod_{i=1}^{n}[A+1-i]_q!} \quad (31.2.1)$$

and

$$\det_{1\le i,j\le n}\left(q^{jL_i}\begin{bmatrix} A \\ L_i+j \end{bmatrix}_q\right) = q^{\left(\sum_{i=1}^{n}(iL_i)\right)} \frac{\prod_{1\le i<j<n}[L_i-L_j]_q \prod_{i=1}^{n}[A+i-1]_q!}{\prod_{i=1}^{n}[L_i+n]_q! \prod_{i=1}^{n}[A-L_i-1]_q!} \quad (31.2.2)$$

and

$$\det_{1\le i,j\le n}\left(\binom{BL_i+A}{L_i+j}\right) = \frac{\left(\prod_{1\le i<j<n}(L_i-L_j)\right)\left(\prod_{i=1}^{n}(BL_i+A)!\right)\left(\prod_{i=1}^{n}(A-Bi-1)_{i-1}\right)}{\left(\prod_{i=1}^{n}(L_i+n)!\right)\left(\prod_{i=1}^{n}((B-1)L_i+A-1)!\right)} \quad (31.2.3)$$

and

$$\det_{1\le i,j\le n}\left(\frac{(A+BL_i)^{j-1}}{(j-L_i)!}\right) = \prod_{i=1}^{n}\frac{(A+Bi)^{i-1}}{(n-L_i)!}\prod_{1\le i<j\le n}(L_j-L_i) \quad (31.2.4)$$

Think About It...

Mathematics serves as a handmaiden for the explanation of the quantitative situations in other subjects, such as economics, physics, navigation, finance, biology and even the arts.

Howard F. Fehr
Mathematics Educator, 1901–1982

FURTHER READING

ABRAMOWITZ, M., and STEGUN, I. *Handbook of Mathematical Functions*, Dover Publications Inc., New York, 1972.

ANDREWS, G.E. and BURGE, W.H., Determinant identities, *Pacific J. Math.* 158 (1993), 1-14.

CARLITZ, L. Some determinants of q-binomial coefficients, *J. Reine Angew. Math.* 226 (1967), 216-220.

KRATTENTHALER, C. Determinant identities and a generalization of the number of totally symmetric self-complementary plane partitions, *Electron. J. Combin.* 4(1) (1997), Article R27, 62 pp.

KRATTENTHALER, C. “Advanced Determinant Calculus” (https://www.mat.univie.ac.at/~kratt/artikel/detsurv.html and arXiv:math/9902004v3), (1999), 67 pages.

KUPERBERG, G., Symmetries of plane partitions and the permanent determinant method, *J. Combin. Theory Ser. A* 68 (1994), 115-151.

KUPERBERG, G., An exploration of the permanent-determinant method, Electron. *J. Combin.* 5 (1998), Article Number R46, 34 pp.

CHAPTER 32

Glossary of Terms

THIS GLOSSARY is not intended to be exhaustive, but to bring a consistency to the language used in this book.

GLOSSARY

Bijective Proof: In combinatorics, bijective proof is a proof technique for proving that two sets have equally many elements, or that the sets in two combinatorial classes have equal size.

Boltzmann's constant: The Boltzmann constant is defined to be $1.380649 \times 10^{-23} m^2 kg s^{-2} K^{-1}$. It is the proportionality factor that relates the average relative kinetic energy of particles in a gas with the thermodynamic temperature of the gas.

Congruences modulo m: Let a and b be integers and let m be a non-negative integer. If $a-b$ is a multiple of m, then we say that a and b are congruent modulo m and write $a \equiv b \, (mod \, m)$. That is, we say

$$a \equiv b \, (mod \, m) \Leftrightarrow a - b = |mq; \; for \; some \; integer \; q.$$

Cube Stack: A plane partition may be represented visually by the placement of a stack of $\pi_{i,j}$ unit cubes above the point (i, j) in the plane, giving a three-dimensional solid as shown in the picture at start of this chapter.

Euclid's Orchard: Euclid's orchard is a two-dimensional array of one-dimensional "trees" of unit height planted at the lattice points in one quadrant of a square lattice.

Euler sum of depth N: The two-place function $\zeta(s,t)$ has a natural generalization

$$\zeta(s_1, s_2, s_3, \ldots, s_N) = \sum_{n_1 > n_2 > n_3 > \cdots > n_N} \frac{1}{n_1^{s_1} n_2^{s_2} n_3^{s_3} \cdots n_N^{s_N}}, \tag{32.0.1}$$

DOI: 10.1201/9781003174158-32

which defines an Euler sum of depth N and weight $\sum_i s_i$. Some authors reverse the order of the variables.

Euler Sum: The double Euler sum is

$$\zeta(s,t) = \sum_{n>m>0} \frac{1}{n^s m^t} = \sum_{n=1}^{\infty} \frac{1}{n^s} \sum_{m=1}^{n-1} \frac{1}{m^t}.$$

Euler's constant γ: The constant γ is defined by

$$\gamma = \lim_{n\to\infty} \left(\sum_{k=1}^{n} \frac{1}{k} \right) - logn = \lim_{n\to\infty} (H_n - logn).$$

Exactly Solved Model: In statistical mechanics this is a lattice point model exhibiting spin charges at each point subject to certain positioning rules. The feature of an exactly solvable model is that there are exact formulas to calculate the free energy and the transition phase temperature. Examples of exactly solved models are: Six-Vertex Model, Eight-Vertex Model, and Hard Hexagon Model.

Farey sequence: The Farey fraction sequence of order n denoted F_n, is the sequence of completely reduced fractions, either between 0 and 1, or without this restriction, which when in lowest terms have denominators less than or equal to n, arranged in order of increasing size.

First quadrant: In the 2D Cartesian lattice, the first quadrant is the set of points with co-ordinates $\langle a,b \rangle$ where a and b are real numbers greater than zero. The two positive axes bordering this are not part of the quadrant, but sometimes we may specify a quadrant and include one or both of the bordering axes.

First hyperquadrant: In the nD Cartesian lattice, for n greater than 2, the first hyperquadrant is the set of points with co-ordinates $(a_1, a_2, 21.1ar, a_n)$ where a_1 to a_n are real numbers greater than zero. The n positive axes bordering this are not part of the quadrant, but sometimes we may specify a quadrant and include one or more of the bordering axes, which has the effect of including 0 as a co-ordinate.

Generating Function: Suppose there is a sequence of numbers or of functions $a_1, a_2, a_3, \ldots$. The generating function for that sequence is the power series $1 + a_1 z + a_2 z^2 + a_3 z^3 + \ldots$. Often the power series is a Taylor series of some closed form function, say $f(z)$.

Greatest Common Divisor (gcd): Let a and b be natural numbers. The greatest common divisor of a and b, denoted by $gcd(a,b)$, is defined to be the largest natural number that is a divisor of both a and b. Similarly, let a_1, a_2, $\ldots, a_n$, be natural numbers. The greatest common divisor of a_1, a_2, $\ldots, a_n$, denoted by $gcd(a_1, a_2, \ldots, a_n)$, is defined to be the largest natural number that is a divisor of all of a_1, a_2, $\ldots, a_n$.

Harmonic Number: The nth harmonic number is the sum of the reciprocals of the first n natural numbers:

$$H_n = 1 + \frac{1}{2} + \frac{1}{2} + \frac{1}{3} + \cdots + \frac{1}{n} = \sum_{k=1}^{n} \frac{1}{k}$$

Starting from $n = 1$, the sequence of harmonic numbers begins:

$$1, \frac{3}{2}, \frac{11}{6}, \frac{25}{12}, \frac{137}{60}, \frac{49}{20}, \frac{363}{140}, \frac{761}{280}, \frac{7129}{2520}, \frac{7381}{2520}, \ldots$$

Hyperpyramid Lattice: In the nD Cartesian lattice, for n greater than 3, the first hyperpyramid is the set of points with co-ordinates $(a_1, a_2, \ldots, a_n)$ where a_1 to a_{n-1} are integers and a_n is a positive integer, such that for every possible $\gcd(a_1, a_2, \ldots, a_n) = 1$, the set of all lattice points in the hyperpyramid are those having co-ordinates $v = (a_1, a_2, \ldots, a_n)$ where if $m = gcd(a_1, a_2, \ldots, a_n)$ then the points given by the scalar product set of all lattice points mv for each positive integer m define the hyperpyramid. The hyperpyramid is also said to be "radial from the origin".

Integer Partition: An integer partition of a positive integer n is a sum of non-increasing positive integers, that sum equal to n. The function $p(n)$ is the number of possible ways this sum can occur for n.

Integrable Function: Yang-Baxter integrable models in condensed matter have long been the traditional setting in which a range of exactly solved models have been realised in experiments. These include the early experiments related to the two-dimensional Ising model. Other integrable (exactly solved) models in statistical mechanics are the hard hexagon model, the Heisenberg spin chain, the quantum Ising chain with transverse and longitudinal fields, and a spin ladder model.

Lozenge tilings of a hexagon: A lozenge tiling is a tiling of a planar domain with three types of lozenges which we refer to as *left*, *top*, and *right* tiles.

Opaque forest: In discrete geometry, an opaque set is a system of curves or other set in the plane that blocks all lines of sight across a polygon, circle, or other shape. Opaque sets have also been called barriers, beam detectors, opaque covers, or (in cases where they have the form of a forest of line segments or other curves) opaque forests.

Plane Partition: In mathematics and especially in combinatorics, a *plane partition* is a two-dimensional array of nonnegative integers $\pi_{i,j}$ (with positive integer subscripts i and j) that are nonincreasing in both subscripts.

Polya's Orchard Visibility Problem: Each lattice point is a tree in a circular orchard of centre at the origin and radius r. If the tree radius is more than 1/r units, one cannot see out of the orchard in any direction. But, if the trees radii are $1/\sqrt{(r^2+1)}$, one can see out at certain angles. The problem formulates scenarios of trees radii and orchard radius.

Polylogarithm: The polylogarithm function is defined by a power series in z, which is also a Dirichlet series in s:

$$Li_s(z) = \sum_{k=1}^{\infty} \frac{z^k}{k^s} = z + \frac{z^2}{2^s} + \frac{z^3}{3^s} + \dots$$

This definition is valid for arbitrary complex order s and for all complex arguments z with $|z| < 1$; it can be extended to $|z|$ greater than unity by the process of analytic continuation. Note $Li_s(1)$ is the Riemann zeta function $\zeta(s)$.

Riemann zeta function: The classical Riemann zeta function is defined by

$$\zeta(s) = \sum_{k=1}^{n} \frac{1}{k^s}, \quad \Re s > 1.$$

Vector Partition: A partition of an n-dimensional vector $v = \langle a_1, a_2, \dots, a_n \rangle$ is a finite sum of n-dimensional (or nD) vectors each having weakly decreasing sequence of 1st vector components, 2nd vector components, 3rd vector components, etc. under vector addition. The vectors that comprise a single sum adding to v_n are called *parts* of the vector partition. A suitable partial ordering is for example, a set of weakly decreasing integers that sum to each component of the vector v_n, such as:

$$\begin{aligned} \langle a_1, a_2, a_3, \dots, a_n \rangle &= \langle b_1, c_1, d_1, \dots, \alpha_1 \rangle \\ &+ \langle b_2, c_2, d_2, \dots, \alpha_2 \rangle \\ &+ \langle b_3, c_3, d_3, \dots, \alpha_3 \rangle \\ &+ \langle b_4, c_3, d_3, \dots, \alpha_4 \rangle \\ &\vdots \quad \vdots \\ &= \langle b_m, c_m, d_m, \dots, \alpha_m \rangle \end{aligned}$$

so we have then clearly that

$$\begin{aligned} a_1 &= b_1 + b_2 + b_3 + \dots + b_m, \quad \text{for} \quad b_1 \geq b_2 \geq b_3 \geq \dots \geq b_m \geq 0; \\ a_2 &= c_1 + c_2 + c_3 + \dots + c_m, \quad \text{for} \quad c_1 \geq c_2 \geq c_3 \geq \dots \geq c_m \geq 0; \\ a_3 &= d_1 + d_2 + d_3 + \dots + d_m, \quad \text{for} \quad d_1 \geq d_2 \geq d_3 \geq \dots \geq d_m \geq 0; \\ \vdots & \quad \vdots \quad \vdots \\ a_n &= \alpha_1 + \alpha_2 + \alpha_3 + \dots + \alpha_m, \quad \text{for} \quad \alpha_1 \geq \alpha_2 \geq \alpha_3 \geq \dots \geq \alpha_m \geq 0. \end{aligned}$$

$\langle a_1, a_2, a_3, \dots, a_n \rangle$ may be construed as the arrow starting at the nD origin ending at the nD first hyperquadrant Cartesian lattice point $(a_1, a_2, a_3, \dots, a_n)$, where each a_i $(1 \leq i \leq n)$ co-ordinate is a non-negative integer.

Visible Lattice Point: A point in n-dimensional Euclidean space whose co-ordinates are $(a_1, a_2, \dots, a_n)$ where $a_1, a_2, \dots, a_n$ are co-prime integers as an n-tuple.

Visible Point Vectors (VPVs): A Visible Point Vector is a vector $\langle a_1, a_2, a_3, \ldots, a_n \rangle$ with each a_i a positive integer in n-dimensional Cartesian space where the greatest common divisor of the co-ordinates $gcd(a_1, a_2, a_3, \ldots)$ is equal to unity. Conceptually this can be thought of as a vector arrow starting at the origin $(0, 0, 0, \ldots, 0)$ with destination point $(a_1, a_2, a_3, \ldots, a_n)$. Since the co-ordinates are n-wise co-prime, the arrow goes unimpeded from origin to the point, so the point is said to be *visible from the origin*. It is well known that the probability that any n-space point is visible-from-origin, is $1/\zeta(n)$.

Yang-Baxter Equation: In physics, the Yang-Baxter equation (or star-triangle relation) is a consistency equation which was first introduced in the field of statistical mechanics. It depends on the idea that in some scattering situations, particles may preserve their momentum while changing their quantum internal states. The Yang-Baxter equation has long been recognised as the masterkey to integrability, providing the basis for exactly solved models which capture the fundamental physics of a number of realistic classical and quantum systems.

Bibliography

[1] ABRAMOWITZ, M. and STEGUN, I. *Handbook of Mathematical Functions.* Dover Publications Inc., New York, 1972. 10th Printing, 1972, with corrections.

[2] ADIGA, C., BERNDT, B.C., BHARGAVA, S. and WATSON, G.N. Chapter 16 of Ramanujan's Second Notebook: Theta Functions and q-Series. In *Memoirs of the American Mathematical Society*, volume 315. Amer. Math. Soc., Providence, RI, 1985.

[3] AHLGREN, S., ALLEN, P. and TANG, S. Congruences like Atkin's for the partition function. *Trans. Amer. Math. Soc. Ser. B*, to appear, 2023. arXiv:2112.09481v2 [math.NT].

[4] AHLGREN, S., BECKWITH, O. and RAUM, M. Scarcity of congruences for the partition function. *Am. J. Math.*, to appear, 2023. arXiv:2006.07645v3 [math.NT].

[5] ALKAUSKAS, G. The beginnings of significant mathematical results in the school children competitions (in Lithuanian). *Alpha Plus Omega*, 1(4):94–98, 1997.

[6] ALKAUSKAS, G. Generalization of the Rödseth-Gupta Theorem on Binary Partitions. *Lith. Math. J.*, April 2003.

[7] ALKAUSKAS, G. and MENDEL-STRASSE, G. Congruence Properties of the Function that Counts Compositions into Powers of 2. *J. Integer Seq.*, 13, 2010. Article 10.5.3.

[8] ALLADI, K. and GORDON H. Partition Identities and a Continued Fraction of Ramanujan. *J. Comb. Theory, Series A*, 63:275–300, 1993.

[9] ALLADI, K., BERNDT, B.C., PAULE, P., SELLARS, J. and YEE, A.J. (Editors). *George E. Andrews 80 Years of Combinatory Analysis.* Springer Nature Birkhäuser, Trends in Mathematics, Switzerland, 2021.

[10] ANDREWS, G.E. A polynomial identity which implies the Rogers-Ramanujan identities. *Scr. Math.*, 28:297–305, 1970.

[11] ANDREWS, G.E. *Problems and prospects for basic hypergeometric functions. The theory and application of Special Functions. (R Askey, Editor).* Academic Press, New York, 1975.

[12] ANDREWS, G.E. *The Theory of Partitions*, volume 2 of *Encyclopedia of Mathematics and its Applications.* Addison-Wesley Publishing Company, Advanced Book Program, Reading, Massachusetts, USA, 1976.

[13] ANDREWS, G.E. Plane partitions II: The equivalence of the Bender-Knuth and MacMahon conjectures. *Pacific J. Math.*, 72:283–291, 1977.

[14] ANDREWS, G.E. Plane partitions I: The MacMahon conjecture. In G.-C. Rota, editor, *Studies in foundations and combinatorics*, volume 1 of *Adv. in Math. Suppl. Studies*, pages 131–150, 1978.

[15] ANDREWS, G.E. An introduction to Ramanujan's "lost" notebook. *Amer. Math. Monthly*, 86:89–108, 1979.

[16] ANDREWS, G.E. Partitions and Durfee dissection. *Amer: J. Math.*, 101:735–742, 1979.

[17] ANDREWS, G.E. *Partitions: yesterday and today.* New Zealand Math. Soc., Wellington, 1979.

[18] ANDREWS, G.E. Macdonald's conjecture and descending plane partitions. In J. G. Williams T. V. Narayana, R. M. Mathsen, editor, *Combinatorics, representation theory and statistical methods in groups*, volume 57 of *Lecture Notes in Pure Math.*, pages 91–106. Marcel Dekker, New York, Basel, 1980. Young Day Proceedings.

[19] ANDREWS, G.E. Totally symmetric plane partitions. *Abstracts Amer. Math. Soc.*, 1:415, 1980.

[20] ANDREWS, G.E. Ramanujan's "Lost" Notebook. III. The Rogers-Ramanujan continued fraction. *Adv. Math.*, 41:186–208, 1981.

[21] ANDREWS, G.E. The hard-hexagon model and Rogers-Ramanujan type identities. *Proc Nat. Acad. Sci. USA*, 78:5290–5292, 1981.

[22] ANDREWS, G.E. q-Series: Ten Lectures. In *NSF-CBMS Regional Conference*, pages 175–182. Arizona State University, 1985.

[23] ANDREWS, G.E. *q-series: Their development and application in analysis, number theory, combinatorics, physics, and computer algebra.* Number 66 in Regional conference series in mathematics. Amer. Math. Soc., Providence, RI, 1986.

[24] ANDREWS, G.E. Plane partitions (V): The t.s.s.c.p.p. conjecture. *J. Combin. Theory Ser. A*, 66:28–39, 1994.

[25] ANDREWS, G.E. and ASKEY, R. Enumeration of Partitions: The Role of Eulerian Series and q-Orthogonal Polynomials. In Martin Aigner, editor, *Higher Combinatorics*, pages 3–26, Dordrecht, 1977. Springer Netherlands.

[26] ANDREWS, G.E. and BERKOVICH, A. A trinomial analogue of Bailey's lemma and N = 2 superconformal invariance. *Comm. Math. Phys.*, 192:245–260, 1998.

[27] ANDREWS, G.E. and BERNDT, B.C. *Ramanujan's Lost Notebook: Part V Paperback.* Ramanujan's Notebooks. Springer-Verlag, New York, 2018. ISBN-13: 978-3030085506.

[28] ANDREWS, G.E. and BURGE, W.H. Determinant identities. *Pacific J. Math.*, 158:1–14, 1993.

[29] ANDREWS, G.E. and ERIKSSON, K. *Integer Partitions.* Cambridge University Press, Cambridge, UK, New York, USA, Port Melbourne, Australia, Madrid, Spain, Cape Town, South Africa, 2004.

[30] ANDREWS, G.E. and GARVAN, F.G. Dyson's crank of a partition. *Bull. Amer. Math. Soc. New Series.*, 18(2):6 pages, April 1988. PDF:www.ams.org/journals/bull/1988-18-02/S0273-0979-1988-15637-6/S0273-0979-1988-15637-6.pdf.

[31] ANDREWS, G.E., BAXTER, R.J. and FORRESTER, P.J. Eightvertex SOS model and generalized Rogers-Ramanujan-type identities. *J. Stat. Phys.*, 35:193–266, 1984.

[32] APOSTOL, T. *Introduction to Analytic Number Theory.* Springer-Verlag, New York, 1976.

[33] APOSTOL, T.M. and ZUCKERMANN, H.S. On the functional equation F(mn)F((m.n))=F(m)F(n)f((m,n)). *Pac. J. Math.*, 14:371–384, 1964.

[34] ATKIN, A.O.L. Proof of a conjecture of Ramanujan. *Glasgow Math. J*, 8:14–32, 1967.

[35] ATKIN, A.O.L. and SWINNERTON-DYER, P. Some properties of partitions. *Proc. London Math. Soc.*, 3(4):84–106, 1954.

[36] AULUCK, F.C. On some new types of partitions associated with generalized Ferrers graphs. *Proc. Cambridge Philos. Soc.*, 47:679–686, 1951.

[37] BAAKE, M. and GRIMM, U. Combinatorial problems of quasi-crystallography. preprint Institut fur mathematik, Universitat Greifswald, December 2002.

[38] BAAKE, M. and HUCK, C. Ergodic properties of visible lattice points. *Proc. Steklov Inst. Math.*, 288(1):165–188, 2015. (English summary) Reprint of Tr. Mat. Inst. Steklova Vol. 288, pages 184-208 (2015).

[39] BAAKE, M., GRIMM, U. and WARRINGTON, D.H. Some remarks on the visible points of a lattice. *J. Phys. A: Math. Gen.*, 27:2669–2674, 1994.

[40] BAAKE, M., MOODY, R.V. and PLEASANTS, P.A.B. Diffraction from Visible Lattice Points and kth Power Free Integers. arxiv preprint, June 1999. http://arxiv.org/abs/math.MG/9906132/.

[41] BAGEMIHL, F. Some opaque subsets of a square. *Michigan Math. J.*, 6(2):99–103, 1959.

[42] BAILEY, D.H. and BORWEIN, J.M. Computation and structure of character polylogarithms with applications to character Mordell-Tornheim-Witten sums. *Math. Comput.*, 85:295–324, 2016.

[43] BAILEY, D.H. and BROADHURST, D.J. A Seventeenth-Order Polylogarithm Ladder: *http://arxiv.org/abs/math.CA*/9906134/. arxiv preprint, June 1999.

[44] BAILEY, D.H., BORWEIN, J.M. and CRANDALL, R.E. Computation and theory of extended Mordell Tornheim Witten sums. In *8th Asia-pacific Complex Systems Conference, International Workshop on Operator Theory and Its Applications.* Australian Mathematical Sciences Institute (AMSI), 2007.

[45] BAILEY, D.H., BORWEIN, J.M. and GIRGENSOHN, R. Experimental Evaluation of Euler Sums. *Exper. Math.*, 3:17–30, 1994.

[46] BAILEY, D.H., BORWEIN, P.B. and PLOUFFE, S. On the Rapid Computation of Various Polylogarithmic Constants. *Math. Comput.*, 66(218):903–913, April 1997.

[47] BAILEY, W.N. Some identities in combinatory analysis. *Proc. Lond. Math. Soc. Second Ser.*, 49:421–435, 1947.

[48] BAILEY, W.N. Identities of the Rogers-Ramanujan type. *Proc. Lond. Math. Soc. Second Ser.*, 50:1–10, 1949.

[49] BATCHELOR, M.T. and FOERSTER, A. Yang-Baxter integrable models in experiments: From condensed matter to ultracold atoms. *J. Phys. A: Math. Theor.*, 49(17), March 2016.

[50] BAXTER, R.J. Hard hexagons: exact solution. *Journal of Physics A: Mathematical and General*, 13(3), 1980. L61-L70, Bibcode 1980JPhA...13L..61B.

[51] BAXTER, R.J. Rogers-Ramanujan identities in the hard hexagon model. *J. Stat. Phys.*, 26:427, 1981.

[52] BAXTER, R.J. *Exactly Solved Models in Statistical Mechanics.* Academic Press, New York, 1982. Reprinted 1989 and 2007.

[53] BAZHANOV, V.V., KELS, A.P. and SERGEEV, S.M. Quasi-classical expansion of the star-triangle relation and integrable systems on quad-graphs. *J. Phys. A: Math. Theor.*, 49:44pp., October 2016.

[54] BEHREND, R.E., DI FRANCESCO, P.D. and ZINN-JUSTIN, P. On the weighted enumeration of alternating sign matrices and descending plane partitions. *J. Combin. Theory Ser. A*, 119:331–363, 2012.

[55] BEHREND, R.E., DI FRANCESCO, P.D. and ZINN-JUSTIN, P. A doubly-refined enumeration of alternating sign matrices and descending plane partitions. *J. Combin. Theory Ser. A*, 120:409–432, 2013.

[56] BELAVIN, A.A., POLYAKOV, A.M. and ZAMOLODCHIKOV, A.B. Infinite conformal symmetry in two-dimensional quantum field theory. *Nucl. Phys. B*, 241:333, 1984.

[57] BENDER E.A. and KNUTH D.E. Enumeration of plane partitions. *J. Combin. Theory Ser. A*, 13:40–54, 1972.

[58] BERKOVICH, A. Fermionic counting of RSOS states and Virasoro character formulae for the unitary minimal series $M(\nu, \nu+1)$: Exact results. *Nucl. Phys. B*, 431(1-2):315–348, 1994.

[59] BERKOVICH, A. and MCCOY, B.M. Continued fractions and fermionic representations for characters of M(p, p') minimal models. *Lett. Math. Phys.*, 37(49), 1996.

[60] BERKOVICH, A. and MCCOY, B.M. Rogers-Ramanujan Identities: A Century of Progress from Mathematics to Physics. *Documenta Mathematica*, Extra Volume ICM 1998 - III:163–172, 1998.

[61] BERKOVICH, A., MCCOY, B.M. and SCHILLING, A. Rogers-Schur-Ramanujan type identities for M(p, p') minimal models of conformal field theory. *Comm. Math. Phys.*, 191(325), 1998.

[62] BERNDT, B.C. *Ramanujan's Notebooks, Part IV.* New York: Springer-Verlag, 1994. ISBN 0-387-94109-6.

[63] BORWEIN, D., BORWEIN, J.M. and BRADLEY, D.M. Parametric Euler sum identities. *J. Math. Anal. Appl.*, 316:328–338, 2006.

[64] BORWEIN, D., BORWEIN, J.M. and GIRGENSOHN, R. Explicit evaluation of Euler sums. *Proc. Edinburgh Math. Soc.*, 38:277–294, 1995.

[65] BORWEIN, J.M. and BAILEY, D.M. *Mathematics by Experiment: Plausible Reasoning in the 21st Century.* Wellesley, MA, 2003. Ed. A K Peters.

[66] BORWEIN, J.M. and BORWEIN, P. *Pi and the AGM.* John Wiley and Sons, New York, 1986.

[67] BORWEIN, J.M., BRADLEY, D.M., BROADHURST, D.J. and LISONEK, P. Special Values of Multidimensional Polylogarithms. *CECM*, 98(106), 1998.

[68] BORWEIN, J.M., BRADLEY, D.M., BROADHURST, D.J. and LISONEK, P. Special values of multiple polylogarithms. *Trans. Am. Math. Soc.*, 353:907–941, 2001.

[69] BORWEIN, J.M., GLASSER, M.L., MCPHEDRAN, R.C., WAN, J.G. and ZUCKER, I.J. *Lattice Sums then and now*, volume 150 of *Encyclopedia of Mathematics and Its Applications.* Cambridge University Press, 2013.

[70] BRENT, R.P. On the Zeros of the Riemann Zeta Function in the Critical Strip. *Math. Comput.*, 33:1361–1372, 1979.

[71] BRENT, R.P.; van de LUNE, J.; te RIELE, J.J. and WINTER, D.T. On the Zeros of the Riemann Zeta Function in the Critical Strip. II. *Mathematics of Computation*, 39(160):681–688, October 1982.

[72] BRESSOUD, D.M. A new family of partition identities. *Pacific J. Math.*, 77(3):71–74, 1978.

[73] BRESSOUD, D.M. On a partition theorem of Göllnitz. *J. Reine Angew. Math.*, 305:215–217, 1979.

[74] BRESSOUD, D.M. Some identities for terminating q-series. *Math. Proc. Cambridge Phil. Soc.*, 89:211–223, 1981.

[75] BRESSOUD, D.M. and ZEILBERGER, D. A short Rogers-Ramanujan bijection. *Discrete Math.*, 38:313–315, 1982.

[76] BRESSOUD, D.M. and ZEILBERGER, D. Generalized Rogers-Ramanujan Bijections. *Advances in Math.*, 78:42–75, 1989.

[77] BRINGMANN, K. and LOVEJOY , J. Ranks and congruences for overpartition pairs. *Int. J. Number Theory*, 4:303–322, 2008.

[78] BRUINIER, J.H. *Borcherds products on $O(2,l)$ and Chern classes of Heegner divisors*, volume 1780 of *Springer Lecture Notes in Mathematics.* Springer-Verlag, 2002.

[79] BRUINIER, J.H. and FUNKE, J. On two geometric theta lifts. *Duke Math. J.*, 125:45–90, 2004.

[80] BRUINIER, J.H. and FUNKE, J. Traces of CM-values of modular functions. *J. Reine Angew. Math.*, 594:1–33, 2006.

[81] BRUINIER, J.H., and ONO, K. Algebraic formulas for the coefficients of half-integral weight harmonic weak Maass forms. *Adv. Math.*, 246:198–219, 2013.

[82] BRUINIER, J.H., JENKINS, P., and ONO, K. Hilbert class polynomials and traces of singular moduli. *Math. Ann.*, 334:373–393, 2006.

[83] CAÑADAS, A.M.; BALLESTER-BOLINCHES, A.; and GAVIRIA, I.D.M. Solutions of the Yang-Baxter Equation Arising from Brauer Configuration Algebras. *Computation*, 11(1), 2023.

[84] CAMPBELL, G.B. Multiplicative functions over Riemann zeta function products. *J. Ramanujan Soc.*, 7(1):52–63, 1992.

[85] CAMPBELL, G.B. Dirichlet summations and products over primes. *Int. J. Math. Math. Sci.*, 16(2):359–372, 1993.

[86] CAMPBELL, G.B. A generalized formula of Hardy. *Int. J. Math. Math. Sci.*, 17(2):369–378, 1994.

[87] CAMPBELL, G.B. A new class of infinite products, and Euler's totient. *Int. J. Math. and Math. Sci.*, 17(3):417–422, 1994.

[88] CAMPBELL, G.B. Infinite products over visible lattice points. *Int. J. Math. and Math. Sci.*, 17(4):637–654, 1994.

[89] CAMPBELL, G.B. *Combinatorial identities in number theory related to q-series and arithmetical functions.* Doctor of philosophy thesis, School of Mathematical Sciences, The Australian National University, October 1997.

[90] CAMPBELL, G.B. A closer look at some new identities. *Int. J. Math. and Math. Sci.*, 21(3):581–586, 1998.

[91] CAMPBELL, G.B. Infinite products over hyperpyramid lattices. *Int. J. Math. and Math. Sci.*, 23(4):271–277, 2000.

[92] CAMPBELL, G.B. An Euler product transform applied to q-series. *Ramanujan J.*, 12:267–293, 2006.

[93] CAMPBELL, G.B. Some n-space q-binomial theorem extensions and similar identities. arxiv preprint, June 2019. arXiv:1906.07526v1 [math.NT]: https://arxiv.org/abs/1906.07526.

[94] CAMPBELL, G.B. An interview with Rodney James Baxter. *Aust. Math. Soc. Gazette*, 47(1):24–32, March 2020.

[95] CAMPBELL, G.B. Fun with numbers: Rational solutions to $x^y y^x = v^w w^v$. *Aust. Math. Soc. Gazette*, 49(5):210–211, November 2022.

[96] CAMPBELL, G.B. Continued Fractions for partition generating functions. arxiv preprint, Jan 2023. arXiv:2301.12945 [math.CO], https://arxiv.org/abs/2301.12945.

[97] CAMPBELL, G.B. Vector Partition Identities for 2D, 3D and nD Lattices. arxiv preprint, Feb 2023. arXiv:10.48550/ARXIV.2302.01091 [math.NT], https://arxiv.org/abs/2302.01091.

[98] CAMPBELL, G.B. Visible Point Partition Identities for Polylogarithms, and Parametric Euler Sums. arxiv preprint, Jun 2023. arXiv:2306.02241v1 [math.CO], https://arxiv.org/abs/2306.02241.

[99] CAMPBELL, G.B. Visible Point Vector Partition Identities for Hyperpyramid Lattices. arxiv preprint, Sep 2023. arXiv:2309.16094v1 [math.CO], https://arxiv.org/abs/2309.16094.

[100] CAMPBELL, G.B. and ZUJEV, A. The series that Ramanujan misunderstood. arxiv preprint, Oct 2016. arXiv:1610.03693v1 [math.NT], https://arxiv.org/abs/1610.03693v1.

[101] CANTINI, L. and SPORTIELLO, A. Proof of the Razumov-Stroganov conjecture. *J. Combin. Theory Ser. A*, 118:1549–1574, 2011.

[102] CARLITZ, L. Rectangular arrays and plane partitions. *Acta Arith.*, 13:29–47, 1967.

[103] CARLITZ, L. Some determinants of q-binomial coefficients. *J. Reine Angew. Math.*, 226:216–220, 1967.

[104] CAUCHY, A. Mémoire sur les fonctions dont plusieurs ... . In *Oeuvres de Cauchy*, pages 42–50. C. R. Acad. Sci. Paris, Gauthier-Villars, Paris, 1893. T. XVII, p. 523, 1re série, T. VIII.

[105] CHAN, H.C. Ramanujan's cubic continued fraction and an analog of his "most beautiful identity". *Int. J. Number Theory*, 6:673–680, 2010.

[106] CHANDRASEKHARAN, K. *Arithmetical Functions*, volume 67. Springer-Verlag, New York, 1970.

[107] CHAPMAN, R. A new proof of some identities of Bressoud. *Int. J. Math. and Math. Sciences*, 32:627–633, 2002.

[108] CHAVEZ, F. and DUHR, C. Three-mass triangle integrals and single-valued polylogarithms. *J. High Energy Phys.*, 2012. article id.114.

[109] CHEEMA, M.S. and MOTZKIN, T.S. Multipartitions and multipermutations. *Proc. Symp. Pure Math.*, 19:37–39, 1971.

[110] CHURCHHOUSE, R.F. Congruence properties of the binary partition function. *Proc. Cambridge Philos. Soc.*, 66:371–376, 1969.

[111] CIUCU, M. Enumeration of perfect matchings in graphs with reflective symmetry. *J. Combin. Theory Ser. A*, 77:67–97, 1997.

[112] CIUCU, M. The equivalence between enumerating cyclically symmetric, self-complementary and totally symmetric, self-complementary plane partitions. *J. Combin. Theory Ser. A*, 86:382–389, 1999.

[113] CIUCU, M. and KRATTENTHALER, C. Plane partitions II: $5\frac{1}{2}$ symmetry classes. In M. Kashiwara, K. Koike, S. Okada, I. Terada, H. Yamada, editor, *Combinatorial Methods in Representation Theory*, volume 28 of *Advanced Studies in Pure Mathematics*, pages 83–103. RIMS, Kyoto, 2000.

[114] COOK II, D. and NAGEL, U. Enumerations of lozenge tilings, lattice paths, and perfect matchings and the weak Lefschetz property. arxiv preprint, May 2013. website: www.arxiv.org/abs/1305.1314.

[115] CORTEEL, S. and LOVEJOY, J. Overpartitions. *Trans. Amer. Math. Soc.*, 356:1623–1635, 2004.

[116] DATA, E., JIMBO, M., KUNIBA, A., MIWA, T. and OKADO, M. Exactly solvable SOS models: local height probabilities and theta function identities. *Nucl. Phys. B*, 290(231), 1987.

[117] DEWAR, M. and RAM MURTY, M. An asymptotic formula for the coefficients of j(z). *International Journal of Number Theory*, 9(3):1–12, 2013.

[118] DICKS, R. Congruence relations for r-colored partitions. *Journal of Number Theory*, 249:377–399, 2023.

[119] DYSON, F.J. Some Guesses in The Theory of Partitions. *Eureka (Cambridge)*, 8:10–15, 1944. ISBN 9780821805619. PDF:www.archim.org.uk/eureka/archive/Eureka-8.pdf.

[120] ECKSTEIN, G. On an Olympiad problem. *Romanian Math. Mag.*, 7:15–16, 1998.

[121] ELIMELECH, R., DAVID, O., De la CRUZ MENGUAL, C., KALISCH, R., BERNDT, W., SHALYT, M., SILBERSTEIN, M., HADAD, Y., and KAMINER, I. Algorithm-assisted discovery of an intrinsic order among mathematical constants, Oct 2023. https://arxiv.org/abs/2308.11829.

[122] ERDÖS P. and RICHMOND B. Concerning Periodicity in the Asymptotic Behaviour of Partition Functions. *J. Austral. Math. Soc. (Series A)*, 21:447–456, 1976.

[123] EULER, L. Introductio in analysin infinitorum. In *Opera Omnia*, volume 1. Marcum-Michaelum, Brousquet, Lausannae, 1748. Chapter 16.

[124] EULER, L. *Institutiones calculi integralis*, volume 1. Marcum-Michaelum, Brousquet, Lausannae, 1768. http://doi.org/10.3931/e-rara-3788.

[125] FADDEEV, L.D. How Algebraic Bethe Ansatz works for integrable model, May 1996. https://arxiv.org/abs/hep-th/9605187.

[126] FEIGIN, B.L. and FUCHS,D.B. Verma modules over the Virasoro algebra. *Funct. Anal. Appl.*, 17(3):241–242, 1983.

[127] FELDER, G. and VARCHENKO, A. Multiplication Formulas for the Elliptic Gamma Function. arxiv preprint, Dec 2002. arXiv:math/0212155v1 [math.QA]. www.doi.org/10.48550/arXiv.math/0212155.

[128] FONSECA, T. and ZINN-JUSTIN, P. On the doubly refined enumeration of alternating sign matrices and totally symmetric self-complementary plane partitions. *Electron. J. Combin.*, 15, 2008. Article Number R81, 35 pp.

[129] FORRESTER,P.J. and BAXTER,R.J. Further exact solutions of the eightvertex SOS model and generalizations of the Rogers-Ramanujan identities. *J. Stat. Phys.*, 38(435), 1985.

[130] FRANKLIN, F. Sur les developpement du produit infiniti $(1-x)(1-x^2)(1-x^3)\ldots$,. *Comptes Rendus*, 82:448–450, 1881.

[131] FRNKA, R.A. *Asymptotic Formulae for Restricted Unimodal Sequences.* Phd thesis, Louisiana State University and Agricultural and Mechanical College, April 2017. LSU Doctoral Dissertations. DOI: 10.31390/gradschool_dissertations.4460.

[132] GARVAN, F.G. Generalizations of Dyson's rank. PhD Thesis, Pennsylvania State Univ., May 1986. https://dokumen.pub/generalizations-of-dysons-rank.html.

[133] GARVAN, F.G. New combinatorial interpretations of Ramanujan's partition congruences mod 5, 7 and 11. *Trans. Amer. Math. Soc.*, 305(1):31 pages, Jan 1988.

[134] GASPER, G. and RAHMAN, M. *Basic Hypergeometric Series*, volume 35 of *Encyclopedia of Mathematics and its Applications.* Cambridge University Press, (Cambridge - New York - Port Chester - Melbourne - Sydney), 1990.

[135] GAUSS, C.F. Disquisitiones generales circa seriem infinitam .... In *Werke 3*, volume II, pages 123–162. Comm. soc. reg. sci. Gött. rec., 1876. article first printed 1813; reprinted in Werke 3 1876.

[136] GIBBS, J.W. *Elementary Principles in Statistical Mechanics.* (Reprinted by Dover, New York 1960), 1902.

[137] GILMORE, T. Three interactions of holes in two dimensional dimer systems. arxiv preprint, Mar 2015. preprint last revised 16 Mar 2015. doi.org/10.48550/arXiv.1501.05772.

[138] GLASSER, M.L. and ZUCKER, I.J. Lattice Sums. *Th. Chem.: Adv. Persp.*, 5:67–139, 1980.

[139] GOLDFELD, D. Beyond the last theorem. *Math Horizons*, 4:26–34, September 1996.

[140] GÖLLNITZ, H. Partitionen mit Differenzenbedingungen. *J. Reine Angew. Math.*, 225:154–190, 1967.

[141] GORDON, B. Two theorems on multipartite partitions. *J. London Math. Soc.*, 38:459–464, 1963.

[142] GORDON, B. Some continued fractions of the Rogers-Ramanujan type. *Duke Math. J.*, 32:741–748, 1965.

[143] GORGON, B. and HOUTEN, L. Notes on plane partitions I. *J. Combin. Theory Ser. A*, 4:72–80, 1968.

[144] GORGON, B. and HOUTEN, L. Notes on plane partitions III. *Duke Math. J.*, 36:801–824, 1969.

[145] GOSPER, R.W. and SCHROEPPEL, R. 2 dimensional visible lattice point invisible polygons. In BEELER, M., GOSPER, R.W. and SCHROEPPEL, R., editor, *HAKMEM: MIT Artificial Intelligence Laboratory, Memo AIM-239*, page 19. Cambridge, MA, 1972. ITEM 48 (Gosper, Schroeppel), Feb 1972.

[146] GUPTA, H. Proof of the Churchhouse conjecture concerning binary partitions. *Proc. Cambridge Philos. Soc.*, 70:53–56, 1971.

[147] HAMMOND, P. and LEWIS, R. Congruences in ordered pairs of partitions. *Int. J. Math. Math. Sci.*, 47:2509–2512, 2004.

[148] HARDY, G.H. *Ramanujan.* Cambridge University Press, London and New York, 1940. (reprinted by Chelsea, New York).

[149] HARDY, G.H. An extension of a theorem on oscillating series. In *Collected Papers*, volume VI, pages 500–506. Clarendon Press, Oxford, 1974.

[150] HARDY, G.H. On certain oscillating series. In *Collected Papers*, volume VI, pages 146–167. Clarendon Press, Oxford, 1974.

[151] HARDY, G.H. and LITTLEWOOD, J.E. A further note on the converse of Abel's theorem. In *Collected Papers of Hardy*, volume VI, pages 699–716. Clarendon Press, Oxford, 1974.

[152] HARDY, G.H. and RAMANUJAN, S. Asymptotic formulae in combinatory analysis. *Proc. London. Math. Soc.*, 2(17):75–115, 1918. Also in Collected papers of S. Ramanujan, pp. 276-309, Cambridge Univ. Press, Cambridge, 1927; reprinted by Chelsea, New York, 1962.

[153] HARDY, G.H. and WRIGHT, E.M. *An Introduction to the Theory of Numbers.* Oxford University Press, Clarendon, London, 1971.

[154] HEINE, E. Untersuchungen uber die Reihe ... . *J. Reine angew. Math.*, 34:285–328, 1847.

[155] HEINE, E. *Handbuch der Kugelfunctionen, Theorie und Andwendungen*, volume 1. Reimer, Berlin, 1878.

[156] HIRSCHHORN, M.D. and SELLERS, J.A. Arithmetic relations for overpartitions. *J. Comb. Math. Comb. Comp.*, 53:65–73, 2005.

[157] INGHAM, A.E. A Tauberian theorem for partitions. *Ann. Math.*, 42:1075–1090, 1941.

[158] ISHIKAWA, M. On refined enumerations of totally symmetric self-complementary plane partitions, I. arxiv preprint, June 2006. preprint 12 Jun 2006.

[159] ISHIKAWA, M. On refined enumerations of totally symmetric self-complementary plane partitions, II. arxiv preprint, June 2006. preprint 6 Jun 2006 : www.arxiv.org/abs/math/0606082.

[160] JACOBI, C.G.J. *Fundamenta nova theoriae functionum ellipticarum*, volume 1. Regiomonte fratrum Bornträger., 1829. (reprinted in *Gesammelte Werke*, Vol. 1, pp 49-239; Reimer, Berlin, 1881).

[161] KAC, V.G. Contravariant form for infinite-dimensional Lie algebras and superalgebras. *Lect. Notes in Phys.*, 94:441, 1979.

[162] KAC, V.G. Modular invariance in mathematics and physics. In *Proceedings of the AMS Centennial Symposium*, volume 337. American Mathematical Society, 1992.

[163] KASTELEYN, P.W. Dimer statistics and phase transitions. *J. Math. Phys.*, 4:287–293, 1963.

[164] KASTELEYN, P.W. Graph theory and crystal physics. In *Graph Theory and Theoretical Physics*, pages 43–110. Academic Press, London, 1967.

[165] KEDEM, R., KLASSEN, T.R., MCCOY, B.M. and MELZER, E. Fermionic quasi-particle representations for characters of $(G^{(1)})_1 \times (G^{(1)})_1/(G^{(1)})_2$. *Phys. Lett. B*, 304:263–270, 1993.

[166] KEDEM, R., KLASSEN, T.R., MCCOY, B.M. and MELZER, E. Fermionic sum representations for conformal field theory characters. *Phys. Lett. B*, 307:68–76, 1993.

[167] KENYON, R. *An introduction to the dimer model*, volume 16 of *Statistical mechanics, IAS/Park City Math. Ser.* Amer. Math. Soc., Providence, R.I., 2009. pp. 191-230.

[168] KING, R.C. and EL-SHARKAWAY, N.G.I. Standard Young tableaux and weight multiplicities of the classical Lie groups. *J. Physics A*, 16:3153–3178, 1983.

[169] KOTĚŠOVEC, V. A method of finding the asymptotics of q-series based on the convolution of generating functions. arxiv preprint, Sep 2015. arXiv:1509.08708 [math.CO]. doi.org/10.48550/arXiv.1509.08708.

[170] KOUTSCHAN, C., KAUERS, M. and ZEILBERGER, D. Proof of George Andrews's and David Robbins's q-TSPP conjecture. *Proceedings of the National Academy of Sciences (PNAS)*, 108(6):2196–2199, February 2011.

[171] KRATTENTHALER, C. *The major counting of nonintersecting lattice paths and generating functions for tableaux*, volume 115 of *Memoirs of the American Mathematical Society*. Amer. Math. Soc. Providence, R.I., 1995. (fourth of five numbers).

[172] KRATTENTHALER, C. Determinant identities and a generalization of the number of totally symmetric self-complementary plane partitions. *Electron. J. Combin.*, 4(1), 1997. Article R27, 62 pp.

[173] KRATTENTHALER, C. Advanced Determinant Calculus. arxiv preprint, 1999. 67 pages arXiv:math/9902004v3 : www.arxiv.org/pdf/math/9902004.pdf.

[174] KRATTENTHALER, C. Another involution principle-free bijective proof of Stanley's hook-content formula. *J. Combin. Theory Ser. A*, 88:66–92, 1999.

[175] KRATTENTHALER, C. Plane Partitions in the work of Richard Stanley and his School. arxiv preprint, May 2015. 33 pages arXiv:1503.05934v2 [math.CO]. www.doi.org/10.48550/arXiv.1503.05934.

[176] KUPERBERG, G. Self-complementary plane partitions by Proctor's minuscule method. *Europ. J. Combin.*, 15:545–553, 1994.

[177] KUPERBERG, G. Symmetries of plane partitions and the permanent determinant method. *J. Combin. Theory Ser. A*, 68:115–151, 1994.

[178] KUPERBERG, G. Four symmetry classes of plane partitions under one roof. *J. Combin. Theory Ser. A*, 75:295–315, 1996.

[179] KUPERBERG, G. An exploration of the permanent-determinant method. *Electron. J. Combin.*, 5, 1998. Article Number R46, 34 pp.

[180] LAI, T. and ROHATGI, R. Enumeration of lozenge tilings of a hexagon with a shamrock missing on the symmetry axis: www.arxiv.org/abs/1711.02818. arxiv preprint, July 2018. preprint last revised 31 Jul 2018.

[181] LANDEN, J. *Mathematical memoirs*, volume 1. Book on Demand Ltd. 112, 1780. Volume 1 Paperback - 11 November 2013.

[182] LEHMER, D.N. An asymptotic evaluation of certain totient sums. *Amer. J. Math.*, 22:293–355, 1900.

[183] LEWIN, L. *Polylogarithms and associated functions.* Elsevier Science Ltd.: North-Holland, New York, 1981. ISBN 0-444-00550-1,(First published London: Macdonald, 1958.).

[184] LEWIN, L. The dilogarithm in algebraic fields. *J. Austral. Math. Soc. Ser. A*, 33:302–330, 1982.

[185] LEWIN, L. (Ed.). *Structural Properties of Polylogarithms.* Amer. Math. Soc. - Providence, RI., 1991. ISBN 0-8218-1634-9.

[186] LITTLEWOOD, J.E. Review of the Collected Papers of Srinivasa Ramanujan. *Mathematical Gazette*, 14:427–428, 1929.

[187] LIVINGOOD, J,. A partition function with the prime modulus $p > 3$. *Am. J. Math.*, 67:194–208, 1945.

[188] The LMFDB Collaboration. The L-functions and modular forms database. web page, 2023. [Online; accessed 18 February 2023]. www.lmfdb.org.

[189] LOSSERS, O.P. The generalized totient function, solution to problem No. 6446. *Amer. Math. Monthly*, pages 434–434, 1985.

[190] MAC LANE, S. and BIRKHOFF, G. *A survey of modern algebra.* N.Y., Macmillan, fourth edition, 1977.

[191] MACDONALD, I.G. *Symmetric Functions And Hall Polynomials.* Oxford:Clarendon Press; New York/London: Oxford University Press, 2nd edition, 1995.

[192] MACMAHON, P.A. Partitions of numbers whose graphs possess symmetry. *Trans. Cambridge Philos. Soc.*, 17:149–170, 1899.

[193] MACMAHON, P.A. *Combinatory Analysis*, volume 2. Cambridge Univ. Press, London and New York (reprinted by Chelsea, New York, 1960), 1916.

[194] MAHLER, K. On a Special Functional Equation. *J. London Math. Soc.*, 15:115–123, 1940.

[195] MAHLER, K. On a Special Function. *J. Number Theory*, 12:20–26, 1980.

[196] MASSER, D.W. *Open problems.* Proceedings of the Symposium on Analytic Number Theory. London: Imperial College, 1985. Chen, W. W. L. (ed.).

[197] MAZURKIEWICZ, S. Sur un ensemble fermeé, punctiforme, qui rencontre toute droite passant par un certain domaine. *Prace Mat.-Fiz. (in Polish and French)*, 27:11–16, 1916.

[198] MEINARDUS, G. Asymptotische Aussagen über Partitionen. *Math. Z.*, 59:388–398, 1954.

[199] MILLS, W.H., ROBBINS, D.H. and RUMSEY, H. Alternating sign matrices and descending plane partitions. *J. Combin. Theory Ser. A*, 34:340–359, 1983.

[200] MILLS, W.H., ROBBINS, D.H. and RUMSEY, H. Self-complementary totally symmetric plane partitions. *J. Combin. Theory A*, 42:277–292, 1986.

[201] MILLS, W.H., ROBBINS, D.H. and RUMSEY, H. Enumeration of a symmetry class of plane partitions. *Discrete Math.*, 67:43–55, 1987.

[202] MILLS, W.H., ROBBINS, D.H. and RUMSEY,H. Proof of the Macdonald conjecture. *Invent. Math.*, 66:73–87, 1982.

[203] MOSSERI, R. Visible Points in a Lattice. *J. Phys. A: Math. Gen.*, 25:L25–L29, 1992.

[204] NINHAM, B.W., GLASSER, M.L., HUGHES, B.D. and FRANKEL, N.E. Möbius, Mellin, and Mathematical Physics. *Physica A*, 186:441–481, 1992.

[205] ODLYZKO, A.M. and WILF, H.S. n coins in a fountain. *Amer. Math. Monthly*, 95:840–843, 1988.

[206] OESTERLÉ, J. Nouvelles approches du "théorème" de Fermat. *Asteérisque, Séminaire Bourbaki exp.*, 694(161):165–186, 1988. ISSN 0303-1179, MR 0992208.

[207] OKADA, S. On the generating functions for certain classes of plane partitions. *J. Combin. Theory Ser. A*, 51:1–23, 1989.

[208] PENNINGTON, W.B. On Mahler partition problem. *Ann. of Math.*, 57:531–546, 1952.

[209] PERRY, D. A polynomial variation of Meinardus' theorem. *International J. Number Theory*, 11(01):251–268, 2015.

[210] PROCTOR, R.A. Shifted plane partitions of trapezoidal shape. *Proc. Amer. Math. Soc.*, 89:553–559, 1983.

[211] PROCTOR, R.A. New symmetric plane partition identities from invariant theory work of De Concini and Procesi. *Europ. J. Combin.*, 11:289–300, 1990.

[212] RADEMACHER, H. On the partition function $p(n)$. *Proc. London Math. Soc.*, 2(43):241–254, 1937.

[213] RADEMACHER, H. On the expansion of the partition function in a series. *Ann. Of Math.*, 44:416–422, 1943.

[214] The RAMANUJAN MACHINE. The Ramanujan Machine parallel computing project. (web page), 2023. [Online; accessed 22 October 2023]. www.ramanujanmachine.com.

[215] RAMANUJAN, S. Some properties of $p(n)$, the number of partitions of n. *Proc. Cambridge Philos. Soc.*, 19:207–210, Nov 1919.

[216] RAMANUJAN, S. *Notebooks (Two Volumes)*. Tata Institute, Bombay, 1957.

[217] RAMANUJAN, S. On certain trigonometrical sums and their application to the theory of numbers. In *Collected Papers of S. Ramanujan*, pages 179–199. Cambridge University Press, Cambridge (1927), 1962. reprinted by Chelsea, New York, 1962.

[218] RAMANUJAN, S. *The Lost Notebook, and Other Unpublished Papers*. Narosa, New Delhi, 1988.

[219] RICHMOND, B. Asymptotic relations for partitions. *J. Number Theory*, 7:389–405, 1975.

[220] RICHMOND, B. Asymptotic relations for partitions. *Canadian J. Math.*, 27:1083–1091, 1975.

[221] RIEMANN, G.F.B. Über die Anzahl der Primzahlen unter einer gegebenen Grösse. *Monatsber. Königl. Preuss. Akad. Wiss. Berlin*, pages 671–680, Nov 1859.

[222] ROBBINS, D.P. The story of $1, 2, 7, 42, 429, 7436, \ldots$. *Math. Intelligencer*, 13:12–19, 1991.

[223] ROCHA-CARIDI, A. *Vertex Operators in Mathematics and Physics*. Springer, Berlin, 1985. ed. J. Lepowsky, S. Mandelstam and I.M. Singer.

[224] RÖDSETH, Ö. Some arithmetical properties of m-ary partitions. *Proc. Cambridge Philos. Soc.*, 68:447–453, 1970.

[225] ROGERS, L.J. Second memoir on the expansion of certain infinite products. *Proc. London Math. Soc.*, 25:318–343, 1894.

[226] ROGERS, L.J. On Two Theorems of Combinatory Analysis and Some Allied Identities. *Proc. Lond. Math. Soc. Second Ser.*, 16:315–336, 1917.

[227] ROTH, K.F., and SZEKERES, G. Some asymptotic formulae in the theory of partitions. *Q. J. Math. Oxf. Ser (2)*, 5:241–259, 1954.

[228] RUIJSENAARS, S.N.M. First order analytic difference equations and integrable quantum systems. *J. Math. Phys.*, 38(2):1069–1146, February 1997.

[229] SANDIFER, C.E. Chapter 32: Who proved e is irrational?. How Euler Did It. *Math. Assoc. Am.*, pages 185–190, 2006. ISBN 978-0-88385-563-8. LCCN 2007927658.

[230] SCHUR, I. Ein Beitrag zur additive Zahlentheorie und zue Theorie der Kettenbruche. *S.-B. Preuss, Akad. Wies. Phys.-Math, Kl.*, pages 302–321, 1973. First printed 1917: Proceedings of the Royal Prussian Academy of Sciences in Berlin; Reprinted 1973: Ges. Abhandlungen, Vol. 2, pp. 117–136.

[231] SCHUR, I. Zur additive Zahlentheorie und zue Theorie der Kettenbruche. *S.-B. Preuss, Akad. Wies. Phys.-Math, Kl.*, pages 488–495, 1973. First printed 1917: Proceedings of the Royal Prussian Academy of Sciences in Berlin; Reprinted 1973: Ges. Abhandlungen, Vol. 3, pp. 43-50.

[232] SIVARAMAKRISHNAN, R. *Classical Theory of Arithmetic Functions.* Marcel Dekker, Inc., (New York and Basel), 1989.

[233] SLATER, L.J. *Generalized Hypergeometric Series.* Cambridge University Press, 1966.

[234] SLOANE, N.J.A. Expansion of $\prod_{m>=1}(1+x^m)$; number of partitions of n into distinct parts; number of partitions of n into odd parts. The On-Line Encyclopedia of Integer Sequences (OEIS), 2020. sequence A000009: *https://oeis.org/A000009*.

[235] SLOANE, N.J.A. Generalized Pentagonal Numbers. The On-Line Encyclopedia of Integer Sequences (OEIS), 2020. sequence A000326 *https://oeis.org/A000326*.

[236] SLOANE, N.J.A. Number of planar partitions (or plane partitions) of n. The On-Line Encyclopedia of Integer Sequences (OEIS), 2022. sequence A000219. (www.oeis.org/A000219).

[237] SLOANE, N.J.A. Expansion of e.g.f. exp(x/(x-1)). The On-Line Encyclopedia of Integer Sequences (OEIS), 2023. sequence A293116. (www.oeis.org/A293116).

[238] SOFO, A. Integral identities for sums. *Math. Commun.*, 13:303–309, 2008.

[239] SOFO, A. Double Integral Representation of Sums. *Int. J. Math. Anal.*, 3(10):483–489, 2009.

[240] SOFO, A. Harmonic numbers and double binomial coefficients. *Integral Transforms Spec. Funct.*, 20(11):847–857, 2009.

[241] SOFO, A. Integral forms of sums associated with harmonic numbers. *Appl. Math. Comput.*, 207(2):365–372, 2009.

[242] SOFO, A. Series associated with Polygamma functions. *Scientia Series A, Mathematical Sciences*, pages 95–104, 2009.

[243] SOFO, A. Closed Form Representations of Harmonic Sums. *Hacettepe Journal of Mathematics and Statistics*, 39(2):255–263, 2010.

[244] SOFO, A. Double Integral Identities and Zeta Functions. *Int. J. Contem. Math. Sci.*, 5(17-20):869–876, 2010.

[245] SOFO, A. Harmonic sums and integral representations. *J. Appl. Anal.*, 16(2):265–277, 2010. pp1869–6082 (online).

[246] SOFO, A. Integrals and Polygamma Representations for Binomial Sums. *J. Integer Seq.*, 13:1–12, 2010.

[247] SOFO, A. Sums in Terms of Polygamma Functions. *Bull. Math. Anal. Appl.*, 2(3):40–52, 2010.

[248] SOFO, A. Triple integral identities and Zeta functions. *Appl. Anal. Discret. Math.*, 4(2):347–360, 2010.

[249] SOFO, A. Harmonic Number Sums in Closed Form. *Math. Commun.*, 16(2):335–345, 2011.

[250] SOFO, A. Harmonic Number Sums in Higher Powers. *J. Math. Anal.*, 2(2):15–22, 2011.

[251] SOFO, A. Harmonic Numbers and Cubed Binominal Coefficients. *Int. J. Comb.*, 2011.

[252] SOFO, A. Integral Identities for Rational Series Involving Binomial Coefficients. *Bull. Malaysian Math. Sci. Soc.*, 34(3):631–637, 2011.

[253] SOFO, A. Summation Formula Involving Harmonic Numbers. *Anal. Math.*, 37(1):51–64, 2011.

[254] SOFO, A. Euler-related sums. *Math. Sci.*, 6(10), 2012.

[255] SOFO, A. Finite Sums in Pascal's Triangle. *Fibonacci Q.*, 50(4):337–345, 2012.

[256] SOFO, A. Harmonic numbers of order two. *Miscolc Math. Notes*, 13(2):499–514, 2012.

[257] SOFO, A. Mixed binomial sum identities. *Integral Transforms Spec. Funct.*, 24(3):187–200, 2012.

[258] SOFO, A. New classes of harmonic number identities. *J. Integer Seq.*, 15, 2012.

[259] SOFO, A. Reciprocal power sums. *Integers*, 12, 2012.

[260] SOFO, A. Generalized hypergeometric function identities at argument ± 1. *Integral Transforms Spec. Funct.*, 25(11):909–921, 2014.

[261] SOFO, A. Shifted harmonic sums of order two. *Commun. Korean Math. Soc.*, 29(2):239–255, 2014.

[262] SOFO, A. New families of alternating harmonic number sums. *Tbil. Math. J.*, 8(2):195–209, 2015.

[263] SOFO, A. Quadratic alternating harmonic number sums. *J. Number Theory*, 154:144–159, 2015.

[264] SOFO, A. Harmonic numbers at half integer and binomial squared sums. *Honam Math. J.*, 38(2):279–294, 2016.

[265] SOFO, A. Harmonic numbers at half integer values. *Integral Transforms Spec. Funct.*, 27(6):430–442, 2016.

[266] SOFO, A. Identities for Alternating Inverse Squared Binomial and Harmonic Number Sums. *Mediterr. J. Math.*, 13(4):1407–1418, 2016.

[267] SOFO, A. Integrals of logarithmic and hypergeometric functions. *Commun. Math.*, 24:7–22, 2016.

[268] SOFO, A. Polylogarithmic connections with Euler sums. *Sarajevo J. Math.*, 12:17–32, 2016.

[269] SOFO, A. Sums of quadratic half integer harmonic numbers of alternating type. *J. class. anal.*, 8(2):99–111, 2016.

[270] SOFO, A. A master integral in four parameters. *J. Math. Anal. App.*, 448:81–92, 2017.

[271] SOFO, A. Alternating cubic Euler sums with binomial squared terms. *International J. Number Theory*, 14(05):1357–1374, 2018.

[272] SOFO, A. Evaluation of integrals with hypergeometric and logarithmic functions. *Open Math.*, 16(1):63–74, 2018.

[273] SOFO, A. General order Euler sums with multiple argument. *J. Number Theory*, 189:255–271, 2018.

[274] SOFO, A. Integrals of polylogarithmic functions with negative argument. *Acta Univ. Sapientiae Math.*, 10:347–367, 2018.

[275] SOFO, A. Families of Integrals of Polylogarithmic Functions. *Researchgate*, 7, 2019. Mathematics 10.3390/math7020143.

[276] SOFO, A. and CIMADEVILLA VILLACORTA, J. New identities for the Arctan function. *J. Math. Anal.*, 3(3):1–10, 2012.

[277] SOFO, A. and CVIJOVIC, D. Extensions of Euler harmonic sums. *Appl. Anal. Discret. Math.*, 6:317–328, 2012. 2.

[278] SOFO, A. and HASSANI, M. Quadratic harmonic number sums. *Appl. Math. E-Notes*, 12, 2012.

[279] SOFO, A. and NIMBRAN, A. Euler sums and integral connections. *Mathematics*, 7:833, 09 2019.

[280] SOFO, A. and SRIVASTAVA, H.M. Identities for the Harmonic Numbers and Binominal Coefficients. *Ramanujan J.*, 25(1):93–113, 2011.

[281] SOFO, A. and SRIVASTAVA, H.M. A family of shifted harmonic sums. *Ramanujan J.*, 37(1):89–108, 2014.

[282] STANLEY, H.E. Spherical Model as the Limit of Infinite Spin Dimensionality. *Phys. Rev.*, 176(2):718–722, 1968.

[283] STANLEY, R.P. Theory and applications of plane partitions: Part 1,2. *Stud. Appl. Math.*, 50:167–188, 259–279, 1971.

[284] STANLEY, R.P. A baker's dozen of conjectures concerning plane partitions. In P. Leroux (eds.) G. Labelle, editor, *Combinatoire énumérative*, volume 1234 of *Lecture Notes in Math.*, pages 285–293. Springer-Verlag, Berlin-New York, 1986.

[285] STANLEY, R.P. Symmetries of plane partitions. *J. Combin. Theory A*, 43:103–113, 1986. Erratum 44 (1987), 310.

[286] STANLEY, R.P. *Enumerative Combinatorics*, volume 2. Cambridge University Press, Cambridge, 1999.

[287] STEMBRIDGE, J.R. Nonintersecting paths, pfaffians and plane partitions. *Adv. Math.*, 83:96–131, 1990.

[288] SVED, M. On the Rational Solutions of $x^y = y^x$. *Mathematics Magazine*, 63(1):30–33, 1990. JSTOR, https://maa.org/programs/faculty-and-departments/classroom-capsules-and-notes/on-the-rational-solutions-of-xy-yx. Accessed 14 Sep. 2022.

[289] SZEKERES, G. An asymptotic formula in the theory of partitions, I. *Quart. J. Math. Oxford Ser.*, 2(2):85–108, 1951.

[290] SZEKERES, G. An asymptotic formula in the theory of partitions, II. *Quart. J. Math. Oxford Ser.(2)*, 4:96–111, 1951.

[291] SZPIRO, L. Propriétés numériques du faisceau dualisant rélatif. *Astérisque.*, 86:44–78, 1981. Seminaire sur les pinceaux des courbes de genre au moins deux. Zbl 0517.14006.

[292] SZPIRO, L. Présentation de la théorie d'Arakelov. *Contemporary Mathematics*, 67:279–293, 1987. Zbl 0634.14012.

[293] TAKEDA, W. Visible lattice points and the extended Lindelöf hypothesis. *J. Number Theory*, 180:297–309, 2017. MR3679799 Reviewed.

[294] TOH, P.C. Ramanujan type identities and congruences for partition pairs. *Discrete Math.*, 312:1244–1250, 2012.

[295] TÓTH, L. Sums of products of Ramanujan sums. *Annali dell'universita' di Ferrara*, 58:183–197, 2011.

[296] WATSON, G.N. Theorems stated by Ramanujan. VII: Theorems on continued fractions. *J. London Math. Soc.*, 4:37–48, 1929.

[297] WATSON, G.N. A Note on Spence's Logarithmic Transcendent. *Quart. J. Math. Oxford Ser.*, 8:39–42, 1937.

[298] WEISSTEIN, Eric W. Visible Point Vector Identity. From MathWorld–A Wolfram Web Resource., July 2012. Article with 2D and 3D identities (https://mathworld.wolfram.com/VisiblePointVectorIdentity.html).

[299] WEISSTEIN, Eric W. Visible Point. From MathWorld–A Wolfram Web Resource., July 2020. Article with references (https://mathworld.wolfram.com/VisiblePoint.html).

[300] WIKIPEDIA. Hard Hexagon Model. On-line encyclopedia article, July 2020. Brief statistical mechanics article on the model and its solution. *https://en.wikipedia.org/wiki/Hard_hexagon_model*.

[301] WRIGHT, E.M. Asymptotic partition formulae I. Plane partitions. *Q. J. Math.*, 1:177–189, 1931.

[302] WRIGHT, E.M. Partitions of multipartite numbers. *Proc. Amer. Math. Soc.*, 28:880–890, 1956.

[303] ZAGIER, D. Traces of singular moduli, Motives, polylogarithms and Hodge theory, Part I. In *Int. Press Lect. Ser., 3*, pages 211–244. Int. Press, Somerville, MA, 2002. first printed Irvine, CA,1998.

[304] ZEILBERGER, D. Proof of the alternating sign matrix conjecture. In *Electron. J. Combin. 3 (no. 2, "The Foata Festschrift")*, 1996. number R13, 84 pp.

[305] ZHAO, H. and ZHONG, Z. Ramanujan type congruences for a partition function. *Electron. J. Combin.*, 18:P58, 2011.

Index